DETROIT PUBLIC LIBRARY
3 5674 00572833 3

P9-ARG-253

Introduction to

Fifth Edition

Harcourt Brace Jovanovich, Inc.
New York Chicago San Francisco Atlanta

Introduction to Geography

Fifth Edition

Front Endpaper

Left page, a "false" infrared Ektachrome photograph of the southern California coastline produced from a multispectral-scanner image transmitted from 567 miles above the earth by ERTS—I (Earth Resources Technology Satellite—I). A. The Oxnard plain. B. The San Andreas fault. C. The city of Los Angeles. The cities of Long Beach and Carpenteria lie to the southeast and the southwest, respectively; Bakersfield and the southern San Joaquin Valley are to the northwest; and the Mojave Desert is to the northeast. The rectangular outline indicates the area shown in greater detail on the opposite page. (NASA Photo)

Right page, an infrared Ektachrome photograph of California's Oxnard plain, as viewed from a NASA RB-57 aircraft at an altitude of 60,000 feet (about 12 miles). A. Wheeler Canyon. B. The city of Oxnard. C. The city of Ventura. D. Oxnard Air Force Base. Field crops in the agricultural areas around Oxnard, mainly vegetables, are indicated in red. Fallow fields or bare soils are represented by a greenish tint that tends toward black where the soil is moist. Citrus and avocado orchards appear in reddish brown, approximately the color of the natural chaparral vegetation in the hills north of Ventura. Also visible north of Ventura is a jumbled pattern of roadways associated with nearby oil drilling. (NASA Photo)

A
C
B
D
N

Geography

Henry M. Kendall
Late of Miami University

Robert M. Glendinning
University of California, Los Angeles

Clifford H. MacFadden
University of California, Los Angeles

Richard F. Logan
University of California, Los Angeles

H. Craig MacFadden

in association with
George Thambyahpillay
Sri Lanka University

Introduction to Geography, Fifth Edition

Cover photos: Nancy Palmer Agency; NASA

The globe on Plate I is reprinted by permission of Rand McNally & Company, © Rand McNally & Company, R. L. 73-GP-14.

The Geologic Time Scale in Appendix E was adapted from that of J. Lawrence Kulp in *Science,* April 14, 1961. Used by permission of *Science* and the author.

ISBN: 0-15-542152-2

Library of Congress Catalog Card Number: 75-30453

Printed in the United States of America

Preface

In recent times, the study of humankind's relationship with the earth's environment has become a matter of increasing practical importance. With human population growth placing ever-greater pressures on natural resources—in some instances, to the danger level—many people are beginning to recognize that this environment is not the inexhaustible gift of nature they had assumed it to be. As they awaken to the reality that the earth is not only a varied and ever-changing planet but a highly fragile and endangered one, people are also realizing that they must obtain a better understanding of their habitat if they are to restore and preserve it. There is, therefore, increasing public awareness of the need for additional research and study in those areas of the physical and social sciences that relate to human use of the environment. In geography, work involving natural environmental processes and the dynamics of culture has always been a major part of the discipline; today, this work is rapidly expanding into new areas and in new directions in quest of solutions to environmental problems.

This edition of *Introduction to Geography* responds to these new directions while carefully reflecting its predecessors' balanced coverage of the total human habitat. As an introductory textbook for today's college student, it addresses the expanding needs of many courses, old and new, in geography and allied fields involved with environmental structure and utilization. The focus of this book is on the relationships between human societies and the natural environment. The introductory chapters of Part

One, which treats physical geography, present an overview of the earth, including such fundamentals as interplanetary relationships and the interactions of the earth's hydrosphere, lithosphere, and biosphere. Chapters 4 through 6 are concerned with the development of landforms, while Chapters 7 through 12 discuss the composition and dynamics of the atmosphere. Chapters 13 and 14 deal with the earth's native vegetation and soils, and with their responses to climate, landforms, and human activity. Chapter 15, the last in Part One, presents an original study that illustrates some of the major interactions of physical features and human population within a specific area of California. The first two chapters in Part Two, on cultural geography, introduce the basic concepts employed in that field. Chapters 3 through 5 trace the origins and diffusions of humankind and describe the present patterns of major institutions—including languages, religions, and political orders. Chapters 6 through 10 deal with the husbanding of the earth, by focusing on human livelihoods through time, and Chapters 11 and 12 are concerned with settlement types, urbanization, and the basic functions of urban societies. Chapters 13 and 14 of Part Two conclude the volume with views on population changes, spatial linkage systems, and the emerging realities of societal impacts on the environment.

Clarity, balance, and conciseness have been the watchwords in the planning and writing of this book; and while naturally presenting several important theoretical works, including those of Christaller, von Thünen, and Weber, we have usually sought to project the reader's thinking beyond the classroom to the facts and problems of the "real world." It is our hope that this approach will encourage people to view the study of geography as more than mere academic training in nomenclature—and thus encourage many more of them to become truly responsible guardians of this planet and of its human societies and cultures.

No extended listing of acknowledgments will be attempted here, but our thanks go to many colleagues and associates and to several staff members at Harcourt Brace Jovanovich (Joanne Daniels, Helen Faye, Harrison Griffin, Carolyn Johnson, Susan Joseph, Jayne Klein, and Harry Rinehart)—who have given us continuing encouragement, direction, and constructive editorial criticism.

Robert M. Glendinning
Clifford H. MacFadden
Richard F. Logan
H. Craig MacFadden

Contents

Chapter 3 *The Earth's Waters, 16*

Chapter 4 *Introduction to Landforms, 46*

Chapter 5 *Landforms Shaped by Fluvial Action, 76*

Chapter 6 *Landforms: Coastal, Karstic, Glacial, and Eolian, 100*

Chapter 11 Air Masses, Fronts, and Lows and Highs: Weather, 214

Chapter 12 Climate and Weather, 234

Chapter 13 Vegetation and the Environment, 274

Chapter 14 *Soils and the Environment, 304*

Chapter 15 *The Physical Environment: A Type-Study, 336*

PART TWO *Cultural Geography*

Chapter 1 *The Cultural Complex, 350*

Chapter 2 *Culture and Nature, 366*

Appendixes

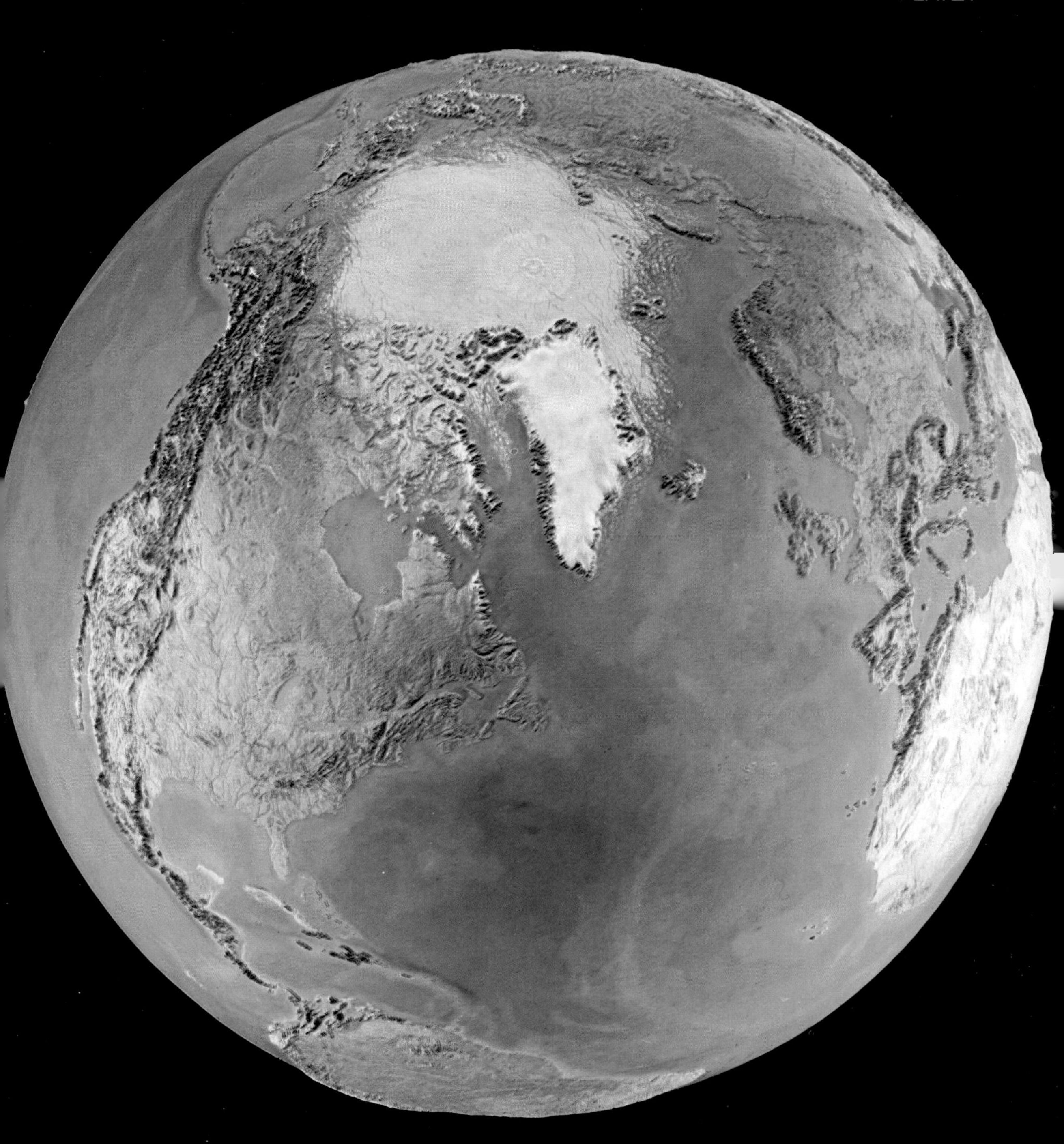

THE PHYSICAL WORLD

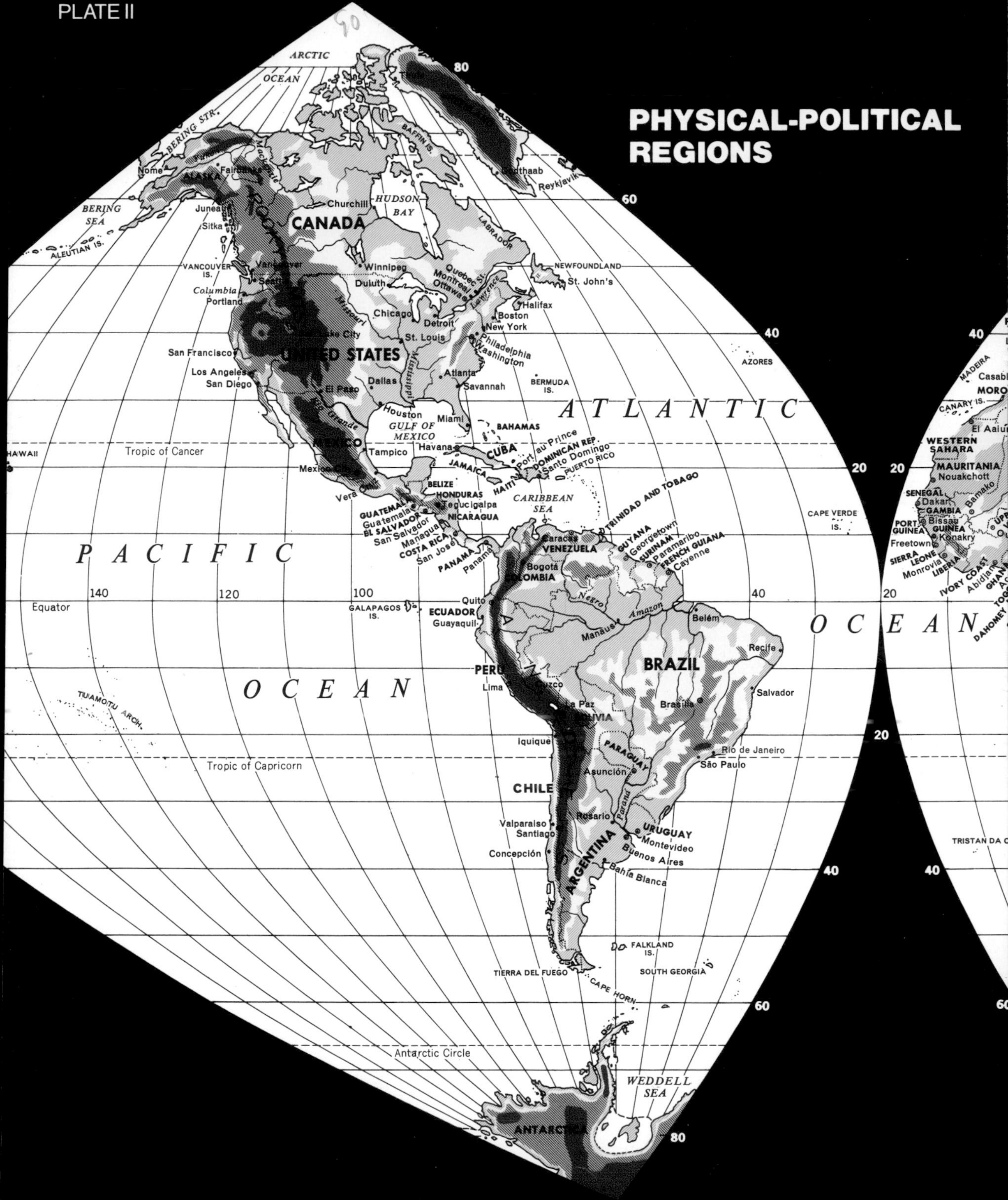
PLATE II
PHYSICAL-POLITICAL REGIONS
ARCTIC OCEAN
BERING STR.
BAFFIN IS.
Godthaab
Reykjavik
Nome
Fairbanks
ALASKA
BERING SEA
ALEUTIAN IS.
Juneau
Sitka
Churchill
HUDSON BAY
CANADA
LABRADOR
VANCOUVER IS.
Vancouver
Seattle
Columbia
Portland
Winnipeg
Duluth
Quebec
Montreal
Ottawa
St. Lawrence
NEWFOUNDLAND
St. John's
Halifax
Missouri
Chicago
Detroit
Boston
New York
St. Louis
Philadelphia
Washington
San Francisco
UNITED STATES
Mississippi
Los Angeles
San Diego
Dallas
Atlanta
El Paso
Savannah
BERMUDA IS.
AZORES
Rio Grande
Houston
Miami
GULF OF MEXICO
BAHAMAS
ATLANTIC
MEXICO
Tampico
Havana
CUBA
Port au Prince
DOMINICAN REP.
Santo Domingo
PUERTO RICO
HAWAII
Tropic of Cancer
Mexico City
JAMAICA
HAITI
Vera Cruz
BELIZE
HONDURAS
Tegucigalpa
NICARAGUA
CARIBBEAN SEA
TRINIDAD AND TOBAGO
GUATEMALA
Guatemala
EL SALVADOR
San Salvador
Managua
COSTA RICA
San José
PANAMA
Panama
Caracas
VENEZUELA
GUYANA
Georgetown
SURINAM
Paramaribo
FRENCH GUIANA
Cayenne
CAPE VERDE IS.
PACIFIC
Bogotá
COLOMBIA
Negro
Amazon
Equator
GALAPAGOS IS.
Quito
ECUADOR
Guayaquil
Manaus
Belém
OCEAN
Recife
BRAZIL
PERU
Lima
Cuzco
Salvador
Brasília
OCEAN
TUAMOTU ARCH.
La Paz
BOLIVIA
Iquique
PARAGUAY
Rio de Janeiro
São Paulo
Tropic of Capricorn
Asunción
CHILE
Paraná
Rosario
Valparaiso
Santiago
URUGUAY
Montevideo
Buenos Aires
Concepción
ARGENTINA
Bahía Blanca
FALKLAND IS.
TIERRA DEL FUEGO
SOUTH GEORGIA
CAPE HORN
Antarctic Circle
WEDDELL SEA
ANTARCTICA
MADEIRA
CANARY IS.
El Aaiun
WESTERN SAHARA
MAURITANIA
Nouakchott
SENEGAL
Dakar
GAMBIA
Bamako
PORT. GUINEA
Bissau
GUINEA
Konakry
Freetown
SIERRA LEONE
Monrovia
LIBERIA
IVORY COAST
Abidjan
GHANA
DAHOMEY
TRISTAN DA C
80
60
40
20
140
120
100
40
20
20
40
60
80
40
20
20
40
60

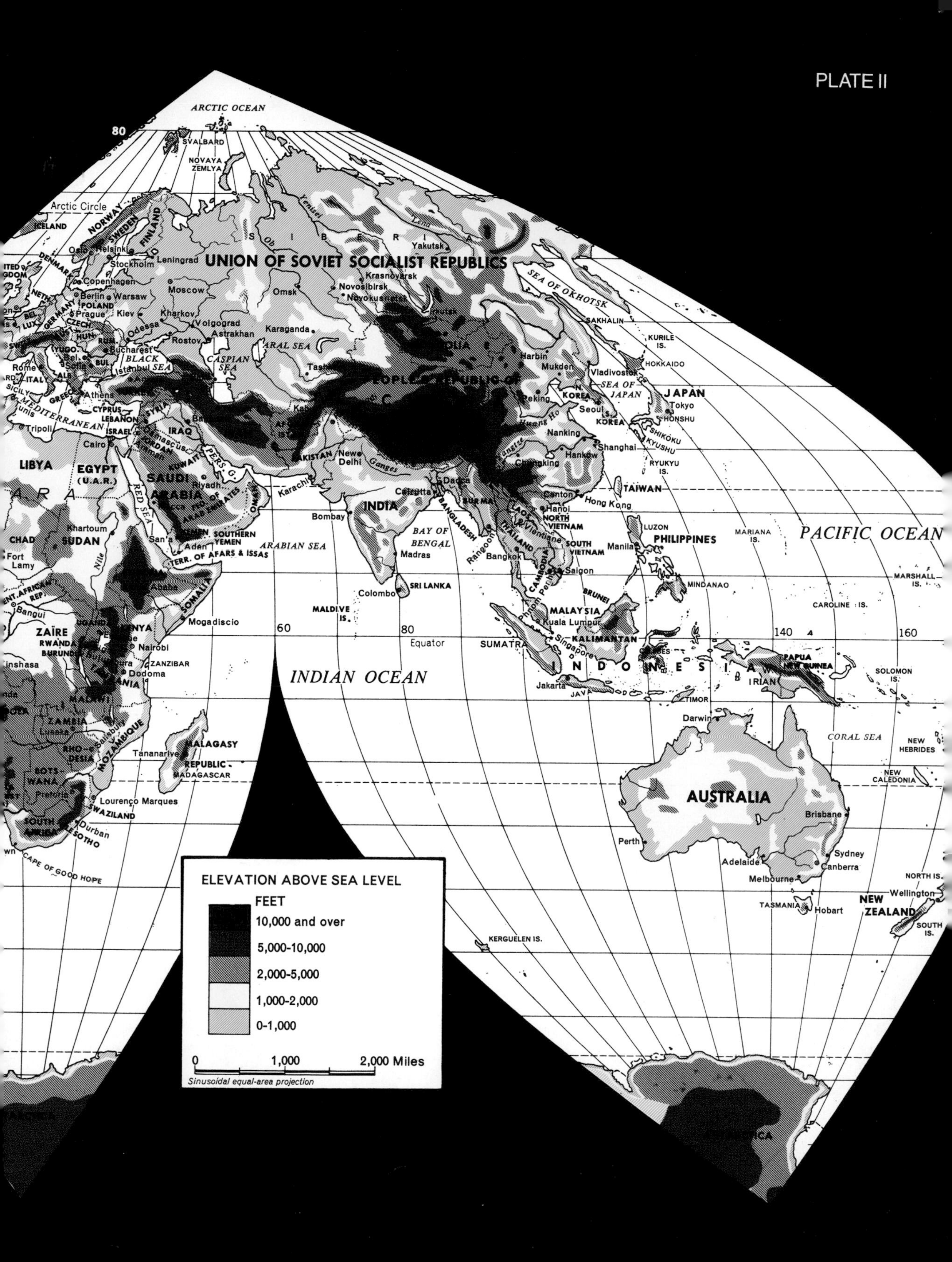
ARCTIC OCEAN
SVALBARD
NOVAYA ZEMLYA
Arctic Circle
ICELAND
NORWAY
SWEDEN
FINLAND
DENMARK
Oslo
Helsinki
Stockholm
Leningrad
UNION OF SOVIET SOCIALIST REPUBLICS
Copenhagen
Berlin
Warsaw
POLAND
Prague
Kiev
Moscow
Kharkov
Volgograd
Astrakhan
Rostov
Odessa
Bucharest
BLACK SEA
Istanbul
Sofia
BUL.
YUGO.
Rome
ITALY
SICILY
GREECE
Athens
MEDITERRANEAN
Tunis
Tripoli
CYPRUS
LEBANON
ISRAEL
SYRIA
Damascus
JORDAN
Amman
IRAQ
KUWAIT
PERS. G.
SAUDI ARABIA
Riyadh
OMAN
RED SEA
Cairo
LIBYA
EGYPT (U.A.R.)
CHAD
Fort Lamy
SUDAN
Khartoum
Nile
YEMEN
SOUTHERN YEMEN
Aden
San'a
TERR. OF AFARS & ISSAS
SOMALIA
Addis Ababa
Mogadiscio
KENYA
Nairobi
ZANZIBAR
Dodoma
RWANDA
BURUNDI
ZAÏRE
Kinshasa
Bangui
MALAWI
ZAMBIA
Lusaka
RHODESIA
Salisbury
MOZAMBIQUE
BOTSWANA
Pretoria
Lourenço Marques
SWAZILAND
SOUTH AFRICA
LESOTHO
Durban
CAPE OF GOOD HOPE
MALAGASY REPUBLIC
MADAGASCAR
Tananarive
CASPIAN SEA
ARAL SEA
Karaganda
Tashkent
Kabul
Baghdad
IRAN
AFGHANISTAN
PAKISTAN
Karachi
New Delhi
Ganges
INDIA
Bombay
ARABIAN SEA
BAY OF BENGAL
Madras
Calcutta
Dacca
BANGLADESH
SRI LANKA
Colombo
MALDIVE IS.
S I B E R I A
Ob
Yenisei
Lena
Yakutsk
Krasnoyarsk
Omsk
Novosibirsk
Novokuznetsk
Irkutsk
MONGOLIA
PEOPLE'S REPUBLIC OF CHINA
Harbin
Mukden
Peking
Huang Ho
Yangtze
Nanking
Shanghai
Hankow
Chungking
Canton
Hong Kong
TAIWAN
BURMA
Rangoon
LAOS
Vientiane
Hanoi
NORTH VIETNAM
SOUTH VIETNAM
THAILAND
Bangkok
CAMBODIA
Phnom Penh
Saigon
MALAYSIA
Kuala Lumpur
Singapore
SUMATRA
KALIMANTAN
BRUNEI
INDONESIA
Jakarta
JAVA
TIMOR
IRIAN
PAPUA NEW GUINEA
PHILIPPINES
LUZON
Manila
MINDANAO
SEA OF OKHOTSK
SAKHALIN
KURILE IS.
HOKKAIDO
Vladivostok
SEA OF JAPAN
N. KOREA
Seoul
S. KOREA
JAPAN
Tokyo
HONSHU
SHIKOKU
KYUSHU
RYUKYU IS.
MARIANA IS.
PACIFIC OCEAN
MARSHALL IS.
CAROLINE IS.
SOLOMON IS.
INDIAN OCEAN
Equator
60
80
140
160
80
Darwin
AUSTRALIA
Perth
Adelaide
Melbourne
Sydney
Canberra
Brisbane
TASMANIA
Hobart
CORAL SEA
NEW HEBRIDES
NEW CALEDONIA
NEW ZEALAND
NORTH IS.
Wellington
SOUTH IS.
KERGUELEN IS.
ANTARCTICA
ELEVATION ABOVE SEA LEVEL
FEET
10,000 and over
5,000-10,000
2,000-5,000
1,000-2,000
0-1,000
0
1,000
2,000 Miles
Sinusoidal equal-area projection

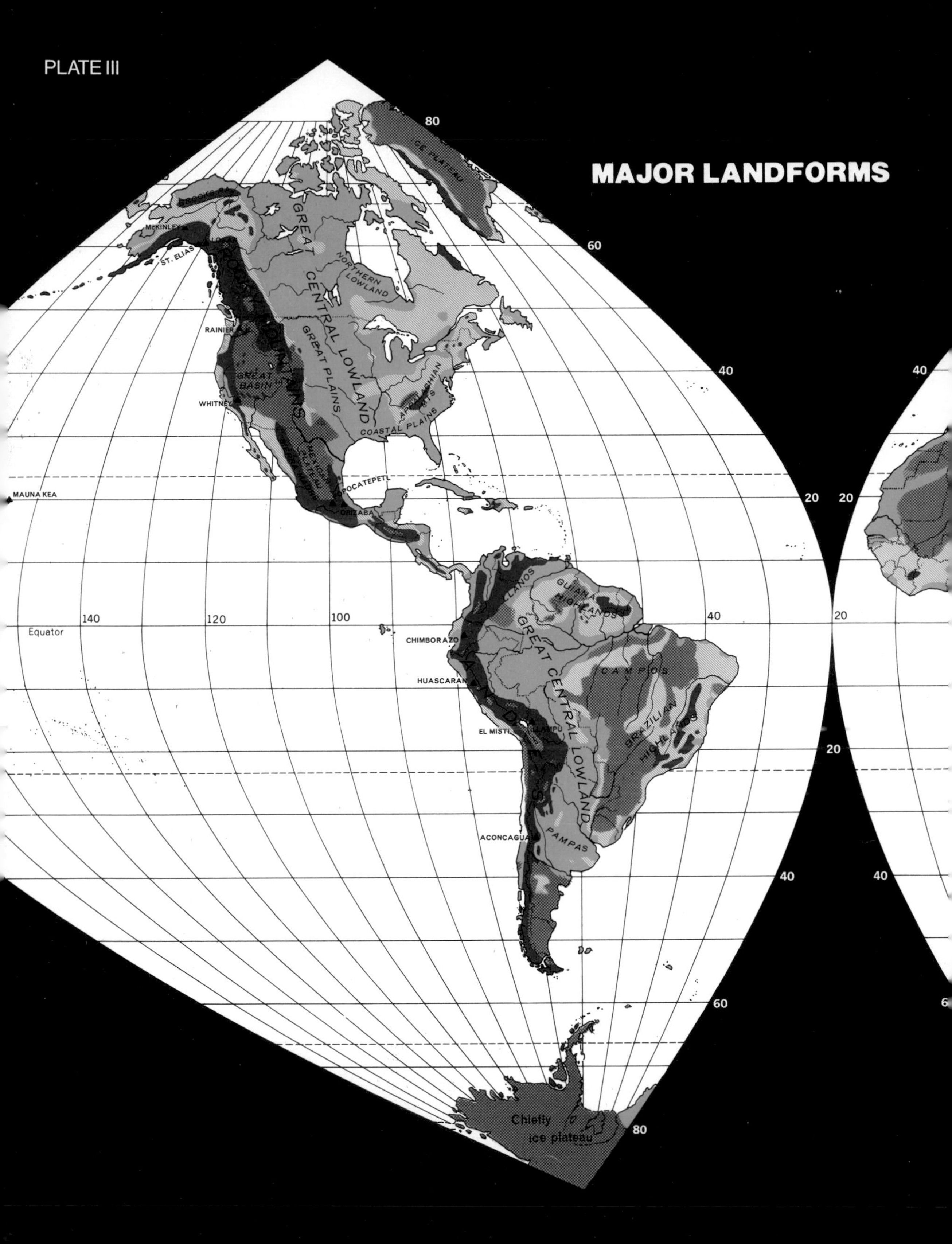
PLATE III
MAJOR LANDFORMS
80
60
40
20
140
120
100
Equator
GREAT CENTRAL LOWLAND
NORTHERN LOWLAND
GREAT PLAINS
GREAT BASIN
COASTAL PLAINS
McKINLEY
ST. ELIAS
RAINIER
WHITNEY
MAUNA KEA
POCATEPETL
ORIZABA
CHIMBORAZO
HUASCARAN
EL MISTI
ACONCAGUA
LLANOS
GUIANA HIGHLANDS
CAMPOS
BRAZILIAN HIGHLANDS
PAMPAS
Chiefly ice plateau

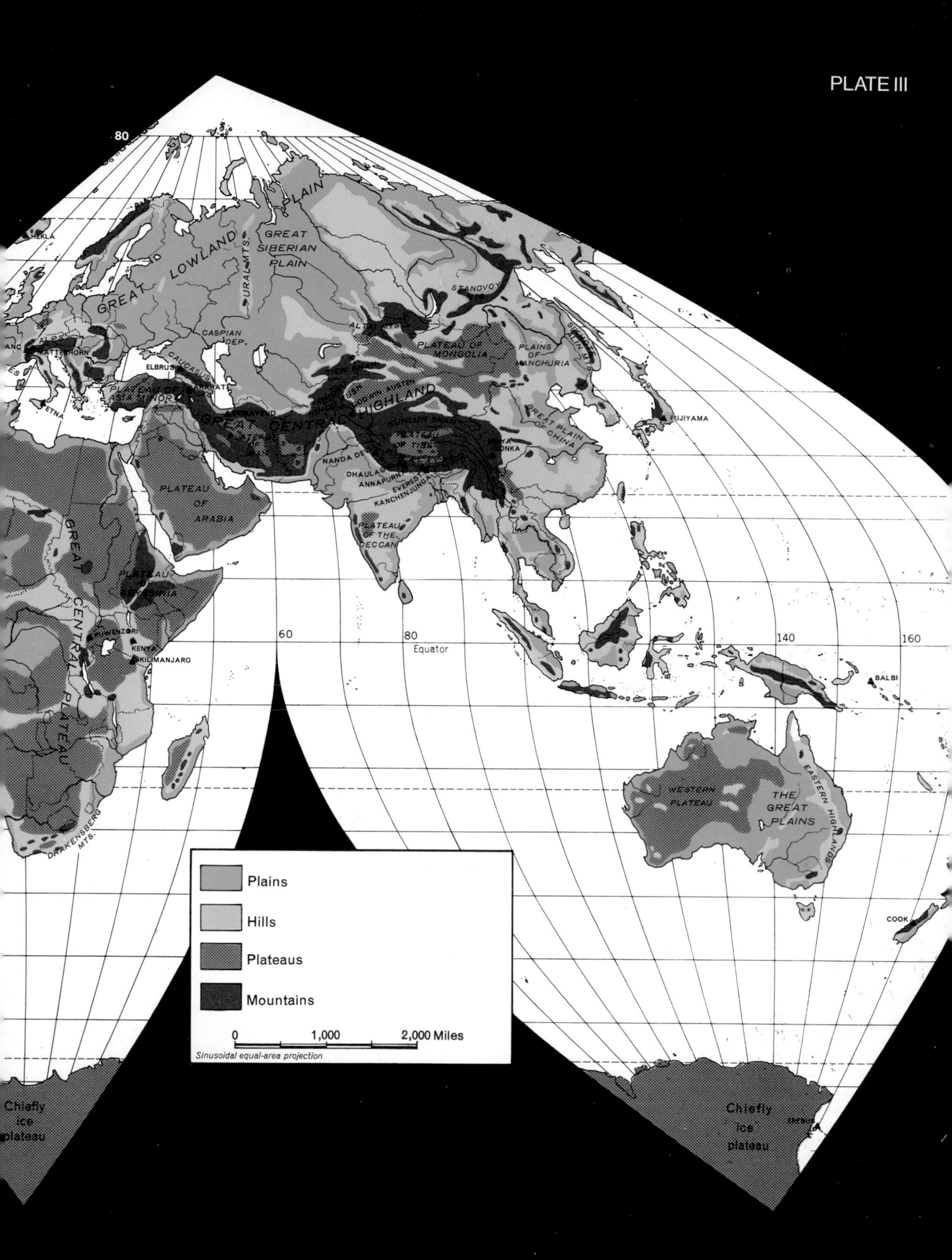
HEKLA
GREAT LOWLAND PLAIN
URAL MTS.
GREAT SIBERIAN PLAIN
STANOVOY MTS.
ALPS
MATTERHORN
CASPIAN DEP.
ALTAI MTS.
PLATEAU OF MONGOLIA
PLAINS OF MANCHURIA
ELBRUS
CAUCASUS
TIEN SHAN
PLATEAU OF ASIA MINOR
ARARAT
ETNA
GREAT CENTRAL HIGHLAND
FUJIYAMA
KUNLUN SHAN
GREAT PLAIN OF CHINA
PLATEAU OF IRAN
PLATEAU OF TIBET
NANDA DEVI
DHAULAGIRI
ANNAPURNA
EVEREST
KANCHENJUNGA
PLATEAU OF ARABIA
GREAT CENTRAL PLATEAU
PLATEAU OF THE DECCAN
PLATEAU OF ABYSSINIA
RUWENZORI
KENYA
KILIMANJARO
60
80
Equator
140
160
BALBI
WESTERN PLATEAU
THE GREAT PLAINS
EASTERN HIGHLANDS
DRAKENSBERG MTS.
COOK
Plains
Hills
Plateaus
Mountains
0
1,000
2,000 Miles
Sinusoidal equal-area projection
Chiefly ice plateau
Chiefly ice plateau

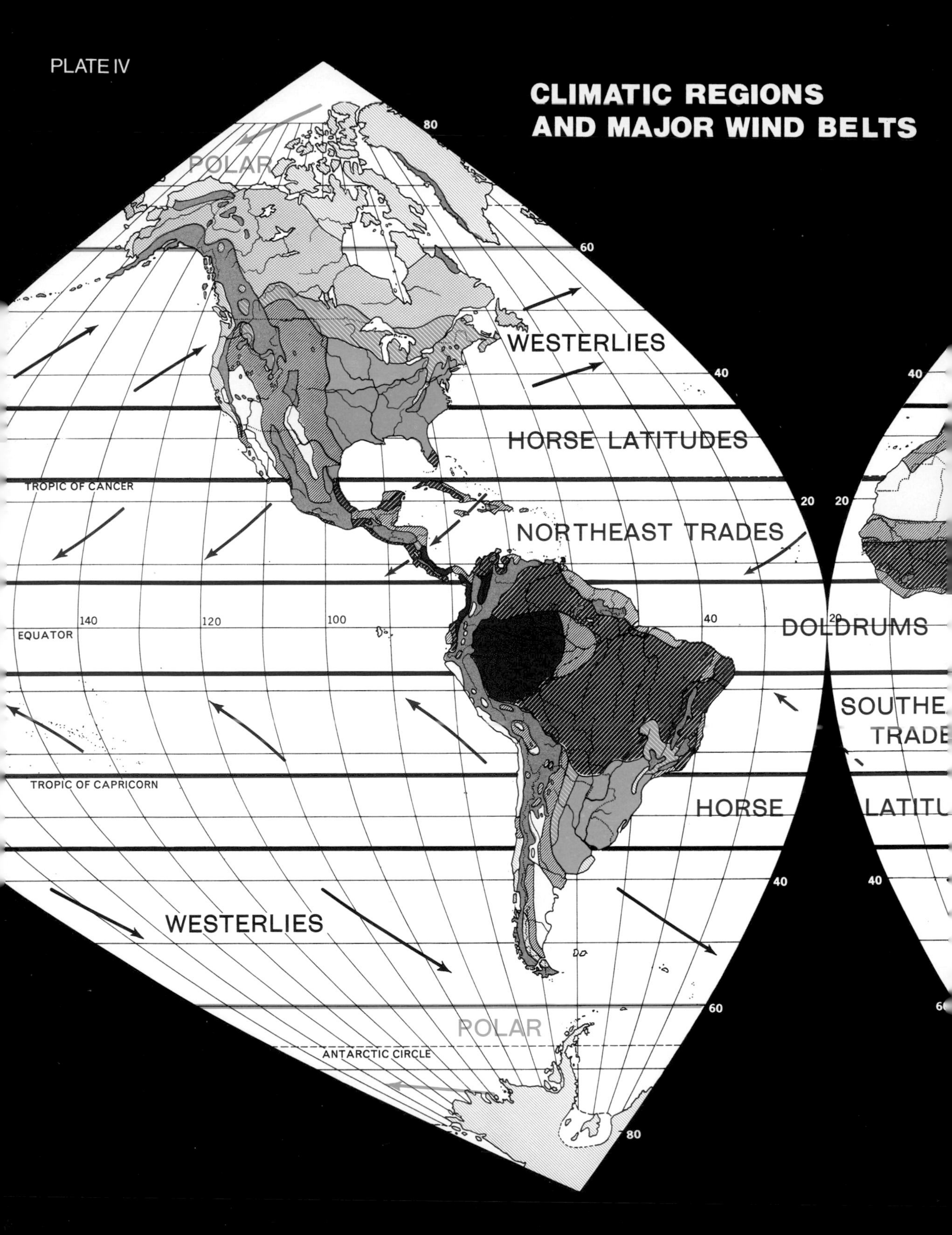
PLATE IV
CLIMATIC REGIONS
AND MAJOR WIND BELTS
POLAR
80
60
WESTERLIES
40
HORSE LATITUDES
TROPIC OF CANCER
20
NORTHEAST TRADES
EQUATOR
140
120
100
40
20
DOLDRUMS
SOUTHE
TRADE
TROPIC OF CAPRICORN
HORSE
LATITU
40
WESTERLIES
60
POLAR
ANTARCTIC CIRCLE
80

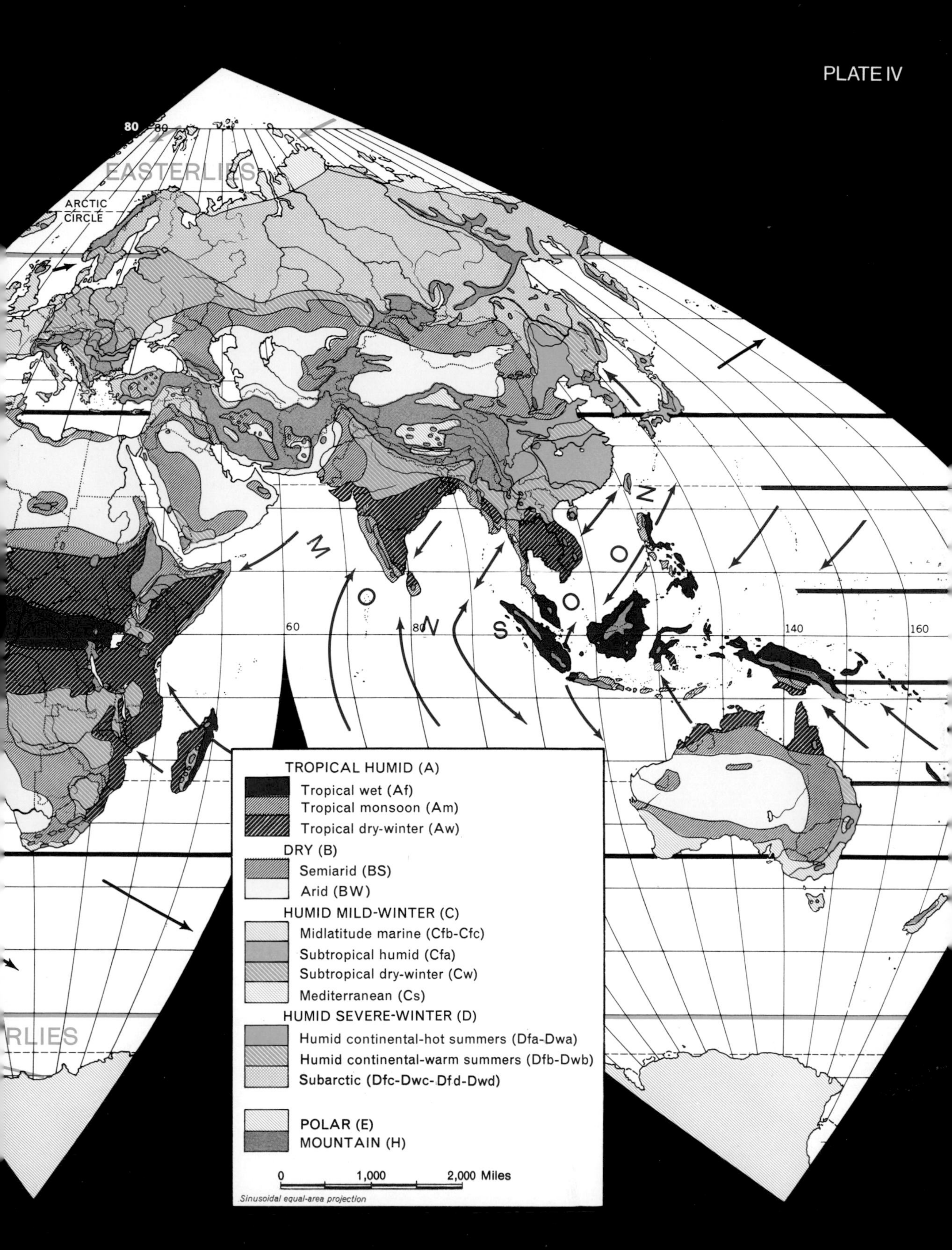
EASTERLIES
ARCTIC CIRCLE
80
80
60
80
140
160
MONSOON
RLIES
TROPICAL HUMID (A)
Tropical wet (Af)
Tropical monsoon (Am)
Tropical dry-winter (Aw)
DRY (B)
Semiarid (BS)
Arid (BW)
HUMID MILD-WINTER (C)
Midlatitude marine (Cfb-Cfc)
Subtropical humid (Cfa)
Subtropical dry-winter (Cw)
Mediterranean (Cs)
HUMID SEVERE-WINTER (D)
Humid continental-hot summers (Dfa-Dwa)
Humid continental-warm summers (Dfb-Dwb)
Subarctic (Dfc-Dwc-Dfd-Dwd)
POLAR (E)
MOUNTAIN (H)
0
1,000
2,000 Miles
Sinusoidal equal-area projection

VEGETATION REGIONS

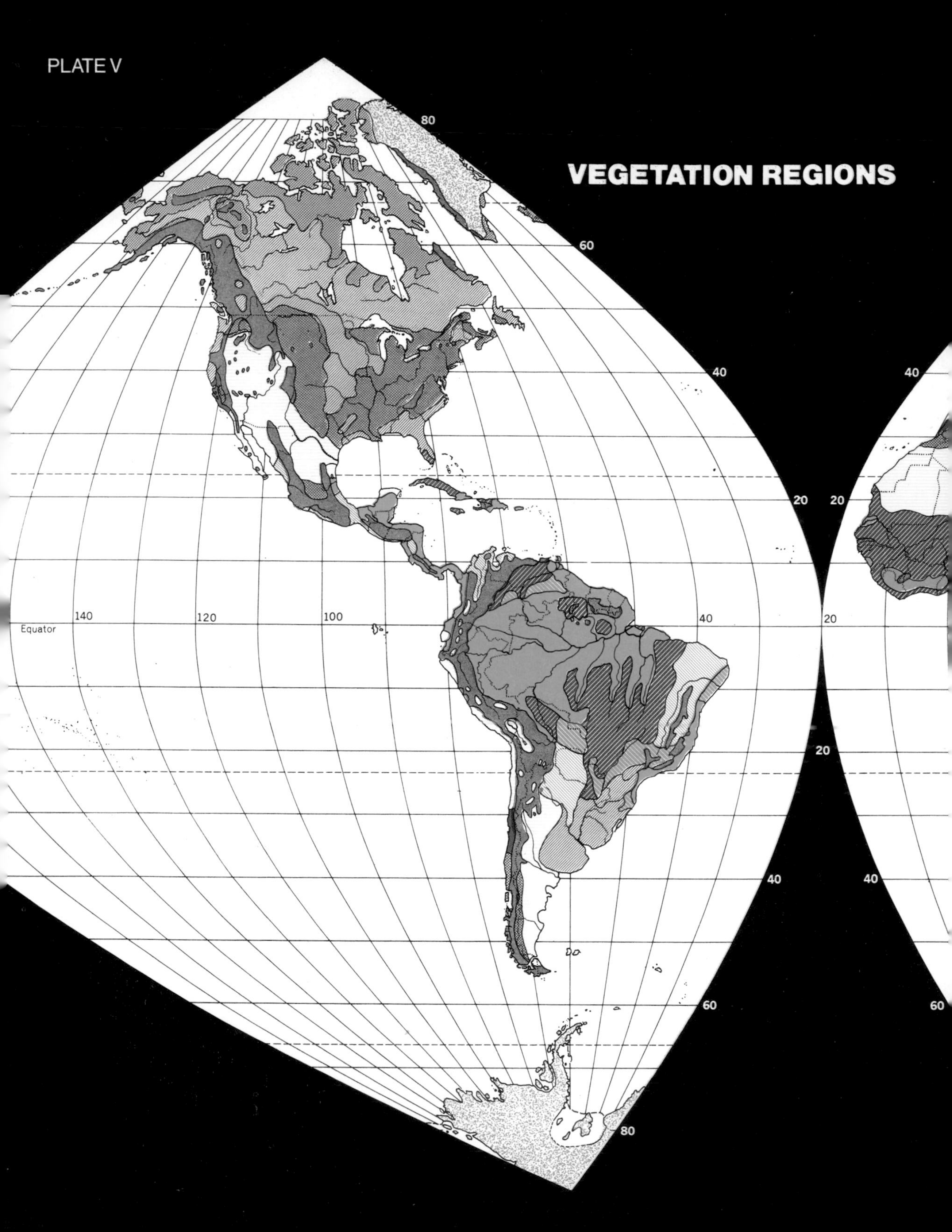

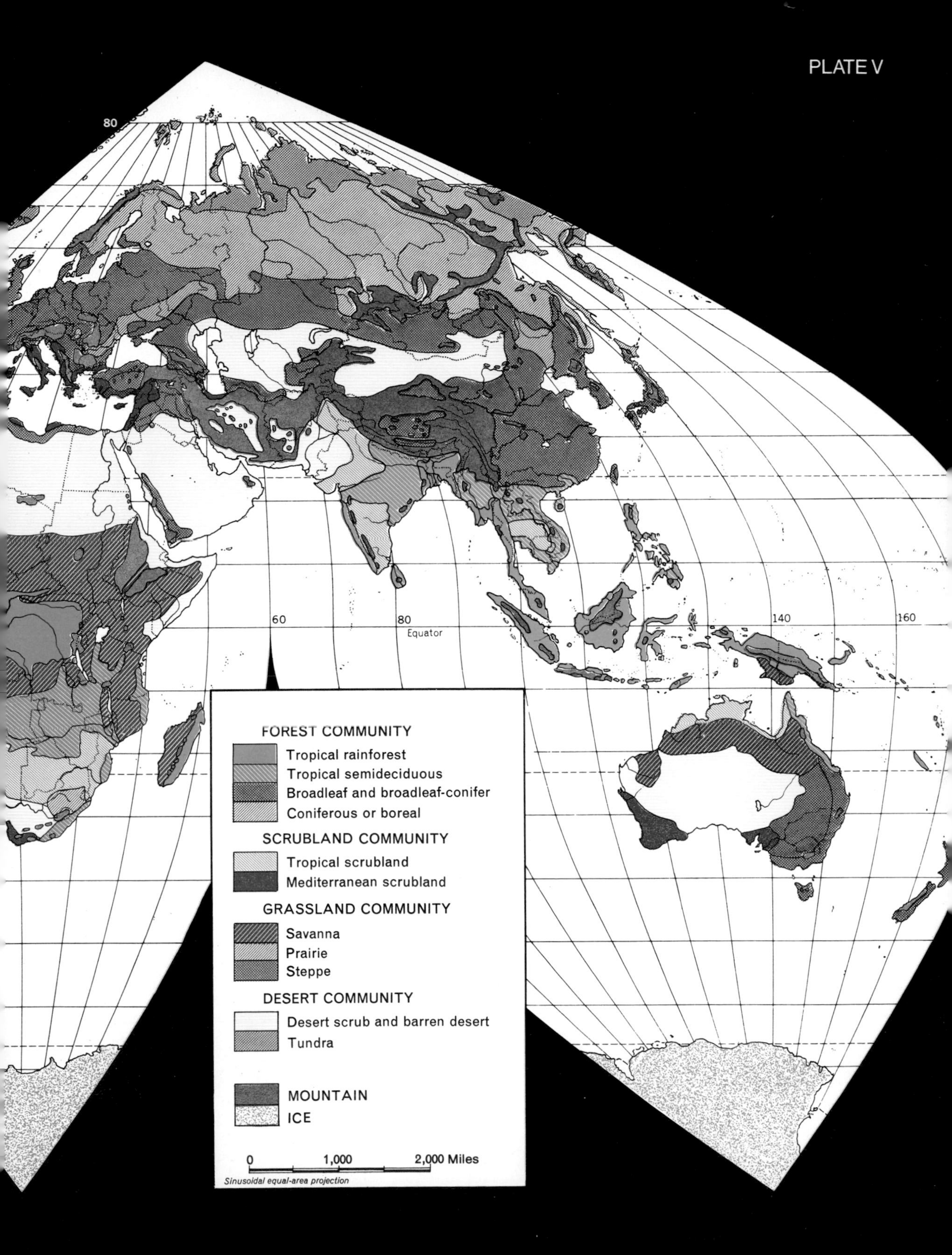
80
60
80
Equator
140
160
FOREST COMMUNITY
Tropical rainforest
Tropical semideciduous
Broadleaf and broadleaf-conifer
Coniferous or boreal
SCRUBLAND COMMUNITY
Tropical scrubland
Mediterranean scrubland
GRASSLAND COMMUNITY
Savanna
Prairie
Steppe
DESERT COMMUNITY
Desert scrub and barren desert
Tundra
MOUNTAIN
ICE
0
1,000
2,000 Miles
Sinusoidal equal-area projection

ZONAL SOILS

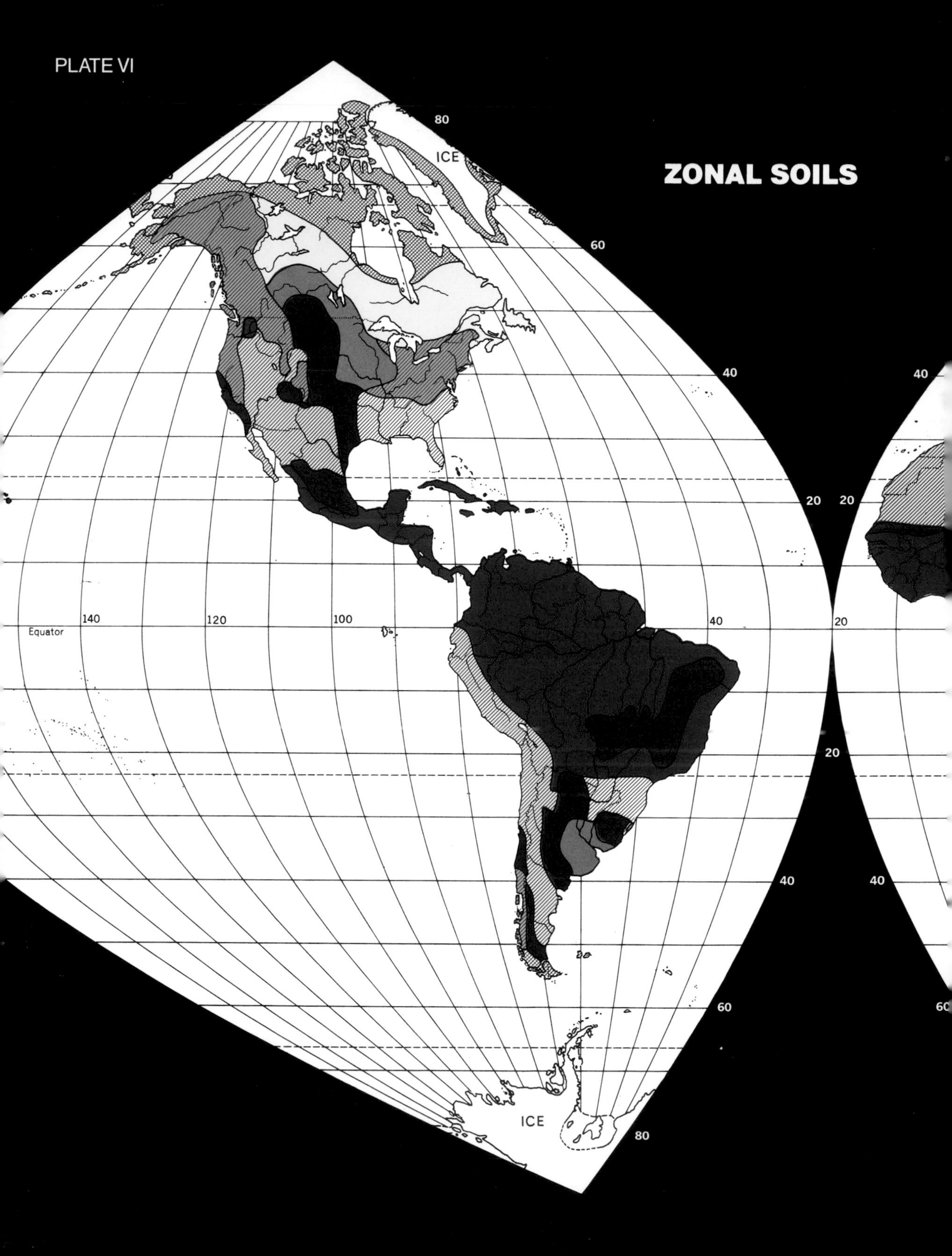

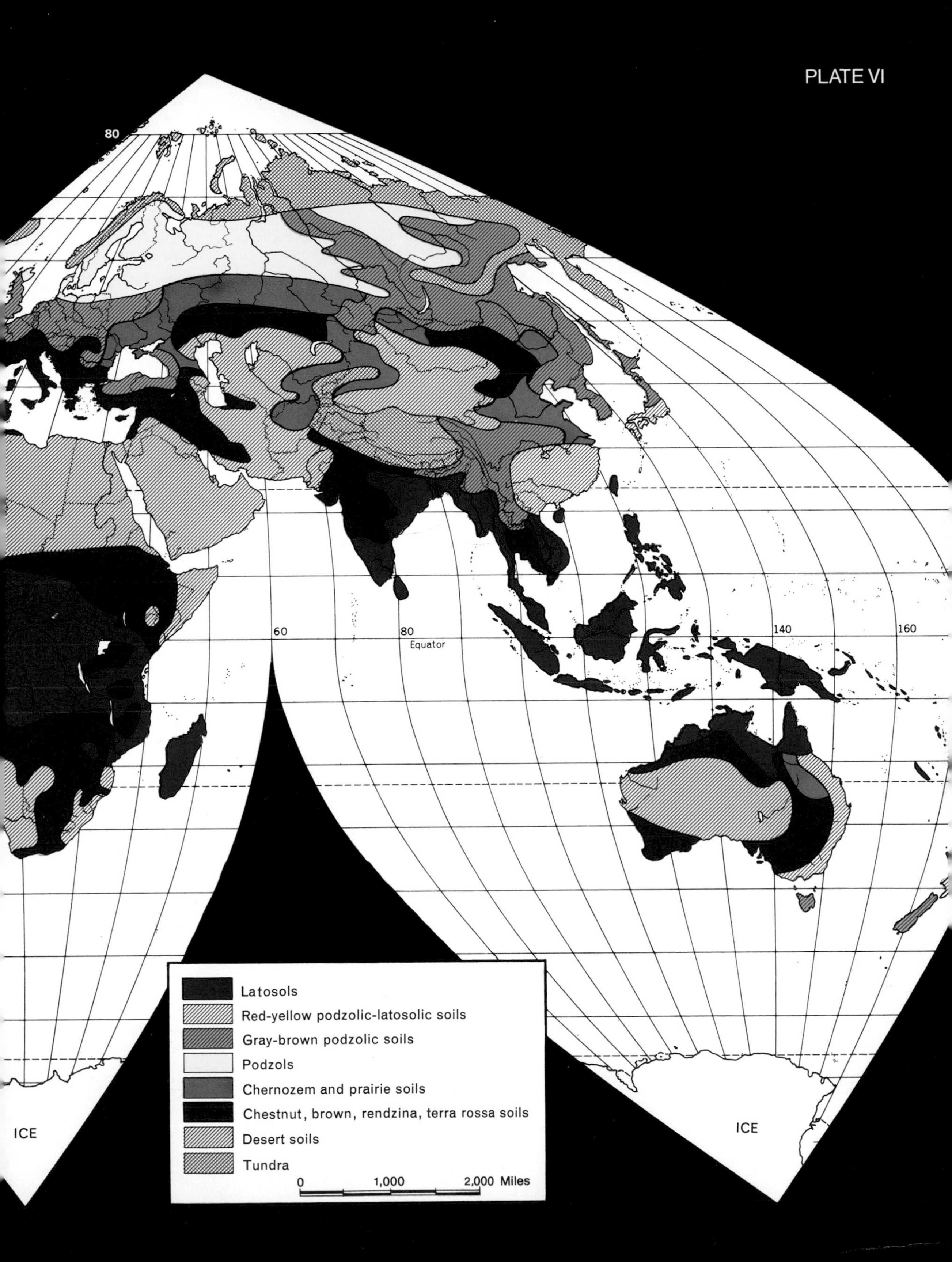
80
60
80
Equator
140
160
ICE
ICE
Latosols
Red-yellow podzolic-latosolic soils
Gray-brown podzolic soils
Podzols
Chernozem and prairie soils
Chestnut, brown, rendzina, terra rossa soils
Desert soils
Tundra
0
1,000
2,000 Miles

POPULATION DENSITY

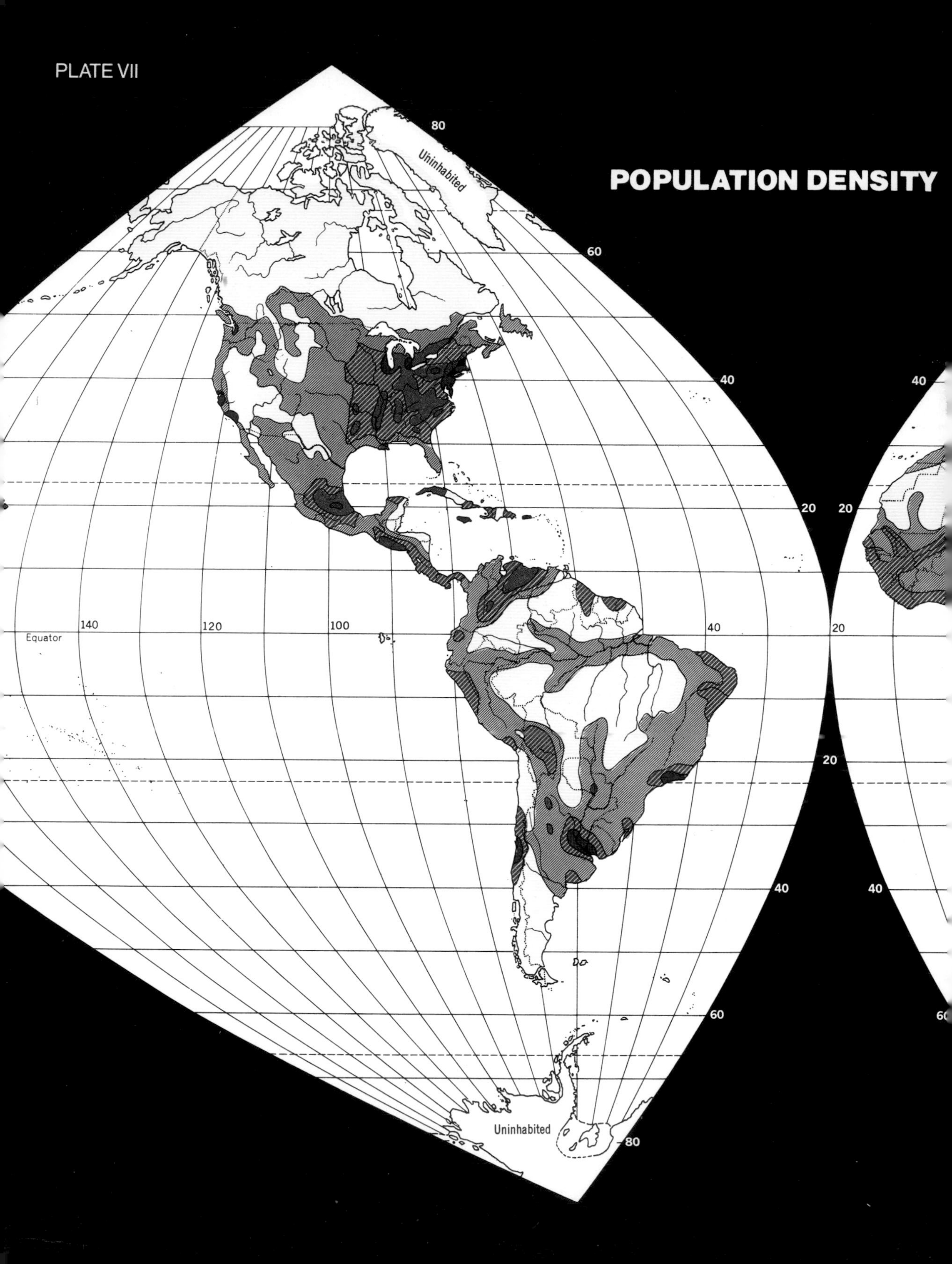

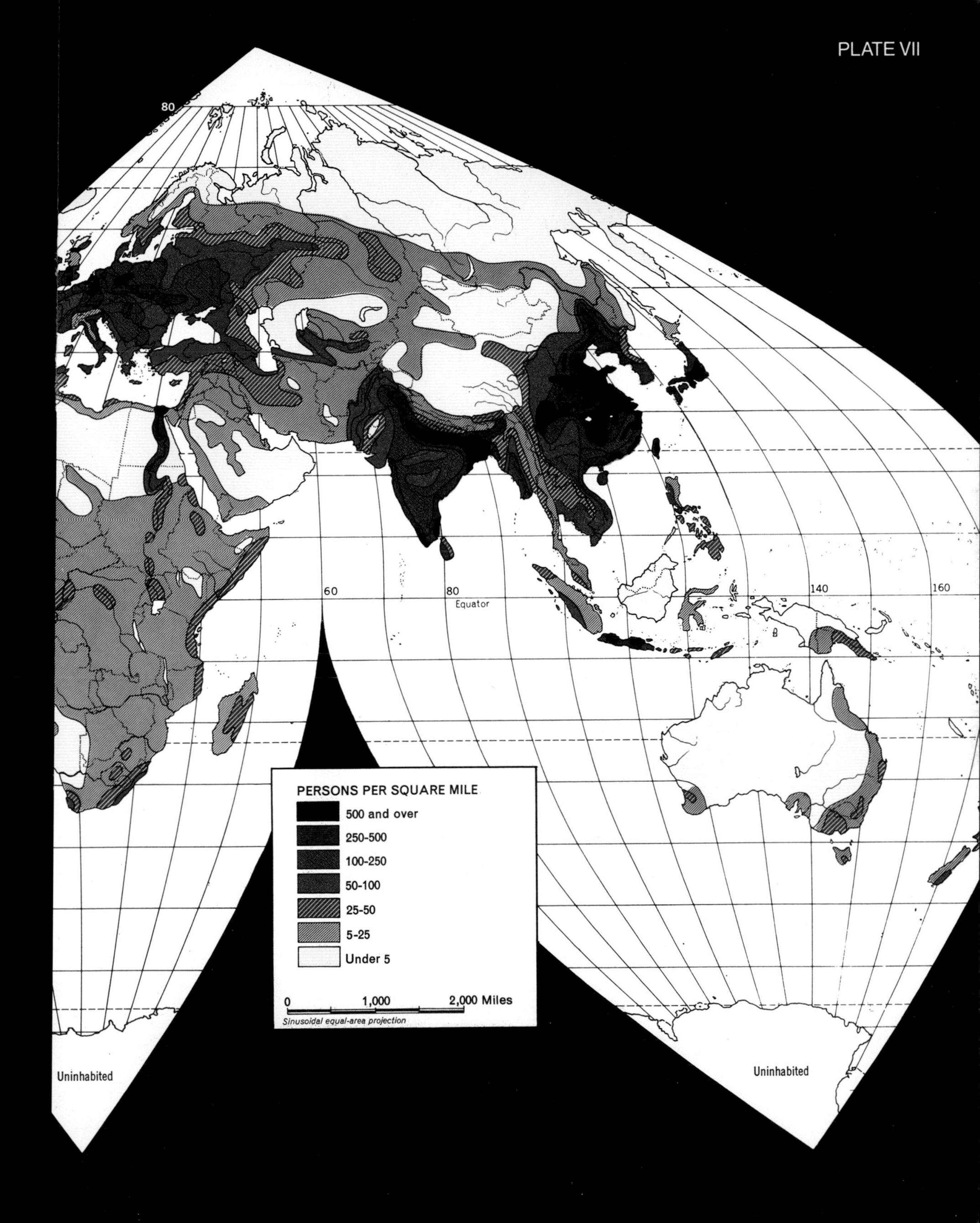
80
60
80
Equator
140
160
PERSONS PER SQUARE MILE
500 and over
250-500
100-250
50-100
25-50
5-25
Under 5
0
1,000
2,000 Miles
Sinusoidal equal-area projection
Uninhabited
Uninhabited

MAJOR PHYSICAL-GEOGRAPHIC REGIONS

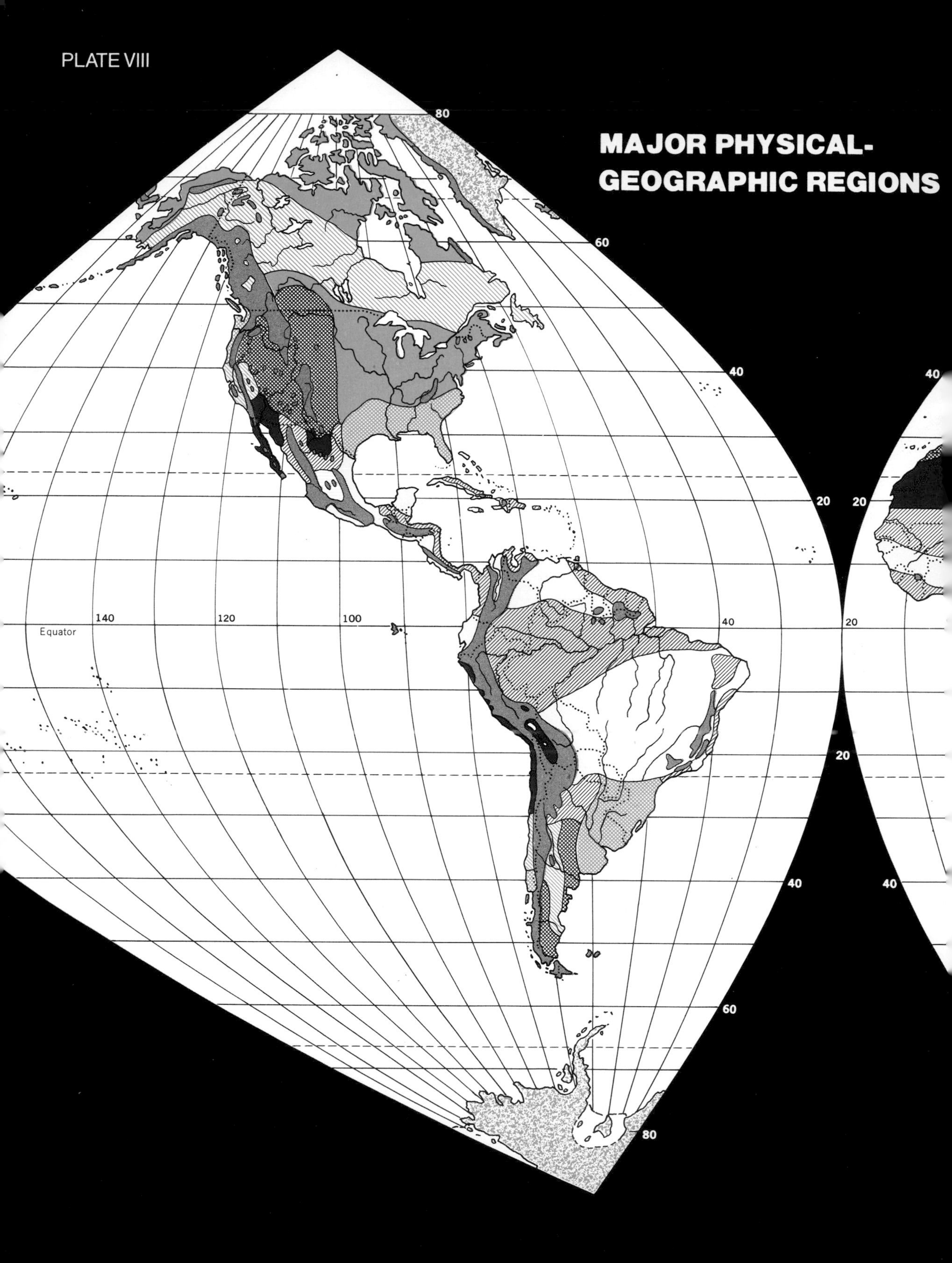

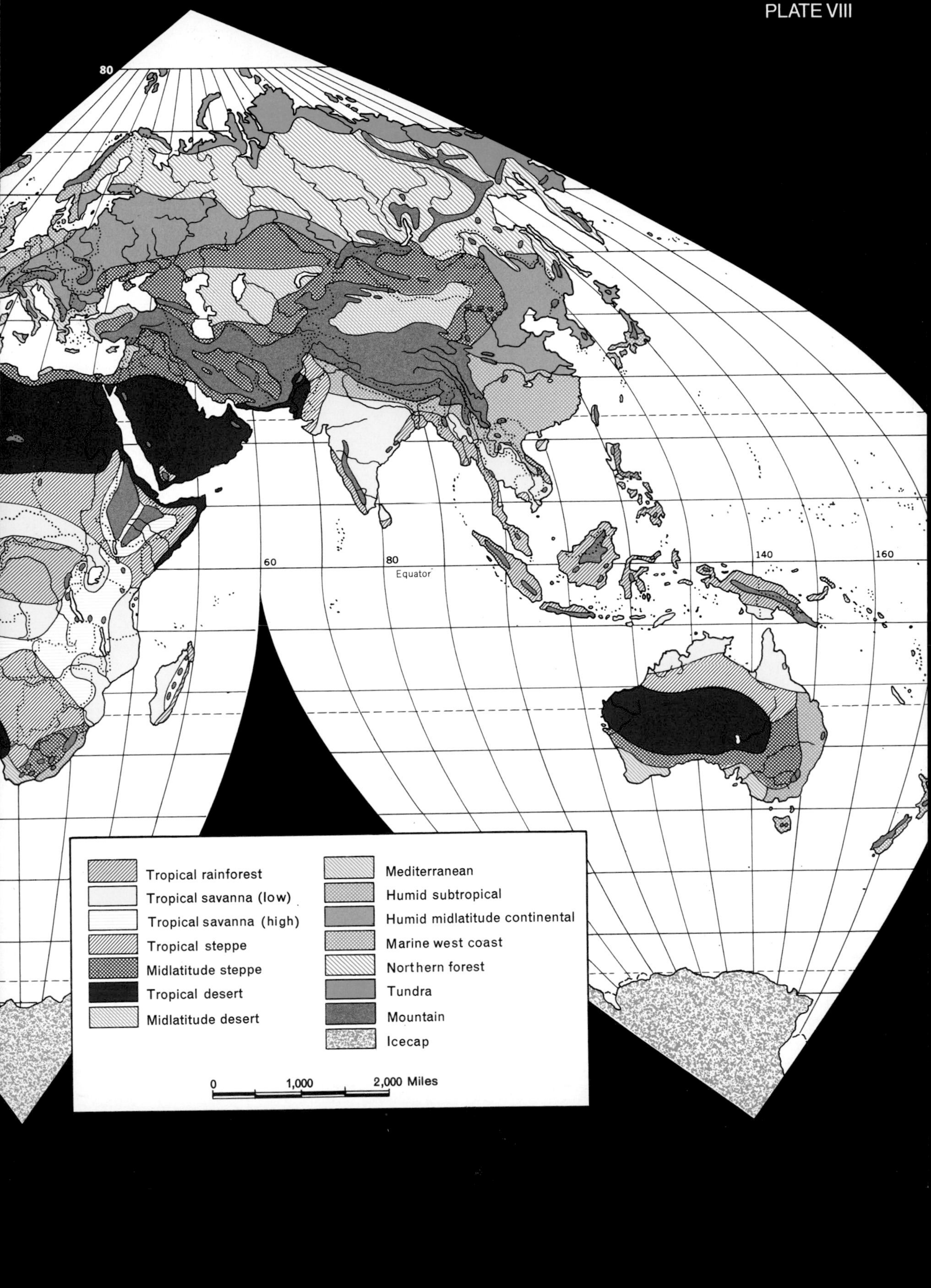
80
60
80
Equator
140
160
Tropical rainforest
Tropical savanna (low)
Tropical savanna (high)
Tropical steppe
Midlatitude steppe
Tropical desert
Midlatitude desert
Mediterranean
Humid subtropical
Humid midlatitude continental
Marine west coast
Northern forest
Tundra
Mountain
Icecap
0
1,000
2,000 Miles

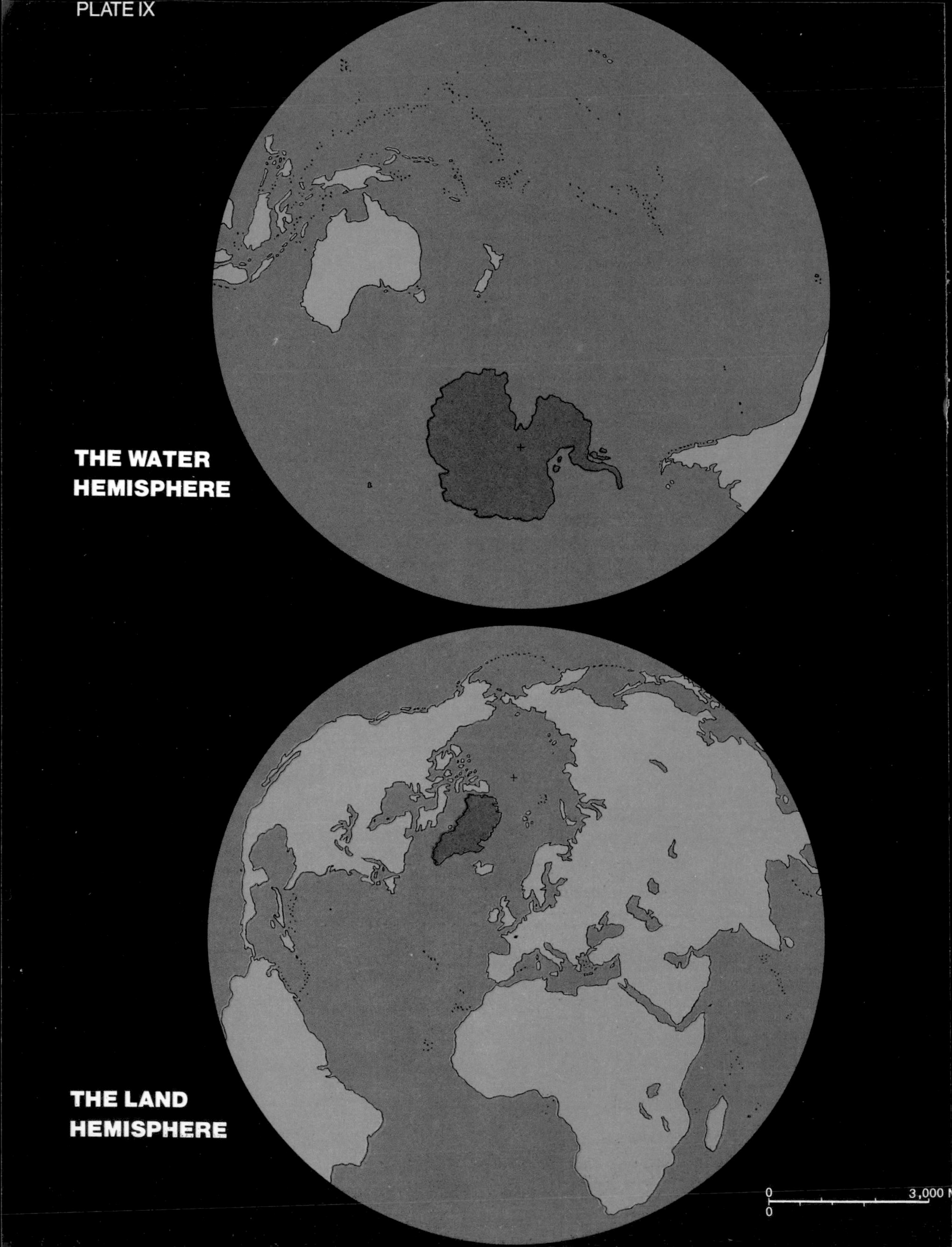
THE WATER
HEMISPHERE
THE LAND
HEMISPHERE
0
3,000
0

Maps

Color Plates

PART TWO *Cultural Geography*

Part One

Physical Geography

Mountains and valleys viewed from Mt. Saentis in northeastern Switzerland. (Swiss National Tourist Office)

1

The Geographic Complex

Physical Geography

The study of geography can be divided into two parts: the study of the earth's environment, or *physical geography,* and the study of the relationships between people and that environment, or *cultural geography.* The next fifteen chapters will deal largely with physical geography, but they will occasionally draw examples from the cultural area in order to emphasize or clarify points set forth in the discussion.

The earth's physical environment—the human habitat—is composed of the *hydrosphere,* the *lithosphere,* the *atmosphere,* and the *biosphere,* or, in other words, water, land, air, and most living things. More specifically, the major physical elements that compose the physical environment are water, landforms, climate, vegetation, and soil (Fig. 1–1).

Each of these elements has great significance in its own right and may be studied separately as the subject of an independent branch of science. Thus there is *hydrology* (the study of water), *geomorphology* (the study of landforms), *climatology* (the study of the atmosphere), *plant geography* (the study of plant locations), and *pedology* (the study of soils); taken in combination, they form a single closely integrated system known as the *physical complex.*

In its most encompassing form, a study of the physical complex involves no less than a study of the entire world; for the earth, with its surrounding envelope of atmosphere, is essentially a closed system operating endlessly within itself, and developments affecting one portion of the complex in one part of the world may bring about marked changes in another portion of the complex in some far-distant region. On a smaller scale, the complex can also be studied by focusing on the environment of a particular geographic area. This may involve an entire continent, such as North America, or a subdivision thereof, such as the Midwest, the Corn

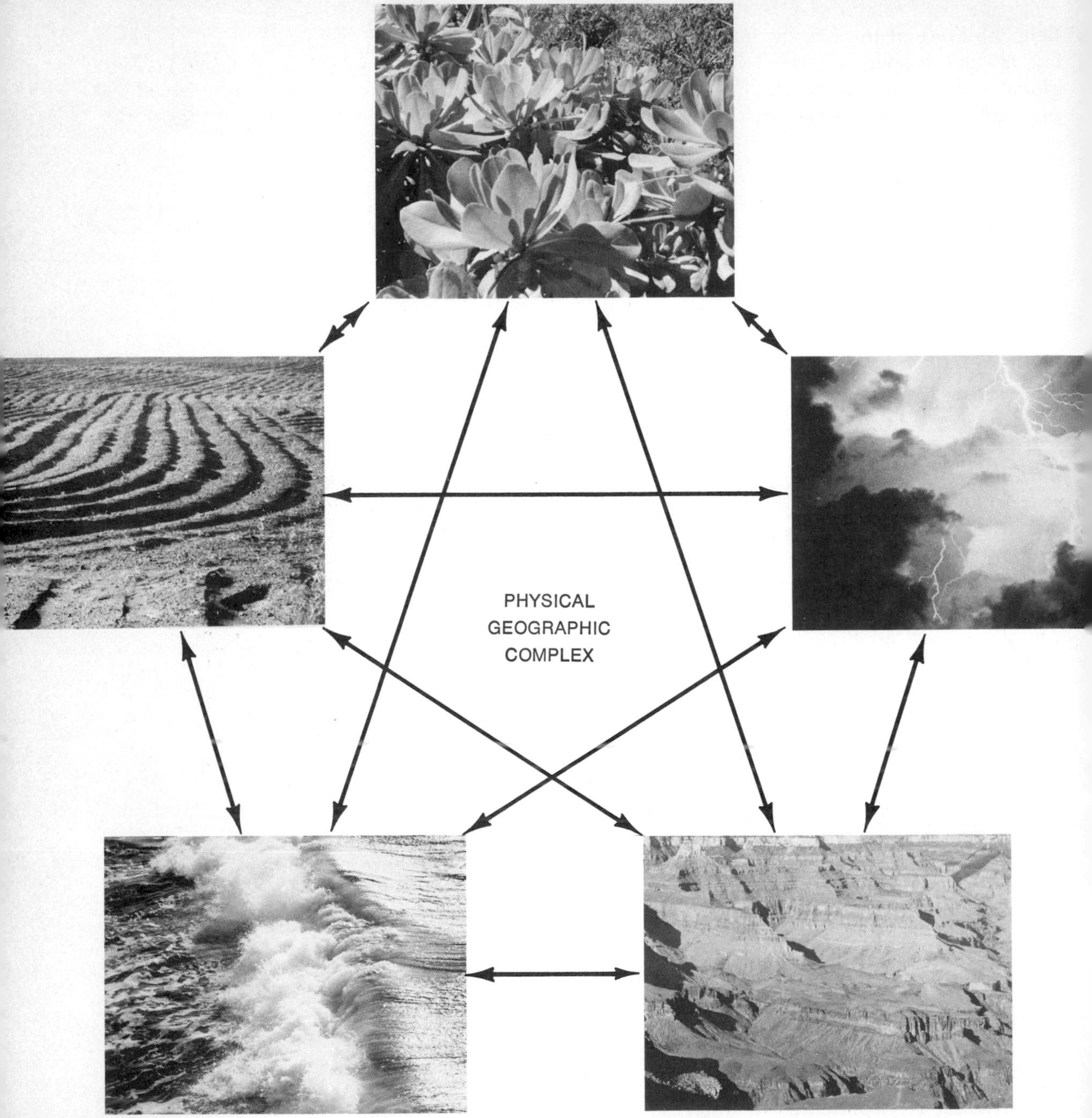

PHYSICAL
GEOGRAPHIC
COMPLEX

FIGURE 1–1

The major physical elements—water, landforms, climate, vegetation, and soil—which, together, form the *physical complex.* (Respectively, Ray Atkeson; Santa Fe Railway; NOAA; Hawaiian Visitor's Bureau; Washington State Department of Commerce)

Belt, the State of Illinois, the Kankakee Valley, or even a specific farm. But whether in global terms or in microcosm, a study of the physical complex makes possible both an understanding of each of the earth's elements and—more important, if the pollution of the environment is to be reversed—an appreciation of their endless interaction.

The Human Significance of Physical Geography

The study of physical geography can be informative, interesting, and rewarding. For most people, however, it is of interest primarily in terms of its direct relationship to their ability to make a living. It is as the basic component of their day-to-day existence that the conditions of the physical environment, especially those of the water, land, and air, strike home.

Until quite recently, the basic physical conditions within the environment were generally ignored by most people, or were simply taken for granted. Water was needed for many domestic and occupational purposes, but there seemed to be water in untold abundance; land was needed to build on and to grow crops and trees on, but there seemed to be land in inexhaustible supply; and air was needed for the support of life, but there seemed to be usable air everywhere.

Today the picture has changed drastically and surprisingly. In the millions of years of human existence, people have increased their numbers, spread across the land, and expanded their living space; at the same time, they have worn out and poisoned soils, used up or rendered nonusable much of the available water, and polluted much of the air (Fig. 1–2).

Much of this use and abuse of the physical environment has occurred within the past century and results primarily from the recently burgeoning population—it more than doubled between 1900 and 1970—and the ever-growing demands of modern industrial society. As more and more people have demanded more food, clothing, housing, tools, machines, luxuries, and the like, the conditions and resources of water, land, and air have been altered to the point that widespread resource deficiencies and critical pollutions either exist or are dangerously imminent.

Now we are confronted with a situation that literally forces us to take stock and formulate plans to insure our continued existence. Such resource inventory and planning, even in its political, social, and economic aspects, must be based primarily upon the present conditions of, and the future possibilities for, the physical environment.

It is not the basic objective of this work to assess, or suggest solutions for, the world's present problems, nor to discuss the manifold and global aspects of human ecology. Rather, the objective is to provide a general portrait and evaluation of the basic background factors and conditions of the physical environment, including, as one important factor, pertinent recognition of the modifications of that habitat by human beings.

A study of physical geography should not be limited solely to the reading of this work, nor to such reading augmented by the usual laboratory experience. Rather, it must

FIGURE 1–2

Human spoliation of the physical environment. Top left, the polluted South Platte River near Denver, Colorado, 1965. (Wide World Photo) Top right, an oil-covered gannet. (G. Komorowski, National Audubon Society) Center left, an eroded farmland in Tennessee. (USCS) Center right, a land area ridged by surface mining near Du Quoin, Illinois. (National Coal Association) Bottom left, smog-enshrouded St. Louis, April 10, 1966. (St. Louis Post Dispatch) Bottom right, logged Douglas fir trees in the State of Oregon. (U.S. Forest Service)

also involve a firsthand analysis of the environment in which we live. Unless you take your study beyond the classroom and printed materials—unless you consciously make it a part of your daily life—it will be little more than an academic exercise. Discipline yourself to observe, comprehend, and evaluate your physical world. In so doing, not only will you enhance your awareness and appreciation of the physical environment but you will also be in a much better position to assess and evaluate intelligently the many aspects and problems of human inhabitation of the earth.

REVIEW AND DISCUSSION

1. What constitutes the physical environment—the human habitat?
2. In relation to your own frame of reference, discuss the advantages of studying the physical and cultural elements of the habitat separately, with only occasional cross references.
3. Are the limits of the habitat fixed permanently, or do they alter? If alterations do occur, how are they accomplished?
4. What do you understand by the statement, "The physical complex embodies the entire world"?
5. On the basis of your previous readings and observations, justify the contention that "much of the use and abuse of the physical environment has occurred within the past century."
6. What occurrences of recent decades helped bring about the situation that is now forcing people to accelerate the study and understanding of the physical environment and to better provide for its preservation and continued functioning?
7. Observe and evaluate the physical environments in which you live, work, and travel. Are they being improved, or impaired? How would their deterioration affect your future?

SELECTED REFERENCES

Ackerman, E. A., Chairman. *The Science of Geography* (Report of the *ad hoc* Committee on Geography, National Academy of Science—National Research Council.) Washington, D.C., 1965.

Bowman, I. *Geography in Relation to the Social Sciences.* New York: Scribner, 1934.

Broek, J. O. M. *Compass of Geography.* Columbus, Ohio: Merrill, 1966.

Freeman, T. W. *A Hundred Years of Geography.* Chicago: Aldine, 1962.

Hartshorne, R. *Perspective on the Nature of Geography.* Chicago: Rand McNally (for the Association of American Geographers), 1959.

James, P. E., and C. F. Jones, eds. *American Geography: Inventory and Prospect.* Syracuse, N.Y.: Syracuse University Press (for the Association of American Geographers), 1954.

Leighly, J., ed. *Land and Life, A Selection from the Writings of Carl Ortwin Sauer.* Berkeley, Calif.: University of California Press, 1963.

From the moon the earth is a very small orb almost lost in the black void of space. (NASA)

2

Planet Earth

The Earth in the Universe

To an individual standing on its surface, the earth appears very large and complex. But to an astronaut looking at the earth from the face of the moon, it seems a very small orb that is almost lost in the black void of space.

Indeed, the earth is a mere speck within the solar system, which in turn is but a minor part of the Milky Way galaxy; and the Milky Way itself is but one of countless galaxies within the universe. Yet, as comparatively minute as it is, the earth is the "life planet" of the universe—at least in terms of the types of life with which human beings are familiar; and it is, for all practical purposes, humankind's total physical habitat.

The Earth in the Solar System

Location in Space

The earth is the third of nine major planets spaced outward from the star that man calls the sun (Fig. 2–1). Its mean distance from the sun is about 93,000,000 miles. Its closest neighbor, averaging about 239,000 miles distant, is its own satellite, the airless, lifeless moon. Its closest neighbors among the major planets are Venus and Mars, which are about 25,800,000 and 48,600,000 miles away, respectively. They, like the moon and the other planets, are lifeless (as far as we know); but, unlike the moon, they are sur-

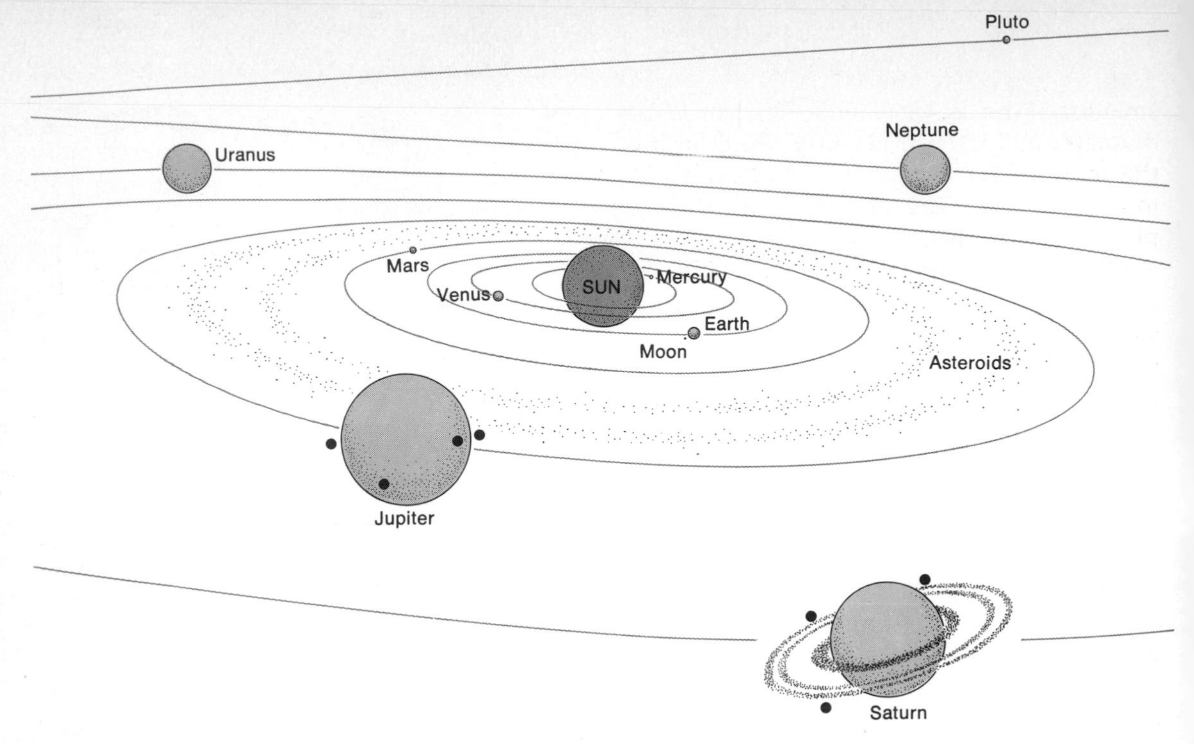

FIGURE 2–1

The nine planets and the asteroids of the solar system are shown here in their relative sizes and orbits around the sun. (The sun is not drawn to scale, but is vastly reduced.)

rounded by "gaseous envelopes," or atmospheres, which though of a different composition than the earth's atmosphere, have prompted speculation that Venus and Mars may possess some very rudimentary (or at least different) life forms.

The earth, in its perennial journey around the sun, is held to its location in space by a fine balance between two opposing forces: *gravitational attraction* and *centrifugal force*. If the sun's gravitational attraction were greater, the earth would be drawn into the fiery gases of the sun and destroyed; if centrifugal force were greater, the earth would literally fly off into space.

The most significant aspect of the earth's placement within the solar system is that the planet is neither too close to nor too far away from the sun. For all practical purposes, the sun is the sole source of heat and light for the earth and for the other planets as well. If the earth were much closer to the sun, the planet would be burning hot; if it were much farther away, it would be frigidly cold. For example, Mercury, the planet closest to the sun, has a "sunny side" temperature estimated at from 500° to 700°F.—high enough to melt tin or lead; and Pluto, the planet farthest from the sun, is thought to have a temperature of some −370°F.

Size and Shape

The earth has a circumference of nearly 25,000 miles, a diameter of nearly 8,000 miles, and a surface area of nearly 197,000,000 square miles. In comparison, Pluto, the smallest of the major planets, has a probable diameter of 3,100 to 3,600 miles, and Jupiter, the largest, has a diameter of about 88,700 miles (Fig. 2–1). The sun, dwarfing all of its planetary satellites, has a diameter of nearly 865,000 miles. If the earth were placed at the sun's center, the distance from the earth's surface to that of the sun would be a bit more than 428,000 miles. Even the orbit of the earth's moon would be some 189,000 miles beneath the sun's surface.

For most study purposes, the earth may be considered a sphere. Actually it bulges slightly in the equatorial region and is slightly flattened in the polar sections. Thus, technically, the earth is essentially an oblate spheroid: the oblateness, or "equatorial bulge," being due to the effect of centrifugal force on a rapidly rotating planet that is not composed of absolutely rigid matter. The departure from true sphericity is shown by the fact that there is a difference of about 27 miles between the polar and the equatorial diameters (Fig. 2–2). If shown to scale on an 18-inch globe, this difference would be only .06 of one inch—too slight for the eye to discern. Another kind of departure from true sphericity is the result of the irregularity of the surface of the earth's crust. However, between the crust's extremes, even as measured from the highest mountain to the greatest ocean depth, there is a vertical difference of only about 66,800 feet (Fig. 2–3)—so slight as to represent no more than the thickness of the paper on a standard 18-inch globe.

FIGURE 2–2

Measurements of the earth.

Surface area (*in square miles*)		
oceans and seas		139,715,400
land (including inland waters and Antarctica)		57,225,000
	Total	196,940,400
Polar measurements (*in miles*)		
diameter		7,899.99
circumference		24,818.60
Equatorial measurements (*in miles*)		
diameter		7,926.68
circumference		24,902.45

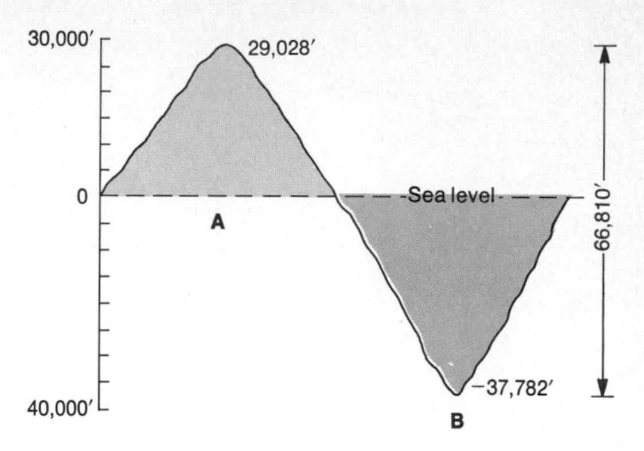

FIGURE 2–3

A schematic profile of the highest and lowest points on the surface of the earth's crust. A. Mt. Everest in the Himalayas. B. The Mindanao Trench just east of the Philippines.

Motions: Revolution, Rotation

The sun, and hence the earth and the other planets of the solar system, is hurtling through space at some 43,000 miles per hour. Although significant astronomically, such motion is of no geographical consequence, for the system moves as an integral whole. Geographically significant, however, are the major motions of the earth within the solar system: its *revolution* and *rotation.*

Revolution refers solely to the earth's continual orbiting around the sun. The time required for each complete revolution, at an average earth velocity of about 66,600 miles per hour, is the period that is known as one year (Fig. 2–4). The earth's *path of revolution,* or its *orbit,* is slightly elliptical, rather than truly circular, and the sun is located at one of

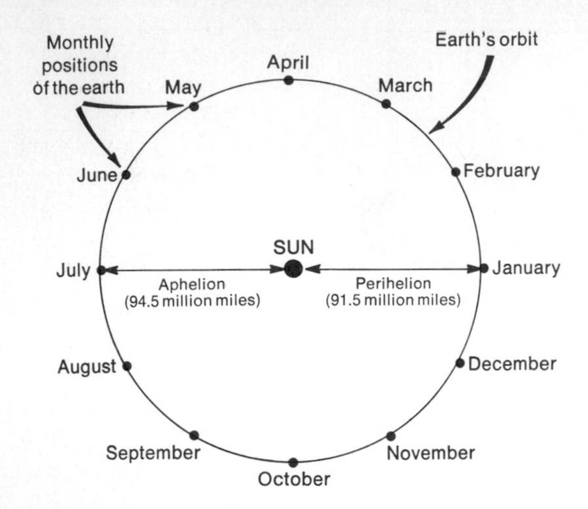

FIGURE 2–4

The revolution of the earth around the sun. (The sun here is not drawn to scale.)

the foci of the ellipse, rather than at the center of a circle. As a result, the distance from the earth to the sun varies somewhat. When at its closest, in the relationship called *perihelion* (occurring about January 1), the earth is about 91.5 million miles from the sun; at its farthest, in the relationship called *aphelion* (occurring about July 1), it is about 94.5 million miles from the sun (Fig. 2–4). The difference in distance, even though it is some three million miles, has no real geographical significance: it is not great enough to have any important effect on the receipt of solar energy by the earth, nor is it enough to upset the basic gravitational relationship between the sun and the earth.

Rotation refers solely to the essentially uniform spinning of the earth on its axis as it revolves about the sun. The time required for one complete earth rotation is designated as one day. The *speed of rotation* is such that the earth presents a constantly changing face to the sun, thereby creating a continual migration of the dividing line between the hemispheres of light and darkness. If the earth rotated at such a speed as to always present the same face to the sun, there would be a permanent hemisphere of darkness (night) and a permanent hemisphere of light (daytime), or, in other terms, a permanently cold hemisphere and a permanently hot hemisphere. The *direction of earth rotation* is toward what has been designated as the east, and thus the sun appears to rise in the east and set in the west; and each day (daylight period) and night marches westward around the earth (Fig. 2–5).

FIGURE 2–5

The rotation of the earth on its axis. A single source of light (the sun) plus the rotational speed of the planet (causing it to present a constantly different face to the sun) create daylight and darkness—one complete rotation being the 24-hour period called a day.

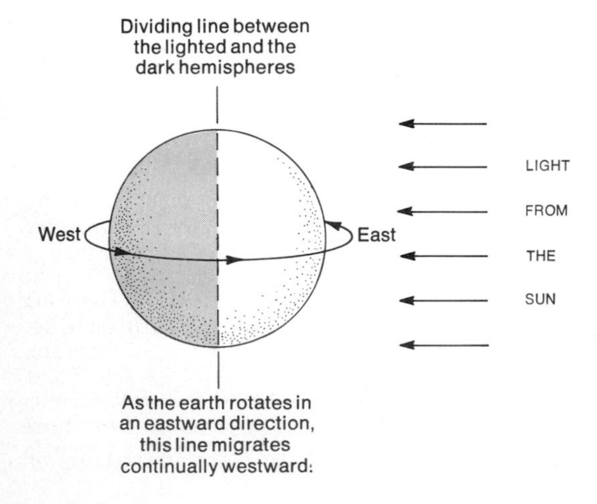

Attitude: Inclination, Parallelism

As the earth revolves around the sun, its axis is inclined at an angle of 23½° to the plane of its orbit (Fig. 2–6). This *inclination* remains constant, and thus the earth's axis is always parallel to any of its former or subsequent positions.

Because of the *parallelism* of the earth's axis, in addition to the facts of inclination

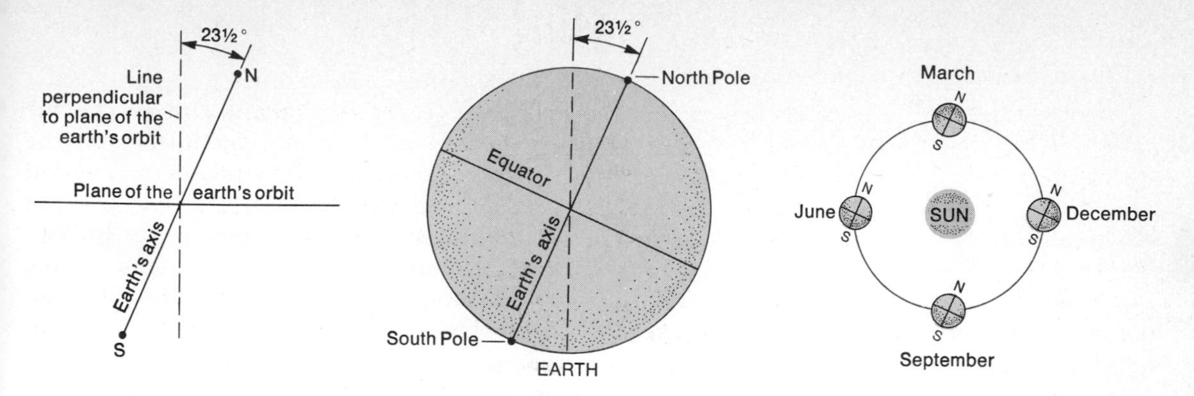

FIGURE 2–6

As the earth rotates, it does not "stand up straight." Rather, its axis is always inclined approximately 23½° from a line perpendicular to the plane of the earth's orbit.

and revolution of the earth around the sun, the Northern Hemisphere is tipped toward the sun for half of the year and away from the sun for the other half (Fig. 2–7). And, in a complementary sequence, the same is true for the Southern Hemisphere. Thus, there is a period of the year when the Northern Hemisphere receives more sun energy than the Southern Hemisphere, and an opposite period when the situation is reversed. The result of this is that the earth has seasons: a summer season in whichever hemisphere is tipped toward the sun, and a winter season in the hemisphere tipped away from the sun. The cause of seasons on the earth may be expressed in a kind of shorthand formula: revolution + inclination + parallelism = the seasons.

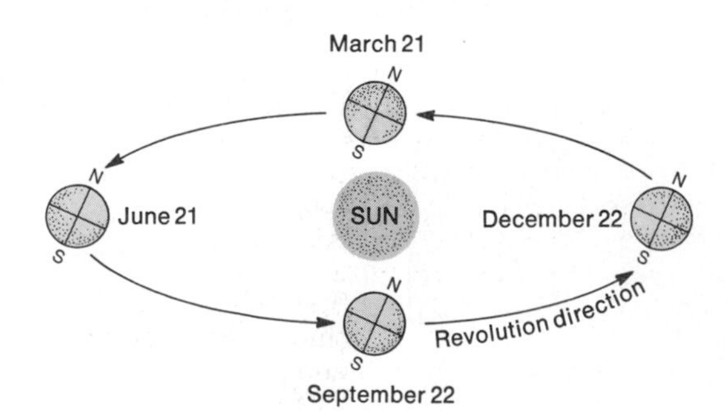

FIGURE 2–7

The Northern Hemisphere is tipped most directly toward the sun on June 21 or 22. This is a relationship called the *summer solstice* of the Northern Hemisphere. At the same time, the Southern Hemisphere is tipped most directly away from the sun. This is a relationship called the *winter solstice* of the Southern Hemisphere. On December 21 or 22, the situation is reversed, with the Northern Hemisphere experiencing its winter solstice and the Southern Hemisphere its summer solstice. Midway between the solstices, on March 21 or 22 and September 22 or 23, the Northern and Southern hemispheres are neither tipped toward, nor away from, the sun. These dates mark the *equinoxes:* the September equinox is the fall, or *autumnal,* equinox of the Northern Hemisphere, and, simultaneously, the spring, or *vernal,* equinox of the Southern Hemisphere; the March equinox is the vernal equinox of the Northern Hemisphere and the autumnal equinox of the Southern Hemisphere.

The Earth's Physical Face: An In-Space View

Physical Variety

The foregoing discussion suggests the nature and variety of humankind's planetary habitat. It notes that the earth is a spherical, ever-curving planet marked by an endless succession of days and nights, yearly solar orbits, and seasonal changes within each period of orbit. Constant in all of this is the vital receipt of solar energy, which provides the earth with light and heat and thus literally fuels its life and nonlife systems, and the existence of a sun mass which holds the earth to its perpetually orbital journey.

If one takes a vantage point in space, far enough away to see the whole earth, yet close enough to distinguish major earth features and conditions, many other aspects of the planet's variety emerge (Fig. 2–8). One notices that the earth is enveloped by a broken and continually changing cloud blanket, and through large and small breaks in the clouds, one can see waters and lands. As the earth turns, it becomes obvious that most of it is covered by extensive oceans, lesser seas, and occasional large lakes. Obvious, too, is the fact that the Southern Hemisphere is far more a "water hemisphere" than is the Northern Hemisphere, and that the Pacific Ocean alone covers nearly half of the earth. The lands seen vary in size from the continental masses to the larger islands, with the Europe-Asia-Africa landmass dwarfing all others. Also, in viewing the landmasses, it is apparent that there is relatively little land area for people south of the equator, and a major portion of that which exists is the bright snow and ice realm of Antarctica—an area about 1.7 times the size of Australia, and about 1.4 times the size of the United States.

FIGURE 2–8

The earth from the Mediterranean Sea area to the Antarctic (south polar) ice cap. Through the broken cloud cover, most of the coastline of Africa and the Arabian Peninsula can be seen. The Asian mainland lies on the horizon toward the northeast. (NASA)

Further viewing of the landmasses from a position in space discloses, however dimly, that there are extensive regions of rough mountains and hills, as well as large areas of smoother plains and plateaus. Furthermore, different colorations from one land region to another indicate different kinds of surface cover—ranging from the darks of the major forest areas, to the yellow-orange-brown pastels of the arid lands, to the whites of the snow cover in polar and near-polar regions.

From space, one does not see the earth's climates and weather conditions, but one is given numerous clues that help to identify them. For instance, the clouds that lie close to the earth's surface suggest the existence of condensed water vapor. The swirling, changing patterns of the clouds indicate the air and air movements that carry and shape them. Characteristically less cloudy regions suggest drier conditions, and the permanently white expanses of Antarctica and Greenland signify glacial conditions.

A Limited Habitat

As the earth is viewed from this "in-space" position it becomes at least partially apparent that, like an astronaut's vehicle, "spaceship earth" has limited resources for human support: most of its surface is water; a large portion of the remainder is arid or semiarid, hilly or mountainous, ice and snow covered, or swampy. The limited supply of readily habitable land would present no problem, however, if the number of people were few and economies simple. But, as this is not the case, the earth's life support systems are already strained, and their burden continually increases.

REVIEW AND DISCUSSION

1. How is Earth unique among the planets of the solar system?
2. Can you prove or disprove the contention that Earth is the only "life planet" in the universe? If so, how?
3. What constitutes our solar system, and of what geographic importance is it that our planet is part of a solar system?
4. What is the most significant effect on the earth of the two opposing forces, gravitational attraction and centrifugal force?
5. That the sun is the only major source of energy in the solar system is of what significance to people and their physical environment on the earth?
6. Discuss and evaluate the various motions and movements of the earth within the solar system, especially with reference to the sun. Of what significance are these motions and movements to human beings?
7. The Southern Hemisphere is far more a water hemisphere than the Northern Hemisphere, and the Pacific Ocean covers nearly half the earth's surface. Of what significance are these facts to the earth and its human inhabitants?

SELECTED REFERENCES

Barnett, L., and the editors of *Life*. *The World We Live In*. New York: Time, Inc. 1955.
Gamow, G. *The Birth and Death of the Sun*. Baltimore: Penguin, 1945.
Harrison, L. C. *Sun, Earth, Time and Man*. Chicago: Rand McNally, 1960.
Kuiper, G., ed. *The Earth as a Planet*. Chicago: University of Chicago Press, 1955.
Mehlin, T. G. *Astronomy*. New York: Wiley, 1959.
Russell, H. N., R. S. Dugan, and J. Q. Stewart. *Astronomy; Vol. I: The Solar System*, rev. ed. Boston: Ginn, 1945.
Strahler, A. N. *Planet Earth: Its Physical Systems Through Geologic Time*. New York: Harper & Row, 1972.

Shoreline waters of Acadia National Park, Maine. (Helen Faye)

3

The Earth's Waters

The prime geographic fact concerning the surface of our planet is that it is mostly water; thus, it has been suggested, had the planet been named by someone viewing it from space, it would probably have been called "Planet Sea," rather than Planet Earth. Only about 25 percent of the planet's surface is earth in the original sense of that word—that is, solid ground, soil, land—the remainder being ocean surface (about 70 percent), and lake, pond, and river surface (nearly 5 percent). In addition, there are some seven million cubic miles of surface water held in the form of ice throughout the five million square miles of Antarctica and most of the 840,000 square miles of Greenland. All in all, there is enough surface water, both liquid and solid, so that if the surface of the earth's crust were smooth, the entire earth would be submerged to a depth of about 8,000 feet.

Water is even more prevalent than is indicated by its surface occurrence, for there is water in the soils and rocks of the earth's crust, as well as some, in vapor form, in the lower atmosphere (Fig. 3–1).

Water is a major constituent, and absolute necessity, of all living things, whether plant or animal. The human body, for example, is about 70 percent water; and its blood is about 90 percent water. Humans are so dependent on potable water that, in all but the rare instance, they can exist without it for only about seven to ten days—and then, not without serious discomfort and injury.

Besides being an element vital to human life, water is also a support medium for myriads of other life forms. This applies predominantly in the oceans, where the great bulk of all life, large and small, land and water, is supported.

Another role of water is the changing of the surface of the earth's crust. By entering their cracks and pores, water is instrumental in causing rocks to decay and literally go to pieces; and moving water is instrumental in wearing away high areas and filling in low ones. If sea level remained constant, and if there were no continuing deformations of the earth's crust, the continents would eventually be leveled off by a combination of stream and wave action.

Chemically pure water: a compound of two parts hydrogen and one part oxygen (H_2O).

"Ordinary" water: H_2O, plus varying amounts of other chemicals and pollutants.

Brackish water: slightly salty water, unpalatable, but usually potable (drinkable).

Salt water: ocean and sea water; nonpotable; considerable content, in solution, of common salt (sodium chloride), plus salts of magnesium chloride, sodium sulphate, calcium chloride, and potassium chloride, plus, in solution, traces of gases found in the atmosphere (especially, nitrogen, oxygen, hydrogen, carbon dioxide, and argon).

Hard water: water high enough in mineral content to interfere with the action of soaps, to form scale in pipes and boilers, to stain sinks and clothing, and even to make bathing difficult—as opposed to water of lesser mineral content ("soft water").

Surface water: water on the earth's surface (ocean, sea, river, lake, pond, marsh water).

Ground water: in common usage, all subsurface water, but, technically, only the water found (and usually considered below the soil mantle) in the cracks and crevices of rocks in the earth's crust, or deep-seated in rock debris.

Soil water: technically, subsurface water in the soil mantle held within a few feet of the land surface.

Meteoric water: only water dropped (precipitated) from the atmosphere (rain, snow, sleet).

Connate water: water trapped in sediments and sedimentary rocks at the time of their deposition.

Magmatic (juvenile) water: water set free during the crystallization of deep-seated igneous (molten) rocks and during volcanic activity.

"Liquid" water: water in the ordinary state.

"Solid" water: water in the form of ice, snowflakes, hail, and so on.

"Gaseous" water: water in vapor form (predominantly in the atmosphere).

FIGURE 3–1

Descriptions of various kinds of water found on the earth.

The Earth's Water Supply

For all practical considerations, the earth's total water supply is constant. Moreover, its various water forms—vapor, liquid, and solid—are maintained in practically constant proportions over many thousands of years. Significant variations in their amounts occur only during periods of widespread glaciation or deglaciation. At such times, the volume of ocean water will either decrease or increase (sea level will fall or rise), depending on how much water is "locked up" in continental ice sheets.

The Hydrologic Cycle

The relative constancy of the proportions of the earth's water forms is the result of a system called the *hydrologic cycle.* As Figure 3–2 shows, this cycle consists of a continuous succession of processes: *evaporation, movement, condensation, precipitation, runoff,* and *percolation.* Simply, water is evaporated primarily from the oceans and is carried, in vapor form, by the winds. Whenever the vapor-carrying air is cooled sufficiently, part of the moisture is condensed. If the condensation is pronounced beyond the point of mere cloud formation, the moisture is precipitated to the earth's surface in the form of rain, snow, sleet, or hail. Some may fall directly into the ocean. If, on the other hand, it falls over the land, the precipitation feeds streams that normally carry it back into the oceans, or else it supplies underground waters that seep into the oceans or into their contributing streams. The result is a sort of continuous water "merry-go-round," or water cycle, which perpetuates the earth's total *water balance.*

Much Salt Water, Little Fresh Water

Considered as a whole, the earth has a comparatively stable "water budget" on which to live. The major problem is that most of its water is overwhelmingly salty; for some 97.2 percent is saline ocean water, and an additional small fraction is the salt water of certain inland lakes, seas, and salt marshes. Another portion is fresh water that is locked up primarily in Antarctic glaciers. The end result is a meager total of fresh water on the earth's land surface and underground.

Another problem is that fresh water is not evenly distributed over the lands—either in terms of regions or in terms of periods of time. Not only is atmospheric behavior such as to create desert or semidesert areas in many regions, but, even in normally humid

FIGURE 3–2

The Hydrologic Cycle. In the longterm global view, the oceans are the ultimate source and destination of all of the earth's waters, which circulate within and between the major physical systems of water, land, and air. A highly idealized depiction of the hydrologic cycle is shown at the lower right. The bulk of the diagram, although more complex, is also a simplified representation of the same cycle. Moisture is *evaporated, carried aloft, moved, condensed,* and *precipitated,* over both the oceans and the land. That which is precipitated over land is returned to the oceans by means of *surface runoff* (through lakes and streams) or *groundwater percolation* (through soils and crustal rocks).

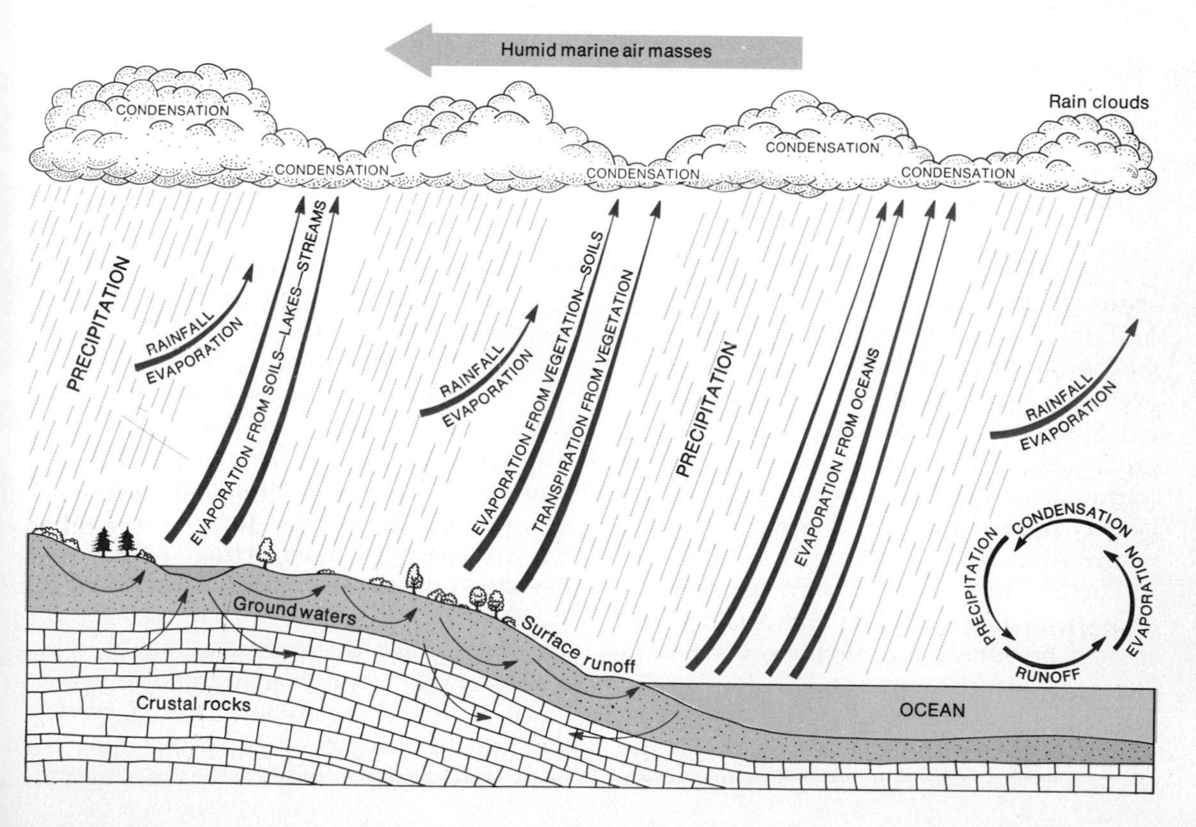

areas there are often noticeably dry years that are part of wet and dry precipitation cycles.

Will We Run Out of Water?

As far as human beings are concerned, there would be little or no water problem if they avoided the desert or semidesert areas and lived in comparatively small numbers in the regions of reliably abundant precipitation; but, such is not the case. Humans are not only inhabiting and overdrawing the subsurface waters of most deserts, but also importing humid-land waters to many of those deserts. Even in many humid regions, the increasing demands of large populations for domestic and industrial water have outstripped local surface and subsurface supplies and forced the importation of water from other, often rather distant, humid areas. When the imported water supplies have also proved insufficient, it has been necessary to curtail water use arbitrarily for the duration of the crisis.

In actuality, there is enough usable water on the earth to meet the needs of a population far greater than that of the present. Full use of that water, however, would entail vast and tightly integrated distributional systems much more elaborate than any in existence today. In addition, the same water would have to be used more than once, or perhaps, recycled over and over again. Eventually, numerous large areas may have to depend on desalted, hence more expensive, ocean water. Beyond all this is the problem of pollution, which already plagues the earth's streams, lakes, and underground waters, and, as recent studies have revealed, even threatens the waters of the oceans.

The Oceans and Seas

The Pacific Ocean is known as the "King of the Oceans" (Plates I and IX; Fig. 3–3). It is so vast that it could encompass all the lands of the earth with enough space left over to nearly equal the size of South America. Moreover, the area of the Pacific is approximately equivalent to that of the other oceans combined. From north to south, the Pacific extends from the Bering Strait, between Alaska and Siberia, to the edge of Antarctica; from east to west, it stretches from the Americas to Asia and Australia, its greatest breadth being roughly equivalent to half the distance around the earth (Plate II). In total, the Pacific covers nearly one-third of the earth's surface (Fig. 3–4). It is also the deepest of the oceans, in regard to both average depth and deepest point. In volume, it contains at least half of all the water on earth (excluding that water held in vapor form in the lower atmosphere).

The Atlantic Ocean, the second largest of the oceans, has an area equaling less than half that of the Pacific. Its latitudinal extent, from the edge of the Arctic Ocean to Antarctica, is about the same as that of the Pacific, but its east-west extent, from Africa and Europe to the Americas, is far less (Plate II; Fig. 3–3).

The Indian Ocean, often thought to be comparatively small, has an expanse nearly equal to that of the Atlantic. It, too, bathes part of the margin of Antarctica, but its northward reach is only to the tropical and

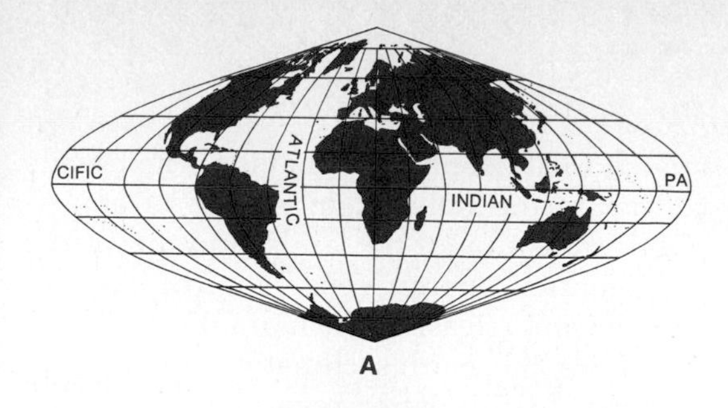

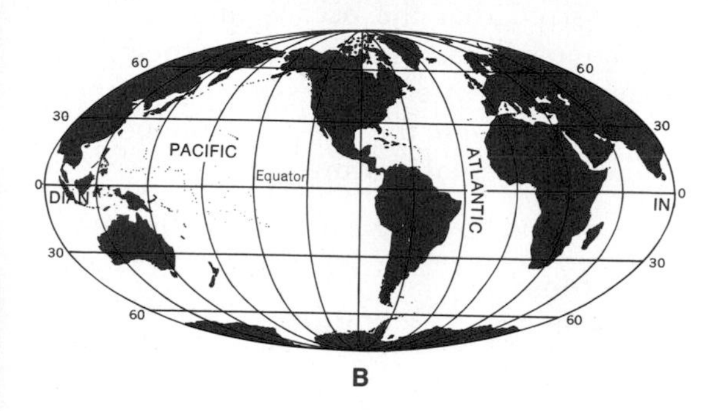

Pacific Ocean	63,985,000
Atlantic Ocean	31,529,000
Indian Ocean	28,357,000
Arctic Ocean	5,541,000
Mediterranean Sea	1,145,000
Bering Sea	878,000
Caribbean Sea	750,000
Gulf of Mexico	700,000
Sea of Okhotsk	582,000
East China Sea	480,000
Yellow Sea	480,000
Hudson Bay	472,000

C

FIGURE 3–3

The vastness of the world's oceans and seas is more than twice that of all the continental areas combined: 71 percent to 29 percent. A. The Sanson Flamsteed (sinusoidal) equal-area projection is one view of the ocean's vastness. B. The Mollweide equal-area projection is a slightly different view of the world's oceanic continuity. C. Sizes of the oceans and larger seas (in square miles).

subtropical shores of India and its adjacent countries. On the west, it touches most of the eastern shore of Africa; on the east, it extends to Australia, extreme southeastern Asia, and the large islands of Sumatra and Java (Plate II; Fig. 3–3).

The Arctic Ocean, centering on the North Pole and nearly landlocked by the poleward margins of Eurasia and North America, is the earth's "little ocean," both in area and depth. Yet, its area is appreciably greater than that of Europe and a bit more than the expanse of Australia. Unlike the other oceans, the Arctic does not extend to any tropical, subtropical, or midlatitude areas, for it is truly a polar sea—a realm of permanent cold and restlessly moving and colliding packs of sea ice (Plate II; Fig. 3–3).

Most of the water bodies referred to as *seas* (or, in some cases, as *gulfs* or *bays*) are actually portions, or *inland extensions,* of the oceans themselves. Some examples are the Bering and South China seas in the Pacific region, the North Sea and the Gulf of Mexico in the Atlantic region, and the Arabian Sea and the Bay of Bengal in the Indian Ocean region. Even the most land-penetrating seas, the Mediterranean Sea and the Black Sea, are nonetheless connected to the Atlantic Ocean. However, some water bodies that are popularly called *seas* are completely sepa-

FIGURE 3–4

The Water Hemisphere. The Pacific Ocean covers nearly one-third of the earth's surface.

rated from any ocean; examples include the Caspian Sea and the Dead Sea, both of which are salty, and even some freshwater lakes, such as the Great Lakes, are sometimes spoken of as *inland seas* (Plate II). Obviously, the term *sea* is an unfortunate one in that it is not precisely defined (it has even been used to signify *ocean,* or *oceans,* as in Rachel Carson's famous book, *The Sea Around Us*). A glance at Figure 3–3 will show that the Mediterranean dominates the world's seas (in the lesser sense) in the same fashion that the Pacific dominates the world's oceans, for this sea alone covers an area equivalent to about one-third that of the United States.

Major Functions

The oceans and seas have many functions, both geographic and nongeographic. Because of their size, they serve as barriers to the overland movement of humans and goods from one land area to another; yet, at the same time, they provide "highways" for the relatively low-cost bulk movement of goods over long distances. From time immemorial they have provided food for shore-dwelling peoples, and, more recently, have supplied food obtained by modern, far-ranging fishing vessels, for inland dwellers as well. In the future, these waters may supply food at a much increased rate, especially if economical methods of obtaining and processing minute sea life (*phytoplankton* and *zooplankton*) can be developed—if overfishing, water pollution, and the like, have not exterminated such species. Most experts, however, do not regard ocean-derived food (whether fish, mollusks, ocean mammals, plankton, or seaweed) as very important in solving the future food problems of a burgeoning world population.

With respect to major "earth systems," the prime functions of the oceans are those related to atmospheric behavior. The oceans constitute the only major source of atmospheric moisture for the lands, and they serve as gigantic "energy cells" for the receipt, storage, and release of the radiant sun energy that fuels the earth's climatic and weather systems. In this latter regard, it should be emphasized that the oceans affect not just the atmosphere directly above them, but, because of earth-encompassing atmospheric circulation, even the most inland portions of the continental landmasses.

Properties of Ocean and Sea Waters

Salinity. The earth's many streams and groundwater systems are constantly dissolving, and flushing into the oceans, large quantities of minerals and mineral salts obtained from the rocks and soils of the earth's crust. And, since the evaporation of water from the oceans and seas through the hydrologic cycle necessarily leaves such salts and minerals behind, their concentration in the oceans is a continuing process. The degree of concentration of these dissolved salts in the oceans is known as *salinity,* or saltiness, and is stated as so many parts per thousand by weight (with oceanic salinity averaging about 35 parts per thousand). About two-thirds of the saline content of the oceans and seas is sodium chloride, or common table salt, present in simple solution. The remaining one-third is largely dissolved salts of magnesium, sodium, calcium, and potassium chlorides, with additional trace quantities of at least half of all the chemical elements known to exist on earth, including all the gases of the atmosphere.

The salinity of the oceans and seas varies considerably from place to place but correlates closely with the amounts of local precipitation, evaporation, and freshwater

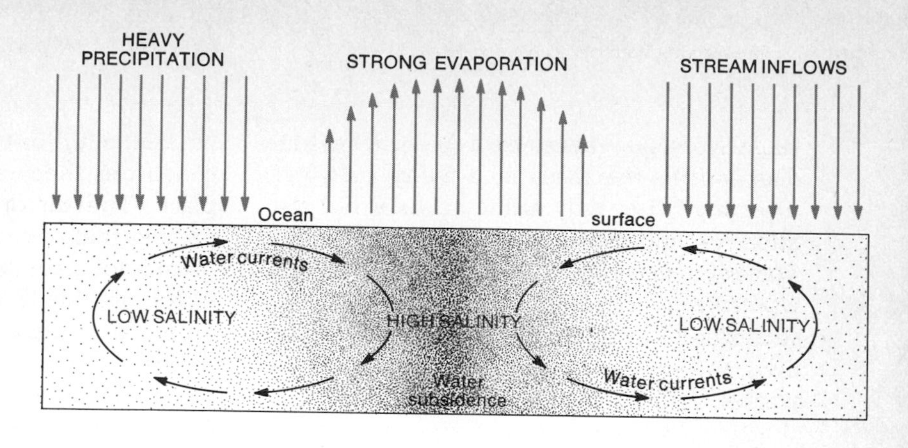

FIGURE 3–5

Heavy *precipitation* and appreciable *stream inflows* induce *low-salinity* conditions that react with adjoining high-salinity areas to produce compensating movements of the waters. *High salinity* is induced by *strong evaporation* of the waters.

inflows (Fig. 3–5). Lowest salinities are found where rainfalls are heavy and near the mouths of major rivers. Highest salinities occur in hot subtropical areas, where there is a great deal of evaporation and very little precipitation, as well as in the two polar regions.

Density. The ratio between the weight and the volume of a substance is known as its *density*. The density of pure water has been arbitrarily given the value of 1.000. Seawater, because of the salts dissolved in it, has a higher density, averaging about 1.027; and the higher the salinity, the higher the density. Water expands and contracts with changes in temperature, and its density changes accordingly: as water cools, it contracts, and more and more can be crammed into the same space; thus there is a greater mass within the given volume, and hence a greater density. In short, the density of water increases with increasing salinity and decreasing temperature.

Water with a density greater than that of the water surrounding it tends to sink; water with a lower density rises. Thus *density currents* are created. In the equatorial regions, the ocean waters warmed by the sun expand and, due to their consequently lower density, rise toward the surface. Continued warming produces continued expansion, which pushes water away to the sides; thus a surface flow develops, from the equatorial regions toward the poles. In the Arctic, the colder conditions cause the water to contract, resulting in its heavier density and gradual sinking. If the earth had no land obstructions to interfere with the free movement of ocean water, a worldwide water circulation pattern would be established, with warm tropical water flowing poleward at the surface, and cold Arctic water moving equatorward at some depth. The continents obstruct such movements, of course, but nevertheless, density differences are among the causes of most ocean currents.

Temperature. Although the land and water surfaces in a given region receive similar amounts of energy from the sun per unit of surface area, the surface temperatures that result are strikingly different. During the day, the soil or rock composing a land surface retains, within a very shallow surficial zone (usually only a few inches in depth), most of the heat it receives from the sun; thus on a sunny summer day the surface becomes relatively hot. Conversely, during the night, heat is rapidly lost by radiation into space, and the surface becomes relatively cool. In a similar way, this retention-radiation sequence also influences the contrast in land temperatures that occurs as summer and winter.

Large bodies of water, on the other hand, show relatively little variation in temperature, either daily or seasonally. This is because: solar energy is able to penetrate deeply into the mass of a body of water, warming it not only at the surface, but also at depth; and heat is distributed throughout the water mass by both horizontal and vertical currents. Thus the body of water warms very slowly in the morning and in the summer, and retains its heat at night and in the winter.

Movements of Ocean Waters

The waters of the oceans and seas are in a constant state of circulation. Moreover, their motion is three-dimensional: horizontal, vertical, and oblique. Some of these circulatory movements result in the transportation of large quantities of water, such as the continuous flows of cold polar waters equatorward and of warm tropical waters poleward, and, to a lesser degree, the upwelling of deep waters and the gradual descent of surface waters. All of these movements are compensatory: that is, for each movement there is a compensating movement in the opposite direction; they are created by differences in the density of waters, by the propulsion of waters by steadily blowing winds, and by the rotation of the earth.

Waves: Wind and Seismic. Wind, blowing over the surface of a body of water, produces several forms of movement in that water. Most of the momentum of surface winds is expended in the generation of waves; a considerably lesser role of winds is the development of water currents. Waves are certainly the most obvious and most common type of water movement within the surface levels of the ocean, but even so, waves do not result in any significant transportation of water from one location to another. The apparent forms of waves move horizontally, but the water particles that constitute them actually remain largely in place. The major importance of waves to man is their effect on his utilization of the oceans for navigation, and even more, their effect on the features of coastal interfaces where many people live and work. As the depth decreases near the shore, oncoming waves strike the sloping bottom and tip their crests forward to eventually crash onto the shore in the form of breakers. This wall of water repeatedly raised and slammed against the shore bottom or, if close enough inshore, against cliffs or artificial walls or dikes, is extremely powerful and can cause erosion or even extensive periodic destruction. Waves also constitute a tremendous form of energy, one which people have never seriously attempted to tap.

There are two major types of waves in the world's oceans and seas: *wind waves*, common to all waters, and *seismic waves*, rare, short-lived, but violent.

Wind waves are formed through two major mechanisms: *direct wave push* and *surface drag*. Direct wave push is a mechanical action in which wind pushes on the windward slope of a wave itself, in the direction of the flow, having the same effect on the water as it would on a boat or similar object. Surface drag is an indirect mechanical action created by the friction between moving air and the water's surface. The magnitude of wind waves is determined by a number of factors, the most important of which are the wind's *velocity*, *duration*, and *steadiness* of blow, and its *extent* of unobstructed sweep.

Wave motions, developed by light or moderate velocity winds in deep waters, are smoothly progressive, or, perhaps more correctly stated, smoothly repetitive—the wave form travels through the water in a progressive oscillatory motion. A simple stylized wave form is depicted in Figure 3–6, in

FIGURE 3–6

A typical wind-generated wave form has a *crest* and a *trough* and may be measured in terms of its length and height.

which the various wave segments and specifications are graphically incorporated. *Wave height,* certainly one of the most obvious features of waves, is measured as the vertical differential between the bottom of a wave's *trough* and the top of its *crest. Wave length,* by contrast, is measured as the horizontal distance between two consecutive crests or two consecutive troughs. The rapidity with which a wave form progresses through the water is referred to as *wave velocity,* which involves movement of the wave form but not any transportation of water. That is, during the wave form's forward movement through wave lengths, each molecule of water in a wave completes a circular vertical orbit, moving slightly forward while on the wave crest and slightly (almost equally) backward while in the wave trough (Fig. 3–7). Since each orbit moves a molecule only a bit more forward than backward, there is only a very slight gain in its forward position, giving a very slight surface drift to the water in the direction of the wind and wave form movement. This slight motion, however, is usually not the same, or as effective, as the movements set up by surface drag.

Tsunamis, or seismic waves, are waves sent out across the open ocean in series, as a result of sudden and violent disturbances of the ocean bottom by crustal movement or volcanic activity. Some seismic waves are as much as one hundred miles in length, in great contrast to wind waves, and may travel halfway around the world, or more. Their height in the open oceans is little more than a foot or two, but they may move at speeds of 250 to 500 miles per hour. When large fast-moving seismic waves strike low unprotected coasts, great coastal flooding and extensive damage and loss of life may result.

Currents: Warm and Cold. The movements that transport relatively large quantities of ocean water over great distances, in fairly constant patterns and amounts, are called *ocean currents.* These are of many types and patterns: horizontal, vertical, circular, and combinations of all types. Some ocean currents are warm and some are cold (Fig. 3–8); some are highly saline and others are but weakly saline; some move largely at or near the surface and some move at intermediate or even great depths. Of them all,

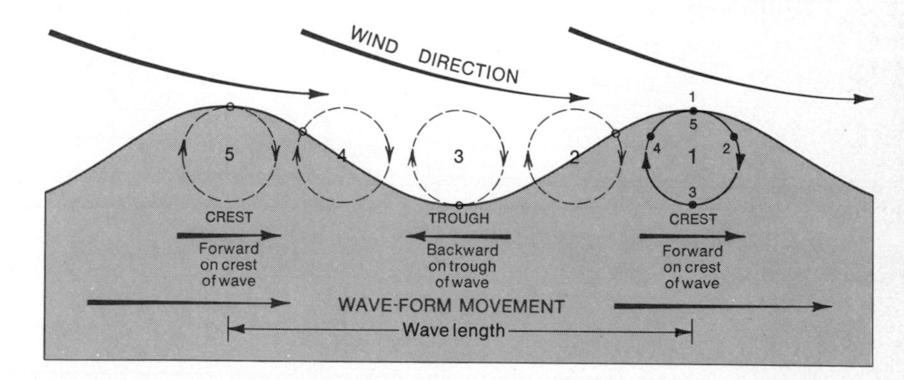

FIGURE 3–7

The stationary orbital motion of a water particle during the passage (#1–5 in diagram) of a wind-blown deep-water wave form of low wave height.

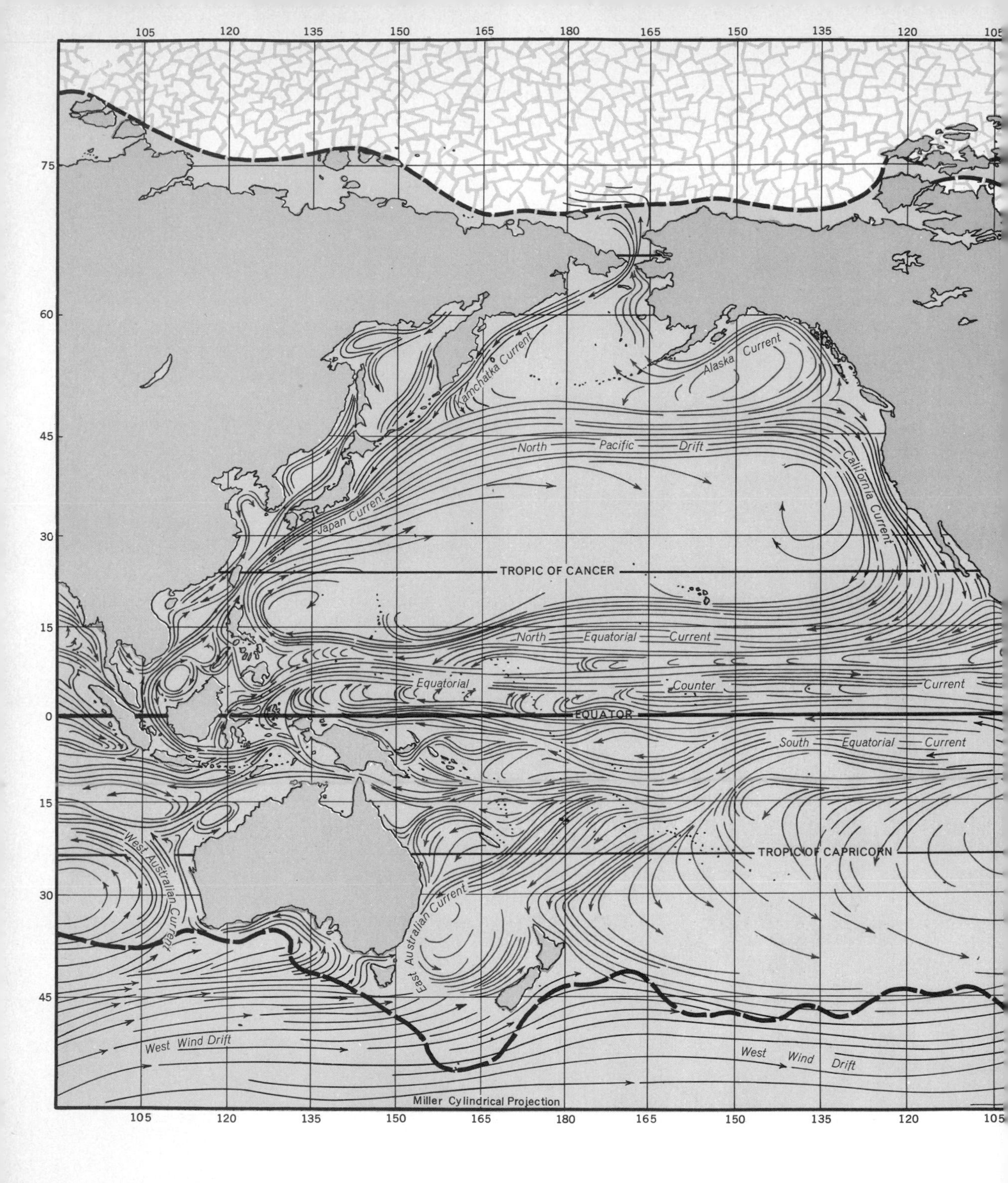

Kamchatka Current
Alaska Current
North Pacific Drift
California Current
Japan Current
TROPIC OF CANCER
North Equatorial Current
Equatorial Counter Current
EQUATOR
South Equatorial Current
TROPIC OF CAPRICORN
West Australian Current
East Australian Current
West Wind Drift
West Wind Drift
Miller Cylindrical Projection

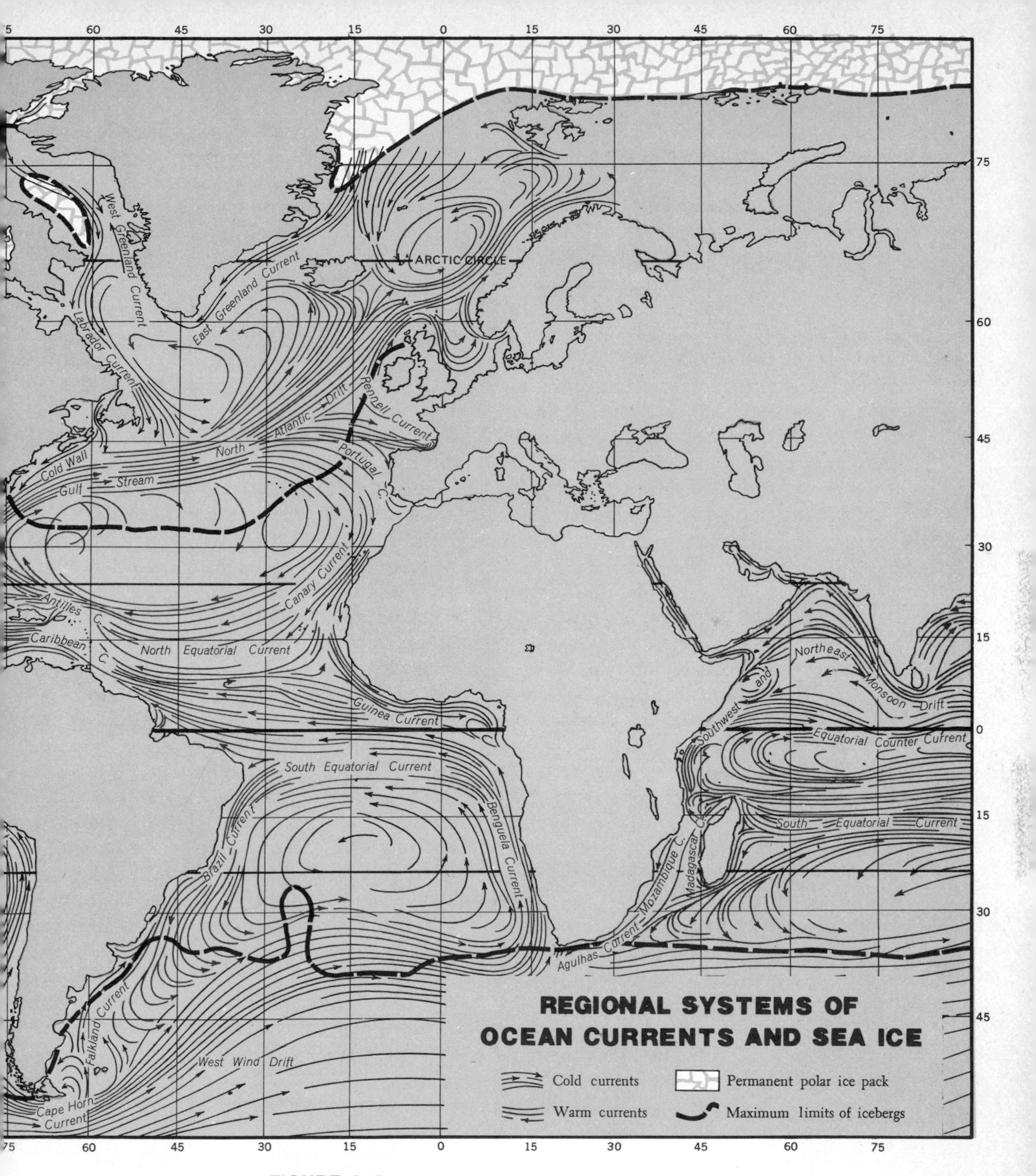

FIGURE 3–8

The world's major cold and warm ocean currents, which form contrasting systems in the Northern and Southern hemispheres, are particularly significant in their effects on climate, ship navigation, and commercial ocean fishing. Note the equatorward limits of polar pack ice and icebergs.

those operating at or near the surface of the great oceans are the most significant geographically. These multiple *surface currents* have considerable direct effects on navigation, economic resource developments, and the utilization of many coastal lands of the world. They also have important indirect influences on societal life styles through their effect on world climate and weather.

Most of the major surface currents originate in the activities of the prevailing planetary winds, which set them in motion through friction, or *push and drag*. As noted earlier, the differences in water density within the oceans also cause current flows, with resultant large-scale transportation of water from place to place; and, as also noted, unequal heating by insolation, unequal cooling by radiation, or a combination of both can promote these differential density conditions. The superchilled surface waters of the polar regions sink to lower levels and displace slightly less chilled waters, which in turn spread equatorward to displace the somewhat warmer tropical waters. In any such transportation of water there must always be a compensating movement—a replacement, a complete circulation—because water is not compressible. Of great importance also is the rotation of the earth; since the earth rotates toward the east, there is a tendency for water to slip back toward the west as the earth turns, which has the effect of creating a west-flowing current in the equatorial region. As it approaches the shores of the continents, this flow of water is deflected poleward by the landmass, thus setting up circular current patterns.

Coriolis Force. Any object moving freely on the surface of the earth is affected by the rotation of the earth, which, moving in an easterly direction, makes a full turn every twenty-four hours. Since the earth's circumference is approximately 25,000 miles, the rotational speed at the equator is a little more than 1,000 miles per hour. At more poleward locations, the speed is less, reaching zero at the poles themselves.

Even in an ordinary room there is a detectable difference between the speeds of the north and south walls. If a weight suspended from the ceiling in the middle of a room is drawn to the south wall, held there until all vibration has ceased, and then released, it will swing across to the north wall and, swinging back, will return to a point slightly removed from the starting point. If the room is located in the Northern Hemisphere, the south (equatorward) wall is moving at a faster rate than the north wall. The weight released from the south wall will have an easterly speed faster than that of the floor it passes over or of the wall it eventually reaches, and it will therefore deflect slightly to the right as it swings. As it pauses before reversing its swing, the weight will reduce its speed to that of the north wall. Returning, it will have a slower eastward speed than the floor it passes over, and the floor and the south wall will move out from under its path. Once again the weight will deflect to the right as it swings, and it will reach the south wall somewhat west of its starting point. This procedure is called "Foucauld's Experiment," and the principle involved is termed the *Coriolis force* (named for the French mathematician who first observed it). The rule can be briefly stated as follows: in the Northern Hemisphere, any freely moving object tends to be deflected to the right; in the Southern Hemisphere, it will be deflected to the left.

Thus, an ocean current tends to flow not in a straight line, but in a gradual curve, and, if space permits, it will eventually return to the point at which it started. Such a curve will be clockwise (to the right) in the Northern Hemisphere, and counterclockwise (to the left) in the Southern Hemisphere.

Surface Ocean Current Systems. Between the tropics of Cancer and Capricorn, most of the surface currents and, hence, the greatest volumes of water, are moving steadily toward the west. This is in accord with the rotation of the earth and the strong frictional drag of the tropical easterly, or *trade winds,* which tend to pile up the surface waters against the eastern shores of the landmasses in low latitudes. This westward movement of tropical water is a twofold one, being composed of the North Equatorial Current and the South Equatorial Current (Fig. 3–8).

Some of the water piled up against the eastern shores of the tropical landmasses finds its escape between the two Equatorial Currents, forming the east-flowing Equatorial Counter Current. Most of it, however, turns poleward in both hemispheres, partly as a result of the Coriolis force. The continuing effect of this force eventually draws these poleward currents away from the continents, and, in the middle latitudes, turns them eastward—back across the ocean traversed in the tropical area. Reaching the eastern shores of that ocean, the currents are likely to divide into two branches, one continuing the rotary pattern dictated by the Coriolis force, and the other (usually smaller) current defying the rule and turning poleward.

The splitting of the ocean currents in middle latitudes is well illustrated by the North Atlantic Drift, which is the offspring of the Gulf Stream. This current, or drift, sends part of its water equatorward as the Canary Current, which in turn becomes the North Equatorial Current mentioned previously. The other part of the North Atlantic Drift flows in a northeasterly direction, past Iceland and Norway and on into the margin of the Arctic Ocean. It brings relatively warm temperatures to these high latitudes and thus accounts for ice-free ports in Norway, north of the Arctic Circle. The North Atlantic Drift is partly responsible for the climatic mildness characteristic of the British Isles and of much of northwestern Europe, despite their northerly latitudes (Fig. 3–8).

On the western side of the North Atlantic, cold waters from the Arctic Ocean, known as the Labrador Current, hug the eastern shores of the landmass as they move southward to meet and slide under the warmer waters of the Gulf Stream. Unfortunately, the Labrador Current brings many icebergs from the great glaciers of Greenland to menace the heavily traveled sea lanes of the North Atlantic shipping route. Particularly during the summer season, when they are freed from the winter ice lock in Baffin Bay, these floating ice mountains cause trouble for ships, and shipping lanes are often shifted southward to avoid the areas of greatest iceberg concentration.

The system of surface circulation in the northern half of the Pacific is strikingly similar to that of the North Atlantic. However, the narrow passage between eastern Siberia and western Alaska does not allow as free an interchange of Pacific waters with the Arctic Ocean waters as does its counterpart in the North Atlantic. Instead, there is only a cold-water current passing from the Arctic to the North Pacific through the narrow passage of the Bering Strait. This current, the *Kamchatka,* is the Pacific counterpart of the Labrador Current. There are, however, no icebergs, for there is no Greenland glacier to furnish them to the ocean's waters. Otherwise the two systems are much the same; the general pattern of *clockwise* circulation occurs in both the North Atlantic and the North Pacific.

The surface currents of the Southern Hemisphere in the southern parts of the Pacific, Atlantic, and Indian oceans, are essentially the same as those in the Northern Hemisphere, but the movement is *counter-clockwise* (Fig. 3–8). There is an interchange of tropical and polar waters: through westward movements across the oceans near the equator, poleward movements along eastern shores of the continents, return drifts across the oceans in the middle latitude positions, splitting of currents in middle latitudes near west coasts, and, on the whole, a generally rotary circulation. Since the continents terminate southward in the middle latitudes, a west-to-east current (the West Wind Drift) sweeps unobstructedly around the world south of Cape Horn, the Cape of Good Hope, Australia, and New Zealand.

Tides: Ocean and Sea. From time immemorial, coastal dwellers and seafarers have been aware of, and have had to contend with, the rhythmic rising and falling of sea level known as the *oceanic tides.* For some 2,000 years, it has been recognized that such tides were somehow caused by the moon and the sun. It was only about 300 years ago, however, that Sir Isaac Newton's discovery and explanation of the law of gravitation led to at least a general understanding of the cause and behavior of tides.

The moon and the sun exert appreciable gravitational forces, or pulls, on the earth, both on its crust and on its water bodies. Their effect on the comparatively solid crust is very minor, but their effect on water bodies, since they are fluid, is appreciable. Because of the huge solar mass, it might be expected that the sun would exert a greater pull on the earth than does the moon. The reverse, however, is the case, for the moon is much closer to the earth, and, in obedience to the law of gravitation, exerts a much greater pull than the sun, despite being so much smaller. Thus, it is the moon that plays the dominant role in the causation and magnitude of the tides.

The gravitational effect of the moon creates twin bulges in the earth's ocean and sea waters. There is always one bulge on the side of the earth that faces the moon and another, a compensatory bulge, on the side away from the moon (Fig. 3–9A). As the earth rotates (eastward) on its axis, these twin bulges move (westward) in a continual procession around the earth. And, because the earth makes a complete rotation each day, for any given place there are normally two *high tides* (two bulges) and two *low tides* (lower water levels between bulges) each day.

The tidal bulges are highest when the moon, sun, and earth are in a straight line, either *in conjunction* (moon and sun both on the same side of the earth) or *in opposition* (moon and sun on opposite sides of the earth). At such times, the combined gravitational pulls of the sun and the moon produce high tidal bulges, or *spring tides* (Fig. 3–9B). When the three bodies are *in quadrature* (the earth-moon line at right angles to the earth-sun line), the gravitational pulls compete and a low tidal bulge, or *neap tide,* occurs (Fig. 3–9C).

The distance between the earth and the moon also causes variations in the height of the tides. The moon describes an elliptical orbit around the earth, coming as close as 221,500 miles at *perigee* (nearest or lowest point) and receding to 252,700 miles at *apogee* (farthest or highest point). Thus its gravitational attraction is stronger at perigee, increasing the tidal bulges markedly at that time. Moreover, due to the combined and ever-varying effects of conjunction-opposition and apogee-perigee, the height of the tides varies from day to day.

In the open ocean, the tidal bulges and the lower water levels between the bulges are scarcely noticeable. Along the coasts, however, and especially where the tides strike straight-on and pile up along shallow water coasts, the *tidal range* (the difference in water level between high and low tides) may be very great—in several places it exceeds 50 feet. One can easily appreciate how such a tidal range, or even a range of

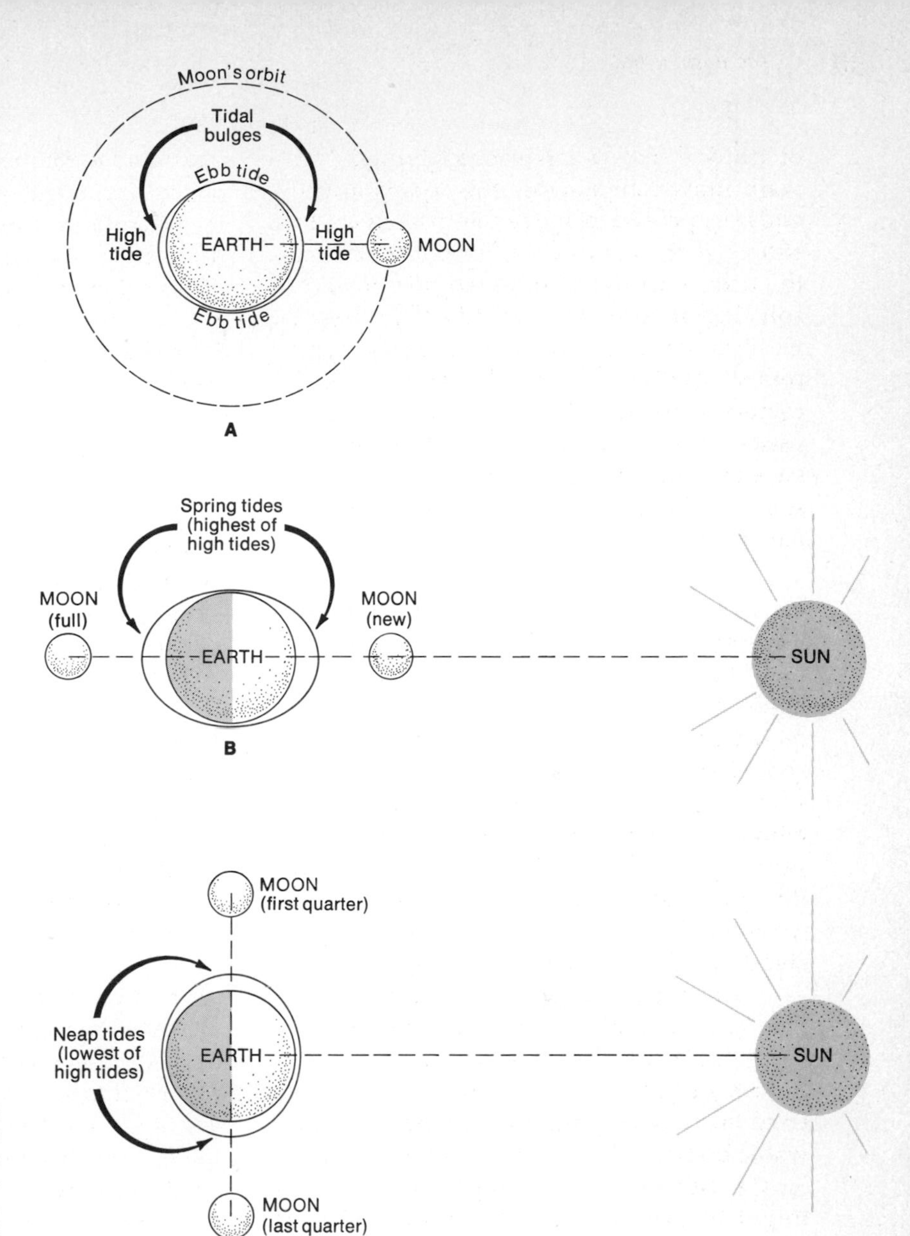

FIGURE 3–9

The creation of the oceanic tides by the moon and the sun. A. The normal *high* (flood) and *low* (ebb) *tide* situations. B. The *spring tide* situation: the moon and sun pull on the earth together. C. The *neap tide* situation: the moon and sun pull against one another.

only a few feet, can affect such things as navigation, ship dockage, fishing operations, and the recreational use of coasts and coastal waters (Fig. 3–10).

Despite the problems they create, tides serve several useful functions. The *tidal currents,* especially those associated with low, or *ebb,* tides help to clear bays and river mouths of channel-filling sediments, to sweep out sewage and other debris, and to prevent accumulations of stagnant water. In some instances, high, or *flood,* tides enable ships to pass over mud, sand bars, and rock ledges that they otherwise could not navigate. Someday in the future, the dream of harnessing the tides to help supply human-

FIGURE 3–10

The *daily tidal range* along many coastal areas is considerable, as here, at Parrsboro, Nova Scotia, in the Bay of Fundy. At the left, low tide; at the right, high tide. (Information Canada Photothèque)

kind's need for more sources of domestic and industrial energy—especially "clean," or nonpolluting, energy—may become a reality.

The foregoing discussion makes no pretense of complete coverage of tides. For such, reference should be made to books cited at the end of this chapter.

Sea Ice and Icebergs. Most of the ice in ocean waters is pack ice—*floes* (cakes) and *fields* of saltwater ice found in the polar regions (Fig. 3–11). Pack ice is monotonously flat, except where these huge cakes have ground against one another and caused their edges to buckle and pile up to form *pressure*

FIGURE 3–11

An ice field at the foot of Mt. Erebus, in McMurdo Sound, Antarctica, seen from the bow of the U.S. Coast Guard icebreaker *Eastwind.* Several pressure ridges are visible across the flat ice in the middle distance. (U.S. Navy) *ridges.* Because the Arctic polar area is an ocean and the Antarctic a continent, there is more pack ice in the Northern Hemisphere than in the Southern Hemisphere. Arctic pack ice has entrapped or crushed many vessels attempting to navigate sea routes and fisheries along the northern continental shores of North America and Eurasia. But despite these dangers, there may still be possibilities for the further use of the Arctic Ocean as suggested, for example, by the under ice voyages of U.S. Navy submarines and the more recent trial runs of gigantic oil tankers from the North Atlantic to Alaska (Fig. 3–12).

FIGURE 3–12
The giant new oil tanker *Manhattan* confronts arctic pack ice during a trial run from the North Atlantic to Alaska. (Exxon)

Some ice of Arctic origin moves into the northwestern portion of the North Atlantic (Fig. 3–8). Fields of ice also constitute navigational hazards in the Bering Sea, Baffin Bay, and the Sea of Okhotsk. However, even in the Arctic Ocean, there is always some open water. Contrary to popular belief, sea ice there never forms a continuous sheet from shore to shore, even at the height of the coldest season. Storms, ordinary winds, and currents keep the ice broken up and shifting from place to place.

Some small icebergs are derived from local valley glaciers along mountainous coasts, like those of southeastern Alaska, but they are short-lived and few in number. The largest and most numerous icebergs come from the Greenland icecap and from the vast ice fields of Antarctica. Tabular Antarctic icebergs, unlike the jagged Greenland type, do not menace any major shipping lanes, nor do they regularly move as far from their points of origin. Only the rare whaling ship or the scientific exploration vessel must be on guard against them, particularly during fogs or storms. The farthest equatorial limits of icebergs sizable enough to be hazardous at sea are shown in Figure 3–8.

Surface Waters of the Lands

Lakes and Their Functions

Considering the enormity of the earth's oceans and seas, the smaller inland water bodies might seem insignificant in comparison. However, geographic significance is not measured by size alone, and many lakes, or group of lakes, have geographic importance far beyond their mere size or location. Lakes serve in the storage and regulation of excess waters prior to their release into the environment, as well as in the protection of the environment from periodically destructive floods and erosion. Thus, lakes may be more directly entwined with the lives of more people, and with the function and preservation of more human habitats, than are the oceans and seas.

Some of the continental landmasses have developed relatively few lakes, except for an occasional artificial one. In South America, natural lakes of appreciable number and size are relatively absent. The same is true of much of Australia and Africa, and parts of Asia. This relative absence of natural lakes is simply a matter of the physical environment having worked against their formation and preservation. The development and continued existence of lakes demand two conditions: first, some sort of natural catchments and basins that have restricted outlets; and second, sufficiently regular inputs of water to keep the basins at least partially filled. Thus, because of the absence of one or both of these, nature rules against lake development throughout the earth's extensive dry-land regions, and even in large parts of the humid lands. Where climates are humid and natural basins are present, lakes readily develop; in time, however, even these humid-land lakes are totally, or at least partially, emptied, as gnawing streams cut channels progressively deeper into their rims and drain away the impounded waters, or as they become filled with inwashed sediments and aquatic vegetation. Thus, lakes (and swamps) are among the most unstable of the major physical features of the earth's environment.

In contrast to those considerable regions of the earth with few or no lakes, there are extensive regions of the planet where they are numerous. Most of the large regions with numerous lakes were once glaciated areas, in most cases affected by continental glaciation rather than by valley glaciation. North America, Europe, and the northern Asian landmasses are well endowed with a variety of natural lake developments. In these regions, the characteristic land surface unevenness, resulting from both glacial scour and glacial deposition, normally results in a variety of natural depressions, or basins, that have no low-level outlet. Most such glaciated regions, with their natural basins and troughs, occur within areas of humid climates, so there is sufficient water input to fill the basins. Streams are constantly engaged in cutting into the lake rims, but by and large, there has not been sufficient time since the glaciers disappeared for the streams to have fully drained the lakes, or for them to have become completely filled with vegetative debris and sediment (Fig. 3–13).

There are several lake regions of the world that were never glaciated, and lakes in such areas owe their origin to other physical processes. For instance, the development of a lake region in equatorial Africa was the result of physical displacements of the earth's crust and of valley damming by lava flows. In northern Florida, lake development is largely the result of soluble rock conditions; and the lakes of the lower Yangtze Basin are due to variations in the deposition of river alluviums and, in part, to some warping of the earth's crust. The biblically famous salt lake known as the Dead Sea lies in a trough formed by the downdropping of part of the land surface (Fig. 3–14). There are other means of lake formation, but the generalization that lakes are most numerous and varied in humid glaciated regions, especially those of recent continental glaciation, is not invalidated by the exceptions.

The Great Lakes System. Of all the continents, North America has the greatest number of lakes, from small and large ones at comparatively low elevations to small ones at mountainous elevations. Of all these lakes, from Canada's Great Bear Lake in the far north to Lake Nicaragua in the tropics, the cluster known as the Great Lakes is the most conspicuous and the most important (Fig. 3–15). Not only is it the earth's chief inland

FIGURE 3–13

Among the extensive lake regions of the world that were once areas of continental glaciation is this region on the north shore of Great Slave Lake, Northwest Territories, Canada. The town of Yellowknife appears at the lower right. (Canadian Pacific Railway)

waterway, but it serves many other purposes as well—from water supply to commercial fishing and recreation.

Lake Superior, the largest of the lakes, is the most extensive body of fresh water on the globe. It is about 350 miles long and 160 miles wide. Its surface is about 600 feet above sea level; its deepest part, about 1,300 feet, places its bottom nearly 700 feet below sea level. With an area of approximately 32,000 square miles, it is large enough to cover all of Belgium and the Netherlands and to ex-

FIGURE 3–14

A classic view of the Dead Sea from its surrounding barren cliffs. This highly saline lake occupies the lowest portion of a huge north-south trench formed by the down-dropping of a block of the earth's crust. (Israel Information Services)

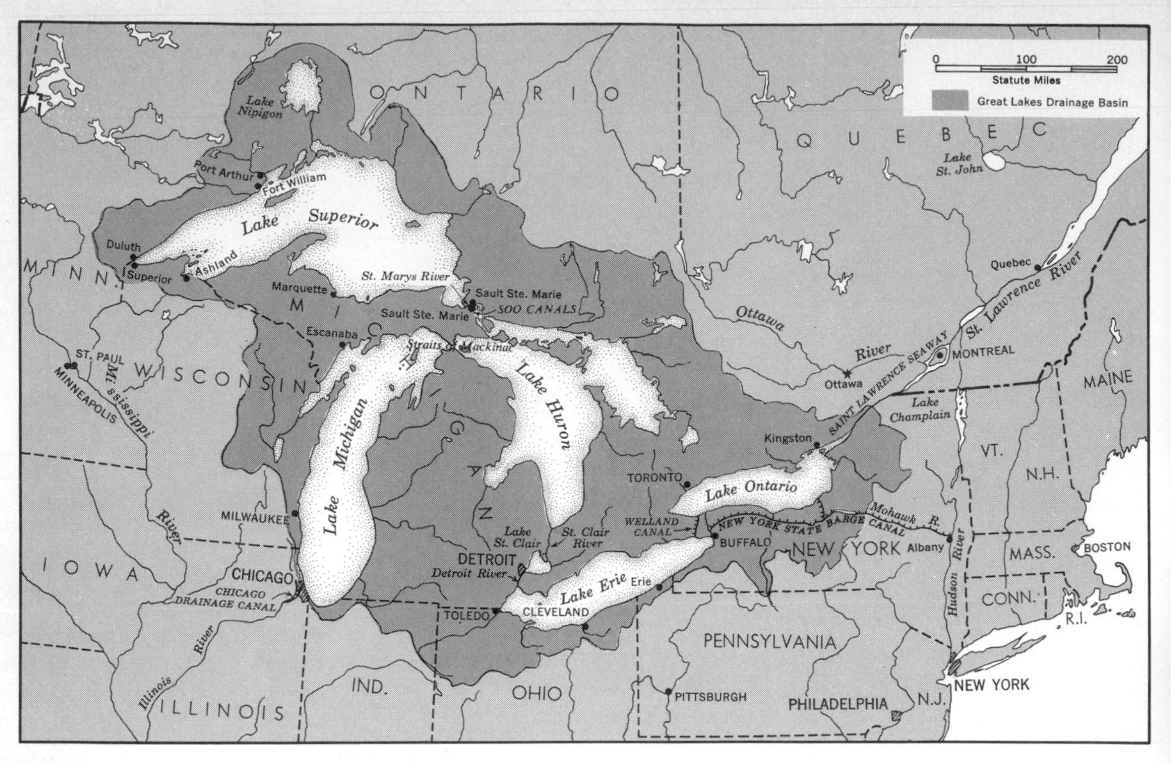

FIGURE 3–15

The Great Lakes and the St. Lawrence River and Seaway constitute the major inland waterway of the world.

tend over half of Denmark as well. A single outlet, the St. Marys River, connects it with Lake Huron. The Soo Canals, paralleling the rapids section of the upper St. Marys, allow vessels to move between Lake Superior and the Lower Lakes.

Lakes Michigan and Huron, which lie about 20 feet lower than Lake Superior, are connected by the Straits of Mackinac. The combined flow of Superior, Huron, and Michigan—except for a minor diversion through the Chicago Drainage Canal to the Illinois River—moves by way of the St. Clair River into small and shallow Lake St. Clair, which drains, by means of the Detroit River, into the western end of Lake Erie. At Erie's eastern end, the water forms the Niagara River, which tumbles over Niagara Falls on its way to Lake Ontario. The Welland Canal in Canada makes it possible for vessels to move between Lakes Erie and Ontario. Although Lake Ontario is the smallest of the Great Lakes (little Lake St. Clair is not considered one of the Great Lakes), it is still nearly as large as Massachusetts. From its eastern end pour the waters of all the lakes, waters that form the full stream of the St. Lawrence River. Together, the Great Lakes cover an area equal to that of the United Kingdom, or an area nearly one-third greater than New England; and the parts of the shorelines that lie entirely within the United States are longer than the distance from New York to London.

The Great Lakes have been singled out for emphasis here because of their size, arrangement, and geographic importance. Since the opening of the Great Lakes–St. Lawrence Seaway, in 1959, the lakes have become an even greater commercial network. More will be said about them in later chapters in this book.

Streams and Their Functions

Streams are the arteries of natural drainage on the earth's surface—the major channels of surface-water flows toward lower levels of land. They vary greatly in size and length, from small tributaries known locally as *rills, brooks, creeks, arroyos,* or *wadis,* for example, to broad, full-flowing rivers. Where landmasses and drainage basins are large and precipitation is abundant, streams of all sizes are numerous and permanent; where drainage basins are small, and especially where precipitation is meager, streams are often small and *intermittent* (Plate II; Fig. 3–16).

The size and closeness of streams affects human activities and settlements in a variety of ways. Primitive people camped at places along or near streams, where water and food were available and where there were highways for rafts or boats. People, primitive or not, have often settled on islands in rivers, where the encompassing water provided a barrier against enemies. River junctions have been strategic points for the control of, and participation in, the trade that moved along their routes. River mouths, joining with the sea, gave access to ocean routes. When industrial centers arose, the riverbank position provided large quantities of water for rapidly growing urban populations and supplied water for power and for the processing of materials. In many areas, the fine dark alluviums of river floodplains attracted agricultural populations. In the eastern United States and, to a lesser extent, in the drier west, the bulk of the westward movement was locally carried and directed by streams. These are but some of the ways in which rivers and people have been, and still are, related.

Humid-Land Streams

In regions of humid climate, all except the smallest tributaries of the drainage systems are normally permanently flowing, but their volumes may vary seasonally in the

FIGURE 3–16

Areas having large drainage basins and abundant precipitation develop streams that are major full-flowing rivers having numerous and permanent tributaries. Here, one such area in the Mekong River Delta, Vietnam. (U.S. Department of Defense)

amounts of water available from precipitation and surface and underground drainage. High and low water stages represent ordinary behavior, but the streams are always present. Some streams may vary only slightly between high and low water levels, particularly where rainfall is of comparatively even distribution throughout the year, and where waters are not partly locked up, in the form of ice, during the winter season. Variations in stream flow are significant physically, but they are also significant in their relationship to man and his activities. Streams that alternately flood and subside are of much less use to humans than those that maintain relatively uniform volumes.

Some Major Systems. All of the continents, except Antarctica, have large and important river systems (Plate II). All of these systems have their sources in humid lands, principally in rainy and snowy mountain areas, and most of them flow within humid regions throughout their courses (Plate IV). Exceptions to this latter condition are such rivers as the Nile, the Tigris-Euphrates, and the Colorado, all of which flow through climatically dry areas for most of their lengths and, as a result, actually lose water volume as they flow toward their mouths. Such rivers are known as *exotic rivers,* though they are exotic only in the portions of their courses that extend through dry regions.

Humid-land and dry-land rivers have both played important parts in human history, because the development of river transportation, the control of floodwaters, the generation of power, and the insurance of sufficient and proper supplies of domestic, industrial, and irrigation waters within their spheres have demanded much cooperative, rather than individual, activity among people.

One of the largest river systems in the world is the Mississippi–Missouri, about 3,800 miles in length from its source in the northern American Rockies to its mouth in the Gulf of Mexico. Among its conspicuous physical features are a large number of tributaries, broad and fertile floodplains in its lower sections, and a wide, flat delta (which it is still continuing to build) into the Gulf of Mexico. Another major river system of great volume is the Amazon. From its source in the high Andes, it flows eastward across the breadth of the Amazon Basin to empty a wide and muddy flow of water into the Atlantic. So great is its volume and so heavy its load of fine suspended materials that the Amazon outflow can be seen for many miles out into the Atlantic, a visibly muddy "river in the ocean." Many islands split its area of discharge into the ocean so that it has several mouths instead of one. The Amazon is the only river that is navigable for ocean vessels for nearly the breadth of a continent, all the way from the Atlantic to Iquitos in Peru. Major river systems of the other continents are the Nile of Africa, the Ob-Irtysh of Siberia, the Yangtze of China, and the Murray of Australia. However, Australia, with its disproportionate area of dry climates and its small drainage basins in its humid sections, has few large permanent streams.

Dry-Land Streams

Stream behavior in regions of arid or semiarid climates is much different from that in humid lands. The streams, except for the few that are exotic, are intermittent rather than permanent and are usually smaller and less numerous. Yet, these intermittent streams of the dry lands present many vexing problems. Periodically, they pour floodwaters onto settlements and play havoc with transportation systems (Fig. 3–17). Extensive flood prevention works are necessary for the protection of towns and cities, as well as of roads and railroads. It is somewhat paradoxical that floods and flood

FIGURE 3–17

Results of a flood on the intermittent Mojave River, in the Mojave Desert, at Barstow, California. Note the breached highway crossing the floodplain and the damaged fields. (Spence Air Photos)

damage may be as severe, or more so, in deserts as in lands of heavy rainfall.

The reasons for these dry-land drainage difficulties lie basically in the combination of amounts and distributions of precipitation; that is, the problems are fundamentally climatic in origin. Because the flow of water is short lived, floodstage sediment loads are dropped, and channels become clogged. Subsequent floodwaters spread out over wide areas and quickly carve new channels that are soon glutted with debris as the waters disappear. Except where stream channels are deeply entrenched, such behavior means that flood areas are usually broad and not well defined; nevertheless, they are the natural drainways, and humans must pay the penalty if they build settlements or transport lines near them. Very often, people seem to know so little about stream behavior under such dry climatic conditions that they fail to recognize that they are in the natural line of water flow—until floodwaters suddenly strike, and then it is too late.

Groundwaters of the Lands

Of the widespread surface waters of the earth's land area, large amounts are continually sinking into the upper parts of the earth's crust to form *groundwater*. Small amounts of the earth's groundwaters are *connate water*, having been resident in rock structures since their formation; and other small fractions are *magmatic water*, having been released as a result of magma emerging at the earth's surface as lava. But nearly all groundwaters are *meteoric water*, water directly derived from precipitation and surface waters and endlessly recirculated in the hydrologic cycle.

Although not usually visible, these groundwaters are as important as surface

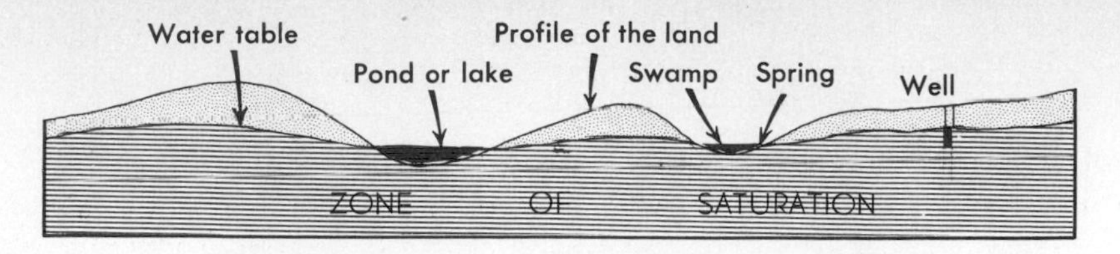

FIGURE 3–18

A section of the upper portion of the earth's crust illustrating the zone of saturation, the water table, the relationship of the water table to the land surface, and some possible developments of lakes, swamps, springs, and wells.

waters, both in the physical processes of the environment and in usefulness to people. From the groundwaters come springs, wells, and most of the moisture so necessary for the growth of natural vegetation and crops. As water moves downward from the surface, it fills the spaces between the soil particles and the crevices and cracks of bedrock as far down as they penetrate into the earth's crust. The zone in which all such spaces and cracks are filled is called the *zone of saturation* (Fig. 3–18), and its top level is known as the *water table.* The water table follows approximately the profile of the surface of the landforms, although it approaches closer to the surface in low-lying places than on hilltops. Where the water table intersects the earth's surface in low-lying spots, swamps, ponds, lakes, or springs develop. Generally, the groundwater table lies nearer to the surface in regions of humid climates than it does in those of arid or semiarid climates. But even in the latter areas, the groundwater table may reach the surface in a few low-lying spots, and in these, salt marshes normally develop.

Soil Waters and Their Chemistry

An extremely critical factor in the growth of all natural vegetation and crops is the availability of soil moisture. Soil moisture is of three kinds: *gravitational water, hygroscopic water,* and *capillary water* (Fig. 3–19). Gravitational water is the water that moves downward, by force of gravity, toward the zone of saturation. If gravitational water is too abundant, soils become waterlogged and less useful for most domestic crops, although such "drowned" soils are still capable of supporting various types of native swamp vegetation. Hygroscopic water forms as an extremely thin film around individual soil particles, somewhat as though each particle were wrapped in moist cellophane. It is often referred to as "unavailable water" because

FIGURE 3–19

The three kinds of soil moisture: *gravitational, hygroscopic,* and *capillary waters.*

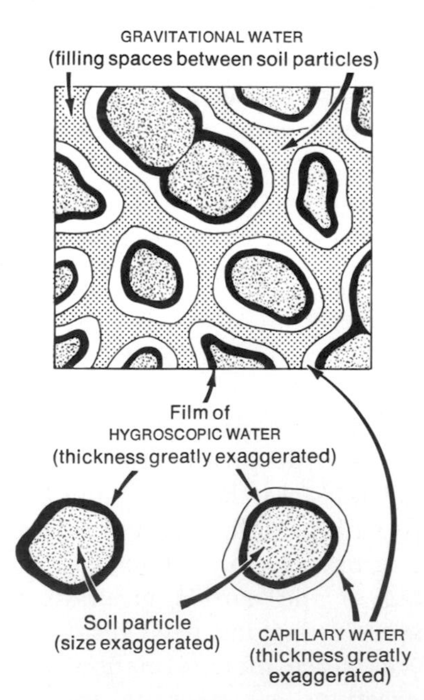

plants cannot utilize it—it is so tightly held to the particle that plant roots are unable to detach it. Capillary water is the water that moves by capillary action, or attraction, in the same fashion that spilled ink is drawn into and along the fibers of a blotter. Its ability to move, even against the pull of gravity, is significant for, as topsoils dry out, capillary water is drawn upward through the soil where it becomes available to the root systems of plants. However, if such action is long continued, particularly if the water is raised all the way to the surface, soils become somewhat saline. This occurs when the water drawn to the surface is evaporated and leaves its minerals behind. If there is insufficient precipitation to wash the minerals, or mineral salts, back down into the lower parts of the soil, permanently saline soils are created. If these are mildly saline, a few crops tolerant of mineral salts may still be grown; but with high salinity, no crops can grow, save perhaps the few types of native vegetation that are specially adapted to salty soil. Such specially adapted plants are called *halophytes.*

Springs and Wells

Springs. When groundwater issues at the earth's surface it creates either *springs* or smaller water sources known as *seeps.* Whether in deserts or in lands of plentiful rainfall, both are of considerable use to humans and animals. If they are hot springs or if they are heavily charged with certain minerals, they may be used for health baths and general resort purposes. A spring in the desert usually makes possible an oasis settlement; a line of springs, as along the base of a range of hills or mountains, allows ranches, or even small towns, to exist. Springs on farms or ranches may constitute their sole water supply; they may be covered by small structures called *springhouses,* which serve as cool places for the storage of milk and other perishables. While some springs are safe for human use, others, especially those that draw their water supply too directly from local surfaces, may be dangerously polluted. The underlying causes of springs are many: Figure 3–20 indicates a few of the more common possibilities.

FIGURE 3–20

Some of the subsurface conditions that lead to the formation of *springs.*

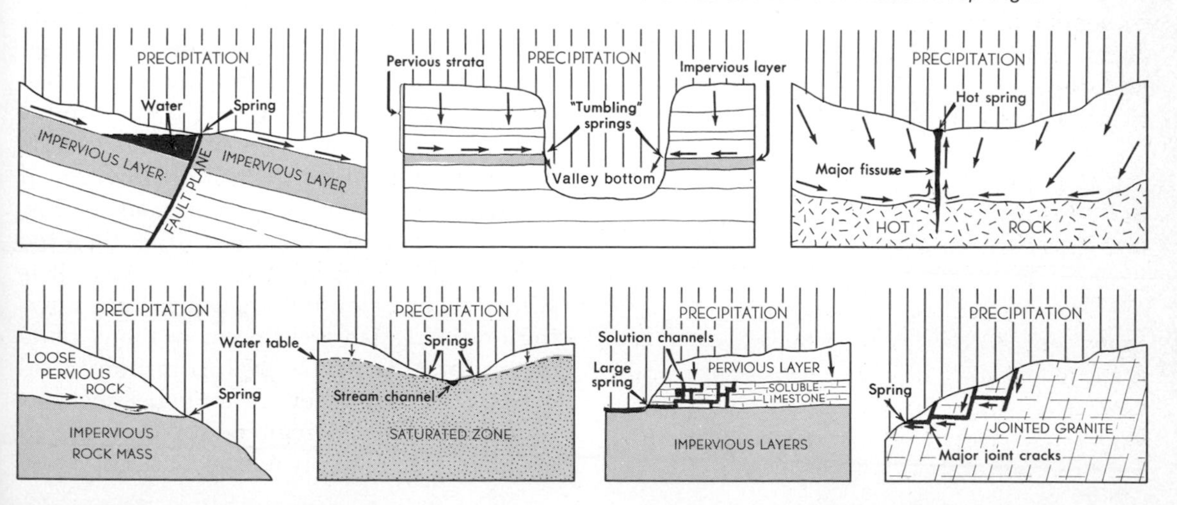

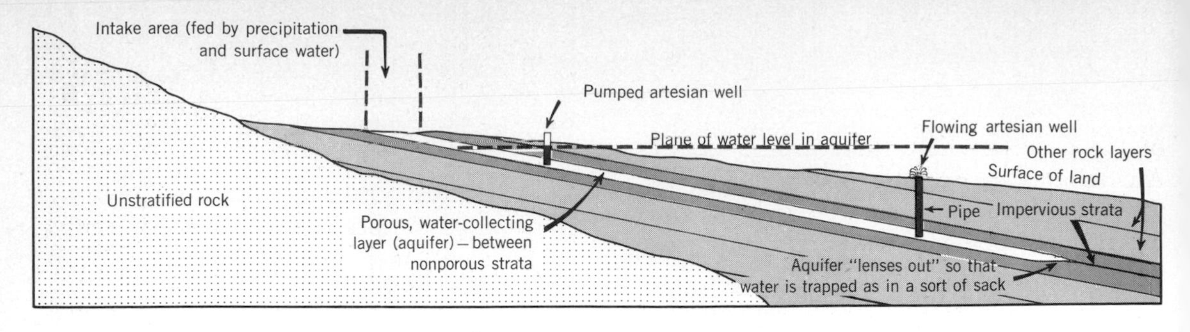

FIGURE 3–21

The basic subsurface conditions necessary for the production of *artesian well water*. In this instance, the *aquifer* (water-collecting rock layer) has not been filled all the way to the intake area, and the aquifer is underlain as well as capped by impervious rock.

Artesian Wells. These are wells of a special and important category. The physical conditions necessary for their development are illustrated in Figure 3–21. There must be sloping, or dipping, layers of rock, the edges of which are exposed at the crust's surface. At least one of the dipping layers must be porous enough to absorb and collect water from the surface, taking the water in at the edge that is exposed to precipitation and surface flow. Additionally, the porous, water-collecting layer must be capped, or overlain, by a relatively impervious rock layer; otherwise its water will be lost into the adjoining rock layers. As water fills the porous sloping layer, or *aquifer,* considerable pressure is built up in that layer, especially in the part that is farthest removed from the intake area and is receiving the push of all the water that has collected on the upslope of the dipping stratum. By drilling into a lower part of the aquifer, where the water is under great pressure, one may obtain a flowing, or *artesian well* (Fig. 3–22). If the pressure is continuous and steady, artesian water continues to flow without the need for pumping. However, if the drilling of a well, or too many wells, causes the gravity pressure to drop, then the water must be pumped to the surface. It is still artesian water, but the well is not a flowing one. Artesian wells occur in many parts of the earth, in dry as well as in humid regions. Obviously, they are apt to be most significant in the dry regions, for they are an excellent water supply for the most arid parts

FIGURE 3–22

High-pressure water spouts from the control valve of an artesian well in the United Arab Republic. (United Arab Republic Tourist Agency)

of the deserts where there are no springs. Australia is particularly famous for its rich resource of artesian water, the bulk of which lies in its steppe and desert regions.

Traditional Wells. Most types of wells do not bring their waters to the surface naturally. People have had to dig or drill past the water table and then contrive some method of lifting the water to the surface (Fig. 3–23). The problem of water supply from wells is relatively simple where the water table is close to the surface, but it is difficult and expensive where the depths to the zone of saturation are great, or where the water table continually drops due to excessive removal of water. The cost of developing a water well can range from a few to many thousands of dollars, and the cost of pumping and maintenance will continue as long as the well is in use.

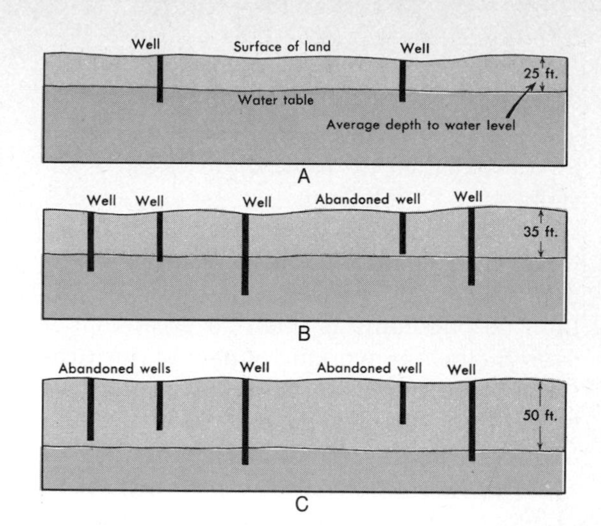

FIGURE 3–23

A sequence of changes in the problem of securing water from wells in an area where the withdrawal of groundwater exceeds its natural recharge. A. A water table little affected by wells. B. Too many wells—withdrawal exceeds the natural recharge. The water table has dropped slightly. C. The water table is greatly lowered, most wells are abandoned, and pumping costs are very high for those wells still in operation. (A period as short as 20 years may mark the change from the condition shown in A to that shown in C.)

Water Pollution: Natural and Human-Induced

There are two sources of pollution of the earth's waters: *natural pollution,* and *human-induced pollution.* Natural pollution is the result of such things as: the inwash of sediments, mineral salts, poisonous chemical elements, organic debris, supercharged solutions from hot springs and geysers, natural oil seeps; and the fallout associated with pollens, dusts, and explosive volcanic activity. Human-induced pollution includes such things as: the discharge of body wastes, garbage and rubbish, household detergents, industrial wastes, pesticides, commercial fertilizers; the fallout from the combustion of fossil fuels (natural gas, petroleum, coal), leaded gasoline, and atomic materials; and thermal pollution incident to the discharge of hot industrial waters, including waters from thermonuclear plants.

Natural pollution is offset to an appreciable degree by natural chemical and mechanical processes that, in time, purify and clarify water. The offset, however, is not complete in some instances. For example, underground water may be permanently poisoned locally by arsenic and copper compounds, or may be too high in unwanted content of lime, sulphur, or iron compounds; ocean water is slowly and constantly being further polluted by the intake of mineral salts, all of which are left behind as evaporation takes place from ocean surfaces.

Human-induced pollution is also offset to some degree by natural processes, especially if the pollutants are biodegradable, but, in far too many cases, the pollutants are not of such a nature as to be effectively modified by natural processes, and the sheer volume of pollutants is often so great that it overwhelms the working of natural purification. Thus, there are many rivers that are running sewers, ponds and lakes that are cesspools, underground waters that are poisonous, and portions of the oceans that are so polluted that they destroy marine life and render coastal areas unsafe for recreational use.

The problem of human-induced water pollution is not a new one: history is replete with accounts of streams and lakes despoiled by human-induced erosion, of foul-smelling urban sewers, polluted harbors, disease-laden wells and rivers, and debris-covered river banks and shores. What is new is that water pollution induced by human beings is now widespread and acute because of the tremendous population increase and the proliferation of industrial, agricultural, and transportational activities. The seriousness of the problem is reflected, at least in many areas of the world, by a general public clamor for new antipollution laws and the stringent enforcement of those already existing. Unfortunately, the chemistry and mechanics of water pollution are so complicated and so little understood that there is widespread disagreement among experts as to its causes, effects, and possible remedies. Humans have been polluting the earth's waters for a long time, and it will take a long time to solve the problems thus created.

REVIEW AND DISCUSSION

1. Water is even more prevalent on the earth than is evidenced by its surface occurrences. What are the other major waters included in the phrase "the earth's waters"?
2. Support the assertion that water is a major constituent and absolute necessity of all living things on earth, whether plants, animals, or people. Also, list as many other functions of water within the physical environment as you can.
3. With a world map in hand, choose several distant localities and try to describe the general structure of the hydrologic cycle in each. (Keep these structures in mind and check their accuracy when you study Chapters 9 and 10.)
4. Discuss the meaning of the phrase "the earth's water budget."
5. Are people likely to experience water shortages in the future? What can they do to prevent them?
6. The oceans and seas have many functions in the human habitat. Discuss their major functions, both on a world basis and on more limited regional bases.
7. Discuss the major properties of ocean waters and their importance in the world environment.
8. Explain the general circulation of major ocean currents and the forces that set them in motion. How does the Coriolis force influence the ocean currents of the Pacific?
9. With respect to occurrence and function, compare and contrast lakes and streams as surface waters.
10. Compare and contrast humid-land streams and dry-land streams in terms of their development, occurrence, and functions.
11. The groundwaters of the land are varied and widely distributed. Discuss their types, sources, and importance in the physical environment.
12. Consider some of the causes and effects of water pollution in the world today. What are some possible solutions for this problem in areas with highly developed agricultural and industrial economies and dense populations?

SELECTED REFERENCES

Addison, H. *Land, Water, and Food.* London: Chapman & Hall, 1955.
Bardach, J. *Harvest of the Sea.* New York: Harper & Row, 1968.
Briggs, P. *Water, the Vital Essence.* New York: Harper & Row, 1967.
Carson, R. L. *The Sea Around Us.* New York: Oxford University Press, 1961.
Chorley, R. J., ed. *Water, Earth and Man.* London: Methuen, 1969.
Coker, R. E. *Streams, Lakes, Ponds.* Chapel Hill, N.C.: University of North Carolina Press, 1954.
———. *This Great and Wide Sea.* New York: Harper Torchbooks, 1962.
Dugan, J. *Man Under the Sea.* New York: Harper & Row, 1956.
Engel, L., and the editors of *Life. The Sea.* New York: Time, Inc., 1961.
Fairbridge, R. W., ed. *The Encyclopedia of Oceanography.* Encyclopedia of Earth Sciences Series, Vol. 1. New York: Reinhold, 1966.
Frank, B., and A. Netboy. *Water, Land and People.* New York: Knopf, 1950.
Furon, R. *The Problem of Water: A World Study.* New York: American Elsevier, 1967.
Hunt, C. A., and R. M. Garrels. *Water: The Web of Life.* New York: Norton, 1972.
King, C. A. M. *An Introduction to Oceanography.* New York: McGraw-Hill, 1963.
Kuenen, P. H. *Realms of Water: Some Aspects of Its Cycle in Nature.* New York: Wiley, 1963.
Lane, F. C. *Earth's Grandest Rivers.* New York: Doubleday, 1949.
———. *The World's Great Lakes.* New York: Doubleday, 1948.
Marx, W. *The Frail Ocean.* New York: Sierra Club–Ballantine, 1969.
Milne, L., and M. *Water and Life.* New York: Atheneum, 1964.
Neumann, G., and W. J. Pierson, Jr. *Principles of Physical Oceanography.* Englewood Cliffs, N.J.: Prentice-Hall, 1966.
Oceanography: Readings from Scientific American. San Francisco: Freeman, 1971.
Richie, C. *Men Against the Desert.* New York: Macmillan, 1952.
Stewart, H. B., Jr. *The Global Sea.* Princeton, N.J.: Van Nostrand, 1963.
Sverdrup, H. U., M. W. Johnson, and R. H. Fleming. *The Oceans: Their Physics, Chemistry, and General Biology.* Englewood Cliffs, N.J.: Prentice-Hall, 1942.
Water Facts. Washington, D.C.: U.S. Dept. of Agriculture. Published occasionally.
Water: The Yearbook of Agriculture, 1955. Washington, D.C.: U.S. Dept. of Agriculture, 1955.
Williams, A. N. *The Water and the Power.* New York: Duell, Sloan & Pearce, 1951.
Weyl, P. K. *Oceanography: An Introduction to the Marine Environment.* New York: Wiley, 1969.
Yasso, W. E. *Oceanography: A Study of Inner Space.* New York: Holt, Rinehart and Winston, 1965.

Northeast across Berthoud Pass in the Rocky Mountains of Colorado. (Lovering, U.S. Geological Survey)

4

Introduction to Landforms

The earth itself, unlike the oceans and the seas, is a huge rock sphere. Yet, like them, the earth has currents, flows, and tidal movements within its physical mass. Particularly within its thin crust, or *lithosphere,* there are continuous movements that result in the warping, folding, and breaking of the earth's rock members, as well as in the movement of its molten rock materials. The surface of the earth's crust, which is overlain by the oceans, seas, and atmosphere, is constantly undergoing modification. The ocean's currents and waves, along with all forms of precipitated moisture, slowly shape and reshape the crustal surface. Where the crust is exposed to the atmosphere, weathering and erosion reshape its surface even more pronouncedly than do the oceans. The end result of these movements is the formation of a wide variety of shapes of the earth's crust that are called *landforms;* those formed beneath the earth's waters are *submarine landforms,* and those formed by crustal exposure to the atmosphere are *subaerial landforms.*

Landforms vary greatly in size: from continents and ocean basins, to mountains, hills, plains, and plateaus, to such small ones as canyons, arroyos, fiords, deltas, alluvial fans, sand dunes, glacial ridges, and volcanoes. For people, the subaerial landforms have the greatest significance, for it is on these landforms that they live, build structures and settlements, establish their lines of transport and communication, and produce most of their foodstuffs and other necessities. No other aspect of the physical environment is, literally, more fundamental to people than landforms.

Because landforms are so intricately related to most daily existence, even cultural geography finds it necessary to study them. Such study is even more essential in the field of physical geography, for the subaerial shapes of the earth's crust are conspicuous parts of the physical landscape and are intimately associated with virtually all of the planet's other physical conditions—from its waters and climate to its vegetation and soil. The specific study of landforms, which involves their nature and origins, constitutes

FIGURE 4-1

In response to geomorphic conditions, human populations often concentrate in small coastal plains areas, as here, in the city of Atami, twenty-five miles southwest of Tokyo, in mountainous Japan. (Rotkin, P.F.I.)

a special branch of geography known as *geomorphology* (Greek: *geos* = earth; *morphos* = form; *logos* = discussion), the discussion of the forms of the earth.

Illustrative of the influence geomorphic features can have on the cultural landscape is the role surface conditions have played in the development of Japan. The major islands of this country are about 85 percent hilly and mountainous, with the remainder being composed of small, discontinuous coastal plains and small mountain basins. In obeisance to the geomorphic conditions, most of Japan's population and associated human activities is concentrated in the limited plains and basin areas (Fig. 4–1). As a result, the landform map of Japan largely mirrors its population map, with the hill and mountain areas having sparse populations, and the plains and basin areas having heavy and very heavy population densities.

Another example of the role of landforms in the human scene is the influence they exerted on the westward course of settlement of the United States. Where hilly or mountainous terrain lay athwart the routes of westward migration, travel was comparatively difficult and was mainly funnelled through a few corridors, gaps, and passes. These included the Hudson-Mohawk route in the northeast and the Cumberland Gap farther south, and, in the far west, such high saddles as Donner Pass in the Sierra Nevada of California (Fig. 4–2).

FIGURE 4-2

Donner Pass (in the foreground), which carries U.S. Route 40 over the high Sierra Nevada between California and Nevada. Donner Lake is shown at center right, and the high mountain country of the Sierras is visible in the background. (State of California Department of Transportation)

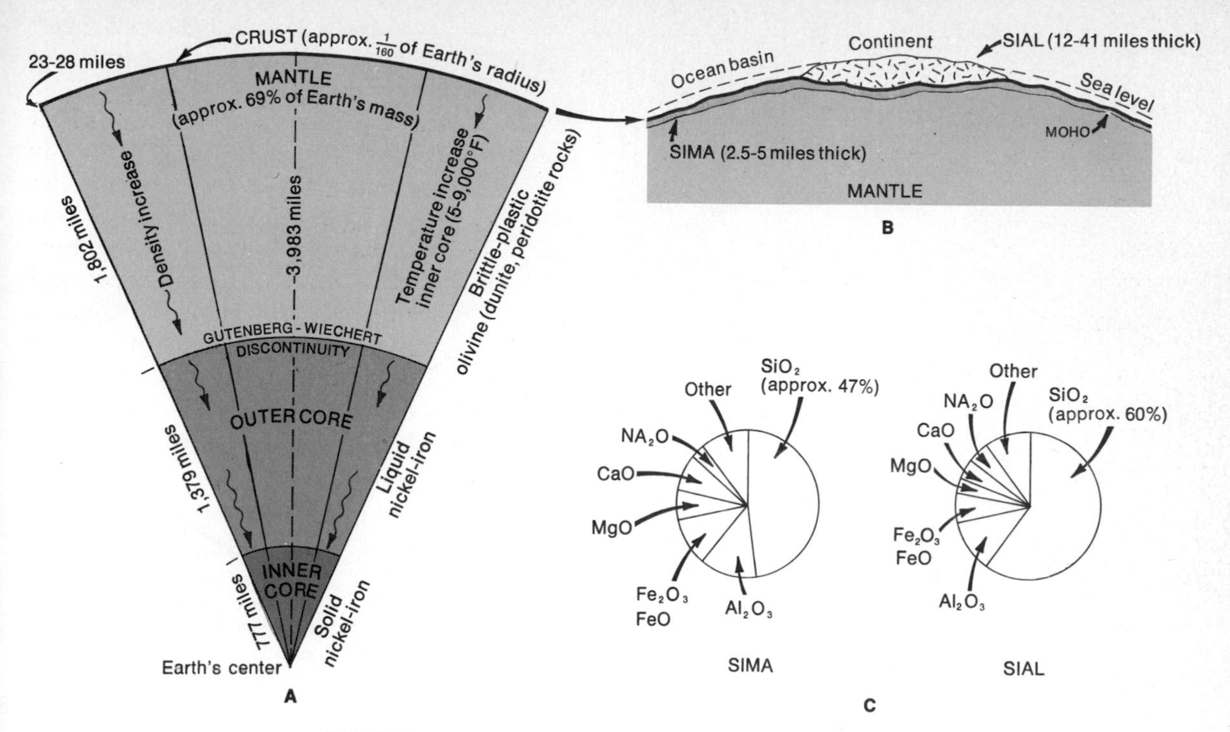

FIGURE 4–3

The general nature of the rock-earth. A. A segment of the earth (surface to center). The Gutenberg-Wiechert Discontinuity is recognized as the division between the mantle and the outer core. B. A segment of the earth's crust (sediments and sedimentary rocks not shown). The *Moho* (Mohorovičić Discontinuity) marks the division between the sima and the underlying mantle. Its existence is predicated on the study of the increase in velocity of earthquake waves from sima to mantle. C. The approximate chemical composition of the crustal sima (mainly basaltic) and the sial (mainly granitic).

The Earth's Crust, or Lithosphere

Landforms, being the shapes of the surface of the earth's crust, are obviously directly related to that crust. Although the study of the earth's crust, including that of its rock components, comprises the field of *geology*, it is also an essential part of physical geography.

Crustal and Subcrustal Zones

The *lithosphere*, or crust of the earth, is relatively thin: its thickness varies from about 2½ miles to 43½ miles (Fig. 4–3A and B). It is thinnest under the ocean basins and thickest under the continents.

Crustal Layers: Sima and Sial. The earth's crust consists primarily of a lower continuous layer, both under the oceans and under the continents, and an upper discontinuous layer comprising the continents themselves (Fig. 4–3B). The lower, thinner, and continuous layer is the *sima;* the upper, thicker, and discontinuous layer is the *sial.* (These terms are derived from the initial pairs of letters in the words *silicon, magnesium,* and *aluminum,* which are the critical

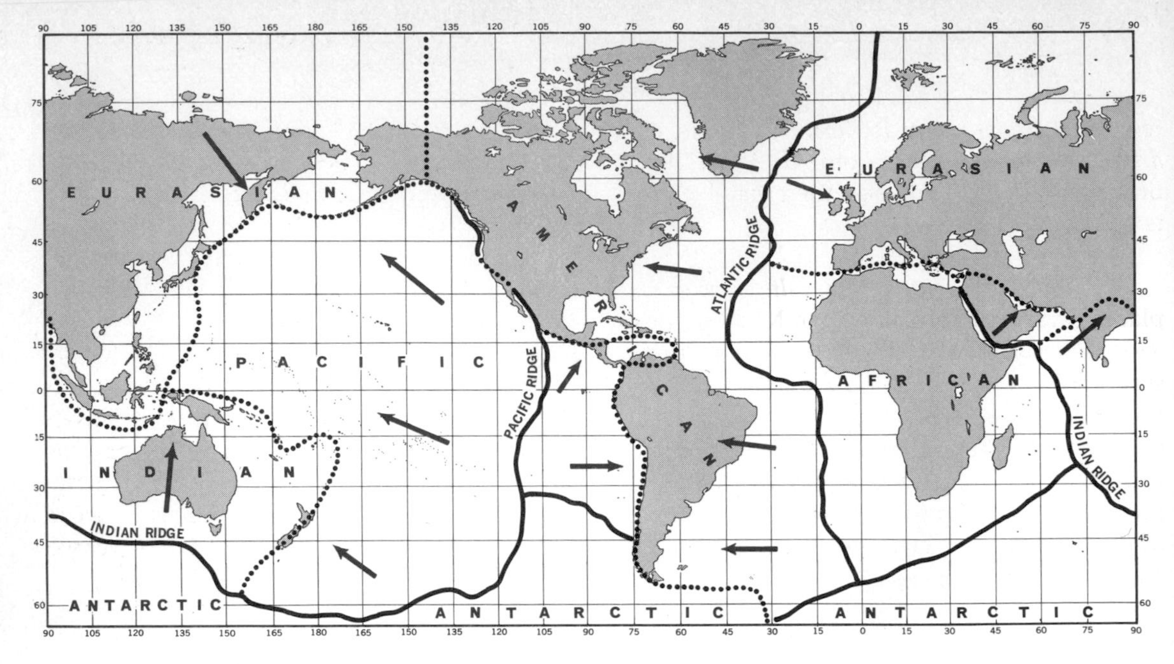

FIGURE 4–4

The global midocean ridge system of the earth's crust (solid lines), and the pattern of crustal plates (solid and dotted lines) composing most of the earth's crustal surface. Arrows show the schematic direction of motion (*continental drift*) of the five major plates (the African plate is apparently stationary at present).

constituents of the rocks involved.) The sima is composed largely of heavy (dense), dark-colored, and fine-grained rocks, mainly *basalt,* while the sial is composed largely of light-weight, light-colored, and coarse-grained rocks, mainly *granite;* consequently, it is common to refer to the sima as essentially *basaltic* and the sial as essentially *granitic.* In a sense, the lighter granitic sial constitutes huge "rock rafts" (the continents) that are floating on the denser basaltic sima. Although the sima and the sial are both composed of the same minerals, the proportion of those minerals within the layers varies appreciably: the sima contains proportionately greater amounts of the dark, heavy minerals, particularly compounds of iron and magnesium (FeO, Fe_2O_3, and MgO); and the sial contains much greater proportions of light-colored, light-weight minerals, especially compounds of silicon and aluminum (SiO_2 and Al_2O_3) (Fig. 4–3C).

Rock Zones Beneath the Crust: Mantle and Cores. Beneath the crust are the major *rock zones,* or layers, of the earth: the *mantle,* the *outer core,* and the *inner core.* Figure 4–3A shows, in a generalized fashion, the position, extent, and composition of these zones. Taken both singly and in combination, they are of prime importance to geology, geophysics, and seismology, but, unlike the crust, they are of little direct significance in the study of the earth's crustal landforms and physical landscape expression.

Crustal Plates

Plates and Marginal Fracture Zones. It appears that the crust of the earth is divided into large horizontal *plates,* which float differentially in relation to one another —immense rock slabs floating on a substratum of semiliquid, or *viscous,* rock matter. This movement is termed *continental drift*. The global pattern of the division lines, or *fracture zones*, between the crustal plates is shown in Figure 4–4.

Plate Movements and Results. In some places, most notably along the Mid-Atlantic Ridge—a submarine ridge running north to south in the middle of the Atlantic Ocean—the plates are slowly pulling apart. In others, as along the western edge of North America, they are grinding horizontally against one another, or, as along the Pacific-facing coast of Japan, meeting head-on. Where they are pulling apart, deep-seated molten rocks rise in the fracture zone to build ranges of volcanic mountains; thus, the submarine Mid-Atlantic Ridge is comparable in height and extent to the Rocky Mountains of western North America. Where the crustal plates are moving horizontally against one another, earthquakes are frequent, as along the famous San Andreas fault zone of California. Where the plates collide head-on, the results are gigantic rock folds, deep ocean trenches paralleling the folds, numerous earthquakes, and marked volcanic activity, as in the Japanese islands and adjacent Pacific Ocean bottom. Figure 4–5 illustrates these movements of the plates of the earth's crust.

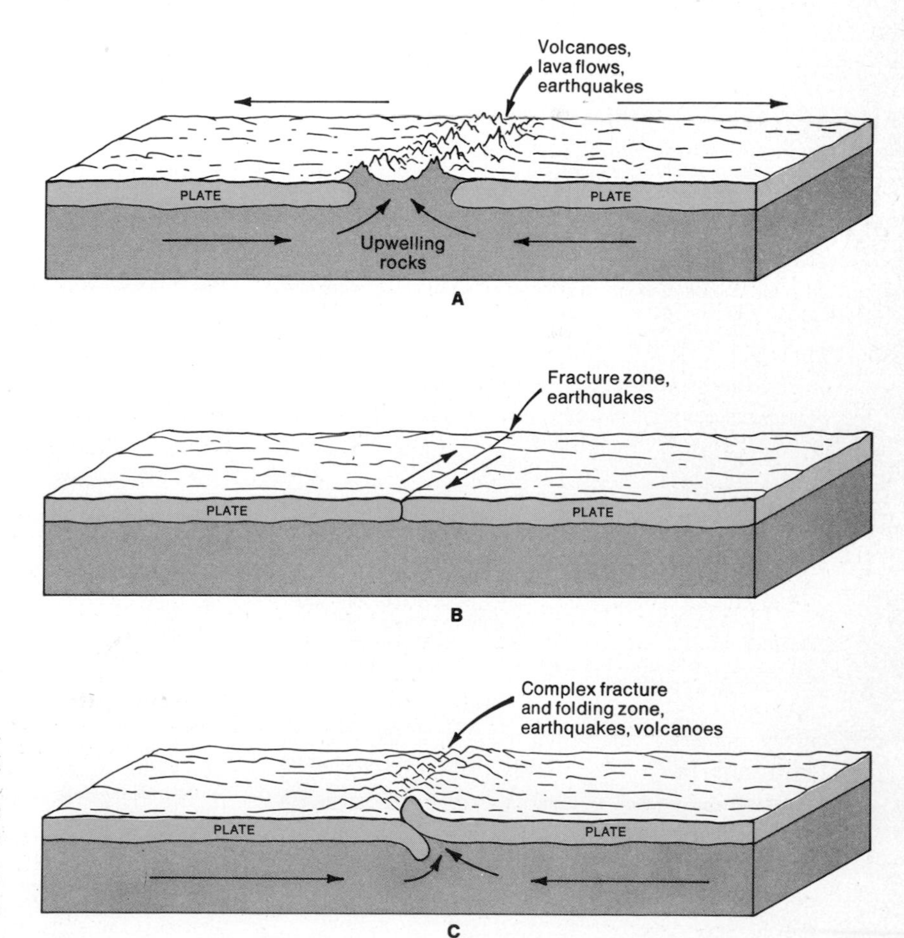

FIGURE 4–5

Movements of the earth's major *crustal plates.* A. Crustal plates pulling apart (as in the Mid-Atlantic Ridge). B. Crustal plates moving horizontally against one another (as in California). C. Crustal plates underriding and/or overriding one another (as along the coast of Japan).

The movements of the crustal plates are most significant in the creation of volcanic features, earthquake zones and features, and the buckling into upfolds and downfolds of major portions of the earth's crust, both beneath the oceans and along the continental margins.

Rocks of the Earth's Crust

Rocks and Landforms. The crust of the earth is composed of rock. Landforms themselves, being the shapes of the crustal surface, are likewise composed of rock, either solid bedrock or loose, broken rock fragments. Therefore, a considerable part of the understanding of what landforms are, how they form, and where they are, depends upon an understanding of the composition and nature of the rocks composing them. For a glossary of rock terms, see Figure 4–6.

Rock Composition. Rocks are mineral solids. The major rock-forming minerals are quartz and the feldspars and ferromagnesians. The major mineral-forming elements are oxygen, silicon, aluminum, and

FIGURE 4–6

A glossary of some fundamental rock terms.

Acidic vs. basic rocks: acidic rocks are high in silica, and low in calcium, magnesium, iron, and so on (as in granite, and as in the sial); while basic rocks are high in calcium, magnesium, iron, and so on, but are relatively lower in silica (as in basalt, and as in the sima).

Bedrock: essentially unweathered solid rock, whether outcropping at the surface or buried under a mantle of weathered rocks (overburden; mantle rock; regolith).

Coarse-textured vs. fine-textured rocks: refers to the size of the grains or particles composing the rock (as coarse-grained sandstone or fine-grained shale).

Consolidated (indurated) vs. unconsolidated rocks: "solid" rock (as a chunk of sandstone) as opposed to a nonhardened (nonindurated) deposit of clay or silt.

Crystalline vs. noncrystalline rocks: crystalline rocks are composed of crystals that formed within the rocks as molten materials cooled (as crystals of quartz and feldspar in a cooling granite mass); as opposed to such a noncrystalline rock as sandstone (the component grains of which are sand grains rather than grains of individual crystals); slow-cooling rocks (time for crystals to grow) are termed *coarse-crystalled* (as granite); fast-cooling rocks (little time for crystals to grow) are termed *fine-crystalled* (as basalt).

Density of rocks: refers to specific gravity (relative density), which, in this regard, is the weight of a particular rock of a given volume compared to the weight of water of the same volume—one cubic foot of water (62.4 pounds) is taken as unity "1;" pumice rock floats on water, and its specific gravity is therefore less than 1; other rocks, being denser than water, have specific gravities greater than 1; so-called light (low-density) rocks have specific gravities in the general order of 2.5 to 2.7 (granite is 2.7), and so-called heavy (high-density) rocks are 3 to 3.5 (gabbro is approximately 3).

Hardness of rocks: a comparative reference to the ease or difficulty with which rocks (more correctly, minerals) may be scratched (as by a diamond point or hardened steel point)—on the *Mohs scale*, talc is softest (#1) and diamond is hardest (#10).

Intrusive vs. extrusive rocks: intrusives are those molten rocks that have forced themselves into other rocks (as a layer of basalt forced between two sedimentary rock strata); extrusives are those rocks that have issued from within the earth's crust to collect and harden on the surface (as lava flows).

Jointing in rocks: refers to cracks, or crack planes, produced in rock layers or rock masses by tension or torsion; in sedimentary strata, the joints are usually vertical or slightly inclined to the bedding planes of the strata; in nonsedimentaries, the joints or joint planes usually develop as parallel sets at right angles to one another, thus outlining blocks of the rock mass (as in jointed granite).

Light-colored vs. dark-colored rocks: light-colored rocks are acidic rocks composed mainly of light-colored minerals (as in granite); dark-colored rocks are basic rocks composed mainly of dark-colored minerals (as in basalt).

Stratified (bedded) rocks: rocks having definite layers (strata) incident to deposition (as superposed layers of sandstone or limestone, or layers within each) as opposed to massive, unlayered masses of, for example, rhyolite or basalt.

Thick-bedded vs. thin-bedded rocks: thick-bedded rocks have thick strata, as opposed to the comparatively sheet-like strata of thin-bedded rocks.

iron (together comprising about 88 percent of the earth's crust), plus lesser amounts of such elements as calcium, sodium, potassium, and magnesium, and far smaller amounts of such chemical elements as gold, silver, platinum, and so on.

Classes of Rocks. All rocks can be divided into three types: *igneous, sedimentary,* or *metamorphic.*

Igneous rocks (Latin: *igneus* = fire) are rocks that have cooled and solidified from an earlier molten state. In some cases, the molten rock, or *magma,* forced its way into beds of other rock where it cooled very slowly. Such *intrusive* igneous rocks sometimes formed large masses at great depths; these rock masses are called *batholiths* (Greek: *bathos* = depth; *lithos* = rock) and are often composed of *granite.* (Today, intrusive igneous rocks may be exposed at the surface as a result of erosion carried on over great periods of time.) *Granite* (Latin: *granum* = grain) is one of the *crystalline rocks,* as it is composed of myriads of interlocking crystals; it occurs as a massive rock, often without any variation in its nature over very large areas. Other intrusive rocks are found in the form of *sills* (thin layers of molten rock formed between layers of other rock) and *dikes* (filling cracks running vertically or at right angles to layers of other rock).

Extrusive igneous rocks are better known and more spectacular than the intrusive ones. They are formed when molten material, forced out (*extruded*) from the interior of the earth onto its surface, cools and hardens. That which flows over the surface and eventually hardens into rock is called *lava;* some is hurled explosively into the air, and, cooling and solidifying in the air, falls as *volcanic ash* and *cinder.*

Sedimentary rocks are composed of particles derived from the breaking up of other rocks. After being transported by water, wind, or ice, the particles are deposited as *sediments,* and are eventually *consolidated* into rock. *Consolidation* is accomplished by either *compaction,* in which the mineral grains adhere tightly to one another as a result of pressure (usually in the case of clayey materials), or *cementation,* whereby grains of sand or other materials are cemented together by lime, iron, or silica.

Sedimentary rocks, as well as the sediments from which they are composed, are classified on the basis of the size of their grains. They range from coarse-grained *gravel,* which is cemented into a *conglomerate,* through finer *sand* (which may become *sandstone*), and *silt* or *clay* (both of which may become *shale*), to the finest-grained *lime* and *marl* (which may later form *limestone* and *dolomite*).

Most sediments are deposited in layers, or *strata,* each individual stratum reflecting a set of conditions existing at the moment of deposition. Thus, a stream flowing into the sea may deposit a layer of coarse gravel during flood periods, a stratum of sand during the waning stage of the flood, and a series of strata of silt and clay during the nonflood periods. These strata naturally accumulate with the oldest on the bottom and the most recent on the top; eventually, if conditions are favorable, they may all be consolidated into *stratified rock.*

Metamorphic rock (Greek: *meta* = change; *morphos* = form) is rock that has been altered from its original form by tremendous heat and pressure. The original rocks, being buried under a great thickness of other rocks, have had their physical structures as well as their mineral components transformed by the great weight of the layers above them and by the high temperatures prevailing deep within the earth. The transformation is often so complete that it is impossible to determine what the original material may have been. A table of such transformations

Unconsolidated sediment		Consolidated sedimentary rock		Metamorphic rock
Gravel	→	Conglomerate	→	Quartzite
Sand	→	Sandstone		
Clay	→	Shale	→	Slate → Schist
Lime or marl	→	Limestone or dolomite	→	Marble

Molten igneous material		Igneous rock		Metamorphic rock
Magma	→	Granite	→	Gneiss
Magma	→	Lava (some types)	→	Schist

FIGURE 4–7

The consolidation and metamorphism of various kinds of rock.

is presented in Figure 4–7. Some metamorphic rocks, especially *gneiss,* are composed of interlocking crystals and thus, like granite, are also crystalline rocks.

Underlying the sedimentary rocks of the continents is the *basement complex,* a mass of very ancient igneous and metamorphic crystalline rocks, chiefly granite and gneiss. In some continental areas, the basement complex is exposed at the surface in the form of a vast low dome called a *shield.* Such shields constitute the most stable portions of the earth's crust. Notable examples include the Canadian, or Laurentian, Shield of eastern Canada, and the Fenno-Scandian Shield of northwestern Europe.

The Shaping of Landforms

Primary Processes

Landforms do not happen by chance: each shape or combination of shapes, whether it be the mass of the Himalayas or a cluster of sand dunes, is the result of much sequential work accomplished over long periods of time. Some landforms are shaped by processes operating within the earth itself, particularly within the lithosphere: these are the *endogenous,* or *tectonic, processes* carried on by *diastrophism* and *vulcanism.* Other landforms are shaped by processes operating at the surface of the lithosphere: these are the *exogenous processes,* carried on by *weathering, erosion,* and *deposition.*

Endogenous Processes

Diastrophism. The earth's crust is under constant strain and stress, and at certain times and places, it yields to those forces. When it does, the result is a modification, major or minor, slow or cataclysmic, of the shapes of the crustal surface. Such modification is called diastrophism.

Under some conditions, parts of the crust are compressed into various types of folds. Such action takes place very slowly, over hundreds of thousands or tens of millions of years. Furthermore, for the rocks to be folded, they must be at a considerable depth in the crust, where the confining pressure of the rock layers above them can prevent them from breaking, as would be the result with unconfined surface rocks. All types of rock may be subject to folding, but it seems best displayed in the stratified sedimentary rocks (Fig. 4–8). Folding may affect such rock formations as granites, but this is very difficult to determine due to the widespread homogeneity of the rock.

Warping is the simplest form of folding, consisting merely of a slight bending of the earth's crust. Though its effects may not be detectable on a local scale, warping is the reason that nearly horizontal beds of marine sediments are now found at considerable elevations above the sea in many parts of the world. It continues today in many areas, perhaps even over most of the earth's surface, and results in the slow drowning of river mouths in areas of coastal subsidence, and elsewhere, in the stranding of seaports and the creation of raised beaches and uplifted marine terraces. A most remarkable example of such warping within historic time is provided by the Temple of Jupiter Serapis at Pozzuoli, west of Naples, Italy. There, a set of stone columns erected by the Romans some 2,000 years ago now stands at the edge of the sea. At a height of 20 feet above the water level is a series of holes bored into the solid stone by a marine-type organism (*Lithodomus*), and the shells of some of these marine animals are still inside some of the holes. After the temple was erected, then, the land was downwarped some 20 feet, thus submerging the columns and making it possible for the borers to attack the stone. Later, it was uplifted by about the same amount. That this was a local warping and not a rise and fall of the surface level of the sea is evident when

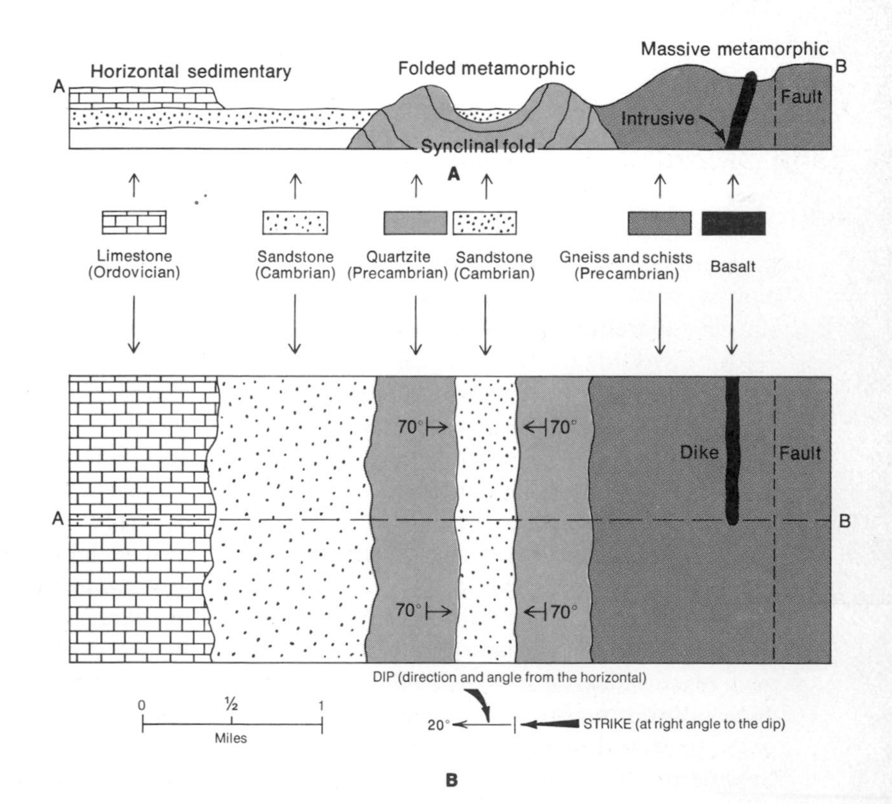

FIGURE 4–8

A sample of a geologic map. A. A land surface profile and geologic cross section. B. An areal geologic map representative of *bedrock geology* (as opposed to surface-mantle materials). (Refer to Appendix E regarding geologic ages as stated under each symbol in the legend.)

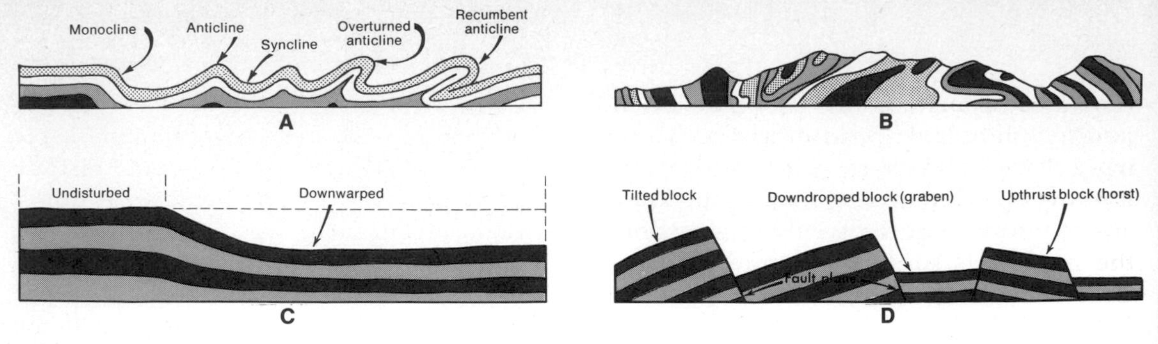

FIGURE 4–9

Diastrophic changes in the earth's surface. A. Relatively simple folding of the earth's crust (eroded). B. Complex folding (eroded). C. Simple warping (uneroded). D. Block-faulting (uneroded). Elevated blocks form hills or mountains; depressed blocks form valleys and trenches.

one realizes that the seas of the world are all interconnected, so that a rise or fall of the sea at Pozzuoli would have to have been duplicated everywhere else in the oceanic world (which did not occur).

Not all warping occurs on so small a scale, however. Marine strata, deposited a few feet below sea level, today constitute the surface rocks across large expanses of northern Arizona's Grand Canyon Country plateaus, often, at elevations of more than 7,000 feet; yet, essentially, they are still in their original horizonal attitude.

In many places, the stratified rocks of the earth's crust have been *buckled* rather than warped, creating several types of folds: simple uparching folds, or *anticlines* (Fig. 4–9); downfolds (or troughs) called *synclines;* one-sided folds, or *monoclines;* and such asymmetrical folds as *overturned* and *recumbent anticlines.* These may all be on very small scales—so small that a rock containing an entire anticline can be held in one's hand—or on such large scales that ridges and even entire mountain ranges are created—an example being the Appalachian Mountains in central Pennsylvania, formed by a succession of anticlines and synclines. Some anticlines and synclines reach such huge proportions that they cover a major portion of a continent, in which case the folds are termed *geosynclines* and *geoanticlines.*

As was indicated above, folding requires that the rock be considerably confined by burial beneath other beds of rock. The absence of such confining pressure results in the breaking rather than the bending of the rock. The rocks of the outer part of the earth's crust have been shattered in every direction by fractures. Those along which no movement of one side relative to the other has occurred are called *joints;* those along which slippage has occurred are called *faults.*

Joints result from the cooling and contraction of igneous rocks, the drying of sedimentary rocks, and the stresses produced by diastrophism in the folding of the local area. They commonly occur in sets running parallel to each other, and, in many places, two or three sets in different planes split the bedrock into blocks.

Fractures along which displacement has occurred are called faults. Most faults have moved largely in a vertical direction, but a few, mainly those associated with the borders of the crustal plates, have had a horizontal-lateral movement. In some rare cases, one block has ridden up over the other and formed a *thrust fault.*

If the movement along the fault is vertical, a cliff called a *fault scarp* is formed

(Fig. 4–10). Sometimes the area between two parallel faults is dropped downward, forming a *graben;* elsewhere, it may be elevated, forming a *horst* (Fig. 4–9). Some fault scarps are of great proportions: the one forming the eastern face of the Sierra Nevada in California, for example, extends well over 10,000 feet high.

Movements occur repeatedly along a fault. Between the movements, tension builds up until it is strong enough to overcome friction. When the tension is suddenly released, the two blocks of the fault slip abruptly into new positions and simultaneously create a jarring earth tremor or earthquake that, transmitted through the crust, may be felt far from the fault.

Vulcanism. The most readily visible landforms produced by vulcanism are *volcanoes.* Some, such as the very conspicuous *conical volcanoes* represented by Fujiyama and Vesuvius, are composed of layers of lava alternating with layers of ash and other material deposited during explosive activity (Fig. 4–11A); others, such as Mauna Loa in Hawaii, are broad *dome volcanoes* created by nonexplosive accretions of lavas (Fig. 4–11B).

Less eye-catching landforms created

FIGURE 4–10

The Hebgen fault scarp that developed at the time of the Montana earthquake of August 17, 1959. (Montana Highway Commission)

FIGURE 4–11

Structural developments of the two major types of volcanoes. A. The *conical volcano*, such as Fujiyama, is usually the result of explosive activity. B. The *dome volcano*, such as Mauna Loa, is usually the result of nonexplosive activity.

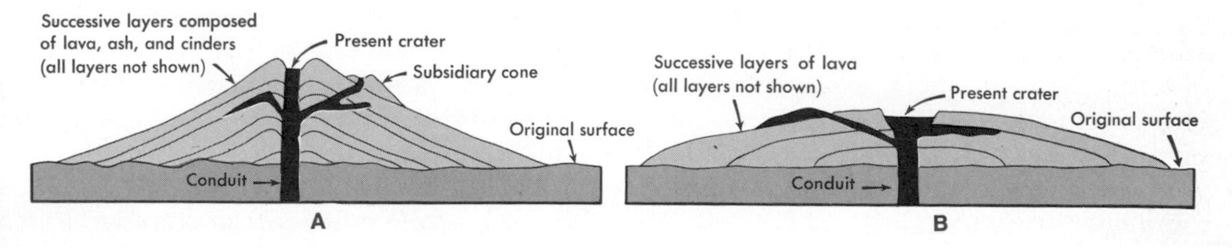

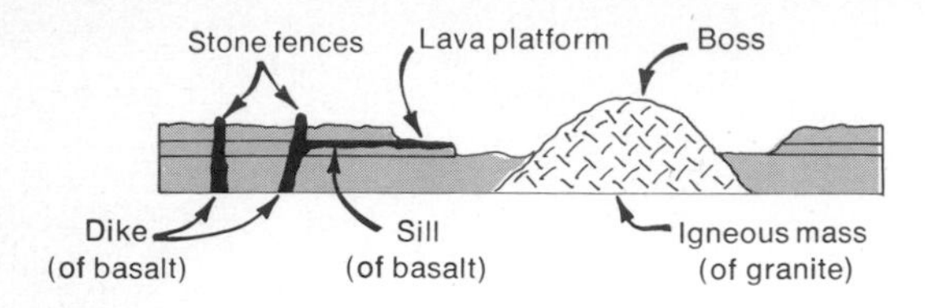

FIGURE 4–12

Such intrusive features as dikes and sills may produce, following weathering and erosion, *stone fences* and *lava platforms;* subterranean igneous masses may be exhumed to form *bosses.*

by volcanic extrusions are *lava plains,* or *lava plateaus,* where lavas have issued from crustal vents and fissures to cover, layer-on-layer, extensive portions of the earth's surface. The Columbia Plateau of the northwestern United States provides an example.

In still other parts of the lithosphere, molten rock may not reach the surface. Instead, it moves as intrusive material in crustal cracks or between rock layers to form *dikes* (vertical), *sills* (horizontal), and *subterranean masses.* But, the fact that dikes, sills, and buried lava masses are not formed at the surface does not mean that they have no landform significance, for subsequent weathering and erosion may expose them as *stone fences* (from dikes), *lava platforms* (from sills), and *bosses* (from igneous masses) (Figs. 4–12 and 4–13). Additionally, masses of molten material may force layers of sedimentary rocks upward to form a sort of crustal blister called a *structural dome* (Fig. 4–14).

Exogenous Processes

Mass Wasting. There is a constant struggle between the processes that tend to elevate, or build up, the earth's surface and

FIGURE 4–13

This huge mass of granite has been exhumed by the processes of weathering and erosion to become a minor landform. Located in the southern part of the Appalachian uplands near Atlanta, Georgia, it is known as Stone Mountain. Its smooth, rounded form is largely the result of the "leafing off," or *exfoliation,* of the granite mass. (Hillers, U.S. Geological Survey)

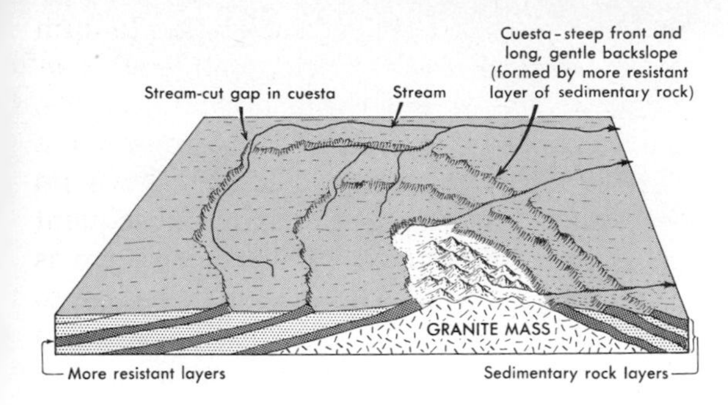

FIGURE 4–14

A diagram of the Black Hills area of South Dakota, an area that typifies an *eroded structural dome.* Whether a subterranean mass becomes exhumed or not, it may still deform overlying rock layers enough to create a structural dome.

those that tend to wear it down. The most widespread of all the downwearing processes is *mass wasting*. It is a direct result of the downward pull of gravity upon rock materials, and it affects all loose materials located on sloping surfaces, especially thick, loose materials lying on very steep slopes that are well lubricated by water. Under certain circumstances, even apparently solid bedrock may slump, avalanche, or drop, as in situations where there is rock overhang or where there are cliffs or near-cliffs.

Over the earth's crustal surface as a whole, the downwearing caused by mass wasting is tremendous, even though the features produced are often far less conspicuous than those caused by running water. The mass wasting processes that play such a major role in moving materials from higher to lower levels, thus helping to wear away the earth's high spots and fill in the low ones, are numerous and complex. For a listing of the major ones, see Figure 4–15.

Weathering. Weathering is the breaking up of surface and near-surface rocks by any chemical and mechanical means (Fig.

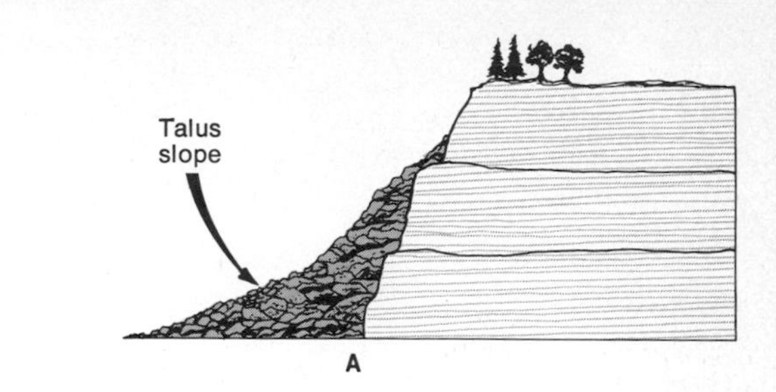

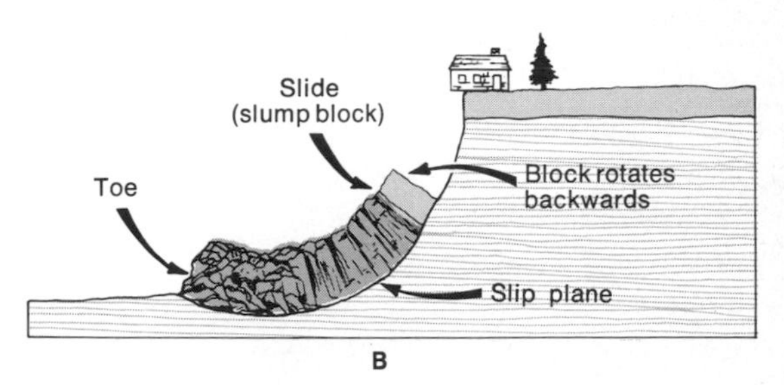

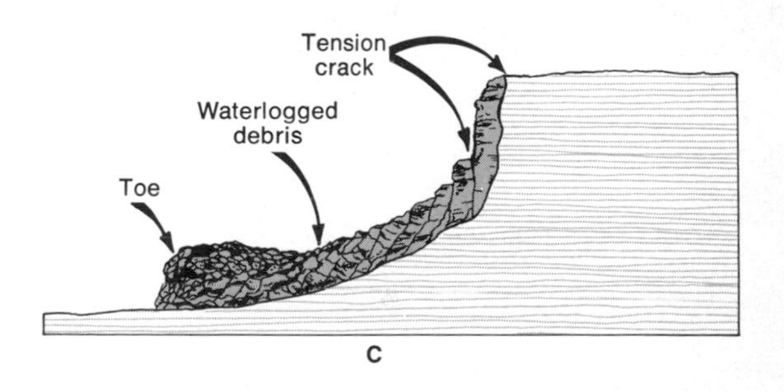

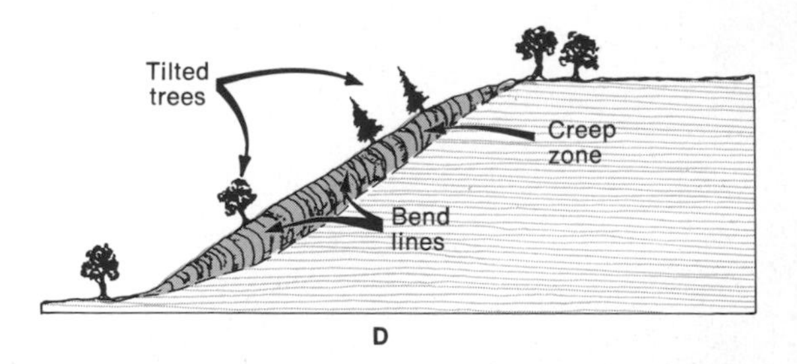

FIGURE 4–15

The most widespread of all the processes of downwear is *mass wasting,* the most common types of which are shown above. A. *Fall and roll* of rock debris from a cliff-face. B. *Slide* (landslide) movement of rock materials on a steep slope. C. *Flow* of earth materials on a slope. D. *Creep* of earth materials on a slope.

FIGURE 4-16

Some results of weathering. Top left, a mantle of *regolith* above weathered limestone bedrock in the Galliluro Mountains of Arizona. (Jerome Wyckoff) Top right, *jointed bedrock* at Donner Pass in the Sierra Nevada, California. (Helen Faye) Bottom left, on Chipmunk Mountain in the Hudson Highlands of New York State, thin sheets of granite separated from the larger mass by the process of *exfoliation*. (Jerome Wyckoff) Bottom right, *woolsacks* produced in the weathering of quartz monzonite at Joshua Tree National Monument, California. (Jerome Wyckoff)

4–16). If accomplished by *chemical actions*, it is termed *decomposition;* if the result of *mechanical actions*, it is called *disintegration*. Working singly, or, as in most regions, in combination, these processes change bedrock to rock fragments, and rock fragments to successively smaller pieces or particles; the end product is a mantle of unconsolidated rocks that is known as *regolith*. In this way, rocks are broken down to provide the basic parent materials for soils and the loose materials for mass wasting and eventual transportation by the agents of erosion.

Decomposition, or chemical weathering,

is carried on by the processes of *oxidation, carbonation, hydration,* and *solution.* Rock decay by oxidation results from the union of rock elements with oxygen, as in the union of iron and oxygen to form the familiar "rusty" iron oxides. Carbonation is the union of such elements as carbon (from carbonic acid solutions) and calcium. Hydration is the taking on of water in actual chemical combination, as in the formation of hydrous oxides of iron. Solution represents the physical breakdown of a substance in a liquid resulting in a homogeneous mixture. Sometimes the soluble material occurs as scattered mineral grains in a mass of otherwise insoluble rock; the solution of such grains has the effect of weakening the whole mass. For example, the cementing agent in sedimentary rocks is frequently highly soluble lime; when it is dissolved, the whole rock mass crumbles. In some cases, as with limestone, the entire mass of the rock is soluble.

In all these ways, decomposition plays its part in the weakening and breakdown of the surficial rocks of the earth's crust. Decomposition is at a maximum in those regions of the earth that are perennially hot and moist, and at a minimum in regions of long-continued cold and ice-locked moisture or in areas of pronounced dryness, as in the deserts.

Disintegration, or mechanical or physical weathering, is the actual fracturing of rocks, whether bedrock or pieces of rock. It is accomplished by several processes, most importantly: *frost action, organic weathering, salt crystal growth, unloading, differential heating and cooling,* and *diastrophic weathering.*

Frost action is a process in which water enters cracks and crevices in rocks and then freezes, with great expansive force, breaking the rocks apart. Where alternate freezing and thawing are common, as in high mountains and portions of the polar regions, frost action is at a maximum, and it normally results in an abundant regolith composed of more or less angular fragments. When collected at the base of slopes, the regolith is called a *talus,* often forming *talus cones* or *talus aprons.* When developed on level or gently sloping surfaces, it is termed a *felsenmeer* (German: sea of rocks).

Organic weathering is the result of animal activities and plant growth. Ants, termites, and burrowing animals are responsible for some mechanical breakage, but that accomplished by the slow, inexorable growth of plant roots is far more important. If one envisions the countless billions of prying and wedging plant roots reaching into the rock mantle, an appreciation of the results is inevitable.

Salt crystal growth is somewhat like the growth of ice crystals formed as water freezes. Wherever land moisture laden with mineral salts moves to or near the surface and evaporates, salt crystals form and enlarge to exert prying and flaking actions on rock materials. This type of weathering is most prevalent along the seeps that occur on and at the bases of cliffs located in arid or semiarid regions. The result is a sort of sapping and flaking action that produces caves, pockets, and alcoves in the rock faces, and accumulations of fine, gravity-impelled rock debris.

Unloading is the exploding off and sheet breakage of rocks due to a decrease in the surrounding pressure. As buried rocks subject to the pressures of overburden are exhumed by erosion, the release of pressure causes them to expand and break. Unloading is especially apparent in rock quarries where the rapid removal of overburden by mechanical means hastens the explosions of rocks from quarry walls and floors.

Differential heating and cooling of rock surfaces results from day-and-night and seasonal variations in temperature. (Heating causes expansion and cooling causes contraction.) Because the rock materials neither expand nor contract evenly, differential stresses are created and rock breakage ensues. The repeated expansion and contraction of individual mineral grains due to changes in temperature cause the grains to become loosened and detached. Water, freezing in the tiny spaces between them, has the same effect. Such removal of grains from their sockets is termed *granular disintegration,* and often results in the accumulation of *grus* (grain-sized detritus—that is, disintegrated but not decomposed material) in pockets on the surface of the rock and about the bases of boulders, slopes, and so on.

Daytime heating causes the expansion of a thin layer at the surface of each mass of rock. At night, the penetration of heat into the mass causes the expansion of deeper layers at the same time that the surface is contracting because of nocturnal cooling. Lines, or zones, of stress are thus created, eventually producing shear planes along which the rock splits into thin sheets; the subsequent breaking off of these sheets is known as *exfoliation.*

Most bedrock is broken by a series of parallel cracks penetrating deep into its mass—the result of either tension or torsion from the bending of that portion of the earth's crust at some time in the remote past. Often, several such sets of planes intersect each other at various angles throughout great areas of bedrock. Along such planes of breakage, or *joints,* heat and cold are able to penetrate well into the bedrock. Disintegration is thus induced, and the rock separates into joint-bordered blocks.

Temperature changes penetrate the individual blocks more deeply on the edges than on the faces, and still more deeply at the corners. Disintegration is therefore hastened at such places, and results in *spheroidal weathering,* the process in which the originally cuboid angular blocks eventually assume rounded forms. Since such blocks resemble sacks stuffed with fleece, the term *woolsack weathering* is sometimes applied to the process (Fig. 4–16). In arid and semiarid regions, where the bedrock is composed of either sandstone or granite, whole mountains may be made up of such boulders, representing the disintegration of a once-solid bedrock mass into separate spheroidally-weathered blocks.

In addition to the weathering of rock materials due to natural decomposition and disintegration, there is that brought on by humans. *Human-induced weathering* is associated with the use of tools, machines, and explosives. The breaking up of bedrock and regolith by the use of picks, mattocks, crowbars, and sledgehammers has proceeded through millenia; more recently, it has involved dynamite, jackhammers, bulldozers, and powershovels; and today, if people choose, they can literally disintegrate sections of the earth's crust in an instant by means of nuclear explosions.

Erosion and Deposition. *Erosion* is predominantly a process of *transportation* of weathered materials followed, inevitably, by the *deposition* of those materials. As the materials are moved, they grind and bang against one another and scrape the surfaces over which they pass, thereby performing *corrasion,* or *abrasion.* At the same time, if water is the erosive agent, some of the transported materials, and pieces of the surface over which they are carried, are dissolved to accomplish the work of *corrosion.*

The major agents of erosion and deposition are *moving water, moving air,* and *moving ice.* The most important element is motion, for without it the agents have no working power: thus, stagnant water cannot cut valleys; still air cannot shift dusts or

FIGURE 4-17

The major agents of erosion and deposition are *moving water, moving air,* and *moving ice.* Top left, an oblique air view shows the results of erosion by moving water in the Rocky Mountain area of the northwestern United States. (Fritz Henle, Cities Service Co.) Top right, a glacially-carved (eroded) valley in the North Tongass National Forest of Alaska, whose steep sides rise to 5,000 feet. (L. J. Prater, U.S. Forest Service) Bottom left, river deposition in the narrow bed of Henson Creek near Lake City, Colorado. (Standard Oil Co., N.J.) Bottom right, the landforms created by eolian deposition are usually local, as in the case of these sand dunes near the center of the floor of Death Valley, in California. (George A. Grant, National Park Service)

sands; and stationary ice cannot move earth debris or grind and scour bedrock (Fig. 4–17).

Moving water is the dominant agent of erosion and deposition over the earth's surface. This is true even in the deserts, as indicated by the typical occurrence of such water-cut and water-deposited features as gullies, arroyos, alluvial fans, and dried lake beds. Water may move in thin sheets (*sheet flow*), in more narrow forms (*stream flow*), or as shore currents (*littoral currents*) and *waves.* The resultant landforms vary from such erosional (*degradational*) forms as gullies, canyons, broad valleys, wave-cut cliffs, and current-scoured channels to such depositional (*aggradational*) forms as deltas, alluvial fans, floodplains, alluvial coastal plains, and sandbars.

Moving air, in the form of wind, has far less ability to transport earth materials than does moving water or moving ice; however, it is often powerful enough to pick up and carry dusts (*loess*) or to move sands. Dusts caught in a wind may be lofted to great heights and scattered over large areas (Fig. 4–18), but moving air seldom lifts sands or gravels more than two or three feet above the surface. In general, the erosional and depositional work accomplished by wind is local; yet, locally, the landforms created by wind, or *eolian,* action may predominate, as in sand dune areas (Fig. 4–17), or where sandblast, or *wind corrasion,* has scoured and etched rock surfaces into often fantastic patterns (Fig. 4–19). In a few instances, as in the North China Plain, the work of the wind is more than local: there, eolian deposition has created a thick, virtually level deposit of loess over many extensive areas.

Moving ice, like moving water, has local effects—as sea or lake ice it grinds against rocky shores or pushes loose earth materials into littoral ridges—but the major erosional (and depositional) effects of moving ice occur in connection with the movement of glaciers. Both valley and continental glaciers create erosional (*scour*) and depositional landforms of considerable variety, such as: the ice-gouged, sea-filled lower valleys, or *fiords,* of Norway; the U-shaped valleys of the higher Colorado Rockies; the

FIGURE 4-18

Wind-lofted dusts may be carried to great heights and over long distances. At left, a dust storm in Baca County, Colorado; at right, large amounts of soil being lifted up and carried away in such a storm. (Soil Conservation Service, USDA)

scoured hills and scour basins of Canada's Laurentian Shield (Fig. 4–20); and the loose-debris depositional forms of glacial outwash plains and morainal ridges in the state of Illinois.

As people are involved in weathering so are they also erosional and depositional agents. *Human-induced* erosions and depositions do not affect the continental surfaces as a whole, but they may dominate local landform expression (Fig. 4–21). Some examples are the "cut and fill" associated with highway and freeway construction, the removal of hills (especially where they constitute barriers in cities), the filling in of marsh lands, the terracing of agricultural land, the step-terracing of residential lots and streets in hilly areas, the debris knolls and ridges left by strip mining, the debris-glutted bottoms of valleys in areas of placer mining and deforestation, and the huge

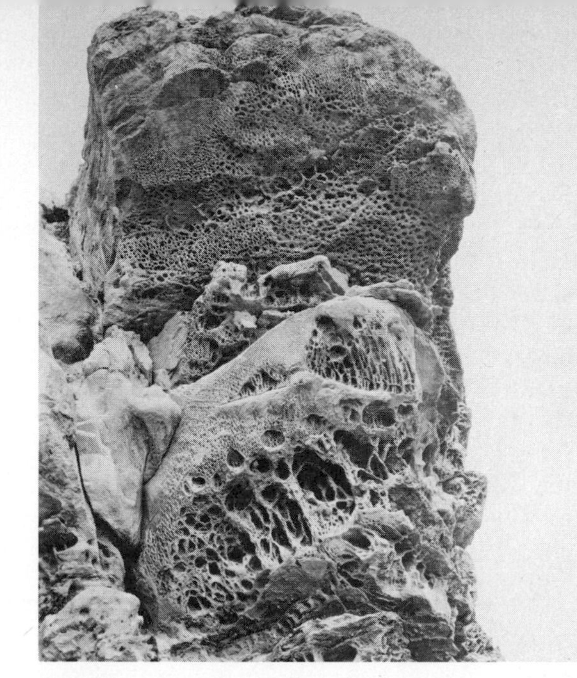

FIGURE 4–19

This twelve-foot high sandstone bluff near Livingston, Montana, has been weathered and sandblasted (*wind corrasion*) into new and intricate shapes. (Darton, U.S. Geological Survey)

FIGURE 4–20

A glacially scoured bedrock area in the Labrador section of the Laurentian Shield. Note the many lakes and ponds in the "scour depressions." The sinuous ridge angling across the area from lower left to upper right is an *esker*, one of the depositional landforms of continental glaciers. (Edward Cabot, from *The Geographical Review*, July 1946, published by the American Geographical Society)

FIGURE 4-21
Adding to the weathering done by natural agents, humans have also become erosional and depositional agents. A typical example of their weathering influence is this "cut and fill" operation near an interchange in California's freeway system. (California Division of Highways)

dumps and open pits associated with the mining of iron and copper ores. Modern tools and machines have given humans a considerable erosional and depositional ability. The machine most representative of this new power is the bulldozer, which is used to such an extent that many local landforms are now said to be "bulldozogenic" in nature. Thus, whether in the end it proves to be productive of good or of ill, "man has truly become a geomorphic agent" to be reckoned with.

Landform Classification

The major subaerial landforms are *plains, plateaus, hills,* and *mountains.* Figure 4–22 illustrates, diagrammatically, the general character and positional relationships of the major submarine and subaerial landforms. The minor subaerial landforms, such as floodplains and sand dunes, will be discussed in the two following chapters.

Major Subaerial Landforms

Plains. A plain is a sizeable area having low *relief* (the vertical distance between ridge crests and valley bottoms) and low *altitude* (the vertical distance above sea level) relative to the surrounding areas (Figs. 4–22 and 4–23).

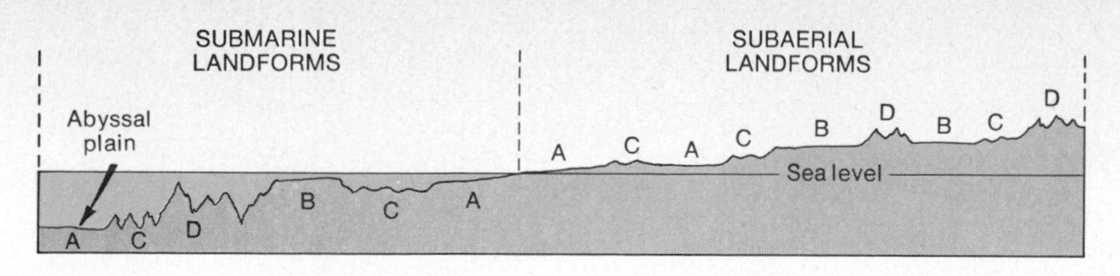

FIGURE 4–22

A profile showing the major types of landforms in the ocean basins and on the continents: (A) plains, (B) plateaus, (C) hills, and (D) mountains.

Plateaus. Plateaus are also known as *tablelands,* or elevated plains. Like plains, they are of low relief, but, unlike them, they lie at considerable altitudes, normally 2,000 feet or more above sea level. The Colorado Plateau (Fig. 4–24), for example, has an average altitude of 7,000 to 8,000 feet above sea level, yet the eye may scan many miles of its surface without discerning any major landform interruptions. Some plateaus or portions thereof have been cut up into numerous deep gorges and knife-edged or flat-topped ridges, as in the Grand Canyon section of the Colorado Plateau, or into V-shaped valleys and rounded hills, as in the Allegheny Plateau. Such erosionally modified tablelands are termed *dissected plateaus,* and often appear to be hill regions.

FIGURE 4-23

Occasionally, plains rise to elevations above 5,000 feet. An example of such *high plains* is the western margin of the Great Plains of the United States, shown here surrounding the city of Alliance, in western Nebraska. The plains in this region rise uninterruptedly from the Gulf Coast. (Fairchild Aerial Surveys)

FIGURE 4–24
The southern part of the Colorado Plateau near Winslow, Arizona, is a flat land that has been slightly dissected. Canyon Diablo, 500 feet wide and 225 feet deep, is one of the few canyons that have been eroded into this type of flat land surface. (Spence Air Photos)

Hills. Hilly country is, in a sense, a miniature version of mountain country (Fig. 4–25). Local relief is marked but is less than 2,000 feet. Slopes may be as steep or even steeper than mountain slopes, but they are shorter; and, despite comparatively small summit areas, there is often relatively more nearly level land on the ridge tops than is characteristic of mountainous lands. Hilly country is markedly different from mountain country, however, in that there is no readily discernible landscape zonation from bottom to top. The Berkshires of western Massachusetts, the South China Hill Country, and the Southern Upland of Scotland are examples of hilly country; Figure 4–25 provides pictorial examples of other hilly regions.

Mountains. True mountain country has nothing in common with plains or plateaus, unless the latter have been dissected to create relief conditions of mountainous proportions (Fig. 4–26). Mountainous territory is marked by comparatively high altitude (somewhere around 2,000 feet at the minimum), relatively high relief (usually at least 2,000 feet locally), long slopes (regardless of local degree of steepness), small summit areas, and pronounced vertical zonation of landscapes. One of the most obvious attributes of mountain country is the existence of vertical vegetation zonation; in mountains, the zonation is conspicuous, whereas in hilly regions it is absent or only weakly discernible. With respect to vegeta-

FIGURE 4-25

Often, hilly country appears to be miniature mountain country. Top, hills near Cashmere, in central Washington. (Washington State Department of Commerce) Center, the eastern foothills of the Coastal Range near Los Banos, in the southern San Joaquin Valley, California. (Helen Faye) Bottom, rolling hills of Emmenthal, Canton Berne, Switzerland. (Swiss National Tourist Office)

FIGURE 4–26
The high eastern front of California's Sierra Nevada thrusts steeply above the Owens Valley, with Mount Whitney (14,495 feet, at arrow) almost lost in the high complex. In the middle foreground, a local line of hills, separate from the Sierra, interrupts the valley floor. (H. L. Kostanick)

tion types and changes, ascending from the bottom to the top of most mountains is comparable to traveling from lower latitude regions to higher latitude regions.

Minor Subaerial Landforms

Variety. The minor landforms comprise a nearly infinite variety of local features. As such, they represent the detailed expressions present within the major landform regions of plains, plateaus, hills, and mountains.

In regions of dry climate, characteristic minor landforms include alluvial fans, arroyos, dunes, and mesas; in regions of humid climate, they include gullies, floodplains, ridges, and hills; and in regions of continental glaciation, ridges and blankets of glacial debris, scoured bedrock hills, and bedrock basins are added.

Significance to Humans. In terms of human activities, minor landforms have a direct and essentially personal relationship. As people cultivate the land, construct their buildings and settlements, establish their lines of transport, and so on, it is the local, minor landform conditions, as opposed to the extensive, major landform conditions, that are of most direct significance to them. Moreover, as regards esthetic and recreational aspects, it is usually a minor landform to which people personally relate, such as a favorite hill or ravine, a breathtaking rocky headland or dune-bordered shore, a favorite ski slope, or a prized camping spot.

Landform Analysis

Major Systems of Analysis

There are two major methods used in the analysis of landforms: the *empirical* and the *genetic*. The empirical method involves the description of their origin and evolution through natural processes. For purposes of illustration, the landform known as a *mesa* will be considered in this regard.

Using the empirical method, a mesa may be described as a landform having a level top, encompassing cliffs, and an encompassing apron of sloping, loose rock debris (Fig. 4–27A). This analytical method describes the form, but provides no concept of how it evolved. In contrast, the genetic method not only describes the mesa, but also explains its origin. As shown in Figure 4–27B, the mesa was once part of the plateau composed of a sequence of soft, easily eroded beds underlying a single hard, resistant layer, known as a *caprock* or *rimrock*. Headward-eroding streams ate back into the plateau's edge in such a manner as to slowly isolate a part of the tableland, while the plateau cliffs retreated because of weathering, gravity fall, and sheet erosion. Finally, the isolated mesa was created as an outlying remnant marking the former extent of the plateau, its cliffs having resulted from the undermining of the hard caprock due to the faster erosion of the comparatively softer underlying beds.

Thus, analyzed genetically, the term *mesa* becomes more than just the name of a form; it becomes a term that describes not only the shape, but the way in which the processes of weathering and erosion proceeded to progressively modify the rock materials to produce the mesa landform.

Structure, Process, and Stage

Using the genetic method, any landform may be understood in terms of *structure, process,* and *stage*. Consideration of these factors provides both an appreciation of a landform as it exists and a comprehension of its evolution.

Structure. The term *structure,* or geologic structure, refers to both the nature and arrangement of the landform materials. In the instance of the mesa, the structure is composed of essentially horizontal sedimentary rock layers with a capping stratum of rock that is comparatively more resistant to weathering and erosion (Fig. 4–27A). Another type of structure can be seen in the Nashville Basin of Tennessee (Fig. 4–28). This region is part of a broad structural dome made up of warped layers of sedimentary rock. As weathering and erosion wore away the top of the dome, the underlying, softer strata (limestone) were exposed to create the basin floor, and the harder strata eroded back to form an encircling basin rim. Other examples of the relationships between geologic structures and landforms may be seen in such illustrations as Figures 4–8 and 4–14.

Process. The term *process,* as used in connection with landform analysis, is a group term that refers to all phenomena associated with the shaping of geologic structures and materials: it includes, then, the endogenous processes (diastrophism and vulcanism) and the exogenous processes (mass wasting, weathering, and erosion). The processes involved in the formation of

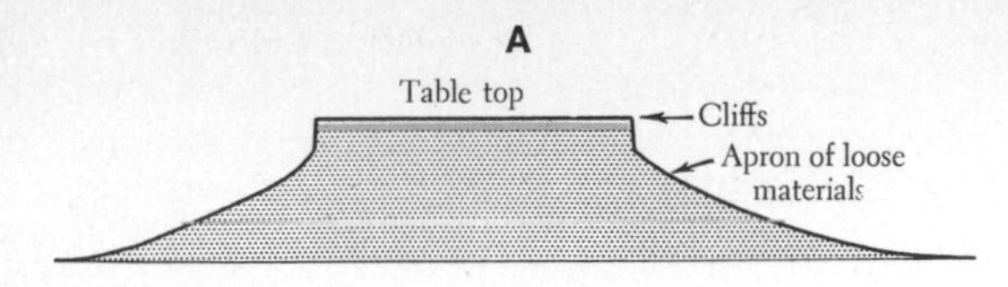

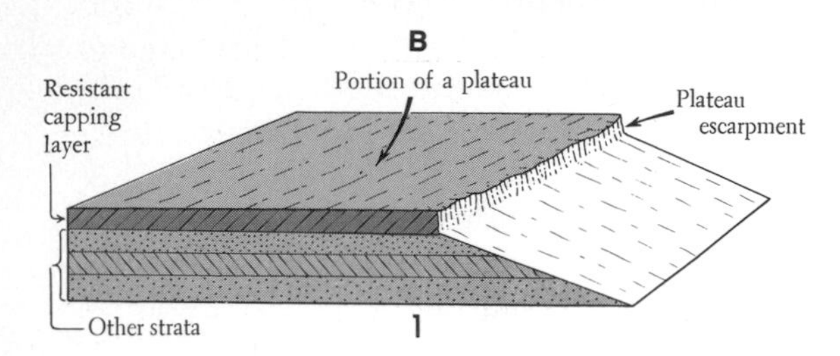

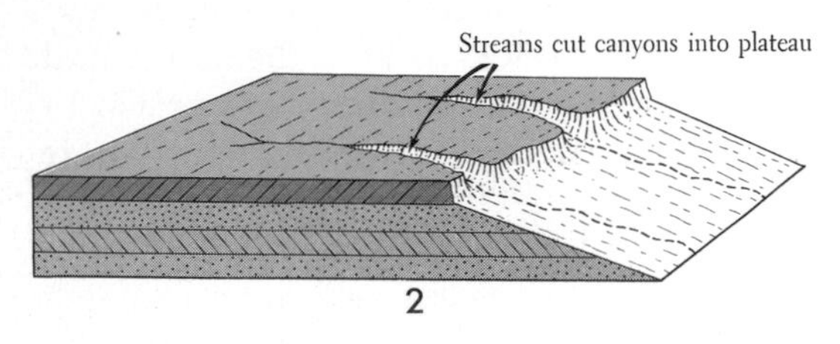

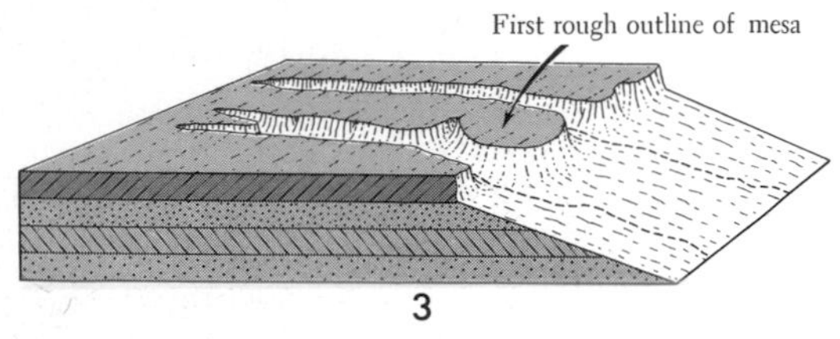

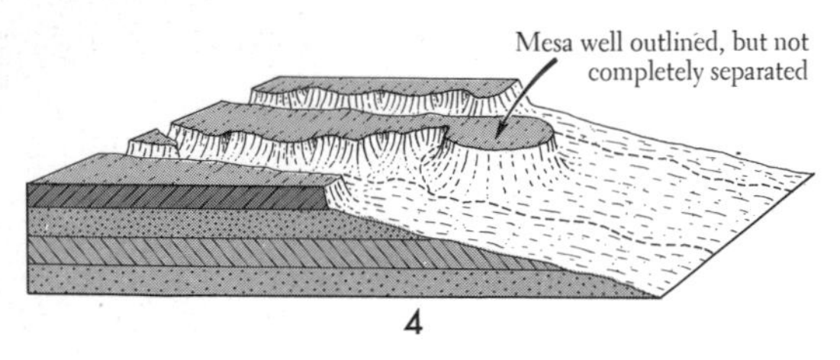

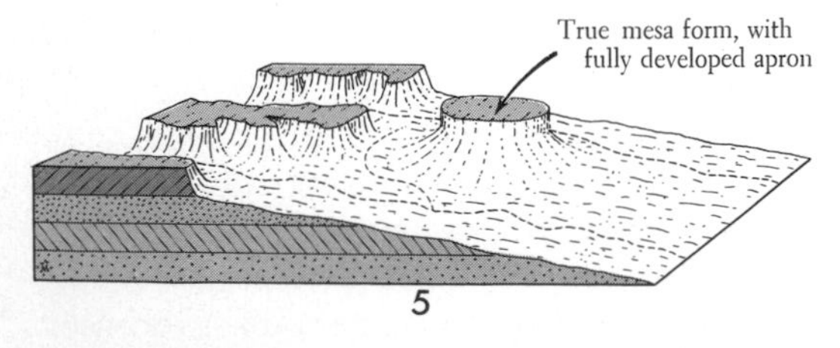

FIGURE 4–27

The structure of a mesa. A. Profile of a typical mesa. B. Stages in the development of one type of mesa.

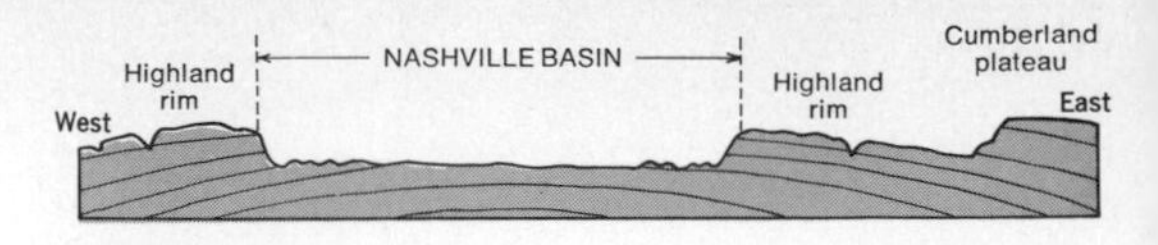

FIGURE 4–28

Geologically, the broad area of the Nashville Basin and vicinity in Tennessee is a large structural dome.

the mesa, for example, were mass wasting, weathering (both decomposition and disintegration), and stream erosion: all of these worked in combination to etch out and ultimately produce the mesa form from the once horizontal geologic structure. Another illustration of process is seen in the creation of the ordinary wave type of sand dune (Fig. 4–29). Such dunes owe their existence to sands that have been collected and shaped by wind. Sand grains are rolled, slid, and bounced across one another to form miniature ridges at right angles to the direction of strongest wind blow. These ridges retain their form as they advance as mobile dunes: sands move up the gently sloped windward side and then fall at an angle to produce the steeper leeward slope.

Stage. Stage refers to the sequential changes accomplished by the other processes as they shape the geologic structures. Stage does not deal with the measurement of time, but rather, with the amount of work that has gone into the evolutionary development of a given landform or landform complex—whether it takes one year, one hundred years, or countless millenia. As noted

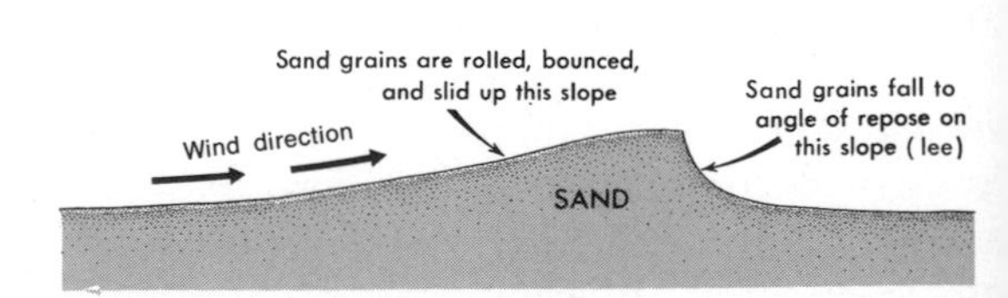

FIGURE 4–29

Profile of a *wave dune.* (The angles of the slopes are somewhat exaggerated.)

earlier, the mesa was not created suddenly, but in stages incident to the degree of break up, transportation, and deposition of rock materials. In similar fashion, a valley reflects in its shape the degree to which the processes have done their work: if it is V-shaped, with steep slopes and no floodplain, it represents a valley form in the initial phase of development; if it possesses alternately steep and relatively gentle slopes and an incipient floodplain, it is a valley in mid-development; if it has a continuous floodplain, it is a valley in the final stage of development (Fig. 4–30).

FIGURE 4–30

The final stage in the genesis of valley forms is marked by the *riverplain valley*, long stretches of which are filled with alluvium. The example shown here is the Connecticut River Valley, in Massachusetts. (Parks, Standard Oil Co., N.J.)

REVIEW AND DISCUSSION

1. From personal experience in your own locality, attempt to justify the statement that "no other aspect of the physical environment is, literally, more fundamental to people than landforms."
2. As the landform map of Japan largely mirrors that country's pattern and density of population, so does the landform map of the world mirror world population patterns and densities. Using Plates II, III, and VII, cite several similar correlations, in countries of dense populations and in those of sparce populations, and amid both the low latitudes and the high latitudes.
3. How is the distinction between major landforms and minor landforms made? What is the difference between major landforms of the earth and major landforms of the continents? Identify some minor landforms in the region in which you live.
4. Compare and contrast the types of rocks that make up the sima and the sial. What significant results occur in the lithosphere due, at least in part, to the compositional differences between the sima and the sial?
5. Consult an appropriate reference work and determine in some detail the nature and results of the earth's crustal plate movements. Give special attention to the plate components themselves, the mid-ocean ridges, and the resulting associated major landforms of the continental margins.
6. Compare and contrast the three types of rock that make up the earth's crust: igneous, sedimentary, and metamorphic. Why are most sedimentary rocks formed in layers, or strata?
7. Some landforms are shaped by processes operating within the lithosphere itself, while other landforms are shaped largely by processes operating at the surface of the lithosphere. Discuss these processes and their resultant landforms.
8. Draw cross-sectional diagrams to represent the differences between an anticline, a syncline, and a monocline. Illustrate a thrust fault, a graben, a horst, and a structural dome.
9. What is mass wasting? How does it operate and what is its comparative world importance in the shaping of landforms? Distinguish between weathering and decomposition, and the end results of each.
10. What are the major agents of erosion and deposition? Which of these is considered to be the dominant agent over the earth's surface? How does this dominant agent operate and what are the landform results? Is it the dominant agent in both humid- and dry-climate regions?
11. Identify in general terms the differences between the world's major subaerial landforms: plains, plateaus, mountains, and hills. Create a map of continental North America that shows the principal occurrences of these major landforms.
12. Identify in general terms the distinctions between several of the minor landforms of a region you know firsthand. Of what significance are they to the human occupance of the region?
13. Which of the above major and minor landforms are the result of endogenous processes, and which are the result of exogenous processes?
14. Discuss the concepts of structure, process, and stage in landform analysis.

SELECTED REFERENCES

Bloom, A. L. *The Surface of the Earth.* Englewood Cliffs, N.J.: Prentice-Hall, 1969.
Bott, M. H. P. *The Interior of the Earth.* New York: St. Martin's, 1971.
Bullard, F. M. *Volcanoes: In History, in Theory, in Eruption.* Austin, Tex.: University of Texas Press, 1962.
Chorley, R. J., A. J. Dunn, and R. P. Beckinsale. *The History of the Study of Landforms or the Development of Geomorphology, Vol. 1: Geomorphology Before Davis.* London: Metheun; New York: Wiley, 1964.
Clark, S. P., Jr. *Structure of the Earth.* Englewood Cliffs, N.J.: Prentice-Hall, 1971.
Continents Adrift: Readings from Scientific American. San Francisco: Freeman, 1972.
Davis, W. M. *Geographical Essays.* New York: Dover, 1954.
Dott, R. H., and R. L. Batten. *Evolution of the Earth.* New York: McGraw-Hill, 1971.
Ernst, W. G. *Earth Materials.* Englewood Cliffs, N.J.: Prentice-Hall, 1969.
Fairbridge, R. W., ed. *The Encyclopedia of Geomorphology.* Encyclopedia of Earth Sciences, Vol. 3. New York: Reinhold, 1966.
Holmes, A. *Principles of Physical Geology*, 2nd. ed. New York: Ronald, 1963.
Jefferies, H. *Earthquakes and Mountains.* London: Methuen, 1950.
King, L. C. *The Morphology of the Earth,* 2nd. ed. New York: Hafner, 1967.
Lobeck, A. K. *Geomorphology.* New York: McGraw-Hill, 1939.
Longwell, C. R., and R. F. Flint. *Physical Geology.* New York: Wiley, 1969.
Meinisz, F. *The Earth's Crust and Mantle.* New York: American Elsevier, 1964.
Putnam, W. C. *Geology.* New York: Oxford University Press, 1964.
Shelton, J. S. *Geology Illustrated.* San-Francisco: Freeman, 1966.
Shimer, J. A. *This Sculptured Earth.* New York: Columbia University Press, 1959.
Spock, L. E. *Guide to the Study of Rocks.* New York: Harper & Row, 1962.
Strahler, A. N., and A. H. *Environmental Geoscience.* Santa Barbara, Calif.: Hamilton, 1973.
Thornbury, W. D. *Principles of Geomorphology.* New York: Wiley, 1954.
———. *Regional Geomorphology of the United States.* New York: Wiley, 1965.
Woolridge, S. W., and R. S. Morgan. *An Outline of Geomorphology: The Physical Basis of Geography.* New York: Longmans, Green, 1959.
Wyckoff, J. *Rock, Time and Landforms.* New York: Harper & Row, 1966.

The lower Rio Grande Valley a few miles south of Weslaco, Texas. (Standard Oil Co., N.J.)

5

Landforms Shaped by Fluvial Action

As noted in the previous chapter, the major landforms of the earth's landmass are plains, plateaus, hills, and mountains. Within these broad categories, however, are the virtually infinite local details that constitute the minor landforms of the continental surfaces (Fig. 5–1). The fact that such local landforms are termed "minor" is in no way indicative of lesser significance; in fact, it is with these minor landforms that people are most directly and intimately concerned and confronted, and it is with the nature and pattern of their arrangement that other features of the physical landscape, whether drainage and soils or natural vegetation and climate, are so vitally related.

With occasional exceptions involving very local volcanic or diastrophic activity, the myriad minor landforms of the continents and the continental edges reflect the shaping actions of moving water, moving ice, and moving air. It is these agents of erosion and deposition, abetted by the processes of weathering and mass wasting, that create the local landform scene. Moving water is considered to be the dominant agent in the formation of these minor landforms on the surface of the earth.

Fluvial Landforms of Humid Areas

Stream Origin and Work Capacity

When a land surface is supplied with water, as, for example, from rain or melting snow and ice, the water moves downslope. Much of it moves as sheet-flow, whose water soon becomes concentrated in the form of rills that join to create creeks; these creeks then join to form small rivers, which, in turn, join to create large, full-flowing rivers (Fig. 5–2). Sheet-flow, including rainwash, has some ability to pick up and transport earth materials, and, with time, through

FIGURE 5-1
The minor, rather than the major, landforms of the earth engage the most intimate concern. This southeastward view shows the Delaware Water Gap, which separates minor landforms of New Jersey (on the left) and of Pennsylvania (on the right). (Pocono Mountains Vacation Bureau)

the process of *sheet erosion,* it will noticeably transform entire slopes (Fig. 5–3). Water's most pronounced erosive ability, however, comes when it is concentrated in linear flows in the process of *fluvial,* or *stream, erosion* (Latin: *fluvis* = river), a process that becomes progressively stronger as the stream increases in size from rill to creek to river.

Velocity is the chief determinant of water's ability to shape the land by erosion or by deposition. Even a slight increase in the velocity of water is accompanied by a marked increase in its ability to carry materials. In view of this, it is understandable that flood damage caused by streams is often quite great. Figure 5–4 provides an explanation of the relationships between stream velocities and water's ability to move different-sized materials.

The velocity of a stream depends on the factors of *slope, volume,* and *load.* Slope is the angle, or *gradient,* down which the water is moving; volume, the amount of water that is moving; and load, the amount of material being carried by the moving water. Water's greatest velocities are attained by large volumes of water that flow with only a slight load down a very steep gradient in a restricted channel. Contrarily, its slowest velocities are characterized by small volumes of heavily loaded water that move on a very gentle gradient across a broad surface. If a stream flows over solid bedrock and, then, without any change in gradient, over loose, easily eroded material, it will pick up more load in the latter stretch than in the former and its velocity will decrease as it progresses. If a stream issues from a steep narrow canyon onto a gentle plain, the change in gradient will check its velocity and cause it to deposit some of its load (unless, of course, it had little or no load to start with). If rain swells

FIGURE 5-2

Stream size and erosional activity increase from rill to creek to small river to great river. Top left, rills etched into a hillside, all but destroying its value as an orchard. (Soil Conservation Service, USDA) Top right, a creek in the Badgworthy Valley in north Devonshire, England. (British Information Service) Bottom left, the Hudson River, an important small river in the state of New York. (New York State Department of Commerce) Bottom right, a great river, the Mississippi, at a point near Natchez, Mississippi. (Standard Oil Co., N.J.)

FIGURE 5-3
In the absence of sufficient ground cover, sheet erosion has led to the devastation of this newly planted apple orchard near Tehachapi, California. (Soil Conservation Service, USDA)

the volume of a stream, gradient and load remaining constant, the stream's velocity will increase. Such changes, by affecting velocity, determine whether a stream will erode or deposit, and hence, whether it will further cut a valley or obscure it with forsaken material.

The Processes of Stream Erosion

No land area is perfectly smooth; consequently, while a runoff may begin as a sheet-flow, it soon converges into a series of definite channels, which, keeping always to the lowest and steepest routes, take it to its ultimate destination in the sea or some interior basin. Thus, running water on the earth's surface is usually in the form of flowing streams, the form in which it carries on fluvial action and in which it erodes most efficiently. The following discussion will be largely concerned with the fluvial processes and will devote special attention to that of fluvial erosion.

It must first be pointed out, however, that weathering is a very critical factor in the whole erosion process: not only does weathering prepare the earth's surface for easy removal of materials, but it also provides streams with tools for the direct corrasion of rock surfaces.

Fluvial erosion consists of several separate processes: *hydraulic action, abrasion, solution,* and *transportation.* While each is distinct in its own right, they usually tend to function together, though with varying degrees of importance.

Hydraulic action, produced by the churning motion of flowing water, lifts up and carries away loose material. This material, coming in contact with the present load of the stream, acts as an agent of *abrasion,* wearing away the stream bed and reducing the sizes of the other fragments in the load. The soluble portions of this new material, particularly limestone and dolomite, are dissolved and carried away in *solution.*

Particles of solid matter are carried away in several ways. That which is very finely divided, such as clay, is carried in *suspension,* while heavier material is rolled or bounced along on the bedrock. The heavy particles can only be rolled, whereas the light ones can be lifted by hydraulic

FIGURE 5–4
The relation between stream velocity and the ability of a stream to move materials of different sizes.

Stream velocity	*Size of materials moved*
1/3 miles per hour	as large as sand grains
3/4 miles per hour	as large as gravels
3 miles per hour	as large as small stones (2 to 3 inches in diameter)
6 miles per hour	as large as large stones (10 to 11 inches in diameter)
20 miles per hour	as large as huge boulders (16 to 17 feet in diameter)

action and swept away more rapidly. Thus, the stream acts as a *selective agent,* carrying light objects the farthest. *Transportation* is really part and parcel of each of the processes mentioned above since, in each case, material must be carried away in order for the process to be completed.

Most of the world's streams flow in valleys of their own making, and, conversely, most valleys were created by the streams now flowing in them. The sizes of most valleys correspond to those of the streams within them. Many hillslopes, especially where denuded of their vegetation, are seamed with *rills* (miniature valleys cut by miniature streams) or *gullies* (larger than rills, and cut by larger streams). Aside from these, there are many names used in different regions to denote the various valleys, but, in most cases, the stream and the valley are commensurate.

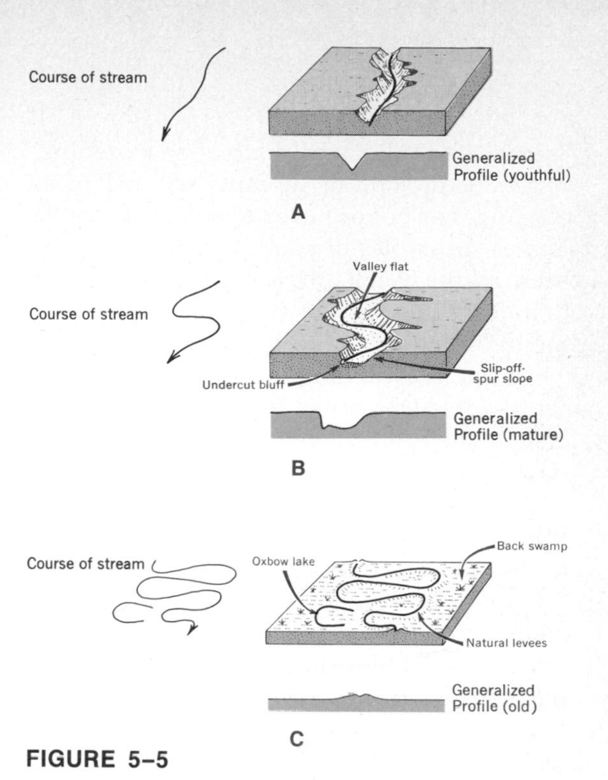

FIGURE 5–5

The erosion cycle of a valley in a region of humid climate. A. *Youthful valley.* B. *Mature valley.* C. *Old valley.*

The Erosion Cycle

Stages in Valley Forms. The work of creating and shaping a valley cannot be accomplished by the stream all at once; instead, it must be done sequentially, resulting in the production of different *valley stages,* each with its own characteristic form (Fig. 5–5). Initially, the stream attempts to cut its bed to the level of the body of water into which it flows. This level, a horizontal plane ultimately coincident with sea level, is called *base level.* In its effort to attain base level, which can be attained only at its mouth, the stream cuts *vertically* as far downward as it can and then begins to erode *laterally.* Vertical erosion ceases when the stream's velocity factors—gradient, volume, and load—are in a state of balance. This state of balance is known as *grade* (not to be confused with gradient or slope, and a stream in this condition is said to be graded. During the stage marked by vertical erosion, a valley has one type of shape (Fig. 5–5A); when lateral erosion becomes dominant, new shapes slowly emerge (Fig. 5–5B and C). Thus, there is an *erosion cycle,* each stage of which is characterized by a typical form and is generally referred to as either *youth, maturity,* or *old.*

The first stage of the normal valley erosion cycle in regions of humid climate is the youth stage, and the valley present is called a *youthful valley* (Fig. 5–5A). A valley in this stage of development can be readily recognized by the V-shape of its transverse profile: its sides are steep; there is no level land in the valley bottom; the stream or the active channel occupies all, or nearly all, of the bottom of the "V;" and the valley and its stream follow a relatively straight course across the land. The youthful valley's profile shows clearly that its stream is still cutting vertically, and that valley widening, by lateral erosion or by rainwash and mass wasting, has not occurred to any appreciable degree. Examples of youthful valleys may be seen in the small tropical gully or, on the grandest scale, in the typical, nonglaciated, mountain gorge.

The second major stage, maturity, is represented by what is called the *mature valley*. The original V-shape has been modified, primarily by lateral stream erosion, so that the valley now has the asymmetrical profile shown in Figure 5–5B. Lateral erosion on the outside of each stream curve has cut away part of the old "V" to form an *undercut bluff*. On the inside of the curve, where the stream's velocity is less than on the outside, part of the stream's load has been deposited to form an arc-shaped *valley-flat* that marks the very beginnings of a river floodplain. As the result of cutting (undermining) on the outside of each bend and deposition on the inside, the stream has widened the bottom of its valley, and its course meanders more and more as time goes on (compare the stream courses shown in Figure 5–5).

The third major stage in the cycle of valley shaping by humid-land streams is represented by what are termed *old valleys* (Fig. 5–5C). The broad, continuous floodplain of a valley in this stage reflects the continuation of lateral erosion and load deposition begun in the mature stage. Bordering the floodplain, but very far from the river, there may be low *valley bluffs* that are steepest where migrating meanders are undercutting, or have recently undercut, the valley margins (Fig. 5–5C). Within the floodplain, the stream twists in free-swinging meander loops. Veneering the valley's erosional floor of bedrock is a considerable thickness of *alluvium* (Latin: that which is washed to a place)—material carried from upstream during flood conditions as a part of the load of the stream and deposited elsewhere as the water's velocity decreased. The banks of the stream are noticeably higher than the surrounding floodplain. When the stream is in its floodstage, it carries a heavy load of fine mud, silt, and clay. As the water from the main channel overflows the bank, its velocity is sharply reduced by frictional drag from the ground and by the impeding action of bushes and grass, and some of its load is dropped immediately. Often, a single flood will add several inches to the height of a bank, which, when it is raised substantially above the floodplain, is called a *natural levee* (French: *levée* = raised). Between it and the border of the valley, the land is lower than the river, poorly drained, and often occupied by *back swamps*. Old meander loops mark former positions of the stream channel; some, still filled with water, are called *oxbow lakes* (Fig. 5–6), while others, partly filled with water and vegetation, are called *oxbow swamps*.

Thus, from the first small notch of the youthful gully to the wide plain of the old valley, humid-land valleys undergo sequential changes in shape: streams etch their valleys into the land first vertically and subsequently in broad meander sweeps. These plane away the land surface horizontally and deposit alluviums to create river floodplains.

Ultimately, continuing erosion by all the streams of a region results in the reduction of virtually every remnant of the uplands to the level of the floodplains, producing an almost perfectly flat area over which sluggish rivers flow at will. Such a landscape is termed a *peneplain* (Greek: *pene* = almost). Even in its most complete stage, however, it is probable that an occasional uneroded remnant will rise above the general level. This remnant is called a *monadnock*, a term derived from the Indian name of a mountain of this nature in southern New Hampshire.

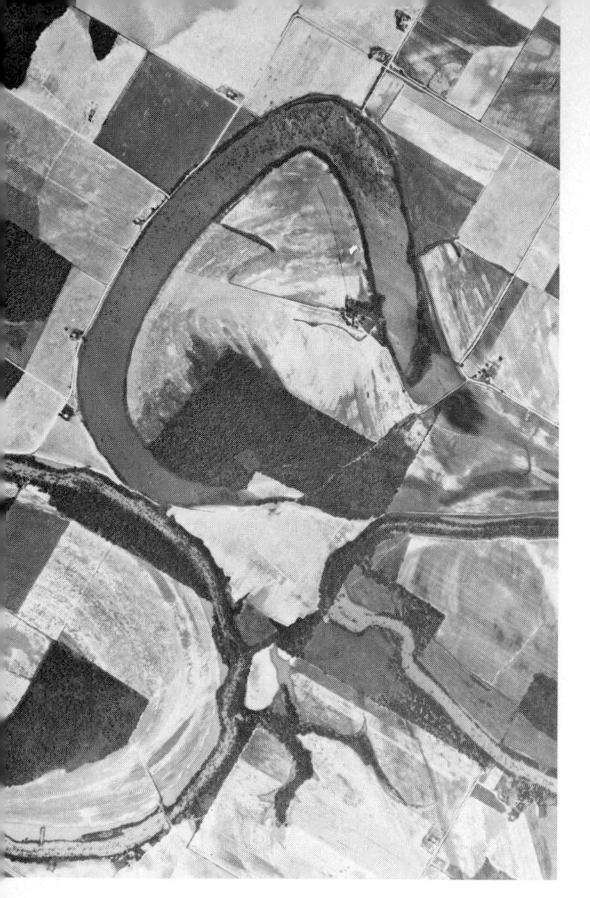

FIGURE 5-6
Soon after they are formed, many small ox-bow lakes are closed in by vegetative growth; but, the larger ones—such as these near Vicksburg, Mississippi—generally remain for long periods. (U.S. Army Corps of Engineers)

In several parts of the world, peneplains have developed and, then, during some later geologic period, have been upwarped to a higher elevation. When this occurs, the whole erosional process starts anew, with streams beginning to dissect the old surface. With the realization of this repetitive situation, the term *erosion cycle* came to be applied to the entire sequence. While some authorities today challenge the validity of the total concept on various grounds, it still remains a useful tool for understanding the development of landforms and a meaningful method of classifying landform types.

In most cases, the erosion cycle has not been allowed to continue to the ultimate stage of peneplanation, but has been interrupted at some point by diastrophic forces that have warped the entire region to new elevations. If the new position is higher than the previous one, the streams acquire a new steepness, hence increased erosive power, and they thus begin to incise new, youthful valleys into the floors of the more mature or old valleys. Because the streams and valleys are again youthful, it is said that *rejuvenation* has occurred.

Significance of Valley Shapes. Whether an area possesses valleys in the youthful, mature, or old stage makes a big difference in its physical landscape expression; and, even more significantly, it makes a great deal of difference to the area's inhabitants. If the area has a large number of youthful valleys, there is little or no level valley land available for human use, and if such valleys lie close together, there is practically no level upland separating them (Fig. 5–7A). In either instance, both land travel and land transport of goods are comparatively difficult and expensive.

If an area has a preponderance of mature valleys, especially if they are deeply entrenched, people will also have difficulties in traveling and in transporting goods (Fig. 5–7B). Overland travelers will be faced with valley crossings and steep valley sides that may necessitate following routes even more devious than those in areas of youthful valleys. This is because the undercut-bluff portions of mature valleys are too steep for roads or railroads; thus, a given route will have to proceed down a more gentle slope onto the valley-flat and then, facing a bluff, turn either upstream or downstream to cross the river to the next valley-flat, and then continue up another comparatively gentle slope. However, mature valleys do provide more level land for human use than

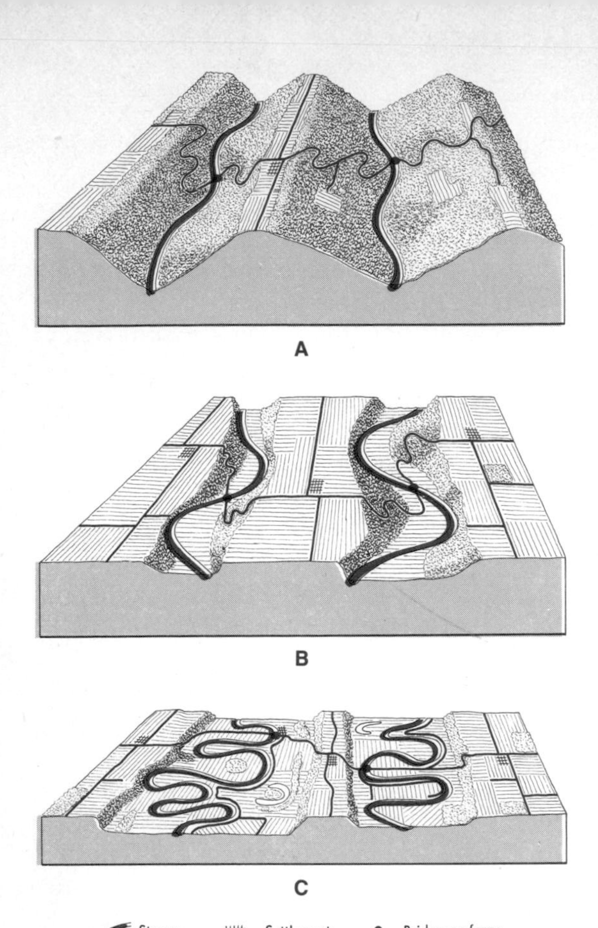

FIGURE 5–7
Some effects of different *valley shapes* on land use in an area of humid climate. A. *Youthful valleys.* B. *Mature valleys.* C. *Old valleys.*

do youthful ones, because there is still some level land in the form of valley-flats on the inside of the meander curves (Fig. 5–7B). Such features are normally the sites of agricultural activity.

In an area where the valleys are old, there is a great deal of nearly level land, particularly where floodplains are very broad and close together (Fig. 5–7C). Valleys of this sort, if the climate is right, ordinarily support large farming populations, for the alluvial soils are usually of high productivity, extending over many square miles with little interruption. Often, however, many sections of the floodplain are too swampy to use for crops, although they may produce swamp forests and their margins may grow lush grasses suitable for feeding domestic animals. In these areas there is also the ever-present danger of floods that may spread from one valley-bluff to another, destroying crops, settlements, roads, railroads, and even the inhabitants themselves (Fig. 5–8). Because of these flood conditions, people often distribute many of their crops and settlements along natural levees in the hope that these slightly higher sections will remain above the floodwaters.

Stages in Interfluve Forms. At the same time that valleys are being etched into the land, areas between the valleys are undergoing changes. These interstream areas are called ridges, or *interfluves.* Like valleys, certain normal interfluve types develop in a definite sequence.

The first stage in the normal humid-land erosion cycle of ridges is the *youthful interfluve.* Interfluves in this stage are characterized by a nearly level top and steeply sloping sides. The top is part of the original upland surface that has only been slightly eroded; the sides are those of the adjacent valleys (Fig. 5–9A). The exact condition of the sides will depend on whether the adjacent valleys are youthful, mature, or old. If they are youthful, then the sides of the youthful interfluves are the steep sides characteristic of youthful valleys.

As weathering, soil creep, rainwash, and stream erosion eat into and sharpen the original youthful ridges, *mature interfluves* are created. At first, such ridges are very narrow and sharply crested, like steeply pitched roofs. Later, the sharp crests are rounded and towered as a result of weathering and rainwash, but they are still relatively narrow (Fig. 5–9B).

FIGURE 5-8

Widespread flooding—which resulted in a great loss of human life, homes, crops, and transport lines—in the Mississippi floodplain at Greenwood, Mississippi, in March, 1973. (U.S. Army Corps of Engineers)

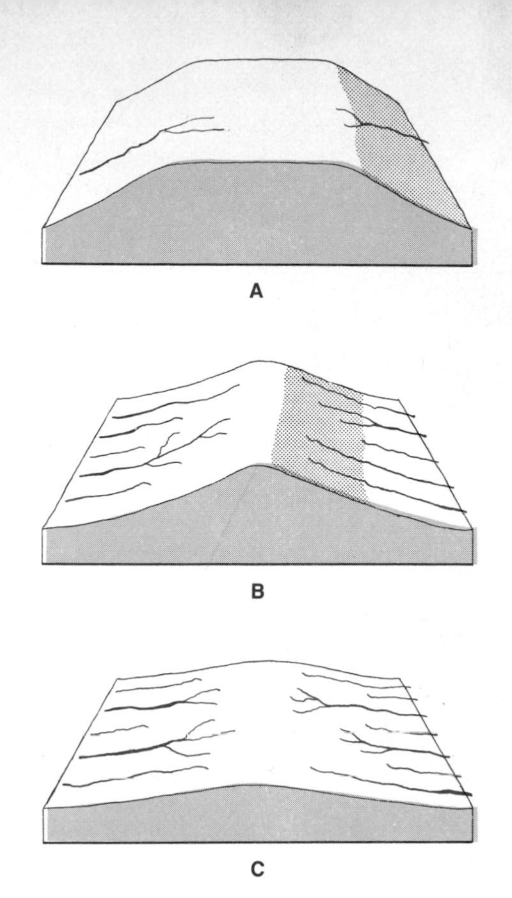

FIGURE 5–9

The erosion cycle of a normal interfluve in a region of humid climate. A. *Youthful interfluve.* B. *Mature interfluve.* C. *Old interfluve.*

With continued wearing away, the *old* stage is reached, wherein the interfluves become so low and comparatively broad that they can hardly be called ridges, but they are still interfluves (Fig. 5–9C). The lowness and smoothness are further enhanced by downwash and by the slow creep of soils until the interfluves become only gentle swells in the land surface (Fig. 5–10).

FIGURE 5–10

Profile of a region in the old stage of development.

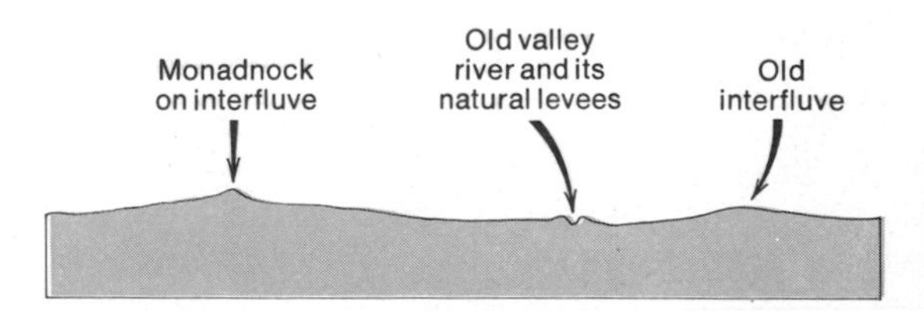

Significance of Interfluve Shapes. As with the typical valley forms produced by running water, the type, or types, of interfluves in a given region affects the general appearance of the land and is important in determining the human use of land. Where broad youthful interfluves are numerous, there is much nearly-level land available for agriculture and settlements, and transport lines can be laid out in almost any fashion desired (Fig. 5–11A).

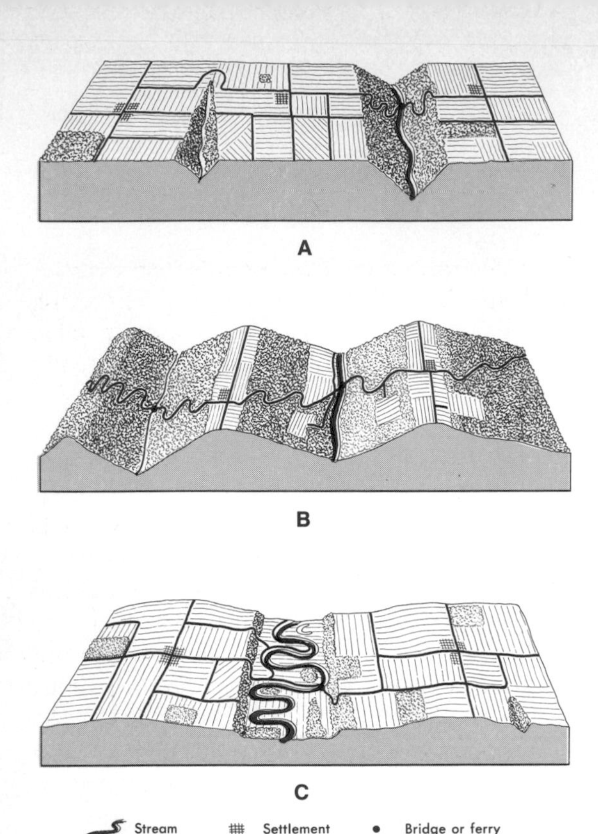

FIGURE 5–11

Some effects of different *interfluve shapes* on land use in a region of humid climate. A. *Youthful interfluves.* B. *Narrow mature interfluves.* C. *Old interfluves.*

In contrast, countrysides made up largely of mature interfluves provide practically no level land. If their slopes are utilized for crops, great care is needed to prevent the soils, loosened by cultivation, from being washed away into the valleys. If necessity demands they be used, such slopes may be terraced to provide level land not available under natural conditions (this is a common practice in densely settled portions of the Orient). In general, however, mature ridges are best suited for forests or for permanent scientifically managed pastures. The same conditions that limit agricultural use of mature interfluves make the laying out of travel routes upon them somewhat difficult. Quite commonly, their major roads, as well as their distribution of dwellings, are obliged to follow the crests of the interfluves' narrow ridges (Fig. 5–11B).

In areas of old interfluves, there is relative freedom from the limitations imposed by landforms. The gently rolling condition allows cropland to be widely spread. There is land suitable for villages, towns, and cities, and travel routes can be structured in almost any pattern. In these respects, areas of old interfluves closely resemble those of youthful interfluves, as both have a considerable amount of nearly level or undulating land (Fig. 5–11C).

Valley Form and Interfluve Form Combinations. Valleys and interfluves are inseparable—they occur in all the possible age combinations imaginable. Through use of the various terms that are applied to valleys and interfluves of different stages, local landforms can be described simply and meaningfully.

For example, there is a portion of the Cumberland Plateau in eastern Kentucky whose landforms may be described simply and concisely as consisting of youthful valleys and mature interfluves. This immediately creates the image of a countryside with a surface profile that resembles the teeth of a saw—the "V's" of the youthful valleys and the inverted "V's" of the mature ridges: "W's." In this instance, the saw-tooth profile is not on a mountainous scale, for the vertical distance from the ridge top to the valley bottom is hundreds, instead of thousands, of feet. Yet nearly all the land is steeply sloping, and the relatively sparse agricultural population is forced to use sidehill fields to supplement the meager level land occurring along the narrow ridges and the equally narrow creek beds.

In similar fashion, the landforms of another area may be described as old valleys and old interfluves. Such a description portrays an entirely different kind of countryside from the one mentioned above: it has very few irregularities; its valleys are nearly level and relatively broad; its interfluves are low and gently sloping (Fig. 5–11C). It is an area whose surface offers people abundant freedom to settle and work. Parts of the inner (and older) portion of the Atlantic Coastal Plain of the United States are examples of areas that are hospitable to humans in this way.

Stream Deposited Forms. Not only does running water help to create landforms by carving valleys and by eating into the modifying interstream areas, but it also produces new forms by actual construction. Some such depositional forms have been noted already in connection with the formation of valley-flats in mature valleys and of floodplains in old valleys. Other common examples, in which the entire form is depositional, are the small cones and fans of alluvial material that are built up at the mouth of almost every small gully.

A large part of a stream's load eventually makes its way to the point where the stream discharges into an ocean or a lake. There, where the stream current is finally stopped, the load is dropped, forming a *delta.* Some deltas have the triangular shape of the Greek letter Δ, with one apex pointing upstream; others form a variety of different shapes. In some cases, the main stream channel divides repeatedly into several *distributaries,* each having its own set of natural levees, which are prolonged seaward far beyond the point where the intervening areas are occupied by water. The pattern formed by this type of development is called a *bird's-foot delta,* an example of which is the delta of the Mississippi River. Where ocean currents are more powerful, deltal extensions are blunted and an *arcuate delta* is built, as in the case of the Nile River. Some deltas form at the heads of bays and are termed *estuarine deltas.*

Stream Patterns

Maps portraying the flow lines of surface waters show many of the patterns that individual rivers and streams have etched into the land. That these patterns are a vital part of geographic expression and not just features that have been unduly emphasized on maps, is illustrated by almost any method of viewing the earth's physical patterns from above—for example, maps, aerial photography, remote sensing, or actual air flight.

The most common drainage pattern is *dendritic*—a treelike pattern, with trunk and branches joining at acute angles. A dendritic pattern develops in a region where the surface waters cut into rocks that are, whether loose or consolidated, relatively uniform in their reaction to erosion—such rocks as granite or homogeneous sandstone—or where these waters flow wholly on loose clays. Figure 5–12 presents some examples of streams that have developed dendritic patterns.

Other major patterns of streams include the *trellis, radial, annular, braided,* and *glacially deranged* patterns. As a whole, these

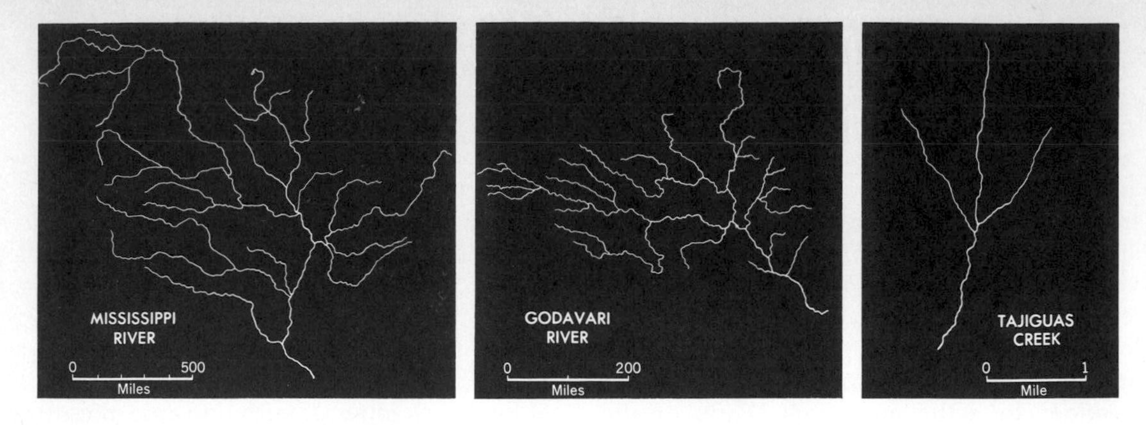

FIGURE 5–12

Examples of the *dendritic* pattern of drainage in streams of large, medium, and small size.

patterns are much less common than the dendritic pattern, but in areas where they do occur, they are equally reflective of environmental factors. The *trellis pattern* consists of relatively straight-lined tributary streams that join their main stream at right angles (Fig. 5–13). This pattern occurs where there is definite banding of the rock structures or where there is structural weakness in the earth's crust. A *radial pattern* develops where streams drain out in all directions from a central area on higher land, like spokes from the hub of a wheel (Fig. 5–14A), as well as in areas where a central low region drains its

FIGURE 5–13

Diagrams of two *trellis* drainage patterns showing the relation of each to landforms and underlying rock conditions. A. Cuestiform plain. B. Ridge and valley area. (See the index for references to *cuestas, synclines,* and *anticlines.*)

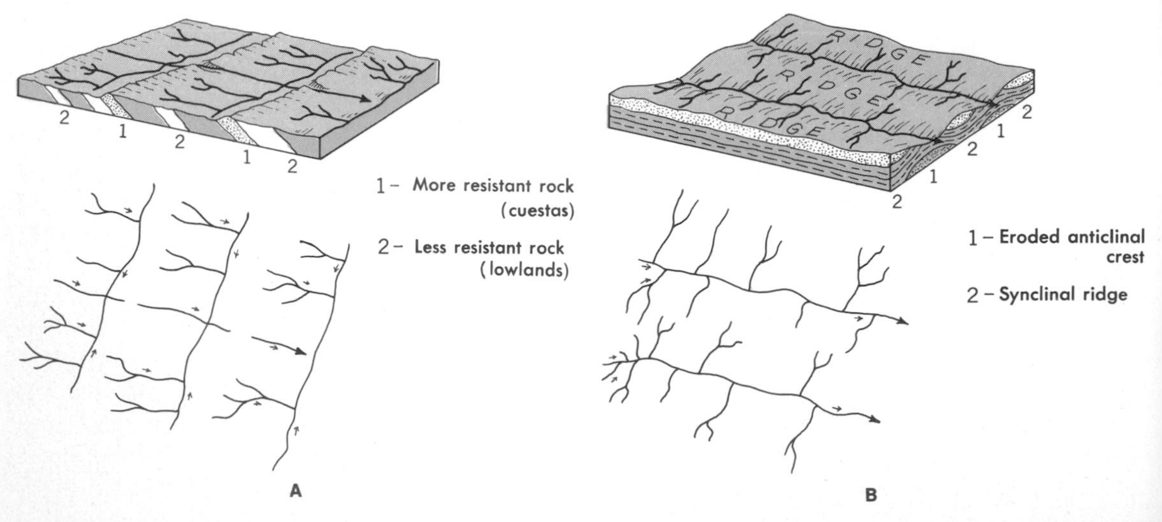

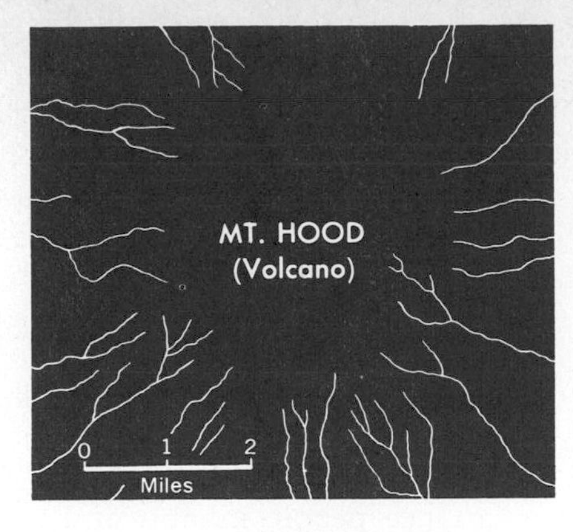

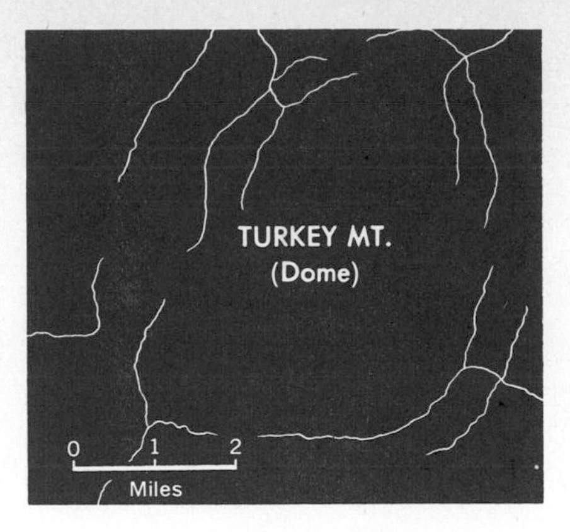

FIGURE 5–14

Some other major stream patterns. Left, *radial* pattern of drainage. Right, *annular* pattern of drainage.

water inward from many or all directions. An *annular pattern,* or *ring,* occurs most frequently as a result of the erosion of structural domes (Fig. 5–14B), and reflects rock structure controls that become operative as soon as the domes have passed the initial stage in their erosion and dissection. In a sense, the annular pattern is a curved or bent trellis pattern. The *braided pattern,* with its numerous interlaced channels, is characteristic of streams that have become so overloaded that they have dropped materials that now clog their original channels (Fig. 5–15).

Stream patterns are important geo-

FIGURE 5–15

A *braided* channel in the aggraded bed of the Komaktorvik River in northern Labrador. Note the many interlocking stream channels. (From Forbes, *Northernmost Labrador Mapped from the Air.* New York, American Geographical Society, 1938)

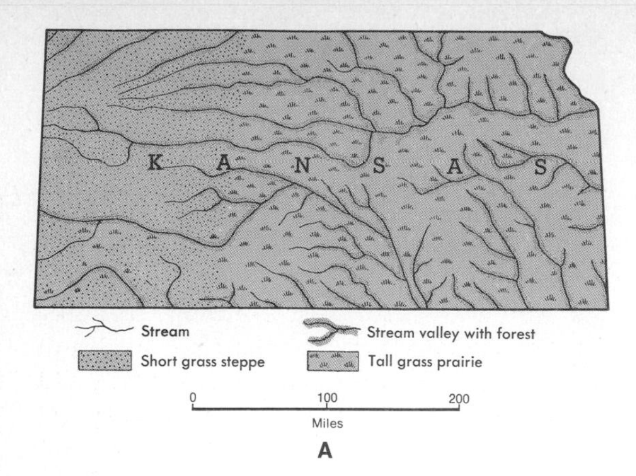

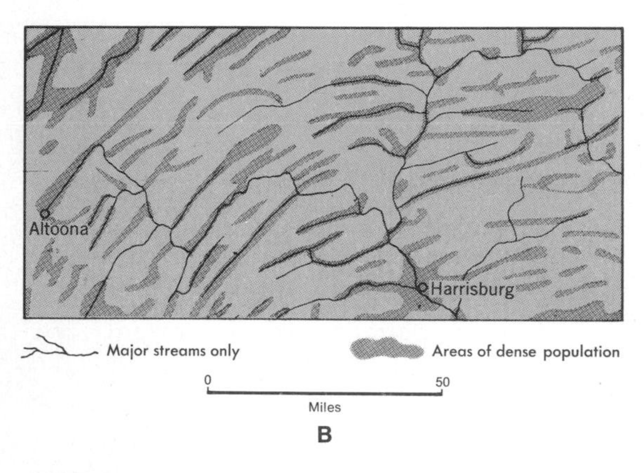

FIGURE 5–16

Some effects of drainage patterns. A. The relation of the dendritic stream pattern to the pattern and type of natural vegetation in a portion of the interior plains of the United States. Forested stream valleys (*galeria forests*) extend through both the short and tall grass areas. B. The relation between the trellis pattern of drainage and the areas of dense population in a ridge and valley area of southern Pennsylvania.

graphically, since they both determine part of the physical plan of an environment and affect the distribution of other physical features within it. One example is seen in the streams of Kansas; their patterns are largely dendritic, as are the patterns of their accompanying floodplains. Water is more abundant in the floodplain soils, enabling trees to maintain themselves there. Thus, lines of trees trace the dendritic pattern of the floodplains, providing a sharp contrast to the drier, grassy interstream areas of Kansas (Fig. 5–16A). In the ridge and valley region of southern Pennsylvania (Fig. 5–16B), the stream pattern is trellised, and this is the key factor in the region's distribution of rich bottomland soils, as well as in the location of those types of regional vegetation that require considerable moisture.

Once people inhabit and utilize a region, stream patterns are reflected in still other ways. In the Amazon Basin, for instance, where the stream pattern is dendritic, many of the towns and small villages are located along the tops of valley bluffs. Thus, the population pattern is also dendritic, though not solidly, for settlements are not continuous; they might be described as forming an interrupted dendritic pattern. In the northern part of Mississippi, the streams also form a dendritic pattern. The best cornlands are found on the fertile, recent alluvium that borders the stream channels; hence the pattern of corn production, at least of heavy yields, is the same as that of the streams. An annular pattern is characteristic of much of the drainage of the Black Hills in South Dakota, and the distribution pattern of ranch lands corresponds closely to it. In a large part of Pennsylvania, chiefly east of the Allegheny Plateau, the stream pattern is trellised, and segments of the pattern mark the positions of the area's most level lands. These lands have the largest population density in that portion of the state; thus the population pattern there is essentially trellised also (Fig. 5–16B).

In addition to its direct influence on population density, the drainage pattern in a region also affects its road pattern. The problems of road alignment and bridge construction in an area of glacially deranged drainage are very different from the same problems in a region of dendritic streams.

Fluvial Landforms of Desert Areas

Because deserts are areas of little rain, their landforms stand in marked contrast to those of humid lands. At first glance, the most conspicuous aspect is the angularity of the area's relief. Another marked difference lies in the drainage pattern: while all streams in humid lands flow to the sea, it is a rare case when a stream in the desert does so, for most terminate locally in "lakes" that have no permanent water. Materials eroded from the headwaters of such streams cannot be carried to the sea but are deposited nearby. The streams of the desert are even more fundamentally different from the permanent streams of humid lands in that they are dry most of the time, and when water does flow in them, it usually flows in violent flood. A final distinction of deserts is that a significant portion of the erosion and deposition that takes place there is done by the wind—an agent of transporation of little importance in the humid lands. Despite these differences, however, it is still primarily fluvial action that is responsible for shaping most of the landforms of the world's deserts.

The deserts of the world can be divided into two geomorphic types: deserts of *mountain-and-bolson* topography, and *plain, plateau,* and *shield* deserts. The two are radically different from one another, their chief common denominator being their essential aridity.

Mountain-and-Bolson Deserts

As the terms imply, a mountain-and-bolson desert is composed of two completely different types of landscape: mountain ranges and intervening *bolsons* (a term, derived from the Spanish word for "purse," that refers to totally enclosed land basins). The mountains are usually upthrust fault blocks and the bolsons are downdropped blocks (Fig. 5–17); in some cases, they are classic horsts and grabens, but, in most in-

FIGURE 5–17

Stages of erosion in a region of dry climate. A. *Youthful stage:* mountains are bold; the depression is only slightly filled. B. *Mature stage:* mountains are considerably worn and dissected; depression-filling is prominent; alluvial fans have coalesced laterally to form continuous *bajadas*, or alluvial-fan piedmont plains. C. *Old stage:* only remnants of mountains remain; the depression is filled.

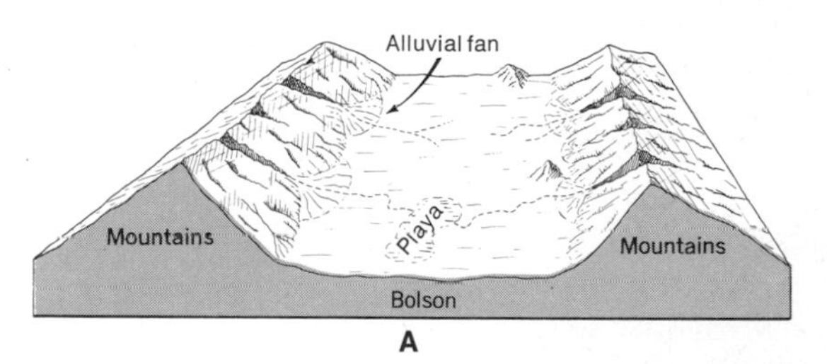

A

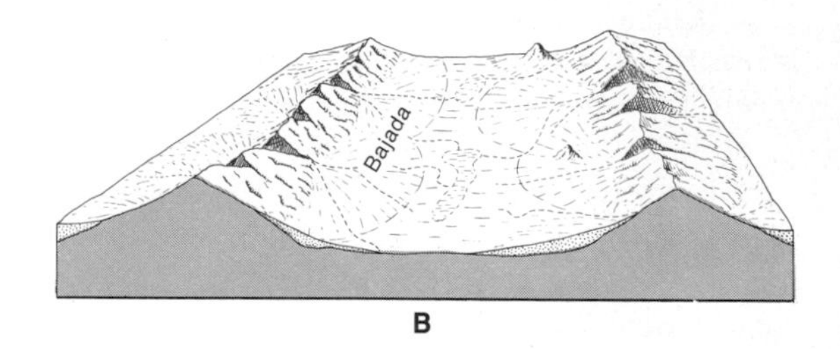

B

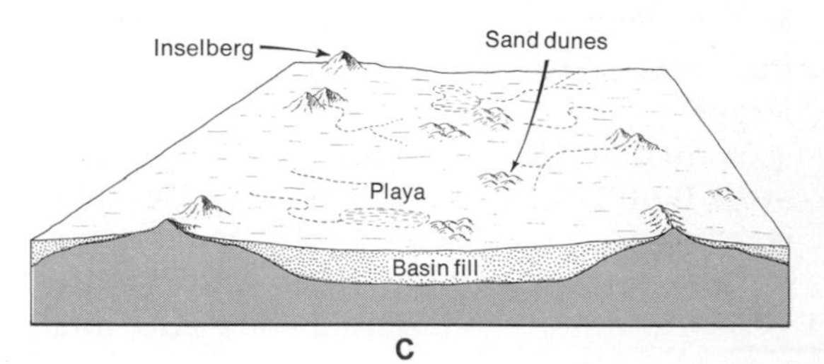

C

FIGURE 5-18
A classic example of a mountain-and-bolson area, as photographed from Dante's View on the east side of Death Valley. (F.E. Dunham, U.S. Forest Service)

stances, the geology is more complex. In the arid climate, the two forms of landscape assume contrasting geomorphic roles: the mountains are the principal areas of erosion and the bolsons are the principal areas of deposition (Fig. 5–18).

Erosional Areas and Activities. Desert mountains support, at best, only a scattered, open pattern of bushes and stunted trees, as soils are thin or nonexistent. The bedrock, being greatly heated during the day and rapidly chilled at night, expands and contracts, causing it to break apart. In short, mechanical weathering results in the disintegration of the surface rock; however, the lack of water slows or prevents its decomposition.

Rain is rare in desert areas, but when it does occur, it takes the form of a torrential downpour. Since there is little or no soil to absorb it, and little or no vegetation to impede it, runoff is instantaneous and total. Sheets of water run down the slopes and sweep along masses of waste and other detritus; sometimes the water becomes so overladen that it acts as a mudflow. The sharp angular debris acts as an efficient cutting tool in these roaring streams, accomplishing more abrasion in minutes than humid-land streams do in months or years. Material, from microscopic particles to huge boulders, is carried in solution or suspension or is merely rolled along the stream bed toward the valley below. Suddenly, the rain ceases, the flood diminishes, and the water's load is deposited, just as suddenly, *in situ* on the bed of the stream. With this, weathering begins again, producing material to be moved in the next rainstorm a month or a decade hence.

The desert's stream beds, then, are generally covered with gravel and blocked by tumbled masses of boulders—most of the year they are bone dry. Its valleys are steep-sided, and surrounded everywhere by bare rock ledges. The mountain ridges are narrow-crested, and flat land is almost nonexistent. *Angularity* is the keynote of desert landforms throughout the world (Fig. 5–17A).

Depositional Areas and Activities. In dry regions, the bolsons are the zones of accumulation. To them, during the brief periods of flood, the intermittent streams carry their load of detritus; in them, that load is deposited; and there, over the millenia, such material continues to accumulate (Figs. 5–17A and 5–18).

Much of the accumulation takes the form of *alluvial fans.* When floodwaters issue from steeply sloping canyons, their velocity is suddenly checked by the change in gradient, thus they drop their load of fine materials, sands, gravels, and boulders. Loss of water by evaporation and by its sinkage into the ground further decreases the streams' ability to carry their load. During a given flood, a stream may flow in a particular channel, but as the flood diminishes, the load of detritus the stream drops partially fills the channel. Succeeding floodwaters seek new and lower lines of flow and then drop their loads along them. This sequence happens over and over again; slowly but surely the alluvial fan takes shape as materials are deposited in a system of channels that eventually spans an arc of nearly 180°, new channels having been made by each flood from the canyon mouth (Fig. 5–19). In general, finer materials are carried to the lower margins and coarser ones are left near the apex. However, some floods are more intense than others, and coarser materials are carried farthest when the water is greatest in volume and velocity. Because of the manner in which their deposition occurs, alluvial soils vary greatly, even within distances of a few feet. Still, most of the finer soil is generally found near the lower margin of the fan and the coarser along the upper portion.

The slopes of alluvial fans are normally quite gentle. They seldom exceed 3 or 4 percent grade (3 or 4 feet of total vertical rise or fall in 100 feet of horizontal distance). As an alluvial fan grows in size, its lateral margins often come in contact with similar fans built at the mouths of adjacent canyons, and, in time, the fans generally merge. Thus, a mountain range may be fronted by a continuous apron of alluvium, a feature variously known as a *bajada* (Spanish: that which is lower down), *coalesced alluvial fan, mountain apron,* or an *alluvial-fan piedmont* (Fig. 5–17B).

At its lower margin, the bajada merges imperceptibly into the floor of the bolson. In the typical case, the lowest point of the bolson is occupied by a *playa,* or *dry lake,* the terminus of the drainage of the entire area—the bolson, the flanking bajadas, and the bolson's tributary mountains. Following torrential summer storms or, in some regions, when the melting of winter snows in the high mountains supplies enough runoff, the "dry lake" may temporarily be wet, but most of the time it is completely dry on the surface. However, the water table is often quite close to the surface, but it will be found

FIGURE 5–19

An *alluvial fan* formed at a canyon mouth. Some of the lines of intermittent water flow are shown on the surface of the fan.

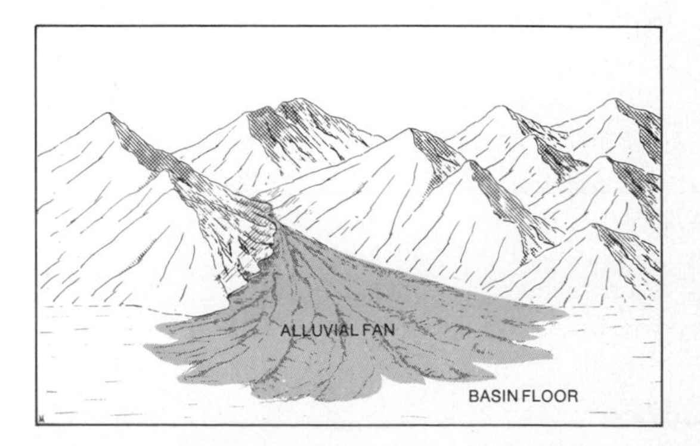

to be brine, at or near the saturation point, with minerals in solution. Such water will have reached the playa in two ways: by flowing across the surface during the wet periods, and by seeping through the alluvium of the bajada from where the streams sank underground, somewhere on the slope. When it flows over the surface, the water transports silts and clays and deposits them in a layer over the surface of the lake. When it seeps underground, it can transport only dissolved minerals—salts it has derived from the mineral constituents of the rock it has passed over or through (almost all kinds of rock yield some form of salt as a weathering product). At the lake, the briny water is drawn to the surface by capillarity (as coffee is drawn up into a lump of sugar) and there, as the water evaporates, the salt is left behind to form a crust or to build up small pinnacles of gleaming white salt crystals (Fig. 5–20). Playas covered with salt layers are called *salars.*

FIGURE 5-20

Salt crystals, deposited when briny water from underground is drawn to the surface and evaporated, stand as miniature ridges, peaks, and pinnacles in the Devil's Golf Course near Badwater on the floor of Death Valley. (Santa Fe Railway Photo)

In the course of time, the continued input of alluvium from the surrounding mountains may raise the floor of the bolson to a level as high as the lowest point in the rim. Thereafter, water flowing into the bolson from the mountains will find an outlet across that low point and make its way into an adjacent basin; thus, one bolson becomes tributary to another, and the drainage systems of the two become integrated. The stream will erode a deeper channel at the outlet of the higher basin and, as that rim is lowered, will begin to erode the materials composing the playa and the bajadas of the higher bolson.

Pediments. *Pediments* are bedrock features with surface configurations almost indistinguishable from alluvial fans, and usually formed on granite (Fig. 5–21). They apparently result from several processes working in combination: lateral erosion by streams as they discharge at the heads of alluvial fans, the retreat of the slopes of the mountain face due to mechanical disintegration of their granite, and sheet-wash during torrential storms. In some cases, the entire mountain has been worn away, forming a *pediment dome,* a smooth dome-shaped area with bedrock everywhere at, or just beneath, the surface.

Pluvial Periods. In the not far-distant past in some of the dry lands of the world, conditions were much more moist. During parts of the Pleistocene Age, the most recent geologic epoch, lasting from one million to about 12,000 years ago, somewhat greater rainfalls over some of today's deserts and

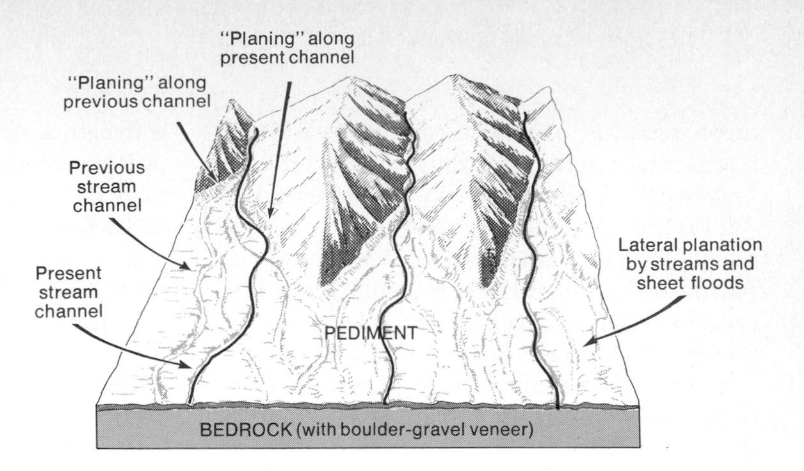

FIGURE 5–21

The development of a *pediment.* Only a few of the former streamflow lines are shown on the pediment surface.

much heavier precipitation over the bordering areas caused many of today's intermittent streams to flow strongly all year round, resulting in the formation of large lakes in many bolsons. Today, old shorelines mark the outlines of these lakes. These periods (there were several of them in the Pleistocene) correspond to the Glacial Periods (or Ice Ages) in other parts of the world, and are termed *pluvial periods* (Latin: *pluvis* = rain).

Distribution. Mountain-and-bolson deserts occupy a large part of the American Southwest—the entire state of Nevada, western Utah, southeastern California, southwestern Arizona, and southern New Mexico—and extend southward into the Mexican states of Sonora and Chihuahua. They are also the typical deserts of Chile and western Argentina; most of the arid region of Iran, Baluchistan, and Afghanistan; the higher portions of Arabia; and the Sinai Peninsula; and they prevail among the great deserts of the interior of central Asia.

Plain, Plateau, and Shield Deserts

Wide, sweeping horizons, endless expanses of featureless plains, and huge accumulations of sand are one's first impressions of the *plateau deserts.* In central and northern Arabia and adjacent Syria and Jordan, across the great expanse of the Sahara, on the high plateaus of southern Africa, or throughout the immense "outback" of Australia, in the colorful Navajo Country, Canyonlands, and Grand Canyon District of the southwestern United States, the majesty, solemnity, and vastness of the scene produce an awesome feeling that affects the very soul.

Geology. Plateau deserts generally occupy relatively stable areas of the earth's crust. Some, composed of almost perfectly horizontal beds of marine sediments, have been uplifted into their present positions thousands of feet above sea level by simple warping of the crust, or by faulting near the borders of the area, but with scarcely any disturbance of the original attitude of the beds. Other areas are underlain by ancient masses of granitic rock that, extremely stable through eons of geologic time, have been exposed by long-continued erosion and worn down to the featurelessness of peneplains.

Surface Features. The *hamada,* or bare rock plain, is a common feature of plateau deserts. In granitic areas, it represents the last stage in the downwastage of the area, most features above a certain base level having been reduced through lateral erosion and mechanical weathering (granular disintegration, exfoliation, and so on). Normally, a thin veneer of waste appears in patches over the surface, but soil or any important regolith development is lacking. Occasional low domes of essentially unjointed granite still rise above the general level in some areas; such an erosional remnant is shown in Figure 5–17C and is called an *inselberg* (German: island mountain).

In areas of horizontal rocks, *stripped plains* occur atop erosion-resistant strata—such as those of sandstone, limestone, or lava—where a former cover of some weaker strata has been stripped away. Occasional remnants of the higher beds may remain, betokening their former widespread occurrence. Where several resistant strata occur at different levels, separated by weak, easily eroded beds, a succession of cliffs and terraces may develop, each terrace being maintained by a resistant stratum. Undermining of the resistant bed by the erosion of the softer underlying members causes occasional collapse of the caprock, and the gradual retreat of the cliff. A terrace remote from the cliff-foot is often surfaced with bare rock, but one in the area near the cliff-foot is likely to be buried under detritus fallen or washed from above. Detached outliers of a terrace often remain in front of the cliff, standing as mesas, flat-topped remnants, or as *buttes,* remnants with no flat top remaining (Figs. 5–22 and 5–23).

Large expanses of plateau deserts are surfaced with *reg,* gravelly plains often topped with *desert pavement,* a concentration of pebbles representing the larger materials in the gravel, the finer particles having been blown away by the wind. Also present is the *erg,* a region covered with sand dunes. While the erg is the popular image of the desert, actually only some 5 to 10 percent of the world's plateau deserts are sand-covered, and even smaller portions of the mountain-and-bolson deserts have that appearance.

When the downcutting of a master stream drains a portion of plateau desert,

FIGURE 5–22

Two common origins of *buttes.* A. Buttes as remnants of mesas. B. Butte formed by the weathering and erosion of a conical volcano (the more resistant plug becomes the core of the butte).

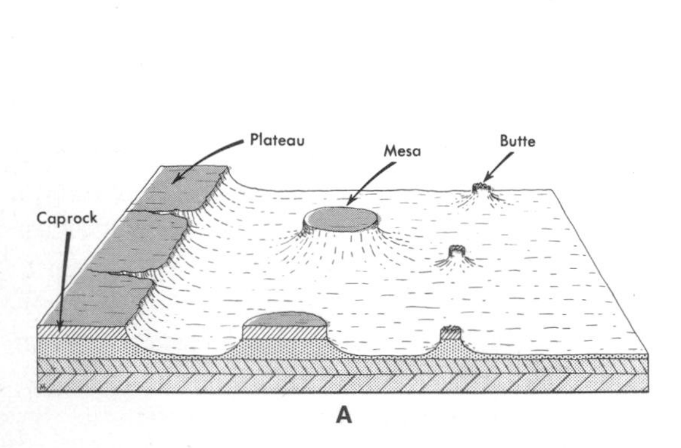

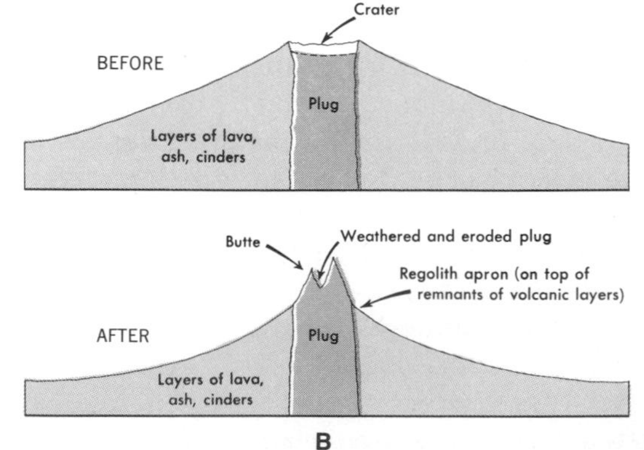

FIGURE 5-23
Some of the excellent examples of mesas and buttes that are found in Monument Valley, Arizona. (TWA)

one result is the dissection of the adjacent area by its rejuvenated tributaries. The type of relief thus produced is dependent upon the nature of the bedrock. Where all beds are soft (as in the case of shale), an intricately dissected landscape is developed—a veritable maze of narrow, steep-sided gullies separated by razor-edged ridges—a type of relief called *badlands.* Where, for the most part, resistant and weak beds alternate, a succession of cliffs and rock terraces occurs, much like the succession of retreating cliffs and terraces described above, but with much narrower terraces. Where the rock occurs in an unbedded, unjointed mass, deep, narrow, vertical-walled canyons are developed.

FIGURE 5-24

A *wadi* near Roswell, New Mexico. Its wide-bottomed, vertical-sided channel shows traces of the intermittent stream that flows from the mountain in the distance. (Jerome Wyckoff)

Most of the drainage of plateau deserts, however, is carried on in a much less spectacular manner. The typical stream course of the plateau desert is the *wadi*—also known as a *wash* in the southwestern United States, or an *arroyo* in the southwestern United States and Mexico. The wadi is a wide-bottomed, vertical-sided trench, from several to perhaps 100 feet deep; although dry most of the time, it is flooded by occasional or rare rainstorms (Fig. 5–24). Its shape is a direct consequence of the intermittency of the stream flow. When running water is present, it carves a more or less V-shaped valley at first, like the youthful valley of humid areas. As water disappears its load is dropped, filling in the lower part of the "V" and creating a level bottom. Succeeding water encounters a channel that is glutted with previously dropped materials; therefore, the flood spreads out over the wadi floor and actively undermines and steepens the valley sides. This procedure is repeated over and over again. Each *freshet*, or stream overflow, relays material farther downstream, leaves other material in its place, and erodes the banks so that they are farther and farther apart. For these reasons, a wadi, large or small, constantly maintains its steep sides and flat bottom, a shape markedly different from that of the normal youthful valley in any humid area.

REVIEW AND DISCUSSION

1. Distinguish between sheet erosion and fluvial erosion—in terms of character, effectiveness, and resultant landform features.
2. Account for the fact that a slight increase in the velocity of a stream greatly increases its ability to erode and carry materials. How does stream velocity depend on the factors of slope, volume, and load? What is hydraulic action and what is its function?
3. The creating and shaping of a valley by a stream is not accomplished all at once, but rather, proceeds in stages. Discuss and diagram the valley form that is typical of each of these stages. Why is the valley-forming process referred to as an erosion cycle? What forces can delay, interrupt, or speed up a normal cycle of erosion?
4. Explain the formation of valley flats in a mature valley in a humid region.
5. What is meant by: base level, undercut bluff, floodplain, back swamp, oxbow lake, peneplain, monadnock?

6. Discuss the significance of valley shapes to human occupance of a region.
7. At the same time that valleys are being created and shaped, the interstream areas are also undergoing change. What are the stages typical of these interfluve forms? How do they relate to the valley stages? Must the valley and the interfluve stages always be the same?
8. Draw a cross-sectional diagram that combines mature valleys and youthful interfluves. Could this combination occur in nature?
9. Discuss the significance of interfluve shapes to human occupance of a region.
10. Why are deltas considered lanforms of stream deposition? How do tributaries relate to deltas?
11. Stream patterns develop in a variety of forms, but the most common pattern is dendritic. Why? Why is the pattern developed not always dendritic? Discuss several other major stream patterns and account briefly for each type. Do stream patterns frequently or seldom change from one type to another? Why? Why do stream patterns mirror population patterns?
12. What will a mountain-and-bolson region look like by the time it reaches the old stage?
13. Why is angularity the keynote of desert landforms throughout the world? Why are the slopes of alluvial fans normally quite gentle? What are playas covered with salt layers called?
14. Explain the development of a mesa and the development and typical characteristics of a badlands region.
15. Why does a wadi, large or small, maintain its steep sides and flat bottom—a shape markedly different from that of the normal youthful valley in any humid area?

SELECTED REFERENCES

Davis, W. M. *Geographical Essays.* New York: Dover, 1954.

Dury, G. *The Face of the Earth.* Baltimore: Penguin, 1959.

Fairbridge, R. W., ed. *The Encyclopedia of Geomorphology.* Encyclopedia of Earth Sciences, Vol. 3. New York: Reinhold, 1966.

Leopold, L. B., M. G. Wolman, and J. P. Miller. *Fluvial Processes in Geomorphology.* San Francisco: Freeman, 1964.

Lobeck, A. K. *Geomorphology.* New York: McGraw-Hill, 1939.

Morisawa, M. *Streams: Their Dynamics and Morphology.* New York: McGraw-Hill, 1968.

Penck, W. *Morphological Analysis of Landforms: A Contribution to Physical Geology,* trans. from German by H. Czech and K. C. Boswell. New York: St. Martin's, 1953.

Strahler, A. N. *The Earth Sciences,* 2nd ed. New York: Harper & Row, 1971.

———, and A. H. *Environmental Geoscience.* Santa Barbara, Calif.: Hamilton, 1973.

Thornbury, W. D. *Principles of Geomorphology.* New York: Wiley, 1954.

Rocky promontories off the coast of the Japanese port of Kushimoto. (Japan Tourist Association)

6

Landforms: Coastal, Karstic, Glacial, and Eolian

Coastal Landforms

In addition to the landforms produced by fluvial action in humid and arid lands, several other types of landforms are widely spread on the surface of the earth. Among the most outstanding of these other landform types are coastal landforms—that is, those landforms produced by waves and currents along the earth's coasts, or shorelines.

A shoreline is far more than just a line separating sea from land: it is a zone, partly above and partly beneath the sea, of constant activity and change. It is in this narrow zone that the sea wages an endless attack upon the land, and usually wins; yet, at times, it inadvertently loses a little, as the land extends long arms and encircles part of the force that is attacking it.

Dynamic Aspects

The forces used by the sea in its attack upon the land come from two dominant sources: *waves* and *currents*. The former are primarily erosive agents; the latter direct their energies toward transportation.

Waves. In the open sea, the water particles involved in producing a wave move in a circular orbit whose vertical axis is equal to the height of the wave. When a wave-form approaches the shore, the friction of the shore bottom decreases the velocity of the lower part of the wave, making the orbital circle become an ellipse whose axis leans shoreward (Fig. 6–1). The wave becomes higher and narrower, and its front side steepens and becomes increasingly concave until, eventually, the crest, which is arching forward, becomes unsupported and collapses, thus forming a *breaker*. At this point, the orbital movement of the wave is replaced by a linear one, and the water rushes up the beach to a point far above sea level.

As long as the wave is in its circular orbital form, it exerts little erosive force: it can be deflected by any solid vertical wall because its water has very little forward mo-

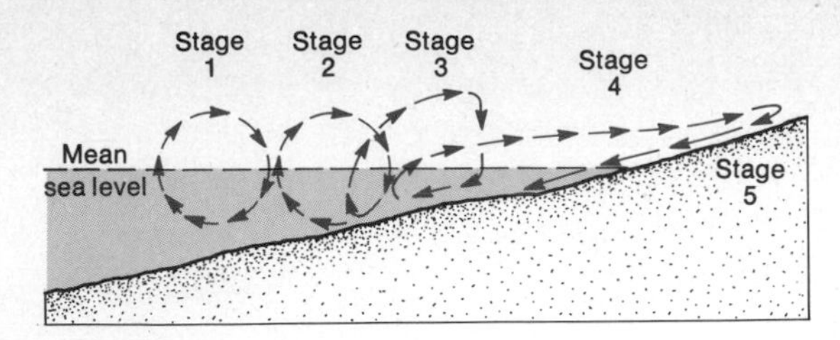

FIGURE 6–1
Water particle trajectories in a wave-breaking sequence: (1) a water particle in circular orbit in the open sea, (2) the wave-form touches bottom, preventing the water particle from completing its orbit, (3) caught up in the next wave, the same particle is thrown forward into an unsupported position, falls, and is carried seaward by the undertow, (4) caught by a third wave, the same particle is thrown forward by the momentum of the wave, and runs high on the beach, and (5) the particle is drawn down the beach by gravity, creating an undertow.

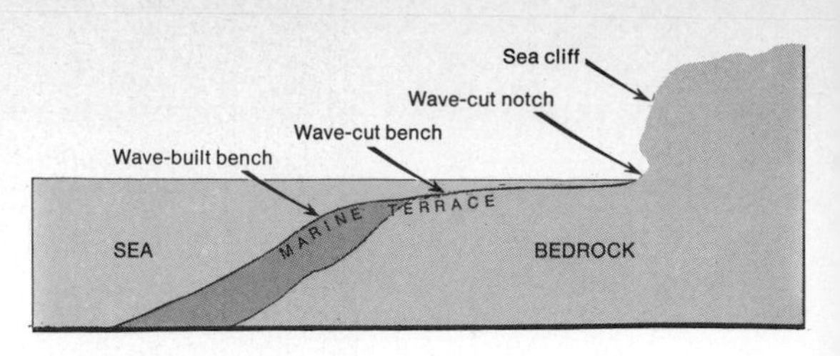

FIGURE 6–2
A cross section of a bold coast, showing a typical *wave-cut sea cliff, wave-cut bench*, and *wave-built bench* (the latter two comprising a *marine terrace*).

tion. A breaker, however, exerts tremendous force as it flings itself shoreward. In addition to this direct force, a great volume of air is compressed between the curling crest and the cliff, shore, or artificial structure against which the wave is breaking. It is this air that makes the heavy booming sound associated with breaking waves; and when it is forced into the joints of the rocks of a sea cliff, it exerts a powerfull wedging force along these fractures.

The surging of water in a breaking wave is a great abrasive force. Grains of sand, pebbles, and even boulders are rolled about, bumped together, and often flung shoreward. This reduces the size of the detritus itself and at the same time serves as a tool in the abrasion of the bedrock of the shore. The waves break with greatest force at high tide, when they are least impeded by frictional contact with the shore bottom. As a result, many sea cliffs are most severely eroded at the high-tide level and, at that level, a hollowed notch is often eroded into the cliff (Fig. 6–2). Such undermining is important in maintaining the steep face of many sea cliffs.

Currents. The water in a breaking wave climbs far up the beach and then, losing its momentum, flows back down the beach under the influence of gravity. In so doing, it passes beneath several subsequent waves, aiding greatly in upsetting them. This seaward-moving current operating beneath the landward-moving waves is called the *undertow*. When waves are large, the undertow is correspondingly powerful and is able to transport seaward great quantities of material.

The *longshore current*, or *littoral drift* (Latin: *littoralis* = shore), is a current that develops where waves break obliquely upon the shore (Fig. 6–3). As this type of wave breaks, its water flows up the beach at an oblique angle in a direction at right angles to the crestline of the wave, but when it re-

FIGURE 6–3
The activation of a *longshore current* (*littoral drift*).

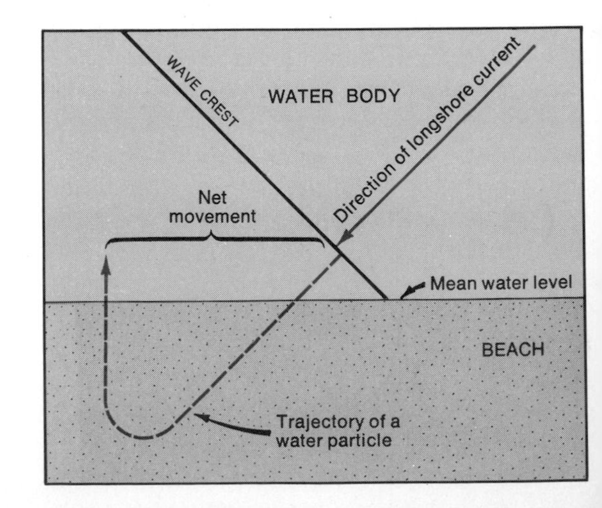

FIGURE 6–4

Stacks standing above a wave-cut bench at Hopewell Cape, at the mouth of the Petitcodiac River, New Brunswick, Canada. (Canadian Government Travel Bureau)

turns to the sea, it moves straight down the beach. Thus, a given water particle moves progressively down the beach in a zigzag fashion and makes a net movement forward with each wave. Any unattached object on the beach moves in the same manner, and thus masses of sand and other detritus move steadily along the shore in a downwave direction. Such movement may be very limited on an irregular coastline, but, in some cases, it may span a remarkably long distance: for example, the sands of the east coast of Florida are derived from the rivers of the Carolinas and Georgia.

In some inlets and straits, *tidal currents* are strong and effective in transporting material, but their total role in shore processes is very minor compared to that of the longshore drift.

Shore Features

As waves erode the land, they form what is called a *sea cliff,* which gradually retreats under the waves' continuous attack. The actual erosion is restricted to the zone between the crests of the highest waves at high tide and the troughs at low tide, and is most intense at the high-tide level. There, a notch develops that undermines and collapses the cliff above, wearing it slowly back. Thus a *wave-cut bench* is eroded into the solid rock. It gradually broadens landward by continued cliff recession, and is smoothed and lowered through abrasion by the detritus that is endlessly swept back and forth across it by the waves and their undertow. The amounts of detritus composing a beach varies. Part of it falls over the seaward edge of the wave-cut bench and accumulates on the slope below, creating a *wave-built bench,* whose surface prolongs the line of the wave-cut bench seaward. Combined, these two benches constitute a *marine terrace* (Fig. 6–2). However, more resistant portions of the sea cliff may temporarily defy erosion and stand above the wave-cut bench as isolated *stacks* (Fig. 6–4).

Abrasion acts upon the beach material, gradually changing it, at its place of origin, from large angular blocks to well-rounded stones and, eventually, farther down the beach, to grains of sand. All such material is in transit, although it may remain in one spot for a long time. Irregularities in the shape of the coastline affect the rate of movement, causing great accumulations of materials on the upcurrent sides of headlands and a paucity of them on the downcurrent sides.

Where a straight coast is suddenly terminated by a bay or inlet, the longshore drift continues straight into the bay (Fig. 6–5). The material moving with the current gradually builds an embankment, or *spit,* into the bay. If not interrupted by currents, this spit may build completely across the bay, forming a *bay-mouth bar,* although a channel is usually maintained through it by the ebb and flow of the tide. In the latter case, the end of the spit may be turned landward by the inflowing tide, forming a *hook* (attempts to construct a hook turned seaward are thwarted by the vigor of the waves and the littoral drift on the outer side of the spit). An accumulation of sand sometimes forms in the lee of an island; if it becomes large enough to connect the island to the mainland, it is termed a *tombolo,* and the island is called a *land-tied island.*

FIGURE 6–5
Some depositional shore features.

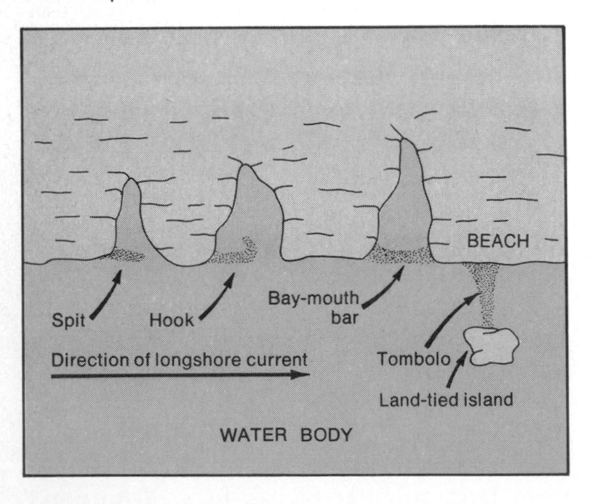

Along a very smooth, shallow coast, waves drag the bottom far seaward, picking up sand and moving it landward. At the line of breakers, they lose their energy and drop the sand, building up a narrow *offshore bar.* Storm waves, having greater power and working at a higher stage of the sea, add to the bar above ordinary sea level, and when the sand dries out after the storm, the wind constructs dunes. The long, narrow, sandy island thus constructed parallel to the shoreline is called a *barrier beach,* and is separated from the shore by a broad, shallow *lagoon* (Fig. 6–6).

Coral Reefs

Many tropical and subtropical shorelines, where water temperatures are in excess of 69° F., are bordered by *coral reefs,* colorful structures formed by *polyps*—small marine animals that secrete a coating of lime (derived from sea water) around the exterior of their bodies. Polyps grow in massive colonies at depths ranging from 180 feet to mean tide level; young polyps develop atop the remains of dead ones, and the whole group grows upward, being gradually cemented into limestone. They thrive best on the outer edges of reefs, where the churning surf provides the plentiful supply of oxygen necessary for their growth; thus, the reefs tend to grow outward from the land.

Coral reefs fall into three groups: *fringing reefs, barrier reefs,* and *atolls.* Fringing reefs occur as platforms built seaward from the shore, their growth extending ever seaward. Barrier reefs form parallel to the shoreline, often miles from it, and are separated from it by a lagoon. Atolls are circular reefs, often formed in midocean, enclosing a large lagoon. While reefs are usually slightly sub-

FIGURE 6–6
A *barrier beach* at Ras Tanura, on the Persian Gulf, Saudi Arabia. (Aramco)

merged, being indicated only by the line of breakers crashing upon them, in some places, coral debris ground by the waves into limy sand may have accumulated in large enough amounts to form small islets.

Types of Shorelines

Shorelines can be classified into four main categories: *submerged, emerged, neutral,* and *compound coasts.*

Submerged coasts are the result of crustal downwarping. Assuming that the area had a relief pattern produced by normal stream erosion, submergence will cause the sea to invade the river valleys, forming deep *estuaries,* the interfluves will become headlands and promontories, and some isolated hilltops may become islands. Initially, the coastline is extremely irregular, but the action of the waves soon begins to blunt the headlands, bars begin to form across the bay mouths, and the coastline is gradually straightened and shortened.

Assuming that the former sea floor was fairly flat, an emerged coast will be quite straight, with depths increasing only very gradually offshore. Along such a coast, an offshore bar and, eventually, a barrier beach will develop, separated from the mainland by a shallow lagoon.

Neutral coasts are dependent upon conditions other than submergence or emergence. Included in this group are coasts fringed with coral reefs or formed by deltas, lava flows, or fault scarps. Compound coasts exhibit more varied combinations than any others. Actually, most coasts are compounds, since few places in the world exhibit the simplicity of geomorphic history necessary to be purely emergent or submergent; but it is often useful to concentrate on the dominant aspect of a shoreline when attempting to explain its history.

Significance of Shore Forms

The general type of coast and the individual coastal forms in an area greatly influence the margins of its landmasses and also have a decisive effect upon human activities there. Coasts composed of prominent rocky cliffs interspersed with only limited beaches, and perhaps offering no bays for the use of ships, make access to the land highly difficult; whereas low, shallow-water coasts, characterized by bars, wide beaches, and lagoons, usually offer many opportunities for recreation and for navigation by small, shallow-draft vessels (but restricted navigational possibilities for large, ocean-going ships). Between these extremes, there are many combinations. Especially useful to humans is a coast with prominent headlands and numerous deep and broad embayments. The headlands provide protection from storms and are often so beautiful that they become resort areas. The bays provide deep anchorages for ships and reach far enough into the land—often into wide rivers—to tie the activities of the sea and land together. Many of these bay areas are ideal for resort purposes also.

Continental Shelves

Extending seaward from the shores of exposed landmasses are gently sloping plains, or platforms, that are actually submerged continuations of the landmasses themselves. These submarine continental margins are the *continental shelves.* Technically, they are not shore forms but are so closely associated with them, that they warrant inclusion in the discussion of such features. Furthermore, they are, geologically, extensions of the continental masses, rather than parts of the deep-ocean basins.

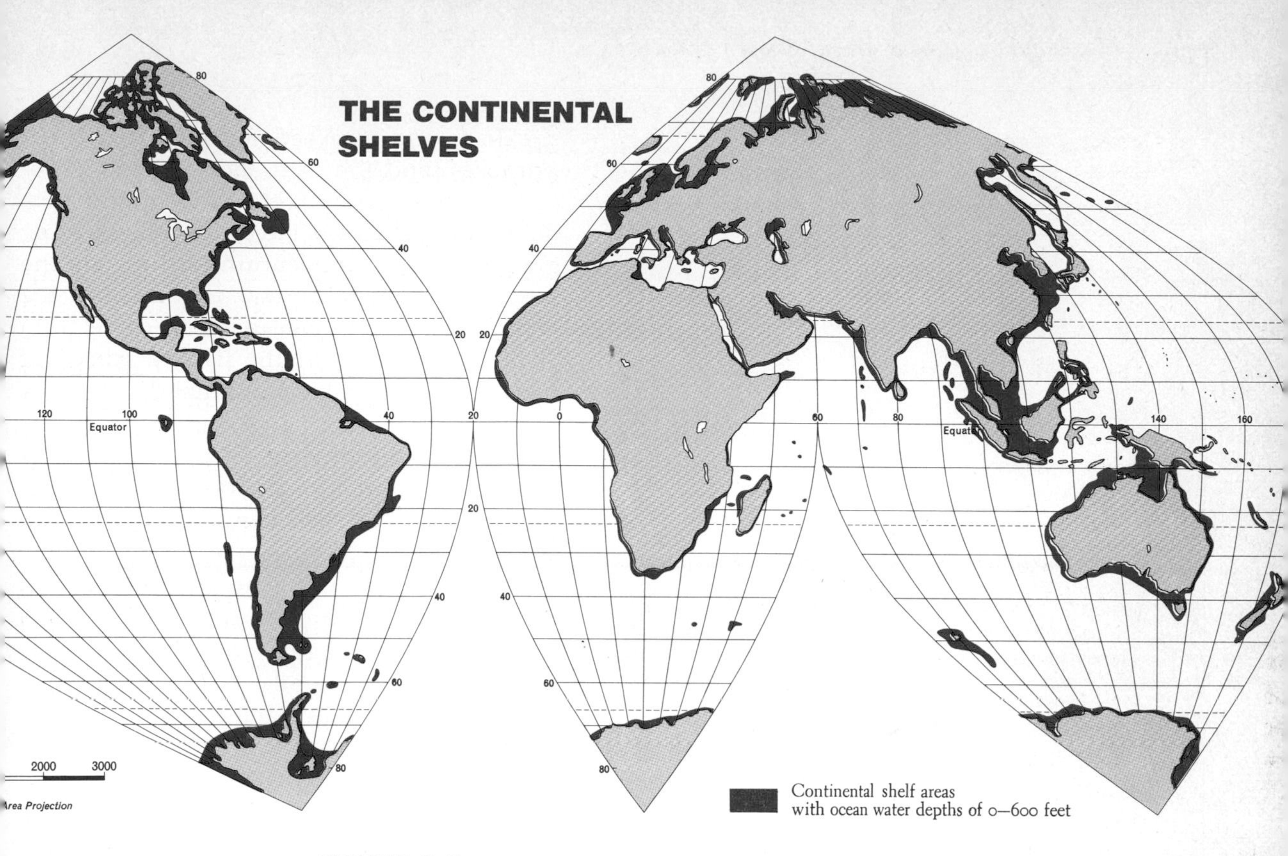

FIGURE 6–7

The world's *continental shelves* are of great human importance, particularly in reference to fisheries, petroleum, gas, minerals, coastal shipping and recreation, and pollution problems associated with shallow, inshore waters. Note the greatly varying widths of the continental shelves.

General Characteristics. Continental shelves vary greatly in horizontal extent, as Figure 6–7 indicates: in some cases, such as most of Hudson Bay and the North Sea, nearly all of the water is underlain by continental shelves; in other places, such as the margins of Africa and the Pacific Coast, the shelves are only narrow ribbons that lie between the edge of exposed land and the ocean deeps.

Floored and built up mainly by the relatively even deposition of sediments derived from eroding landmasses and of the limy remains of marine life, continental shelves are normally smooth and very gently sloping toward their seaward edges. The waters of these shelves generally vary in depth from 0 to 600 feet before increasing rapidly to depths of thousands of feet, where the base of the *continental slope* (Fig. 6–8) joins the ocean-basin floor.

Present-day continental shelves are largely old coastal plain surfaces that have become submerged. These surfaces were created, for the most part, as subaerial landforms during the time of extensive continental glaciation, the Pleistocene Epoch, which terminated some 12,000 years ago (see Appendix E). During this period so much of the world's water became locked up in the form

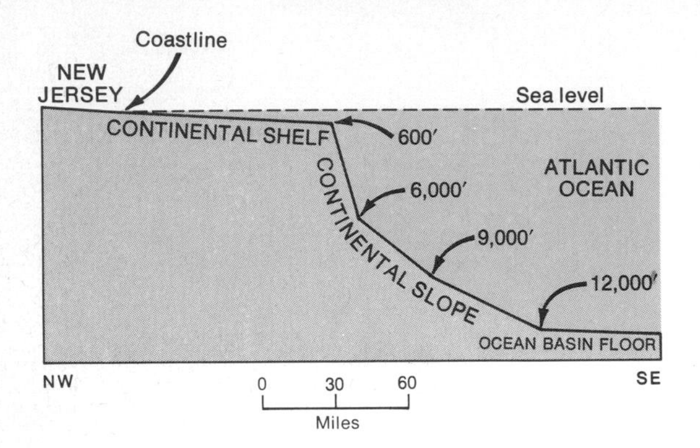

FIGURE 6–8

A generalized profile of the *continental shelf* and the *continental slope* off the low, sandy coast of New Jersey. In general, the sharp angle between the world's continental shelves and the abruptly descending continental slopes occurs at depths of roughly 600 feet below sea level. (The vertical has been greatly exaggerated here in order to emphasize the difference in gradient between the surfaces of the continental shelf and the continental slope.)

of ice that sea level dropped several hundred feet. Then, as the glaciers melted away, sea level rose to its present position, drowning much of the former coastal plain surfaces, and also submerging the lower parts of former subaerial valleys to create estuaries—such as those of the River Thames in England and the Hudson River in the United States—thereby greatly aiding ocean navigation. In some cases, the drowning of former valleys has created *submarine canyons;* however, most canyons of this sort have other origins. Significant among these are landslides and mudflows, and the turbidity currents that enable sediment-laden ocean waters to scour canyons into the continental slope.

Human Utilization. From time immemorial, coast-dwelling peoples have utilized the continental shelf areas. The shallow waters have furnished such products as fish, shellfish, and seaweed for food and other purposes; and the near-shore waters, especially where coasts were not continuously bold and harborless, have been the training ground for humans learning to operate the first small boats. Today, the bulk of the world's fish catch continues to come from shelf waters, such as the fishing banks off New England and Newfoundland and in the North Sea.

Within recent years particularly, greatly increased international rivalries and disputes have developed over control of ocean waters and ocean-bottom areas. Such disputes reflect not only the age-old struggle for commercial and naval control of the high seas, but also the active competition for the fisheries and mineral resources of shelf areas. Despite many international conferences and a few compacts incident to them, the problems are increasing in number and magnitude. Some nations claim offshore rights to a 3-mile limit, others to a 12-mile limit, and some to as much as a 200-mile limit, the latter often extending far beyond the given continental shelf, as is the case off Ecuador and Peru. Thus, the media continue to mention such items as the seizure of United States fishing vessels in coastal waters, the poaching by Japanese and Russian fishing fleets within coastal waters claimed by the United States, disputes between nations over the ownership of petroleum and natural gas deposits that lie beneath the North Sea, and so on.

In the United States, there is also much public awareness of the existence and partial exploitation of petroleum deposits located in the continental shelves off California and the Gulf Coast. This interest and concern revolves mainly around the past and potential pollution of nearby shores and coastal waters as a result of oil spills from offshore wells.

Karstic Landforms

Some land areas, whether humid or arid, and whether coastal or interior, owe the nature of their landforms to the solution of rock materials by moving water. As noted earlier, in the discussion of chemical weathering, solution occurs in practically all earth materials found both at the surface and at considerable depths. However, the creation of definite landforms by solvent action requires large-scale removal of underlying rock materials, for enough subsurface material has to be removed to affect the shape of the surface of the land itself. The conditions that permit large-scale erosion by solution occur in several scattered parts of the world, but the Karst, a sizable limestone plateau in northwestern Yugoslavia, is perhaps the classic example; thus, the term *karst* has been applied to all areas exhibiting similar solution-erosion processes and the resultant landform features.

Solution Processes and Forms

Limestone (calcium carbonate) and dolomite (calcium-magnesium carbonate), which will hereafter be referred to jointly as limestone, are soluble in water containing carbonic acid, a weak acid derived from decaying vegetation on or in soil, and therefore plentiful in the groundwater of humid areas. In compact limestone, subsurface water follows along the joints and contacts between adjacent strata, forming small *solution chan-*

FIGURE 6–9
Some typical features of a *karst*. A. Doline. B. Cenote. C. Uvala. D. Magote.

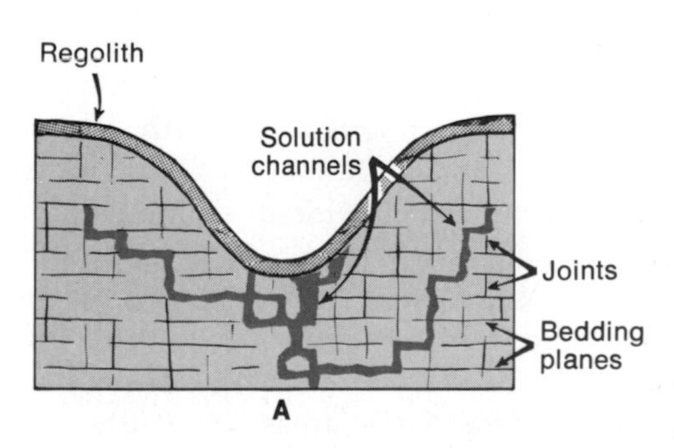

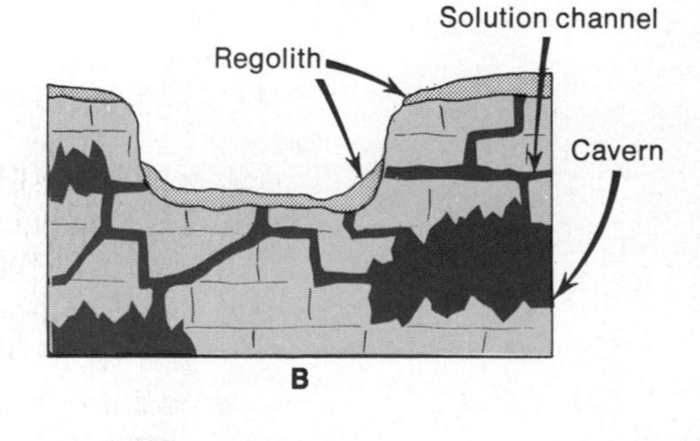

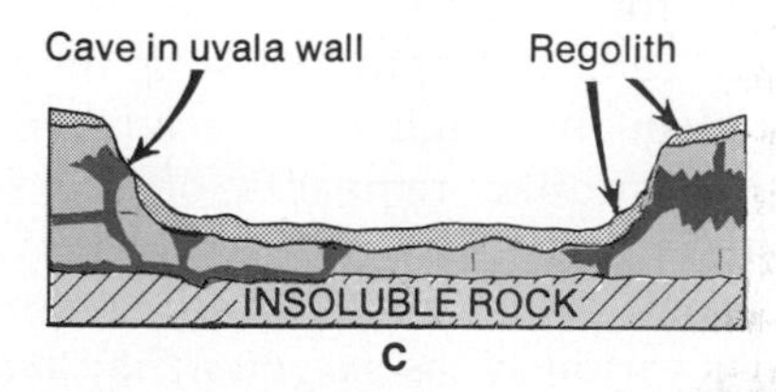

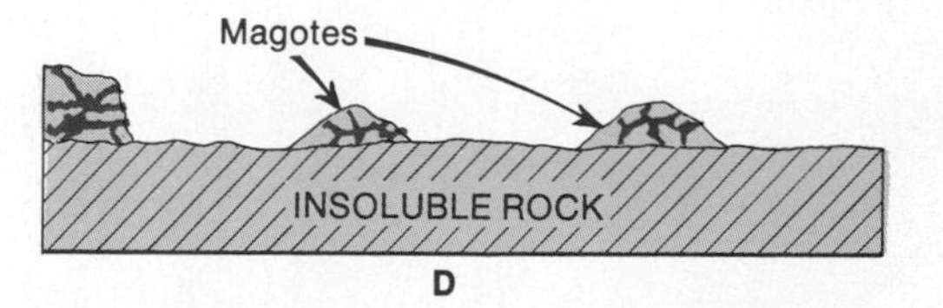

nels (Fig. 6–9A). As further solution takes place, these small solution channels become large ones, and the rock members are honeycombed with a completely developed, three-dimensional system of passageways. Where the rock is more soluble, these channels are expanded into *caverns* (Fig. 6–9B). When caverns are numerous and occur in lines at different levels, they are referred to as *galleries.* Drip water from cavern ceilings creates limy icicle-like depositional forms known as *stalactites.* When the same drip water builds up on cavern floors, it forms blunt, inverted limy cones called *stalagmites* (Fig. 6–10). If stalactites and stalagmites join, they form *pillars.* Other common features of caverns are shallow pools of cold, clear water that are rimmed and dammed by limy deposits to form staircased *travertine terraces.*

FIGURE 6–10

Stalactites and *stalagmites* in a cavern at Luray, Virginia. (Luray Caverns Photo)

Surface Forms. By the time an area has numerous well-formed caverns, the work of solution-erosion is reflected in its landforms. As near-surface cavern roofs collapse, either slowly or suddenly, landform depressions called *sinkholes* are formed. Sinkholes are circular or oval, and vary in size from a few feet to perhaps hundreds of feet across. There are two major types of sinkholes: one is funnel-shaped and is called a *doline* (Fig. 6–9A); the other is shaped like a shallow well and is called a *cenote* (Fig. 6–9B). Where sinkholes are numerous, entire areas are pockmarked with large and small depressions. If several caverns collapse along the same line, or if the roof of an underground stream falls in, a trenchlike *uvala* is produced. Like sinkholes, uvalas may be relatively small or may extend for thousands of feet (Fig. 6–9C): many of the uvalas in the karst region of Yugoslavia are sizable enough, and have sufficient soil accumulation on their floors, to contain entire agricultural settlements. Some karst areas have been stripped of their soils or regolith, by either natural erosion or human-induced erosion, particularly that incident to deforestation and overgrazing, to the point that they have become exposed bedrock areas, or *naked karsts,* laced with the intricate patterns of exposed vertical solution channels that form a *lapies surface.* Such areas are quite barren and unproductive.

If the work of solution continues long enough, the limestone in an area is almost entirely removed, producing a new level, lower than the original surface, marked by hilly, moundlike remnants of the soluble limestone rock. Such residuals, networked with solution channels and caverns, are known variously as *magotes, haystacks,* or *hums* (Fig. 6–9D).

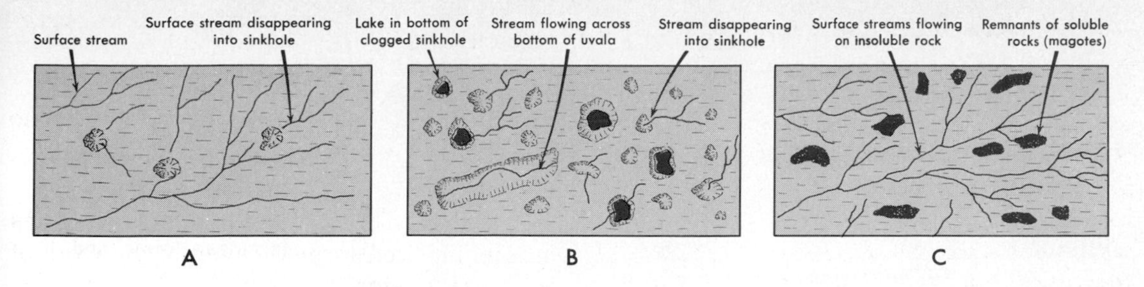

FIGURE 6–11

Drainage conditions in karst areas in different stages of development (compare with Figure 6–12). A. Drainage (chiefly on the surface) in the early stage of karst development. B. Drainage (chiefly underground) in a well-developed karst. C. Drainage (again on the surface, but at a lower level than before) in the last stage of karst development.

Karst Drainage. In the initial phases of karst development, drainage conditions are normal. Some precipitation sinks into the regolith, some evaporates, and some collects and runs along low lines on the surface to form streams. As soon as solution channels, caverns, and sinkholes appear, drainage is greatly modified. Water, sinking into the regolith and the bedrock, moves downward very rapidly through the honeycombed strata. Surface waters travel only very short distances before they disappear into sinkholes and the underground channels that drain them (Figs. 6–9A and B, 6–11A and B). The lack of definite streams and the rapid removal of water to lower levels may, even in humid regions, produce droughty soils; in lands with arid and semiarid climates, the karst condition produces such extreme dryness that it often leaves the surfaces completely barren of vegetation.

Depending largely on their degree of development, some karst regions possess numerous small lakes and ponds. These usually reflect the blocking of sinkhole outlets, so that each sinkhole acts as a reservoir to entrap local surface flow. Such ponds or lakes may drain away overnight if the outlets into the subterranean channels are freed of the earth materials that block them. Thus, aged maps of karst regions may show ponds and lakes that no longer exist and will not indicate those that have come into existence relatively recently.

By the time the limestone of a karst has been largely removed, drainage is again at the surface, and more or less normal drainage patterns are present (Fig. 6–11C).

The Karst Erosion Cycle. This discussion of karst landforms and karst drainage illustrates a definitely sequential development. It indicates three major stages in what may be termed the *karst erosion cycle:* youth, maturity, and old (Fig. 6–12). *Youthful karst* is characterized by a few sinkholes and by the predominance of surface drainage. *Mature karst* is marked by the presence of numerous sinkholes, a few or many uvalas, and almost entirely subterranean drainage. In the third major stage, drainage has returned to the surface, and about all that remains to mark the area as karst are the scattered, knobby and hilly remnants of the soluble rock; these are known as *old karst.*

The Significance of Karst Conditions

The fact that an area is a karst region makes a great deal of difference to the area's inhabitants. In youthful karst, assuming a humid climate, water is abundant.

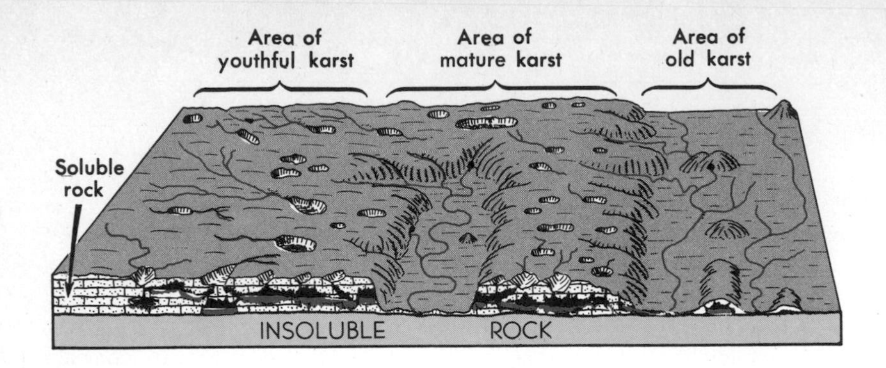

FIGURE 6–12
Typical surface forms, subterranean forms, and drainage conditions in different stages of the *karst erosion cycle.*

Streams are on the surface and underground channels have not developed to the degree that they quickly rob the soils of necessary moisture. In addition, sinkholes are few, so that there are only minor interruptions in the surfaces of fields and only slight restraints placed on the pattern of land routes.

If, however, sinkholes are numerous, living conditions are different—even in humid climates, water becomes a problem. There are few or no surface streams, and soils may be deficient in moisture. Although these conditions may be partially offset by the use of sinkhole ponds and lakes as sources of water for domestic animals, such water is still of little use to humans, since it is usually polluted and stagnant. In most karst areas, rainwater has to be collected from the roofs of dwellings and stored in surface containers or in underground cisterns. Soils may be so dry that they support only drought-resistant crops or demand special dry-farming methods. Moreover, the countryside is greatly interrupted by depressions, which affect the amount of land available for crops and greatly limit the alignment of routes. Where sinkholes are of the doline type, more land is available for agricultural use; for example, the sides of the dolines may be used as pastures and the ponds that may collect in their bottoms, as water for stock. If the sinkholes are cenotes, much land is wasted. The sides, often cliffed, are too steep to use, and any impounded water is inaccessible to livestock. In addition, it often is necessary to construct fences around cenotes to prevent roaming livestock from falling into them.

Areas that have reached the stage of old karst are little affected by karst conditions, as only remnants of the original soluble rocks remain. Streams are on the surface, and soils are not losing their moisture to subterranean solution channels. However, in places, magotes may be so numerous as to interfere with the agricultural use of fields and to cause marked deviation in the alignment of roads and railroads.

Glacial Landforms

Ice, like water, must be moving in order to take part in the formation of landforms. And as with moving water, ice in motion results in two kinds of work: erosion and deposition, each creating its own particular set of landform features. Large ice masses can eradicate or drastically modify previously existing landforms and substitute for them new and very different ones.

Large stationary accumulations of snow are called *snowfields;* similar masses of ice constitute *icefields.* A mass of ice that is moving, even if the motion is so slow as to be almost imperceptible, is termed a *glacier.*

Continental Glaciation

Glacier Origin and Movement. Glaciers are formed when more snow falls during the colder season of the year than melts during the warmer season, resulting in an annual increment of snow crystals. As the older layers of snow become buried and pressed down by newer ones, they slowly change into a coarse, granular snow, called *névé,* or *firn.* In time, under still greater depth and pressure, the granular snow itself is transformed into solid glacial ice. A view of the vast ice plateau of Antarctica (Fig. 6–13) reveals little more than mile after mile of snow, but a few feet under the snow is the névé and, below that, the ice mass itself.

When an appreciable thickness of ice has accumulated, movement begins. In the central area of the glacier, the ice may attain a thickness of more than a mile. The great weight of the upper part of the glacier exerts tremendous pressure on the lower portions. Seeking relief from that pressure, the ice flows out from the center in all horizontal directions. The movement is very slow, and although ice cannot flow like water, it does possess sufficient plasticity to acquire a ponderous creeping motion. Continental ice sheets that are still "alive" and advancing push their edges forward only a few feet each year. Even then, there are times when the rate of melting equals the rate of advance, and the line of the "front" remains stationary. In other periods, the front retreats, even though the ice itself is still moving forward. This condition reflects a melt rate that is greater than the rate of forward motion of the mass. Still another situation arises when the supply of new snow, in the center of the

FIGURE 6–13

The stark whiteness of snow-covered ice plateaus is exemplified in this view of Antarctica's Antenna Plateau and McMurdo Sound. (U.S. Navy)

accumulation area, is no longer sufficient to cause forward motion; then, the ice merely lies in place or stagnates, until melting destroys it. Its last vestiges are the places where it was thickest, or whatever great chunks of it were well buried and insulated under earth debris.

If ice moves into the sea, its front breaks off, or *calves,* to form *icebergs.* Where it descends into the sea from considerably higher and rougher lands, as in Greenland, the resultant icebergs are sharply jagged in appearance; where it moves into the sea waters, in a thick, nearly horizontal sheet, as it does along much of the Antarctic shore, the icebergs are flat and tablelike (Fig. 6–14). The flat icebergs, moved by winds and currents, menace only the occasional whaling vessel and the ships supplying Antarctic scientists, but the Greenland type, carried southward by the Labrador Current, menace the shipping of the world's busiest oceanic trade route—the North Atlantic.

Landforms of Glacial Erosion. When a continental glacier moves its billions of tons of ice across a landmass, the resultant effects are almost unbelievable. Loose soil, stones, and other regolith are pushed along by the ice and are incorporated into the glacier's underside, which then acts as a scouring and grinding machine, planing away and gouging into the solid bedrock it passes over. Since the ice is often a mile or more in thickness, the weight and pressure exerted by it makes it an extremely effective agent of abrasion. However, because rock varies in hardness from place to place, and because the local velocity and thickness of the ice also varies, the stripping and scouring action

FIGURE 6–14

Icebergs. Above, the jagged Greenland type. (U.S. Coast Guard) Below, the tabular Antarctic type, shown here as it breaks away from the shelf ice. (U.S. Navy)

is very uneven. In some areas the ice gouges deeply; in others it merely scrapes lightly at the uppermost materials.

The most conspicuous landform formed by glacial scour is the *roche moutonnée* (Fig. 6–15), a type of hill that has been shaped from solid bedrock by the erosive action of moving ice. On the side from which the ice approached (the *stoss,* or *struck,* side) it has a smoothed and relatively gentle slope; on the opposite side (the *lee* side) it is rough and steplike. The steplike condition is the result of a plucking action, whereby the lower ice carried away blocks of bedrock of different sizes, blocks that later became glacial boulders. Roches moutonnées occur in many sizes; some are small enough to fit into a small room, while others are hundreds of feet in height and several miles in length. Regardless of size, their frequent occurrence gives a unique expression to a whole region.

Less readily visible is another ice-shaped landform known as a *scour basin.* It has been scooped out of solid bedrock, and its surface is just as hard and sometimes as barren, as that of the roche moutonnée. However, because it is a depression rather than a hill, its surface is often partially buried by a thin veneer of loose glacial debris or covered by the water of a lake or the thick growth of a swamp.

The third of the more common landforms of glacial scour is the *glacial trough,* a gigantic groove that may extend cross-country for miles. It usually marks the course of a former streamcut valley that ran along the axis of ice motion.

Landforms of Glacial Deposition. The materials that are picked up and carried in the lower portion of the ice are eventually spread over the land in many forms. In some areas, the debris is laid down so thickly that all previously existing landforms are completely smothered, and bedrock lies at considerable depths; in others, the debris blanket is thin and patchy, and bedrock, smoothed and grooved by ice scour, sticks through the unconsolidated materials. In general, the debris is thinnest toward the centers of ice accumulation and thickest near the margins of the farthest advance of the massive ice.

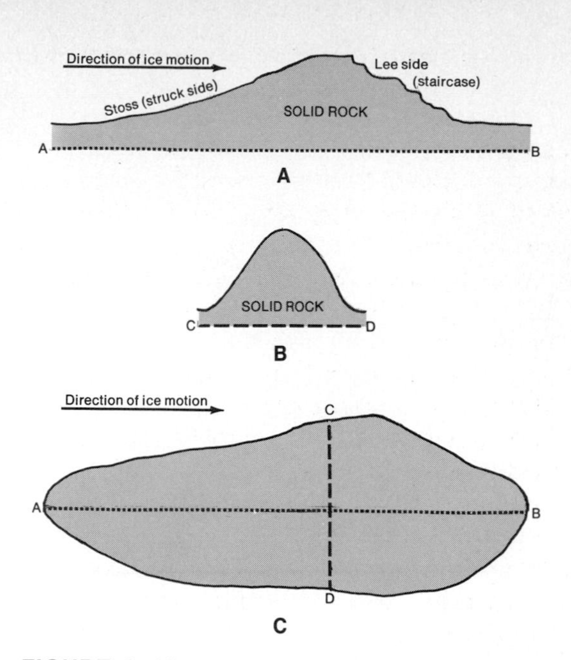

FIGURE 6–15

A *roche moutonnée.* A. Longitudinal section. B. Transverse section. C. Typical shape when viewed from directly overhead.

Known as *till,* or *glacial drift,* this debris is a heterogeneous mixture of material: huge stones and boulders, as well as fine clays derived from preglacial regolith; boulders plucked from hillsides or detached from scoured plains; and *rock flour* produced by the glacier's abrasion of the bedrock it traversed.

Glacial deposits may be divided into two general classes: those deposited directly by the ice, or *moraines;* and those reworked by water or wind before their final deposition. Moraines are the surface expressions of deposits of till, and because the conditions of deposition vary greatly, the ap-

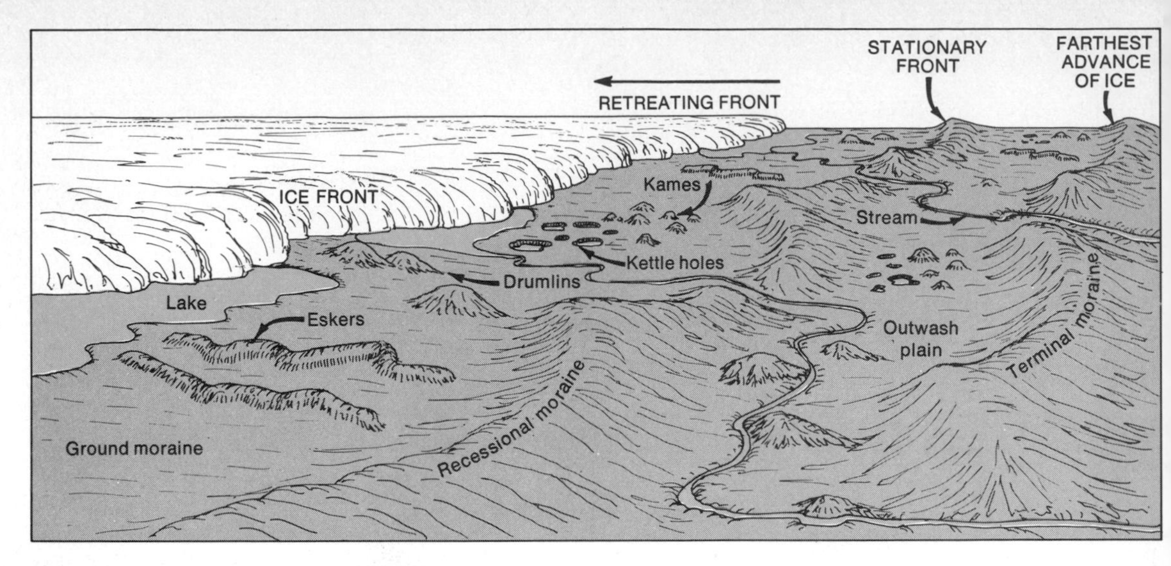

FIGURE 6–16

The kind and characteristic location of typical *landforms of glacial deposition*, as deposited by a continental ice sheet. The features shown in the pattern of arcs and lobes mark the successive positions of the stationary ice front.

pearance of moraines also varies. A *morainal ridge* marks the line along which the glacial front remained stationary for some time. Forward-moving ice carries debris to such lines and piles it up to form ridges ranging from 20 to 200 feet in height (Fig. 6–16). Moraines of this type, which were laid down at the line of farthest advance of the ice, are known as *terminal moraines;* those laid down elsewhere, each marking a stage in the retreat of the line of the glacial front, are called *recessional moraines.* The only difference between them is their geographical position, for they were formed in the same manner and from the same types of materials. Between the ridges of terminal and recessional moraines, or between two recessional moraines, the retreating ice front drops a gently rolling carpet of *ground moraine.* Like the ridge moraines, it is comprised of materials ranging from silts to huge boulders. A special feature of ground moraine is the *drumlin,* a type of hill shaped like an egg that has been sliced longitudinally in two. Composed of till that is relatively high in clay content, drumlins are formed underneath the ice sheet and streamlined by the ice passing over them. They commonly occur in groups with their long axes always in the line of glacial flow, and range up to 200 feet in height and to about 2 miles in length (Fig. 6–17).

The melting of a glacier produces vast amounts of *meltwater,* and much of the detritus transported by a glacier is carried away in floods of this meltwater, sorted by it, and eventually deposited as some form of water-borne sediment; either stream-deposited (*glaciofluvial*) or lake-deposited (*glaciolacustrine*).

The *outwash plain* is the most common form of glaciofluvial deposit. It is well named, for it consists of debris that was washed out from the glacial front by meltwater and deposited, layer on layer, to form a plain of very low relief (Fig. 6–16). Such plains may extend as far as several city blocks or several square miles. They are notably sandy: the meltwater that flows in a sheet from the ice front is not powerful enough to handle gravels and boulders, but it does have the ability to transport silts, clays, and sands. As its force lessens, it drops the sands and builds the outwash plain, while the finer silts and clays are carried farther.

Continental glaciers push into a region with complete disregard for its topography and drainage systems. Hence, not infrequently, an ice sheet acts as a temporary dam, impounding waters in large but short-lived lakes. Meltwater, flowing into such lakes, deposits in them whatever sediments it is carrying. Thus, *glaciolacustrine deltas* are built of these sandy materials, deltas that are left high and dry when the ice melts and the lakes drain away. On the floors of such lakes, the fine silts and clays are deposited to form a *glaciolacustrine plain.* A very large and famous plain of this sort occupies most of southern Manitoba province in Canada and extends in an ever-narrowing finger across the border into northwestern Minnesota and the eastern edge of North Dakota. This area was once covered by the cold waters of glacial Lake Agassiz, waters that were caught between the wall of the ice front to the north and the slightly higher land to the south (Fig. 6–18). As long as the ice dam existed, materials, particularly fine ones, were washed into the lake's shallow waters and distributed very evenly over the lake floor. Later, when the ice dam disappeared, Lake Agassiz drained away into Hudson Bay, leaving behind a lake bed that is nearly as flat as a table top and that constitutes some of the most productive soils of the world. These soils, fine-textured and fertile, have proved capable of growing a great variety of high-grade wheat in large quantities; they comprise the basic resource of much of the major agricultural region known as the Spring Wheat Belt of the United States and Canada. Remnants of old Lake Agassiz can still be found today in the cluster of lakes that includes Lake Winnipeg and Lake Winnipegosis.

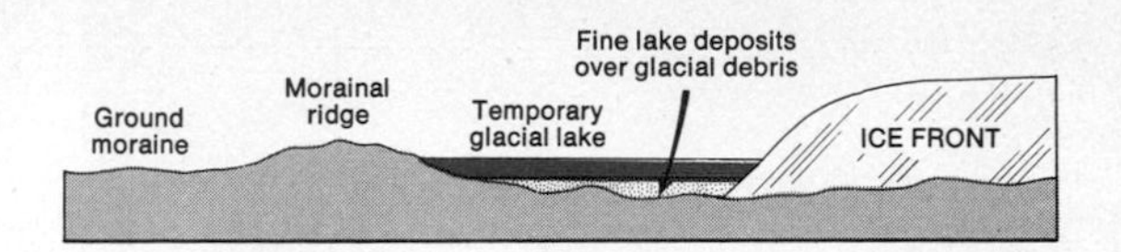

FIGURE 6–18

A cross section representing one type of origin of a *glaciolacustrine plain.* Following the disappearance of the ice dam and the glacial lake, the lake-bed deposits are exposed as a plain of extremely low relief.

FIGURE 6–17

A *drumlin* formed by the Green Bay glacier in Fond du Lac County, Wisconsin. The steep front end (toward which the ice pushed) has been left wooded, while the gentler slopes have been cultivated. (USGS)

Leading out from temporary glacial lakes are the ribbonlike forms known as *glacial spillways.* These were once the outlets for surplus lake waters; today some of them are stream valleys, but others are simply elongated depressions floored with alluvial material. Because they form low corridors, their courses are often used as routes for highways, railroads, and canals.

Other minor glacial landforms include *eskers, kames,* and *kettle holes* (Fig. 6–16). An esker is a low, sinuous ridge that winds cross-country for a few hundred feet or several miles (Fig. 6–19). It represents a portion

FIGURE 6–19

An *esker* traces its sinuous course across an area of ground moraine near Fort Ripley, Minnesota. (W. S. Cooper)

of a subglacial tunnel that was blocked with stream-borne materials; after the ice melted away, the mold of the tunnel was left behind as a low ridge. Because it is well drained, an esker often serves as a route for a trail or road, and because it is made of gravel, it is often the site of a gravel pit. Kames are small knobby hills; in single occurrence they are seldom conspicuous, but when they occur in tightly packed groups, they create a sort of miniature hill country. Kettle holes are small funnel-shaped depressions. They mark places where blocks of ice were buried in glacial debris. As the blocks melted, the materials covering them dropped lower and lower until pits were created. They may be found almost anywhere in glacial debris, as in *pitted outwash plains, pitted morainal ridges, pitted ground moraines,* and as pits among clusters of kames (Fig. 6–20). Kettle holes, because of their shape, are often mistaken for sinkholes, but there is no connection between a glacial kettle hole and a karst sinkhole.

Broad outwash plains, dried out during the long glacial winters, were fine sources from which the wind could derive great clouds of dust. Settling eastward somewhere to the leeward, the film of dust from each storm, added to those from other storms, amounts to considerable deposits of loess in some places. Today, many feet of loess cap the bluffs at Vicksburg, Mississippi: the eastern edge of the Mississippi River floodplain was the site of much outwash deposition during the Ice Age, even though it is far outside the glaciated area.

Glacial Drainage. When continental ice masses invade a region, they change more than just the landforms. From the preceding discussion, it is evident that they have a vital effect on soils and on the pattern of routes. Even more important, however, is the effect on conditions of drainage, for glaciation plays complete havoc with preexisting stream patterns and stream behavior.

Whether in areas of scour or deposition, a *glacially deranged pattern* of drainage is created. The pattern is well named, for it is one without any apparent system or reason (Fig. 6–21). Lakes and ponds are numerous, and they lie in the depressions created by unequal scour and uneven deposition. Scattered among the lakes and ponds are many

FIGURE 6–20

A *kame* and a *kettle pond* near Franklin Lake, New Jersey. (Jerome Wyckoff)

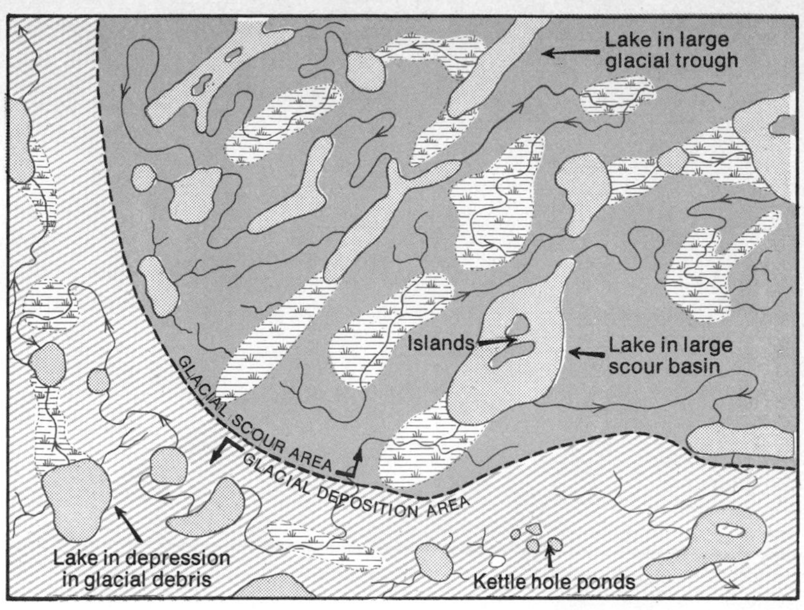

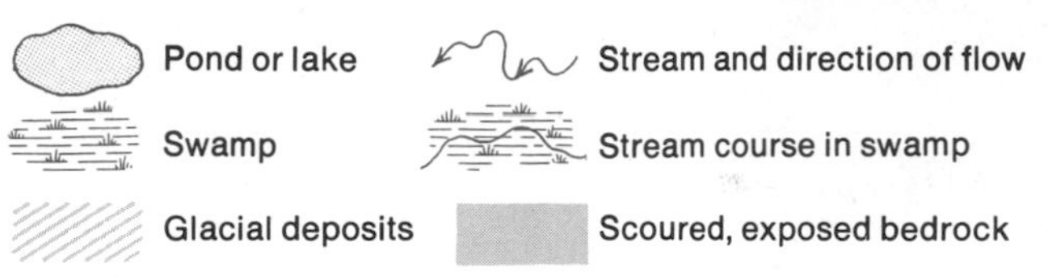

FIGURE 6–21
A *glacially deranged* pattern of drainage in an area of glacial scour (above dashed line) and in an area of glacial deposition (below dashed line).

swamps, some of them representing former lakes and ponds that have been filled with inwashed materials and by the growth of dense swamp vegetation. Connecting the lakes, ponds, and swamps are many large and small streams. They wander in a random fashion, for running waters always seek the low lines in a glaciated surface and follow the varying slope of the land. Their wanderings reflect the lack of sufficient time since the ice disappeared in which to carve definite and coordinated channels and patterns of drainage. Falls and rapids are the rule rather than the exception, although they are shorter-lived where streams are eroding into loose glacial drift than where they are crossing solid bedrock.

The Significance of Continental Glaciation. The achievements of the great ice sheets affect the appearance and nature of millions of square miles of the earth's surface. Figure 6–22 shows that the glaciers of Pleistocene times (the most recent Great Ice Age) covered truly huge areas and that today, largely in Antarctica and Greenland, their shrunken remnants are still present.

In Europe, the center of ice accumulation was at the edge of the Fenno-Scandian Shield, chiefly in what is now the mountainous backbone of Norway. From this elevated area the ice pushed out in all directions: the British Isles were glaciated as far south as London; and the continent of Europe was covered as far south as a line extending from the mouths of the Rhine River through Germany and Poland, and on across the Russian Plain to northwestern Siberia.

In North America, a more extensive area was affected. There were four centers of ice action: the Labrador center, lying between the eastern shore of Hudson Bay and the coast of Labrador; the Keewatin center, between Hudson Bay and Great Slave Lake;

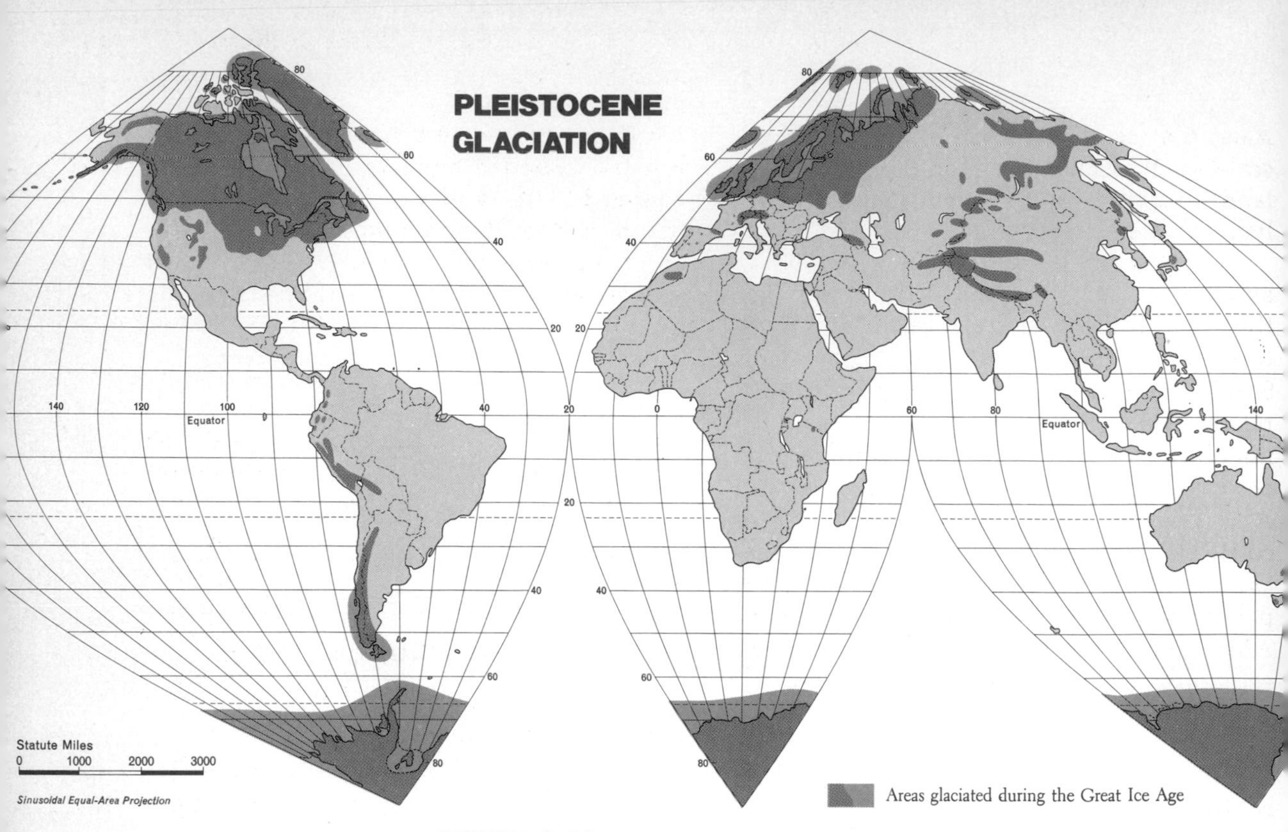

FIGURE 6–22

Appreciable portions of the earth's surface were affected by continental glaciers during the Great Ice Age of Pleistocene times, especially in North America, Europe, and Antarctica. Large remnants of these continental glaciers still cover most of Greenland and Antarctica.

the Patrician center, between Lake Superior and Hudson Bay; and far to the west, in the crest areas of Canada's high mountains and on into Alaska, the elongated Cordilleran center. From these four centers, the ice moved out in all directions until, at one stage, it covered practically all of Canada, parts of Alaska, and a considerable portion of the coterminous United States. The latter was glaciated as far south as a line extending across Long Island to the Ohio and Missouri rivers and on to Montana, Idaho, and Washington. By the time the ice wasted away, it left, to nearly half of North America, an inheritance of glacial landforms, glacial soils, and glacially deranged drainage. Only at the outermost margins of the oldest deposition has there been sufficient time for normal stream erosion to restore ordinary drainage conditions and create new, nonglacial landforms.

Elsewhere in the Northern Hemisphere, ice sheets covered most of Greenland and Ellesmere Island, the Brooks Range of northern Alaska, and smaller areas in the American Rockies, the Cascades, and the Sierra Nevada. In Europe and Asia, they covered parts of the Alps, the Pyrenees, the Caucasus, and many of the ranges that stretch from the Pamir mountain hub to northeastern Siberia. (It is interesting to note that most of Siberia was not then, and never has been, glaciated, chiefly because the site was so far removed from a source of moisture that it did not receive sufficient snowfall.) Off the shores of parts of Europe and

Asia, the ice affected, on more than a local scale, such islands and island groups as Iceland, Svalbard (Spitzbergen), Novaya Zemlya, and the New Siberians.

In the Southern Hemisphere, the ice smothered Antarctica and pushed on into the sea, creating a glacial mass at least twice the size of the coterminous United States. South America's ice sheets, affecting only a relatively small part of that continent, lay mainly in the southern Andes, with smaller occurrences in the central and northern Andes. Elsewhere in the world, small masses covered much of South Island, New Zealand, portions of Tasmania, and the southern part of Australia's Great Dividing Range. In Africa, there were minor glaciers in the highest mountains of the central highland and in the High Atlas.

Even this brief discussion should make it clear that the modifications wrought by continental glaciers are much more than just matters of local geographic concern. In addition, the present geography of Greenland and Antarctica, areas representing more than six million square miles of the earth's surface, is essentially an "ice geography;" in these regions, there remains a "laboratory" in which the nature and behavior of continental ice sheets and glacial climates like those that affected so much of the world during Pleistocene times are studied.

Valley Glaciation

In great contrast to the extensive ice blankets of continental glaciers are the much smaller masses and tongues of valley ice. Each is born in a high valley head in those parts of mountains where more snow falls in winter than melts in summer. As soon as the accumulating snow has been transformed into ice and the ice has acquired sufficient mass to move, the glacier begins its slow journey down the valley. By the time the work is finished and the ice is gone, the typical V-shaped mountain valley has been changed to the typical U-shaped form of valley glaciation, the crestline has been sharpened and made more irregular, and much of the whole preglacial surface has been greatly modified (Fig. 6–23).

FIGURE 6–23

A portion of a mountain region before, during, and after *valley glaciation*. Notice how greatly the process of valley glaciation has modified the landforms and the drainage conditions in the region.

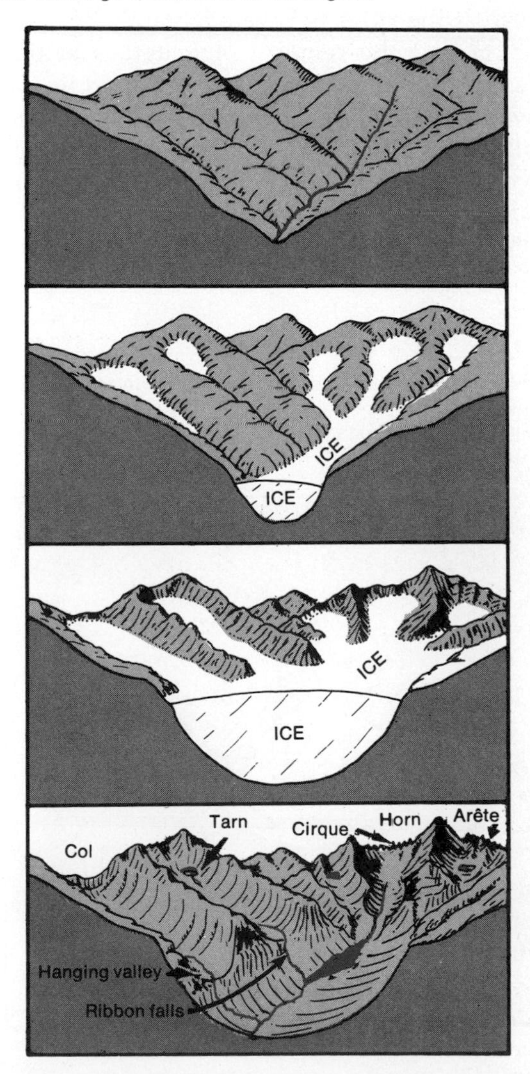

Typical Landforms and Drainage. The first landform created by the work of valley ice is the *cirque*. This form is shaped in the valley head, where the glacier is formed and where its ice is thickest. By a combination of scour and frost action, the valley head is carved into a type of small basin that has the form of an amphitheater: it is very steep on three sides, while the fourth is open in the downvalley direction (Fig. 6–23). Where two cirques are formed close together in adjacent valley heads, the divides between them are sharpened into a knife-edged ridge, or *arête*. If the ridge is lowered, mainly by frost action and the removal of materials at the base by ice plucking, it becomes a saddle, or *col*, which makes a low place in the mountain crests and may be used as a pass route. Where three or more cirques have been carved into the sides of a single mountain, it may be sharpened into a pyramidal peak, known as a *horn*, such as the famous Matterhorn of the Swiss Alps.

As the glacial tongue moves down the valley from the cirque basin, its strong erosive action steepens the valley sides and somewhat deepens the valley bottom, thus changing the "V" to a wide and irregularly shaped "U." Streams of ice in tributary valleys create U-shaped valleys of their own, but they may also add their ice to the main glacial tongue. As a rule, the ice in tributary valleys, having less volume and power, does not cut as deeply as the main ice flow, and, hence, *hanging valleys* are formed.

Valley ice also builds landforms of deposition (Fig. 6–24). If the ice travels the full length of the valley and emerges to fan out on a piedmont plain, a *terminal morainal ridge* is created. With a retreating front, *ground moraines* are laid down, some out on the piedmont and some in the valley floor. Successive states of a stationary front are marked by the damlike forms of *recessional moraines*. Along the valley sides, glacial debris accumulates in irregular heaps called *lateral moraines*. Where two valley glaciers join, the lateral morainal material that unites at their junction will appear later as a *middle*, or *medial*, *moraine*, which will be found in the midst of the valley floor.

A great part of the beauty of a glaciated valley results from its glacially deranged drainage. The uppermost drainage feature is the *tarn* — a small lake that occupies the lowest portion of the cirque basin and sends its waters tumbling over the rocky edge of the cirque into the valley below (Fig. 6–23). Farther downvalley, a series of small, shallow lakes normally lie in local scour basins or are dammed behind deposits of morainal debris. Out on the piedmont, if the glacier reached that far, there is usually a marginal lake impounded by the curve of the terminal moraine (Fig. 6–24). Connecting all the lakes, from tarn to piedmont, are the alternately

FIGURE 6–24

The characteristic landform and drainage features in the lower portion of a completely glaciated mountain valley (in this instance, the ice fanned out on the piedmont).

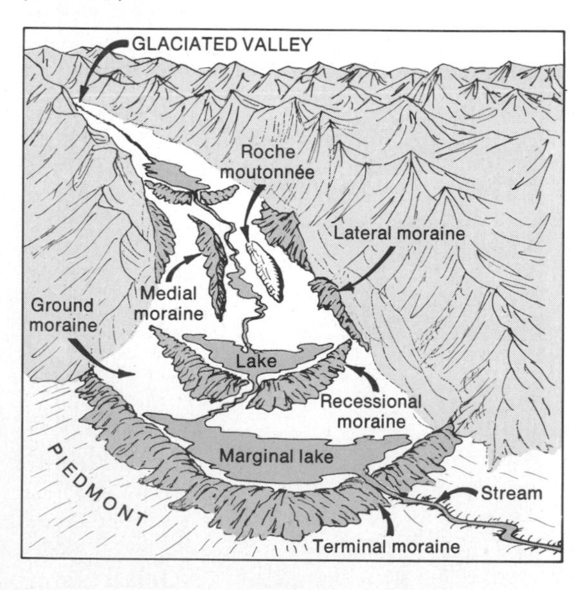

swift and sluggish waters of the valley stream. Where hanging valleys are present, the streams from them leap into space as misty ribbons of white water. A famous example of such *ribbon falls* is Bridalveil Falls in the glacial trough of Yosemite (Fig. 6–25).

The Significance of Valley Glaciation. Although valley glaciation has affected much smaller portions of the land surface of the earth than has continental glaciation, it still occurs in thousands of places: from the high mountain valleys of South Island, New Zealand, and the southern Andes, to those of the Brooks Range of Alaska and the Stanovoi Range of Siberia; and in such well-known alpine regions as the Rockies and the Sierra Nevada of North America, and the Alps, Pyrenees, and Caucasus of Europe.

Many of the earth's glaciated valleys, as in Switzerland, are used intensively by humans. The valley ice created more level or

FIGURE 6–25

Bridalveil Falls, one of several *ribbon falls* that plunge from tributary *hanging valleys* into Yosemite Valley, in the western flanks of California's Sierra Nevada. (Union Pacific Railroad)

FIGURE 6–26
Roseg Valley, one of Switzerland's many fertile and picturesque glaciated valleys. (Swiss National Tourist Office)

nearly level land than was present before, thus providing more land suitable for crops, pasture, and settlements (Fig. 6–26). The glacially deranged drainage provides many natural reservoirs as well as falling waters that aid in the generation of hydroelectric power. The total condition, marked by such features as tarns, arêtes, horns, hanging valleys, and ribbon falls, provides a scenic beauty that is the chief resource for recreational and resort industries. It should also be pointed out that glaciated mountain valleys are important to lands that lie below them, and, sometimes, to lands miles away. The valleys, with their many natural reservoirs, collect waters and release them slowly. This not only minimizes the dangers from floods in lower areas, but it also insures a steadier supply of water for piedmont settlements, or, if the piedmont areas are in dry lands, for irrigation.

Eolian Landforms

Wind (eolian action) is the least potent agent of erosion and deposition. However, in any one spot, wind-created landforms may dominate the land surface as, for example, in the Sand Hills of central Nebraska or the Erg Oriental of the Sahara. Wind accomplishes most of its work in arid lands, where the regolith is normally dry, transportable by the wind, and little-protected by vegetation. There are, however, some parts of humid lands in which wind-created forms are conspicuous, particularly along the margins of water bodies where loose beach sands are available and exposed to wind action. Examples are the many dune areas along the leeward (in this case, eastern) shore of Lake Michigan, the sand ridges of the Landes region of southwestern France, and the dune hills located in low pockets along the bold and rocky coast of Oregon (Fig. 6–27).

Erosional Forms

Only in rare instances are wind velocities sufficient to move materials as large as gravels, but they are often strong enough to move dust and sand. Dust (loess) is picked up and transported in suspension, and may be carried to heights of thousands of feet and to distances of many hundreds of miles before it settles out of the air. Sands are moved mainly by the processes of *creep* and *saltation*. The term *creep*, as used in this particular connection, refers to the movement of sand grains as a result of bombardment by other sand grains. *Saltation*, the paramount process of sand movement, is a kind of bouncing action whereby wind plucks sand grains upward at a steep angle, and then transports them forward and downward on a more gentle trajectory until they impact the surface.

FIGURE 6–27
Dunes along the eastern shore of Lake Michigan, near New Buffalo, Michigan. (Henry Mayer, National Audubon Society)

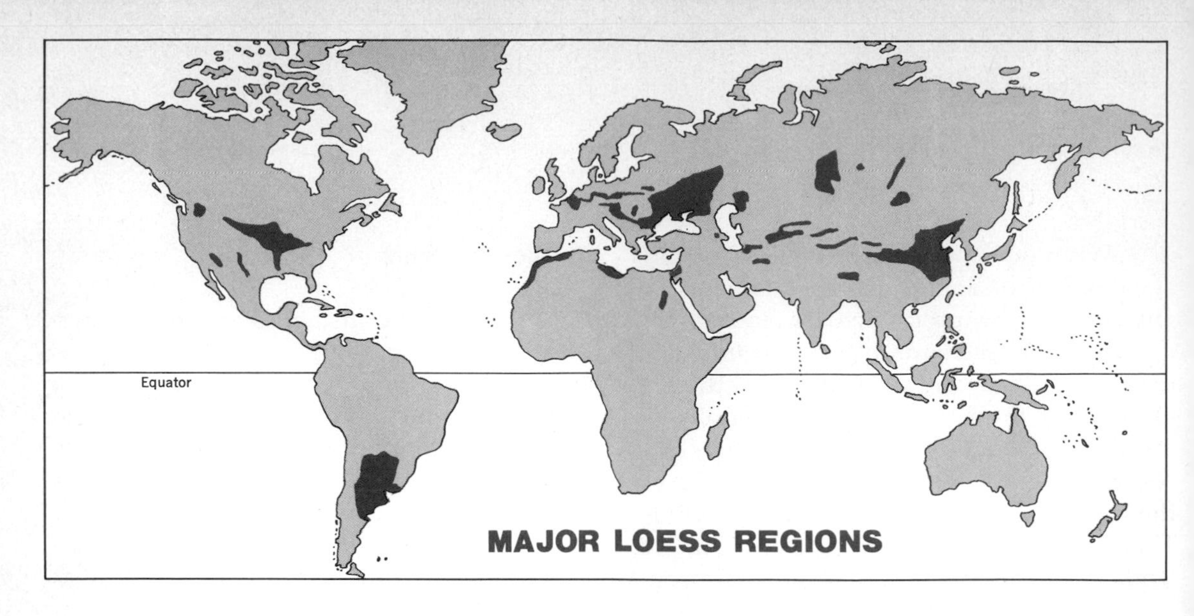

FIGURE 6–28
The distribution of the world's major loess regions. (After A. K. Lobeck, *Geomorphology*, © McGraw-Hill Book Co., Inc., 1939. Used with permission.)

This bouncing type of transportation is carried on to a height of up to two or three feet above the surface (sand is rarely lifted more than some three feet above the surface over which it is moving).

The general process of removal of dusts and sands from a given place is called *deflation*. If deflation dominates an area long enough, the surface is lowered and tends to become veneered with the gravels or other comparatively large rock materials left behind as the dusts and sands are removed. This often occurs on old, inactive, alluvial-fan surfaces and also on large and small reg or desert-pavement surfaces. In some places, deflation may create depressions, called *blow-outs,* which may be a few feet deep and a few yards across or scores of feet deep and hundreds of yards across. The lee sides of such depressions are commonly marked by streamer dunes of dusts and sands.

Where relatively flat bedrock surfaces are exposed to pronounced sandblast, the rock may be scoured out to form *abrasion basins,* or the surface of the bedrock may be shaped into the solid wavelike forms known as *yardangs.* Sandblast may also etch intricate patterns into exposed rocks. These do not constitute landforms in the ordinary sense, but they do provide an interesting sculptural detail on the exposed face of certain landforms. The abrasive action of wind-blown sand may also buff rock surfaces to produce shiny *desert polish.* Desert polish is particularly associated with *desert rind,* a rind, or thin surface layer, that is created when dry air draws mineral-laden moisture to the rock's surface and the subsequent evaporation of the moisture leaves the minerals behind as a sun-baked crust.

Depositional Forms: Loessal

Loess, readily transported by strong winds, is moved across dry lands in relays. Ultimately, most of it is carried on into humid lands, or, as off the western edge of the Sahara, is spread over the ocean. If deposited in humid areas, it may simply form a mantle on the landforms already present rather than creating landform features of its own. In some regions, however, the loessal accumulation has been so great as to ob-

FIGURE 6–29
Loessal badlands in the Chinese province of Shenhsi, near the Ching Ho and Huang Ho rivers. (United States Air Force)

scure the original surface and create extensive *loessal plains* (Fig. 6–28). Some loessal plains have been dissected by running water to the point of creating considerable local relief, as in the Loess Hills west of the North China Plain. Parts of China's Loess Hills have been so heavily cut by water that they are now *loessal badlands* (Fig. 6–29). Where dissected, loess produces very steep, often cliffed, slopes, as illustrated in the Council Bluffs area of western Iowa, or in the badland sections of China's Loess Hills.

All loess is not derived from dry lands. Some of it may be obtained from large glacial spillways, where an abundance of very fine glacial debris is deposited and subsequently transported by winds to adjacent areas, as in much of the loess lands along and near the present courses of the Missouri and Mississippi rivers of the central United States.

Whether loess creates landforms or not, it is still of considerable importance, since it either forms rich loessal soils or adds to the enrichment of other soils. Furthermore, loess is comparatively retentive of moisture, so that loessal soils are less "droughty": this is especially important where "normal" humid-land agriculture is being practiced in semiarid margins of dry lands, as in eastern Nebraska and the southern part of the Russian plains, or in the subhumid margins of humid lands.

Depositional Forms: Sandy

The most extensive sandy, or sand-dune, regions of the world stretch in a much-broken belt from the western Sahara to central Asia (Fig. 6–30). The huge deserts of northern Africa and Arabia together possess at least half of the world's dune regions. The Great Australian Desert has several large sandy areas, while in the Americas, sand-dune regions are small, few, and far between, even in the driest parts of the dry lands (Fig. 6–31).

Blown sand may be shaped into an almost infinite variety of variously sized dune forms—far more than can be discussed here. Some of them are simple *sheet,* or *veneer dunes,* which blanket flattish areas and lower hillslopes, and which often extent upslope to the crests of rocky saddles. Others are *hillock dunes,* formed around large boulders or around clumps of vegetation (Fig. 6–32). Where the local surface is hard, the sand supply meager, and the wind unidirectional, crescent-shaped dunes, called *barchans,* are formed, with the horns of each crescent pointing downwind. Where sand supply is relatively abundant and winds of about equal strength blow alternately from different directions, sand becomes heaped into peaklike forms called *pyramidal dunes.*

The most common dunes in large, markedly sandy areas are the *ridge dunes.* Some of these are wave dunes: either *transverse dunes,* whose waves and intervening troughs form at right angles to the direction of the stronger winds, or streamerlike forms, called *longitudinal dunes,* in which the ridges are aligned parallel to the stronger winds. Regardless of its type of dune, any area completely covered by sand dunes forms the typical erg condition mentioned previously in connection with the landforms of the world's dry lands.

If dunes are "on the move," they are called *live dunes;* if they have become stabilized by vegetation (natural or planted by humans), then they are *dead dunes.* Live dunes may generate a great deal of trouble (Fig. 6–33): they may invade roads, railroads, and landing fields, and overwhelm croplands and settlements. In the Sahara and parts of Arabia, for example, many oases have been occupied and developed, then overwhelmed by marching dunes, and then reoccupied by man as the dunes moved on.

FIGURE 6–30

The distribution of the principal sandy regions of the world. (After Trewartha *et al*, *Elements of Geography,* © McGraw-Hill Book Co. Inc., 1967. Used with permission.)

FIGURE 6–31

An aerial view of the central portion of New Mexico's White Sands National Monument, one of the few sand-dune regions in the Americas. The dunes shown average thirty to fifty feet in height. (Bill Belknap, Rapho-Guillumette)

FIGURE 6–32

Hillock dunes in Death Valley, California. Tough desert shrubs trapped moving sand, and new shrubs grew above the accumulations. (Dave Packard)

FIGURE 6–33

Live sand dunes that have encroached upon both an area's natural vegetation and its constructions. (C. Walinsky, Stock-Boston)

REVIEW AND DISCUSSION

1. Explain the contention that a shoreline is not just a line but a narrow zone partly above and partly beneath the sea. Give examples of the dynamism of this zone in the shaping of landforms.
2. What two major forces does the sea use in its attack upon the land? Which is an agent of erosion, and which is an agent of transportation? Do they work independently, or together? How?
3. How do the waters in a breaking wave exert their great abrasive force? How do high tide and low tide influence this force?
4. Compare longshore currents and tidal currents in terms of their roles in the shore-forming processes.
5. Describe the coastal landforms, or shore features, that are normally found along subdued shallow-water coasts. What, if any, general developmental sequence characterizes these coastal landforms? Are coral reefs coastal landforms? Explain.
6. Discuss the similarities and differences between submerged coasts and emerged coasts. Cite examples of each from various parts of the world.
7. What unique effects do different types of coastal landforms have upon human activities?
8. Discuss the location, nature, and characteristics of continental shelves. How are they important?
9. Discuss fully the forms and solution processes that characterize the development of karstic landforms. Explain the development of galleries. What is karst drainage? Compare and contrast the processes that produce karstic landforms with the fluvial processes that create landforms.
10. If you were in a typical karst area, what criteria would you use to determine whether the area was young, mature, or old?
11. Discuss the uses and problems of a karst area for human occupance. Define: stalactites, stalagmites, doline, cenote, uvala, naked karst, magotes, haystacks, and karst erosion cycle.
12. Consult a good geology reference for discussions of the theories of ice motion in glaciers. Discuss the origins and movements of continental glaciation. Define: glacier, iceberg.
13. Landforms shaped by continental glacial erosion differ from those formed by continental glacial deposition. Discuss the formation processes and appearance of both types of landforms. How do the two types also differ from landforms shaped by running water?
14. List and discuss the ways in which whether a region has been glaciated or not is significant to its inhabitants.
15. Distinguish between continental glaciation and valley glaciation, discussing scope, origin, typical landform processes, landform types, drainage patterns, and human significance. Define: cirque, col, horn, and U-shaped valley.
16. Distinguish between: terminal moraine and ground moraine; drumlin and roche moutonnée; outwash and glaciolacustrine plain; eskers, kettle holes, and kames.
17. Winds are often strong enough to move dust and sand. How does the transportation of these two types of materials differ?
18. Distinguish between the various depositional eolian landforms. Of what human importance are the world's great loess plains? Define: blow-outs, sandblast, desert polish, loessal plains, dune ridges, live dunes, and ergs.

SELECTED REFERENCES

Antevs, E. *The Last Glaciation*. New York: American Geographical Society, 1928.
Bagnold, R. A. *The Physics of Blown Sand and Desert Dunes*. London: Methuen, 1941.
Bascom, W. *Waves and Beaches*. Garden City, N. Y.: Doubleday Anchor, 1964.
Davis, W. M. *Geographical Essays*. New York: Dover, 1954.
Dury, G. *The Face of the Earth*. Baltimore: Penguin, 1964.
Dyson, J. L. *The World of Ice*. New York: Knopf, 1962.
Fairbridge, R. W., ed. *The Encyclopedia of Geomorphology*. Encyclopedia of Earth Sciences, Vol. 3. New York: Reinhold, 1966.
Flint, R. F. *Glacial and Pleistocene Geology*. New York: Wiley, 1957.
Guilcher, A. *Coastal and Submarine Morphology*. New York: Wiley, 1958.
Johnson, D. W. *Shore Processes and Shoreline Development*. New York: Hafner, 1965.
King, C. A. M. *Beaches and Coasts*. New York: St. Martin's, 1960.
Lobeck, A. K. *Geomorphology*. New York: McGraw-Hill, 1939.
Longwell, C. R., and R. F. Flint. *Physical Geology*. New York: Wiley, 1969.
Morisawa, M. *Streams: Their Dynamics and Morphology*. New York: McGraw-Hill, 1968.
Putnam, W. C. *Geology*. New York: Oxford University Press, 1964.
Schultz, G. *Glaciers and the Ice Age*. New York: Holt, Rinehart and Winston, 1963.
Sharp, R. P. *Glaciers*. Eugene, Oreg.: University of Oregon Press, 1960.
Shepard, F. P. *Our Changing Coastlines*. New York: McGraw-Hill, 1971.
———. Submarine Geology, 2nd ed. New York: Harper & Row, 1963.
Strahler, A. N. *The Earth Sciences*, 2nd ed. New York: Harper & Row, 1971.
———. *Physical Geography*, 3rd ed. New York: Wiley, 1969.
Thornbury, W. D. *Principles of Geomorphology*. New York: Wiley, 1954.

The atmosphere over the southwestern United States at about 18 miles high. (Official Navy Photo)

7

The Earth's Atmosphere

Humans live, and pursue varied activities, on the earth's surface, at the bottom of an "ocean of air" that we call the *atmosphere* (Greek: *atmos* = vapor; *sphaira* = sphere). The atmosphere envelops and is intimately related to all aspects of the earth's surface and to all the varied phenomena thereon. This atmospheric mass presses down upon the earth's surface with tremendous weight, and, as it does so, it exerts equal pressures on the human body—both externally and internally. Thus the body can move with relative physical freedom, almost as though the mass of air did not exist. The body has become acclimatized to this environment, but in spite of this, when we ascend even comparatively short distances into the atmosphere, we cannot adjust to the rarer and colder atmospheric conditions there without constant protection. In short, whenever they move appreciably above their accustomed earth-bound habitat—whether in high-altitude aircraft, in spaceships, or even in climbing high mountains—humans must take with them portable, miniaturized atmospheric life-support environments. When humans finally achieved their most remarkable feat of extraterrestrial travel—walking on the moon—they carried in their backpacks complete earth-atmosphere duplicating facilities (Fig. 7–1).

The atmosphere enters into nearly every facet of daily human life and affects, directly or indirectly, virtually every element of the environment. These include such diverse aspects as weather and climate, soils and vegetation, minerals, fuels, and even the waters and the lands. Thus, it may be said that the atmosphere is largely the sculptor of our environmental complex and the director of our manifold activities.

The study of the atmosphere can be approached in various ways. The discussion that follows will first consider the chemical composition of the atmosphere's constituent gases and the various zones associated with them; it will then examine and classify the atmosphere on the basis of the temperatures existing at its various levels; and finally, it will survey and classify the dynamics of the atmosphere and its electromagnetic properties at various levels. Each of these aspects has its own profound effects upon the earth's surface and upon all humankind as well.

FIGURE 7–1

Human use of miniaturized atmospheric life-support environments. Left, Apollo 17 astronaut Harrison H. Schmidt wearing a portable life-support pack as he collects samples of the lunar surface. (NASA) Right, Dr. Richard Emerson, at the 10,000-foot level of Washington's Mt. Rainier, wearing an oxygen mask designed especially for mountain climbing. (Wide World Photo)

Composition of the Atmosphere

Vertical Characteristics and Zones

In terms of its constituent gases, the atmosphere presents a notably uniform chemical composition up to altitudes of about 50 miles above the earth. Above this altitude, the atmosphere becomes layered and nonuniform in chemical composition. The lower atmospheric level, of uniform composition, is called the *homosphere* (Greek: *homo* = same); the upper atmospheric level is called the *heterosphere* (Greek: *heteros* = different, or other) (Fig. 7–2). Both of these levels, or *zones*, in which the phenomena of weather and climate are manifest, are of particular importance to physical geography.

The Homosphere. The homosphere, with its uniform-ratio composition of nonvariant gases, and its practically total domination over the earth's weather and climate, is generally considered to be the *earth's atmosphere.* It extends above the earth's surface to about the 50-mile level (Figs. 7–2 and 7–3), where a narrow transitional zone called the *homopause* separates it from the vast expanse of the heterosphere, the atmospheric

FIGURE 7–2

The earth's atmosphere may be classified into various systems of vertical layers, or zones. Here, the vertical *temperature zones* are emphasized, and comparisons with the vertical *composition* and *dynamic* systems are indicated.

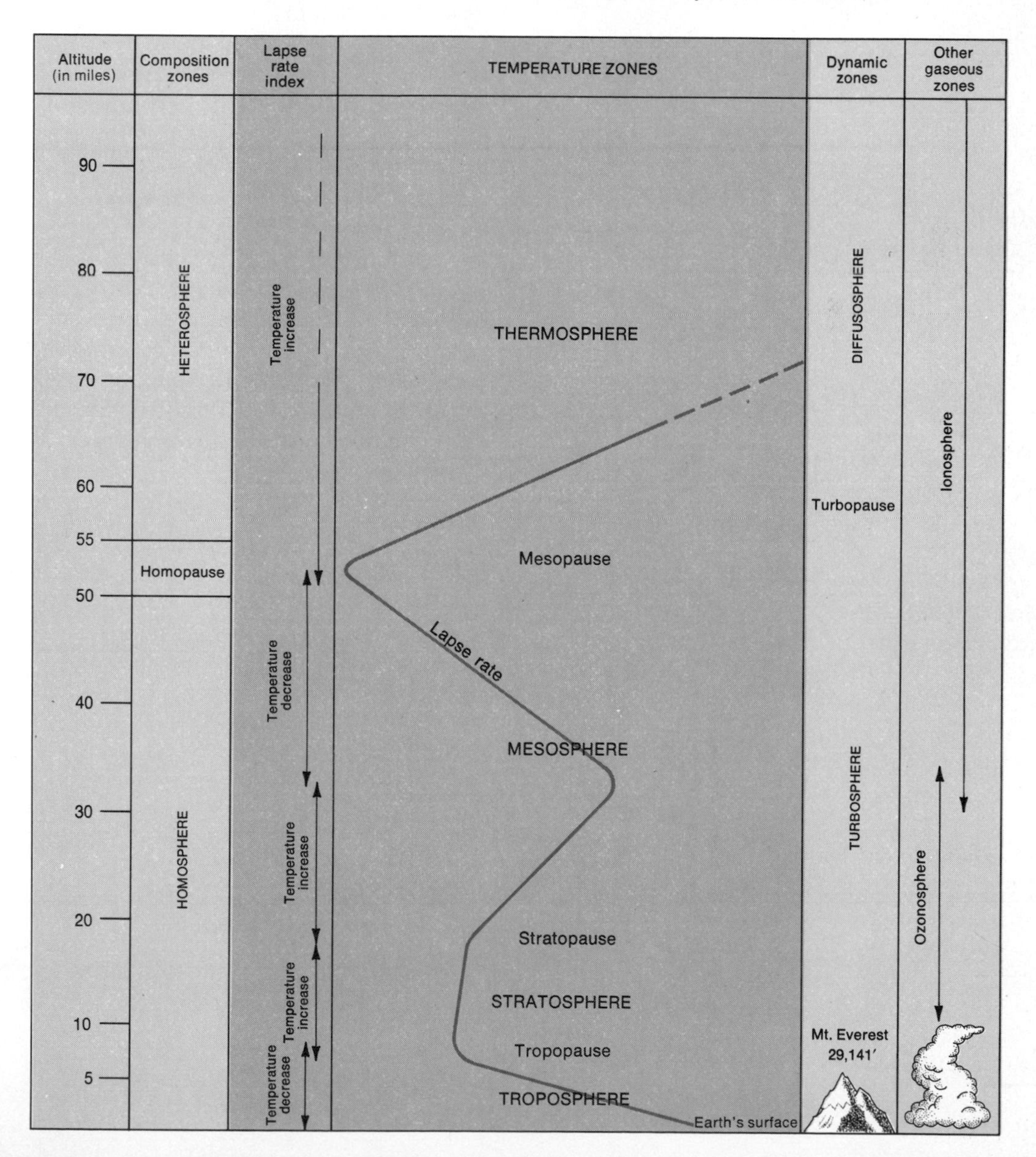

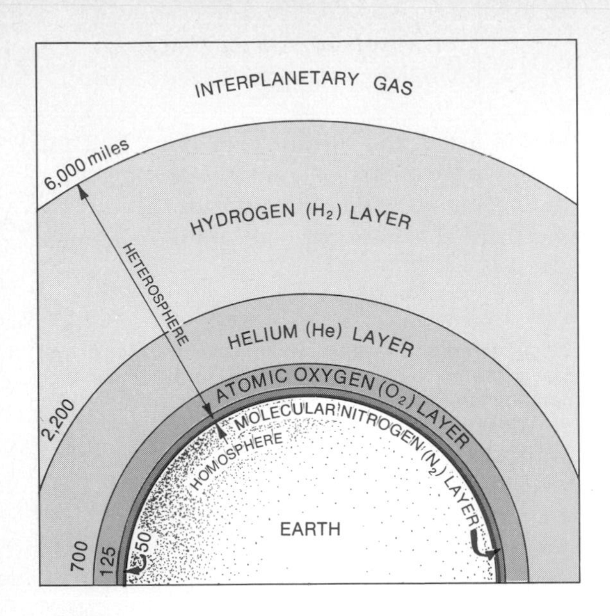

FIGURE 7–3

The lower atmospheric zone, of uniform composition, is known as the *homosphere;* the upper zone, of nonuniform composition, is known as the *heterosphere.*

zone that begins at about the 55-mile level.

Since the total atmosphere (homosphere and heterosphere) is drawn toward the earth's surface by strong gravitational forces, the lowest portions of the homosphere are most heavily compressed against the earth's surface, while the vast extent of the atmosphere above (upper homosphere and heterosphere) becomes rapidly thinner with increasing altitudes. The extreme compression of the lowest portions of the homosphere causes about 50 percent of the total atmospheric mass to be concentrated within about 3.5 miles of sea level, 75 percent within 7.5 miles, and as much as 97 percent within 18 miles of the earth's surface (Fig. 7–4). Thus, as much as 97 percent of the total atmospheric mass is compressed within the lower one-third (18 miles) of the homosphere, leaving only about 3 percent to be spread throughout the upper two-thirds of the homosphere and the entire extent of the heterosphere.

From a geographical point of view, it is especially significant to note that within the lowest and densest layers of the atmosphere —layers in close and intimate contact with the earth's surface—practically all weather and climatic activities develop. And it is here, too, that the forces of weather and climate have their greatest effects upon the physical features (landforms) of the earth and upon all of its multitudinous life forms and systems.

The Heterosphere. Beginning at about 55 miles above the earth's surface, the gases of the atmosphere assume a definite layered arrangement based on their respective molecular weights (Figs. 7–2 and 7–3). The lowest layer, at an approximate altitude of between 55 miles and 125 miles, is the molecular *nitrogen* layer, in which the heavier nitrogen molecules are the dominant constituent. Above it is the atomic *oxygen* layer (*monatomic* layer), which extends upward to an altitude of about 700 miles. Above this is the gaseous layer composed largely of *he-*

FIGURE 7–4

Excessive compacting of the earth's atmosphere puts about half of it within 3.5 miles of the earth's surface, and 97 percent within 18 miles.

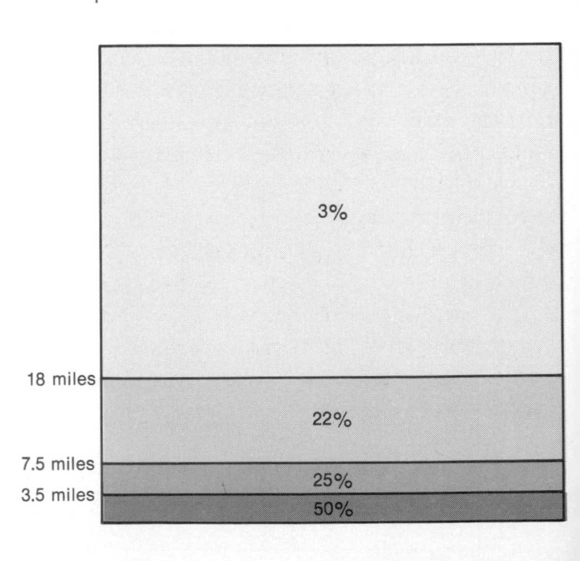

lium; it is dominated by the lighter helium atoms and extends to about the 2,200-mile level. The fourth and uppermost layer, between the 2,200- and 6,000-mile levels, is the *hydrogen* layer, which is dominated by very light hydrogen atoms. Outward, beyond 6,000 miles, the rarified atmosphere merges into the near vacuum of the interplanetary gases and of the occasional minute dust particles of the solar atmosphere.

Nonvariant Gases

Exclusive of water content, the lower portion of the homosphere is composed almost entirely of *nonvariant gases.* From random samplings of the "dry air" in this region, it may be assumed that more than 99 percent of the atmosphere (by volume) is composed of *nitrogen* (over 78 percent) and *oxygen* (nearly 21 percent). The remainder of the nonvariant gases, totalling less than 1 percent, is chiefly *argon* (.93 percent) and *carbon dioxide* (.03 percent), with several trace elements present. In order of importance by volume they are: *neon, helium, methane, krypton, nitrous oxide, xenon,* and *hydrogen.* While it is known that these nonvariant gases always occur in the proportions indicated in Figure 7–5, it must be recognized that human activities bring about, at varying places and times, some changes in this natural balance; this is especially so over large and densely populated cities and over areas of heavily concentrated industry. The most significant changes brought about in the atmosphere over areas of this kind are slight increases in carbon dioxide and very slight decreases in oxygen, as well as very small, but still harmful, increases in the percentages of ozone, nitrous oxides, and carbon monoxide.

Nonvariant gases		*Percentages (by volume)*
Nitrogen (N_2)		78.088
Oxygen (O_2)		20.949
	Total	99.037
Argon (A)		0.930
Carbon dioxide (CO_2)		0.030
	Total	0.960
Neon		0.0018
Helium		0.000524
Methane		0.00014
Krypton		0.000114
Nitrous oxide		0.00005
Xenon		0.0000086
Hydrogen		0.00005
	Total	0.0026866

FIGURE 7–5

The nonvariant gaseous composition (*by volume*) of the lower (dry) atmosphere (the homosphere).

The Nitrogen Cycle. There are several specific physical processes that operate to maintain the delicate balance among the gaseous constituents of the atmosphere. For example, as nitrogen is lost continually from the atmosphere, either by direct escape into space or by combining with other elements or compounds, nature compensates for this loss by a continuing process of replenishment: a loss-gain balancing process known as the *nitrogen cycle.* In this cycle, the multitudes of microorganisms (bacteria) that live in the soil or in the nodules and roots of leguminous plants (peas, clover, beans, alfalfa, and the like) function as *nitrogen-fixing agents.* These bacteria take nitrogen from the atmosphere and convert it into the nitrogen compounds that are essential for the growth of the plants. These compounds are then incorporated into the plant tissues. When these plants or their by-products are consumed by people or animals, the nitrogen is eventually returned to the soil, either in body excretions or in corporeal remains. In the soil, the decaying processes that affect such animal tissue release nitrogen back into the atmosphere; the decay of plants also involves a release of nitrogen into the atmo-

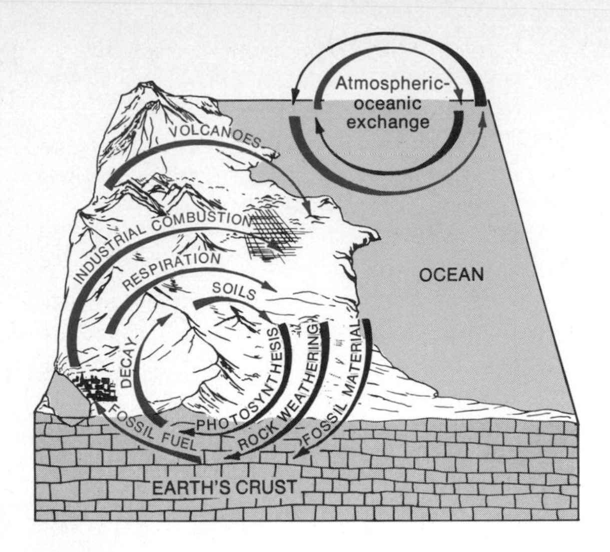

FIGURE 7–6
The carbon dioxide cycle.

sphere. Additionally, lightning discharges in moist air result in the production of nitrogen compounds that are then carried to the earth by falling rain or snow.

The Carbon Dioxide Cycle. A second process, the *carbon dioxide cycle,* also functions to maintain the lower atmosphere's gaseous balance (Fig. 7–6). Although carbon dioxide constitutes only a minute portion of the atmosphere, its occurrence is still essential. Carbon dioxide is the raw material that, in combination with solar energy (sunlight) and water, maintains the process of *photosynthesis,* that is, the manufacture of plant food. In the presence of sunlight, carbon dioxide and water combine and are converted by the process of photosynthesis, with the aid of the green chlorophyll of plants, into compounds of carbon, hydrogen, and oxygen. Thus, plants are almost continuously releasing excess oxygen into the atmosphere; living animals absorb this oxygen when they inhale, and release carbon dioxide back into the air when they exhale. Carbon dioxide is also removed from the atmosphere through the formation of carbonates by organisms such as shellfish, by the continuous chemistry of bacterial action, and by the decay (or weathering) of rocks. It is also replenished by such things as volcanic eruptions and emanations from gas wells and springs. Certain industrial processes, such as the burning of lime (calcium carbonate: $CaCO_3$), smelting, and fermentation, also release carbon dioxide into the atmosphere. It is also carried by streams into the ocean, which, in turn, releases it back into the atmosphere. The ocean is both capable of absorbing excess carbon dioxide from the atmosphere and of eliminating atmospheric deficiencies of it (Fig. 7–6). In fact, it is estimated that the ocean contains almost fifty times more carbon dioxide than the atmosphere, and billions of tons of carbon dioxide constantly enter into the atmospheric-oceanic exchange.

Carbon dioxide is also a vital "balance constituent" with respect to weather and climate. It functions as a heat-balance regulator, in that it helps to trap the heat that emanates from the earth's surface. Various theories pinpoint carbon dioxide variations as being responsible for many major climatic changes. It is believed that a fifty-percent decrease in atmospheric carbon dioxide content would result in a worldwide drop in mean temperatures of as much as 7° F., which, in turn, would help to induce another ice age.

Variant Gases

The atmosphere also contains a number of variant gases. Of these, the most important is *water vapor* (invisible moisture). It is not only essential to all earth-life systems, but it is also highly variable in its world distribution, both in time and space. The other variant gases are comparatively negligible, as they occur in only minute quantities. Among such gases are *ozone, ammonia, sulfur*

Constituent	Formula
Water vapor	H_2O
Ozone	O_3
Ammonia	NH_3
Hydrogen sulphide	H_2S
Sulfur dioxide	SO_2
Sulfur trioxide	SO_3
Carbon monoxide	CO
Radon	Ra
Sodium chloride	NaCl
Soot	
Meteoric dust	

FIGURE 7–7

The variant gases and the solids of the atmosphere.

constituents, and *carbon monoxide* (Fig. 7–7).

Water vapor is the primary source of atmospheric moisture and, hence, plays a vital role in cloud and fog formation through the processes of *condensation* and *precipitation* (rain, snow, hail, and sleet). It also functions as an important agent in the absorption of radiant energy (both direct solar energy and reradiated terrestrial energy), thus helping to maintain the important heat balance of the earth's atmosphere. Atmospheric water vapor is mainly concentrated within the lowest atmospheric levels, very close to the earth's surface—more than half is concentrated within a scant mile-and-a-half of the earth and is scarcely detectable beyond altitudes of five miles (Fig. 7–8).

Water vapor is introduced into the atmosphere, from the varied land, water, and biotic systems of the earth, through the process of *evapotranspiration* (the combined processes of *evaporation* of moisture from land and water, and the *tranpiration* of moisture from vegetation). In the evaporation process, which converts liquid water into a gaseous or vapor form, a change in the water's molecular structure takes place, causing heat energy to be expended. Water vapor is also injected into the atmosphere by the process of *sublimation,* which occurs when matter in a solid state, such as ice or snow, is transformed into a gaseous state without passing through an intermediate liquid state.

Ultimately, water vapor is returned to the earth's surface, mainly through various forms of precipitation, to be reinjected into the atmosphere and returned to the surface again and again, as a part of the continuing global hydrologic cycle (see Figure 3–2). In this important water recycling process, evapotranspiration may be thought of as the earth's continuing "water loss," and precipitation, as the earth's continuing "water charge." Evapotranspiration returns an estimated two-thirds of the precipitation that falls on the land directly to the atmosphere, while the remaining one-third goes into runoff and groundwater storage, and is eventually transported to the oceans.

Solids or Impurities

Another essential component of the atmosphere is the group of substances known as *impurities,* a term that refers to all of the atmosphere's minute solid particles, including terrestrial dust and even meteoric dust, found suspended therein. Some impurities

FIGURE 7–8

The widely varying water-vapor content at various atmospheric elevations (it can be as much as 3 percent). Note its concentration near the earth's surface.

Altitude kilometers	miles	feet	Water-vapor content percent
8	4.97	26,248	0.05
7	4.35	22,967	0.09
6	3.73	19,686	0.15
5	3.11	16,405	0.27
4	2.48	13,124	0.37
3	1.86	9,843	0.49
2	1.24	6,562	0.69
1	.62	3,281	1.0
0	0.00	0	1.3

have their sources in such natural occurrences as volcanoes and forest and grass fires, which inject ash and smoke particles into the atmosphere. Apart from the general activity of volcanoes, occasional dramatic explosions, such as the 1883 explosion of Krakatoa off Java-Sumatra, cause unusual additional injections of dust into the atmosphere. Following the 1883 eruption, Krakatoa's dust spread from the equatorial region into the higher latitudes and had worldwide effects on weather conditions for several years. Deserts, as well as dirt roads and cultivated fields, even in humid regions, contribute considerable amounts of terrestrial dust to the atmosphere. The vast air-ocean interface is another source of impurities, as there is a constant injection of salt particles into the atmosphere (by means of salt sprays), which are carried aloft by vertical air currents and then spread over large areas by the winds. Plant life is also a source of impurities, for it contributes large quantities of minute pollen grains and spores to the atmosphere.

Air Pollutants

Today, it is becoming increasingly apparent that many daily human activities inject myriads of pollutants of many types into the atmosphere. Smoke and soot from domestic heating and cooking, gases and chemicals from industrial plants and refineries, and exhaust gases from internal-combustion engines in growing numbers of automobiles and aircraft, are all being injected into the atmosphere on a global basis, in tremendous amounts and with little human control or even understanding. These countless pollutants, in combination with certain natural atmospheric conditions—such as impeded air movements, fog conditions, and thermal variations—create smog of varying kinds and intensities, an increasingly common feature, particularly in industrial centers and in most of the world's urban centers. These smog conditions are beginning to cause grave concern, at both domestic and international levels, because of their disastrous effects on humans and the total physical environment.

Hygroscopic Nuclei

The various solid-particle impurities present in the earth's atmosphere serve as *hygroscopic,* or *water-attracting, nuclei* around which water vapor condenses and forms minute droplets. Without such nuclei, natural water vapor could not condense, and, therefore, there could be no clouds, fog, rain, or snow. Thus, there is a direct causal relationship between hygroscopic nuclei and the moisture aspects of the atmosphere. Moreover, hygroscopic particles introduced by humans, and present in increasing numbers, now have at least some small effect on condensation and precipitation: humans purposely introduce such particles into the atmosphere, under certain conditions, such as in modern rainmaking experiments with cloud seeding. But it must be noted that in order for cloud seeding to be successful, the natural cloud conditions must be nearly right for natural precipitation. Solid-particle impurities, both natural and human-induced, also cause some diffusion of solar rays, creating haze and varicolored skies.

Other Gaseous Zones

There are two other meteorologically significant gaseous zones of the atmosphere: the *ozonosphere* and the *ionosphere* (Fig. 7–2). The ozonosphere is the layer in which ozone (O_3) is concentrated. It normally includes the vertical region between 12 and 21 miles of altitude, but at times it expands to altitudes of 35 miles. Ozone is a photochemical product resulting from the reaction of ultraviolet radiation upon ordinary (diatomic) oxygen (O_2). In the ozonosphere, these gases act as absorbents of the sun's ultraviolet radiation which, if allowed to enter the lower atmosphere unimpeded, would prove lethal to all forms of life on the earth's surface. The gases do allow a sufficient amount of ultraviolet light to filter through into the atmosphere to maintain life forms. Since the ozone layer absorbs intense ultraviolet radiation, it also becomes a "heat layer." The significance of this may be observed with respect to its upper limit being coincident with the temperature curve noted in Figure 7–2. Within the lower atmosphere, ozone may be formed when lightning discharges act upon ordinary oxygen. Ozone is also one of the natural smog-causing pollutants.

The second of these superimposed, or intermingling, gaseous zones is the ionosphere, which ranges between the 30- and 250-mile levels of the atmosphere. This means that the lower level of the ionosphere is nearly coincident with the upper limit of the ozonosphere. The basic feature of the ionosphere is the occurrence of ionized particles produced by the atomization of gas molecules during ultraviolet radiation. This process of atomization also tends to greatly increase temperature and to produce the temperature curve (1,000° F. to 2,000° F.) found in the higher atmosphere. The upper limits of the ionosphere overlap the molecular nitrogen layer and the atomic oxygen layer. The role of the ionized layers in reflecting radio waves, especially as demonstrated by the lowest (or Kennelly-Heaviside) layer, is well known. It is certain that long-distance radio communication would not be possible if these layers were absent.

Atmospheric Temperature

Temperature characteristics, especially the changes in temperature relative to elevations, constitute the principal criteria used in most geographic study today for distinguishing various layers in the atmosphere. Four such vertical layers are readily recognizable: the *troposphere, stratosphere, mesosphere,* and *thermosphere;* they are separated from each other by transitional zones (Fig. 7–2). Together, the three lower temperature layers correspond very closely to the homosphere, and the upper temperature layer relates in like manner to the heterosphere (Figs. 7–2 and 7–3).

It is a commonly observed phenomenon that as one ascends a high mountain or takes off in a plane, the air becomes progressively colder. This decrease in temperature with increased elevation is called the *normal lapse rate,* or sometimes, the *environmental lapse rate.* The rate of change is about 3.5° F. per 1,000 feet of ascent. Under certain circumstances, one can observe, very close to the earth's surface and occasionally at varying heights within the lower atmosphere, situations in which temperature conditions are reversed, in which, in fact, temperatures are found to increase with ascent. These situations are considered abnormal and are thus called *inversions;* their reversed rates of temperature change are termed *inverted lapse rates.*

The Troposphere

At the turn of the century, the French meteorologist Teissernec de Bort observed that at a certain altitude the normal lapse rate ceases to operate (Fig. 7–2). The term *troposphere* (Greek: *tropos* = turn) was suggested by de Bort to designate the lower region of the atmosphere, directly above the earth's surface, where the normal lapse rate prevails. The upper limit of the troposphere, on the suggestion of the British meteorologist Sir Napier Shaw, was then called the *tropopause* (Greek: *pauein* = to cause to change), designating the level in the atmosphere where the normal lapse rate ceases to operate. The altitude of the tropopause, and therefore the thickness of the troposphere, varies both according to season and latitude. The troposphere is thickest over the equatorial and tropical parts of the earth and shallowest over the poles, indicating clearly that this layer is conditioned by local temperature situations. Over the equatorial zone the troposphere extends about eight or ten miles above the earth, while over the poles it reaches only about five or six miles. (These are mean heights; obviously, because the troposphere is temperature conditioned, during the winter season it descends to somewhat lower heights than during the summer season.) Since the troposphere is not uniform, the tropopause appears in cross section as a convex curve. In addition, the tropopause occurs in several segments, often exhibiting paradoxically overlapping situations, as shown in Figure 7–9. It may also be observed that temperatures over the equator at the tropopause level are colder than those at corresponding tropopause levels over the poles, simply owing to the effective operation of a constant rate of temperature decrease on the normal lapse-rate basis.

It is within the troposphere that the world's weather is largely concentrated—where it develops, lives its short life, and dissipates—in close physical contact with the earth's surface.

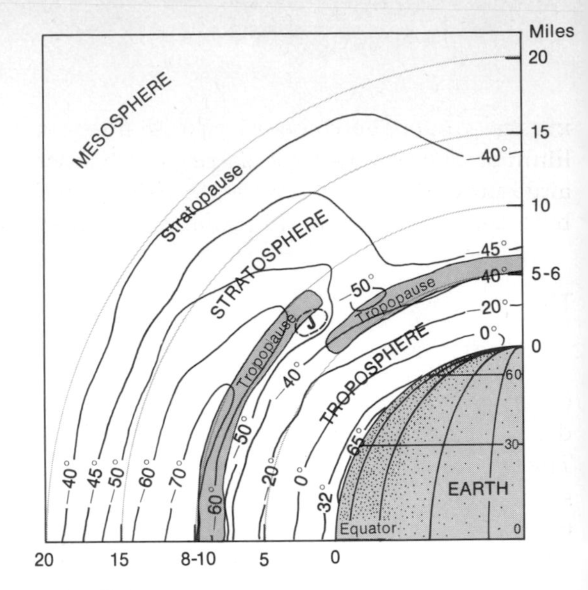

FIGURE 7–9

The disrupted structure of the *tropopause.* The mean-temperature profile is for July along 80° west longitude, and shows the relative temperatures over the tropics and over the poles. The jet stream **(J)** is shown within the disrupted tropopause-overlap between 40°-50° N.

The Stratosphere

In the layer above the troposphere, the lapse rates are constant, but there is an almost imperceptible increase of temperatures that gives the layer a slightly inverted lapse rate. Nevertheless, for all practical purposes, constancy of temperature can be said to be one of its characteristics. This layer, which extends upward from the tropopause to about 18 miles, is called the *stratosphere,* (Latin: *stratus* = bands or stratified), or the *isothermal layer.* In the stratosphere (Figs. 7–2 and 7–9), weather phenomena common to the troposphere are virtually nonexistent, hence it is also known as the "weatherless region." Today's long-range aircraft make use of this weatherless region and thus avoid any turbulent air conditions present in the troposphere below. At the upper limit of the stratosphere there is a marked altitudinal increase in temperature, indicated by a transitional zone called the *stratopause.*

The Mesosphere

In the layer above the stratosphere, called the *mesosphere* (Greek: *meso* = middle), there are two lapse rates in operation. The mesosphere extends upward from the stratopause, through increasing temperatures, to a maximum temperature at about 30 miles, then through rapidly decreasing temperatures to the 50–55-mile level. This 50–55-mile level constitutes a third transitional zone, called the *mesopause,* which coincides closely with thc homopause of the "atmospheric composition" system.

The Thermosphere

Regarding the *thermosphere* (Greek: *thermos* = heat), the region that extends from the mesosphere to the 100-mile-level, it is known only that temperatures continue to increase very rapidly through its expanse (providing a steep inverted lapse rate). Temperatures on the order of 2,000° F. to 3,000° F. are commonly attained in the thermosphere, but, since the density of air there approaches a vacuum, these temperatures are not readily conducted through the air.

The Exosphere

The zone that lies beyond the 700-mile-level is called the *exosphere,* or *magnetosphere* (Fig. 7–10). From this distance on, the high temperatures and very low particle densities permit the escape of some atoms and molecules from the earth's gravitational field. It is the area where ionized particles are trapped within the earth's magnetic fields. The positively-charged protons and the negatively-charged electrons are believed to play a significant role in the transfer of solar energy to the earth and in the occurrence of the *aurora borealis* (northern lights) and the *aurora aus-*

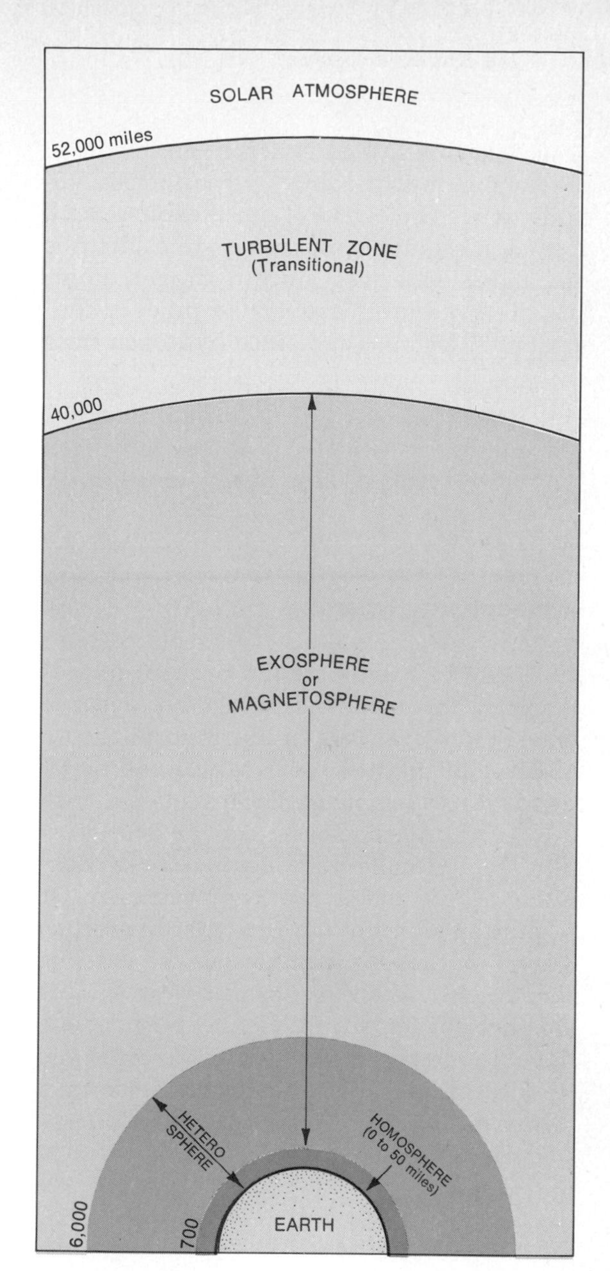

FIGURE 7–10

The zone beyond 700 miles is called the *exosphere*, or *magnetosphere*, and is where ionized particles are trapped in the earth's magnetic fields.

tralis (southern lights), each an electrical phenomenon that occurs within the 50–700-mile zone. The extent of the exosphere and, hence, the outer limit of the atmosphere is debatable. However, since hydrogen atoms are observed up to about 6,000 miles in concentrated fashion, and since hydrogen trace elements are found within the earth's immediate atmospher, in a general sense, the 6,000-mile-level may be accepted as the outer limit of the general atmosphere. A diffuse zone prevails beyond this level and eventually becomes the "solar atmosphere" at some 52,000 miles above the earth.

Other Aspects of the Atmosphere

Atmospheric Dynamics

Although they are not of great importance to geographic study, the *dynamic processes,* or *dynamics,* of the atmosphere are relevant in that they serve as an additional basis for its assessment. The term *turbosphere* has been adopted to designate the lower portion of the 60-mile-high region of the atmosphere where the dynamic processes result in atmospheric turbulence. It is this turbulence that causes the mixing of the gaseous constituents that results in the earth's weather phenomena. The atmosphere above the 60-mile-level, in which the dynamic processes are nonturbulent and essentially diffusive, is generally known as the *diffusosphere* (Fig. 7–2).

Gaseous Properties

Radiant solar energy is the underlying cause of all weather processes. All weather phenomena observed in the lower atmosphere are the result of thermodynamic processes that occur in the atmosphere because it possesses such gaseous properties as multidirectional *mobility, expandability,* and *contractability*. Because the atmosphere is a mixture of gases, it behaves like a gas and thus conforms to the natural physical laws that control all gases. One important property of gases is their ability to expand and contract. They are made up of large numbers of minute molecules that are in constant motion; as the molecules collide, there are resultant changes in both temperature and pressure. Since gas is compressible, there is a contraction of its volume and an increase in its density when pressure is applied to it. Contrarily, gas readily expands in volume when pressure is decreased. Under constant pressure, an increase in temperature results in an expansion of air; a decrease in temperature results in air becoming denser. The pressure-density-volume relationship, innate to all gases, is fully expressed and analyzed in several Gas Laws, the basic meanings of which may be demonstrated by a simple illustration using three closed containers that have different cubic capacities (Fig. 7–11). The air in container *A* occupies a volume corresponding to the cubic capacity of its container, and it will exert equal pressure on all sides of the container. Hence, the volume and pressure in container *A* bear a distinct relationship. If all of the air in container *A* is transferred to container *B* (which is larger than *A*), it becomes obvious that the air from container *A* now occupies a larger volume by being in container *B;* thus, volume has increased, and, in compensation, pressure has decreased,

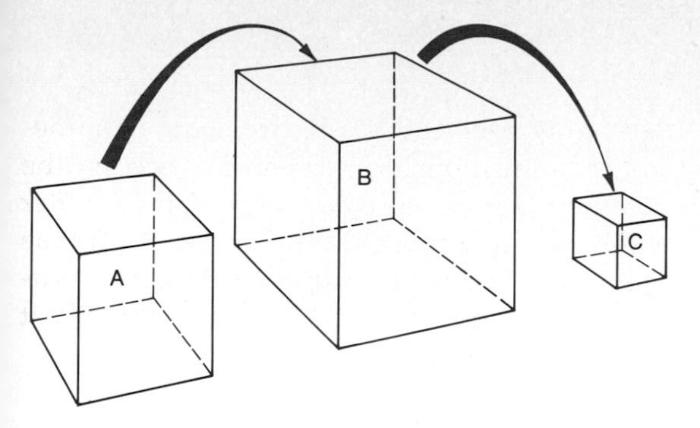

FIGURE 7–11

An illustration of the gas law principle enunciated by Boyle. The pressure-volume relationship of an air mass in container *A* remains constant even when that air mass is transferred to containers *B* (which is larger in cubic capacity than *A*) and *C* (which has a smaller cubic capacity than container *A*). This is so because of the inverse relationship and adjustment made possible by the volume and pressure parameters: greater volume is compensated for by lesser pressure, and lesser volume by greater pressure.

for molecules are colliding less than when the air was confined in the smaller container *A*. In effect, then, an inverse relationship exists between volume and pressure. Thus, although between containers *A* and *B*, the volume has changed in respect to the same mass of air, the combined volume-pressure relationship is maintained as a constant (as long as the temperature is assumed to be stable). Next, the air in container *B* is transferred to container *C*, which is even smaller than container *A*. Now the same air mass occupies a smaller volume than it did in either container *A* or container *B*, and this means that the air molecules now confined within the smallest container will collide more often, so pressure will naturally increase. Again, an inverse relationship is established, with volume decrease being compensated for by pressure increase. The air has now become compressed. Since density refers to the ratio of mass to volume, then density varies inversely with volume. It also follows that the density of a gas varies in direct proportion to its pressure.

It is thus seen that, despite the varying volume-pressure relationships in the three containers, the overall result in each experiment is a constant relationship of the two properties, volume and pressure.

In various important ways, this illustrates the situation prevailing in the atmosphere when a unit, or body, of air undergoes vertical movement. Less dense (lighter), heated air rises, and as it rises, it expands in volume and the pressure decreases. In the reverse situation, when air descends, it becomes compressed, volume decreases, and pressure increases. Both these situations are simplifications, however, for in the real situation in the atmosphere, temperature variations that may produce other results also have to be taken into account.

Weather and Climate

It is mainly the phenomena of weather and climate that draw people's attention to the atmosphere around them. The atmosphere is where the daily changes in weather occur and where the full interplay of the thermal and dynamic parameters operates.

Weather may be defined as the momentary, or instantaneous, state of the atmosphere above a particular place or region. *Climate* is the average, or normal, state of the atmosphere above a particular place. Thus, weather and climate are both related to time and place. In addition, observed weather phenomena may occur repeatedly, even rhythmically, over the same place, and when they do, they eventually become recognized as definite *weather types* (Fig. 7–1).

The concept of climate, as opposed to that of weather, is not always readily understood. Because one cannot sense it directly, as one can such weather elements as wind, rain, or air temperature, climate takes on an abstract quality. Despite this, climate is nevertheless a "real" phenomenon, one that is probably best defined as the sum total of weather conditions in an area over an extended period of time.

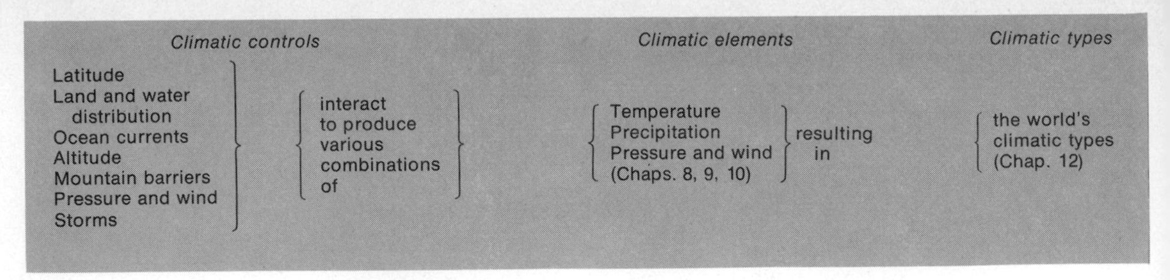

Climatic controls
Climatic elements
Climatic types
Latitude
Land and water distribution
Ocean currents
Altitude
Mountain barriers
Pressure and wind
Storms
interact to produce various combinations of
Temperature
Precipitation
Pressure and wind
(Chaps. 8, 9, 10)
resulting in
the world's climatic types
(Chap. 12)

FIGURE 7–12
The relationship between climatic controls, elements, and types.

Elements of Climate. Although the earth's atmosphere is mobile and constantly changing, a recognition of certain of its basic elements, of their controls, and of the nature and distribution of the effects of these controls is nonetheless essential to a full appreciation of the real world in which man lives (Fig. 7–12). Why is one area forest-covered and another grass-covered? Why are the trees typical of eastern Canada, such as pine, hemlock, and spruce, not present in the Great Plains of the United States? Why are red soils found in Georgia but unknown in Iowa? Why do house builders in the tropics keep their dwellings open to the winds, while those in the north strive to make their houses as windtight as they can? These are only some of the more obvious questions whose answers reflect the atmosphere, weather, and climate, in different parts of the world.

The most immediately obvious *climatic elements* are heat and moisture, that is, *temperature* and *precipitation*. The less apparent ones are *atmospheric pressure* and *wind*.

Controls of Climate. There are many controls that act upon such climatic elements and, in so doing, produce characteristic combinations known as *climatic types* (Fig. 7–12). Of the many *climatic controls*, there are seven that have worldwide significance. They are: latitude, land and water distribution, ocean currents, altitude, mountain barriers, pressure and wind, and storms. The temperature and precipitation conditions of an area are affected by each of the seven, and pressure and wind, as elements, are acted upon by the other six controls. The resultant combinations are reflected in the forms of the land surface, the soil, the vegetation cover, and even in the features of the human scene. These combinations, or climatic types, and their distribution thus offer clues or guides to many other earth phenomena. But before the full meaning of these types can be understood, the individual climatic elements of which they are constituted, and the climatic controls to which they are subject, must be examined in their broader aspects.

REVIEW AND DISCUSSION

1. Explain several ways in which the atmosphere affects the environment and conditions much of all life on earth.
2. Discuss the distinguishing characteristics of the homosphere and the heterosphere. Why is the chemical composition of the atmosphere important in the study of weather and climate?
3. What would you judge is the significance of the fact that 50 percent of the total "atmospheric mass" is compressed within 3.5 miles of the earth's surface? Within what levels of the atmosphere do practically all weather and climate activities develop?

4. Of the numerous nonvariant gases that compose the lower level of the atmosphere, how do nitrogen and oxygen compare with the others by volume? In which of the atmospheric gases are important changes brought about by human activities? How do these changes occur, and to what extent?
5. How does the carbon-dioxide cycle function? Why is carbon dioxide a vital "balance constituent" with respect to weather and climate?
6. Explain why, of all the variant gases in the atmosphere, water vapor is singled out for special emphasis in the study of weather and climate. What is the meaning of evapotranspiration? Of sublimation?
7. Consult a good reference book for a detailed discussion of atmospheric pollutants and pollution. What are the major air pollution problems in your community? What are their major causes? How can they be controlled?
8. Discuss the distinguishing characteristics of the troposphere, stratosphere, mesosphere, and thermosphere. Why are the temperature characteristics of the atmosphere important in the study of weather and climate?
9. What are the meanings of lapse rate, normal lapse rate, and inverted lapse rate? How does lapse rate relate to vertical atmospheric temperature zonation? Explain temperature inversions.
10. The world's weather is largely concentrated within the troposphere. Thus, it is generally below what heights in the lower latitudes, and what heights in the higher latitudes?
11. Although weather phenomena are virtually nonexistent in the stratosphere—which is therefore known as the "weatherless region"—knowledge of the stratosphere is of great and growing importance to humanity. Explain modern society's increased interest in this region.
12. Discuss the meanings of atmospheric mobility, expandability, and contractability. Why and how are these properties of the atmosphere important in the study of weather and climate?
13. Explain the statement that "the atmosphere is the medium of all weather and climate, wherein thermal and dynamic forces operate according to the laws of nature."
14. Define and distinguish between weather, climate, and weather types.
15. Discuss the differences between climatic controls and climatic elements. Why does physical geography concern itself with climatic controls and climatic elements in the study of weather and climate?

SELECTED REFERENCES

Bach, W. *Air Pollution*. New York: McGraw-Hill, 1972.

Barry, R. G., and R. J. Chorley. *Atmosphere, Weather and Climate*. New York: Holt, Rinehart and Winston, 1970.

Blair, T. A., and R. C. Fite. *Weather Elements*, 5th ed. Englewood Cliffs, N.J.: Prentice-Hall, 1965.

Blumenstock, D. I. *The Ocean of Air*. New Brunswick, N.J.: Rutgers University Press, 1959.

Fairbridge, R. W., ed. *Encyclopedia of the Atmospheric Sciences and Astrogeology*. Encyclopedia of Earth Sciences Series, Vol. 2. New York: Reinhold, 1967.

Hare, F. K. *The Restless Atmosphere*. New York: Hillary, 1961.

Lobsack, T. *Earth's Envelope*. New York: Pantheon, 1959.

Massey, H. S. W., and R. L. F. Boyd. *The Upper Atmosphere*. London: Hutchinson, 1958.

Meetham, A. R. *Atmospheric Pollution*. London: Pergamon, 1960.

Murchie, G. *Song of the Sky*. Boston: Houghton Mifflin, 1954.

Petterssen, S. *Introduction to Meteorology*, 3rd ed. New York: McGraw-Hill, 1969.

Riehl, H. *Introduction to the Atmosphere*, 2nd ed. New York: McGraw-Hill, 1972.

The corona of the sun and the moon's silhouette in the midst of a solar eclipse. (High Altitude Observatory, Boulder, Colorado)

8

Temperatures and the Heat Balance

Usually the most immediately obvious, and perhaps the most important, of the climatic elements is atmospheric temperature. Basically, *temperature* indicates the amount of heat, or *atmospheric energy,* present in the air at any given time and place. The temperature of the atmosphere, especially when linked with the atmosphere's moisture content, has a profound effect on the survival, lifestyles, and occupations of humans. Such temperature extremes as a hot, sticky summer day in Illinois and a frigid, snowy winter day in New England are conditions typical of that portion of the atmospheric blanket closest to the earth's surface.

Without actually seeing it, we are, and have always been, continually affected by the atmosphere around us—especially its temperature. For example, in matters of clothing and shelter, people of primitive cultures were governed by the major climatic elements; and prehistoric people sought shelter in caves as protection against the frigid blasts sweeping over them from retreating ice sheets. Similarly, their placement of primitive settlements near sources of water and their avoidance of waterless areas suggest early people's complete dependence on natural moisture. Today, people attempt to regulate both temperature and moisture, mostly inside homes and other buildings: air conditioning, insulation, and central heating are all attempts to control the effectiveness and actual distribution of heat and moisture.

Solar Energy

The energy that emanates from the sun, or *solar energy,* is the earth's only significant source of atmospheric energy. The earth, a relatively small body in the solar system at a considerable distance from the sun, is only able to intercept a tiny fraction—one two-billionth—of the total solar energy output (Fig. 8–1). This energy is absolutely fundamental to the development of the earth's atmospheric temperatures, upon which all the planet's living things depend, and to the earthly presence of sunlight, which is so essential to the growth and reproduction of these life forms.

The sun's phenomenally high temperatures and vast solar energy outputs are the results of continuous *thermonuclear fusion* within its vast interior: a process that involves the conversion of hydrogen to helium with a resultant generation of unmeasurable quantities of heat energy. The sun, then, is actually a huge incandescent nuclear furnace—a giant solar energy plant—emitting the solar radiant energy that heats and fuels the earth's atmosphere and surface, and that sets in motion the intricate thermodynamic processes that result in the earth's weather and climate systems.

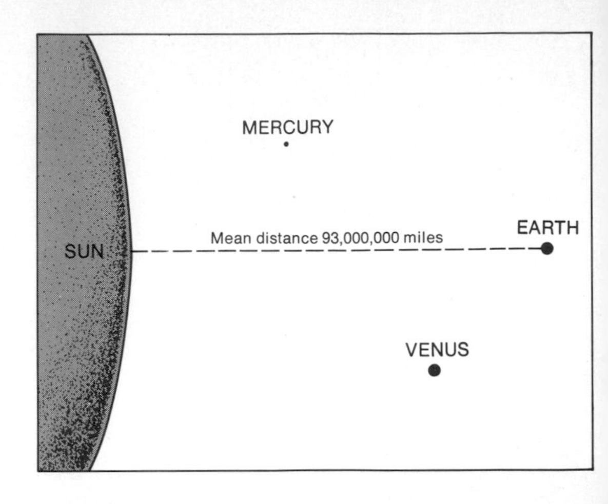

FIGURE 8–1

The sun radiates nuclear energy in all directions into space. The earth receives only a minute amount of this vast total energy output, yet the sun is the earth's only source of energy. (The diameter of the sun is 864,000 miles; that of the earth is 7,926 miles.)

The Solar Spectrum

Solar energy is emitted from the surface of the sun in the form of *electromagnetic radiation,* which takes on a widely varying range of wavelengths. These wavelengths, which combined comprise the *solar spectrum* (Fig. 8–2), range from very *short* (*gamma rays*) to very *long* (*radio waves*). It is important to note here that these wavelengths and their corresponding radiation intensities are inversely proportional. Thus, the shorter the wavelength, the greater the radiation intensity, and *vice versa.* Among the many intermediate-wavelength solar energy rays are those that constitute "visible light," or the *visible spectrum.* This "visible light" is that portion of the total radiant spectrum to which the human eye is sensitive, and which breaks down into the familiar colors violet, indigo, blue, green, yellow, orange, and red.

The solar spectrum is generally considered to include two broad component spectrum ranges, separated by the narrow visible spectrum. Of far greater significance to the physical geographer, however, both in terms of the heating of the atmosphere and with regard to weather processes, is the *shortwave ultraviolet-visible-infrared spectrum range.* The preponderance of all relevant radiant energy is concentrated within this small range, and

FIGURE 8–2

A chart of the *solar spectrum*, representing the solar radiation *wavelengths.* Wavelengths, indicated here in three common scales, are *short* at the *gamma* end of the spectrum and *long* at the *radio* end.

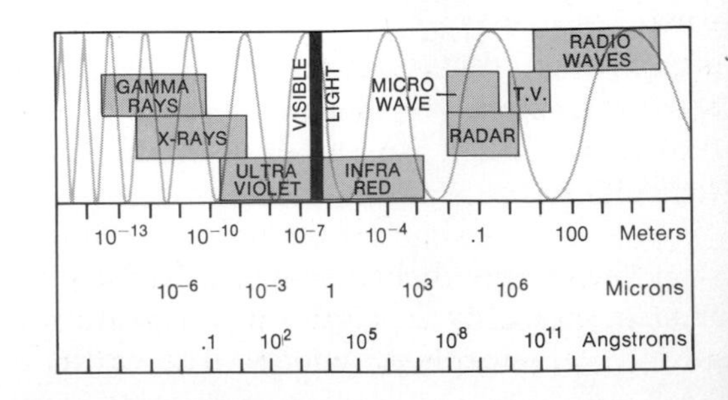

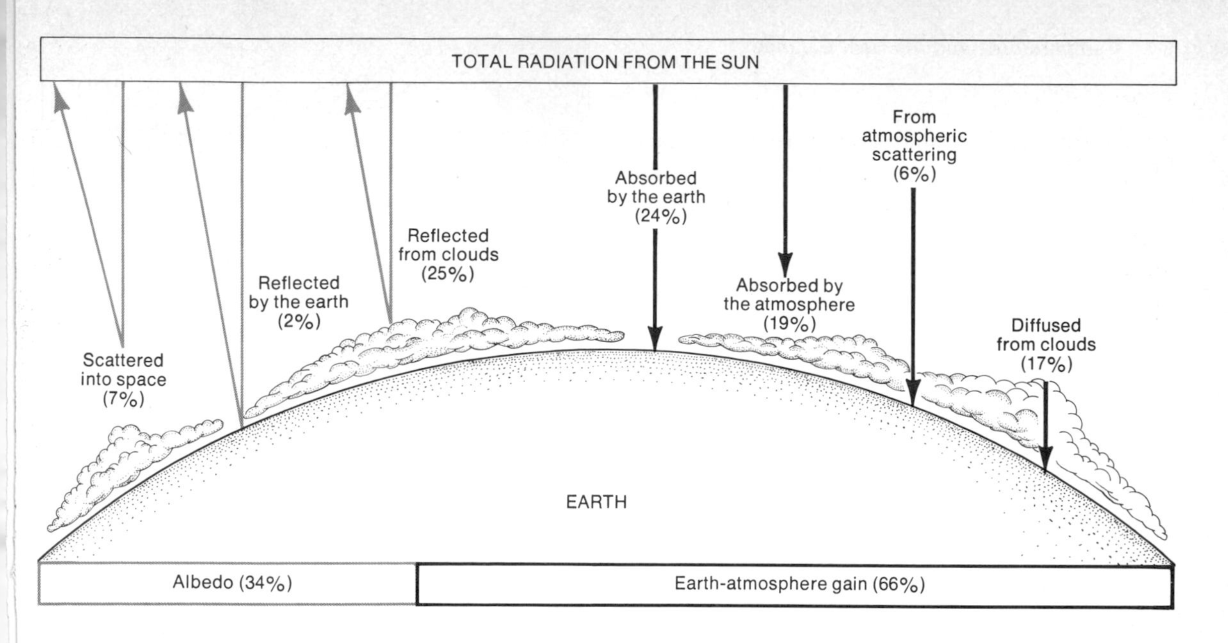

FIGURE 8–3

The dispersion of incoming solar radiant energy in the earth's atmosphere (in the Northern Hemisphere at 45° N.).

most of it emits from the narrow slit that is the visible spectrum (Fig. 8–2). Hence, all solar radiation received at the earth's surface is referred to as *shortwave radiation;* and conversely, all that emitted by the earth is referred to as *longwave radiation.* Thus, from the point of view of the earth's surface, these two kinds of energy are referred to as the *incoming shortwave solar radiation* and the *outgoing longwave terrestrial radiation.*

Insolation

The solar radiation received at the surface of the earth is called *insolation.* It varies in amount from place to place over the earth's surface and from time to time at any given place. Temperature differences themselves and the strong influences of temperature on the other climatic elements combine to create great climatic variations.

While the amount of insolation varies slightly with the output of energy by the sun, the changes resulting from this factor are relatively unimportant, as the greatest variation so far observed is barely more than one percent of the total usually received by the earth. The variations induced by the seasonal changes in the earth's distance from the sun are also of minor consequence—although in January, the earth is some three million miles nearer the sun than it is in July, the resultant difference in the intensity of radiation received is only about one percent (as computed with regard to the outer edge of the atmosphere).

The constant flow of solar energy is unmodified in space and moves freely through space before it is intercepted by the earth's atmosphere. When this unmodified energy approaches the earth, the atmosphere itself acts as a *selective moderator,* or screener, preventing the creation of overly excessive temperature extremes—much as the panes of glass in a greenhouse moderate the temperature within. Much of the solar radiation directed to the earth as shortwave energy is prevented from reaching the earth's surface by the action of *direct reflection, scattering,* and *absorption* that take place within the atmosphere (Fig. 8–3).

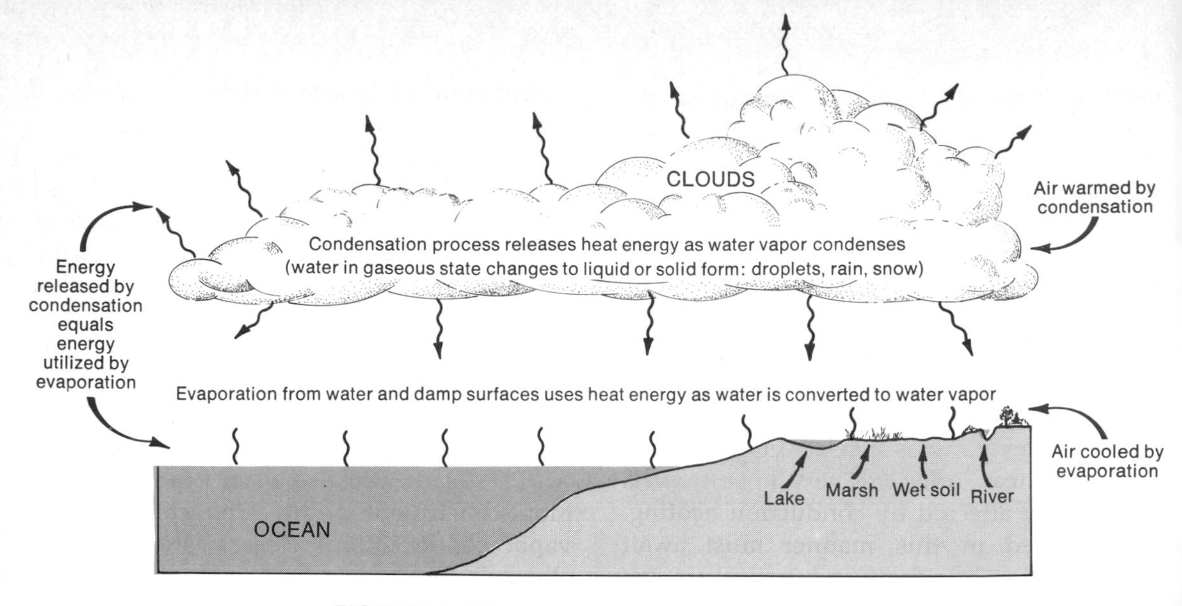

FIGURE 8–14

The heating and cooling of air incident to the *evaporation-condensation* process (evaporation = heat energy utilized and "stored;" condensation = heat energy released as latent heat of condensation).

movement, principally the processes of *convectional, advectional,* and *adiabatic heating.*

Convectional heating is a part of what is known as a *convectional system.* As engendered in fluids or gases, a convectional system is a sort of "closed circulation" system characterized by the rising of matter in a central column, its spreading and outflow aloft, its descent around the margins, and its return to the base of the central column. Figure 8–15 represents the convectional systems present in both water and air.

In the atmosphere's convectional system, the lower atmosphere is heated over any comparatively warm surface, such as a field that is barren and warm surfaced in relation to its surrounding woods. When the low-level air over such a field has been heated, it expands, becoming less dense and therefore lighter, and is displaced from the bottom by inflows of cooler, denser, and thus heavier air from the surrounding wooded area. By means of this circulation, the warmed air from the surface and near surface rises to varying, and often considerable, altitudes within the atmosphere—to as much as 75,000 feet in extreme cases. As the warm air rises, it expands and cools until, at a considerable height, it becomes sufficiently cooled and dense, and thus heavy enough to outflow and descend to take its place once more at the bottom of the heated central rising column (Fig. 8–15B). Thus, convection actually entails both heating and cooling. But, note that prior to the cooling, the heated air has been transported to considerable atmospheric altitudes.

Advectional heating is associated with horizontal or nearly horizontal movement rather than with essentially vertical movement as is convection. Horizontal movements of air, which are referred to as winds, accomplish advectional heating by literally importing warm air from one area to another. For example, in summer, the American Midwest is subjected, for considerable periods of time, to strong, hot, and humid importations of air from warmer latitudes lying to the south. And, in winter, the British Isles are usually strongly affected by warm southwesterly winds moving steadily in from over the comparatively warm waters of the north Atlantic (Fig. 8–12).

Adiabatic heating occurs whenever air descends. Regardless of the cause of the descent, as air descends to lower altitudes, the consequent increase in air pressure and density—as well as the increase in molecular bombardments—causes a steady rise in temperature at the rate of 5.5°F. per thousand feet. The heating is strictly mechanical and internal, with no addition of heat energy from outside scources. In many instances,

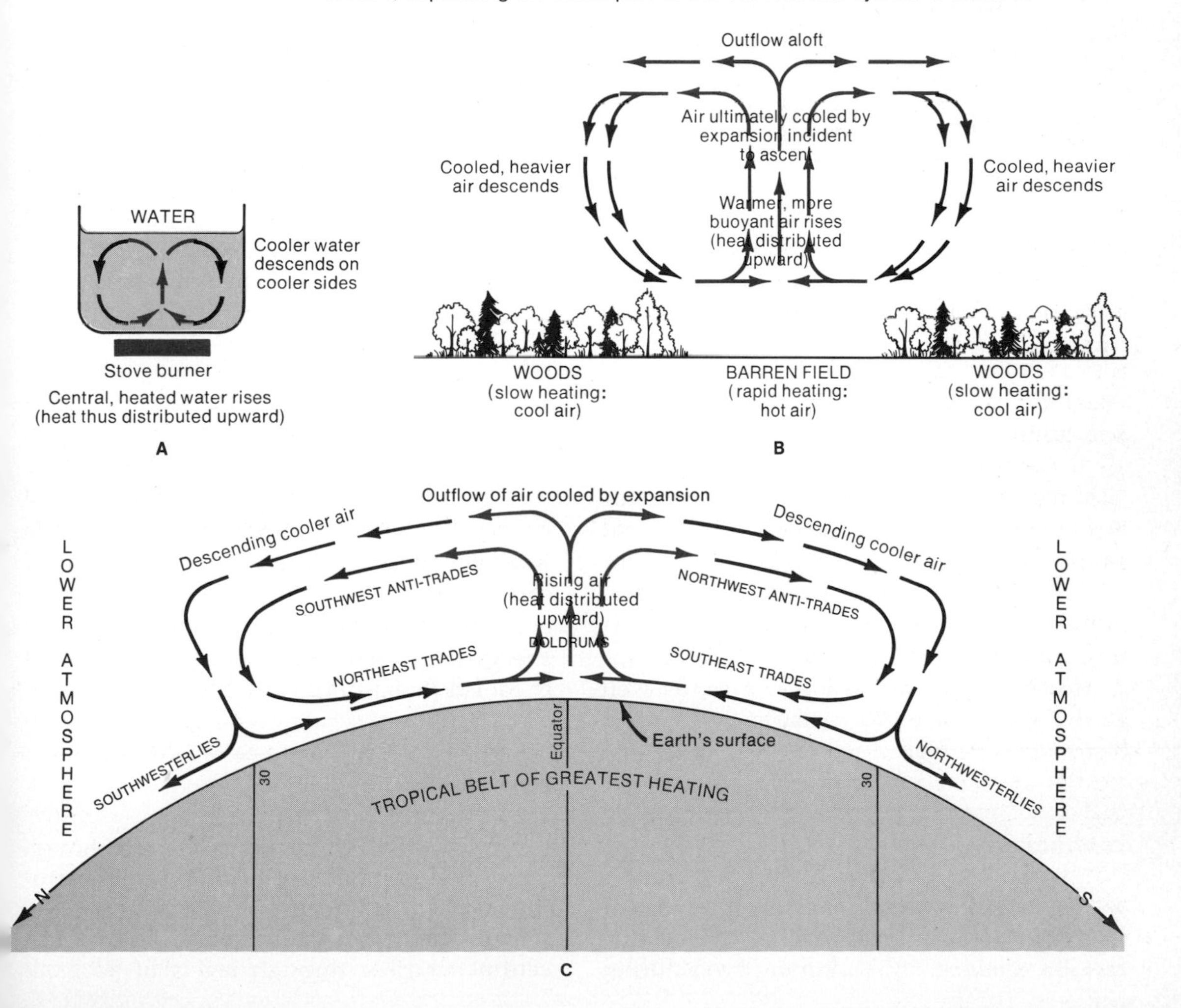

FIGURE 8–15

Cross sections of some *convectional systems.* A. In water. B. In local air. C. In the air of the earth's tropical and near-tropical regions (the vertical here is exaggerated). In a convectional system in air, heat is distributed upward in a central rising column, but, ultimately, the rising air is cooled by expansion and the cooled descending air is distributed downward. Thus, air is both heated and cooled by convection, depending on which part of the convectional system is entailed.

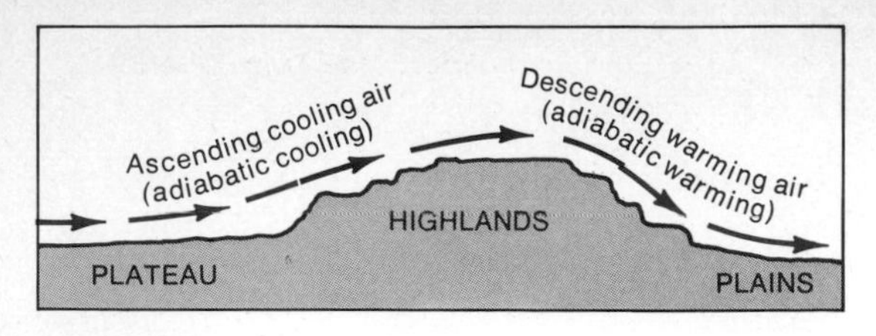

FIGURE 8–16

Adiabatic heating and cooling of air moving over highland regions (such as the Rocky Mountains).

adiabatic heating results from the descent of air on the leeward sides of highlands (Fig. 8–16), but it also occurs in many other situations in the open air.

Measurement of Atmospheric Heat

Figure 8–17 illustrates and explains some of the more pertinent facts concerned with the amount of heat in the atmosphere and the methods by which this heat (temperature) is measured and expressed. Further detailed treatment is deemed beyond the scope of this book, but is readily available in any of the numerous meteorology source books.

Cooling of the Atmosphere

In a general sense, air is cooled by, or in connection with, the same processes that are involved in its heating. Thus, there is cooling by *radiation*, by *conduction*, by *evaporation* (preceding heating by condensation in the evaporation-condensation process), by *convection*, by *advection*, and by *adiabatic* means.

Radiation

Although the earth's surface receives insolation only during the day, it loses heat energy during both day and night by *radiation* to the atmosphere and to outer space. Thus, at night, when the surface temperature becomes lower than that of the air, radiation from the air to the surface, as well as the loss of heat by radiation into space, begins to take place.

Conduction

At night, air in contact with a cooler land (or water) surface loses heat to that surface by *conduction*. In addition, even during the day, air will lose heat to the surface it touches if that surface is comprised of cold-current waters, snow or ice, or landmasses that have not been appreciably day-warmed (perhaps, for example, because seasonally low-angle solar rays supplied meager insolation, or because there was a thick, daytime cloud blanket).

Evaporation

It has been pointed out that in the evaporation-condensation process air is warmed by the release of the latent heat of condensation. Prior to condensation, however, during the process of *evaporation*, heat energy is utilized, and, consequently, the air is cooled (Fig. 8–14).

Convection

As mentioned previously, even though atmospheric *convection* results in air-warming, it eventually causes air-cooling as well (Fig. 8–15B and C). This is because the warm central air rises through levels of progres-

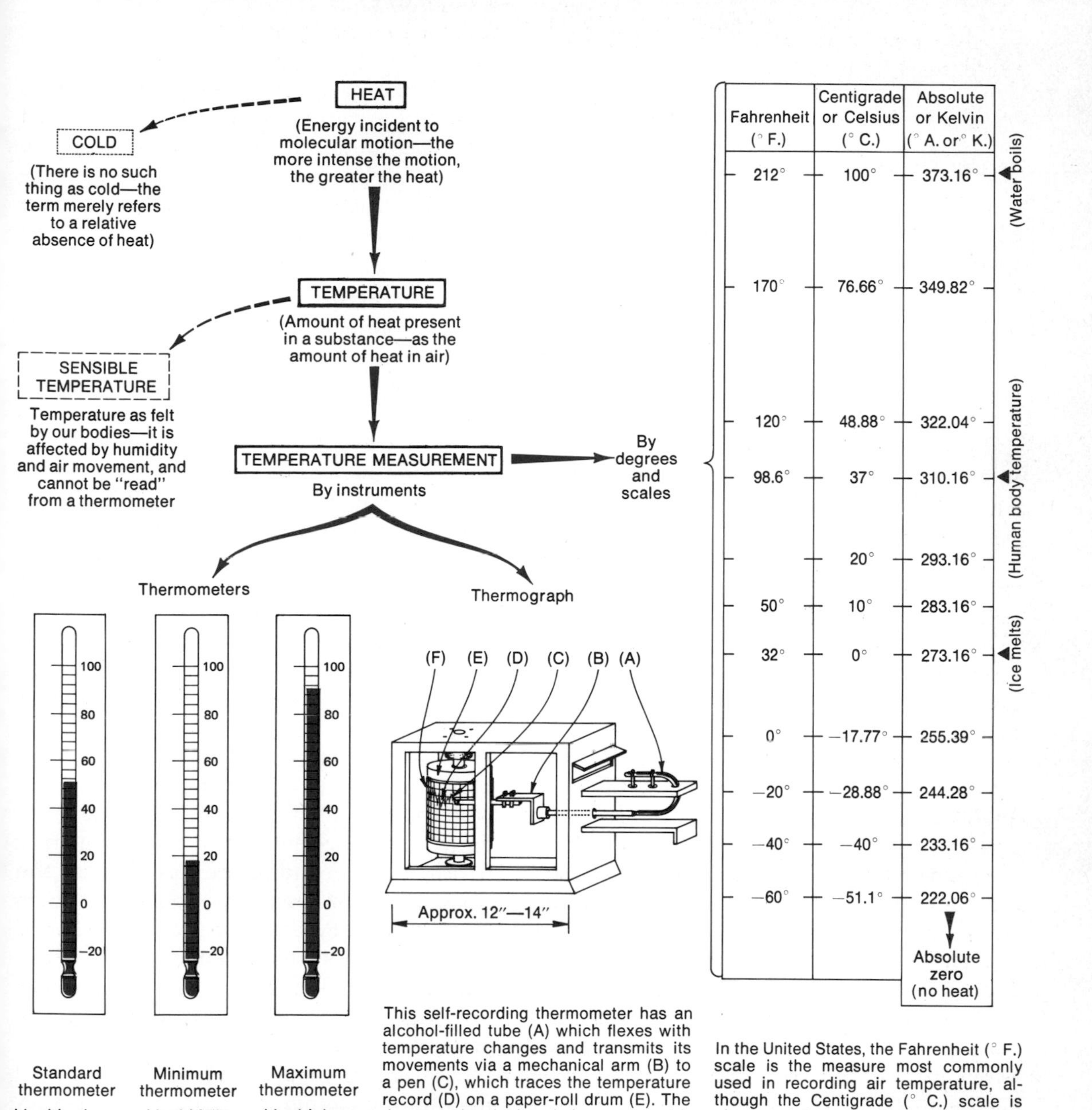

Standard thermometer
Liquid column (mercury or alcohol) rises or falls according to temperature variation

Minimum thermometer
Liquid falls to its lowest temperature and stays there until thermometer is "reset"

Maximum thermometer
Liquid rises to its highest temperature and stays there until thermometer is "reset"

This self-recording thermometer has an alcohol-filled tube (A) which flexes with temperature changes and transmits its movements via a mechanical arm (B) to a pen (C), which traces the temperature record (D) on a paper-roll drum (E). The drum, run by clockwork, has paper gridded by lines indicating days and hours (F). As the drum turns, the minute-by-minute temperature is recorded on the gridded paper. When the paper is removed, a continuous *thermogram* is available.

In the United States, the Fahrenheit (° F.) scale is the measure most commonly used in recording air temperature, although the Centigrade (° C.) scale is characteristic in this regard throughout the rest of the world. The absolute (° A.) scale is a "long" and very finely divided scale that is used for extremely detailed heat measurements, especially by physicists. (For a conversion of ° C. to ° F. or ° F. to ° C., see Appendix D.)

FIGURE 8–17

The measurement of atmospheric heat.

sively decreasing atmospheric pressure, causing it to expand and, with its diminishing density, to cool.

Advection

Just as air may be warmed by the *advection* of warmer air masses, so may it be cooled by the importation of colder air masses. For example, hot July air over downtown Chicago may be cooled by a breeze blowing from comparatively cool Lake Michigan; and, illustrating the same phenomenon on a somewhat larger scale, the British Isles are cooled in July by the westerlies that move steadily in from off the comparatively cool summer waters of the nearby north Atlantic (Fig. 8–12).

Adiabatic Cooling

Descending air, as has been mentioned, is warmed adiabatically. Through its effect on ascending air, the *adiabatic* process plays a part in atmospheric cooling as well. Here again, it involves the rising of air to higher levels resulting in decreasing air pressure. This decrease in air pressure results in expansion of the rising air, which causes decreases in the violence of its molecular motion and, hence, in its temperature. In the example of the Rockies (Fig. 8–16), air descending on the lee (eastern) side of the landmass was adiabatically warmed; however, prior to such warming, air on the windward side had been forced to rise over the highland barrier prompting, as is typical of ascending air, a temperature decrease.

Temperature Distribution and Climatic Controls

Isothermal Maps

The best method for showing geographical distribution of temperatures over the earth is the *isothermal map:* an isotherm is a line connecting points of equal temperature. The temperatures shown may be absolute temperatures for any specific time, or they may be mean temperatures for any designated time period (a commonly used time period being the month). Means for January and July are usually most informative, since they offer a measure of monthly temperatures at the extremes of opposite seasons of the year.

Mean Monthly Temperature Distribution: January and July

World isothermal maps showing actual mean temperatures for January and for July are detailed in Figure 8–18. Both of these maps show the isotherms following roughly east-west directions. In other words, temperatures generally occur in latitudinal belts, with the world's highest means located near the equator in Central and South America, Africa, Arabia, India, the East Indies, and Australia. From these areas poleward—both to the North Pole and to the South Pole—mean temperatures become progressively lower. Essentially, then, *latitude* is the control that determines a particular region's amount of insolation.

A notable departure from the east-west trend of the isotherms occurs wherever they pass from the land to the sea. This is primarily a reflection of the differential heating and cooling of these two surfaces.

The highest mean temperatures during January are found mainly south of the equator. During this month, the greater part of

Australia lies within the area whose mean temperature exceeds 80° F. (that of a small northwestern portion of the continent exceeds 90° F.). Two areas in the interior of South America and a small area on the nose of that continent have mean temperatures over 80° F. In Africa, a narrow strip along the Guinea coast and another along the central eastern coast likewise have monthly means over 80° F.

During this same month, extremely low mean temperatures are found over the Northern Hemisphere. Northeast Asia exhibits the lowest mean temperatures, having a considerable area where the temperature is actually below −40° F., and in January, over the Greenland icecap and the northern portion of North America, subzero mean temperatures are the rule.

Notable differences are evidenced on the July map as well. The highest mean temperatures in July are found mainly in the Northern Hemisphere. Over the land, northwestern Mexico and adjacent sections of the southwestern United States stand out as a hot zone in the Americas; the southeastern United States, the Caribbean, and the east coast of Central America are also a part of this zone. Throughout the Sahara, and eastward, through lowland Arabia, most of India, the coastal districts of southeastern Asia, and the nearby island groups, stretches a similar zone of great heat. By July, regions of intense cold have disappeared from the Northern Hemisphere—that is, except for ice-covered Greenland, and Antarctica has become the center of the cold zone, now located in the Southern Hemisphere.

Still another factor in temperature distribution is illustrated by the paths followed by the January and July isotherms, as seen in Figure 8–18. For the Northern Hemisphere, the January map shows the distinct equatorward bend of the isotherms over the cold landmasses, and the July map shows the distinct poleward bend over the then warmer land; in the Southern Hemisphere these patterns are reversed. Extremes of mean temperature, no matter what the season, always occur over landmasses rather than over water. This is a further demonstration of the difference in the heating qualities of land and water, and one of the major effects the distribution of land and water has on the environment.

Ocean Currents

The effects of ocean currents upon mean air temperatures are also demonstrated in Figure 8–18. Off the western coasts of the continents, currents bring cold water from more poleward positions into the middle latitudes, causing marked equatorward bends in the isotherms locally. Such situations occur along the Peruvian and Chilean coasts, where the Humboldt Current brings cold water from the Antarctic; along the coast of southwest Africa, where the Benguela Current acts similarly; and off the California coast, where the California Current brings cold water southward from the North Pacific. Conversely, off the British Isles and the coast of Norway, the Gulf Stream (also called the North Atlantic Drift) carries relatively warm water well northward of its usual latitude; the warm air that circulates above this portion of the Gulf Stream is reflected in the poleward swing of isotherms in the vicinity.

Finally, regardless of season, there is less deviation from the latitudinally arranged belts of temperature in the Southern Hemisphere than in the Northern Hemisphere. Since the Southern Hemisphere is mostly water, it exhibits a greater uniformity of temperatures. This fact again illustrates the great influence exerted by the earth's land and water distribution.

ern Hemisphere. Over the continents it bends southward, or equatorward; over the oceans it bends northward, or poleward. This characteristic is explained in part by the differential heating of land and water: large landmasses in these latitudes of the hemisphere become very cold during the period of least insolation, while the water bodies, cooling more slowly, remain comparatively warm. The very pronounced poleward bends off the west coasts, as along the western coast of Europe, are in large part accounted for by the poleward flow of warm ocean currents. These currents have a considerable warming effect upon the air that moves over them; the higher temperatures of the low latitudes are thus carried well northward along the courses of the currents.

The individual isotherms in the Northern Hemisphere show quite a different pattern during the summer. The isotherm of 68° F. for July is characteristic of the conditions prevailing in this hemisphere at that time. Beginning in the mid-Pacific near 40° N., the isotherm swings southward almost to the Tropic of Cancer, thus reflecting the cold California current off the west coast of Mexico. Over the North American continent, it pursues a most irregular course, reaching the border between the United States and Canada over the heated lowlands and looping far southward over the high, cold, Sierra Nevada and the Rocky Mountains.

After passing near New York City, the isotherm follows an east-west course across the Atlantic to the coast of Spain. Over ocean-tempered western Europe it stays in the southern portion of the landmass, passing south of the Alps; over eastern Europe, beyond the tempering effects of the Atlantic, it swings sharply northward to the plains of Hungary, and then moves still farther northward to the lowlands of Siberia. From this northernmost position it takes a southward swing to the Sea of Japan and passes over the Japanese Islands in a northward swing that, due to the warmth of the Japanese Current, extends well offshore over the western Pacific.

Thus, over each of the continental landmasses, the summer isotherms of the Northern Hemisphere bend well northward, toward the pole, while over the oceans they are relatively straight; the pattern over the land is the reverse of that of the winter period. Over the oceans, where there is any summer departure from an east-west line, that departure is toward the equator. Again, this is the reverse of the winter tendency. The sharp equatorward bends of the isotherms over the oceans, such as those off the New England shore, result primarily from the presence of cold ocean currents. These cold currents are found off east coasts poleward of 40° latitude, and off west coasts in the latitudes between 20° and 40°.

In the Southern Hemisphere, regardless of season, isotherms have a more nearly constant east-west trend. Tendencies toward an equatorward bend over land in winter and a poleward bend over land in summer are best evidenced in Australia. To a lesser extent, South Africa and South America show the same trend, especially in July, their winter period. Neither of those continents imposes any considerable land breadth south of the equator to interrupt the great expanse of water in the Southern Hemisphere. There is more uniformity of heating surface in that hemisphere, and hence the ordinary contrasts of differential heating of land and water are somewhat masked there by other controls of temperature.

Temperature Range

The difference in degrees between the mean temperatures of the warmest and those of the coldest months of the year is called the *mean annual temperature range*. The smallest ranges occur mainly near the equator and at high latitudes (Fig. 8–19). Largest ranges generally are encountered in the interiors of great landmasses in the higher latitudes; thus the largest seasonal temperature differences occur in the Northern Hemisphere. The smallest temperature differences on land occur along windward coasts, near the equator, and in tropical high mountains. In the latter locations the effects of nearly constant insolation throughout the year and of high elevation are both operative. The mean annual temperature range over large water bodies is also relatively small, thus moderating the temperature ranges of nearby shores. Where winds blow onshore from the oceans, the tempering effect of large water bodies is found to exist not only along the immediate shores but also for varying distances inland.

Ranges very from less than 5° F. on the equator to over 110° F. in the interior of Asia. For example, the temperature range at Manaus, on the Amazon in Brazil, is about 3° F.; that at Verkhoyansk in northeastern Siberia is nearly 119° F.

FIGURE 8–19

Mean annual temperature ranges are generally smallest near the equator and over the oceans at higher latitudes; the largest ranges are found in the interiors of the great landmasses of the Northern Hemisphere. Where the range is 5° F. or less, the belt is set off by dashed lines. (After Haurwitz and Austin)

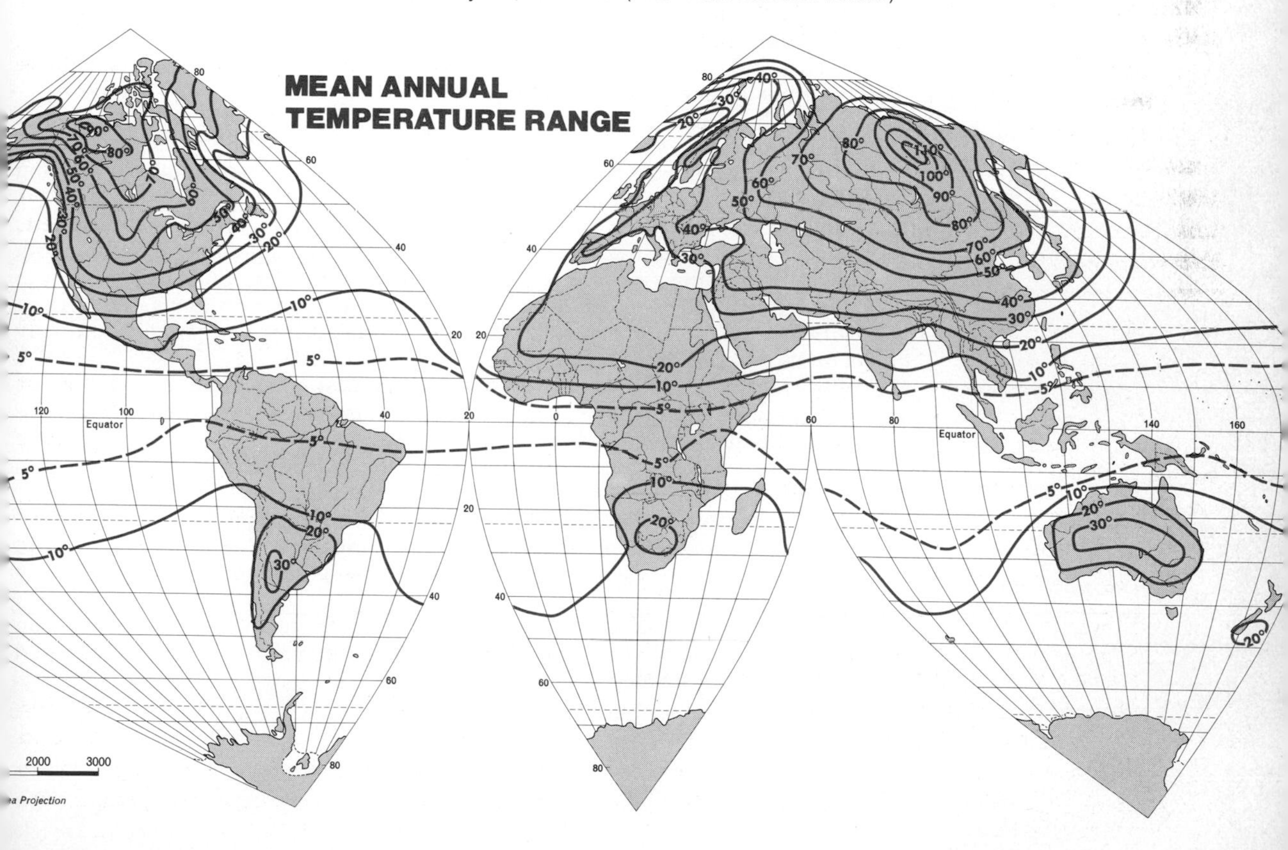

REVIEW AND DISCUSSION

1. Temperature is basically an indicator, or measure, of the amount of atmospheric energy present in the air at a particular time and place. Why is it perhaps the most important of the climatic elements? What other climatic elements combine with temperature in the development of weather and climate?
2. Why does solar-atmospheric energy vary in amount from time to time and from place to place over the earth's surface? What is insolation?
3. Discuss solar energy and the solar spectrum, especially the "visible light" portion. Explain why it is correct to refer to the "incoming shortwave solar radiation" and to the "outgoing longwave terrestrial radiation."
4. Outline and discuss the processes involved in heating the earth's atmosphere. How do direct reflection, scattering, and absorption affect the earth's receipt of solar shortwave energy? Explain what is meant by the "greenhouse effect" of the atmosphere; by terrestrial radiation; by albedo. Discuss insolation as "fuel."
5. Explain the concept of the global heat balance. What is the solar constant?
6. Temporary heat nonbalances do occur repeatedly throughout the earth's atmospheric system, mainly because of latitudinal factors. Explain (with the help of diagrams). Also, explain angle of strike and duration of strike.
7. Of land and water, which one is both heated and cooled more rapidly? What differences result in the temperatures of adjacent water-heated atmospheres and land-heated atmospheres? In this context, what is meant by penetrability and mobility?
8. Discuss and compare the processes involved in the heating of the troposphere by means of energy radiated from water and land surfaces.
9. Explain how additional heating of portions of the troposphere takes place in connection with varying types of atmospheric circulation. Explain conductional, convectional, and advectional heating. How does adiabatic heating differ from these other kinds of atmospheric heating?
10. Outline the processes involved in the cooling of the earth's atmosphere; discuss each of these processes. Compare and contrast adiabatic cooling with adiabatic heating (include references to Figure 8–17 as illustration).
11. How are temperature differences over the earth shown on maps? What is an isotherm? Why are the mean monthly temperatures for January and July commonly used for isothermal maps?
12. Why do isotherms normally follow roughly east-west directions? Do mean temperatures become progressively lower poleward, or equatorward?
13. Notable departures from the normal east-west trends of isotherms occur where they pass from land to water or from water to land. Explain. Discuss, also, the effects ocean currents have on air temperatures.
14. For a clearer understanding of season-to-season changes in mean temperatures, trace the isotherm of 32° F. for both January and July, as indicated in Figure 8–18. Also, trace some other isotherms as they occur on the January and July maps.
15. Explain mean annual temperature range. What effect does this range have on humans and the physical environment? Where on earth is it the greatest, and where is it the smallest?

SELECTED REFERENCES

Bach, W. *Air Pollution*. New York: McGraw-Hill, 1972.

Budyko, M. I. *The Heat Balance of the Earth's Surface,* trans. by N. A. Stepanova. Washington, D.C.: Office of Technical Services, U.S. Dept. of Commerce, 1958.

Gates, D. M. *Energy Exchange in the Biosphere*. New York: Harper & Row, 1962.

Kondratyev, K. Y. *Radiation in the Atmosphere*. New York: Academic Press, 1969.

Miller, D. H. *The Heat and Water Budget of the Earth's Surface*. Advances in Geophysics, Vol. 2. New York: Academic Press, 1965.

Petterssen, S. *Introduction to Meteorology,* 3rd ed. New York: McGraw-Hill, 1969.

Riehl, H. *Introduction to the Atmosphere,* 2nd ed. New York: McGraw-Hill, 1972.

Sellers, W. D. *Physical Climatology*. Chicago: University of Chicago Press, 1965.

the oceans near the margins of low latitudes is due, only partially, to the lower temperatures in these regions, for they are also partially formed by earth rotation. However, the connecting wide belts over the land areas in winter are essentially the result of continental cooling during that season, and the lows that separate the oceanic highs during summer are essentially the result of continental heating in that season.

Still farther toward each pole, there are other low-pressure belts. They are much more intense in the Southern Hemisphere than they are in the Northern Hemisphere, largely because there is little land-water alternation at equivalent latitudes in the Southern Hemisphere.

There are thus five general belts of low and high pressures: an *equatorial belt of low pressure;* the *northern* and the *southern subtropical belts of high pressure;* and the *northern* and the *southern subpolar belts of low pressure.* In addition, there are subsiding eddies around the poles that create high-pressure centers whose area expands in winter (in January in the Northern Hemisphere and in July in the Southern Hemisphere) and contracts in summer. Along any one meridian, the lows can be thought of as *pressure troughs* and the highs as *pressure ridges.* The gradient from any high to an adjacent low brings about a flow of air from the highs into the lows, and thus, recognizable wind belts have become established over the earth's surface.

Pressure Gradients and Winds

While vertical air movement involves the transfer of heat energy from the earth's surface into the atmosphere and its eventual recycling (back to the surface), the horizontal movement of air is primarily a transfer of air from one place to another along or above the earth's surface. This does not mean, however, that vertical and horizontal movements are mutually exclusive. In fact, both types of movement take place within the same closed system and create a three-dimensional pattern of movement known as the *global atmospheric circulation.*

There are three principal forces governing horizontal air transfer: the *pressure gradient force,* the *centrifugal force,* and the *Coriolis force.* The first of these forces comes about because winds tend to flow from areas of higher atmospheric pressure to areas of lower atmospheric pressure along what is called a *pressure gradient.* The pressure gradient in an area creates a natural gradient flow, somewhat like air flowing from a pressurized container (though this analogy has somewhat limited application). *Centrifugal force* is that which tends to pull any particle away from the axis of a rotating body. Since the axis of the earth's rotation is a line that goes through the center of the earth and the poles, there is a tendency for air to move equatorward from the polar regions. The third, the *Coriolis force,* is responsible for the fact that a moving body deflects to the right in the Northern Hemisphere and to the left in the Southern Hemisphere.

When all of these forces are balanced, in a state called *strophic balance,* ideal horizontal air movement is achieved. However, this ideal movement, known as the *geostrophic wind,* is just an approximation; it indicates wind flowing parallel to the isobars, a circumstance that rarely occurs.

Within the lower atmospheric levels, nearer the earth's surface, a still further conditioning force comes into operation. This additional force is called the *friction effect* and is the noticeable drag of air being moved from one place to another.

In general, however, it is the atmospheric pressure gradient forces—of high and low magnitude—and the gradients they establish, that are chiefly responsible for the development of the major atmospheric movements and for the major wind systems of the world.

Measurements of Winds

There are two measurements commonly used in wind observation—wind *direction* and wind *velocity*. Wind direction is generally measured by a *weather vane*, consisting of a bar, with a pointer at one end and a vertical tail surface at the other, mounted so that it can rotate freely on a perpendicular shaft. As the wind blows, the pressure it exerts on the vertical tail surface of the vane causes the pointer to swing around into the direction from which the wind is blowing: thus, a west wind is a wind that is blowing from the west.

Wind velocity is measured by an *anemometer*. One of the most common types consists of three or four light metal cups mounted on a horizontal frame that rotates at the end of a vertical shaft. As the wind blows, the difference in the effect of the pressure on the inside and on the outside of the cups causes the frame to rotate. By counting the rotations, the speed of the wind can be determined, in miles per hour. As with atmospheric pressure, there is a standardized scale of wind velocities—the *Beaufort scale*—named after its originator, Admiral Sir Francis Beaufort (1774–1857), a naval officer and hydrographer.

The World Wind Circulation System

The wind systems of the Northern and Southern hemispheres are strikingly similar—almost mirror images of each other. From the subtropical high-pressure belts, at about 30° latitude in each hemisphere, surface air flows both toward the equator and toward the subpolar lows. Surface air also flows from the high-latitude polar eddies in each hemisphere, toward their respective subpolar lows. In this manner, three of the major wind belts in each of the hemispheres are formed: the *trade wind belt*, or the *tropical easterlies;* the *westerlies wind belt*, or the *surface westerlies;* and the *polar wind belt*, or the *polar easterlies* (Plate IV; Figs. 9–5 and 9–6). Between the facing trade wind belts of the two hemispheres, and associated with the equatorial low-pressure belt, lies a belt of light and variable winds and calms known as the *intertropical convergence zone* (ITCZ), or the *doldrums*. Between the trades and the westerlies in each hemisphere, and associated with the subtropical high-pressure areas, lies a belt of light, variable winds and calms known as the *horse latitudes*. There are thus nine generally distinct wind belts in the world wind system, as is graphically indicated in Figure 9–6.

Major World Wind Belts

Since the rotation of the earth, due to the Coriolis force, exerts a deflective force on all moving bodies on and above its surface, including the air, the flow of winds tends to be deflected above its immediate surface (to the right in the Northern Hemisphere and to the left in the Southern Hemisphere.) Figures 9–5 and 9–6, depicting the various wind belts, should be studied in comparison with the atmospheric pressure

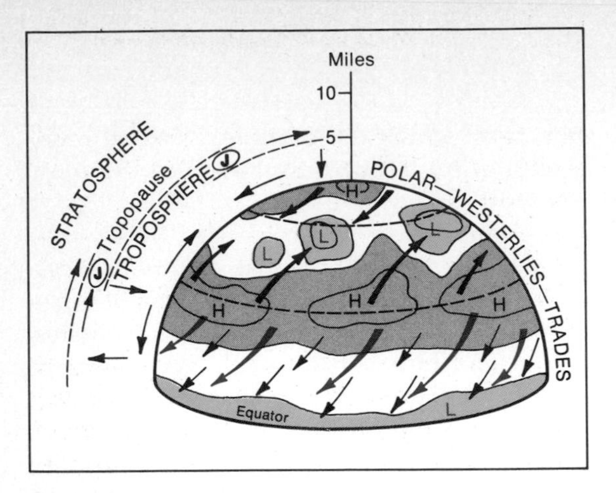

FIGURE 9–5
An approximation of atmospheric conditions north of the equator in the winter, showing pressure distributions, planetary wind belts, and vertical circulation. On the left, the Ⓙs amid the vertical temperature regions indicate the jet streams.

maps in Figures 9–3 and 9–4, and with the ocean wind maps in Figures 9–7 and 9–8. Such a study will clarify the relationship between the generalized theoretical wind belts and the generalized "real world" wind belts—and will make the latter much more comprehensible than they may now be.

Trade Wind Belts. In the Northern Hemisphere, the winds that flow toward the equator deflect to the right and thus become northeasterly or easterly. In the Southern Hemisphere, since the winds are deflected to the left, the equatorward-moving winds are dominantly southeasterly or easterly. Thus, there are actually two belts of equatorward-moving winds—one in each hemisphere. These belts are known as the *trades, trade winds,* or *tropical easterlies,* and depending on the hemisphere, are individually referred to as either northeast trades or southeast trades (Fig. 9–6).

Westerlies Wind Belts. The winds flowing from the subtropical highs to the subpolar lows in each hemisphere are the *prevailing westerlies;* they become southwesterlies in the Northern Hemisphere and northwesterlies in the Southern Hemisphere (Fig. 9–6). In the Southern Hemisphere, where oceans nearly circle the globe in the higher latitudes, the westerlies persist throughout the year and blow from an almost due west direction. Many literary descriptions of "rounding the Horn" around the southern tip of South America attest to the wildness and storminess of this persistent wind belt, where it is somewhat constricted by the landmasses of South America and Antarctica. In the Northern Hemisphere, where large landmasses break the continuity of the subpolar low and impose frictional blocks to the open sweep of these winds, the westerlies are much more variable, both in direction and in strength. During the winter months, however, when the subpolar lows are best developed over the north Atlantic and the north Pacific, storminess on those bodies approaches the ferocity of the Southern Hemisphere's belt of westerlies.

Polar Wind Belts. The *polar winds* of each hemisphere are less well known than are the trades or the westerlies. Since they flow from the poles toward the equator and are also subject to rotational deflection, they are "easterly" winds (Fig. 9–6). They have great strength and blusteriness, as attested to by frequent references in accounts of polar explorations as well as in extensive instrument and photographic records.

Secondary World Wind Belts

Intertropical Convergence Wind Belt (Doldrums). Lying between the two tropical trade wind belts, somewhat astride the equator, is a narrow intertropical wind belt of very light and irregular surface winds and calms; it is known as the *doldrums,* or the *In-*

tertropical Convergence Zone (ITCZ). In this belt the air is generally rising, partially as a result of the high temperatures near the equator, and partially as the result of slight temperature differences between the masses of air flowing in from the northeast and from the southeast, that is, between the northeast and southeast trades. Slight inequalities in heating, and, hence, in atmospheric density, cause one mass to be forced somewhat upward over the other. Where there is an active rising, or a pushing upward, of the air, there exists what is known as the *intertropical front* (Fig. 9–6).

The term *front* is used here to designate the contact surface between bodies of air that are observably different in regard to temperature, moisture, or other properties. In a sense, any front in the atmosphere can be compared to the interface between oil and water when they are poured into the same container—they do not mix, but an observable contact surface (front) is formed, with water on one side and oil on the other. Whether the rising of air in the ITCZ between the trade wind belts is convectionally induced or brought about by action along an intertropical front, the end result is usually an upward movement of air, leading to atmospheric expansion, cooling, and condensation, with attendant cloud developments and precipitation.

Horse Latitudes Wind Belts. Two other similar world wind belts—one in each hemisphere—lie between the trade wind belts and the westerlies belts. Because of the lack of appreciable pressure gradients, these

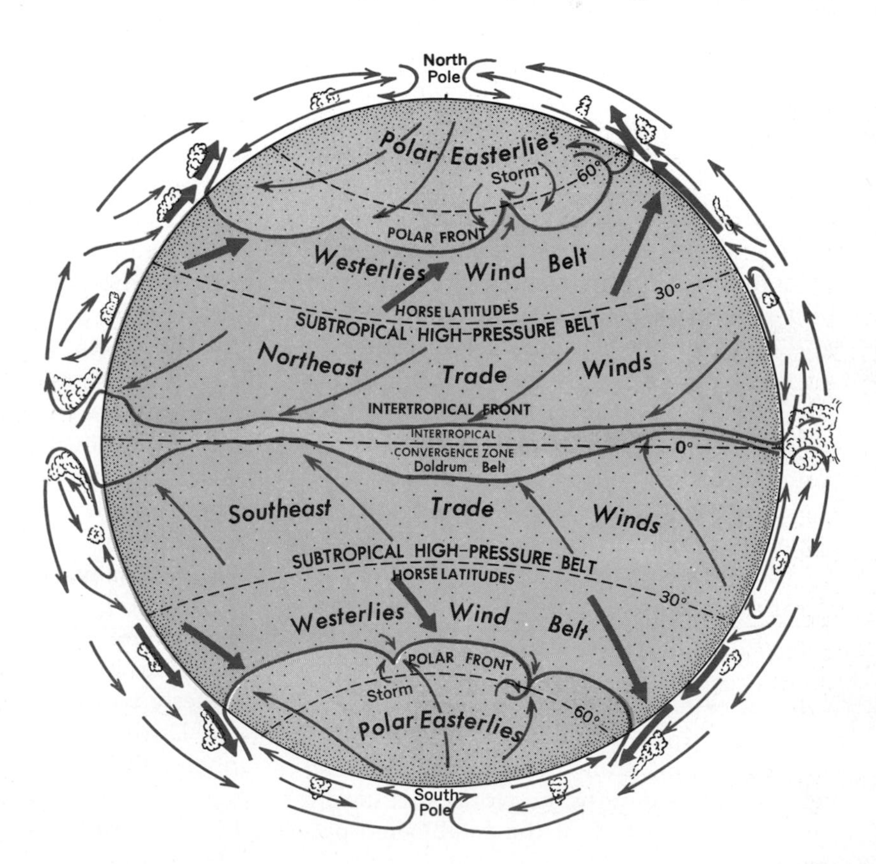

FIGURE 9–6

The world's major wind belts, the frontal zones, and the vertical circulation within the troposphere.

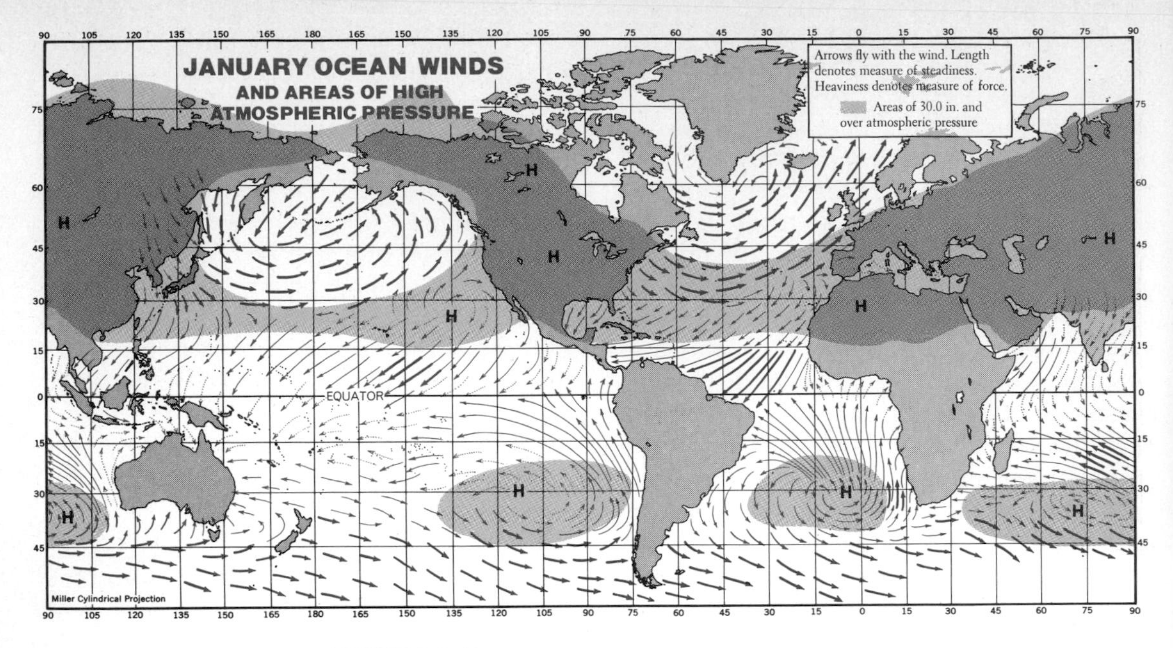

FIGURE 9–7

The actual prevailing winds over the oceans in January show a well-developed counterclockwise circulation over the oceans in the Southern Hemisphere and a strong development of the prevailing westerlies in the Northern Hemisphere.

belts of subtropical high pressures are regions of very weak, shifting, irregular winds or calms, and are known as the *horse latitudes* (Fig. 9–6). Generally, their characteristic calms are more pronounced in mid-ocean, where higher pressures dominate throughout the year. During the winter in each hemisphere, the oceanic highs extend landward over the continents, and as they do so, there is a consequent spreading of the calms of the horse latitudes. As the highs retreat to their oceanic positions during the summer, and as lows begin to form over the continents on either side, flows of air become noticeable on the continental fringes. At those times, these belts of calms are again confined to the oceanic centers.

Oceanic Wind Circulation

Deviations from the simple, idealized belted arrangement of the world's winds are so numerous that all of the established patterns are nearly obscured; still, an understanding of the real situation is easier with such a preface. Under actual conditions, in the summer in each hemisphere, the subtropical belt of high pressure consists of separate oceanic high-pressure cells and separate continental low-pressure cells, which transform it from a continuous belt into a "broken" one (Figs. 9–3, 9–4, and 9–6). From the centers of the oceanic highs, winds blow outward in all directions, are deflected, and rotate in their respective basins—in clockwise directions in the Northern Hemisphere and in counterclockwise directions in the Southern Hemisphere (Figs. 9–7 and 9–8). Actually, then, the portion of each swirl that lies nearest the equator is that hemisphere's belt of trade winds, while the poleward half is the hemisphere's belt of westerly winds.

During the summer in each hemisphere, there is a continuity in both high-pressure belts that is absent in the winter (Figs. 9–3 and 9–4). On the western sides of

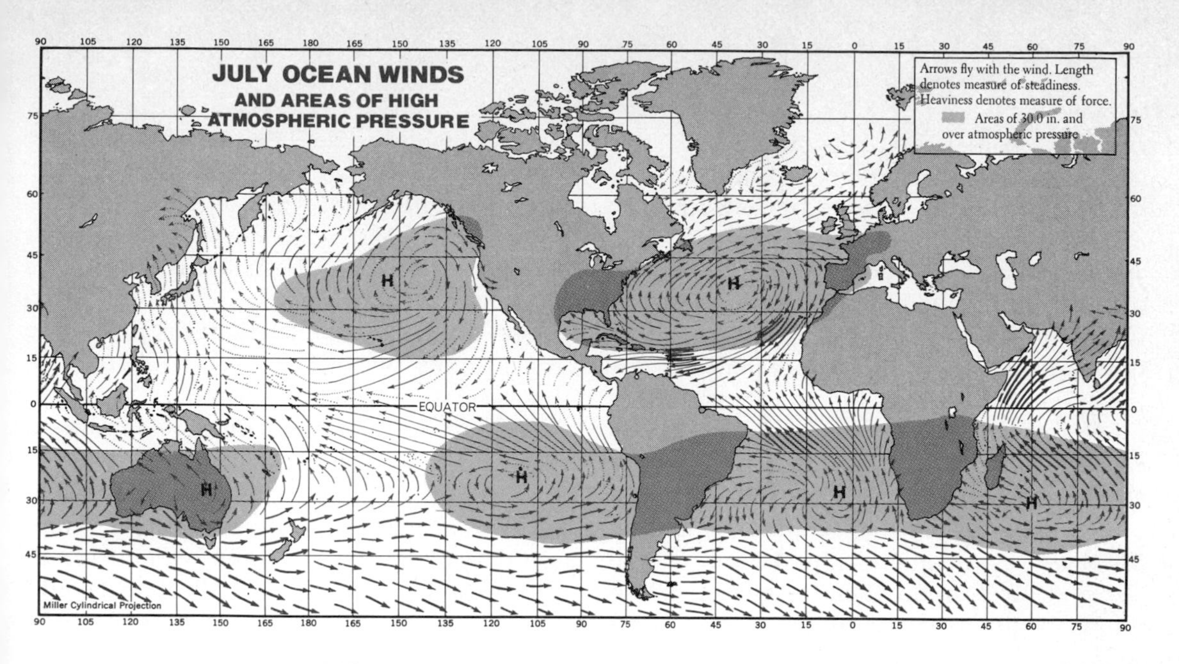

FIGURE 9–8

The actual prevailing winds over the oceans in July show a well-developed clockwise circulation over the oceans in the Northern Hemisphere.

the ocean basins (western sides of the oceanic highs), winds are directionally deflected, so that the trades eventually become the westerlies (Fig. 9–9). On the eastern sides of the ocean basins, they are deflected to eventually become the trades. Continental lows that develop during the summer in each hemisphere have the effect of bringing about an indraft of air toward the continents and, thus, of strengthening the continuity that exists between the trade winds and the westerly winds.

The actual prevailing winds over the oceans in January and July are charted in Figures 9–7 and 9–8. In these charts, each group of arrows points in the direction in which a wind is blowing; the length of the arrows indicates the relative persistence of the wind, and the thickness indicates the relative strength. It should be observed that the trade winds are generally persistent in both seasons and that the prevailing westerlies are strongest in the winter season of each hemisphere.

Monsoons

The large continental landmasses of Asia and North America develop independent wind systems that largely override the general belted world wind system described in Figure 9–6. The winds in Asia offer the best example of such independent systems. Over the southern and eastern parts of the massive Asian continent, several low-pressure cells develop during the April-to-September period (summer in the Northern Hemisphere) (Figs. 9–4 and 9–8). During the October-to-March period (winter in the Northern Hemisphere), the process depicted above is reversed (Figs. 9–7 and 9–10A): high-pressure cells build up over Asia, opposed by an intense low over the Australia-Indonesia region. This reverses the pressure gradient, directing it away from Asia, southward across the equator. Along the new gradient, the outblowing northeasterly winds from India move into the Southern Hemisphere. There, their deflection is reversed so that they become northwesterlies, where, ideally, the belt of southeasterly trades would be expected. Concurrently, across the equator in the Southern Hemisphere (in its winter period), an intense subtropical oceanic high develops. Winds flowing from this winter oceanic high toward the equator start as southeasterly trades (Figs. 9–8 and 9–10B), but the pressure gradient toward India is sufficiently great to cause these trade winds to continue on across the equator to Asia. When these winds cross from the Southern into the Northern Hemisphere, the action of the Coriolis force upon them is reversed: the winds begin to curve to the right toward Asia, from the southwest, into the zone where the northeasterly trades would ideally be expected.

FIGURE 9–9

The rotation of subtropical high-pressure cells over the oceans: clockwise rotation in the Northern Hemisphere and counterclockwise rotation in the Southern Hemisphere. The diagram also illustrates the way in which the outflows of air from the highs are deflected to "feed" the *westerlies* on the poleward sides and the *trades* on the equatorward sides. (H denotes the center of the subtropical high.)

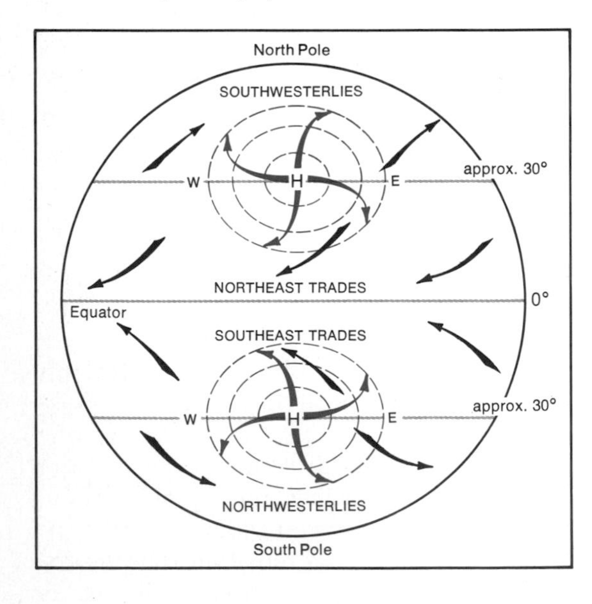

Seasonal inflows and outflows of winds on a continental scale, such as those over southern and eastern Asia, are known as *monsoonal circulations,* or *monsoons.* Monsoons are most completely developed along the littorals of southern and eastern Asia, Indonesia, northern Australia, and the westward bulge of Africa (Figs. 9–7 and 9–8). Elsewhere, monsoonal circulations are not sufficiently intense, or sufficiently persistent, to override the general belted world wind system, and so do not warrant recognition as true monsoons. All of the continents create some monsoonal tendencies, but they seldom dominate the air movements for a whole season or for the entire year, as they do in the Asia-Pacific monsoonal region.

Seasonal Migrations of the World Wind Belts

The wind belts of the world migrate latitudinally with the cycle of the seasons, due largely to the seasonal changes in insolation; these changes cause shifts in the latitudinal positions of the pressure belts and, consequently, in the positions of the wind belts.

During the winter in the Northern Hemisphere and the summer in the Southern Hemisphere, the zone of maximum insolation lies south of the equator (Fig. 8–18A). Therefore, the accompanying equatorial belt of low pressure is somewhat south of the equator. In response to this, the trades of the Northern Hemisphere move closer to the equator than at any other season. All other wind belts of the Northern Hemisphere shift equatorward also, while all of the wind belts in the Southern Hemisphere shift poleward.

During the summer in the Northern Hemisphere and the winter in the Southern Hemisphere, maximum insolation is received north of the equator (Fig. 8–18B). Hence, the equatorial belt of low pressure is developed somewhat north of the equator. All of the wind belts in the Southern Hemisphere then migrate equatorward, while those in the Northern Hemisphere move poleward. In brief, the wind belts of each hemisphere migrate poleward in their respective summers and equatorward in their respective winters (compare Figures 9–7 and 9–8). In addition, the global system of north-to-south and south-to-north migrations corresponds to the similar migrations of the sun's vertical rays (Fig. 8–8).

Over the ocean areas, this latitudinal migration of wind belts involves some 10° to 15° of latitude; over the landmasses it may involve as much as 35°. The shifts over the oceans usually lag behind the movement of the sun's vertical rays by a month or two, while over the land, there is generally little time lag. Both of these time effects correlate closely with the different heating qualities of land and water.

The regions most affected by these migrations lie near the margins of the various wind belts. This is well illustrated by the Mediterranean region of Europe and Africa. During the Northern Hemisphere summer, when that region's wind belts move poleward, the Mediterranean area comes under the marginal influence of the trade winds belt. In the Northern Hemisphere winter, when the wind belts move equatorward, the Mediterranean lies just within the sphere of

FIGURE 9–10

The monsoonal wind circulation over India and its environs in opposite seasons. A. The *outblowing* monsoon in January. B. The *inblowing* monsoon in July. (In January, the monsoonal circulation is much weaker and much less constant than it is in July.)

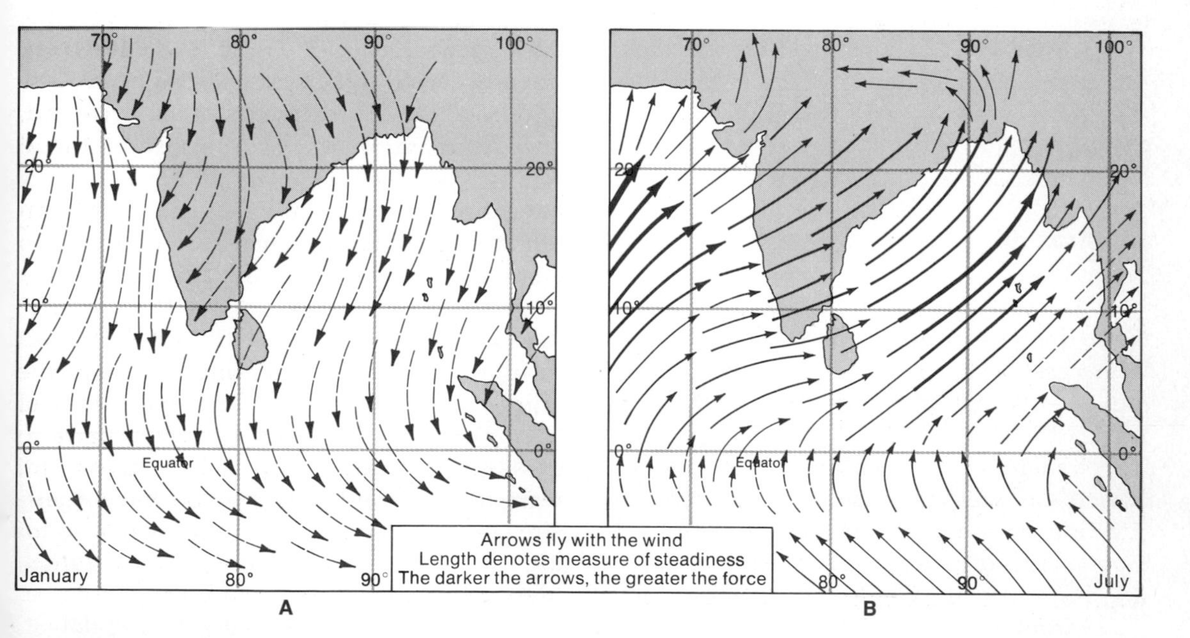

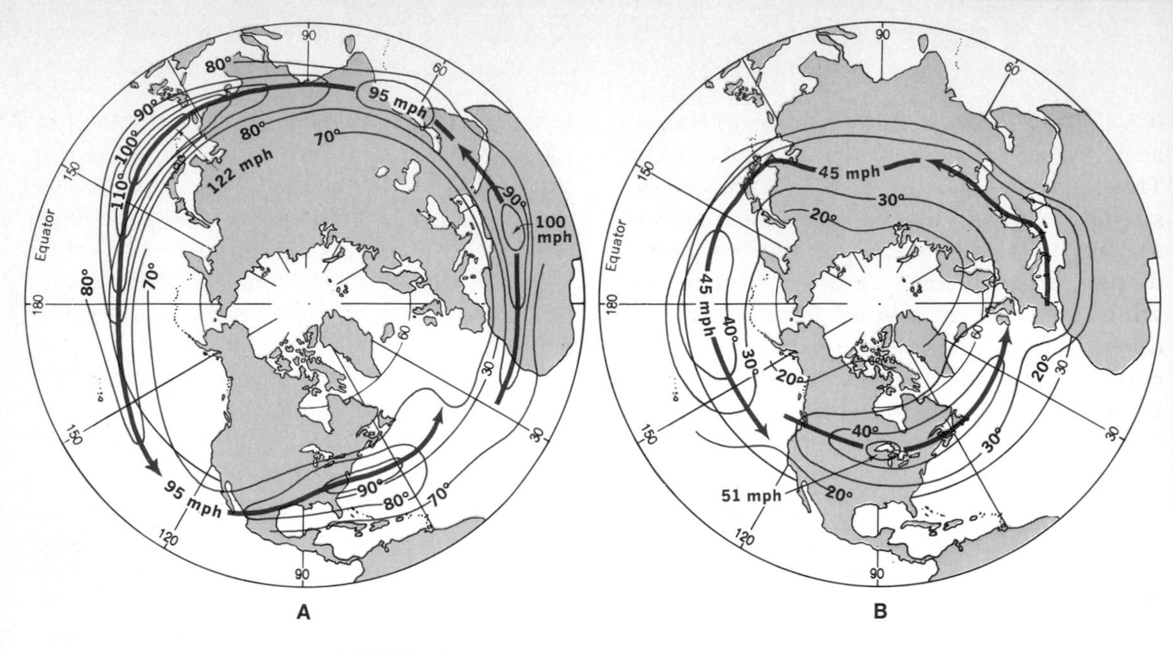

FIGURE 9–11

Circumpolar vortex circulations in the Northern Hemisphere (with wind speeds shown in miles per hour). A. The position of the *expanded* CVC in January. B. The position of the *contracted* CVC in July. (After Namias and Clapp)

the westerlies (compare Figures 9–7 and 9–8). Since the trades generally blow off the land toward the Atlantic, they transport little moisture and result in summer drought. The westerlies, flowing in off the Atlantic, carry both the moisture and the mechanism —cyclonic storms—that combine to make the winters rainy.

Jet Streams

To complete an understanding of the world pattern of atmospheric circulation, it is necessary to consider the atmospheric movements that take place somewhat above the earth—at least in those levels of the troposphere with which humankind is concerned—as well as those at the surface. In addition to the poleward and equatorward movements of air, there are various shifting, high-velocity wind flows that generally move from west to east at high speeds at or near the tropopause (the narrow zone between the troposphere and the stratosphere).

During the Second World War, aircraft flying in the upper troposphere over the Asia-Pacific area, at 35,000 to 40,000 feet, frequently encountered unusually high-velocity winds often approximating their own air speeds. In view of the intensity of these winds, they were named *jet streams,* and have since come to be vitally significant in efforts by humans to increase their understanding of atmospheric dynamics.

It was first suggested that these newly discovered jet streams constituted the linear cores of broader upper-air flows of the westerlies, centering around the North Pole. Continuing studies have provided certain evidence that these flows do prevail—from west to east—and that they do have a jet-stream "core," with wind velocities of up to 250 miles per hour. These high-altitude, high-velocity air flows are now known to be part of the *circumpolar vortex circulation*

(CVC), and they prevail in both Northern and Southern hemispheric circulations. These circumpolar circulations exhibit seasonal migrations of latitudinal expansion and contraction, with the westerlies being disposed to expand equatorward during the winter (Fig. 9–11A) and to contract poleward during the summer (Fig. 9–11B). The actual correlative relationship, however, is yet to be rationally explained.

The main jet streams have two branches: the polar offshoot, or branch, is the *subpolar jet,* and the tropical offshoot is the *subtropical jet.* It is relatively well accepted that the subtropical jet in the Northern Hemisphere has a strong correlation with the southwest monsoon of India. The height of the CVC varies considerably, but it usually persists at high altitudes, within the troposphere or the tropopause, or even within the lower portion of the stratosphere (refer to Figure 7–9 for a profile of the tropopause).

FIGURE 9–12
The meandering cycle of the upper westerlies circulation in the Northern Hemisphere. A. The beginning of the formation of the jet stream, (J). The first of the jet stream waves are forming. C. The waves are more strongly developed. D. The cells or "pools," of cold air (C) and warm air (W) are well formed.

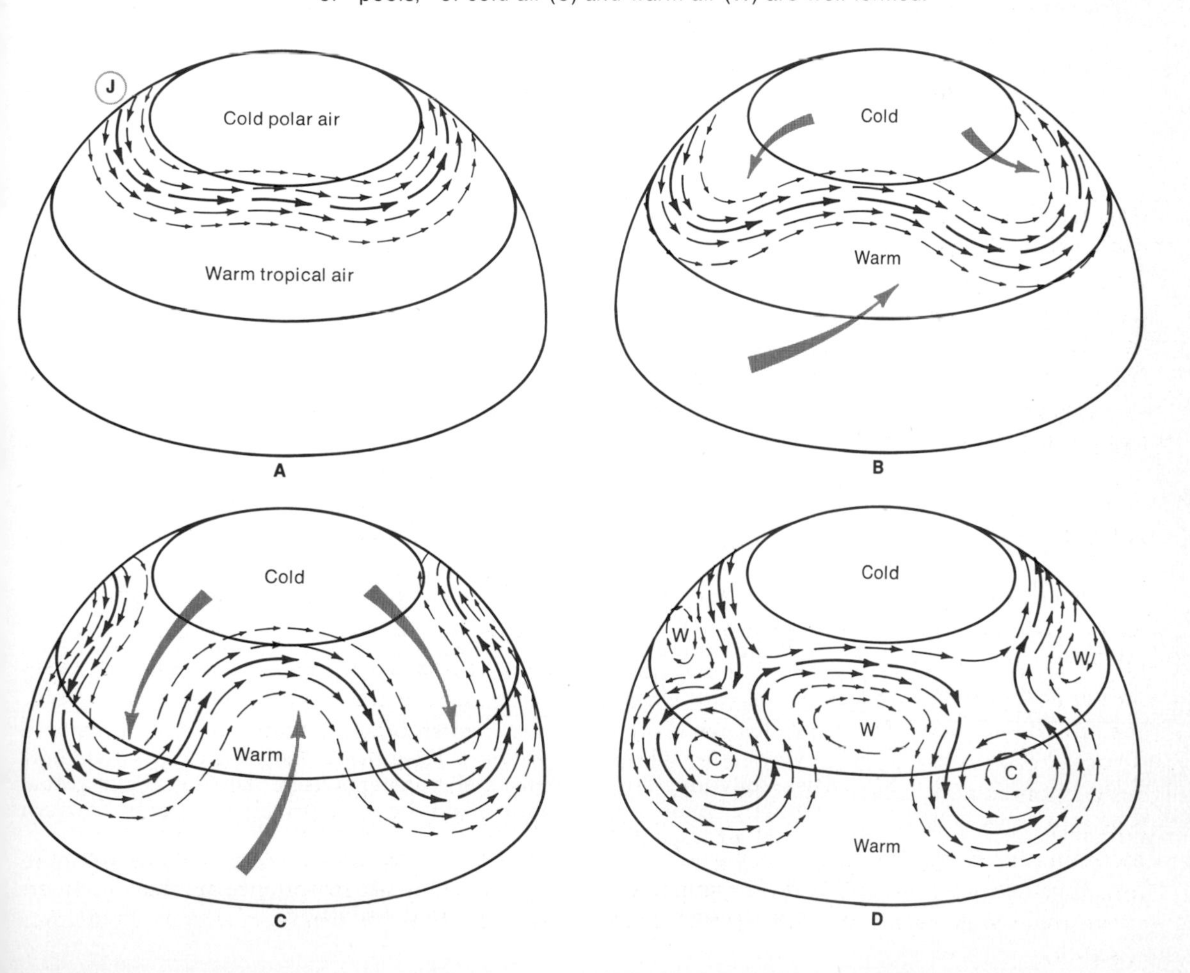

It has also been determined that the CVC tends to follow meandering, or *serpentine,* patterns around the poles and, therefore, it tends to introduce more tropical air into poleward locations and more polar air into equatorward locations than would otherwise develop. The meandering normally takes the form of slight undulations, which usually leave the tropical air and the polar air untouched in their respective geographical and thermal locations (Fig. 9–12A). However, as the meanders deepen and begin to take on stronger wavelike forms (Fig. 9–12B and C), alternate "pools," or "tongues," of polar air form intrusions equatorward, and pools of warmer tropical air form intrusions poleward (Fig. 9–12D).

If, then, one were to visualize the effect of the CVC three-dimensionally, the air near the polar surface would be clearly overlain in the stratospheric levels by intrusions of warmer tropical air. Similarly, the air close to the tropical surface would be overlain by intrusions of colder polar air.

Gradually, over a period of several days or weeks, the heat balance is restored; the pools of cold and warm air, left stagnant, become mixed, and there is gradual reversion to the regular, simple undulation pattern of the CVC. "Pool dispositions" as a result of meanderings may develop in cycles lasting several days, or even weeks. A total explanation for the existence of the CVC and the definite role played by the jet streams in the general circulation of the atmosphere are yet to be worked out in full and definitive detail: but the effects of these phenomena are considerable.

REVIEW AND DISCUSSION

1. How does atmospheric pressure play an important role in the development of weather and climate? Is it as apparent in everyday life as the other major climatic elements? Explain.
2. What is an area of high pressure; an area of low pressure; a pressure gradient?
3. Does atmospheric pressure decrease at a constant rate with increased altitude? What does atmospheric pressure actually mean? How is it measured?
4. Horizontal differentials in the atmospheric pressures of adjacent regions are common occurrences. Basically, what usually causes these pressure differentials to develop?
5. How are horizontal pressure differentials represented on a map? What exactly is an isobar? Are there close relationships between the patterns on isobaric maps and those on mean monthly temperature maps? What are they, and why do they exist?
6. Discuss the world pressure belts—what do they represent; where do they exist; and why do they form where they do? Name and explain the five general global belts of low and high pressure.
7. Technically, what is a wind? Distinguish between winds and air currents. In what ways must winds be considered essential to the physical environment?
8. Discuss fully the three-dimensional transfer system known as the global atmospheric circulation system. Explain the three principal forces that govern horizontal air transfer (wind).
9. Direction and velocity are two principal measurements in wind observation. What do these measurements mean? Of what importance are they in understanding weather and climate types and distributions?

10. Map and explain the major patterns of the world wind circulation system. Are the winds in the Northern and Southern hemispheres similar? Why?
11. In what way does the Coriolis force exert a deflective influence on all winds? Is the effect the same in both hemispheres? Why? Does the Coriolis force affect the hydrosphere as well as the atmosphere?
12. Compare and contrast the three major wind belts. Identify the doldrums (ITCZ) and the horse latitudes. Where are they located? Why are they both important? How?
13. Discuss oceanic wind circulation as a deviation from the simple idealized belted arrangement of world wind circulation (see Figures 9–8 and 9–9).
14. Explain the atmospheric conditions under which a monsoon develops. How does a monsoon affect the general world wind belts?
15. Suggest the reasons for, and the major results of, the seasonal migrations of the world wind belts.
16. What are the jet streams? Where and how do they form? How do they operate? In what ways are they of major importance in weather and climate development? in weather and climate study?
17. Mention as many ways as you can in which a knowledge of the world atmospheric pressure system and the world wind circulation system contributes to an understanding of the world's climatic types and their world patterns.

SELECTED REFERENCES

James, P. E. "A New Concept of Atmospheric Circulation." *Journal of Geography*, Vol. 63 (September 1964), 245–50.

MacDonald, J. E. "The Coriolis Effect." *Scientific American*, Vol. 186 (May 1952), 72–78.

Namias, J. "The Jet Stream." *Scientific American*, Vol. 187 (October 1952), 27–31.

Palmén, E., and C. W. Newton. *Atmospheric Circulation Systems: Their Structure and Physical Interpretation*. New York: Academic Press, 1969.

Petterssen, S. *Introduction to Meteorology*, 3rd ed. New York: McGraw-Hill, 1969.

Ramage, C. S. *Monsoon Meteorology*. New York: Academic Press, 1970.

Reiter, E. R. *Jet Streams*. Garden City, N.Y.: Doubleday, 1967.

Riehl, H. *Introduction to the Atmosphere*, 2nd ed. New York: McGraw-Hill, 1972.

Wexler, H. "The Circulation of the Atmosphere." *Scientific American*, Vol. 193 (September 1955), 114–25.

thermodynamic conditions, it is condensed and precipitated upon the land surface. Of the moisture thus supplied to the land, some runs off directly into lakes and oceans; some is intercepted by plants and animals; some sinks into the ground to become soil moisture and groundwater; and some is carried back into the atmosphere through evaporation. In short, the moisture that falls on the land supplies the major water needs of plants and all living animals, and it provides the waters of lakes and streams as well. The moisture that, in turn, is evaporated from lakes and streams and from vegetation and soils, may again be precipitated over the lands nearby, or it may be carried over great distances by the winds and fall over the oceans. In that case, the interchange of moisture between the oceans and the landmasses are affected even more directly than they would be otherwise (for a description of the *hydrologic cycle,* see Figure 3–2).

Atmospheric Humidity

Absolute Humidity

The measurement of the actual amount, or weight, of water vapor in the atmosphere is referred to as *absolute humidity.* It should be noted that although absolute humidity is expressed as an absolute quantity—the amount, or weight, of water vapor per unit volume of air—it varies considerably if the given air mass expands or contracts. For example, if one cubic foot volume of air containing twenty grains of water vapor were to expand in volume to two cubic feet, then, since the new dimension of that specific air mass would still contain the absolute amount of twenty grains of water vapor, the absolute humidity per cubic foot would change to ten grains.

At any given temperature, the atmosphere has a certain maximum *water-vapor-holding capacity,* as shown in Figure 10–1. If the atmosphere contains as much water vapor as is possible at its given temperature, the air is said to be *saturated.* The capacity of air to hold water vapor increases with any increase in its temperature; moreover, this holding capacity becomes greater at an increasing rate as the temperature rises. Thus, at higher temperatures, air requires greater amounts of water vapor to be considered saturated. It should also be noted that, for every arithmetic-interval increase in temperature, the corresponding holding capacity increases more nearly in a geometrical progression. For example, the holding capacity

FIGURE 10–1

Absolute humidity values for saturated air. (Adapted from meteorological tables of the Smithsonian Institution.)

Temperature (in ° F.)	*Maximum water-vapor-holding capacity (in grains per cu. ft.)*	*Capacity variations (in successive intervals of 10° F.)*
0	0.479	
10	0.780	0.301
20	1.244	0.464
30	1.942	0.698
40	2.863	0.921
50	4.108	1.245
60	5.800	1.692
70	8.066	2.266
80	11.056	2.990
90	14.951	3.895
100	19.966	5.015

increase from 0° to 10° F. is only 0.301 grains per cubic foot, but the increase from 90° to 100° F. is 5.015 grains per cubic foot.

In general, absolute humidity is highest over humid equatorial areas and over the oceans; it decreases toward the poles and inland over the landmasses. It must be recognized, however, that the hot atmosphere over the Sahara, or any other arid region, can have an absolute humidity that is higher than the cooler atmosphere over the British Isles or other humid regions. Absolute humidities in summer are higher than those in winter for the same basic reason: more moisture can be held at high summer temperatures than at low winter ones.

Relative Humidity

Even more significant geographically than absolute humidity in the understanding of weather dynamics is the concept of *relative humidity*. Relative humidity (RH) is the ratio of the amount of water vapor actually present in a unit of air to the maximum amount that unit can hold at the same temperature. This ratio is always expressed as a percent: saturated air is said to have a relative humidity of 100 percent; half-saturated air has a relative humidity of 50 percent.

Figure 10–1 shows that air at a temperature of 70° F. has a total water-vapor-holding capacity of 8.066 grains per cubic foot. If, however, that same air actually contained only 4.033 grains, it would be only half-saturated, and the ratio would be 4.033:8.066, or one-half (50 percent) relative humidity. If it were now possible to either increase the amount of water vapor from 4.033 to 8.066 grains by actually introducing water vapor into the air, or to decrease the temperature of the air to about 50°F. by cooling, for example, then the relative humidity would become 100 percent. That is, the air would now be saturated, since its absolute humidity would equal its maximum water-vapor-holding capacity. The formula for determining relative humidity may be expressed thus:

$$\text{RH } (\%) = \frac{\text{absolute humidity}}{\text{water-vapor-holding capacity}} \times 100$$

If, for example, a given air sample has an absolute humidity of 2.5 grains per cubic foot and, at the same temperature, can hold up to 10.0 grains per cubic foot, then its relative humidity is figured as:

$$\text{RH} = \frac{2.5}{10.0} \times 100 = .25 \times 100 = 25\%$$

When air is described as "dry," the implication is that it does not fulfill its maximum water-vapor-holding capacity. By this criterion, the air sample in the example above would be considered very dry air.

Condensation

When air is cooled, its ability to hold water vapor is reduced, or, in other words, its relative humidity is increased. Continued cooling will eventually bring the air to 100 percent relative humidity (saturation). If the cooling continues beyond this point, the amount of water vapor in excess of that which the air can hold at the new temperature forms minute particles of water (if the temperature is above 32° F.) or ice (if it is below 32° F.). If the moisture changes from a water vapor (gas) to a liquid (water), the process is called *condensation;* if the moisture changes directly from a water vapor (gas) to a solid (ice), it is called *sublimation.*

Dew Point

The temperature at which relative humidity becomes 100 percent, and saturation is achieved, is referred to as the *dew point.* It is often reached on clear, cloudless nights, when the rapid loss of terrestrial radiation to the atmosphere may cool the air to temperatures at which its absolute humidity value corresponds with its saturation value. If further cooling ensues, some of the vapor has to be removed from the air, either by condensation on some cooler object with which the air is in contact or by precipitation. If the dew point is above 32° F., the condensation appears as tiny droplets of water called *dew* (Fig. 10–2). Dew is often erroneously thought of as having fallen, but actually, it forms as a result of *static cooling,* in an *in situ* condition. If the dew point is below 32° F., moisture removed from the air appears as minute ice crystals known as *rime, hoar frost,* or *white frost.* Both dew and white frost collect on cool objects on the surface of the earth. The term *frost* is also used for below-freezing conditions not associated with condensation. For example, *black frost,* or *killing frost,* is a condition of *dry freeze* in which vegetation is damaged and turns black, but in which the air does not reach saturation; *ground frost* is a condition injurious to vegetation in which the temperature just above ground level (grass temperature) is 30.4° F. or below.

FIGURE 10–2

Grape leaves covered with *dew.* (Ralph J. Donahue, National Audubon Society)

Another common form of condensation is *fog.* Fog is a mass of fine water droplets suspended in the air close to the earth's surface; in a sense, it is a cloud near the ground. Although the water particles are condensed, they are too light to fall to the earth; rather, they float in the air and are moved about by whatever light winds occur. The formation of fog by condensation is effected by the lowering of the temperature below the dew point or by the addition of water vapor so that the dew point is reached. The cooling processes that lead to the development of fog are radiation, conduction, and the mixing of warm and cold air.

Another term used in respect to humidity is *specific humidity,* which represents the weight of water vapor per unit weight of moist air (that is, saturated air) and is usually expressed as grains of water vapor per pound of air or as grams of water vapor per kilogram of air. This property of the air is most distinctive in that it is not influenced by volume or temperature variations (as are absolute humidity and relative humidity). The weight of water vapor in a given weight of air remains the same regardless of any change in either the temperature or the volume of the air.

Condensation and Clouds

Essentially, clouds are the visible expression of condensation, for their bases mark the levels above which condensation is occurring in the atmosphere. The saturation of the air with water molecules, it is now known, does not automatically lead to the condensation process and to the formation of clouds. Studies clearly indicate that, in addition to high relative humidity, condensation requires the presence of nuclei in the atmosphere, such as particles of dust, soot, salt, and other impurities, around which the water droplets may condense. These nuclei are specifically termed *hygroscopic nuclei* to indicate that they are constituted of matter that has an affinity for water. In the absence of such hygroscopic nuclei (for example, in "pure air" obtained by a filtering process in the laboratory), it is possible to achieve a situation in which relative humidity values reach well over 100 percent. This atmospheric condition is called *supersaturation.*

As any body of air rises, it cools steadily, at the adiabatic rate. However, when condensation begins, each molecule of water vapor releases a unit of latent heat, thereby heating the air internally. As this happens, the normal *dry adiabatic rate,* operative in clear, dry air, is superseded by a much lower *wet adiabatic rate,* within a cloud. The altitude to which the air must rise before its internal temperature is as low as that of the surrounding air is thus greatly increased, and the speed of its vertical rise is greatly accelerated. This entire condition, termed *convective instability,* is the cause of the rapid vertical growth of various types of clouds, as, for example, transformation of cumulus clouds into towering cumulonimbi, and the development of vertical cumulonimbus patterns on the upper parts of clouds formed by the rise of air over the windward sides of mountains. Such a cloud grows by convection until it reaches a level where its top begins to flatten, and it thus becomes *anvil shaped* (Figs. 10–3 and 10–4). Figure 10–3 portrays three basic stages in cloud growth: the simple cumulus stage, produced by essentially updraft turbulence; the cumulonimbus stage, in which both updraft and downdraft are evidenced; and the cumulonimbus stage in which the anvil shaped top has developed.

Cloud Types

Although they exist in an almost infinite number of forms and are constantly changing, clouds are good indicators of the state of the atmosphere. With increased human use of the lower levels of the atmosphere, an understanding of their nature and function has taken on great significance. Basically, there are three major cloud forms —*cirrus, stratus,* and *cumulus* (Fig. 10–5).

Cirrus clouds (Latin: *cirrus* = a lock of hair) are the filmy or feathery, wispy clouds that appear at great altitudes. Usually composed entirely of ice crystals, when scattered irregularly over the sky—where, in the sunlight, they appear dazzlingly white against a deep blue—they are fair weather clouds. When they are arranged in regular bands and appear to be heavier, they usually presage the coming of stormy weather.

Stratus clouds (Latin: *stratus* = a spreading out) are sheetlike clouds that have almost no individual shapes and are usually dull gray in color; they cover the sky in a uniform layer. As typical winter clouds of the middle latitudes, they often remain as a leaden-gray cover over the whole sky for days at a time. They are largely low- to middle-altitude clouds and are remarkably uniform in thickness: neither the base nor the top of the cloud layer shows much unevenness.

Cumulus clouds (Latin: *cumulus* = a heap, a piling up) are massive, rounded clouds typical of middle latitude summer days. Their base is usually quite even and of low altitude, while their top is cauliflowerlike and may reach even greater heights than those typical of cirrus clouds (Fig. 10–4). The cumulus cloud is usually a fair weather form; but, when on a hot, sultry, summer afternoon, it grows to a great height and sends up great bulbous projections or flattens off at its top and bottom to assume the shape of an anvil, it may well indicate sharp showers, squalls, or even severe thunderstorms. When it appears in this form, the cumulus is commonly referred to as a "thunderhead" (Fig. 10–4).

There is an international classification of clouds that recognizes ten distinctive combination forms divided into four major families (Figs. 10–3, 10–4, and 10–5). Each of these cloud forms gives a definite indication of the condition of the atmosphere. For example, the cirrocumulus and cirrostratus forms typically occur ahead of a storm center and denote its approach. Nimbostratus is associated with steady rain or snow; it produces low ceilings, poor visibility, and ice conditions. Cumulonimbus, the thunderhead, or overgrown cumulus, is associated with violent vertical air movements that often exceed 100 miles per hour within the cloud; it produces tremendous downpours of rain with sharp, brilliant lightning, heavy thunder, and oftentimes, brief but destructive hailstorms.

FIGURE 10–3

Examples of Family D clouds, which extend from the 1,600-foot level to the 50,000-foot level. A. The cumulus stage. B. The cumulonimbus stage. C. The cumulonimbus with anvil head stage.

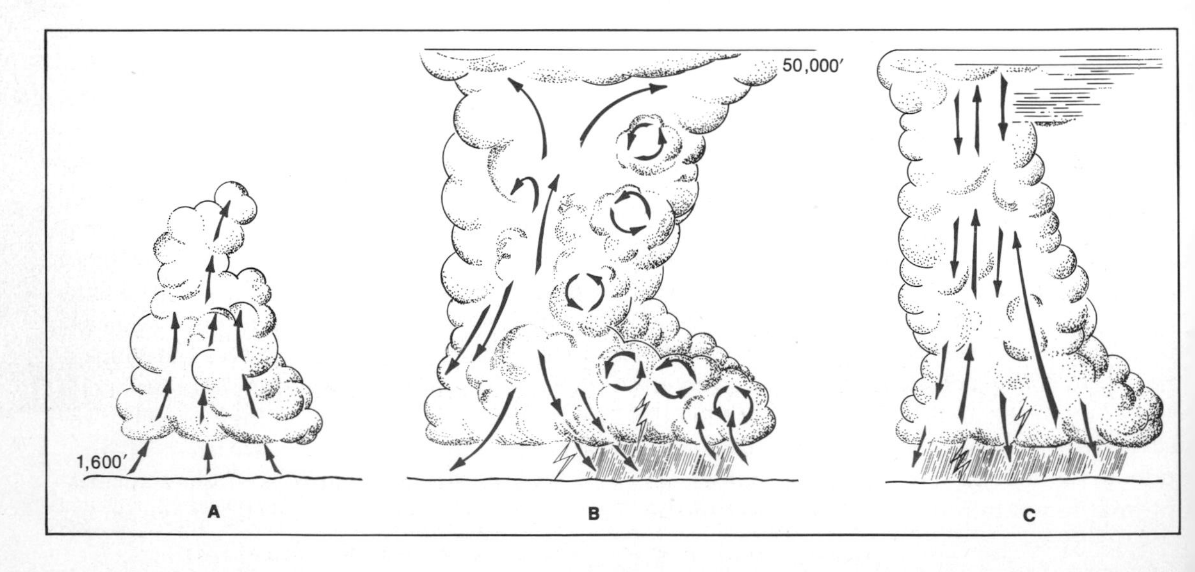

FIGURE 10–4

Vertical developments of the cloud types in family D. Above, fair-weather cumulus—each puffy cumulus cloud indicates the top of a single convection current. (Standard Oil Co., N.J.) Below, this anvil-shaped cloud, a variety of cumulonimbus, is the typical "thunderhead." (E. Fontseré, "Atlas Elementaire des Nuages")

FIGURE 10–5

Horizontal developments of some cloud types in families A, B, and C. Top left, cirrus in parallel trails (with ice particles). (NOAA) Top right, cirrocumulus (with ice particles). (NOAA) Center left, altostratus with cumulus below (with ice particles). (National Film Board of Canada) Center right, altocumulus. (NOAA) Bottom left, stratus. (Library of Congress) Bottom right, stratocumulus from above. (BOAC)

Precipitation Forms and Formation Processes

The term *precipitation* applies to the forms of moisture that issue from the clouds and fall to the ground—such forms as rain, snow, hail, and sleet. The condensation process does not lead directly to precipitation. Indeed, the means by which precipitation is achieved have long been a controversial issue, laboratory conditions having shown that there is not just one development process but rather a number of possible processes. The fact that most clouds do not produce precipitation is still not completely understood: but through recent research and attempts to stimulate artificial precipitation, the complexity of rain formation is becoming more evident.

The process of raindrop growth is vital to precipitation. Raindrops must be great enough in size and weight to counteract the atmosphere's strong turbulent updraft and

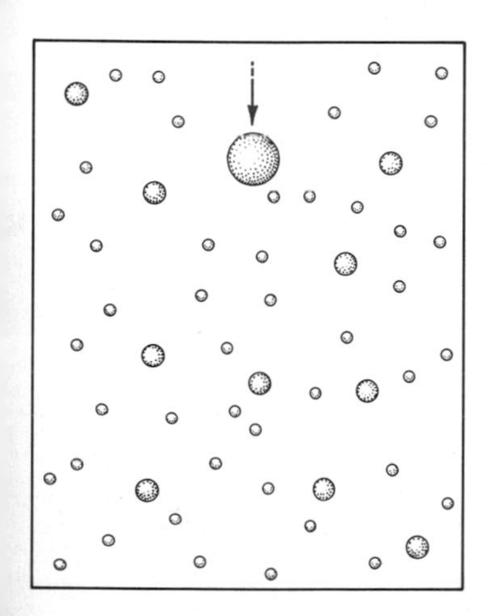

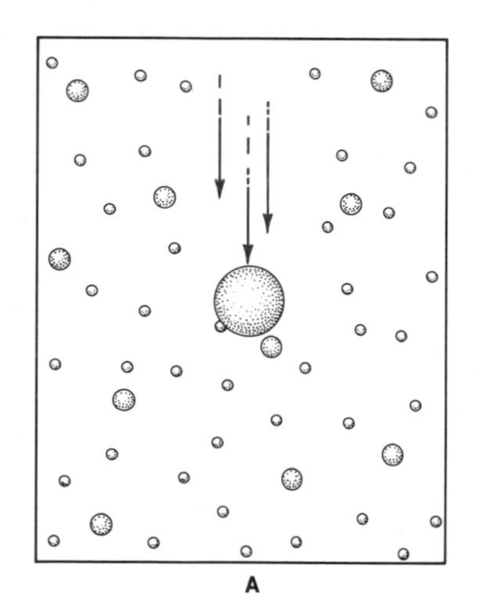

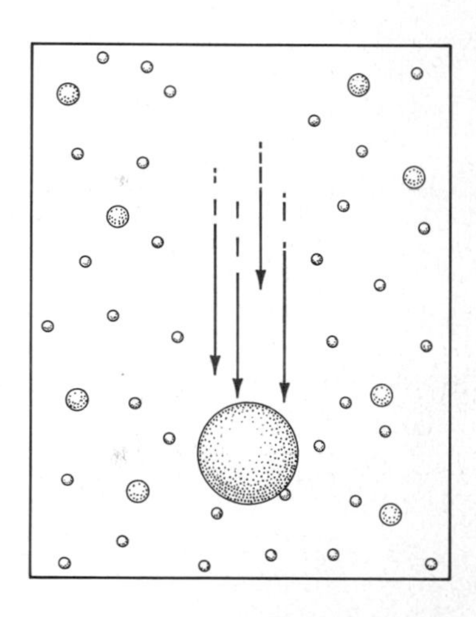

A

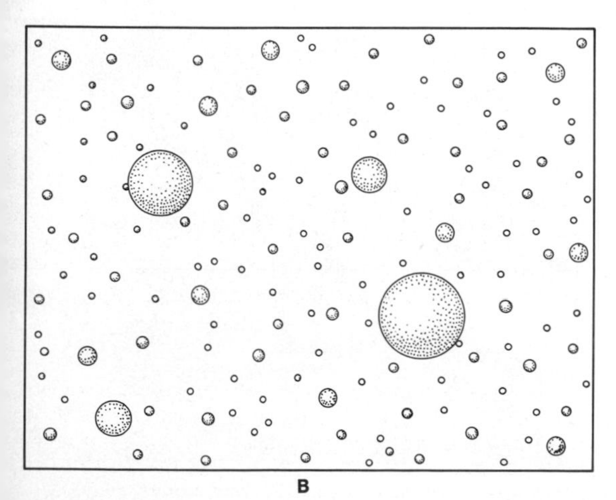

B

FIGURE 10–6

The coalescence of raindrops of different temperatures. Through condensation, the large raindrops continue to acquire moisture evaporated from the warmer, smaller raindrops. A. Coalescence by collisions as a raindrop falls through a cloud. B. Coalescence by evaporation.

fall eventually to earth. If condensation were the only means of raindrop growth, it is estimated that it would take several days for a raindrop to develop to a size sufficient to allow it to reach the earth. Among the suggested mechanisms of raindrop growth, three are generally recognized: the coalescence of drops as a result of the constant collisions they experience in turbulence (Fig. 10–6A); the coalescence of drops of different temperatures (colder drops from the upper part of the cloud and warmer drops from the lower part), whereby moisture evaporated from the warmer and larger-surfaced drops condenses upon the smaller and colder drops (Fig. 10–6B); and the coalescence of drops with opposite charges (which usually develops when the drops divide). But, a cloud may be affected by each of these mechanisms and may still produce no precipitation. Confronting this situation, scientists have found that the introduction of ice particles into such a cloud immediately results in precipitation. As the ice crystals fall through the cloud, they grow larger through the condensation of water droplets upon them and eventually become heavy enough to counteract the atmospheric turbulence and reach the ground. If temperatures in the lower atmospheric layer are sufficiently low, the ice crystals fall as snowflakes; otherwise, they melt while falling through that layer and reach the earth as rain.

Though much surrounding their production is uncertain, raindrops are undoubtedly the source of other forms of precipitation. When caught in an updraft and carried into the freezing cloud top, raindrops become frozen and then, falling back through the relatively warmer lower layers of air, they are slightly melted and accumulate additional moisture. If they are caught in an updraft again, they become frozen again, and this process of acquiring layer after layer of ice may continue within the cloud until, eventually, the accumulation may be heavy enough to fall through to the ground as *hail*. Another type of precipitation based on raindrops is *sleet*, formed when falling rain passes through a frigid layer of air and itself becomes frozen (Fig. 10–7). *Glaze* forms when raindrops fall upon objects and surfaces where temperatures are well below freezing; the water is instantly transformed into a sheet of ice.

FIGURE 10–7

Linemen in Illinois restringing electrical lines damaged during a sleet storm. The fallen branches tore the lines as they gave way under the weight of the sleet. (Gerard, Monkmeyer)

Convectional, Orographic, and Frontal Types

In terms of the kind of air movements that caused it, every instance of precipitation can be described as one of three basic types: *convectional, orographic,* or *frontal.*

Convectional precipitation is associated with strong vertical movements of air. Originally induced by extreme local heating of some portion of the earth's surface, such movements are perpetuated, and often intensified, by the release of latent heat within the rising column of air during condensation. The result of these movements is a cumulonimbus cloud, then a thunderstorm, and often a downpour of rain and/or hail.

Orographic precipitation (Greek: *oros* = a mountain) occurs when moist air is forced to rise over a barrier (usually a mountain) lying athwart its path. The rise in altitude causes the air to cool, which, in turn, results in the condensation of moisture, in cloud formation, and in precipitation. Once over the mountain crest, the air descends, is warmed adiabatically, and quickly reevaporates any cloud particles still present. Hence there is a marked contrast between the cloudy, well watered windward slope of a mountain and the clear, dry leeward side.

Frontal precipitation is produced when a relatively warm, moist mass of air is forced to rise over a mass of colder, heavier air. The resultant cooling and condensation produces precipitation that falls to earth through the lower air mass.

Thunderstorms and Tornados

Whenever masses of warm, moist air are forced to rise very rapidly and to very great heights, they are cooled rapidly and excessively. Water vapor in the cooled air is condensed to form massive vertically extensive clouds (Fig. 10–8) that give rise to lightning, thunder, heavy rainfall, and sometimes hail—in other words, a *thunderstorm.* The large-scale release of latent heat that takes place during rapid condensation warms the air to such an extent that its vertical rise is greatly accelerated. Vertical velocities of over 100 miles per hour are characteristic, and the currents and their accompanying clouds build to great heights—commonly 25,000 feet, and occasionally as much as 75,000 feet.

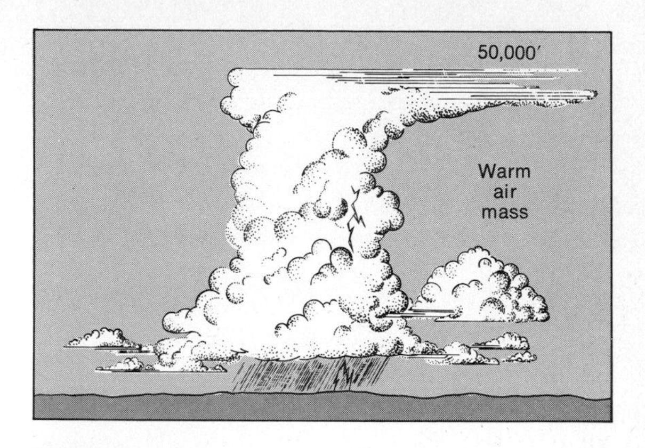

FIGURE 10–8
A profile of the cumulonimbus cloud of a thunderstorm.

The moisture that is condensed as the air rises forms raindrops that, growing steadily in size, are carried up in the vertically rising current. When that current suddenly ceases, or when it slants at an oblique angle or is diverted by a descending current, the raindrops are precipitated earthward. If, as often happens, their accumulated numbers are great, the resultant deluge is usually what is known as a *cloudburst.*

Basically, the thunder and lightning characteristic of the thunderstorm (Fig. 10–9) are induced by the existence of raindrops with electrical charges. A tremendous electrical discharge may develop between posi-

FIGURE 10–9

Brilliant, and often destructive, lightning accompanies the heavy downpour of rain in a thunderstorm. (Wide World Photos)

tively charged raindrops in the lower level of a cloud and the negatively charged earth surface or, between oppositely charged raindrops within the same cloud or adjacent clouds (Fig. 10–10). Although a thunderclap is generally heard on the earth after a lightning flash has been visible, the two phenomena originate simultaneously and are manifestations of the same electrical discharge. However, since sound travels at a slower speed than light, the thunderclap moves more slowly through the atmosphere than does the lightning flash and, in some instances, may seem to follow it by as much as several seconds.

Thunderstorms are seldom more than a few miles in diameter, but frequently, there are several following one another in rapid succession. They always occur within a rising mass of warm air, but the force that starts the upward movement of the air may be the result of a variety of factors. Occasionally, thunderstorms are convective—the result of local differences in atmospheric heating. At other times they are orographic—the result of upward movements of air on a mountain slope. At still other times they may result from the passage of air mass

FIGURE 10–10

A schematic representation of a typical convectional thunderstorm cloud, showing its anvil head (at the top) and the distribution of its positive and negative electrical charges. These charges induce the electrical discharges known as *lightning*. Lightning can take place between a cloud and the earth, between parts of a cloud, or between adjacent clouds.

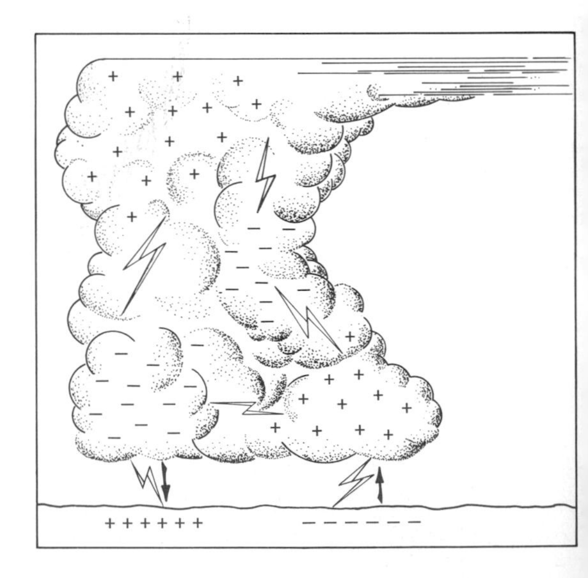

fronts. But, regardless of the generating force, or trigger action, the air that is displaced upward must be warm and moist if a thunderstorm is to develop.

The other important type of storm, usually occurring in continental interiors of the middle latitudes, and mostly in summer, is the *tornado*. Like the thunderstorm, it is generally small in scope: the tornado's diameter at the earth's surface varies from 10 to 50 yards. What occurs over this limited range is seldom insignificant, however. A rapidly spinning funnel shaped cloud reaches down from a very low, heavy cumulonimbus cloud and twists erratically along the ground surface (Fig. 10–11). This funnel shaped cloud represents the vertical core of an extreme low-pressure column of air that is spiraling inward at speeds of up to 300 miles per hour and updrafting at velocities of 100 to 200 miles per hour: the tornado is thus the most violent of all storms. The extreme destructiveness of tornados is limited to their immediate paths, however, since they fortunately have little force beyond their immediate funnel width. When such a storm occurs over water, it is known as a *waterspout,* as water is drawn into the vortex.

Cloud-Seeding

Attempts to produce rain artificially, through *cloud-seeding* (Fig. 10–12), have been numerous but not completely successful. The most effective mechanisms in this effort have been the injection or discharge of silver iodide or dry ice into groups of clouds. The complete success of cloud-seeding requires much more detailed study into the mechanism or, more probably, mechanisms of precipitation.

FIGURE 10–11

The development of a tornado near Gothenburg, Kansas. Note how the funnel cloud reaches down and eventually makes destructive contact with the ground. (U.S. Weather, ESSA)

FIGURE 10–12

A twenty-mile long pattern formed by the seeding of a stratus cloud with dry ice. Dropped into the cloud in this pattern from an airplane, the dry ice has caused supercooled water droplets to change into snow and ice crystals, which, in turn, have fallen out of the cloud or evaporated. (U.S. Army Photo)

Geographical Distribution of Precipitation

By the use of *isohyets* (Greek: *isos* = equal; *hyetos* = rain), or lines, connecting points of equal precipitation, maps are made that distinguish areas of heavy precipitation from areas of lesser precipitation. The basic world pattern of mean annual precipitation (Fig. 10–13) reveals that equatorial and tropical regions are the wettest areas in the world, while the polar regions and the subtropical deserts are the driest.

Equatorial and Tropical Regions

When air, rising in a broad belt along the equator, contains a large amount of water vapor, precipitation results. Since air at the high temperatures common to equatorial regions is capable of holding larger amounts of water vapor, and since the trades and monsoons of those regions flow over large expanses of warm ocean, the moisture that they carry produces more precipitation than anywhere else in the world.

Where onshore winds and rising air combine most effectively in this zone, tremendous amounts of rainfall occur. The highest mean annual rainfall in the world is recorded on the windward slopes of mountains in the trade wind belts. For example, the mean rainfall recorded over seven consecutive years for Mount Waialeale in the Hawaiian Islands is 476 inches per year; Cherrapunji, on the windward southern slope of Assam's Khasi Hills in India, due largely to Asia's summer monsoon, has recorded 366 inches in one month; Baguio on Luzon in the Philippine Islands has received 46 inches in one twenty-four hour period.

FIGURE 10–13

A generalized pattern of mean annual precipitation over the land and water bodies of the world. (After Meinardus and Schott)

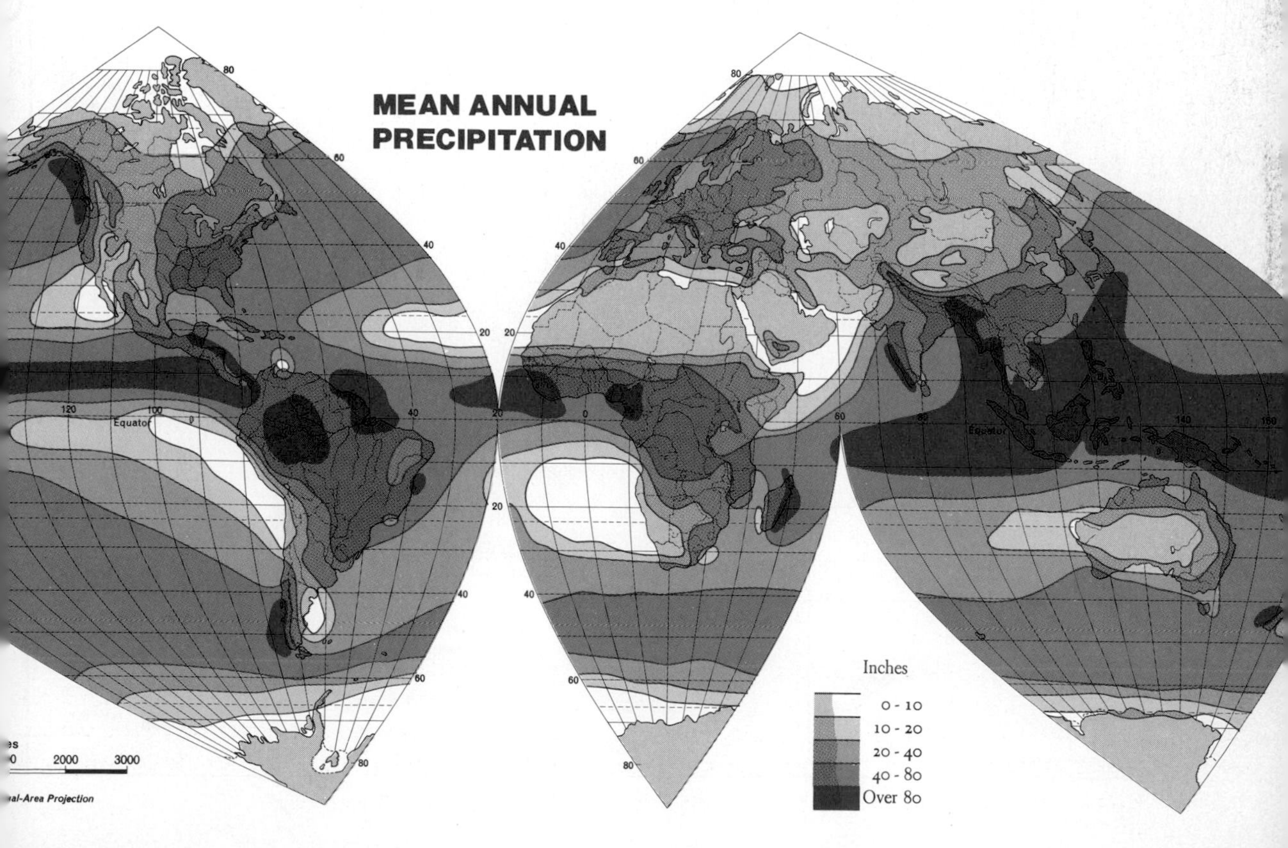

Low Latitude West Coast Regions

That only small amounts of precipitation occur along the west coasts of continents at low latitudes is a fact related to the existence of the subtropical high pressure belt. Winds in these latitudes strike the east coasts, while the west coast areas are regions of offshore winds. In addition, air in these latitudes is settling toward the surface. There is, therefore, no adequate source of moisture, and the mechanism necessary for condensation and precipitation is lacking. While there are no positive records that any one part of the earth's surface is absolutely free of rain, there do exist, for the low latitude west coast regions, several reports of rain gauges that have seldom received rain or snow: one such area is Iquique in northern Chile. In these and similar locations, as in the coastal desert of southern Peru, dew and fog provide about the only appreciable amounts of usable moisture. Usually, as throughout the Sahara, the precipitation is

FIGURE 10–14

The United States well illustrates the contrast between continental margins and their interiors with respect to mean annual precipitation.

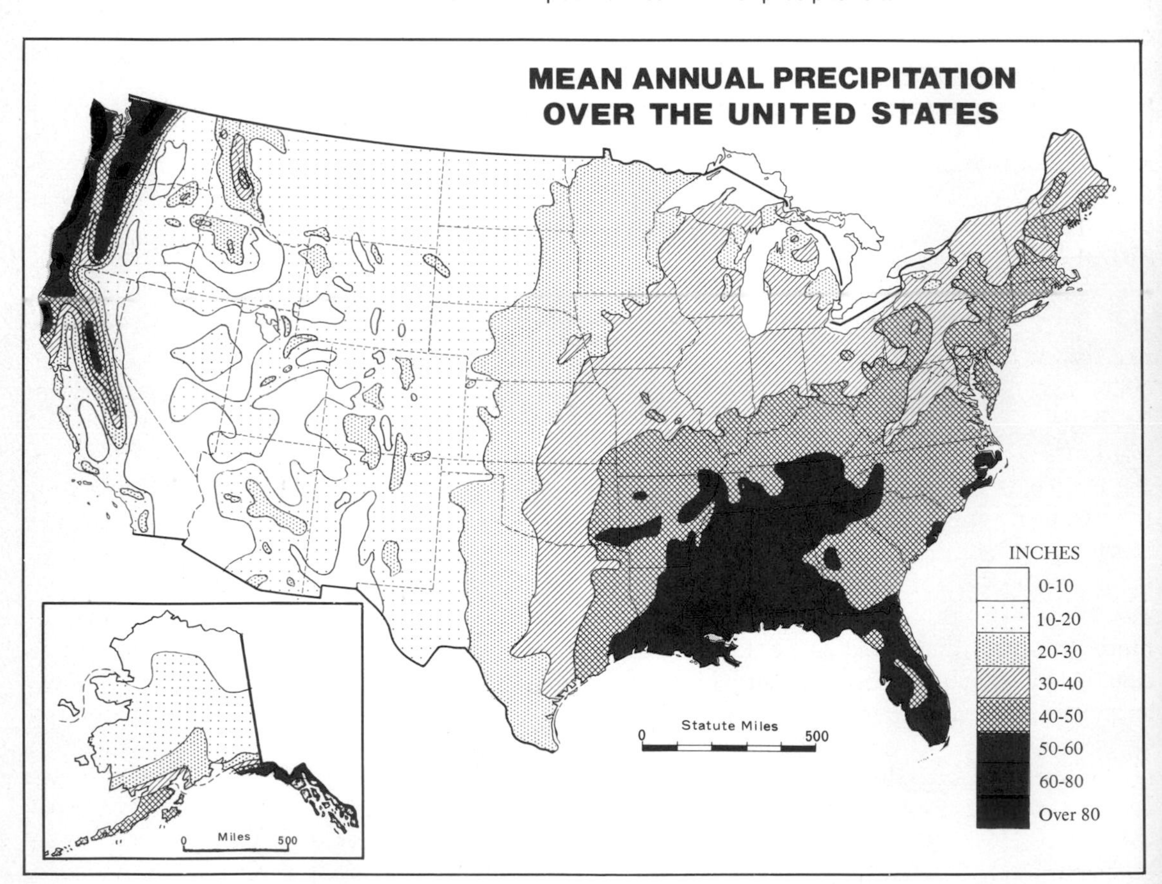

below ten inches per year, and what rainfall there is occurs at irregularly spaced intervals—months or years apart.

Middle Latitude Regions

Throughout the middle latitudes, a moderate amount of precipitation is common, though parts of the continental interiors are characteristically dry. This is the zone in which frontal activity is linked to the zone of interaction between the westerlies and the polar winds. Here, the amount of precipitation depends largely upon an area's distance from the major source of moisture—the oceans. Onshore winds carry much moisture toward the continents, and orographic, convectional, or frontal conditions bring about its precipitation. In either event, much of the precipitation occurs near the continental margins, while smaller amounts are generally received in the interiors (Fig. 10–13).

The map of mean annual precipitation over the United States (Fig. 10–14) illustrates the contrast between continental margins and interiors. There are large amounts of precipitation over the Pacific coast, especially coastal Oregon and Washington. Inland, the decrease in precipitation is very pronounced. In the western United States the control is in large part orographic. Along the Atlantic coast, precipitation is moderately heavy. Decrease toward the interior is again noticeable though, due to the absence of strong orographic features, it is more gradual than that near the Pacific coast. The Great Plains are relatively dry largely because rain-bearing winds from the Gulf of Mexico do not regularly penetrate northwestward into the continental interior, and those from the Pacific have lost most of their moist, marine characteristics before reaching that far inland.

High Latitude Regions

In the polar regions air settles toward the earth's surface, and, in addition, it contains little moisture. The ability of cold air to contain water vapor is very low, and the mechanics by which that small amount might be condensed and precipitated are usually lacking. Hence, the rainfall is light (Fig. 10–13). In this instance, the actual amount of precipitation is of little significance, for temperatures are always low. Few measurements have been made for this reason. Upernavik, Greenland, has an annual mean precipitation of only 9.2 inches, despite its marine position, and Point Barrow on Alaska's north coast has only 5.3 inches annually.

Seasonal Distribution of Precipitation

Along with the total yearly precipitation, its seasonal distribution is also important in any geographic discussion of climate. This distribution is represented by the isohyetal maps for January and July (Fig. 10–15). These maps indicate shifts of rainy and dry zones comparable to the shifts of the world wind belts. For example, almost all those parts of central Africa receiving over two inches of rain during January—when the zone of greatest heating and the intertropical front occupy their farthest southward position—lie south of the equator. In the opposite period of the year, in July—when the vertical strike of the sun as well as the intertropical front are well north of the equator—most of the area receiving over two inches of precipitation lies north of the equator. Similar shifts may be seen in the middle and high latitudes and, almost without exception, the precipitation shifts parallel the shifts of temperature and winds.

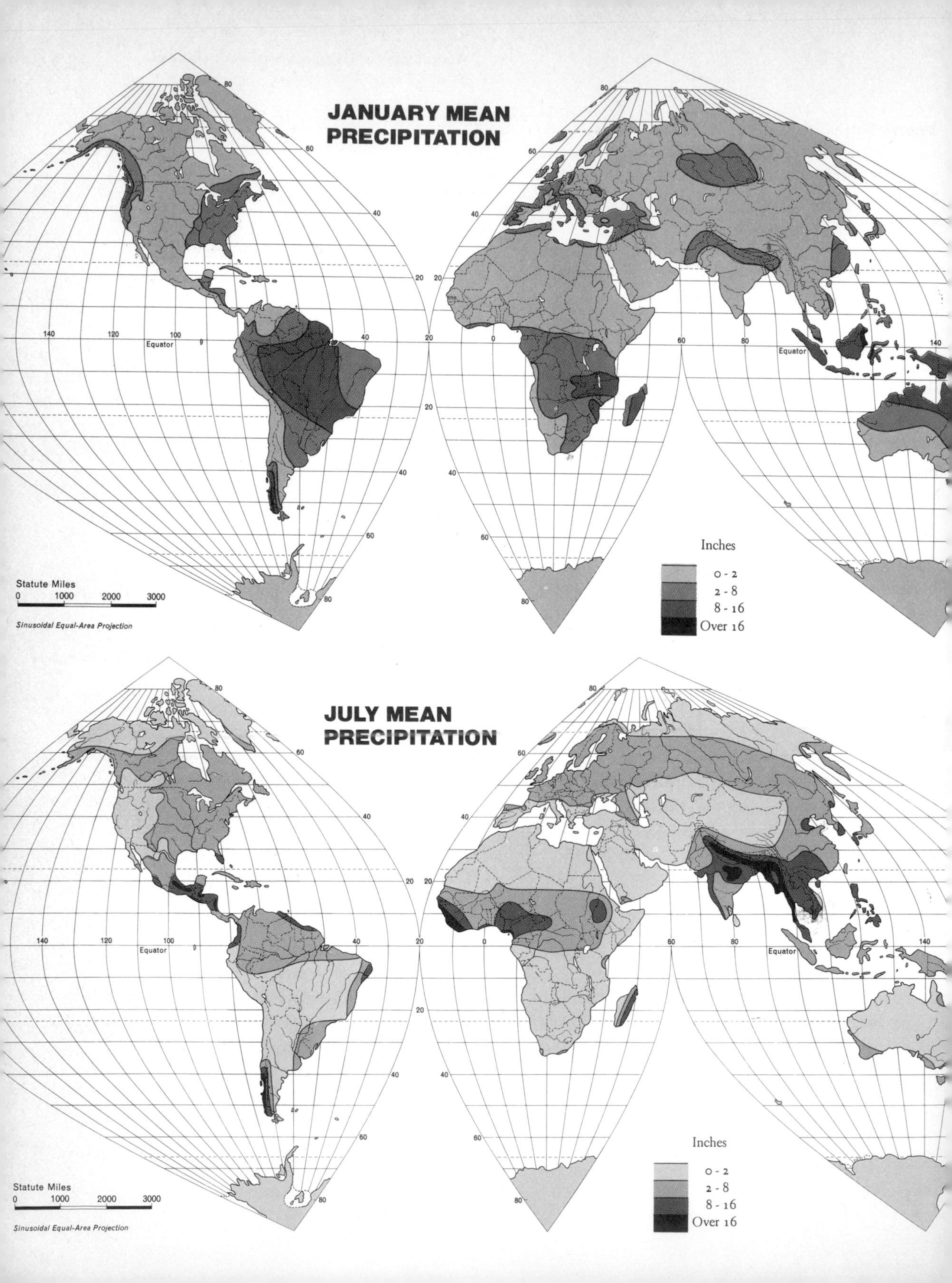

JANUARY MEAN PRECIPITATION
Equator
Statute Miles
0 1000 2000 3000
Sinusoidal Equal-Area Projection
Inches
0 - 2
2 - 8
8 - 16
Over 16
JULY MEAN PRECIPITATION
Equator
Statute Miles
0 1000 2000 3000
Sinusoidal Equal-Area Projection
Inches
0 - 2
2 - 8
8 - 16
Over 16

World Precipitation Variability

Despite the prevalent use of maps showing mean precipitation values for seasons or full years, it is well recognized that variation from the mean value is the rule rather than the exception. Thus, in the study of geography, it is appropriate to indicate the degree of precipitation variability from the mean as well.

It is found that regions of normally heavy precipitation are also areas of least variability (Fig. 10–16). Thus, the areas of the Amazon and Congo basins, the equatorial zone in general, and the Indonesian islands within southeastern Asia stand out as regions exhibiting the world's lowest precipitation variability, between 0 to 15 percent. On the other hand, areas of very high variability (over 40 percent) occur within the lowest rainfall areas of the world. These include the Sahara and Kalahari deserts in Africa, the Arabian, Thar, and Gobi deserts in Asia, the Chilean desert in South America, and the great interior deserts of Australia.

FIGURE 10–15
Mean precipitation over the land areas of the world in January and July.

FIGURE 10–16
The pattern of precipitation variability in the world demonstrates that greater variability is usually more common in areas of light precipitation than in areas of heavy precipitation.

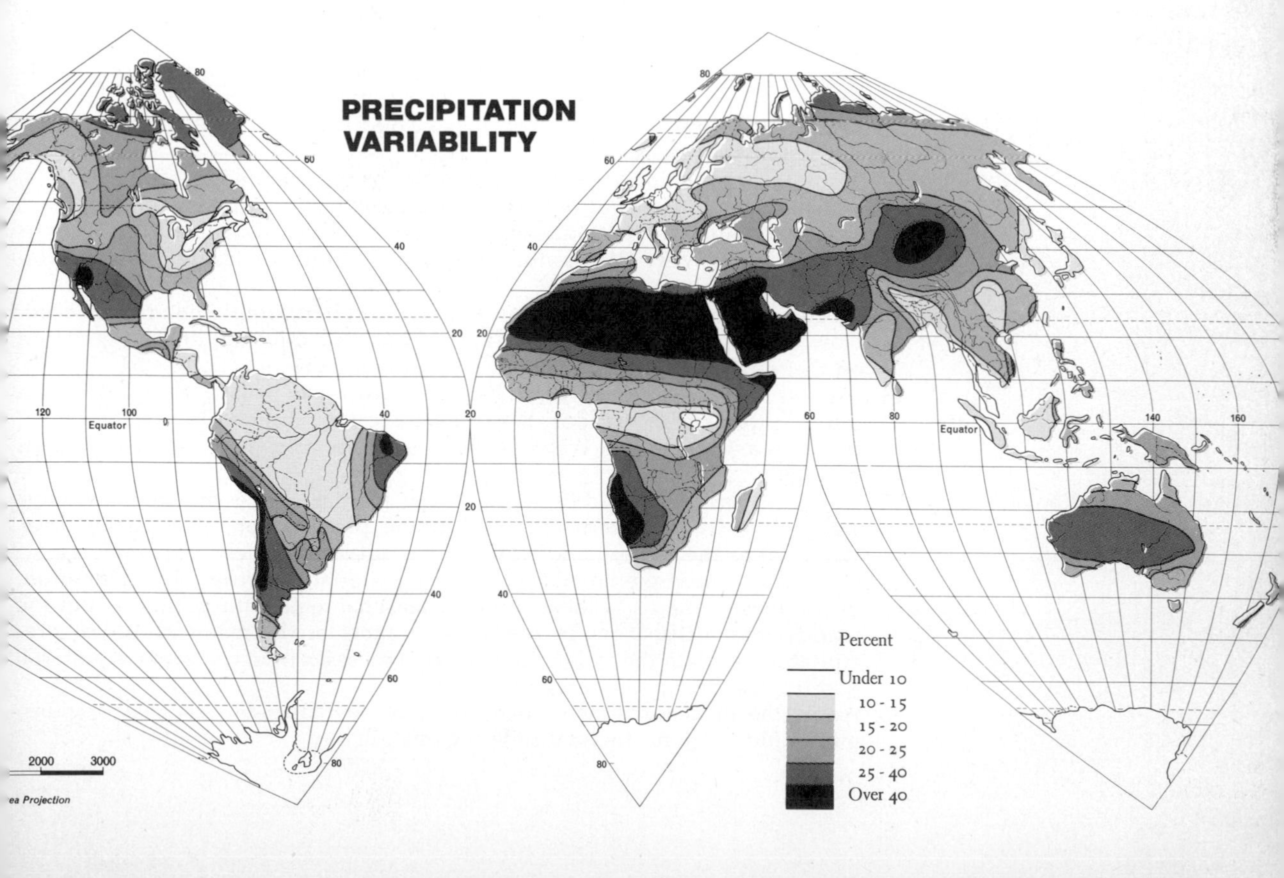

REVIEW AND DISCUSSION

1. At what distance from the earth is most of its atmospheric moisture concentrated? in what temperature zone? What two climatic elements are the principal forces in the "making" of weather and climate?
2. From what global source is the major part of the moisture present in the atmosphere as water vapor derived? Make a sketch of the hydrologic cycle and then turn to Figure 3–2 to check your drawing's accuracy and completeness.
3. Define absolute humidity. What is your understanding of the statement that "although absolute humidity is expressed as an absolute quantity . . . it varies considerably if the given air mass expands or contracts"? When is air saturated? Is the absolute humidity over the Sahara normally higher than that over the British Isles? Explain.
4. Discuss the meaning of relative humidity. Explain why relative humidity is of even greater significance in weather dynamics than absolute humidity. Why does the term "dry" air not imply a total lack of water vapor? What does partial pressure mean?
5. Define and discuss fully the condensation process. What is sublimation?
6. Describe each of the following, including the processes involved in its development: dew, rime, ground frost, and fog.
7. What are clouds? Discuss the process of cloud formation. How are hygroscopic nuclei related to this process? What, visually, marks the level above which condensation is occurring in the atmosphere? Are clouds present only in the troposphere?
8. How are clouds good indicators of the state of the atmosphere? Explain the "reading" of cloud forms with respect to short term weather predicting.
9. Although cloud forms are almost infinite, practically all clouds can be classified into three basic types. What are they? Describe them in several of their variations. Compare and contrast these cloud forms in terms of the weather conditions they presage.
10. Study the international classification of clouds reflected in Figures 10–3, 10–4, and 10–5. With what sort of weather is a nimbostratus cloud associated?
11. In a sequential sense, condensation should lead to precipitation. Does it always do so? Why? Discuss the subject of precipitation forms and formation processes as fully as you can.
12. Discuss the three basic types of situations in which precipitation is produced. How would orographic precipitation normally be mirrored in the vegetation on the windward and leeward sides of a mountain?
13. Discuss the atmospheric conditions and processes usually associated with thunderstorms and tornados. What are the causes of lightning and thunder? What is a waterspout? What is cloud-seeding, and how successful has it become?
14. How are precipitation differentials from place to place shown on a map? Within the world pattern of mean annual precipitation, what major areas stand out as the wettest, and which stand out as the driest? Explain.
15. On a map, locate the broad regions of equatorial and tropical precipitation. Relate these patterns to those of mean seasonal temperatures throughout the world. From discussions in previous chapters of the major causes of precipitation, explain the major atmospheric processes involved in the equatorial and tropical precipitation regions.
16. Apply the procedures mentioned above to: low latitude west coast regions; midlatitude regions; high latitude regions.

17. What are the major reasons and explanations for the differences in seasonal distribution of precipitation from place to place over the earth?
18. What is the meaning of world precipitation variability? How is it, along with measures of seasonal precipitation, of great importance to human beings?

SELECTED REFERENCES

Bach, W. *Air Pollution*. New York: McGraw-Hill, 1972.

Battan, L. J. *Cloud Physics and Cloud Seeding*. New York: Doubleday, 1962.

———. *The Unclean Sky*. New York: Doubleday, 1966.

Mason, B. J. *Clouds, Rain and Rainmaking*. Cambridge: Cambridge University Press, 1962.

Petterssen, S. *Introduction to Meteorology*, 3rd ed. New York: McGraw-Hill, 1969.

Riehl, H. *Introduction to the Atmosphere*, 2nd ed. New York: McGraw-Hill, 1972.

Thornthwaite, C. W., and J. R. Mather. *The Moisture Balance*. Princeton, N.J.: Princeton University Press, 1949.

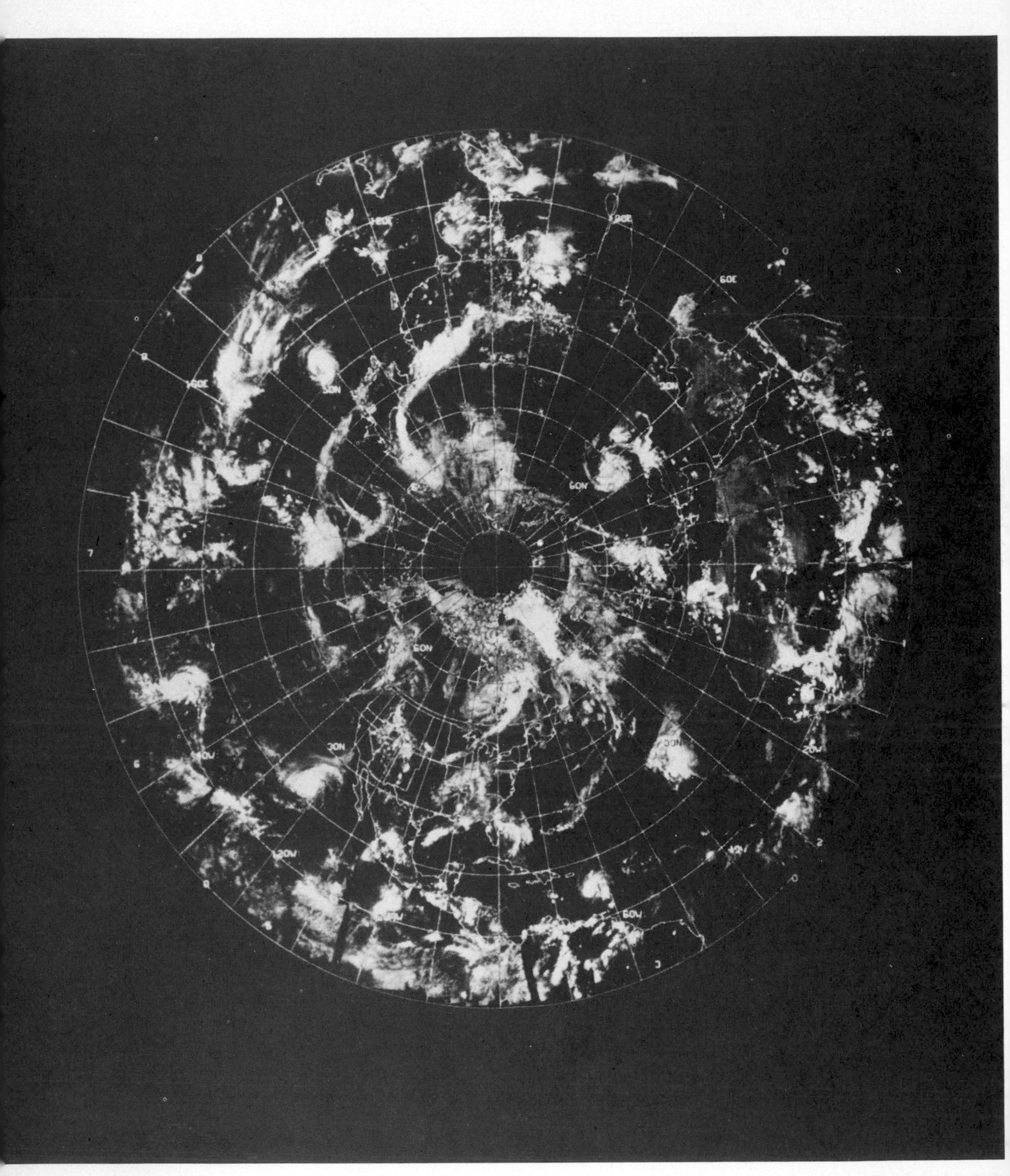

A mosaic of cloud-cover pictures taken by a weather satellite in which more than a dozen storms are visible. (NASA)

11

Air Masses, Fronts, and Lows and Highs: Weather

The dynamic conditions of the earth's atmosphere—the median of climate and weather—strongly influence the earth's total physical environment. That these conditions are so subject to change—locally and globally as well as daily and seasonally—is largely because they are the products of highly variable and unstable weather elements that are interrelated in the weather-making processes.

From the point of view of the physical geographer, the major weather systems at the earth's surface result, ultimately, from the interaction of many factors: most important, the unequal heating of the earth's surface from the equator poleward, the rotation of the earth, the relationship between the oceans and the atmosphere, the friction between the atmosphere and the earth's surface, the differences between land heating and water heating, the unequal geographical distribution of the oceans and landmasses, the earth's topography (landforms), the general circulation of the earth's atmosphere, and the global heat balance. More specifically, the world's weather results from such factors as: air masses, zones of wind convergence and divergence, the polar fronts, the intertropical fronts, the jet streams, and the systems of lows and highs. The occurrence and functions of these more specific factors, considered the "building blocks" of weather, will be the subject of the following discussion.

Air Masses

Sources and Properties

The atmospheric blanket that surrounds the earth is not simply one uniform mass of air—either horizontally or vertically. Quite to the contrary, it is an extremely variable complex of parcels, or cells, of air, each notably distinct in its nature. These cells are called *air masses,* and their two most apparent distinguishing characteristics are their differences in temperature and in moisture content.

More specifically, an *air mass* may be defined as a finite portion of the atmosphere, wherein the horizontal temperature and humidity properties are homogeneous but distinct from the surrounding air conditions. Thus, any sizable horizontal extent of the atmosphere reveals the same variety in physical properties that is apparent in any sizable vertical extent of the atmosphere. In both instances, vertical and horizontal, the dissimilar masses are in constant motion.

An air mass, when in contact with any part of the earth's surface for any appreciable length of time, is gradually affected by some of that area's characteristics. Thus, the temperature-humidity properties of any air mass are determined, or at least considerably conditioned, by the surface area over which it forms—that is, by the *air mass source region* (Fig. 11–1).

FIGURE 11–1

The source regions and physical properties of the various air masses within North America. (Adapted from Haynes, U.S. Department of Commerce)

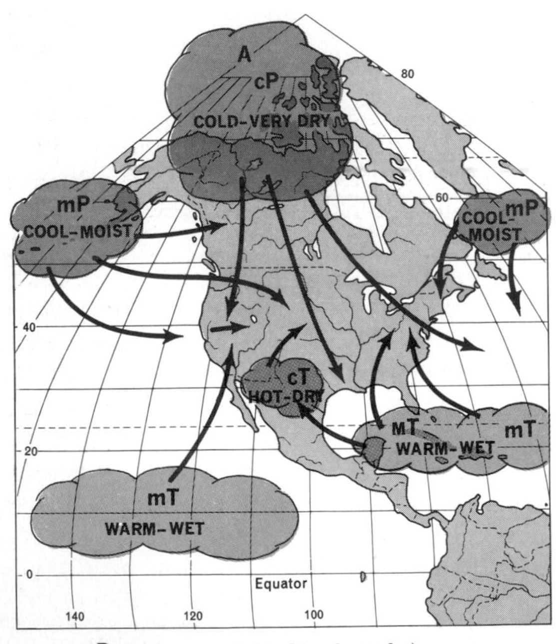

cP Polar continental (cold and very dry)
mP Polar maritime (cool and moist)
mT Tropical maritime (warm and wet)
cT Tropical continental (hot and dry)
A Arctic (very cold and very dry)
E Equatorial (not on this map) (hot and very wet)

When air masses are moved away from their source regions by the planetary wind circulation system, modifications of their original temperature-humidity properties begin immediately. The slower an air mass is forced to move, the closer it remains to its source region, and the longer it will tend to retain the characteristics of that region. The original temperature (heat) and humidity (moisture) characteristics of an air mass, then, are subject to changes in proportion to its distance of travel and time of migration over dissimilar earth surfaces.

Such changes occur in a variety of ways. One quite frequent change occurs by means of a transfer of heat and moisture between the air mass and the portion of the earth's surface over which it lies. The heating or cooling of the air mass may be effected through conduction or radiation, while the moisture content of the air mass is usually altered through evapotranspiration or condensation. Major modifications of air masses, however, are not brought about exclusively at the earth's surface. At higher levels, alterations may involve the horizontal convergence of separate air masses or the divergence of a single air mass.

Air masses often extend over hundreds or thousands of square miles horizontally, but vertically, they are usually restricted to the lower half of the troposphere. This relative proximity to the earth allows them to be very distinct and important climatic factors: any air mass, migrating or stationary, has a direct influence on the weather conditions manifested along its path of migration. In addition, the movement and juxtaposition of the varied air masses in different regions of the world ultimately determine global climatic types and patterns (Fig. 11–1).

Polar and Tropical Air Masses

The world's air masses, most of which originate between latitudes 10° and 70°, may be classified, for general study purposes, on the basis of their source regions and, consequently, their temperature and moisture properties. Air masses that originate in the higher latitudes are classified as *polar* (P) and have relatively cold temperatures; those that originate in the lower latitudes are classified as *tropical* (T) and are generally warm. Air masses originating over large landmasses are considered *continental* (c) and have relatively dry air; those originating over the oceans are designated *maritime* (m) and have relatively moist air. Thus, there are four possible air mass combinations: polar continental, polar maritime, tropical maritime, and tropical continental.

Because the oceans act as moderating influences on the atmosphere above them, maritime air masses tend to have less extreme temperatures than continental ones, and because cool air holds less moisture than warm air, polar air masses tend to be drier than the tropical variety. Consequently, using two relative progressions—cold-cool-warm-hot and very dry-dry-moist-wet—the following classification of air masses of North American origin can be made:

Air mass	*Characteristics*	*North American source regions*
polar continental (cP)	cold, very dry	central Canada, winter
polar maritime (mP)	cool, moist	northern Pacific, all year
tropical maritime (mT)	warm, wet	central Pacific, Gulf of Mexico, all year
tropical continental (cT)	hot, dry	deserts of northern Mexico, summer

Arctic and Equatorial Air Masses

Besides the four predominate types of air masses just discussed, there are also the types typical of the true arctic (very high) latitudes and the true equatorial (very low) latitudes. Of these, the most important are the *arctic* air masses (A), composed of very cold, dry air from the arctic regions, and the *equatorial* air masses (E), composed of hot, very humid equatorial air of oceanic origin (Fig. 11–1).

North American Air Masses

The North American source region for *polar continental* air is the interior of central Canada. During the winter, it becomes extremely cold: temperatures fall far below zero, all water bodies freeze, snow covers all the land, and there is very little warmth from the sun (which rises above the horizon for only a few hours each day). Skies are clear, heat loss by radiation is very great, and, because of the very low temperatures, the water-holding capacity of the air is extremely low.

At times, large segments of this air mass move southward, carrying conditions of extreme cold and dryness far beyond their usual bounds and producing "cold waves" in the Midwest and "northers" in Texas. As long as the air is above a snow-covered surface, its temperature remains very low. As it passes over snow-free ground farther south, however, its temperature rises rapidly (due to the radiation of heat from that warmer surface), although its relative humidity remains very low. In winter, such *modified polar continental* air frequently reaches as far south as southern California, bringing warm, bright, and clear weather during the day, with rather rapid decreases in temperature at night.

Polar maritime air develops in North America over the Gulf of Alaska, the Bering Sea, and the Pacific south of Alaska. Aloft, it is dry and cold—much like the polar continental air of central Canada—but its lower part is greatly modified by contact with the cool sea. Consequently, the temperatures in this part are above freezing, and the air is nearly saturated. (Due to the still relatively low temperatures, the amount of water vapor necessary for saturation is not very great.)

Frequently, this air moves southward and then eastward to the western coasts of southern Alaska, British Columbia, and the coterminous United States. There, it produces clouds, fog, and long-lasting drizzle—especially when the land is colder than the sea (as in winter), or when the incoming air is forced to rise over coastal mountains.

Tropical maritime air develops over the central Pacific, in the general vicinity of Hawaii. Its warmth allows it to hold great amounts of water vapor. On occasion, it moves northeastward toward the North American continent. In passing over the cold current flowing southward along the coast, the lower portion of this tropical air mass is often lowered in temperature to the point of condensation; thus fog forms. The rest of the air mass, unchanged in this respect, brings warm moist air to the land. Where it is forced to rise, as over the coastal mountains, tropical maritime air produces much orographic precipitation; in some areas it also becomes a major source of frontal precipitation.

Tropical maritime air also forms over the Atlantic Ocean, east of Florida and the Bahamas and south of Bermuda, and over all of the Gulf of Mexico and the Caribbean Sea. Very warm and heavily laden with moisture, this air is drawn northward into the central lowlands of the United States, at intervals in winter, and at all times in the summer. In summer, rising convectional currents initiated by the warmer land surfaces produce scattered but heavy showers almost daily, and in winter, frontal situations produce widespread and varied precipitation. As a result, tropical maritime air is the major source of precipitation for the eastern two-thirds of the United States.

Tropical continental air masses are very important over the Sahara, southwestern Asia, and Australia, but they are of lesser significance over North America, because the land area in the latitudes where they might be expected to develop is quite limited. During the summer, over northern Mexico, western Texas, and the southern parts of Arizona and New Mexico, there does occur some development of continental tropical air characterized by very high temperatures and very low relative humidity, but such development is quite weak.

Polar Fronts

Air masses in juxtaposition do not tend to mix, unless they are very similar in temperature and humidity. The terms *front* and *surface of discontinuity* are used to designate the sloping border zones, or surfaces of contact, that are formed between adjacent air masses that are observably different in these qualities. Fronts in the atmosphere, however, are not merely narrow lines, but rather, they are large areas with mean vertical depths of half a mile and horizontal widths that vary from 1 to 50 miles. Fronts that are deep

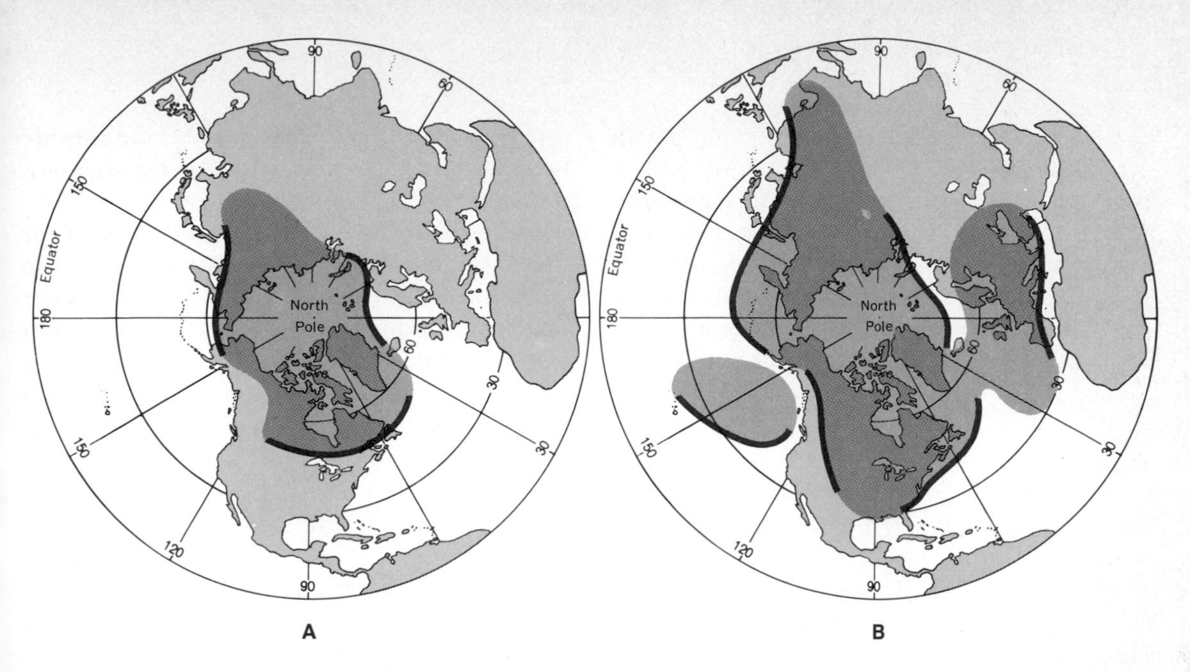

FIGURE 11–2

The seasonal migration of the north polar front. A. In summer, the front is much weaker than in winter, and it does not bulge down over the United States. B. In winter, the most active sections of the polar front in the Northern Hemisphere form in the northwest Pacific and Atlantic.

(high) and well developed are frequently associated with tropospheric jet streams—that is, the fronts are considered parts of their higher level wind-flow patterns.

In the middle latitudes, there are two belts of opposing winds: the westerlies, composed largely of mP and mT air, and the polar easterlies, which contribute the air movement about the periphery of the cP air mass. The westerlies, because they come from the lower middle latitudes, are relatively warm, while the polar winds, having their origins in the higher latitudes, are cold. The convergence of these two types of winds presents ideal conditions for the formation of a permanent front. The front formed in this manner is called the *polar front* (Fig. 9–6). In the Northern Hemisphere, this most important weather-producing zone forms a wavelike, or undulating, belt that, with some interruptions, extends around the world in the higher midlatitudes; the same phenomenon is largely duplicated by the polar front in the Southern Hemisphere's higher middle latitudes.

This latitudinal change in the polar fronts is due, in part, to the seasonal migration of the sun's vertical rays from the Tropic of Cancer to the Tropic of Capricorn. When the sun is over the Tropic of Cancer, in the Northern Hemisphere, the north polar air is heated and expanded by the warmed continents, the pressure near the North Pole is greatly lowered, and thus the northern polar front retreats poleward (Fig. 11–2A). When the sun is over the Tropic of Capricorn, in the Southern Hemisphere, the north polar winds as well as the westerlies become colder, the pressure builds up near the North Pole, and, as a result, the northern polar front is pushed equatorward (Fig. 11–2B).

Thus, the lack of constancy in the positions of both the westerlies and the northern polar easterlies causes an undulating fluctuation in the position of the polar front; when the polar winds are stronger, a mass of cold (cP) air pushes well equatorward into the westerlies; when the westerlies are stronger, a mass of warm (mP or mT) air pushes well poleward. These alternating masses of cold or warm air that seasonally cross North America in north-to-south fluctuating tracks are strong determinants of weather and climate, especially in the central and eastern sections of the continent. On a worldwide scale, the seasonal fluctuations in the position of the polar fronts are closely associated with the various weather and climate conditions found in the middle latitude regions of both hemispheres.

Cyclones and Anticyclones

The polar front, rather than being a smoothly curved line, normally has a number of pronounced irregularities. These are caused by outpourings of polar continental (cP) air that create great south-trending bulges in some areas, and by invasions of tropical maritime (mT) or polar maritime (mP) air that create embayments reaching far poleward in other places. Where a large cell of warm mT or mP air penetrates deeply into the cP mass, a zone of low pressure is created, since the warmer air has a greater tendency to rise than does the adjacent cP air. Conversely, each bulge of colder cP air, because it has a much greater density than the warm air, creates a zone of high pressure. Moved by the prevailing westerlies, these opposing pressure cells travel eastward, in tandem. The low-pressure cell tends to draw more air to itself, while the high-pressure cell tends to expel air; thus, a continuous series of eastward-moving eddies of air are formed. (As a general rule, in the Northern Hemisphere the pattern of wind-flow about a low-pressure cell is always counterclockwise and the pattern about a high-pressure cell is always clockwise; in the Southern Hemisphere these wind-flow patterns are reversed.) The low-pressure cells are called *cyclones,* and they are the main sources of precipitation in the middle latitudes. The tongues of cold, settling polar air are called *anticyclones* and are commonly associated with clear, cloudless weather. Together, cyclones and anticyclones are largely responsible for continuing weather changes, especially throughout the Northern Hemisphere.

Thus, the convergence of cold polar and warm tropical air masses (Fig. 11–3A) forms waves in the polar front (Fig. 11–3B). On the eastward side of each wave, the relatively warm air from the lower latitudes is pushing poleward, trying, as it were, to push the colder polar air out of its way. Since warm air is the force acting upon them, these frontal segments are known as *warm fronts* (Fig. 11–3B). On a weather map, a warm front is represented by a line with a series of small semicircles attached in the direction of the frontal movement (poleward). On the western side of each wave, cold polar air from higher latitudes is pushing in behind the warm air. Here, the cold air is the active force, and the frontal segments are known as *cold fronts* (Fig. 11–3B). A cold front is always represented on weather maps by a line that bears small triangles pointing in the direction of frontal movement (equatorward).

Frontal Weather

Most weather changes and atmospheric disturbances are brought about by the formation of warm fronts and cold fronts. Because of the condensation incident to rising and cooling air, weather conditions along both types of front are associated with cloud formation and precipitation.

Warm Front Weather

Since middle latitude cyclonic storms in the Northern Hemisphere migrate eastward, and since the warm front is the eastern side of a wave in the polar front, it is the warm front that passes over a specific geographic location first. Along a warm front, warm air glides up over cold air along a relatively gentle slope, with gradients of 1 in 300 to 1 in 100. This means that 100 miles ahead of the line along which the front meets the ground, the frontal surface lies from one-third to one mile above the ground (Fig. 11–4A). The angle of contact between the warm and cold air along a warm front is far smaller than that along a cold front. As a consequence, the areal extent of the resulting cloud formations and precipitation is likely to be more extensive than in a cold front. The warm air slides gently but steadily up this surface, forming cirrus clouds well ahead of the front—these clouds can usually be seen over several hundred miles often days in advance of the front's arrival. These are soon followed by lower altostratus clouds and, possibly, by a few very low stratocumulus clouds. As the front approaches a specific location on the earth's surface, light precipitation (rain or snow) will begin, and as the front arrives, the precipitation usually intensifies considerably and extends over a wide area. The period of precipitation is usually relatively long and covers a wide area, but the rain or snowfall is frequently gentle. Just before the front passes, fog or very low black clouds are often observed. As it finally passes, the clouds become higher, precipitation ceases, temperatures rise, and the winds usually shift from easterly or northerly directions to southerly or southwesterly ones.

Weather conditions along a warm front are determined by the degree of stability or instability of the warm air mass. If the warm air is comparatively stable, the storm will be relatively mild; if the warm air is unstable, weather conditions are normally more violent and will be characterized by great ascending air currents that will sometimes create thunderstorms (Fig. 11–4B and C).

Cold Front Weather

Along the cold front, warm air is forced up over the cold air along a relatively steep slope, ranging from 1 in 100 to 1 in 25; that is, 100 miles from the line where the front meets the ground, the frontal surface may be anywhere from one to four miles high (Fig. 11–5A). Although the horizontal extent of a cold front may be several hundred miles in width, the interface between the warm and cold air is so steep that the cold front weather is no more than a narrow zone of advancing violent storms that usually take the form of squalls, thunderstorms, or tornados (Fig. 11–5B and C). Cloud formation is delayed almost to the time of the cold front's actual

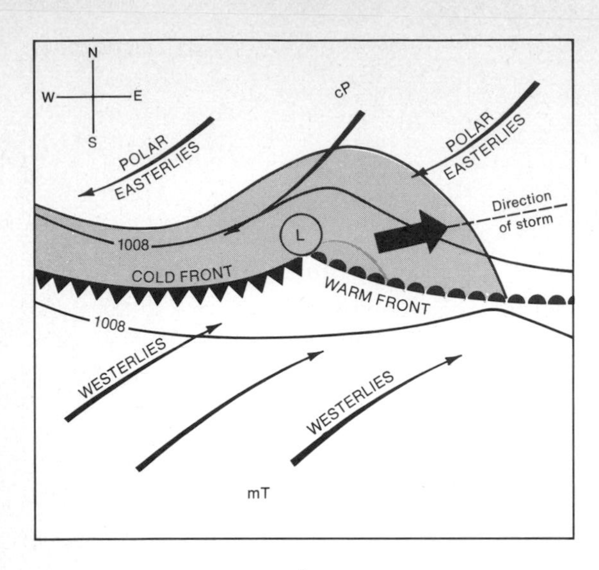

A

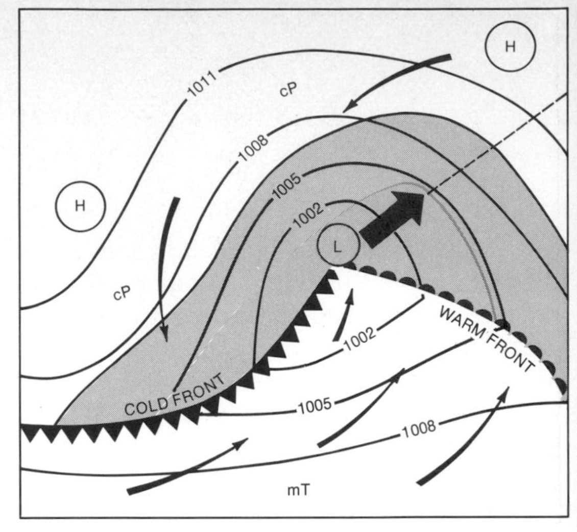

B

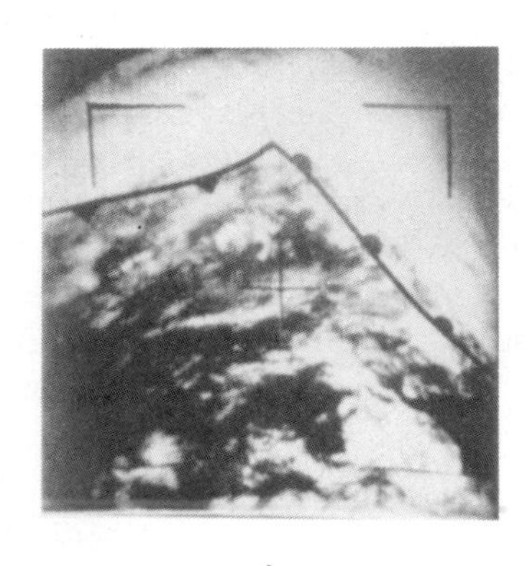

A

B

High-pressure cell

Low-pressure cell

COLD FRONT

WARM FRONT

Occluded front

Section of the polar front

Wind directions

1008 Pressure in millibars

Areas of clouds

Areas of rainfall

FIGURE 11–3

Ground plans of the formation and development of a cyclonic storm along a section of the polar front in the Northern Hemisphere (as would be shown on a series of weather maps). A photograph taken from space appears below each ground plan and shows the corresponding stage in the life cycle of North Atlantic cyclones. A. The *warm front* forms the eastern wing and the *cold front* forms the western wing of this small part of the polar front. Photograph: open wave (52° N. 37° W.; 1004 mb.), April 5, 1962. B. The northeastward-moving maritime tropical air (mT) rises up the inclined plane of the warm front (which, it must be remembered, is three-dimensional) and, cooling by expansion, begins losing its water vapor through condensation (thereby forming clouds) and precipitation. Along the western flank, the active and forceful polar continental air (cP) pushes the cold front rapidly southward and then southeastward. Photograph: closing wave (46° N. 59° W.; 996 mb.), April 10, 1962. C. The cold front, backed by the strong cold air mass, is advancing, and the polar front now assumes a shape like that of closing jaws. Along both the warm front and the cold front, the warmer air is gradually forced above the colder air, and, chilled by expansional cooling, it produces clouds and precipitation. The resulting convergence and rising of warm air along the

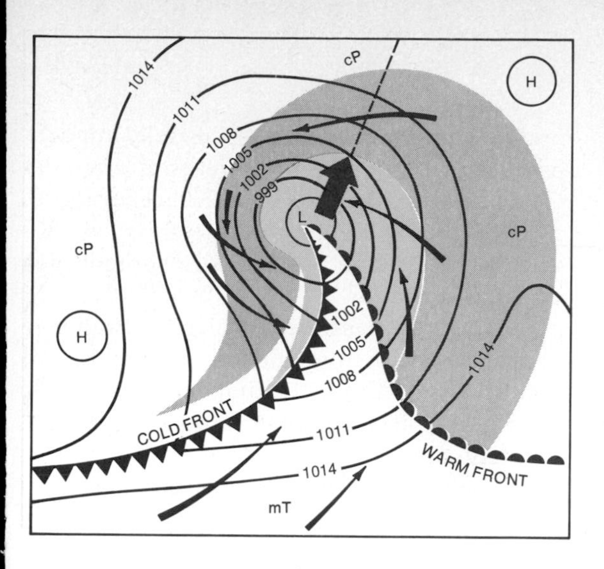

C

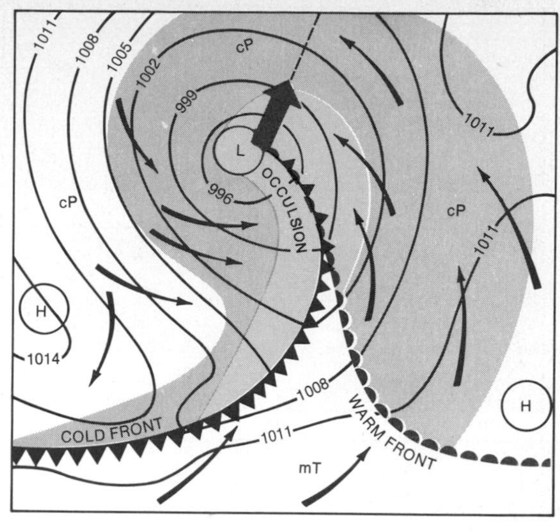

D

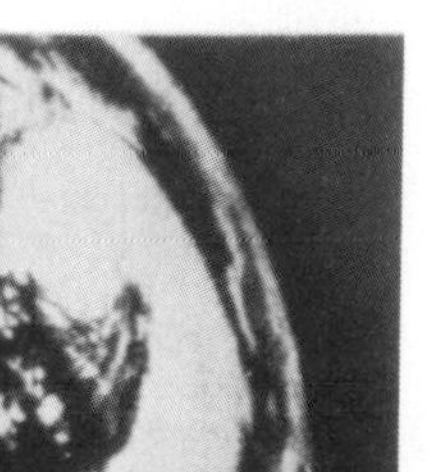

C

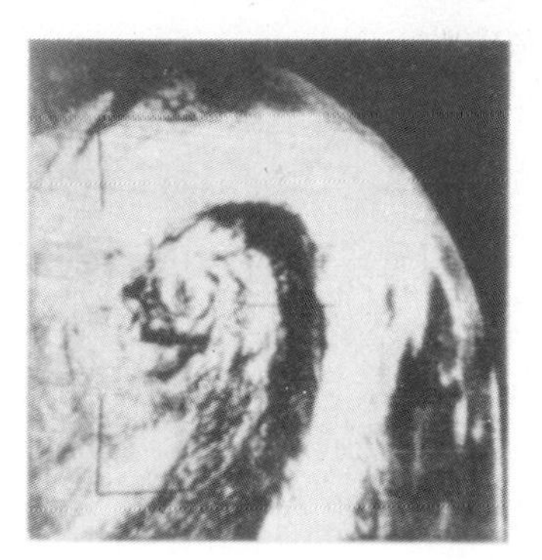

D

polar front causes a low-pressure cell of rising warm air to form. This "tongue" of warm air advancing from the south is called a *warm sector.* Now, the jaws of the polar front are almost completely closed, and the cold front has just about overtaken the warm front, with the low-pressure cell of rising warm air being forced toward the center. One should realize that the Coriolis force deflects the winds to the right (at least in the Northern Hemisphere) and that, as a result, as the jaws of the polar front close more tightly, small eddies of clockwise-whirling air develop on the periphery of the jaws. These clockwise-whirling eddies of cold air are associated with high-pressure cells, or *anticyclones.* Here, these clockwise-whirling eddies and the main high-pressure cell have almost completely surrounded and trapped the low-pressure cell of rising warm air. Air from the high-pressure cells is flowing into the low-pressure cell. This entrapped low-pressure cell is referred to as a *cyclone.* Photograph: 12–18 hours after the occlusion's beginning (48° N. 45° W.; 1006 mb.), September 4, 1961.   D. The cold front has overtaken the warm front, and the warm air has been lofted above the cold air to form an *occluded front.* The eventual dissipation of the warm air will mark the life cycle of this characteristic cyclonic storm. Photograph: deep occluded cyclone, 30–36 hours (51° N. 24° W.; 978 mb.), August 28, 1961.

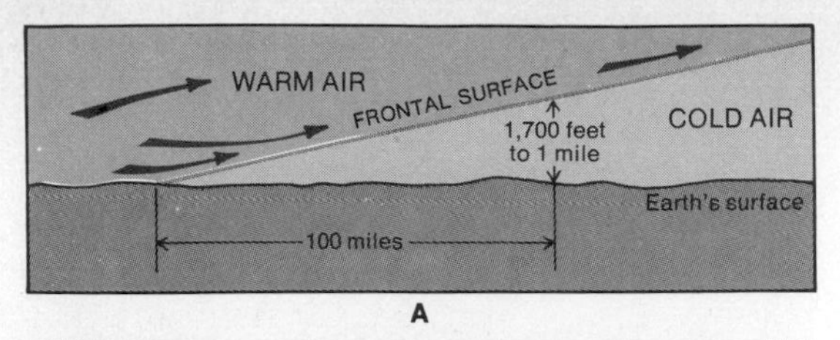

A

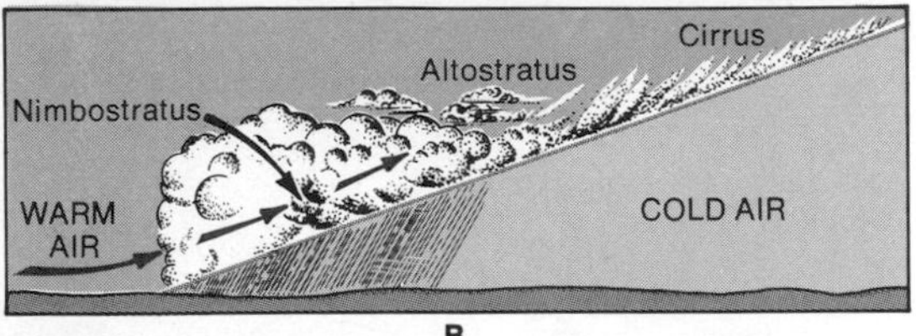

B

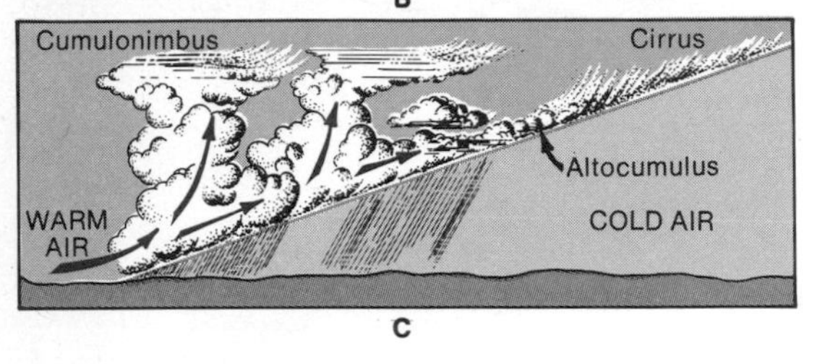

C

FIGURE 11–4

The movement of air along a warm front. A. Stylized. B. Stable. C. Unstable. (Refer, also, to Figure 11–3.)

passage over the earth's surface. Before the passage of the front, warm moist air flows in from a southerly or easterly direction (in the Northern Hemisphere) and, being forced to rise rapidly, forms cumulonimbus clouds almost directly above the ground location of the front. With the formation of clouds, the wind shifts to a westerly direction and there is an accompanying drop in temperature. Precipitation begins either just prior to or immediately after the passage of the front on the ground and, though heavy, is usually of short duration and followed by relatively rapid clearing. After the passage of the rain or snow activity, there is a marked drop in both relative humidity and temperature.

Squall Line Weather

A *squall line* is a line of high winds generated by a cold front. Always preceding the front, it may be hundreds of miles wide and, to the ground observer, usually appears as a single band of black, threatening clouds. Often reaching heights of 40,000 feet or more, these clouds generally produce rain or hail, and the squall line's strong winds sometimes produce tornados.

The evenness with which a squall line storm progresses depends upon the rate and continuity of the front that propels it (Fig. 11–5), and it is almost impossible to accurately determine the pace with which a front will move. Any long-range forecast of its progress, then, is subject to many qualifications. Only if the front moves steadily eastward can the time of its passage be calculated accurately and used to forecast the storm sequence. If the front stagnates, be-

FIGURE 11–5

The movement of air along a cold front. A. Stylized. B. Stable (the slow-moving cold front is lifting the stable air). C. Unstable.

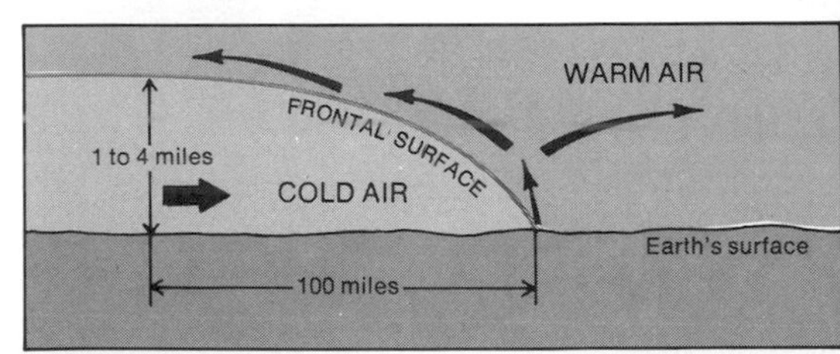

A

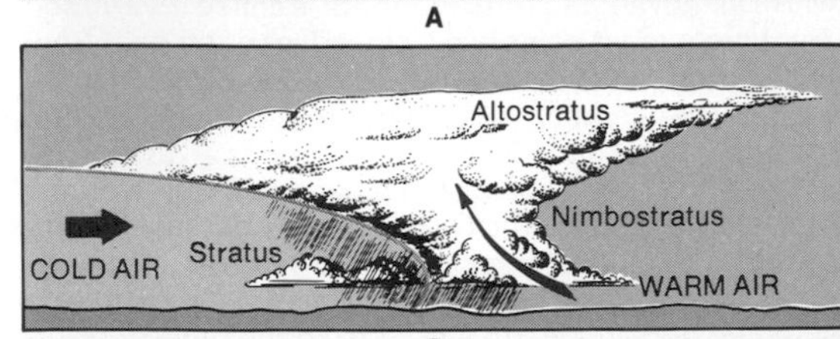

B

C

coming a *stationary front,* or, if it accelerates its timing, weather forecasts must be altered accordingly. Thus, they are often revised with considerable frequency.

Occluded Front Weather

Cold fronts usually move with greater regularity and speed than do warm fronts. Thus, though it starts as the "back jaw" of a cyclonic storm (Fig. 11–3C), a cold front eventually closes in upon the warm front preceding it. When it finally overtakes the warm front, what is known as an *occluded front* is formed (Fig. 11–3D).

If the temperature of the air behind a cold front is lower than that of the air ahead of the warm front, the cold front will push in under the warm front, forcing it aloft. This is known as a *cold front occlusion* (Fig. 11–6B), and it is most common on the eastern sides of continental areas. If the air behind a cold front is of a higher temperature than that ahead of the warm front, the cold front will be forced aloft in a *warm front occlusion* (Fig. 11–6C).

The weather that is associated with occlusions combines the features of both types of frontal storms. Since the occlusion takes place near the center of the cyclone, bad weather is widespread, particularly at the time the occlusion begins. The length and intensity of the period of bad weather depend largely upon the rate at which the warmer mass of air is forced upward. In addition, if the air is forced high enough so that temperature differences are cancelled, the low-pressure center will disappear: air that was relatively warm and light will be replaced by colder and heavier air, and the cyclone will disappear.

The major portion of the precipitation in the belt of the westerlies derives from these cyclonic storms (frontal weather). The moisture for this precipitation is provided by the warm moist air from the low latitude oceans, and the mechanism for condensation is provided by the colder, drier air from the polar regions and the continental interiors.

Because there is a seasonal migration of the zone of contact between warm and cold air masses, some areas are always under the influence of middle latitude cyclonic storms, while others experience their influence during only certain parts of the year. Moreover, it appears certain that shifts in the altitude, latitude, or intensity of the polar front jet stream have profound effects upon the detailed nature of the cyclonic system in the middle latitudes.

FIGURE 11–6

Air mass occlusions. A. The beginning of the occlusion. B. A cold front occlusion. C. A warm front occlusion.

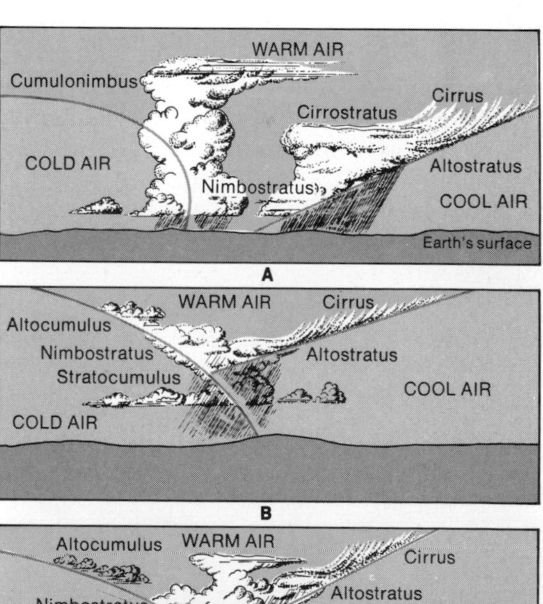

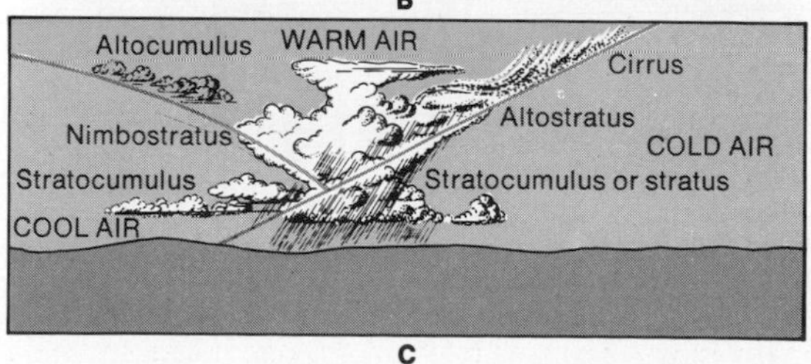

Midlatitude Cyclonic Storms

Within the interfacial zone of convergence between the westerly and polar wind belts (between 35° and 65° of latitude in both hemispheres), there is a constant variation of weather because of a succession of cyclones (lows) and anticyclones (highs). Situated in this zone of convergence, and influenced by the planetary wind circulation systems, these extratropical (midlatitude) pressure cells are put in horizontal motion by the prevailing westerlies and thus travel slowly in an easterly direction. Though frontal weather may develop in the zone of low pressure wind convergence between two anticyclones, the cyclone is more likely to have fronts, or definite wind-shift lines, associated with it and, thus, is the main source of precipitation in the middle latitudes.

Extratropical cyclones are responsible for four major types of weather: warm front weather, cold front weather, occluded front weather, and air mass weather (weather occurring away from fronts). These weather types can represent storms that range in severity from a weak disturbance, producing only light rain or snow, to a highly destructive display of extremely strong winds accompanied by either torrential rain or heavy snowfall.

FIGURE 11–7

An extratropical cyclone centered over the south-central United States, and its associated airflow pattern.

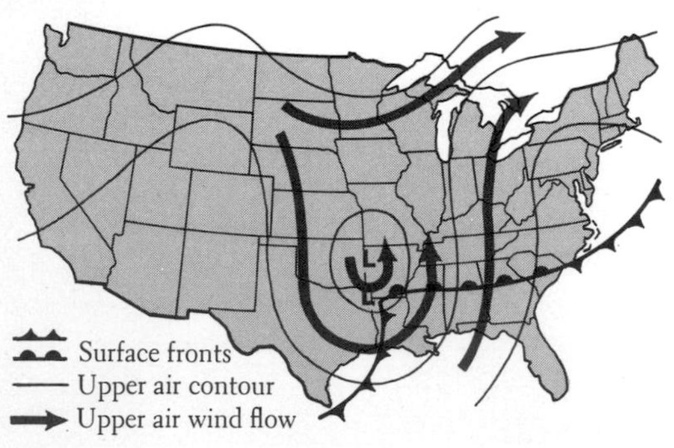

Anticyclones are usually associated with dry cloudless weather, while cyclones are associated with overcast skies and drizzly rains of several days' duration. Where a succession of these alternating highs and lows is the rule, as across most of North America, the weather is a comparable pattern of fair days and stormy days.

Both these pressure zones vary tremendously in size. Cyclones may be anywhere from 100 to 1,500 miles horizontally, although most range from 500 to 1,000 miles (Fig. 11–7). Their shape, when viewed from above, is usually elliptical or oval. Anticyclones are also generally oval in shape (some may be rounded) but, ranging evenly from 100 to 2,000 miles long, they tend to be larger than cyclones.

Since extratropical cyclones and anticyclones are formed along the undulating polar front, their specific routes of travel are almost entirely dependent upon the north-to-south seasonal migration of this front. Though their general direction of travel—being inspired by the westerly winds—is always eastward, the pressure zones take individual tracks that often vary and sometimes become very erratic.

In general, extratropical cyclones travel along certain standard tracks. As shown in Figure 11–8, they tend to approach the United States from the northwest and travel in a southeasterly direction toward the Midwest before they curve toward the northeast. Anticyclones follow roughly the same tracks across the United States, except that they do not normally veer northeastward near the

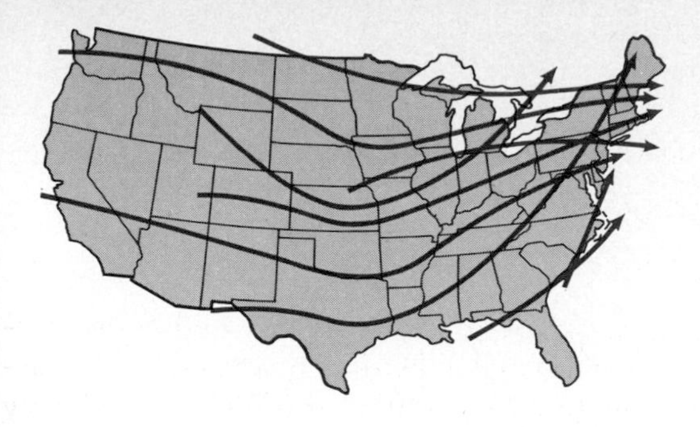

FIGURE 11–8

The paths or storm tracks generally followed by extratropical cyclones as they cross the United States.

eastern seaboard, but proceed more directly out over the Atlantic. The ground speed at which cyclones travel varies, but the average is from 20 to 30 miles per hour, with the rate of speed being greater in winter than in summer. On the average, anticyclones travel at speeds that are much slower than those of cyclones.

During winter in the Northern Hemisphere, when the north polar front is farthest south, both cyclones and anticyclones tend to be more numerous, better developed, and faster traveling than in summer, when the polar front has receded to its most northerly position. Cyclones entering the United States in winter bring moist air and rain or snow to the western states and the Rocky Mountains, and they often effect severe snowstorms as they bear southeastward over the middle of the country and then northeastward over Appalachia and the Atlantic seaboard. These winter cyclones are accompanied by the low temperatures brought by the polar winds that, in winter, extend as far south as the Gulf of Mexico and Florida. Weather reports label this combination a "cold wave," thus emphasizing the important role played by the cold air masses of the polar front (Fig. 11–9).

During summer in the Northern Hemisphere, when the polar front is farthest north, cyclonic systems are less numerous and less vigorous than in winter, and the general atmosphere over the United States is usually less active. The emphasis of summer weather reports is generally on high temperatures; for, during this season of the year, warm, humid tropical air masses tend to move far northward and subject most of the United States to "heat waves" marked by hot, humid air, heavy precipitation, and frequent thunderstorms (Fig. 11–10).

Although cold and warm air masses are both present in summer and in winter, the seasonal fluctuation in the position of the polar front makes the United States subject to the effects of one side of the front in one season and the other side in the opposite season. Twice yearly the front lies midway between its extreme positions and the more equal influence of its two sides results in the moderate temperatures characteristic of spring and fall.

Intertropical Front Weather

Near the equator, between the two trade wind belts, is the zone of very light and irregular surface winds and calms referred to previously as the intertropical convergence zone (ITCZ), or, more popularly, as the doldrums (Fig. 9–6). In this zone, the air is rising, partially as a result of the high temperatures near the equator and partially as a result of slight differences between the converging masses of air flowing in from the

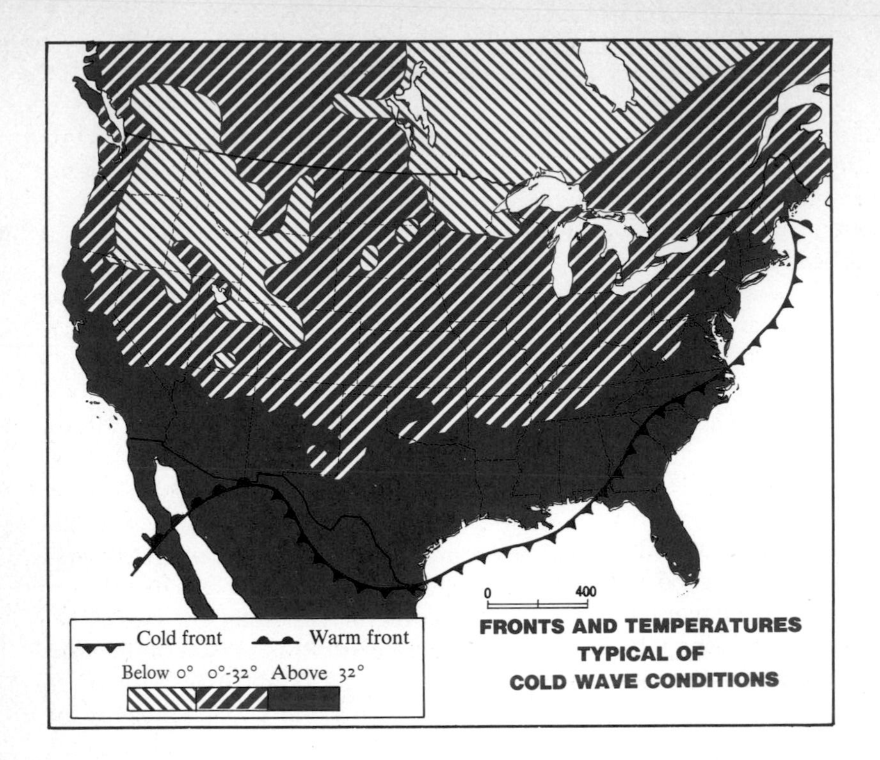

FIGURE 11–9

During the winter season, when a cold front pushes its way southward across Canada and the United States into the Gulf of Mexico, the low temperatures behind the front produce what is called a "cold wave."

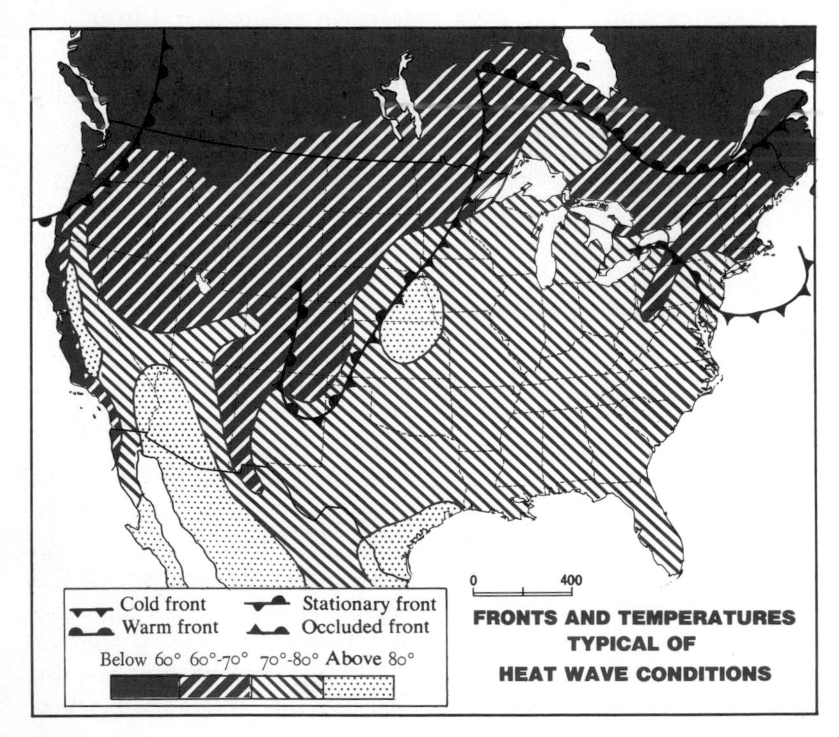

FIGURE 11–10

During the summer season, when a warm front pushes its way northward into southern Canada, the high temperatures behind the front produce what is called a "heat wave."

northeast and from the southeast—the northeast and the southeast trade winds. Slight inequalities in heating, and hence, in density, may cause one of these air masses to be forced upward over the other, thereby producing thunderstorms and severe squall lines. This line of contact between the air masses is known as an *intertropical front* (Fig. 9–6). Commonly, there are two such fronts, marking the two poleward boundaries of the ITCZ. Since their positions are continually changing, the width of the ITCZ is continually changing as well.

The latitudinal position of the ITCZ varies with the seasons. Over the Indian Ocean, for example, the effect this has is one cause of monsoons. In some of its extreme seasonal swings—over China and the Philippines, for example—the ITCZ may actually become fragmented or, in some cases, merge with the polar front.

In contrast to the polar fronts, the intertropical fronts are neither strong nor constant weather producers, for the temperature and humidity characteristics of air masses converging near the equator are very similar. But, during late summer in the Northern Hemisphere, when the ITCZ moves north of the equator, the temperature and humidity characteristics of the two trade wind belts are sufficiently dissimilar to create distinct and relatively active intertropical fronts. This seasonal activation of the fronts accounts, in part, for the formation of the various tropical cyclonic storms that occur during summer.

Tropical Cyclonic Storms

The cyclonic storms that occur in the tropics originate along the intertropical fronts, where the convergence of the belts of the trade winds is most highly developed. Most numerous in each hemisphere during its summer season, they are formed over warm and humid tropical waters, which generally have a surface temperature of at least 80°F. These storms tend to be of a mild and convective nature, and are characterized by cloudiness, scattered showers, and thunderstorms. Their movement is generally westward, owing to the drift of the trade winds. As they move over land areas, and thus increase their distance from a source of moisture, they tend to become less intense and, finally, disappear entirely.

If these tropical storms develop wind speeds over 75 miles per hour, they are then formally referred to as *tropical cyclones.* Such severe tropical storms are more commonly known as: *hurricanes,* when they occur east and southeast of North America; *typhoons,* when they occur in the same latitudes off the east coast of Asia; *cyclones,* when they occur in the northern Indian Ocean area; or *willy willies,* when they occur in northern Australia.

Like other cyclones, tropical cyclones are broad inward and upward spirals of air about a low-pressure center; their diameters usually vary from 100 to 400 miles. Since the whirling motion they represent is brought on by the Coriolis force, they are encountered only at latitudes of at least 5° (there is no Coriolis effect at positions closer to the equator). The degree of development of these cyclones is determined by the amount of moisture contained in the warm air along the intertropical convergence zone. Thus, when

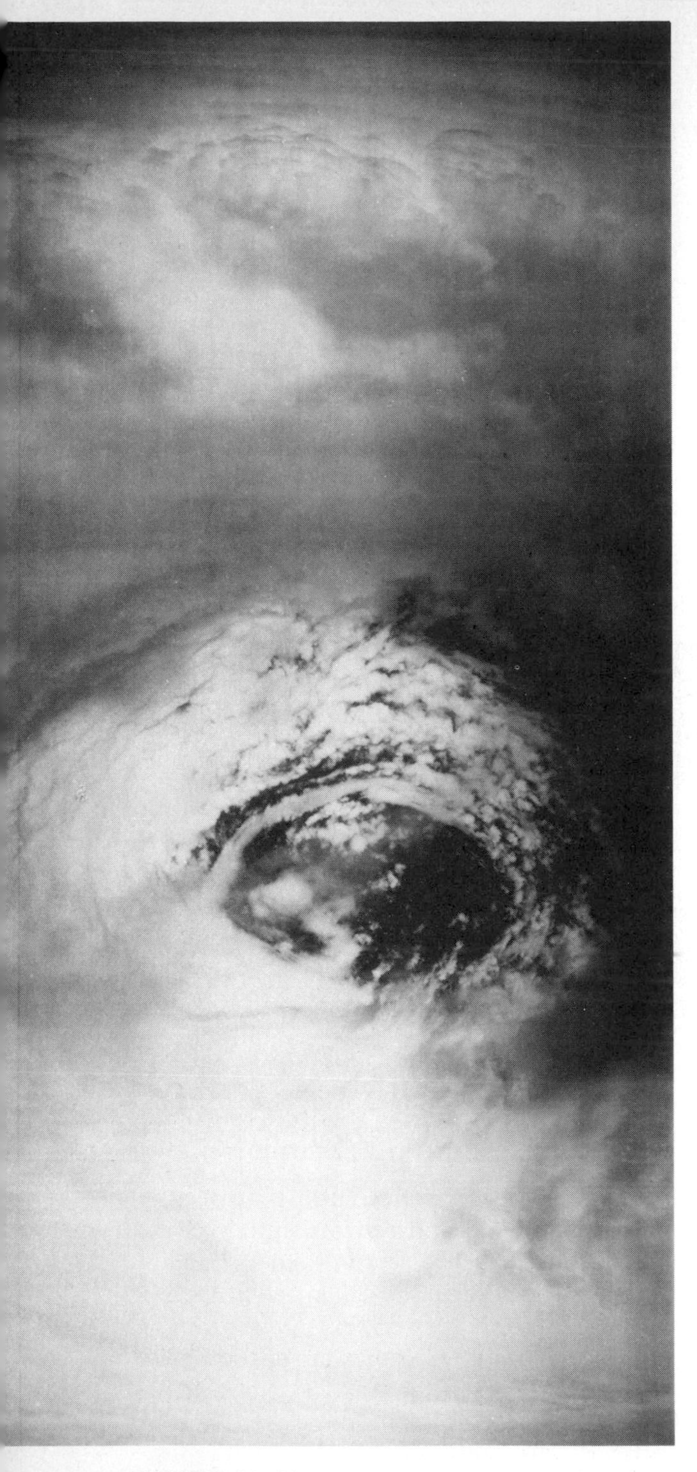

FIGURE 11–11
The *eye* of Hurricane Betsy (September, 1965), as viewed from an airplane. (U.S. Air Force Photo)

the ITCZ migrates—north of the equator in summer, and south of the equator in winter—to regions where the seas are warmer and the winds are more deflected, the development of tropical cyclones increases. An interesting difference between tropical and midlatitude cyclones is that, at the center of the tropical whirl there is a relatively calm, cloudless, and rainless zone known as the *eye*, which may range from 5 to 30 miles in diameter and is wider at the top than at the bottom. Often in excess of 120 miles per hour, the high wind velocities around the edge of the eye account for the destructiveness attributed to tropical cyclones (Figs. 11–3 and 11–11). Since these storms originate over the oceans, they cause serious damage to shipping as well as to any coastal districts they may encounter.

From the standpoint of physical geography, one of the most important aspects of tropical cyclones is their regions of occurrence. The primary examples of such regions are eight in number: the western central Pacific including coastal portions of the Philippines, China, and Japan (where more typhoons occur than any other place on earth); the Bay of Bengal east of India; the Arabian Sea west of India; the central Indian Ocean south of the equator; the Timor Sea northwest of Australia; the Pacific northeast of Australia and New Zealand; the eastern Pacific off the coast of Mexico and Central America; and the triangle formed by the western Atlantic, the Caribbean, and the Gulf of Mexico (Fig. 11–12).

Frequently, in their westward migrations, cyclones that develop into severe storms in the western Atlantic-Caribbean-Gulf region, the western central Pacific, or the Pacific northeast of Australia and New Zealand are drawn into the general wind and oceanic circulation patterns and, as a result, are turned northeast or southeast (depending on the hemisphere) at about 20° to 35°

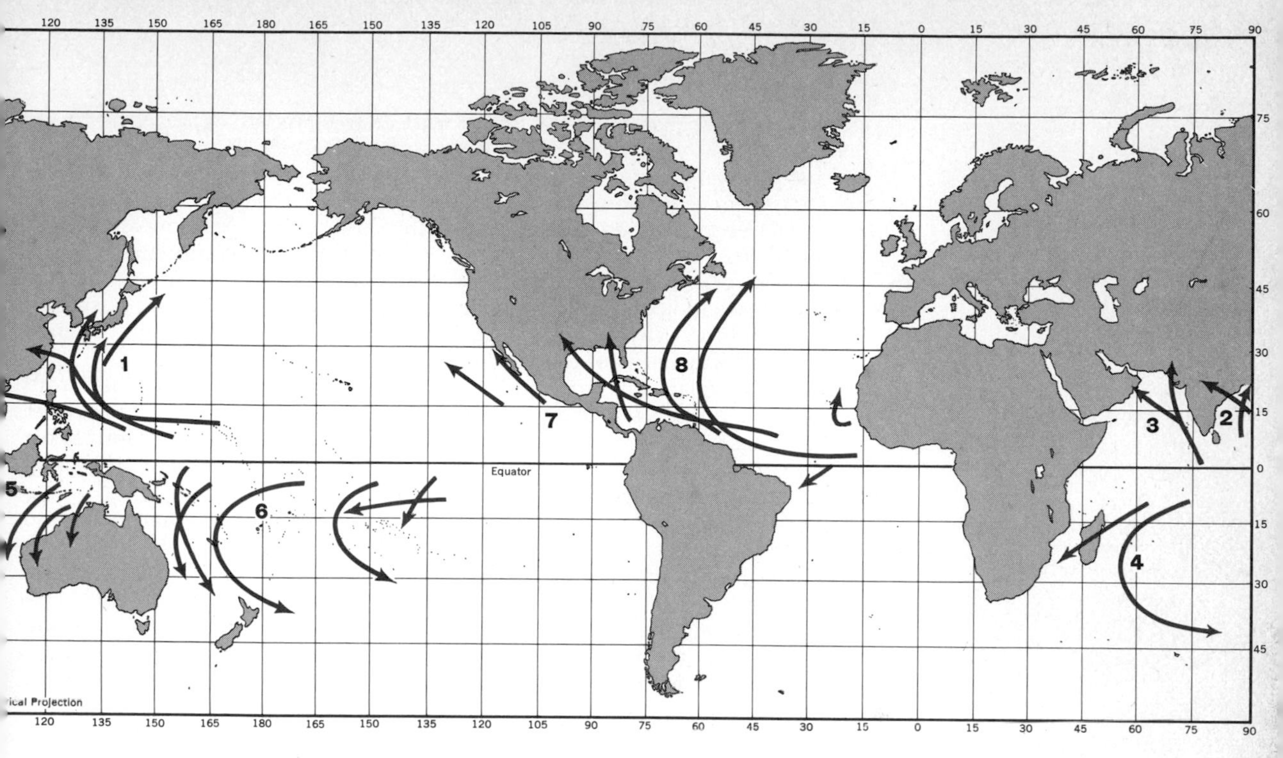

FIGURE 11–12

The principal tropical cyclone regions of the world and the mean trajectories of their cyclonic storms: 1) the western central Pacific (typhoons); 2) the Bay of Bengal east of India (cyclones); 3) the Arabian Sea west of India (cyclones); 4) the central Indian Ocean south of the equator (cyclones); 5) the Timor Sea northwest of Australia (willy willies); 6) the Pacific northeast of Australia and New Zealand (hurricanes); 7) the eastern Pacific off the coasts of Mexico and Central America (hurricanes); 8) the triangle formed by the western Atlantic, the Caribbean, and the Gulf of Mexico (hurricanes).

latitude into the belt of the westerlies. As such a tropical cyclone approaches the middle latitudes, it may join the series of extratropical or, midlatitude, cyclones in the belt of the westerlies and continue at a much abated tempo or, it may retain formidable power and cause enormous damage to life and property as it impinges upon whatever shipping vessels or coastal areas that lie within its path.

REVIEW AND DISCUSSION

1. Assess the following statement: any evaluation of atmospheric phenomena that is made with weather-making processes in mind involves the evaluation of all weather elements. In making your assessment, explain as many intricacies of weather elements as you can, and discuss the significance of each to human beings.
2. Give a succinct definition of an air mass. What is an air-mass source region? How do these source regions influence the air masses within them? When air masses migrate, do their general characteristics change? How?

3. Outline and discuss fully the types and characteristics of the world's air masses: cP; mP; mT; cT; A; E. Identify the source region of each.
4. Discuss fully, with the help of a map, the North American complex of air masses. Account for the development of fog along the west coast of North America. Why is the air that is drawn into the southcentral lowland of the United States usually warm and moist (or wet)? Why are continental tropical air masses of little significance over North America?
5. Do air masses in juxtaposition tend to mix readily? What is a front or surface of discontinuity? Why are fronts considered three-dimensional? How do they relate to the wind belts of the westerlies and the polar easterlies?
6. Thoroughly explain the formation and nature of a polar front. Is a polar front stationary monthly, seasonally, or annually? Is there a polar front in both hemispheres? If a polar front fluctuates latitudinally, what is the reason? What different sorts of weather and climate would result from a fluctuating polar front and from a nonfluctuating polar front?
7. After a careful study of Figure 11–3, sketch its series of weather maps and then add the labels from memory. Also, discuss the sequential developments that take place in such a time-series—that is, when mT or mP air penetrates a cP air mass.
8. Define a cyclone and an anticyclone; a cell; a warm front and a cold front. Which move in a clockwise direction? How are these related to polar fronts? What is an occluded front? What exactly do all of these have to do with weather and climate?
9. Why do cyclonic storms migrate eastward in the middle latitudes?
10. What is meant by frontal weather? Define warm front weather and explain its: development, life span, movement, cloud types, precipitation and temperature norms, winds, and degrees of stability or instability.
11. Compare warm front weather and cold front weather with respect to the characteristics mentioned above. Create diagrams that show the contrasting features of the two weather types. With regard to weather in your own locality, how would one distinguish between the passing of a warm front and the passing of a cold front?
12. Describe squall line weather and occluded front weather. With what type of front is each associated? Do cyclonic storms supply much of the precipitation in the belt of the westerlies? Why? Do the jet streams have any effect upon such storms?
13. Explain the ways in which cyclones and anticyclones vary in size, extent, and rapidity of horizontal movement. From what direction do these storms tend to approach the United States, and what tracks do they take across the country? What is the reason for this tendency?
14. The major weather–climate patterns are influenced greatly by the seasonal migration of the polar fronts. Discuss as many illustrations of this fact as you can.
15. Discuss intertropical front weather, contrasting it with polar front weather.
16. Tropical cyclonic storms differ considerably from those of the middle latitudes. Compare and contrast these two types of storms.

SELECTED REFERENCES

Barry, R. G.,and R. J. Chorley. *Atmosphere, Weather and Climate.* New York: Holt, Rinehart and Winston, 1970.

Battan, L. J. *The Nature of Violent Storms.* Garden City, N.Y.: Doubleday Anchor, 1962.

Blair, T. A., and R. C. Fite. *Weather Elements,* 5th ed. Englewood Cliffs, N.J.: Prentice-Hall, 1965.
Dunn, G. E., and B. I. Miller. *Atlantic Hurricanes.* Baton Rouge, La.: Louisiana State University Press, 1964.
Flohn, H. *Climate and Weather.* New York: McGraw-Hill, 1969.
Flora, S. D. *Hailstorms of the United States.* Norman, Okla.: University of Oklahoma Press, 1956.
———. *Tornadoes of the United States.* Norman, Okla.: University of Oklahoma Press, 1954.
Helm, T. *Hurricanes: Weather at Its Worst.* New York: Dodd, Mead, 1967.
Lehr, P. E., R. W. Burnett, and H. S. Zim. *Weather: A Guide to Phenomena and Forecasts.* New York: Simon and Schuster, 1957.
Lowry, W. P. *Weather and Life: An Introduction to Biometeorology.* New York: Academic Press, 1969.
Namias, J. "The Jet Stream." *Scientific American,* Vol. 187 (October 1952), 27–31.
Petterssen, S. *Introduction to Meteorology,* 3rd ed. New York: McGraw-Hill, 1969.
Reiter, E. R. *Jet Streams.* Garden City, N.Y.: Doubleday, 1967.
Riehl, H. *Introduction to the Atmosphere,* 2nd ed. New York: McGraw-Hill, 1972.
Stewart, G. R. *Storm.* New York: Modern Library, 1947.
Thompson, P., R. O'Brien, and the editors of *Life. Weather.* New York: Time, Inc., 1965.

A desert area in Lebanon. (Silvester, Rapho-Guillumette)

12

Climate and Weather

Ancient Classification of Climates

From early times, major studies of the physical environment have attempted to classify the basic climatic elements in order to provide a means of comparison between one region and another. The early Greeks established such a classification largely upon the single element of temperature. They understood the climate of the area in which they lived—with its hot, dry summers and mild, rainy winters, and its clear summer skies and cloud-filled winter skies. The climate of the area in which they lived and worked posed few problems of sustenance and fewer still of clothing and shelter. To them, theirs was the ideal climate—thus, they called it the *temperate zone.*

Their knowledge of the hot deserts to the south led the Greeks to believe that even hotter lands lay beyond, lands where the sun was so hot that it could turn human skin a darker color, hot enough to be uninhabitable by people like themselves. This was their concept of the *torrid zone.* In a similar manner, they deduced the opposite extreme existing to the north. The occasional cold, north winds of winter, and the stories of the hardships of travelers to the north, led the Greeks to imagine that lands to the north were too cold for human habitation, at least by civilized people like themselves. This was the *frigid zone.*

Almost on the basis of temperature alone, they erected their threefold climatic classification that remains, even to this day, in frequent references to torrid, temperate, and frigid climatic zones. However, today this classification is neither adequate nor fully correct, as it recognizes only one of the basic climatic elements. While there is some reality in such a system, it does not reflect the full combination of climatic elements that have since been found to exert their strong influences on the climates of the world (Figs. 12–1 and 7–12).

FIGURE 12–1

Some of the world's major climatic types. Top left, tropical wet climate (Af): an area of forests, rice paddies, and stilted houses in Tambunan, Malaysia. (Paul Conklin, Monkmeyer) Top right, arid climate (BW): the tiny settlement of Khafs Daghrah, site of an experimental farm, in Nedj, Saudi Arabia. (Standard Oil Co., N.J.) Center left, subtropical humid climate (Cfa): a variety of palm trees, broadleaf trees, and flowering shrubs, in Orlando, Florida. (Florida State News Bureau) Center right, Mediterranean climate (Cs): olive groves, live oak trees, and hillside villages in Andalucía, Spain. (Helena Kolda, Monkmeyer) Bottom left, humid continental with hot summers climate (Dfa): cornfields and meadows surrounding the village of Templeton, in Iowa. (USDA) Bottom right, icecap climate (EF): the USS *Glacier* on iceberg patrol, and some of the penguins who live on Balaena Island, Antarctica. (U.S. Navy)

Modern Classification of Climates

Purposes of Classification

Climatic classification is a tool, not an objective in itself. Many classification systems have been devised for the orderly grouping of varied climatic phenomena into major complexes: mainly to provide means of systematic, meaningful spatial comparison, and to be used as tools in the determination of the basic world climatic pattern, which influences human use of the physical environment so markedly.

Systematic Comparison. While there is great variety in the possible combinations of climatic elements, certain broadly similar ones have been observed to recur frequently throughout the world. Such combinations constitute the basic climatic types, some of which are depicted in Figure 12–1. Each element can be given an exact quantitative, rather than mere qualitative, definition. On the basis of these quantitative definitions, a system, or classification, can be erected. Such a system not only defines each of the climatic types, but indicates, in terms of observable and measurable amounts, the relationships between them. Through the use of such a classification, a descriptive statement is made exact. Thus, no matter who may use the classification, or to what region of the world it may be applied, the meanings are standard. It is then possible to state whether or not one region exists under essentially the same combination of climatic elements as another.

Determination of a Basic World Pattern. Climatic classification also makes possible the determination of the pattern of climate over the world. For any chosen region, observed records of as many locations as possible can be classified. The location of each place, with the climatic type represented by its record, can be plotted on a map. Then lines can be drawn separating the places with one type of climate from those with another. The result is a *climatic map* of the region. The lines on such a map are primarily indicators of direction of climatic change, such change being greatest at right angles to the lines and least along the lines or in a direction parallel to them. If the same process is followed for the whole earth, a basic world climatic pattern (of climatic types) is obtained (Plate IV).

The world pattern of major climatic types is basic to an understanding of the physical environment as a whole, because of the close relationship that exists between climate and the many other elements of the environment. Visible forms of the physical environment, such as landforms, vegetation, and soil, are all modified greatly by the climate in which they exist. For example, a hot, rainy climate produces particular characteristic forms of the land surface, of the plant cover on that surface, and of the soil in which the plants grow: quite unlike those produced in a cold, dry climate.

Similarly, while local soil types may reflect certain differences in the rock materials from which they are formed, on the whole, given the same basic materials, the soil developed in one climate is vastly different from that formed in another. The reddish to yellowish soils of Georgia, for example, are not at all like the deep black to dark brown soils of Iowa.

The ways in which people live in the physical environment are also greatly affected by climate: the type of building best suited for human use in the constantly hot, wet climate of an equatorial forest is hardly adequate protection against the severe winters of northeastern Siberia. The fact that cotton is important in the agriculture of the southeastern United States and absent from that of the New England states is primarily a reflection of climatic difference also. The contrast between the intensive use of the lowlands of Japan and the near emptiness of the lowlands that face the Arctic Ocean is, in large part, the result of the respective climates of these two areas. In short, climate is a force that limits and conditions, either directly or indirectly, much that a society does in its particular environment. Consequently, the distribution of climatic types over the earth is the key to, and explanation for, the arrangements of many other features of the physical environment.

System of Classification

Today, the climates of the world are grouped into categories based upon the two elements most significant to human beings: temperature and precipitation. Long-term statistical records of these elements are available for numerous and widely scattered locations on the earth, and, when analyzed, it becomes apparent that certain outstanding combinations occur regularly. For example, some locations show a combination of hot and humid conditions, others hot and dry, while still others are cold and humid or cold and dry. From these basic combinations, five major categories of specific climatic types are discernible: tropical humid climates (A); dry climates (B); humid mild-winter climates (C); humid severe-winter climates (D); and polar climates (E) (Plate IV; Fig. 12–26). In addition, there is a general category of undifferentiated climates—known as the mountain climates (H).

Each of these broad climatic categories is subdivided to indicate further significant climatic differences from place to place within any major climatic type. For example, within the humid mild-winter climates (C), there are some places that have a summer drought; other places in the same broad type have a characteristic winter drought; and still other places are humid throughout the year.

The classification used above and throughout this book is essentially the widely used one known as the *Köppen system* (Appendix C). The Köppen system uses specific letter symbols, rather than merely descriptive names, to represent the various climatic types, with each letter having a spe-

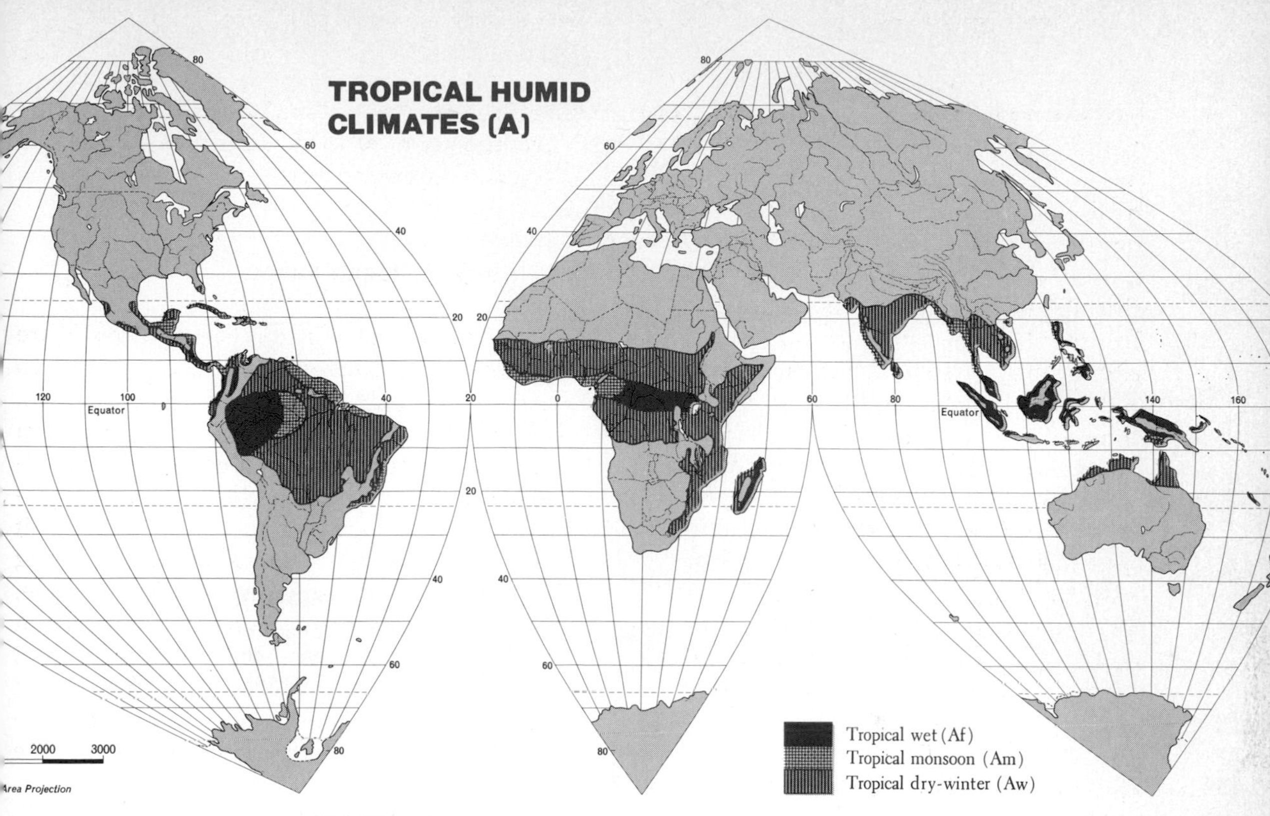

FIGURE 12–2

The *tropical humid climates (A)* of the world are concentrated largely within the equatorial and tropical latitudes and occur in all the continents except Europe. (Refer, also, to Plate IV.)

cific definition. Thus, a humid mild-winter climate in which the summers are characteristically droughty is known in the Köppen system as a Cs climate; if such a climate also has hot summers, it is known as a Csa climate; if the summers are warm rather than hot, it is known as a Csb climate. Each of these terms—such as hot, warm, droughty, or mild—is defined in exact temperatures or amounts of precipitation. (See Appendix C for a detailed discussion of the classification and data.)

Tropical Humid Climates (A)

Climatic Characteristics

Throughout the year, the climates that prevail in the tropical low latitude regions are hot and humid. These tropical humid climates (A) are controlled predominantly by tropical and equatorial maritime air masses and tropical continental air masses (Fig. 11–1). Within this broad framework of similarity, however, there are many significant differences (Plate IV; Figs. 12–1, 12–2, and 12–3).

Types	Areas
Tropical wet (Af)	Central America; equatorial South America; Africa; southeastern Asia; Indonesia; Melanesia
Tropical dry-winter (Aw)	Central America; South America; middle Africa; south and southeastern Asia; Australia
Tropical monsoon (Am)	Central America; coastal South America; coastal Guinea; coastal southern India; Bay of Bengal coasts

Characteristics	
Latitudes	equatorial and tropical
Temperatures	warm to hot (all months); no winter; mean temperature of 64° F. for the coolest month
Precipitation	heavy to moderate rainfalls; convectional
Wind belts	trade wind belts; doldrums; monsoons
Air masses	equatorial; tropical maritime; tropical continental

FIGURE 12-3
Chart of the principal areas and characteristics of the tropical humid climates (A).

Temperature Characteristics. The most common characteristic of all tropical humid climates (A) is constant high temperatures. Climatic types are defined according to the mean temperature during the *coolest* month, because that temperature indicates whether there is a season in which truly tropical vegetation cannot flourish—a period of plant rest induced by low temperatures. The lower limit of the A climates is a mean of 64°F. for the coolest month. This temperature limit closely corresponds to a notable distinction in vegetation type: where the coolest month averages below 64°F., the palm and similar tree types are absent; the forests in these areas are made up of such trees as the oak.

In the heart of the regions of tropical humid climates (A), monthly means are in the 70's and 80's, and the range between the mean for the coolest month and that for the warmest month is very small. The graph of climatic statistics for Singapore, in southeastern Asia (Fig. 12–4A), indicates these two conditions. Here, the mean for the coolest month is 78°F., while the mean for the warmest month is 81°F.—a difference of 3°. This is typical of the most pronounced of the tropical humid climates. Often, the differences between day and night temperatures are greater than those between the coolest and warmest months: a daily range of as much as 15° is not unusual. In Singapore, the average daily range varies between 12° in July and 15° in March. Monotony characterizes such a temperature condition and prevails throughout the heart of all tropical humid climatic (A) regions.

Graphs of climatic statistics for Corumbá, in Brazil (Fig. 12–4B), illustrate another condition. Here, the mean for the coolest month is 70° F., while that for the warmest month is 80° F.—a difference of 10°. The average daily range is approximately 20°. Other factors, particularly moisture differences between these two places, tend to emphasize the differences wrought by temperature. Thus the monotony so pronounced at Singapore is modified at Corumbá.

Precipitation Characteristics. The greatest contrasts between the various subregions of the tropical humid climates (A) are found in the amount and distribution of annual precipitation. All of these subregions are sufficiently humid to be included within the wet climates, and, in many places, there is even excessive wetness, but the similarity ends there. Some of the tropical humid climates (A) have rainfall distributed evenly throughout the year, or, if they do show any period of appreciable dryness, that period is relatively short. In some places, the precipitation during the greater part of the year is very high, and there is no true season of drought. When there is no drought, the climate is of the *tropical wet* type (Af), as in Singapore (Fig. 12–4A).

In other areas of tropical humid climate (A), definite drought during one season alternates with heavy rainfall during another. Where such a drought condition prevails, as in Corumbá (Fig. 12–4B), the climate is of the *tropical dry-winter* type (Aw). It is obvious that the transition from one extreme to the other is gradual and, consequently, that the definition of any dividing line between the tropical wet (Af) and the tropical dry-winter (Aw) types is arbitrary.

Between these extremes of drought and no drought is the tropical rainfall type of climate that is prevalent along parts of the southern and southeastern coast of Asia (and elsewhere), where an excess of precipitation during one season of the year compensates for a lack of it in another. In these areas, where monsoons are most typically devel-

FIGURE 12–4

Graphs of climatic statistics. A. Singapore, in southeastern Asia (tropical wet climate at 1° 18′ N.). B. Corumbá, Brazil (tropical dry-winter climate at 19° 01′ S.). C. Akyab, Burma (tropical monsoon climate at 20° 10′ N.).

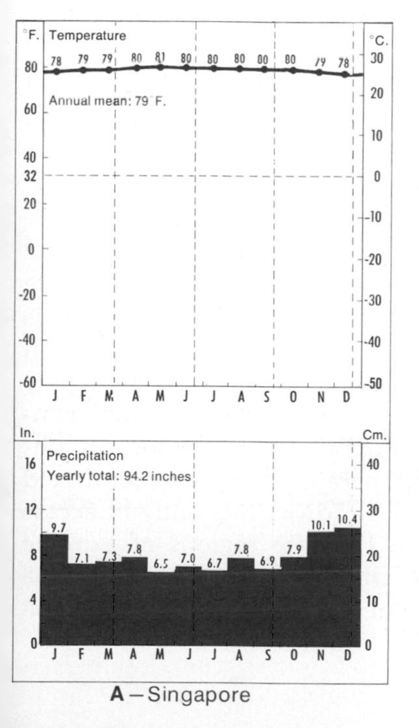

A – Singapore

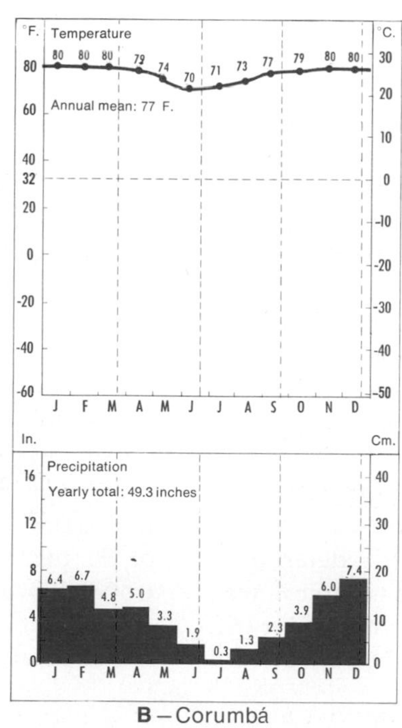

B – Corumbá

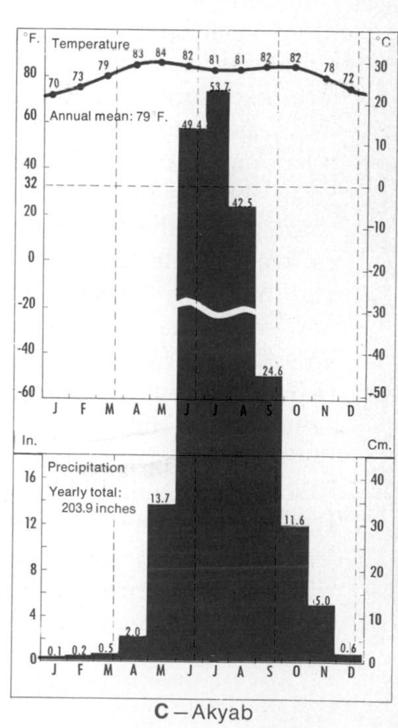

C – Akyab

oped, such a characteristic distribution of rainfall throughout the year occurs. This distribution is shown on the graph of climatic statistics for Akyab, Burma (Fig. 12–4C). It is similar to that in regions of tropical dry-winter climate (Aw)—there is a rainy season and a dry season—but the rainy season is generally longer and the dry season shorter than those in tropical dry-winter regions (Aw). In addition, the amount of rain that falls during the rainy season is excessive. While this rain, coming during a relatively short period, cannot be as effective as it would be if spread throughout the year, it is still effective enough to prevent a complete cessation of vegetative growth processes during the ensuing dry season and, hence, to produce a setting more nearly like that encountered under a tropical wet climate (Af) than that of a tropical dry-winter climate (Aw). This type is known as the *tropical monsoon* climate (Am).

This intermediate type of rainfall is commonly associated with areas in which a monsoonal circulation is present. In some situations, however, other causes, such as local exposure to rainbearing winds during one season and shelter from them during another, may produce this rainfall pattern. Since there is no precise line along which a change from the year-round rain of the tropical wet climate (Af) to the alternating rainy and droughty seasons of the tropical dry-winter climate (Aw) is experienced, in some areas, the so-called tropical monsoon climate (Am) indicates a transition between extremes rather than a monsoonal circulation, as does the Gulf of Guinea coast.

The relationship between the total precipitation for the year and the amount that falls during the driest month determines which of the three varieties—tropical wet, tropical monsoon, or tropical dry-winter (Af, Am, or Aw)—is represented by any one set of precipitation statistics. This precipitation relationship is mirrored in the native vegetation. The tropical wet climate (Af), in which there is no season of rest for plant growth, supports a heavy forest. The tropical dry-winter climate (Aw), in its drier phases, supports a mixed cover of sparse grass and thorny bush that, though lush and green during the rainy season, is dull, brown, and dormant during the height of the dry season. Again, in the intermediate position, the tropical monsoon climate (Am) supports a light forest with heavy undergrowth, rather than the heavy forest of the tropical wet climate (Af), or the grass and bush cover of the drier tropical dry-winter climate (Aw). In tropical monsoon (Am) forests, relative dryness rather than absolute drought is characteristic. In addition, there are no sharp lines between one type of vegetation and another—there is simply a gradual merging.

Subtypes of the Tropical Humid Climates (A)

Tropical Wet Climate (Af). For a direct comparison of the extremes found within the tropical humid climates (A), we may refer again to the climatic graphs for Singapore, Corumbá, and Akyab (Fig. 12–4). Singapore has an average annual precipitation of 94.2 inches. The driest month is May, during which 6.5 inches fall; the wettest month is December, during which 10.4 inches fall. It would be unrealistic to say that Singapore has a dry season. Accompanying such constantly high rainfall in the region of constant high temperatures, there is continued high relative humidity, frequently in excess of 90 percent. In these regions of extreme tropical wet climate (Af), humidity is a greater problem to most human beings than is temperature (Fig. 12–5).

FIGURE 12–5

Workmen bathing in the Ramoi River near Sorong amid the hot, humid, tropical rainforests of western New Guinea. (Standard Vacuum Oil Co.)

Tropical Dry-Winter Climate (Aw). At Corumbá, Brazil, the average annual precipitation is 49.3 inches. The driest month is July, with an average of 0.3 inches; the wettest month is December, with an average of 7.4 inches. The graph (Fig. 12–4B) indicates the pronounced nature of the dry season. When this is compared with the temperature curve, it is evident that the period of greatest dryness comes when the temperatures are the lowest. The doubled effect of seasonal difference is thus made even more noticeable. Relative humidity is less than it is in Singapore and shows considerable fluctuation throughout the year. The relative humidity average for December, the rainiest month, is 84 percent, and that for August, one of the drier months, is 68 percent. The wet season approximates the dominant characteristics of the tropical wet climate (Af); the dry season approximates the climate found along the margins of the moisture-deficient, or dry, climates (B) farther away from the equator.

Tropical Monsoon Climate (Am). The climatic graph for Akyab (Fig. 12–4C) shows a temperature curve similar to that of the other tropical humid climates (A), with the slight exception that the highest temperatures are in the two months just preceding the maximum rainfall. The precipitation distribution throughout the year suggests the characteristics of the tropical dry-winter climate (Aw). The amount of rainfall that comes during the rainy season, however, is much greater than that at Corumbá. The average yearly total at Akyab is 203.9 inches; of this, over 25 percent—53.7 inches—comes during the single month of July. This amount is sufficient, along with the excessive amounts during the other months of the rainy season, to compensate for the long drought period that is comparable in intensity to that of the tropical dry-winter climate (Aw). This is true despite the fact that each of the four months from December through March receives less than one inch of rain. As in the tropical dry-winter climate (Aw), seasonal change is quite pronounced. Temperatures throughout most of the dry season are appreciably lower than those during the rainy season. It is only at the end of the dry season that high temperatures do occur. Relative humidity averages about 70 percent through most of the dry season; it averages in excess of 90 percent from May through September. In terms of human sensibility, there are really three seasons in such a climate: the cooler dry season, the hotter dry season, and the hot rainy season. This climate represents a kind of merging of the extremes of the tropical wet (Af) and the tropical dry-winter climates (Aw), as it combines certain features of both.

Global Distribution

Figure 12–2 shows the global distribution of the tropical humid climates (A): in distinguishing among the three subdivisions it should be noted especially that, throughout most of the region, the tropical monsoon (Am) type frequently forms a transition between the tropical wet (Af) and the tropical dry-winter (Aw) types. Closer observation of the tropical humid region would probably show that this transition actually exists between all areas of tropical wet (Af) and tropical dry-winter (Aw) types. However, such information is lacking, and, consequently, there are some places, such as the northern edge of the Congo Basin in Africa and the inner part of the Amazon Basin in South America, where tropical wet (Af) and tropical dry-winter (Aw) types are shown abutting directly upon one another.

Dominant Controls

Temperature Controls. The constantly high temperatures experienced in regions of tropical humid climate (A) are primarily the result of the location of these regions near the equator. It is in equatorial regions that insolation is greatest and most nearly constant throughout the year. The angle of the sun's rays and the length of the daylight period are the factors primarily responsible for this condition.

Along the equator, the noon sun is vertically overhead twice during the year, about March 21 and September 22. Hence, the heat received along the equator at the earth's surface is most intense at these two times. Along each of the Tropics, the noon sun is vertically overhead on only one day of the year, about June 21 at the Tropic of Cancer and about December 22 at the Tropic of Capricorn. Thus, there is only one period of greatest surface heating in each instance and, hence, only one period of maximum atmospheric heating.

At all locations between the Tropics, the noon sun is overhead twice during the year; as the Tropics are approached, these periods come closer together until they finally merge. At no time during the year within this tropical region is the noon sun far from the overhead position. Because of this, there is no period when heating is greatly reduced. The range is smallest along the equator, and it increases slightly as one moves north or south from it.

Equally important in the control of temperature is the length of the daylight period. Along the equator, the daylight period is of equal length throughout the year (Fig. 8–9). In both directions away from the equator, this length increases during one season and decreases during the opposite season. At the Tropics, the longest daylight period is about 13⅓ hours, and the shortest is about 10⅔ hours. Likewise at the Tropics, the period of the longest day corresponds to that of highest sun, and the period of the shortest day to that of lowest sun. Hence, along the equator, surface heating is essentially constant throughout the year, but at the Tropics, it varies seasonally.

Between the equator and the Tropics the longest day comes between the two high-sun periods that are closest together. At either of the Tropics, the period of longest daylight is also the time of highest sun. Thus, at the Tropics, both controls of insolation—the angle of the sun's rays and the length of the daylight period—operate to produce the greatest heating at one season and the least at the opposite season. At the equator, the angle of the sun's rays varies less than at either Tropic, and the length of the daylight period is constant. Hence, the variation in the amount of insolation throughout the year in the low latitudes (latitudes between the

FIGURE 12-6

The trade wind coast of St. John Island, in the Virgin Islands. Note the heavy growth of tropical forest that reaches to the cliffed shores and the beach. (National Park Service)

Tropics) is greatest at the Tropics and least at the equator. This means that the average annual temperature range decreases from the Tropics to the equator.

Despite this condition, the annual temperature range at the Tropics is relatively small. There is no season when insolation is so reduced as to produce low air temperatures and there is, hence, no winter season.

In addition, a position on or near a windward coast, as in Hawaii or the West Indies (Fig. 12–6), tends to diminish the temperature range during the year. Conversely, a position in a continental interior tends to exaggerate the range. These effects result from the different rates of heating and cooling of land and water masses.

Since temperature ranges are notably small within all tropical humid climates (A), monotonously uniform temperatures are produced, to an even greater extent, in tropical mountains, where the elevation above sea level is increased. The extreme of this condition is illustrated by Quito, in Ecuador (Fig. 12–7), located within 1° of the equator, at an altitude of 9,350 feet. This elevation lowers all temperatures so much that the climate is not of the tropical humid (A) type, but rather, of the tropical mountain type. The warmest month in Quito is September, with an average of 54.9°F. The lowest average, occurring in February, March, April, July, and November, is 54.5°F. Thus, the average annual range is only 0.4°—too small a fraction to be shown on the graph.

Precipitation Controls. The precipitation in all regions of tropical humid climate (A) is associated with a relatively small number of atmospheric conditions. These are: the

FIGURE 12–7

Graphs of climatic statistics for Quito, Ecuador (tropical mountain climate at 0° 17′ S.).

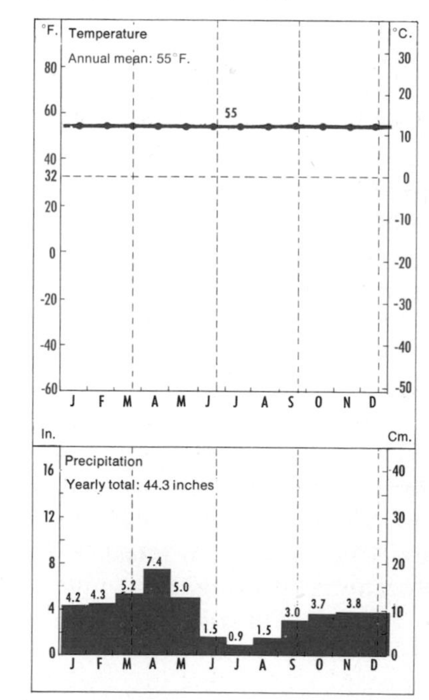

zone of rising air between the belts of trade winds; onshore winds where landmasses intercept the flow of the trade winds; local convectional storms; and monsoons.

Within the doldrums, there is a zone of rising air, induced either thermally or by the presence of the intertropical front (Fig. 9–6). Cooling accompanies the general updraft and brings about condensation of moisture and heavy precipitation. Migration of the trade wind belts with the seasons is accompanied by a similar shift of the zone of rising air. As a consequence, precipitation in the doldrum belt is effective from about 10° N. to about 10° S. during the course of the year. At no one time, however, is this condition effective over the entire 20° belt.

Since the trades blow from an easterly direction, any east-facing coast within the regions where these winds blow provides a means of forcing air to rise, and, hence, a means of producing rainfall. For at least part of the year, the seasonal migration of the trades brings onshore winds nearly to the margins of the low latitudes. The influence of these winds decreases away from the equator and inland from the east coasts.

The tropical wet climate (Af) occurs characteristically where these three forces are most strongly developed. The farther an area is from the zone of convergence of the trades, the farther it is from an east coast, and the farther it is from the zone of maximum heating in which local convection is at a maximum, the smaller the amount of rainfall and the more seasonal its distribution. Thus one passes from the tropical wet (Af) to the tropical dry-winter climate (Aw).

Monsoons upset the simplicity of this pattern. A monsoon typically develops only where a tremendous landmass, through intense summer heating or winter cooling, forms the center of a huge indraft of moist air or a great outdraft of dry air. Even with the existence of monsoon circulation, the rising of air and the resultant precipitation are accomplished primarily by movement onshore at an angle to the direction of the coastline or by the presence of a considerably displaced intertropical front. Monsoons are best developed on low latitude coasts where the trade wind circulation is interrupted. Such a condition exists in southern and southeastern Asia. The presence of a huge landmass north of the equator and an open ocean opposite it in the Southern Hemisphere allows the existence of a well-developed trade wind belt in the Southern Hemisphere, but not in the Northern. The intense heating of the interior of Asia in the period from May through September makes possible an indraft of air from over the Indian Ocean, with resultant heavy rains. During the period from November through March, the accumulation of intensely cold air over the interior of the continent not only prevents any indraft, but causes a strong movement outward, with a consequent lack of rain. A similar summer indraft occurs on the Guinea coast of Africa, but there, the winter outdraft is far less pronounced.

Throughout some areas of tropical humid climate (A), there is much irregularity of precipitation largely because of the variation in the intensity of the precipitation controls. In any one year, the strength of the trade wind circulation may increase or weaken beyond the average. The intertropical front may be more or less well developed than usual. The frequency of convectional storms may change. Monsoon winds may blow with varying strength. In regions where several of these controls act together to produce the total precipitation, the failure of any one may be accompanied by the strengthening of another; thus, there is less uncertainty and less irregularity of precipitation in such places than there is where one force alone is the primary control. Hence, it is within the marginal areas of tropical dry-winter climate (Aw), farthest removed from the equator, from an east coast, or from a monsoon coast, that uncertainty and irregularity of precipitation are most pronounced.

Dry Climates (B)

Climatic Characteristics

Proceeding from the equator toward the poles along continental west coasts, precipitation becomes markedly less. Increasing dryness is reflected in the natural vegetation by the change from forest to grassland and then to desert. Here, the second of the specific climatic types takes over—the *dry climates* (B); see Figures 12–8 and 12–9. These

FIGURE 12–8

The *dry climates (B)* of the world are scattered widely through both the tropical and middle latitudes and encompass portions of all the continents, even a small area in western Europe. (Refer, also, to Plate IV.)

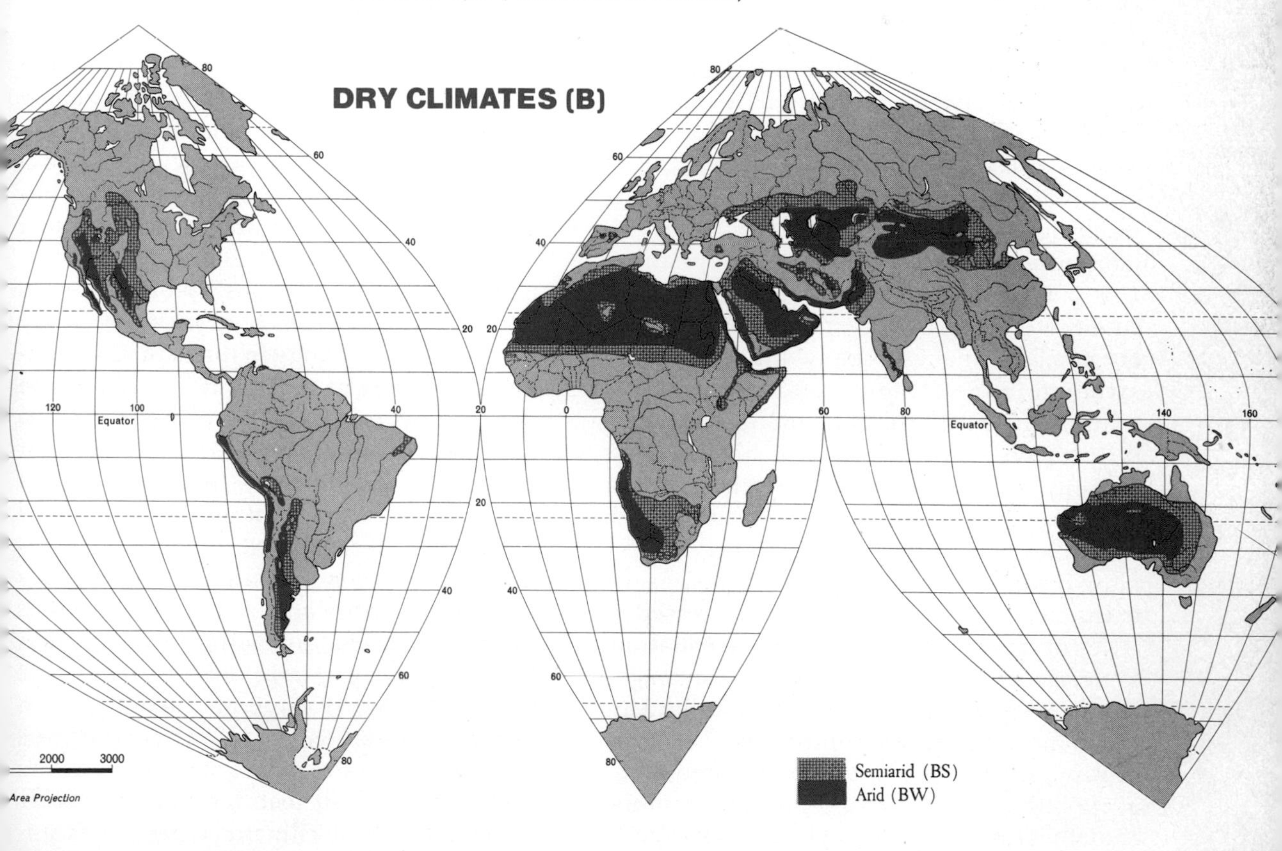

Types	Areas
Semiarid (BSh–BSk)	Northern Hemisphere; western United States; Mexico; northern Africa; Pakistan; central Asia Southern Hemisphere; Argentina; southern Africa; Australia.
Arid (BWh–BWk)	Northern Hemisphere; western United States; Mexico; northern Africa; Pakistan; central Asia Southern Hemisphere; Peru; Chile; Argentina; southern Africa; central Australia.

Characteristics	
Latitudes	tropical and middle
Temperatures	great ranges, both daily and annually, for their latitudes
Precipitation	meager and uncertain, both seasonally and annually—less than potential evaporation
Wind belts	trade wind belts; westerlies wind belts
Air masses	tropical continental; polar continental

FIGURE 12-9
Chart of the principal areas and characteristics of the dry climates (B).

dry climates (B) are influenced largely by tropical and polar continental air masses (Fig. 11–2).

Precipitation Characteristics. The outstanding feature of all dry climates (B) is the meagerness and uncertainty of precipitation. The amount of rainfall is far from being equal to the potential evaporation, and even the small normal average is subject to great variation from year to year. Under such conditions, the moisture available for plant life is incapable of supporting a heavy forest vegetation. No permanent streams can rise within such areas, though large ones may cross them.

The characteristic moisture deficiency of the dry climates (B) cannot be expressed in terms of one specific amount of precipitation, as yearly and seasonal temperature averages affect the evaporation rate profoundly. Under some temperature conditions, 25 inches of rain is sufficient to exceed the amount of evaporation and, thus, to produce a moist climate with its accompanying heavy vegetation. With higher temperatures, the same amount produces a dry climate with sparse vegetation. The division between the dry and the humid climates can therefore be expressed only in terms of the relationship between temperature and precipitation, not in terms of either element alone. A temperature–precipitation relationship that attempts to define the condition of dryness necessary to bring about a change in the natural vegetation from forest to grassland is taken as the boundary between the dry climates (B) and the humid ones. Actually, in most parts of the world, the boundary thus defined lies within the grasslands rather than at the dry edge of the forest; thus, it is close to the median line between the humid and the dry limits of the grassland zones of the world.

The primary divisions of the dry climates are the semiarid (BS) and the arid (BW) types (Fig. 12–8). Both are deficient in moisture but the semiarid is more moist than the arid type. As a consequence of this, the vegetative and soil features found in areas having these two climatic types differ appreciably. A sparse cover of drought-resistant grasses, bushes, and thorny plants takes the place of the continuous cover of the semiarid (BS) areas. No part of the world is known to be absolutely rainless, though some are almost so; hence, except for areas of moving sand dunes and stretches of bare rock with no cover of soil material, areas of arid climate (BW) do have some vegetation.

Rain within regions of arid climate comes at infrequent and irregularly spaced intervals. As a rule, there is no definite rainy season nor is widespread rain a common occurrence. Storms are local, and the effects of any single storm are not felt over any considerable area. Several months or years may pass between successive rains. When they do occur, however, they may be in the form of violent cloudbursts during which a large amount of water falls in a very short time. Yet, it is these storms that replenish the meager underground supplies of water and thus make life possible in these regions. The more gentle showers that occur occasionally barely wet the surface, and the moisture from them is soon returned to the air by evaporation. Clouds often form, with rain falling from them, but none of it ever reaches the ground—high temperatures and dry air cause the rain to evaporate during its descent. Under such conditions, the figure of average yearly rainfall fails to present a complete and true statement, as it suggests only meagerness. To that, the concept of great uncertainty and irregularity must be added. Precipitation in areas of semiarid climate (BS) is not only greater than in arid (BW) areas, but it is also more dependable, as it occurs at more frequent intervals and is more widespread in its effects.

Temperature Characteristics. Temperatures in dry climates (B) represent the whole range found from the low latitudes to the equatorward margins of the high latitudes. The common feature of dryness implies no specific set of temperature values. Yet, what the temperatures of the dry climates do have in common is the element of exceptionally great range, both annual and daily, for the latitudes in which they occur.

In the middle latitudes, the range between the coolest and warmest months is usually between 40° and 50° F. Summers are notably hot and winters are definitely cold. This condition is illustrated by Figure 12–10A, which graphs climatic statistics for Denver, Colorado, which has a mean January temperature of 30°F. and a mean July temperature of 72°F. The mean annual range is 42°. Thus, there is a definite summer season and a definite winter season.

In the low latitudes, the annual range—usually between 15° and 30°—is still very large compared to that of the tropical humid climates (A) or to the climates of tropical uplands or mountains. At Kayes, Mali, in western Africa (Fig. 12–10B), the mean for the coolest month is 77°F. and applies to both December and January. May is the warmest month, with a mean of 96°F. The average annual range is 19°. Thus, even though one season is markedly cooler than the other, no true winter can be said to exist.

It is largely on the basis of the presence or absence of a winter season that further distinctions within both the semiarid (BS) and arid climates (BW) are made. The lack of a cold winter season is characteristic of all dry climates within the low latitudes or on their margins, as in Kayes. Within the middle latitudes, the dry climates (B) have true winter and summer seasons, as in Denver.

Low latitude dry climates (B) have shown the highest temperatures ever recorded on the face of the earth. On the margin of the low latitudes, Azizia, just southwest of Tripoli on the northern coast of Africa, has registered 136° F. in the shade.

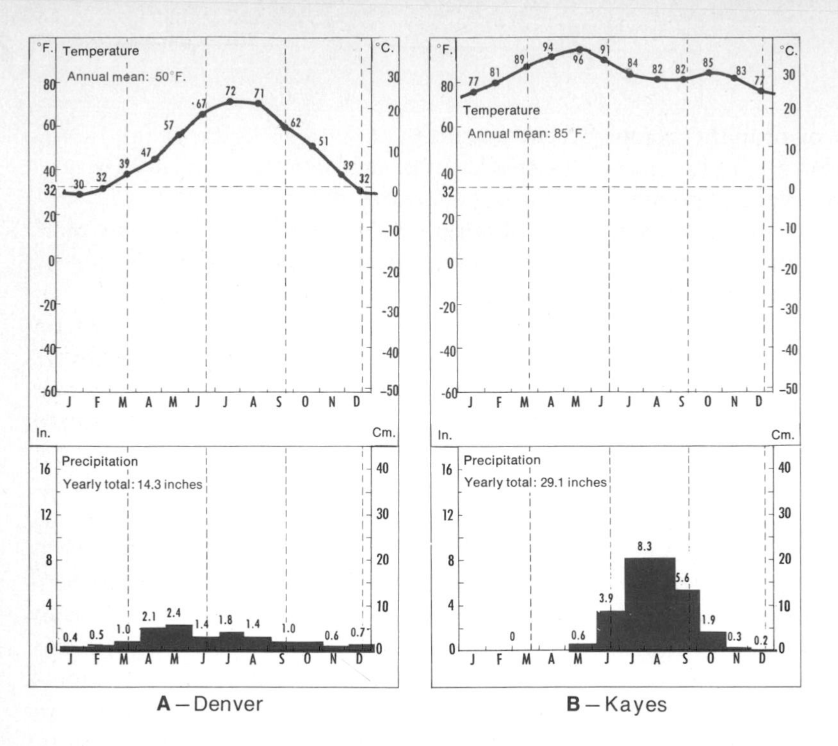

FIGURE 12–10

Graphs of climatic statistics. A. Denver, Colorado (mid-latitude semiarid climate at 39° 44′ N.). B. Kayes, Mali (tropical semiarid climate at 14° 20′N.).

FIGURE 12–11

The pattern of the mean annual amount of cloud cover over the world, expressed in percentages of the sky ordinarily covered by clouds, shows that most dry climates (B), in particular, are areas of comparatively little cloudiness.

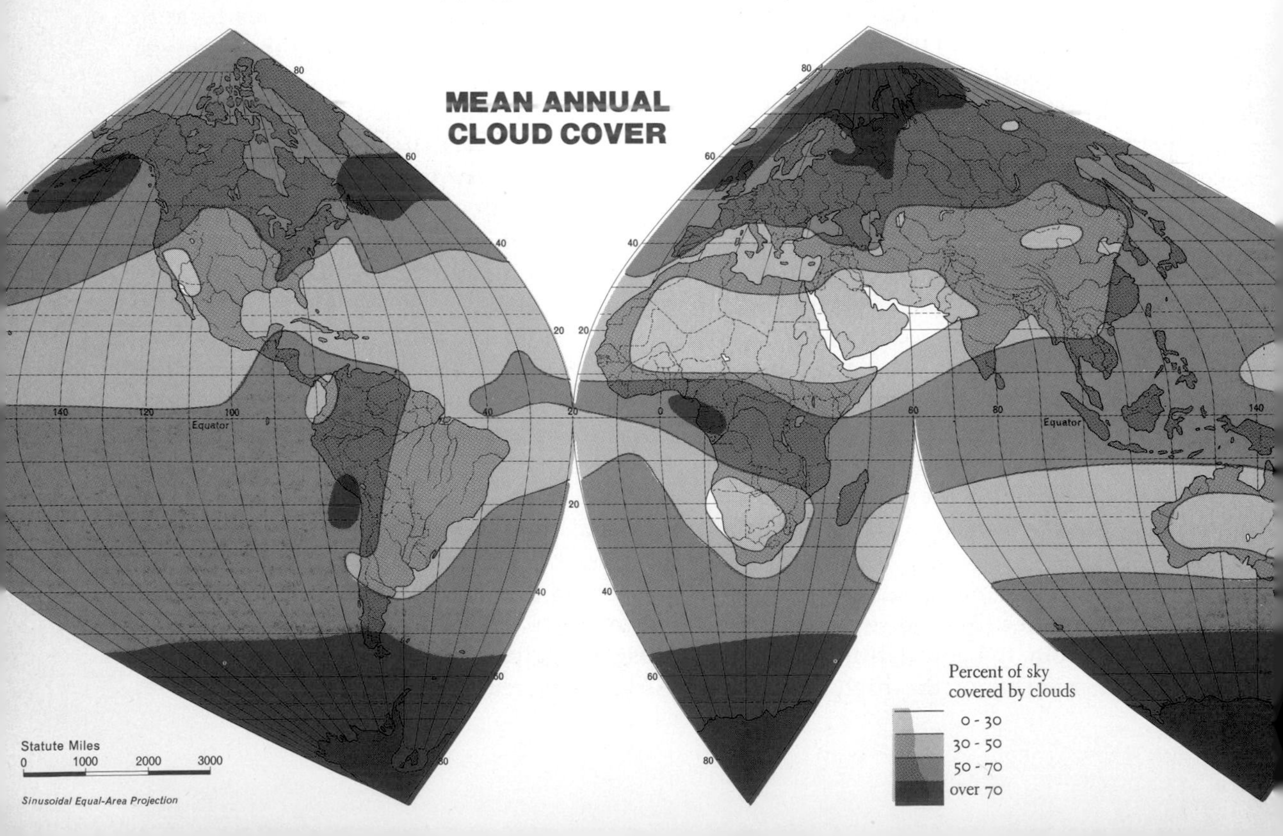

Maximum temperatures of over 100° F. have been recorded in every month except January at Wadi Halfa in northern Sudan, and records indicate that, in that month, a temperature of 99°F. has been attained. In the north-central Australian desert, shade temperature is reported to have exceeded 100°F. for more than sixty consecutive days.

Despite such excessively high temperatures during the day, night readings sometimes are painfully low, often below freezing. The mean daily range in regions of desert climate (BW) in the low latitudes is usually 25° F. to 30° F. Extreme daily ranges often are as great as 70°; at Azizia, which holds the record, there was a 100° range recorded—from 126°F. to 26°F.—within one 24-hour period. Ordinarily, the daily ranges are greater in the warmer part of the year than in the cooler part, but, at all times in most dry climates (B), the contrast between day and night is startling.

Other Features. Two other outstanding features in areas of dry climate (B) are insolation and wind. Both are at a maximum in the low latitude dry climates, but they are significant elements in the other dry climates as well. Throughout the hotter and drier sections, skies are seldom obscured by clouds (Fig. 12–11), and over 80 percent of the possible sunlight is received at the earth's surface. Cloudiness increases to about 50 percent in the cooler and moister sections, such as those that border the ocean or adjoin humid climates in the middle latitudes. High winds are common, especially during the daytime. As a result, the air is frequently laden with dust particles. Dust storms are often sufficiently heavy to blot out the sky, and their force is great enough, though they seldom last for long periods, to make them among the most destructive, hazardous, and uncomfortable phenomena of the world's dry climatic (B) regions.

Subtypes of the Dry Climates (B)

Semiarid Climate (BS). Kayes, in Mali (Fig. 12–10B) has a representative low latitude semiarid climate (Fig. 12–8). As in the tropical humid climates (A), there is no true winter season, but the range of 19° between the warmest and coolest months exceeds that found in the tropical humid types. Precipitation is small in amount, yet it is much larger than that in arid climates (BW). Rain is distributed through the year in a manner similar to that in the tropical dry-winter climate (Aw) (Fig. 12–4B), in which drought occurs during the low-sun period. The precipitation at Kayes is, however, too small in amount to equal the potential evaporation at the temperatures there. There is less dependence upon additional sources of moisture than in areas of arid climate, but any continued intensive cultivation of the soil must usually be supported by irrigation.

In the middle latitudes, the semiarid climate (BS) is characteristically represented by Denver, Colorado (Fig. 12–10A). There is a marked winter season, during which most of the precipitation is in the form of snow. The average annual temperature range is high—42°. High summer temperatures are common, with 105° F. as the maximum recorded; low winter temperatures are indicated by the minimum record of −29° F. As in the low latitudes, total precipitation is greater in amount and is much more dependable than in desert areas. While relative humidity is still low compared to that of the humid climates, the average is much higher than that in arid climates (BW). Consequently, the temperatures, especially those of winter, appear to be even more extreme than they actually are. Intensive agricultural use of land with this type of climate demands irrigation, though pastoral use and extensive cultivation are possible without it.

Arid Climate (BW). The general characteristics of the arid (BW) type (Fig. 12–8) as it occurs near the margins of the low latitudes are illustrated by Yuma, Arizona (Fig. 12–12A). Precipitation is insignificant in amount, and what little there is, is unevenly distributed throughout the year. Though a yearly average of 3.3 inches is shown, the rainfall is uncertain and variable in amount; during one year, Yuma received less than one inch, while during another, it received over eleven inches. January is the coolest month, with a mean temperature of 55° F., and July is the hottest, with a mean of 91° F. Thus, the yearly range is 36°. A maximum of 120° F. and a minimum of 22° F. have been recorded. There is, on the average, bright sunshine 90 percent of the year. The excessively high temperatures caused by the blazing sun in July and August are made bearable by the dryness of the air. Relative humidity is low, averaging 55 percent in the early morning and only 27 percent in the late afternoon. Human utilization of an area with this type of climate is strictly limited to the regions that can be supplied with water in addition to that received as rain.

The climate of Albuquerque, New Mexico (Fig. 12–12B), is more truly representative of the middle latitude arid climate (BW) than is that of Yuma. In contrast to Yuma, the hot season is less intense, and there is a pronounced winter. The highest monthly mean temperature, that for July, is 77°F., 14° less than that at Yuma; the coldest monthly mean, that for December and January, is 34°F., over 20° lower than that at Yuma. The maximum in summer reaches 104°F., but the winter minimum drops to –10°F. Killing frosts may be experienced at any time from the end of October to the middle of April, and snow is not unknown as a form of winter

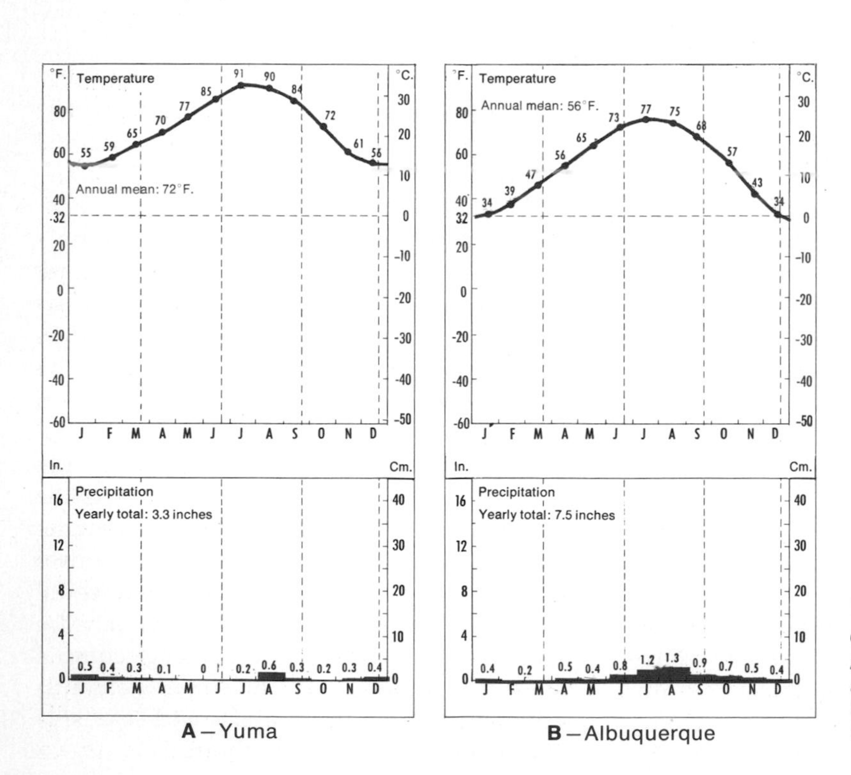

FIGURE 12–12

Graphs of climatic statistics. A. Yuma, Arizona (arid climate at 32° 40′ N.). B. Albuquerque, New Mexico (arid climate at 35° 05′ N.).

precipitation. Bright sunshine occurs nearly as much of the time as it does at Yuma, and the air has the same characteristic dryness. Human utilization of the land, as in all arid climates, tends to be limited by the availability of water.

Global Distribution

The distribution of the dry climates (B) throughout the world is shown in Fig. 12–8. The dry climates extend inland from the low latitude west coasts, swinging poleward as they penetrate into the hearts of the landmasses. The areas of arid climate (BW) are flanked by regions of semiarid climate (BS). On the equatorward side of the areas of arid climate, the semiarid areas provide the transition from the arid (BW) to the tropical dry-winter climates (Aw); on the poleward side, the semiarid areas indicate the transition from the arid (BW) to the humid mild-winter (C) or the humid severe-winter (D) climatic types.

Dominant Controls

Precipitation Controls. Distance from sources of moisture, orographic barriers, or the lack of winds moving in the proper direction to transport moisture are responsible for the meager rainfall in all dry climates (B). The total rainfall in steppe (BS) areas is greater than that in arid (BW) ones, because the former are either closer to moisture sources or are so located that rainbearing winds are experienced more frequently. Such precipitation as there is in areas of arid climate usually results from local convectional storms, whereas that in areas of semiarid climate is usually produced by causes effective in adjacent humid climatic regions. Within the low latitudes, the controls are those of the tropical humid climates (A); in the middle latitudes, cyclonic storms cause the greater part of the precipitation.

Temperature Controls. The great annual and daily ranges of temperature experienced within areas of dry climate (B) are directly related to the dryness of the air and to the sparsity of the vegetation cover. Dry air permits a high proportion of solar energy to reach the surface during the daylight period. Clear, cloudless skies, resulting from the relatively low moisture content of the air, allow heating to continue uninterruptedly over long hours. Little of this energy is absorbed by the sparse plant cover or used in the evaporation of water. Thus, surfaces are often heated to such high temperatures that they are painful to the touch. Rapid radiation from the earth's surface produces very high air temperatures. With the setting of the sun, however, the receipt of solar energy by the earth's surface stops, and the dry air and cloudless skies allow a high loss of heat to outer space. Consequently, air temperatures drop with startling rapidity, so that the nights often seem very chilly after the glaring heat of the days.

Maximum daytime heating occurs when the noon sun is most nearly overhead. Though the noon sun is never at a very low angle in these latitudes, the difference in heating between the high- and low-sun seasons is more marked than in humid areas in the same latitudes.

In the middle latitudes there is no period in the year when the noon sun is directly overhead. But, in the season when the sun is highest, the daylight period is much longer than in the low latitudes (Fig. 8–10). Hence, total insolation in summer is relatively large in amount, and that season is relatively hot. In winter, the sun is at a much lower angle than it ever is in the low latitudes, and the daylight period is much shorter. The receipt of solar energy is thus greatly reduced. The process of radiation, however, is as little interrupted in the middle latitude areas of dry climate (B) as it is in those of the low latitudes. Consequently, the low-sun season is notably cold. The range between the seasons in the middle latitudes is thus even greater than the range between the seasons in the low latitude areas of dry climate.

Humid Mild-Winter Climates (C)

Other humid climates, by general groupings, are the humid mild-winter climates (C), the humid severe-winter climates (D), and the polar climates (E). It should be recalled that the seasons in regions of tropical humid climates (A) are defined primarily in terms of precipitation conditions. In these other regions of humid climate, however, it is temperature conditions that produce the outstanding differences between the seasons.

FIGURE 12–13

The *humid mild-winter climates (C)* of the world are concentrated largely within the subtropical low latitudes of both hemispheres and occur in all the continents. (Refer, also, to Plate IV.)

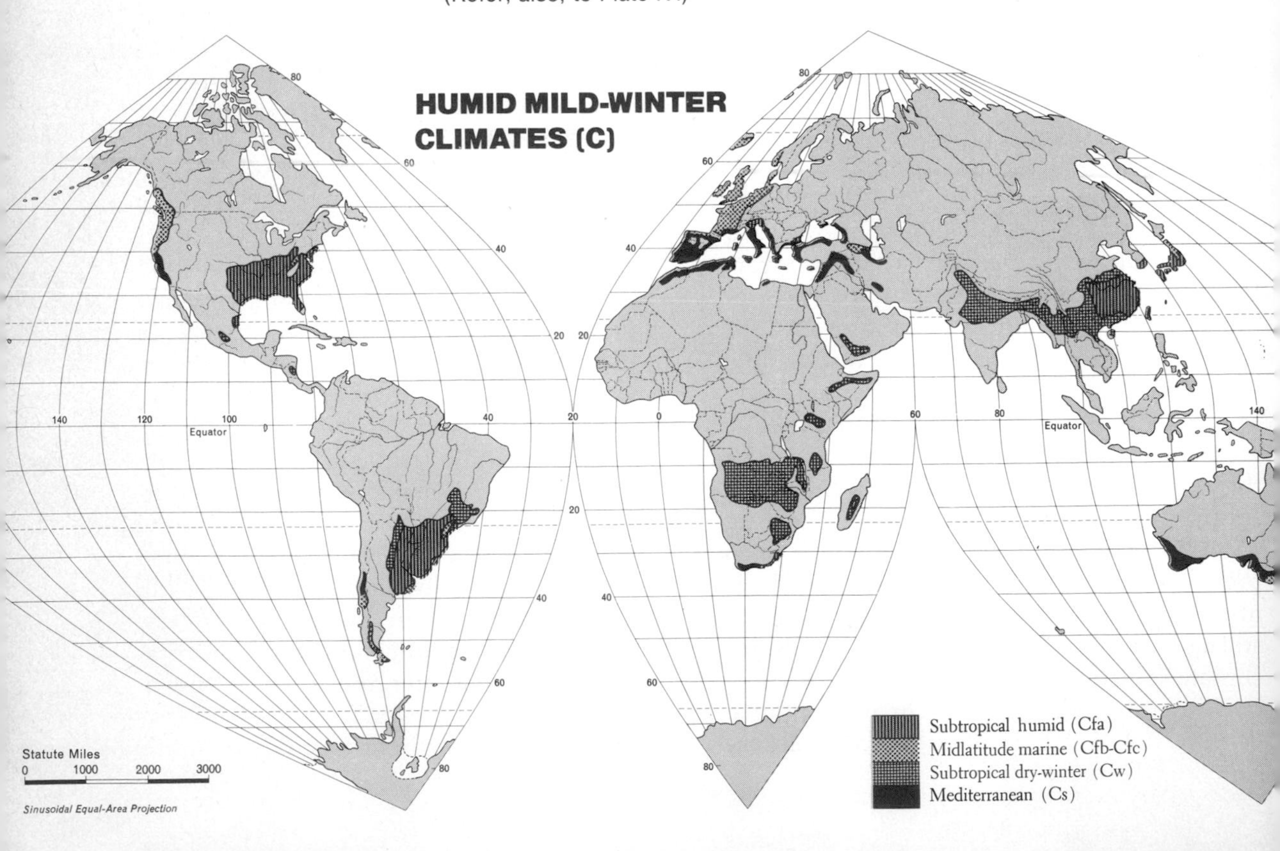

Types	Areas
Subtropical humid (Cfa)	Northern Hemisphere: southern United States; central China; Korea; Japan Southern Hemisphere: South America; Australia
Midlatitude marine (Cfb–Cfc)	Northern Hemisphere: western United States; western Europe—United Kingdom to Iceland; northern Japan Southern Hemisphere: southern Chile; southern Africa; eastern Australia; New Zealand
Subtropical dry-winter (Cw)	Northern Hemisphere: India; Bangladesh; Burma; southern China Southern Hemisphere: eastern Brazil; central Africa; eastern Australia
Mediterranean (Cs)	Northern Hemisphere: California; Mediterranean littorals Southern Hemisphere: middle Chile; southern Africa; southern Australia

Characteristics	
Latitudes	subtropical low latitudes of all the continents
Temperatures	mild winters; no mean monthly temperature below 32° F.
Precipitation	humid, considerable variation, both seasonally and annually
Wind belts	westerlies wind belts
Air masses	tropical maritime; tropical continental; polar maritime

FIGURE 12–14
Chart of the principal areas and characteristics of the humid mild-winter climates (C).

Climatic Characteristics

The humid mild-winter climates (C) occur poleward of both the tropical humid climates (A) and the tropical portions of the dry climates (B). They lie sufficiently poleward to possess mild-winter seasons, but not far enough poleward to have severe winters (Figs. 12–13 and 12–14). These humid mild-winter climates (C) are controlled largely by the interactions of tropical maritime and continental air masses and polar maritime air masses (Fig. 11–2).

Temperature Characteristics. In the humid mild-winter climates (C), temperatures during part of the year drop below those in the tropical humid climates (A). The lower temperatures are reflected in shortened growing seasons for both natural vegetation and crops, and they also effect different individual plant species from those characteristic of the humid tropics and introduce different seasonal rhythms into the cycle of plant life.

Along the line of contact of the mild-winter climates (C) with the tropical humid climates (A) and the low latitude dry climates (B), the seasonal rhythm is poorly developed; many of the forest trees remain leaf-covered and green throughout the year. Along the poleward margins of the humid mild-winter climates (C), the seasonal changes are marked by budding, flowering, and leafing in spring and by the loss of leaves in autumn.

The presence of a mild-winter season is reflected in many other features, such as the kinds of crops grown, the systems of cultivation used, and the types of buildings in which people live.

On the tropical margins, the winter monthly mean temperatures are in the low 60's, as at Hong Kong (Fig. 12–15A). Here, the mean for the coolest month, February, is 59° F. Farther poleward, the winter monthly mean temperatures are in the low 50's, as at Sacramento, California (Fig. 12–15B), where the mean for December and January is 46° F. In the continental interiors and near the poleward margins of the humid mild-winter climates (C), low monthly means are in the 40's, as at Nashville, Tennessee (Fig. 12–15C), with a low mean of 39° F. for January. Winter temperatures are considered to become truly severe where the mean temperature approximates 32° F. for the coldest month. It is this figure that may be taken as the cold limit of the humid mild-winter climates (C).

From the point of view of human activity, the nature of the summer is as significant as that of the winter. On the equatorward margins and in the continental interiors, the summers are long and hot, frequently with means in the high 70's or the low 80's. Hong Kong has a July mean of 82° F., and Nashville one of 79° F. In coastal locations and where marine influences penetrate well into the interior, the summers are warm rather than hot and are somewhat shorter. San

FIGURE 12–15

Graphs of climatic statistics. A. Hong Kong (subtropical dry-winter climate at 22° 15′ N.). B. Sacramento, California (Mediterranean climate at 38° 35′ N.). C. Nashville, Tennessee (subtropical humid climate at 36° 10′ N.).

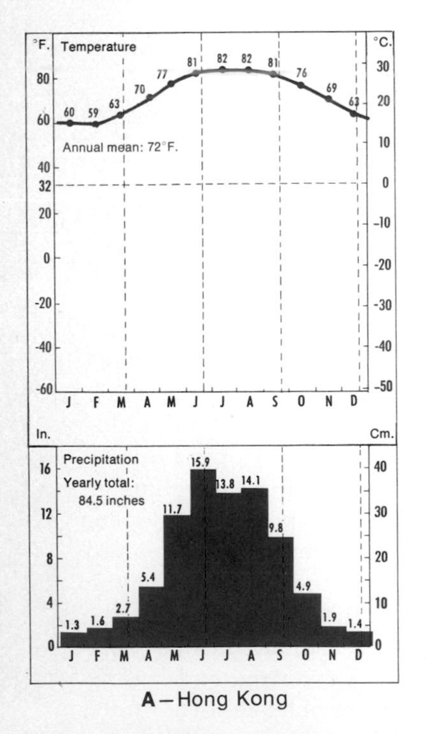

A—Hong Kong

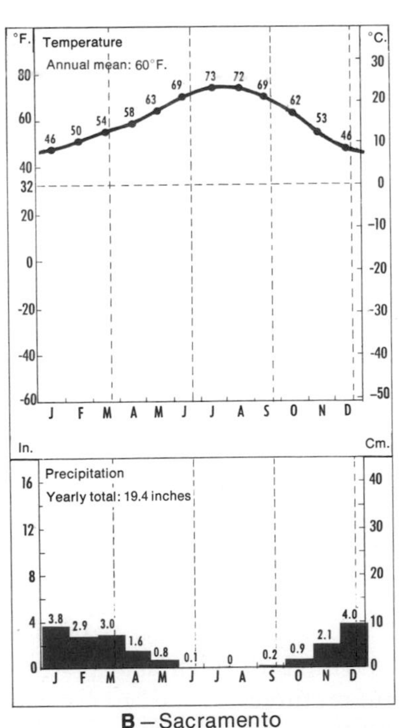

B—Sacramento

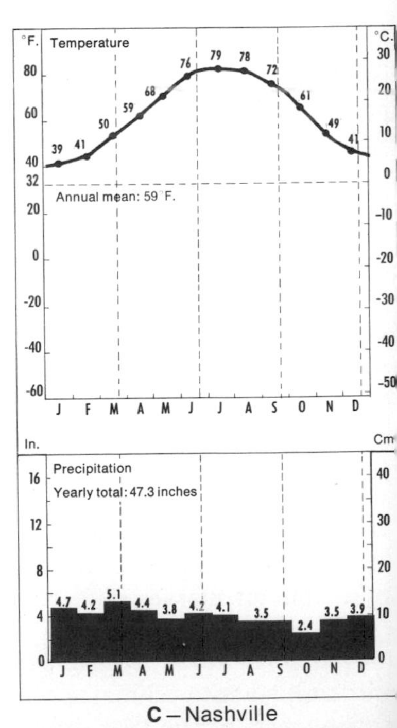

C—Nashville

Francisco's warmest month (Fig. 12–16C) has a mean of only 60° F., and Greenwich, England (Fig. 12–16A) records a mean of only 63° F. Some parts of west coasts in the higher middle latitudes have a very short, very cool summer. At Lerwick, in the Shetland Islands off the northern coast of Scotland (Fig. 12–16B), the warmest months are July and August, with means of 52° F., and there are only three months with means of 50° F. or higher.

Precipitation Characteristics. None of the humid mild-winter climates (C) is deficient in moisture for the year as a whole, but some of them show seasonal deficiencies. There are three possible explanations for rainfall distribution through the year: the rain may be concentrated in the summer season, as at Hong Kong (Fig. 12–15A); it may be concentrated in the winter period, as at Sacramento (Fig. 12–15B); or it may be evenly distributed throughout the year, as at Nashville (Fig. 12–15C).

In at least one respect, an area with summer rains is like one in which rainfall is evenly distributed throughout the year: in both, rainfall is usually plentiful during the warmer part of the year, at which time plant growth is greatest. The advantage of these two kinds of rain distribution for crops and, hence, the food supply, is obvious. That the winter is dry in one instance only serves to emphasize the period of plant rest induced by lower temperatures.

FIGURE 12–16

Graphs of climatic statistics. A. Greenwich (London), England (midlatitude marine climate at 51° 28′ N.). B. Lerwick, Scotland (midlatitude marine climate at 60° 08′ N.). C. San Francisco, California (Mediterranean climate at 37° 45′ N.).

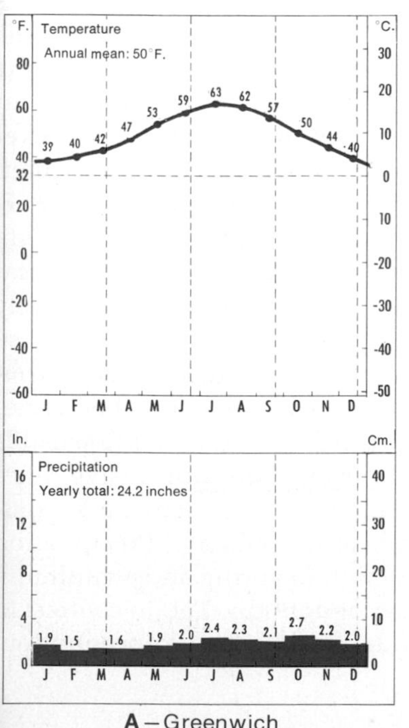

A – Greenwich

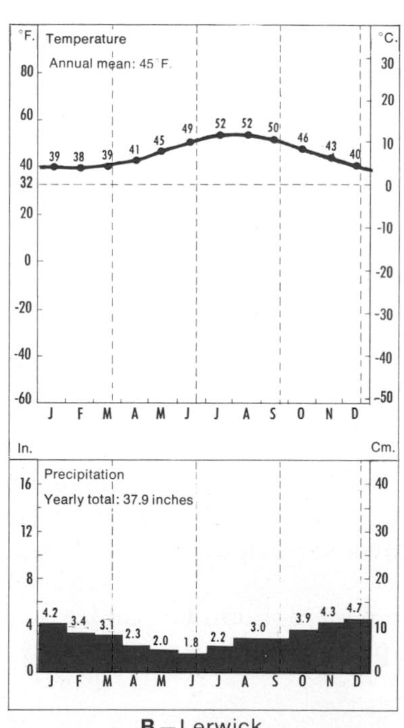

B – Lerwick

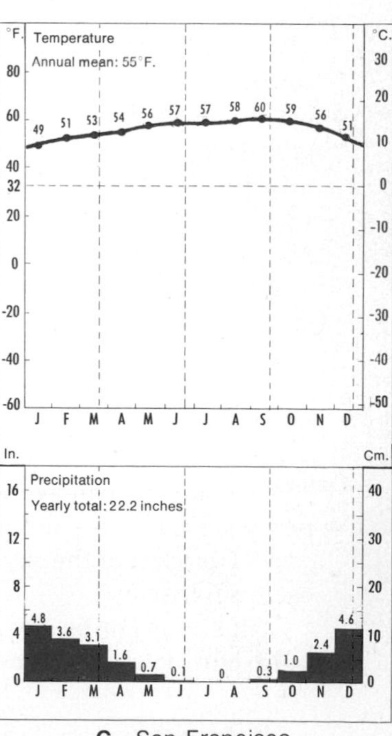

C – San Francisco

The condition represented by winter rain is quite different. The rain of winter is followed by summer drought of an intensity equal to that of the dry climates (B). Plant growth is greatly restricted, and intensive cultivation is dependent on irrigation during the growing season. Fortunately, the winters in areas where rainfall is concentrated in that season are usually mild, or even warm, as, for example, in southern California or along the Mediterranean shores of Europe and Africa, so that cessation of plant growth is not characteristic.

Precipitation varies greatly in amount and in form throughout the humid mild-winter climates (C). On a windward coast, especially if it is hilly or mountainous, such as that of northwestern Scotland, and on the equatorward margins, as along the coasts of Mississippi, Alabama, and western Florida, the amount is often excessive—80 to 100 inches a year is not uncommon. The total is much less in continental interiors, as in eastern Kansas and Oklahoma, or in the lee of highlands, as in southeastern Scotland. The specific controls of precipitation vary widely, but areas within this climatic region are subject to both the kind of rain that continues gently over relatively long periods of time and the kind that falls torrentially for short periods.

Local convectional storms are almost unknown in this climatic region. In terms of the number of rainy days per year, some of the cloudiest and rainiest areas in the world are located here. Yet the amount of rain is often remarkably small. Lerwick, though recording an average yearly rainfall of only 37.9 inches, has an average of 260 rainy days per year, and, on the average, between 70 and 80 percent of the sky is covered by clouds each month.

General rains are common in humid mild-winter areas where precipitation is concentrated in the winter. They may be heavy or light, and they may vary in duration from a few hours to several days. Local convectional storms, however, are infrequent, and the amount of rain they cause is negligible.

Subtypes of the Humid Mild-Winter Climates (C)

Differences in the occurrence of precipitation through the year and in the relative warmth of the summer in the humid mild-winter climates (C) are sufficient to create four distinct subtypes of this climate. These are shown on the map in Figure 12–13 as: subtropical humid climate (Cfa); midlatitude marine climate (Cfb and Cfc); subtropical dry-winter climate (Cw); and Mediterranean climate (Cs).

Subtropical Humid Climate (Cfa). The world distribution of this climate is shown on Figure 12–13, and its major characteristics are illustrated by Figure 12–15C, the climatic graph for Nashville, Tennessee. On this graph, note that the long, hot summer—four months with a mean temperature of over 70° F.—contrasts sharply with the winter, when the lowest monthly average is 39° F. Daily temperatures in midsummer are high, usually in the 80's or 90's, and they occasionally exceed 100° F. This heat is normally accompanied by high relative humidities, so that the temperature one feels (*sensible temperature*) is much higher than the thermometer temperature indicates. Often, the sensible temperature is even more oppressive, for a short time, than that produced by the steaminess of a tropical wet climate. Precipitation shows a maximum in spring and a minimum in late summer or autumn. Yet, in contrast to areas of well-defined seasonal concentration, the rainfall is rather evenly spread throughout the year (Fig. 12–15C). Though the average figures show adequate moisture throughout the year, droughts of short duration are common. Snow occurs in winter, but it seldom remains on the ground for more than a few days at a time. The ground seldom freezes deeply and never remains frozen throughout the winter.

Midlatitude Marine Climate (Cfb and Cfc). The world distribution of this climate is shown on Figure 12–13. Though there are areas with these types of climate on the west coast of North America, better examples occur in northwestern Europe. Consequently, Greenwich, England (Fig. 12–16A) is chosen to represent the midlatitude marine type (Cfb). In contrast to those of the subtropical humid climate (Cfa), the summers are warm rather than hot and the winters are milder. Nashville and Greenwich have the same low mean temperature of 39°F. for January, yet Greenwich is 15° of latitude farther poleward than Nashville. The warmest month in Greenwich is 16° cooler than that in Nashville. In the midlatitude marine climate (Cfb), precipitation is distributed throughout the year in a manner similar to that in the subtropical humid climate (Cfa), but it is rarely of the local convectional type. Relative humidity is higher over the year period, and there is greater cloudiness.

The extreme phase of the midlatitude marine type, the Cfc phase, is illustrated by the climatic graph for Lerwick, Scotland (Fig. 12–16B). The summers are so cool that agricultural use of the land is almost impossible, yet the winters are only slightly more severe than in the subtropical humid climate. Lerwick is nearly 25° of latitude farther north than Nashville, yet its coolest winter month averages only a fraction of a degree lower. On the other hand, freezing temperatures have been recorded at Lerwick in every month except July and August. Although the average annual precipitation is only 37.9 inches, that amount is excessive, for, at the prevailing temperatures, evaporation is low.

Subtropical Dry-Winter Climate (Cw). The distribution of this climate is shown on Figure 12–13. It occurs, for example, in Hong Kong (Fig. 12–15A), where a very hot and humid summer contrasts with a cooler and drier winter. Except for the low temperatures at one season, weather conditions are similar to those in the tropical monsoon climate (Am), as at Akyab (Fig. 12–4C). The winter dry period is seldom completely without rain, and usually enough rain falls during the summer to maintain soil moisture through the year. Since the winter is not cool enough to experience more than rare frosts, it is possible to carry on agriculture throughout the year.

Mediterranean Climate (Cs). The global distribution of areas with a Mediterranean climate (Cs) is shown on Figure 12–13. It is unique among all the world's climates, in that precipitation is concentrated in winter while the summers are droughty. Despite this unique feature, there are two distinct phases of the Mediterranean climate: the *coastal phase,* which has shorter and warmer summers, and the *interior phase,* which has longer and hotter summers.

In coastal locations, which are fully exposed to marine influences, summer temperatures are reduced and winter ones increased. Relative humidity is high throughout the year. Cloudiness is characteristic, and fog is a common phenomenon. Summer high temperatures rarely attain 100° F., and freezing days are likely to occur only during a very brief period in midwinter and, then, chiefly on the poleward margins of these areas. This type, the coastal phase of the Mediterranean climate (Csb), is represented by San Francisco (Fig. 12–16C).

In interior locations, which are away from marine influences, summer temperatures are higher and winter ones are lower; relative humidity is lower, especially in summer; and cloud and fog are greatly reduced. This type, the interior phase of the Mediterranean climate (Csa), is represented by Sacramento, California (Fig. 12–15B), where burningly dry, sunny days succeed one another almost without interruption through the summer. Temperatures in that season frequently approach those of the low latitude semiarid (BS) and arid climates (BW), with daily maxima between 85° and 95° F. Absolute maxima are usually well over 100°F. in Sacramento, and a temperature of 114°F. has been recorded there. Freezing temperatures are more common in such an inland Mediterranean climate area than they are in Mediterranean climate (Csb) areas along coasts, and may occur over a two-month period in winter.

Global Distribution

Between the coasts and the dry interiors, humid mild-winter climates (C) are found (Fig. 12–13). West of the dry interiors, Mediterranean (Cs) types occur between 30° and 40° latitude in both hemispheres. They are succeeded poleward by midlatitude marine types (Cfb and Cfc) that continue to approximately 60° latitude in both hemispheres. Small areas of subtropical humid climate (Cfa) are found in the interior adjacent to the dry margin, particularly in parts of southeastern Europe. East of the dry interiors, subtropical humid (Cfa) or subtropical dry-winter types (Cw) extend poleward to about 40° latitude from areas of tropical wet (Af) and tropical dry-winter climates (Aw).

Dominant Controls

Temperature Controls. The presence of a winter season is a direct reflection of distance from the equator. In addition, distance from the equator produces greater mean annual temperature ranges. The mean annual range is controlled basically by seasonal changes in insolation. Since these changes increase away from the low latitudes, mean annual temperature ranges also increase in that direction. Marine influences counteract this tendency, and continental influences amplify it. Since the middle latitudes, where the humid mild-winter climates (C) occur, are the regions of prevailing westerlies, it is along the continental west coasts that marine influences are strongest. Marine influences are weaker along east coasts and weakest in continental interiors.

Precipitation Controls. Numerous controls of precipitation are operative in the different types of humid mild-winter climates (C). The prevailing westerlies are rainbearing onshore winds throughout the year in midlatitude marine areas (Cfb and Cfc). Cyclonic storms, following the general path of the westerlies, add to the total precipitation. Areas of Mediterranean climate (Cs) likewise receive most of their rain from the westerlies and from cyclonic storms, but these two controls are effective only during the winter in the latitudes where the Mediterranean climate (Cs) occurs. This is primarily a reflection of the seasonal shift of wind belts. In areas of subtropical humid climate (Cfa), cyclonic storms are the major source of rain in all seasons, but that derived from local convectional storms, which occur in summer, is another important source. Within areas of subtropical dry-winter climate (Cw), the major controls are the same as those in tropical dry-winter (Aw) and tropical monsoon climates (Am); they are convection, movements of the intertropical front, onshore winds in the belt of trades, and monsoons. The slight rains of winter in subtropical dry-winter areas (Cw) are generally associated with the passage, at irregular intervals, of weakly developed cyclonic storms.

Humid Severe-Winter Climates (D)

Climatic Characteristics

Poleward from the humid mild-winter climates (C) are the humid severe-winter climates (D); see Figures 12–17 and 12–18. Here, again, there are distinct characteristics for the category as a whole, as well as individual climatic types within the general

FIGURE 12–17

The *humid severe-winter climates (D)* of the world are found only within the middle and higher latitudes of the Northern Hemisphere—within North America and Eurasia. There are no climates of this type in the Southern Hemisphere. (Refer, also, to Plate IV.)

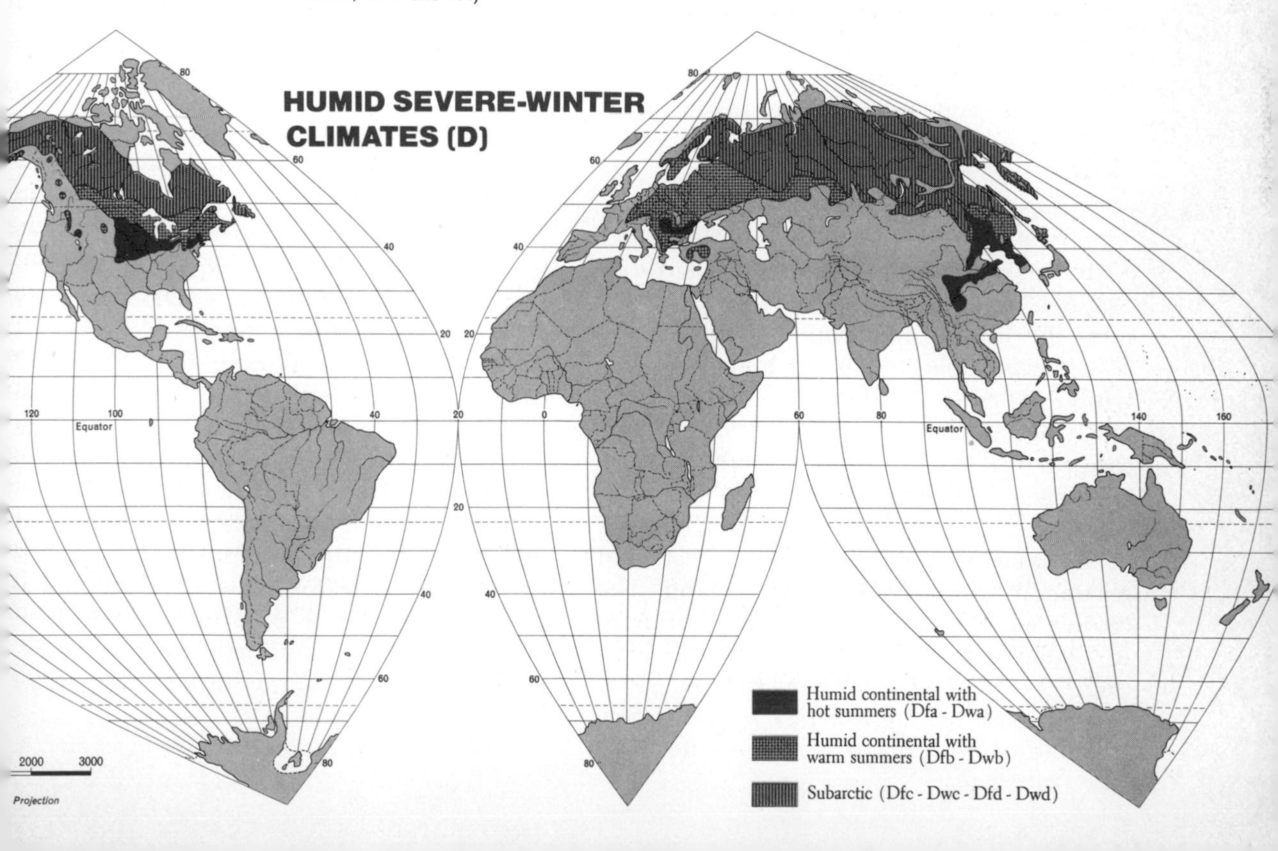

Types	Areas
Humid continental with hot summers (Dfa–Dwa)	Northern Hemisphere: central United States; eastern Europe; central China Southern Hemisphere: none
Humid continental with warm summers (Dfb–Dwb)	Northern Hemisphere: northern United States; southern Canada; eastern Europe; western Siberia; Manchuria Southern Hemisphere: none
Subarctic (Dfc–Dwc–Dfd–Dwd)	Northern Hemisphere: northern Canada; Sweden; Finland; northern Siberia Southern Hemisphere: none

	Characteristics
Latitudes	middle and high latitudes, from North America to Eurasia (Northern Hemisphere only)
Temperatures	great extremes, from hot summers to severe winters—greatest seasonal ranges in the world
Precipitation	moderate to light; summer rains and winter snows
Wind belts	westerlies wind belts; polar wind belts
Air masses	polar continental; polar maritime; tropical continental; tropical maritime

FIGURE 12–18

Chart of the principal areas and characteristics of the humid severe-winter climates (D).

category. These humid severe-winter climates are controlled largely by the interactions of polar continental with maritime air masses and tropical continental with maritime air masses (Fig. 11–2).

Temperature Characteristics. Colder, longer, and more continuous winters, and a greater range from season to season distinguish the humid severe-winter climates from the humid mild-winter ones. On the equatorward margins of the humid severe-winter climates (D), the coldest monthly mean is 32°F. Poleward from there, and toward the continental interiors, winters become increasingly severe. The extreme is reached in northeastern Siberia, where continental influences are strongest and marine influences are totally lacking. The coldest winters in the world, except for those in Antarctica, occur there. Verkhoyansk (Fig. 12–19A) has an average of −59° F. for January; and Sredne Kolymsk (Fig. 12–19B), has one of −41° F. On winter nights, temperatures well below those figures are regularly experienced. At Oimekon, in the same general area, an absolute minimum of −108° F. has been reported.

Extreme continental position in high latitudes is necessary to the production of such low temperatures. In the Southern Hemisphere there are no large landmasses in similar latitudes. Consequently, the humid severe-winter climates (D) occur *only* in the Northern Hemisphere.

Frost rules the winter; and the ground freezes to great depths, and remains frozen, for several months. In the most extreme areas, the subsoil is always frozen; only the topsoil ever thaws completely. The length and continuity of the winters are made more pronounced by the suddenness of the transition to and from summer. Some additional seasonal contrast is provided by the long daylight hours of summer and the correspondingly long hours of darkness in winter (Fig. 8–10).

There is great variation in the character

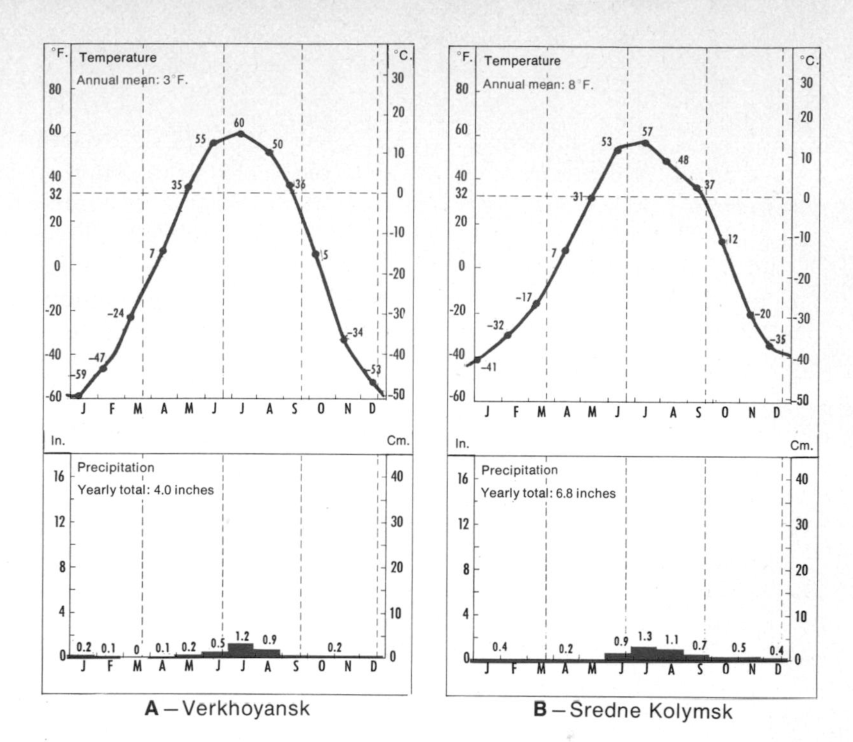

FIGURE 12–19

Graphs of climatic statistics. A. Verkhoyansk (subarctic climate at 67° 43′ N.). B. Sredne Kolymsk (subarctic climate at 67° 49′ N.). (Both locations are in the northeastern portion of the Soviet Union.)

of the summers. Comparatively long, hot summers characterize the southern margin in the continental interiors. Averages in the 70's are usual, and not infrequently during the summer, 100°F. is exceeded for a short period. More frequent daytime fluctuations, regularly lower night temperatures, and lower relative humidity make these high summer temperatures less oppressive than they are in the humid mild-winter climates (C). Summers rapidly become cooler in a northward direction (Fig. 12–20A) and toward coastal areas (Fig. 12–20B). Long, hot summers grade into short, warm summers and, finally, into very short, cool summers. The coolness and shortness of the summer in the higher latitudes are partly compensated for by the longer daylight hours. Close to the Arctic Circle, and within it, there is no real darkness in midsummer, and, as the season progresses, the days are as long—in hours—as those in regions much farther south.

The average annual temperature ranges within the humid severe-winter climates (D)

FIGURE 12–20

The effect of latitude and location on mean summer temperatures in humid severe-winter climate regions (D). A. The decrease of mean summer temperatures northward. B. The decrease of mean summer temperatures from interior areas toward the coasts.

Place	*Approximate latitude (°N.)*	*Mean temperature (July)*
Keokuk, Iowa	40°	76.1°
Cedar Rapids, Iowa	42°	75.6°
Charles City, Iowa	44°	73.2°
St. Paul, Minnesota	45°	72.2°
Bemidji, Minnesota	48°	68.9°
Winnipeg, Manitoba	50°	66°

A

Place	*Approximate latitude (°N.)*	*Approximate distance from ocean (in miles)*	*Mean temperature (July)*
La Crosse, Wisconsin	43–44°	1,050	72.8°
Mt. Pleasant, Michigan	43–44°	730	71.0°
Rutland, Vermont	43–44°	130	69.0°
Plymouth, New Hampshire	43–44°	100	68.1°
Portland, Maine	43–44°	0	67.8°

B

are the greatest in the world. Verkhoyansk has a warm month mean of 60° F. and a cold month mean of −59°F. (Fig. 12–19A). Its range—119°F.—is thus quite extreme. In coastal locations and along the southern margins, the range is between 40° and 50°, and it increases northward and toward continental interiors.

Precipitation Characteristics. Precipitation varies greatly in amount and in distribution throughout the year in the humid severe-winter climates (D). It is heaviest in coastal areas, as at Portland, Maine (Fig. 12–21A), where 40 to 50 inches per year may be expected. It decreases toward continental interiors, where 25 to 35 inches per year is normal, as at Chicago, Illinois (Fig. 12–22A). The amount also decreases northward, to an extreme of less than 5 inches, as at Verkhoyansk. In the humid severe-winter climates (D) there is little variation from year to year. That so small an amount should suffice to create a moist climate is a measure of the effectiveness of moisture in areas of relatively low temperatures.

Throughout most of the humid severe-winter climates (D), the snows of winter and the rains of summer are rather evenly distributed. In northeastern Asia, however, snowfall is meager, and most of the precipitation is received as summer rain. For many purposes, the presence of adequate moisture during the season of greatest heat is of major importance. Hence, whether the precipitation is evenly distributed throughout the year or concentrated in summer, the results are very similar.

Winter precipitation is almost entirely in the form of snow, which accumulates to form a blanket that lasts through the winter. The thickness of this snow blanket decreases

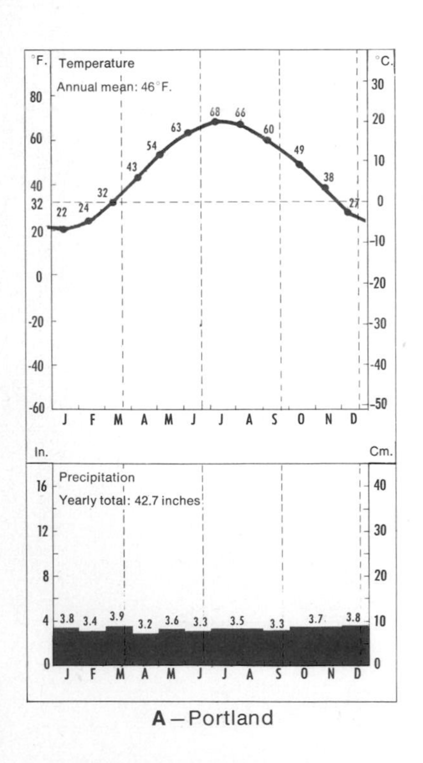

A—Portland

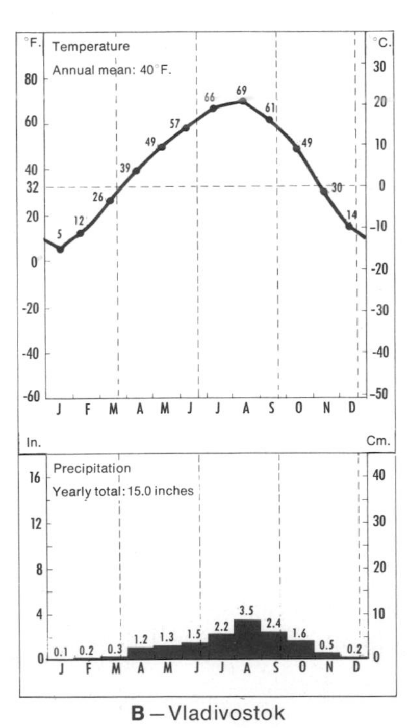

B—Vladivostok

FIGURE 12–21

Graphs of climatic statistics. A. Portland, Maine (humid continental with warm summers climate at 43° 40′ N.). B. Vladivostok, in the eastern Soviet Union (humid continental with warm summers climate at 43° 06′ N.).

away from the coasts and toward the poleward margins.

Subtypes of the Humid Severe-Winter Climates (D)

From the standpoint of significance to humans, there are three subdivisions of the humid severe-winter climates (D). This distinction is made primarily on the basis of the relative warmth of the summer season. The three types, as shown on the map in Figure 12–14, are: humid continental with hot summers climate (Dfa and Dwa); humid continental with warm summers climate (Dfb and Dwb); and subarctic climate (Dfc, Dwc, Dfd, and Dwd).

Humid Continental with Hot Summers Climate (Dfa and Dwa). Within areas of this type of climate (Fig. 12–17), there is variation in the distribution of precipitation through the year. In some areas, no part of the year is dry; in others, there is drought during the winter season. However, the summer, or growing season, is humid in both.

Representative of these phases are Chicago, Illinois (Fig. 12–22A), and Mukden, China (Fig. 12–22B). The essential difference between them is in the distribution of precipitation during the year. At Chicago (Dfa), no season is markedly drier than another; at Mukden (Dwa), there is a notable rainfall contrast between the seasons. In addition, the seasonal contrast at Mukden is marked by a high relative humidity in summer and a low one in winter. There is little change in relative humidity between the seasons in Chicago. Clear, cold winter days succeed one another at Mukden, while cold, often windy and snowy, days alternate with days of

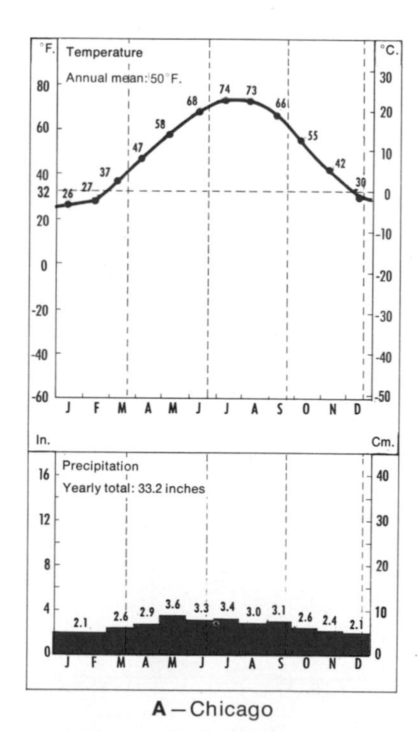

A – Chicago

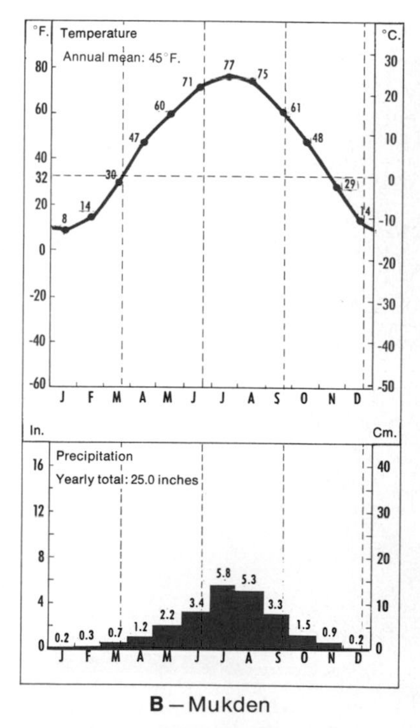

B – Mukden

FIGURE 12–22

Graphs of climatic statistics. A. Chicago, Illinois (humid continental with hot summers climate at 41° 49′ N.). B. Mukden (Shenyang), China (humid continental with hot summers climate at 41° 45′ N.).

mild, drizzling rains in Chicago winters. The summers are much more similar. Both places are subject to prolonged periods of humid heat, with frequent, sudden downpours of rain alternating with cool, dry periods of clear weather.

Humid Continental with Warm Summers Climate (Dfb and Dwb). The distribution of this climate is shown on Figure 12–17. Here, as in the humid continental climates with hot summers (Dfa and Dwa), two phases may be recognized: one in which precipitation is even throughout the year, and one with winter drought. This distinction, however, is not of major importance; the character of the summer is a feature of much greater significance.

Portland, Maine (Fig. 12–21A) and Vladivostok, on the east coast of Siberia (Fig. 12–21B), illustrate the temperature and precipitation features of this type of climate. Compared with the hot summer type of climate, the summers in this type are only warm and are generally short. While two months average over 70° F. at Chicago, no month averages as high as 70° F. in Portland, Maine; at Mukden, three months average over 70° F., but none averages as high as 70° F. at Vladivostok.

Subarctic Climate (Dfc, Dwc, Dfd, and Dwd). Longer and more severe winters, shorter and cooler summers, almost no spring and autumn, and a decreased amount of precipitation are the features that distinguish the subarctic climate (Fig. 12–17) from the humid continental types. Dawson, Canada (Fig. 12–23A) and Okhotsk, in eastern Siberia (Fig. 12–23B) are, respectively, good representatives of the Dfc and Dwc phases of subarctic climate. At both locations, average temperatures are below freezing for over half the year; yet summer temperatures often

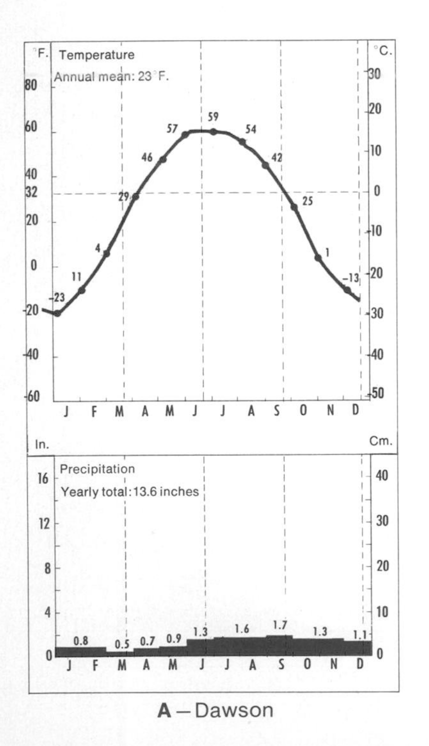

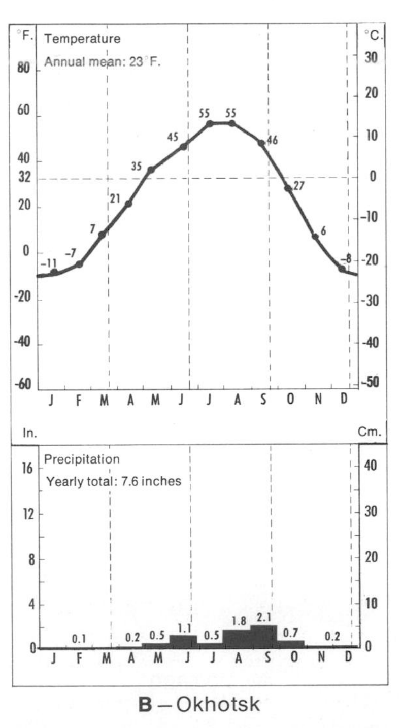

FIGURE 12–23

Graphs of climatic statistics. A. Dawson, Canada (subarctic climate at 64° 04′ N.). B. Okhotsk, in the eastern Soviet Union (subarctic climate at 59° 28′ N.).

reach 80° F. Frosts may occur in nearly every month, and thus, the growing season is short and uncertain. Only the long sunlight hours of the short summer tend to encourage the pursuit of agriculture.

Where continentality is extreme, as in northeastern Siberia, subarctic climates develop some of the most severe winters (Dfd and Dwd) on earth, outside of Antarctica, as at Sredne Kolymsk (Fig. 12–19B) and Verkhoyansk (Fig. 12–19A). Winter air is very dry and clear through the long hours of darkness at that season. This dryness, accompanied by very little, if any, wind, helps to make the low temperatures less severe in their human effects than might be expected. The maximum daily temperature rarely rises to even 0° F. during the winter.

Dominant Controls

Temperature Controls. The summer sun, though never vertically overhead, is above the horizon for many hours of the day. Surface heating is small, per unit of time, as compared with areas farther equatorward, but it continues over a longer period of time in a single day. Total heating is therefore considerable. In addition, there is only a short time during each summer day when radiation from the earth continues without the receipt of some solar energy. At the opposite time of year, the sun is at a lower angle than during the summer. It is above the horizon for only a few hours a day throughout most of the humid severe-winter climate (D) areas (Fig. 8–11), and is actually below the horizon for many days in those areas within the Arctic Circle. Hence, receipt of solar energy is very small at that season, but earth radiation continues uninterruptedly. Thus, air temperatures become very low. Marine and continental influences likewise cause variations in the degree of heat and cold.

Precipitation Controls. Precipitation in summer results both from the passage of cyclonic storms and from local convectional storms, while precipitation in winter is due, almost exclusively, to cyclonic storms. The usual smallness of the amount is mainly the result of great distance from moisture sources.

The pronounced dryness of winter in northeastern Asia results from the intense cooling of that continent in winter. A tremendous mass of cold, dry air forms over the continent and prevents the movement of cyclonic storms across it. In summer, the mass of cold air does not exist; in fact, it is replaced by relatively warm air that results in a wind movement inland from off the cooler fringing oceans. This is the means by which the moisture that becomes the source of precipitation, either from cyclonic storms or from local convectional systems, is moved onto the continent.

Polar Climates (E)

Subtypes of the Polar Climates (E)

The polar climates (E) are at the opposite extreme from the tropical humid climates (A); the latter have no winter and the former have little, if any, summer. A mean temperature of 64° F. for the coolest month is specified as the limit of the tropical humid climates; a mean temperature under 50° F. for the warmest month is specified for the polar climates. These polar climates (E) are

Types	Areas
Tundra (ET)	Northern Hemisphere: arctic fringe of Alaska; Canada; Siberia Southern Hemisphere: none
Icecap (EF)	Northern Hemisphere: Greenland Southern Hemisphere: Antarctica

	Characteristics
Latitudes	high polar latitudes
Temperatures	cold to cool all months; little or no summer; mean temperatures below 50° F. for the warmest month
Precipitation	meager; light, wind-driven snows
Wind belts	polar easterlies wind belts
Air masses	polar; arctic

FIGURE 12–24

Chart of the principal areas and characteristics of the *polar climates (E).*

controlled predominantly by polar and arctic air masses (Fig. 11–2).

There are two phases of polar climates: where the mean for the warmest month is between 32° F. and 50° F., the climate is the *tundra* type (ET); where the mean of the warmest month is below 32° F., the climate is the *icecap* type (EF). For all practical purposes, the icecap climate (EF) is limited to Greenland and Antarctica (Plate IV; Figs. 12–1 and 12–24).

Tundra Climate (ET). Poleward from the areas of humid severe-winter climate (D)

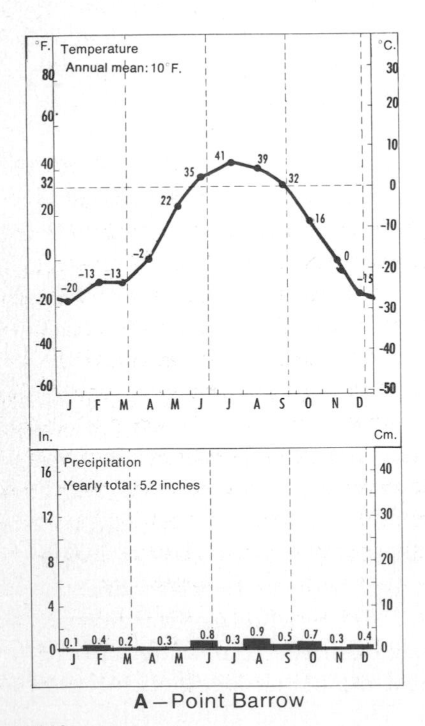

A – Point Barrow

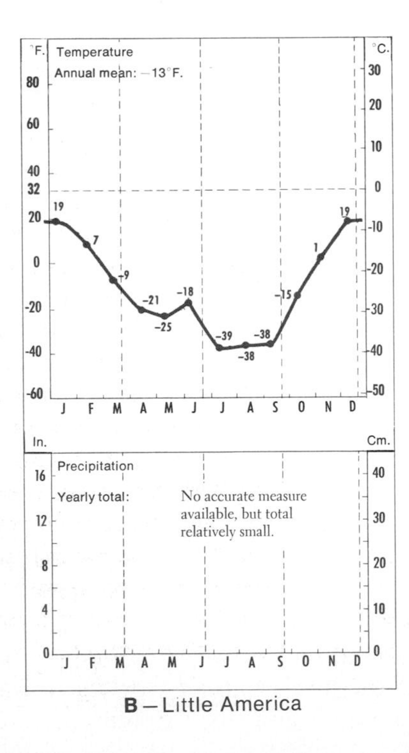

B – Little America

FIGURE 12–25

Graphs of climatic statistics. A. Point Barrow, Alaska (tundra climate at 71° 20′ N.). B. Little America, Antarctica (icecap climate at 78° 30′ N.).

in the Northern Hemisphere, there is a region in which there is no true summer, but in which temperatures do rise above freezing during at least part of the year. Severe frosts occur even during this warmer period, and temperatures are too low to permit agriculture. Mosses, lichens, flowering herbs, and coarse grasses cover the surface, but trees are lacking or exist only as stunted caricatures. This is the *tundra*. Although the warmer season is hardly a summer, it is warm enough to allow some plant activity: temperatures rise into the high 60's. Then, since the boggy ground supplies excessive moisture for evaporation, the air becomes sticky and oppressive. Point Barrow, Alaska (Fig. 12–25A), indicates the major temperature and precipitation features of the tundra type (ET) of climate. Five months average below 0° F., eight months average below freezing, and even the mean of the warmest month is only 41° F. There is also a meagerness of precipitation.

For all practical purposes, the tundra climate (ET) is absent from the lands of the Southern Hemisphere, except for limited occurrences in such areas as Tierra del Fuego, some ice-free fringes of Antarctica, and small patches in high mountains.

Icecap Climate (EF). In areas of icecap climate (EF) monthly average temperatures never rise above freezing. Snow and ice cover the surface throughout the year (Fig. 12–1). Even though the sun remains above the horizon continuously for four to six months of the year, it circles above the horizon so low in the sky that little heating results. Only a steep slope facing the sun for a few hours may be warmed sufficiently to experience a slight melting of snow and ice. In exceptional instances, the temperature may rise into the 70's for a few minutes, but it soon drops back to below freezing as the sun swings around and leaves the slope in shadow.

An equally long period at the opposite season, during which the sun is below the horizon, brings continued hours of darkness and unrelieved cold. Temperature records over long periods are not available, but those from polar expeditions suggest the general conditions (Fig. 12–25B). Eight months of the year average below 0° F. at Little America. Far inland from Little America, at a scientific station on the polar ice plateau near the South Pole, a minimum temperature of −127° F. was recorded. Because of the difficulties of measurement in regions of drifting snow and violent winds, precipitation data for such areas are lacking or, at best, very spotty. It is probable that annual snowfall is equivalent to somewhere between 5 and 10 inches of water.

Mountain Climates (H)

Characteristics

The mountains of the world, in whatever latitude and continental position, present so many variations in their climatic features that it is impossible to fit them into any of the broad, general climatic regions into which the rest of the world's climates have been divided. (For a summary of these, see Figure 12–26.) The mountains are composed of an exceedingly complex mosaic of miniature climatic regions, determined by a wide spectrum of local controls so complex that a thorough study of them constitutes a subscience known as *microclimatology*.

Controls

Elevation. Increased altitude is normally coincident with decreased temperature, at a normal lapse rate of 3.5° F. per thousand feet. The largest range in temperature associated with altitudinal change occurs in high mountains located on or near the equator, as, for example, in the Andes of Ecuador. Because such mountains have tropical lowland climates in their foothill areas and their peaks and crests in a zone of perpetual cold, they run the gamut from climates typical of the equatorial lowlands to those of the polar regions.

Exposure. Exposure factors relative to mountain areas are wind direction and insolation. Mountains oriented across the paths of wind movements have heavier rains on their windward sides (where air is being forced to rise and is therefore cooling and tending to drop moisture) and lighter rains on their leeward sides (where air is descending, being compressed, warmed and tending to pick up moisture, thereby creating a *rain shadow* condition).

Mountains within the midlatitudes receive markedly contrasting amounts of solar radiation on their variously oriented slopes. Equatorward-facing (*adret*) slopes (south-facing in the Northern Hemisphere) receive much greater inputs of insolation than do the poleward-facing (*ubac*) slopes. Thus, adret slopes have relatively greater warmth and a longer growing season, but lose moisture more rapidly and may suffer from aridity even in areas of considerable precipitation. In the Southern Hemisphere, adret and ubac

TROPICAL HUMID CLIMATES (A)
- Tropical wet (Af)
- Tropical dry-winter (Aw)
- Tropical monsoon (Am)

DRY CLIMATES (B)
- Semiarid (BS)
- Arid (BW)

HUMID MILD-WINTER CLIMATES (C)
- Subtropical humid (Cfa)
- Midlatitude marine (Cfb–Cfc)
- Subtropical dry-winter (Cw)
- Mediterranean (Cs)

HUMID SEVERE-WINTER CLIMATES (D)
- Humid continental with hot summers (Dfa–Dwa)
- Humid continental with warm summers (Dfb–Dwb)
- Subarctic (Dfc–Dwc–Dfd–Dwd)

POLAR CLIMATES (E)
- Tundra (ET)
- Icecap (EF)

FIGURE 12–26
A summary of some distinct climatic types

positions relative to the directions north and south are reversed. In tropical areas, no such contrasts between slopes occur.

Areal patterns

Because of the many possible combinations of these factors, as well as a number of other minor, more local controls, mountain climates vary greatly within very short horizontal and vertical distances. Mountain regions, therefore, do not have one climate, but rather a complex of local climates, markedly contrasting, yet occupying only tiny areas in close and irregular juxtaposition. The result is a climatic situation too detailed, areally, for full consideration in this introductory text; thus, for present purposes, the complex is simply mapped under the general heading of Mountain Climates (H), as shown on Plate IV.

REVIEW AND DISCUSSION

1. Discuss the purpose of a classification of climatic types. Why is the ancient Greek classification of climates inadequate today? What does a modern map of climatic types actually depict? How is such a map made?
2. A climate map uses a line to indicate the boundary between one type of climate and another. What does such a boundary line show? If you were traveling across the United States, how could you distinguish the boundaries between the climatic types?
3. What two climatic elements are most significant to humankind? Upon what, then, should a climatic classification be largely based? Describe and discuss the climatic classification used in this book. Why do such people as geographers, agriculturists, foresters, and biologists find it advantageous to use a simple climatic classification that is based upon a standard classification such as the Köppen system? (See Appendix C.)
4. The tropical humid climates (A) are defined as having an average temperature of at least 64°F. for the coldest month of the year. Why? Why not 54°F? How do the many tropical humid climates differ from one another? Compare Af, Aw, and Am climates in terms of forest growth, agriculture, and human comfort.
5. What are the major temperature controls in tropical humid areas? What are the controls of precipitation in these areas? Compare the climate of southwestern India with that of the northern Congo Basin.
6. What are the dry climates? Define the symbols B, BS, and BW as they are used in the Köppen system. What do *h* and *k* signify when added to such symbols? How is the presence of a dry climate reflected in human use of an area?
7. From the description of the dry climates (B), it is apparent that temperatures vary considerably within these areas. Why is temperature a secondary factor in these areas? Is it more, or less, important in tropical humid climates (A)? Explain.
8. Why is there a greater temperature range, both diurnally and seasonally, in the dry climates (B) than there is in the tropical humid climates (A)?
9. Account for the generally meager precipitation in all dry climate (B) areas.
10. Why are temperature ranges along semiarid (BS) and arid (BW) continental west coasts less than those in other areas that are examples of these two climatic subtypes?
11. What characteristic most clearly distinguishes the humid mild-winter climates (C) and the humid severe-winter climates (D) from the tropical humid climates (A)? Account for this characteristic.

12. In what ways may a winter season in a climatic region be of significance to people—both as individuals and as elements of the total environment? Account for the seasonal deficiency of precipitation in certain of the humid mild-winter climate (C) areas.
13. What place in northwestern Europe has a type of C climate that is similar to one found in the western part of North America? Why is it not surprising that both locations are on west coasts?
14. Define the following climatic symbols and describe the climates they represent: Cfa, Cfb, Cfc, Cw, Cs. Cite a good example of each of these climates in North America and in Asia.
15. Why is there no subtropical dry-winter climate (Cw) in the southeastern United States? Why is there a large area of this climate in southeastern China?
16. Explain the absence of humid severe-winter climates (D) in the Southern Hemisphere. In the severe-winter climates (D), spring and autumn are very short seasons. Explain this phenomenon.
17. Account for the fact that summer temperatures in the humid severe-winter climate (D) areas frequently exceed 100°F., whereas in areas of tropical humid climate (A), similar temperatures seldom occur.
18. Define the following climatic symbols, and describe the climates they represent: Dfa and Dwa; Dfb and Dwb; Dfc, Dwc, Dfd, and Dwd. Cite a sizable example of each in Europe and in North America.
19. Define and describe the climate represented by the symbol *E*. Where is a good example of such a climate found? Define ET and EF climates.
20. In general, the earth's climates are arranged in latitudinal belts. Why are the mountain climates (H) something of an exception to this rule? Why are they so complex in their temperature patterns? Is mountain vegetation as complex as mountain climates?

SELECTED REFERENCES

Baird, P. D. *The Polar World.* New York: Wiley, 1964.
Barry, R. G., and R. J. Chorley. *Atmosphere, Weather and Climate.* New York: Holt, Rinehart and Winston, 1970.
Brooks, C. E. P. *Climate in Everyday Life.* New York: Philosophical Library, 1951.
Climate and Man: The Yearbook of Agriculture, 1941. Washington, D.C.: U.S. Dept. of Agriculture, 1941.
Critchfield, H. J. *General Climatology,* 2nd ed. Englewood Cliffs, N.J.: Prentice-Hall, 1966.
Gentilli, J. *A Geography of Climate.* Perth: University of Western Australia Press, 1958.
Gourou, P. *The Tropical World,* trans. by E. D. LaBorde. London: Longmans, Green, 1966.
Hills, E. S., ed. *Arid Lands: A Geographical Appraisal.* London: Methuen, 1966.
Kendrew, W. G. *The Climates of the Continents.* Oxford: Clarendon, 1961.
———. *Climatology,* 2nd ed. New York: Oxford University Press, 1957.
Kimble, G. H. T., and D. Good. *Geography of the Northlands.* The American Geographical Society, Special Publication No. 32. New York: Wiley, 1955.
Lee, D. H. K. *Climate and Economic Development in the Tropics.* New York: Harper & Row, 1957.
Matthews, W. H., W. W. Kellogg, and G. D. Robinson, eds. *Man's Impact on the Climate.* Cambridge Mass.: Massachusetts Institute of Technology Press, 1971.
Miller, A. A. *Climatology,* 8th ed. London: Methuen, 1955.
Money, D. C. *Climate, Soils and Vegetation.* London: University Tutorial Press, 1965.

Peattie, R. *Mountain Geography.* Cambridge, Mass.: Harvard University Press, 1936.
Petterssen, S. *Introduction to Meteorology,* 3rd ed. New York: McGraw-Hill, 1969.
Riehl, H. *Introduction to the Atmosphere.* New York: McGraw-Hill, 1972.
Rumney, G. R. *Climatology and the World's Climates.* New York: Macmillan, 1968.
Stamp, Sir L. D., ed. *A History of Land Use in Arid Regions.* Arid Zone Research No. 17. Paris: UNESCO, 1961.
Thornthwaite, C. W. "An Approach Toward a Rational Classification of Climate." *Geographical Review,* Vol. 38, (January 1948), 55–94.
Trewartha, G. T. *The Earth's Problem Climates.* Madison, Wis.: University of Wisconsin Press, 1961.
______. *An Introduction to Climate,* 4th ed. New York: McGraw-Hill, 1968.

A view of the Vermont countryside. (Newell Green)

13

Vegetation and the Environment

The earth's vegetative mantle, consisting of plants in their varied forms, is a factor vital to the entire life complex on the earth, interacting with, and inseparable from, the total physical environment, or *ecosystem*. A large portion of the insolation received at the earth's surface is first received and used by green plants (the *producers*), then passed on directly, as vegetative foodstuffs, to living animals (the *consumers*); and then, eventually, it is released back into the atmosphere and solar space, through the slow process of decay by bacteria (the *decomposers*).

The study of vegetation here, being within the context of physical geography, will not concern itself with the evolution of plant life nor with the history of the spread and decline of plants. Rather, emphasis will be placed largely on the *bioclimatological* aspects of vegetation geography and on the interrelations of native land vegetation systems with the total environment.

On the basis of their vegetation, its size and complexity, the land areas of the earth can be divided into four major systems, or communities: *forests, scrublands, grasslands,* and *deserts* (Figs. 13–1, 13–2, and 13–3). The development of these distinctive global communities reflects the responsiveness of all natural vegetation to the earth's varied climatic controls, particularly those of moisture availability and temperature (Plates IV and V; Figs. 13–1 and 13–3). Any world map of these vegetation communities must, however, be recognized as being highly generalized, and the plotting of their boundaries must be considered somewhat arbitrary. The major plant communities do not occupy sharply defined regions: they merge imperceptibly, through zones of transition, where the species of one community intermingle and compete with those of another. Furthermore, each community is itself composed of a complex of greatly varied subdivisions, some of which may actually be far-removed outliers of another community. Yet, forest landscapes, for instance, are distinct from all the other vegetation communities—whether the region in which they are located is flat or mountainous, warm or cold. Likewise, grass-covered landscapes exhibit unique properties and present vastly different possibilities

FIGURE 13–1

Some examples of the world's major *vegetation communities. Forest* in North Carolina. (North Carolina Department of Conservation and Development) *Scrubland* in Kenya. (Kenya Information Office) *Grassland* in Oklahoma. (USDA–SCS) *Desert* in Nevada. (U.S. Forest Service)

to people than do similar land surfaces, whether heavily forested or barren. The forests of the tropics appear and respond to human efforts differently than do those of Mediterranean Europe. When the Portuguese and Spanish encountered the Amazon forests during their exploration and colonization of South America, they found them almost insurmountable barriers and therefore of little value as a human habitat. On the other hand, the woodlands of Mediterranean Europe have been used by humans for millenia and are an integral part of the environment in which present-day Western civilization had its beginnings.

Native plant and animal life, once every human community's source of food, clothing, and shelter, have both been profoundly altered by human use of the land, in agrarian areas as well as in large urban ones. Where the inroads have not been extensive, how-

FIGURE 13–2

The global vegetation communities.

FOREST COMMUNITY
- Tropical forests
  - Tropical rainforest, or selva
  - Tropical semideciduous, or monsoon, forest
- Middle and high latitude forests
  - Broadleaf and mixed broadleaf -coniferous forest
  - Coniferous, or boreal, forest

SCRUBLAND COMMUNITY
- Tropical scrublands
- Mediterranean scrublands

GRASSLAND COMMUNITY
- Savanna
- Prairie
- Steppe

DESERT COMMUNITY
- Desert scrub and barren desert
- Tundra

MOUNTAIN VEGETATION

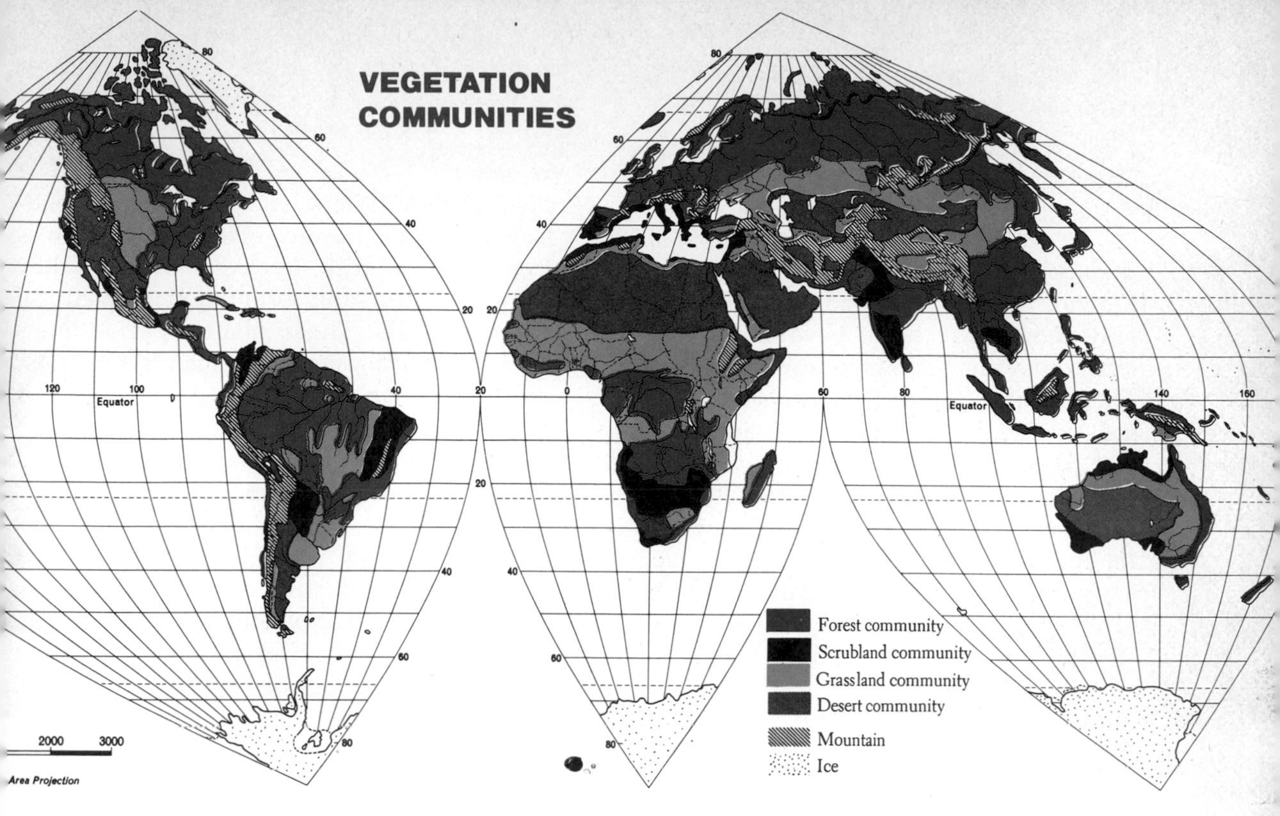

FIGURE 13–3

The distribution of the four major systems, or global communities, that encompass the land vegetation of the world. (Compare with Plate V.)

ever, the native plant and animal complexes stand as visible and direct mirrors of the physical environment. Since plant distribution is determined largely by the combination of elements across the physical environment, it offers considerable information about the contrasts between land areas within that environment—both in the past and at present. Much about each region's precipitation and temperatures, as well as its soils and drainage, can be interpreted from its native vegetation cover. This information proves useful in such environmental tasks as evaluating the reforestation potential of exploited timberlands and planning ecologically sound settlements in undeveloped areas. It should also be especially recognized that many of our classifications of the world's climates and soils make extensive use of native vegetation as the visible terrestrial indicators of their general types and distributions.

Vegetation Distribution Controls

Complexity is a major characteristic of the earth's vegetation distribution (Plate V), and it is equally typical of the multiple factors, or controls, that account for that distribution. The four main environmental controls affecting vegetation distribution are *climatic*, *geomorphic* (landforms), *edaphic* (soils), and *biotic* (living organisms).

Climatic Controls

Climate is definitely the principal influence on the distributional limits of most life forms (Plates IV and V). When plants and animals are taken beyond the general limits of their climatic milieu, they are often unable to survive, especially if they are unable to make certain adjustments. The way in which vegetation relates, and responds, to the total climate is as yet only partly understood, but much is known about its response to individual climatic factors, especially moisture (and humidity), temperature, sunlight, and winds.

Moisture. To most vegetation, an effective moisture supply is probably the dominating feature of the physical environment. It is particularly important to trees and other leaf-bearing plants since, even during periods of moisture stress, they continuously release large amounts of moisture into the atmosphere through the process known as *transpiration*. The rates of transpiration vary tremendously with specific plant types and with specific atmospheric conditions. As they do in human perspiration, high temperatures and low humidities accentuate the loss of moisture by plants, and cold temperatures and high humidities lessen the loss. Leaf structure also determines moisture loss to a considerable extent, with broad, thin, textured leaves giving off moisture at higher rates than small, thick, waxy surfaced, or needle, leaves (Fig. 13–4). Some plant types develop a great deal of spongy tissue throughout their surface and underground structures and are therefore able to collect and store quantities of water for use during periods of low moisture availability. These types of plants are known as *succulents.* Plants that are structurally adapted to extremely limited water budgets are known as *xerophytes* (Greek: *xeros* = dry; *phytos* = plant); while others, such as certain broadleaf trees of the humid regions, are known as *hygrophytes* (Greek: *hygros* = wet), for they show a contrasting adaptation to plentiful water budgets. Those plant types of intermediate tolerances and demands for moisture are known as *mesophytes* (Greek: *mesos* = middle). All vegetation types have some minimum moisture demand below which they either become dormant or die.

Temperature. Temperature is a second climatic control of major importance, since it acts directly and almost instantly upon the physiological processes of most plants: the processes of budding, flowering, fruiting, and seed germination are influenced directly by temperature conditions. In fact, the entire growth, reproductive, and dormant periods of plants are largely cycled by temperature conditions to such a marked degree that the world's major vegetation regions coincide closely with the world's major temperature zones (Figs. 8–18 and 8–19). Each plant type seems to exist under some optimum annual temperature conditions. For each specie there appears to be three critical temperatures: an *optimum temperature,* under which it flourishes best; a *minimum temperature,* below which it ceases to reproduce and exist; and a *maximum temperature,* above which it also fails to survive. Different plant species respond differently to frost conditions: some are killed by even a slight frost, while others require at least some degree of frost to continue their life cycles. Temperature differences also alter the moisture requirements of plants through changes in their transpiration processes. Adaptation to low temperatures is made in different ways by different plant species. Some types become dormant and await warmer temperature conditions;

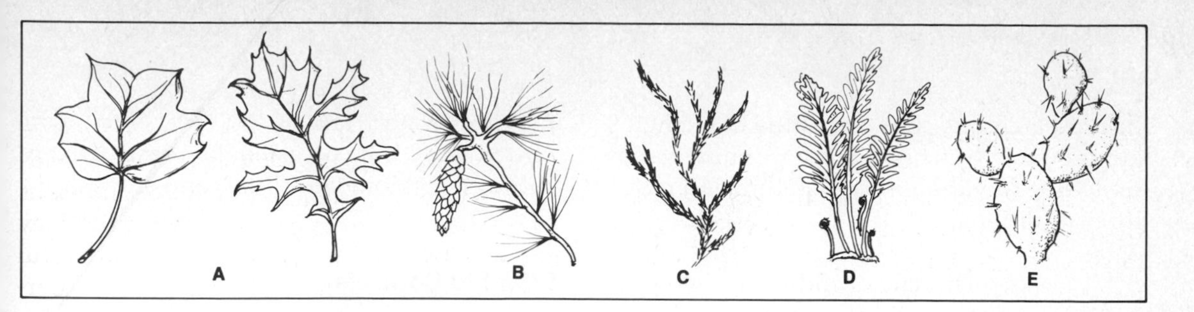

FIGURE 13–4

Some types of leaf structures. A. Broad, thin, textured leaves (left, yellow poplar; right, scarlet oak). B. Needle leaves (western white pine). C. Spiny leaves. D. Fern. E. Succulent leaves (cactus).

during such suspensions of growth, leaf-shedding often takes place, as in the dormant periods of some broadleaf trees and shrubs of the middle latitudes. Other plant types of the middle and higher latitudes, such as coniferous trees, simply become dormant without experiencing any noticeable loss of leaves or undergoing other external modifications.

Some plant species survive protracted low temperature periods, or seasons, by means of seed carryover. The entire life cycle of the plant is completed within the warm temperature season, including the complete disappearance of the vegetative structure. Seeds that are capable of resisting the prolonged low temperature period (often accompanied by a low moisture availability period) propagate the plant with the occurrence of the next warm temperature season. These seed perpetuation plants are known as *annuals,* while plants that survive vegetatively through contrasting temperature periods are known as *perennials.*

As a general rule, the lower the temperatures experienced in a region (both annually and periodically) the fewer the species of plants that will survive (Plates IV and V). The greatest number of plant species exists in the tropical regions (and predominately where warm temperatures and high rainfalls combine). The smallest number of plant species exists in the severely low temperature regions of the arctic and alpine environments of high latitudes and high altitudes. Based upon their various tolerances to low temperature conditions, vegetation types are classified as: *megatherms,* requiring high temperatures; *mesotherms,* favoring moderate temperatures; and *microtherms,* requiring low temperatures.

Sunlight. Sunlight or, more properly, insolation (since certain wavelengths outside the visible spectrum are relevant), is the energy source for *photosynthesis,* the process in which plants combine carbon dioxide from the air with water to produce carbohydrates (starches and sugars), and the basis of the complex food chains upon which most animals, including humans, are dependent. Consequently, the many variations in the intensity and duration of sunlight in most regions have strong influences on this important developmental process, most of which are not yet well understood. Another major role of sunlight is the regulation of the life cycles of plants, since, when a change in day-length occurs (either lengthening in spring or shortening in fall; see Figure 8–9), another stage of a plant's life cycle is begun. This triggering is not yet well understood, but research has shown that the day-lengths are probably basic, with minor timing variations due to variable temperature and moisture conditions.

Winds. Directly or indirectly, other climatic elements, such as winds and clouds, have varying influences of lesser importance on the growth of vegetation and its geographical distribution. Winds frequently have a drying effect on vegetation and, where strong and persistent, may even cause it to become structurally deformed. Winds also play important roles in the distribution of plant pollen, in the scattering of seeds, and in the daily and seasonal shifting of the earth's cloud cover.

Geomorphic Controls

Landforms, or geomorphic factors, variously influence the growth and geographic distribution of vegetation types. Plains and rolling hills are relatively neutral as control factors, while high plateaus and mountains are strong influences (Plate III). Indeed, terrain slopes of all types influence vegetation through their angle, or steepness, and their solar attitude, or orientation to the sun. This is especially so in the middle latitudes, for their equatorward-facing slopes receive vastly greater amounts of insolation, are much warmer, and lose their moisture much more rapidly, than adjacent poleward-facing slopes. The attitude of a slope with respect to the normal wind flow in its area is also a significant part of its influence on vegetation. Windward slopes tend to receive much greater precipitation than leeward ones, especially in areas of predominantly unidirectional air movements, or where the moisture source lies only in one direction. In a broader view, landforms exert their influences through the agents of moving water, snow, and ice, often constructively, but occasionally destructively, through erosion, landslides, and snowslides.

Edaphic Controls

Since pronounced soil characteristics—temperature, water supply, and chemical composition—are usually somewhat localized, soils tend to have highly specific, or localized, influences on plant developments and distributions. Despite their territorial limitations, however, the influence of soil types on vegetation often counteracts comparable climatic influences. While a region may have a climate suitable for particular types of plants, its soils may be so porous or so infertile as to prevent the growth of these expected types and allow the growth of others in their stead. For example, a region of semiarid climate that receives enough precipitation to support a continuous grass cover may have such porous soil that it retains only a small portion of the total precipitation. If this amount proves too little to actually support the grass cover, desert shrubs will become the region's principal vegetation. In such instances, the presence of plants that demand less moisture is said to be edaphically, rather than climatically, determined.

Biotic Controls

Other organisms affect vegetation in a multitude of ways: earthworms help to loosen and aerate the soils; bees and other insects serve in the process of plant cross-pollination; and birds and land animals serve as transporters of plant seeds. Less constructively, insects, birds, and animals also serve as carriers of plant diseases and, in the case of animals, as destroyers of plant cover through overgrazing. Additionally, as consumers of dead tissue and humus, bacteria are endlessly engaged in the elimination of vegetative materials. But on a global scale, people are probably the most potent biotic influences on vegetation. As firemakers, lumberjacks, and cultivators, they have contributed enormously throughout history to the elimination and severe alteration of native vegetation. Their positive influences on plant life, on the other hand, have only recently gone beyond matters of their immediate self-interest—controlling fires, protecting water sources, improving plant species—to include the protection or renewal of vast areas of native vegetation.

Vegetation Classification and Regionalization

Since the number of specific plant types to be found in any one of the vegetation communities—forests, scrublands, grasslands, and deserts—is practically infinite, each community is usually defined only in terms of its most dominant plant types. In a forest community, the dominant vegetation includes not only a dense growth of trees, but also, legions of woody shrubs, grasses, and other low groundcover. Scrublands, usually having some identity with both forests and grasslands, are characterized by shrubs of varying heights and densities but commonly incorporate grasses and trees as well. Grasslands may consist solely of spans of grass, or they may be characterized by various combinations of grasses, shrubs, and trees. And though some deserts are barren, most support meager and discontinuous arrays of bunch grasses, shrubs, and even a few stunted trees (Fig. 13–1). As listed here, these four major plant communities represent descending orders of vegetative size, complexity, and groundcover envelopment—each community's position corresponding directly with its rank, in terms of available moisture (Figs. 13–3 and 10–13). With little doubt, the occurrence and distribution of most of these plant communities over the face of the earth are largely environmentally induced.

Forest Community

The most complex of the major plant communities, forests, are also the most widespread, for they dominate vast stretches of the tropical and middle latitudes on all the continents (Plate V; Fig. 13–3). The tall trees found in most forest areas are usually so close together that their branches touch or even overlap; thus, their foliage shades the ground to a very large extent. This shading effects a distinctly different ground climate in forests than is typical of the world's other major plant communities. A forest's large trees also require more moisture than most of the vegetation types that dominate the scrublands, grasslands, or deserts. Water must be present in forest soils in quantities large enough to satisfy the growth needs of these highly demanding plants; this requires relatively heavy mean annual precipitation. In addition, if the tree roots are to make use of it, the soil moisture must be available in a liquid, rather than a frozen, state; this often severely limits the effective moisture of forests in the higher latitudes. If the annual available moisture of a forest is reduced below the minimum requirements of its resident tree types, the forest will be greatly altered and may even disappear. Conversely, if a scrubland, grassland, or even a desert area experiences so great an increase in its yearly soil moisture that it can thus meet the moisture requirements of extensive tree growth, it will generally develop the vegetative range and complexity typical of forests.

Forest growth usually demands mean monthly air temperatures above 42° F. for at least a short growing season. Where lower temperatures persist throughout most of the year, forests are usually limited, both because the moisture available in areas of continually low temperatures tends to be too limited for tree growth and because the growing period above 42°F. is often too short. Still, the forest community spans a remarkably broad thermal range—from the tropical climates (A) to the severe winter climates (D).

Moisture and temperature variations are reflected in certain outstanding differences among trees. Some trees retain most of their foliage at all times, since each drops its leaves (including needles) serially rather than all at once. Such trees are known as *evergreens.* In contrast, other trees lose all of their leaves in one season and acquire new ones in another season. Trees that show this seasonal rhythm are known as *deciduous* trees. Where temperatures are always high, and precipitation is heavy throughout the year, trees are evergreen, such as the live oak and the red bay in the southern and southwestern United States. Where there is a deficiency of moisture for an appreciable part of the year, either through lack of precipitation or through lowering of temperature so that the moisture is ice-locked and hence not available to the plant roots, a period of rest from growth is introduced into the annual cycle. Under these conditions, trees are usually (but, due to spongy plant structures, not always) deciduous, such as the maples and elms of the northeastern United States.

The manner in which a tree sheds its leaves is often indicated by its leaf form: either needle shaped, like pine leaves, or broad, like maple leaves. Most needleleaf trees are evergreens, while broadleaf trees may be either evergreen or deciduous.

On the basis of the characteristics described above, four main kinds of trees may be distinguished: *broadleaf evergreen, broadleaf deciduous, needleleaf evergreen,* and *needleleaf deciduous.* Entire forests may be dominated by each of the first three types. For example, the forests in areas of tropical wet climates (Af) are made up predominantly of broadleaf evergreen trees; those in areas with severe winters and hot summers (Dfa and Dwa) are predominantly deciduous trees; and those in areas of subarctic climates (Dfc and Dfd) are largely needleleaf evergreens. Trees of the fourth type, needleleaf deciduous, do not commonly occur as a dominant forest type, except in places in eastern upland Siberia, where deciduous pines, or larches, form pure stands.

Scrubland Community

Like a tree, a *shrub,* or *bush,* is a woody plant; however, it lacks the single lengthy trunk of a tree, being composed, instead, of several stems that branch out at or very near ground level. The scrublands of the world are dominated by such plants that, ranging in height from less than one foot to more than twenty feet, grow sparsely in some places and to almost impenetrable densities in others. Most scrublands are found in the lower and middle latitudes generally midway between well-watered lands and deserts. Each represents great seasonal variations in terms of moisture available for plant growth, demonstrating both a pronounced season of water shortage and one of water abundance.

Grassland Community

The world's grassland community is dominated by herbaceous plants and ranges widely through the tropical and middle-latitude subtropical portions of all the continents (Figs. 13–1 and 13–3). For the most part, grassland areas are composed of tall to short grasses, with varying secondary inclusions of *legumes* (plants of the pea family whose roots support bacteria capable of converting nitrogen from the air into a form usable by plants), *forbs* (any nonwoody plant other than grass), and scattered shrubs; however, a smattering, or even thin lines, of small trees may be found along many grassland stream beds. The grasses range in height from a few inches to about a dozen feet, varying directly with climatic conditions (particularly precipitation) and edaphic features. Among the individual plants of the grasslands, most are perennials (which continue their life cycles over a period of years); only a few are annuals (which complete their life cycles in a single growing season and leave seeds that germinate at the start of the next). The grasses and their varied plant associates require less moisture than the dense tree populations of forests but more than most of the plant types found in deserts. Hence, grasses constitute the normal vegetation of the moderately dry regions of the earth's surface: the humid severe-winter climates, their neighboring semiarid climates, and the tropical dry-winter climates.

Desert Community

The desert community includes the world's severely arid regions as well as regions in the cold subarctic latitudes (Figs. 13–3, 10–13, and 8–18). The vegetation cover of a desert—whether an arid desert or a cold one—is light to almost nonexistent. In desert areas, there is no continuous cover of plants such as that formed in the relatively well-watered grasslands, and certainly no crowding of plants as in the copiously watered forests. Yet, the desert community, except for certain icy wastes of the polar regions, is seldom devoid of plant life. When examined closely, even sand-and-rock areas that had appeared to be barren of plant life, normally reveal at least a small number of specially adapted plants. Individual desert plants are widely scattered and many are dormant throughout most of the year, though they may enliven for a short period following one of the occasional rainstorms that characterize the dry deserts of the world.

Most of the plants of the world's desert community are xerophytic—that is, they are structurally adapted to resist drought. Like regions of heavier plant cover, deserts include both perennials and annuals. Particularly striking features of the desert community are the many *succulents,* or watery-tissued life forms, such as the cactus, which are capable of storing moisture to sustain themselves through long periods of dryness, and the *ephemeral vegetation,* a form that completely vanishes between rainy periods.

The cold deserts of the high latitude polar and subarctic climates are known as the *tundra.* Their patchy groundcover includes such plants as mosses, lichens, sedges, some grasses and bushes, and even stunted trees. Many of these plants have had to adjust to long winters without liquid water and thus, in a sense, have become xerophitic. It is mainly for this reason that tundra vegetation is classified in this book as part of the desert community (some classifications consider the tundra as a part of the grassland community—for which there also is considerable justification).

Forest Community

Tropical Forests

Probably the heaviest forest covers in the world exist in the regions of tropical wet climate (Af). Abundant precipitation and continuously high temperatures, two characteristics of the tropical wet climate (Af), create a vast "natural greenhouse" in which the forest is evergreen and broadleaf. As one proceeds away from the areas of tropical wet climate (Af) through tropical monsoon and tropical dry-winter climates (Am and Aw), optimum conditions for forest development decline markedly. The increased prominence of dry seasons over rainy ones soon eliminates the preponderance of the broadleaf evergreen forest. Lesser amounts of precipitation result in a gradual thinning of the forest and a decrease in the size of its individual trees until, finally, a limit beyond which it is too dry for trees is reached. Even in sufficiently humid areas, the existence of definite cool seasons beyond the margins of the tropical humid climates (A) brings about a change in the plant species that make up the forest, and thus its entire aspect is altered. In such areas, the tropical forest is supplanted by some type of middle latitude forest.

Two major forest types may be recognized in tropical regions. The transition from one to the other is gradual, and the types are often intermingled along broad zones rather than grading sharply; nevertheless, marked differences set them apart from one another. The two major contrasting types of tropical forest are: *tropical rainforest* and *tropical semideciduous forest*. In addition to these two types, two associated secondary types have been recognized: *jungle* and *mangrove*. The two major forest types (Plate V; Fig. 13–5) are treated separately in the following discussions, while the secondary types, usually occurring in areas too small to be illustrated with accuracy on maps of the scale used in this book, will be treated only in terms of their specific relationships to the major ones.

Tropical Rainforest, or Selva. Most luxuriant of the various forest types is the *tropical rainforest,* or *selva* (Fig. 13–6), which, from the air, resembles a rolling green sea. Tall trees crowd so closely together that their crowns form a dense, irregular canopy through which little sunlight penetrates to the ground. In profile, the crowns are seen to be arranged in layers, or stories: the lowest is about 50 feet above the ground and the second lowest, about 100 feet. Above the second story extend the upper trunks, branches, and foliage of the forest "giants," which rise to 125 feet or more. Except where the forest has been removed, the canopy of leaves remains unbroken (Fig. 13–7). As a result, the interior of the selva is dark, and the air in it is still and moist. Trunks rise straight and branchless to the crowns, which are always green, for water is continuously available to the tree roots. The moisture supply is so plentiful that the trees are usually shallow rooted and must depend largely on buttresslike ridges at the base of the trunks for their support. The trees are broadleaf evergreens. Although a large number of species are present, there are rarely any sizable numbers of any particular one: where the forest is undisturbed by humans, it is unusual to find more than a few representatives of any one specie in an acre of forest.

Although the tropical rainforest is evergreen, there is a constant loss of leaves

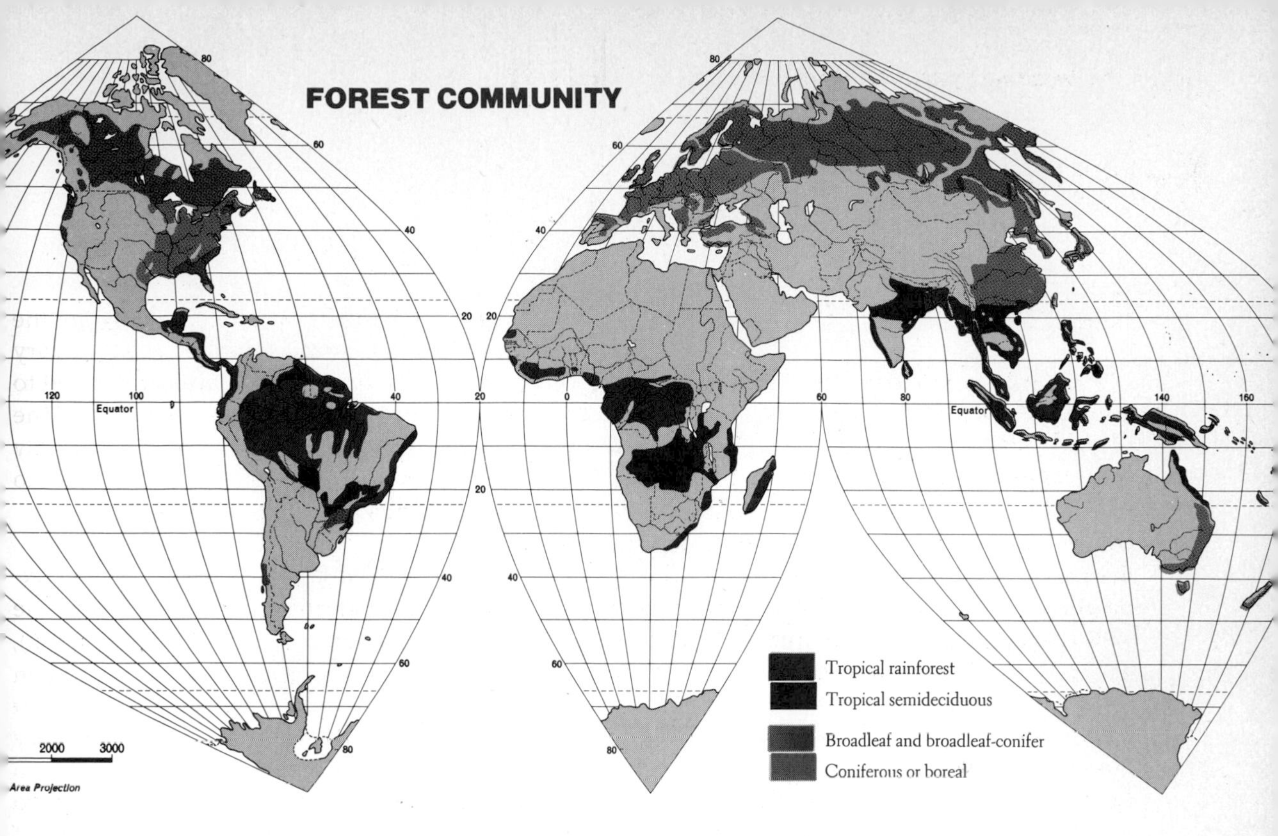

FIGURE 13–5

Forests are the most widespread of the world's native vegetation, and they dominate vast stretches of the tropical and middle latitudes of all the continents. Only in areas of dense, or comparatively dense, populations have the original forests shown on this map been removed. (Compare with Plate VII.)

from all the trees throughout the year; thus, a steady supply of organic matter falls to the forest floor. It might be expected that it would accumulate in large amounts; however, the combination of abundant bacterial life (causing rapid decay) and the activities of a host of termites and beetles result in relatively rapid destruction of the forest litter. Consequently, the floor of the tropical rainforest is remarkably clear. In addition, the absence of sunlight prevents the growth of much underbrush, and thus allows for relatively unrestricted movement through the rainforest.

Characteristically, the tropical rainforest has large numbers of plant types known as *epiphytes* and *parasites*. Epiphytes grow on other plants but do not secure their food from them; parasites grow on and do secure their food directly from other plants. In great profusion, both plant types hang ropelike from the crowns of trees or jut out from tree trunks. Many epiphytes, such as orchids, are brilliantly flowering. Others, such as lianas, also characteristic of the tropical rainforest, are woody-stemmed vines that start from the ground and lace themselves around other plants. Together with the density of the trees, these and many other plant types convey the impression of superabundant forest plant life, which is why this kind of forest is often, though incorrectly, called jungle.

One of the secondary tropical forest types, *jungle*, is actually a dense, relatively low-growing tangle of trees, shrubs, and lesser plants, often laced together with vines, and occurring where sunlight can penetrate

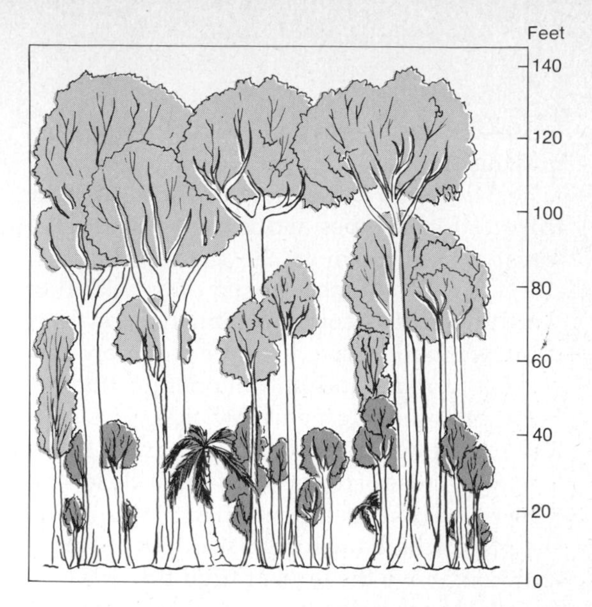

FIGURE 13–7

A schematic profile of a *tropical rainforest* in Trinidad, with several distinct stories apparent. (After J. S. Beard. *The Natural Vegetation of Trinidad*, Clarendon Press, Oxford, 1946.)

to the forest floor. In most tropical rainforests, the shading of the forest floor is so complete that only removal of the tree crowns will admit the sunlight necessary for heavy ground plant growth. This occurs characteristically in three situations. The first is where the selva has been destroyed or damaged: as a result of its having been cleared, fire (caused by lightning or humans) having burned it, or severe local winds having damaged its crown. In such areas, dormant plant seeds germinate very rapidly, and soon, there are thousands of plants in a tangled thicket vying with one another to occupy the space left by the fallen giants that preceded them. Because the plants rise from relatively dry and firm ground, this is ordinarily called *dry jungle.* Eventually, the faster growing and larger specimens crowd out the others, and the patch of jungle disappears. The second characteristic location of dry jungle is on steeply-sloping land where the sun's rays can penetrate through the tree trunks and thus to the forest floor. This is also dry jungle, but it is more permanent than that which springs up in clearings, for the trunks of the trees provide less shade than do the interlocking crowns.

The third location of jungle is in the wet spots of the tropical rainforests: along the banks of streams too wide to be completely shaded by the crowns of the forest giants on either side, on the floodplains of mature and old valleys, on shelving lake shores, or near the seaward edges of low coastal plains. In this *wet jungle,* the dense tangle of vegetation rises directly from shallow water or from deep mud, making it difficult to tell where the land begins and the water leaves off. Wet jungle is much more permanent than dry jungle, for the areas in which it occurs are usually too wet to support most of the species of tree found in the tropical rainforest.

The idea of jungle being the usual vegetation type in equatorial regions came from scant knowledge of the wet jungle. As early explorers penetrated the interiors, largely along the great rivers, they had to travel between river banks densely covered with vegetation. This vegetation seldom extended back from the river for more than a few hundred yards, but to most explorers attempting to find a route away from the stream, it seemed to continue without end. From this experience there came the inference that jungle persisted throughout in these latitudes. Actually, once out of the jungle fringe and into the forest, undergrowth usually becomes relatively light. Such riverbank alignments of dense vegetation are called *galeria,* or *gallery forests,* be-

FIGURE 13–6

The *tropical rainforest,* or *selva,* such as this one on the foothills of the Andes Mountains in Venezuela, is the most luxuriant of all the forest types. (Karl Weidman, National Audubon Society) cause the trees often arch over the stream, making it seem like a gallery, or tunnel.

Mangrove is the other of the secondary tropical forest types and is found where the forest comes down to the sea or to a tidal inlet. It is sometimes spoken of as a kind of wet jungle. It becomes established on very shallow, muddy coasts where there are few strong currents and only moderate fluctuations between high and low tide. The stilt-like roots of one individual mangrove interlock with those of the next to form a close-knit wall rising out of the mudflats. Well above the place where the roots rise out of the mud, and just above the level of high tide, heavily leafed branches bush out. At high tide, the fringe of green may stand out from the shore at a considerable distance. The roots are then entirely covered by water. The barrier formed by mangrove is so thick that sediments brought from the land by the rivers are held close to the shore. A mangrove fringe thus acts as a "land builder" along shallow tropical shores.

The tropical rainforest constitutes one of the most extensive types of forest remaining in the world. In a few localities, such as Java and India, it has undergone much destruction and alteration, but elsewhere, much of it remains; even where clearings have been made, unless measures are taken to restrict it, regrowth is rapid. Rubber, cacao, such cabinet woods as mahogany and ebony, gums and resins, and edible nuts are all derived from trees native to the rainforest. Many tropical fruits have local use, and some, such as the banana, are significant items of world trade. There is, however, one great handicap to intensive utilization of the rainforest—the trees of any one species are widely scattered. Unless some device, such as the creation of an artificial forest similar to those of the Malayan rubber plantations, is resorted to, small-scale forest industries cannot operate efficiently or economically. In lumbering, for instance, selective rather than complete logging must be practiced. Distances from world markets and from large supplies of labor add to the difficulties of exploitation. Many foresters and lumberers have suggested that, with large-scale operations, it is economically possible to exploit the tropical forests by assembling many different kinds of wood instead of carrying on selective logging. Despite this suggestion, rainforest products, except for a few items such as Brazil nuts and chicle, remain luxury goods.

The largest areas of continuous tropical rainforest occur in equatorial South America and Africa. Less extensive areas of this type of forest are to be found in Central America, southeastern Asia, and the East Indies. There is a close relationship between the distribution of tropical rainforest vegetation and that of the tropical wet climate (Af) (compare Plates IV and V), but this correspondence is not absolute. In Liberia, for example, there is rainforest vegetation along with a tropical monsoon climate (Am); here, even though there is a period of less rain, soil moisture never falls below the needs of the forest. Minor discrepancies of this sort result from many causes, most prominently: edaphic (soil-related) factors, a lack of numerous long-term climatic observations, and a lack of exact information regarding the detailed distribution of vegetation types. Despite such minor discrepancies, the similarity of climate and vegetation patterns on a worldwide scale is strongly pronounced. This applies not only to the tropical wet climates, but to all of the world's natural vegetation types and the climates in which they characteristically occur.

Tropical Semideciduous, or Monsoon, Forest. Beyond the tropical wet climate (Af) and the more moist portions of the tropical monsoon climate (Am), an increasing intensity of the dry season more and more strongly enforces a period of rest in plant growth: the forest becomes lighter, individual trees are more widely spaced, and low undergrowth is heavier (Fig. 13–8). As in the teak forests of Burma, pure stands of some species are common. The forest is semideciduous, and the dry season prevents many of the trees from being evergreen. Some of the trees do retain their foliage, but the greater part of the forest is dry and brown throughout the nearly rainless period. Many of the species common to the rainforest are found in the semideciduous forest, though a large number of those that are evergreen in the rainforest are deciduous in the monsoon, or semideciduous, forest.

Jungle is more truly typical in tropical semideciduous forests than in the rainforest because of the wider spacing of the large-crowned trees. As a result of the dense undergrowth, passage through this forest is much more difficult than it is through rainforest. Only here and there, particularly on low river floodplains or on low-lying coastal strips, is there a sufficient number of broadleaf evergreens to cut off sunlight from the forest floor and thereby eliminate jungle. Mangrove is a characteristic coastal fringe, and narrow lines of galeria forest, growing on the stream banks, point out the river courses across low-lying plains.

Generally, semideciduous forest is more easily cleared by primitive groups because of the smaller size of the ground plants, and because a dry season permits clearing by fire; consequently, fewer and less extensive areas of it remain. Commercial operations to recover forest products are easier than they are in the rainforest, for the more frequent stands of individual species make economic utilization of the resource possible. Among the trees of the semideciduous forest that have great commercial significance is *teak,* which provides a wood that is particularly desirable because of its strength and durability.

The tropical semideciduous, or monsoon, forest type is generally found in areas of tropical monsoon climate (Am), and in the transition zone into the wetter parts of the tropical dry-winter climate (Aw). Its greatest extent is in southern and southeastern Asia and adjacent islands (Fig. 13–5), where the forest stretches, with some interruption by other forest types, across central and northeastern India, through southeastern Asia to the Philippines, and on to the northeastern coast of Australia. Other small areas of tropical semideciduous forest occur in eastern Africa and northern South America.

Middle and High Latitude Forests

From the interiors of the continents outward, there is a definite sequence of vegetation types, ranging from desert to

FIGURE 13–8

Teak logging in a *tropical semideciduous,* or *monsoon,* forest in central Burma's Myinbyin Forest Range. (United Nations)

grasslands to forest (Plates IV and V). Within the tropics, where the temperatures are constantly high, the sequence is almost solely determined by the availability of moisture. In the middle and high latitudes, the sequence still depends on the availability of moisture, but it is further modified by marked seasonal fluctuations of temperature. Except on their drier margins, all of the humid climates of the middle latitudes have ample moisture to provide for forest covers. There is, however, considerable variety in the forest cover, due to differences in the seasonal distribution of precipitation or in the severity of the winters. The two principal forest types in the middle and high latitudes are *broadleaf and mixed broadleaf-coniferous forest* and *coniferous, or boreal, forest* (Fig. 13–5).

Broadleaf and Mixed Broadleaf-Coniferous Forest. Climatically, these areas have sufficient moisture for tree growth, and the winters are not cold enough to enforce a significant rest period. The forest is normally broadleaf evergreen. The individual trees belong to different species from those of the tropical forest, they stand somewhat farther apart, and the forest undergrowth is usually more dense. Epiphytes and parasites are common. The whole forest somewhat resembles a miniature tropical rainforest.

Where sandy soils occur, cone-bearing trees, or *conifers,* replace the broadleaf evergreens. In swampy lowland areas, dense forest, dominated by cypress trees, frequently rising to a height of 100 feet from their buttress-rooted bases, is common (Fig. 13–9). Within these swamp forests, a dense tangle of low bushes and trees, interlocked with epiphytes and parasites, creates a gloom and dankness reminiscent of the wet jungle.

FIGURE 13–9

A typical *cypress swamp* on Weeks Island, Louisiana. (Standard Oil Co., N.J.)

Where the winter season is more severe, as in the higher middle latitudes, the forest changes to one of broadleaf deciduous trees with scattered conifers (Fig. 13–10). A seasonal rhythm becomes well established, with budding, flowering, and leafing in spring, a full greenness in summer, and the brilliant change of leaf color with the coming of the first frost. Later, the bare branches of the broadleafs, contrasting sharply with the dark green of the conifers, form a distinctive winter scene. Undergrowth is usually heavy but largely composed of small specimens of the trees that dominate the forest. There is a great accumulation of litter on the forest floor from the seasonal dropping of leaves and the occasional falling of branches and trunks. This litter is not rapidly destroyed, for, under the climatic conditions that exist, the rate of decay is slow; hence, much of the litter remains to furnish soil-enriching material.

In this forest type, pure stands of broadleafs, such as maple, oak, and hickory cover

FIGURE 13–10

A forest of *broadleaf deciduous trees* and *scattered conifers* in western Pennsylvania. Most prominent here are the hemlock, beech, and hickory trees. (Helen Faye)

the areas of better soil, while similar stands of conifers, such as pine, fir, spruce, and hemlock are to be found on poorer soils. Occasionally, grassy openings break the continuity of the forest. Edaphically dry localities—called *heath*—produce a low, bushy cover in which plants such as heather form a woody mantle. Edaphically wet sites are clothed in a similar low vegetation, and many of the ground mosses combine with heather and similar plants to form the *moors*, or *moorlands* (Fig. 13–11). *Meadows*—areas of tall grass—and *canebrakes*—thickets of large jointed-stemmed grasses—edge the rivers in places where wide floodplains have developed.

Of all the major forest types in the world, this mixed broadleaf-coniferous has suffered the most destruction by humans. Broadleaf deciduous woodland once covered much of the British Isles; today, less than five percent of it remains or has been replanted. In the eastern and central United

FIGURE 13–11

A *moor* in the northwest highlands of Scotland near Loch Shin. As in this instance, much of the world's moorlands provide pasture for domestic animals. (British Information Services)

States, less than one-tenth of the original forest has been preserved. In China, only vestiges give evidence of the luxuriant forest that once must have spread over much of the whole country. After the supply of broadleafs was exhausted, scattered conifers provided a convenient source of softwoods, but they, too, have been largely removed.

Such ruthless destruction has given rise to a host of problems. One of the greatest of these is soil erosion. Forest removal creates little, if any, change in precipitation; yet it does cause more moisture to become available as an agent of erosion: since less moisture filters into bare soil than into forest-covered soil, runoff is increased. Where remedial measures are not taken, or, too often, taken too late, gullying and sheet erosion produce surfaces unfit for cultivation: a large proportion of the rainfall flows over the surface of the ground to the streams and carries an increased load of sediment with it. Formerly ample stream channels are now overloaded; hence, they can no longer handle flood loads, and control works are necessary to protect the adjacent valley lands from destruction.

World distribution of the broadleaf and mixed broadleaf-coniferous forest type ranges widely through the middle latitude climates (C and D) of the Northern Hemisphere (Plates IV and V; Fig. 13–5). The main Northern Hemisphere areas correspond roughly to regions of high population density; it is within these areas that forest removal has been most intensive during recent centuries.

Coniferous, or Boreal, Forest. The mixed broadleaf-conifer forest continues poleward, approximately to the boundary between the humid continental severe-winter–warm summer climates (Dfb-Dwb) and the subarctic climates (Dfc-Dwc) (Plates IV and V). Here, the high latitude coniferous forest, ordinarily called the *boreal forest,* or *taiga,* takes over (Fig. 13–5). In these latitudes, the increased coolness and shortness of the growing season provide a climatic environment in which few of the broadleafs thrive. Although some, such as birch, poplar, alder, and willow, are numerous, most of the trees are needleleafs. Needleleaf trees, such as cedar, spruce, hemlock, and fir, do not predominate in this environment because it is ideal for them, but rather, because the broadleafs adapt less easily and thus provide less competition. The forest is tall, dark green, and close-packed along its equatorward margin, but it becomes more and more stunted and sparse toward its poleward edge.

FIGURE 13–12

A *muskeg* in the northern coniferous forest of Canada. Stunted and misshapen trees, many of them dead, rise out of a waterlogged mat of grassy and bushy vegetation, filling in what originally was a pond. (H. M. Laing, National Museum of Canada)

Frequently, there are extensive swampy interruptions in which tamarack, cedar, and spruce dominate, or in which an excess of water produces a quaking bog of low growing, leathery-leafed bushes, mosses, and grasses that form a spongy mattress between the stunted trees—this is the northern *muskeg* (Fig. 13–12). In some sections, such as in Siberia, the amount of precipitation is so slight that forest, as a complete cover, is ruled out. In such places, the cover of steppe-like grass often extends over wide areas among thin, scattered patches of conifers. Also, in regions of pronounced glacial scour, as in the Laurentian Shield, extensive and numerous outcrops of bedrock, as well as many glacial ponds and marshes, make the northern forest very patchy. Where the original cover has been removed, broadleaf deciduous trees are most numerous in the second growth. Low, thin-trunked saplings, rising from a thicket of stunted brush, form an almost useless covering that the conifers very slowly replace.

Scrubland Community

The world's scrublands occur largely in two types of locations: between the tropical forests and the tropical grasslands—that is, along the borders between the tropical wet (Af) and tropical monsoon (Am) climates or the tropical dry-winter (Aw) and semiarid (BS) zones; and in the areas of Mediterranean climates (Cs); see Plate V and Figure 13–13.

Tropical Scrublands

Poleward from the tropical rainforests, a progressive decrease in amounts of precipitation occurs, culminating, ultimately, in the nearly rainless subtropical deserts. Actually, the decreased precipitation results from the increased lengths of the dry season at greater distances from the equator; during the rainy season, even at points quite close to the desert border, the daily rainfall is generally heavy.

In regard to vegetation, the result should be a progression from the tropical rainforest through the semideciduous forest, to thornbush scrubland, to grassland, to desert. This ideal, however, is seldom achieved in nature—too many other features, such as local relief and edaphic and hydrologic conditions enter into the picture. Instead, there are scattered areas of tropical scrublands occupying large expanses of South America, India, and Australia, but not really participating in any orderly progression (Figs. 13–13 and 13–14).

Very commonly, the tropical scrublands are composed of thorny bushes and stunted thorny trees. In the rainier areas, or where the hydrologic conditions supply adequate water, the growth is very dense, with trees and bushes interlacing and making the scrub almost impenetrable. In many places, grass is plentiful and covers the ground thickly. Where the scrubland is drier, as along its borders with grasslands, the shrubs are smaller and more widely spaced, and, in many places, either grass or bare ground consumes more space than does the brush.

Mediterranean Scrublands

The great subtropical deserts are limited on their poleward sides by the occurrence of winter rains brought by midlatitude frontal situations associated with cyclonic

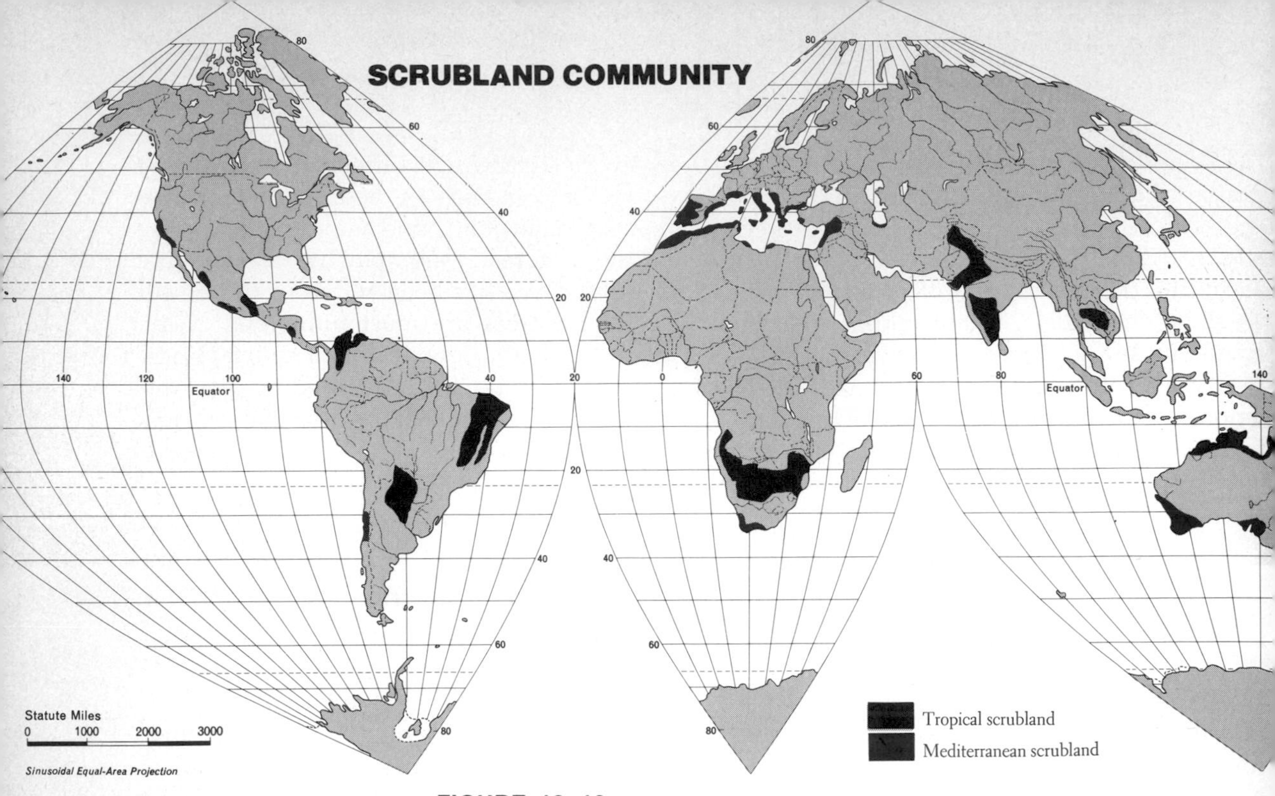

FIGURE 13–13

The world's *scrublands* generally occur between the tropical forests and the tropical grasslands, along the margins between the tropical wet (Af) and tropical monsoon (Am) climates, and in areas having Mediterranean climate (Cs).

storms. Thus, the area comes to be characterized climatically by winter frontal precipitation and a long summer drought. This situation is known as the Mediterranean climate (Cs), because it is best developed in the areas around the Mediterranean Sea.

When precipitation is concentrated in the cooler winter season and heat is concentrated at the opposite time of year, vigorous plant growth is not possible; consequently, a very distinctive, and somewhat unusual, type of vegetation characterizes all such "Mediterranean" lands.

To endure the long summer drought, most Mediterranean plants have acquired at least some xerophytic characteristics. Leaves are small, hard, thick, frequently "varnished," or waxy, or covered with hairs—all mechanisms designed to reduce water loss by transpiration. The barks of the trees, also in an effort to reduce water loss, are thick. Sap is highly resinous, and, when exposed to the air, congeals quickly in order to seal off wounds and prevent excessive "bleeding." Root systems are large, with both great lateral spread and deep penetration into the ground. Many plants are evergreen, but those that are deciduous tend to lose their leaves in summer and get new ones in midwinter, conforming to the periods of drought and moisture rather than to the times of heat and cold.

The characteristic Mediterranean landscape has a fairly dense cover of low to moderate-sized brush on the hills and a scattering of moderate-sized trees on the valley floors and the lower slopes (Fig. 13–15). The brush is highly xerophytic, shows marked seasonal changes, and, due to its resinous sap and the absence of water during the dry season, is highly flammable. Fires are common, and the brush seldom has

FIGURE 13–14

Examples of the wide variety of *tropical scrublands* to be found in Brazil. Left, a *moist* phase, known as *campo cerrado*, north of São Patricio in the state of Goiás. Right, a *dry* phase, known as *caatinga*, west of Campina Grande in the state of Paraiba. (Preston E. James)

a chance to reach its climax stage before another fire occurs. Many plants, however, are highly resistant to fire and quickly resprout from undestroyed root crowns or from vast subterranean root systems; some actually require the heat of fire to release their seeds and ready them for germination.

While there are several different names for this brush (for example, *maquis* and *gerigue* in southern France, *macchia* in Italy and South Africa, *chaparral* in Spain and California), and, while floristically, it is quite different in the various areas in which it is found, the way in which it has adjusted to the similar climates is so similar that,

FIGURE 13–15

Mediterranean scrublands, such as this area on the western foothills of the Sierra Nevada, east of Fresno, California, are characterized by a fairly dense cover of low shrubs on their hillsides and scatterings of medium-sized trees on their valley floors and lower hillslopes. The tallest plants shown here are digger pines; the others are mixed oaks or chaparral shrubs. (Verna R. Johnston, National Audubon Society) physiognomically, examples of it from different regions are almost identical: the vegetation of the Cape Region of South Africa can scarcely be distinguished from the vegetation in the Santa Monica Mountains of southern California or the Serra de Moncique of Portugal.

In favorable situations, the hillside brush grows so densely that it forms an almost hopelessly tangled barrier. It is commonly 15 to 25 feet high. In less favorable situations, the bushes are thinner and stand apart from one another. Grass and other herbaceous plants commonly grow beneath the shrubs and, where accessible to larger animals, provide good grazing. In the Mediterranean area of Europe, the Near East, and North Africa, severe destruction by people and their domesticated animals through the millenia has severely damaged the natural vegetation cover, and thus today, entire areas are almost totally denuded. In other areas, due to the relative recency of the arrival of "civilization," more of the natural vegetation remains.

The valleys and lower slopes of many Mediterranean lands are (or once were) occupied by oak woodlands. While the trees are not tall (usually under 60 feet), they are massive, with huge trunks, widely spreading branches, and a dense cover of foliage.

Mediterranean scrublands occur around the Mediterranean Sea, in California, the tip of South Africa, in central Chile, and in two southern areas of Australia (Plate V; Fig. 13–13).

Grassland Community

The world's grasslands are usually divided into three basic types: *savanna, prairie,* and *steppe*. Savanna is a dense, tall growth of coarse grass that, in the tropical dry-winter (Aw) climatic areas, often has scattered trees. Prairie is a continuous tall-grass cover in the humid continental-hot summers (Dfa-Dwa) and subtropical humid climates (Cfa). Steppe is a short-grass cover associated with semiarid (BS) climates (Fig. 13–16).

Savanna

The great savannas, largely coincident with the tropical dry-winter climates (Aw), are growths of tall grasses, ranging from two feet to as much as a dozen feet in height. They are also characterized by widely-spaced trees—sometimes in clumps, sometimes solitary—of low to medium height, and often include a scattering of moderately tall bushes. Most of the trees and bushes have protective thorns, and many of the grasses have spiny tips or barbs. The grasses are often very coarse, and despite their apparent lushness, are usually relatively unpalatable to grazing animals.

Three seasons typify the savanna lands: the warm and rainy summer, when the grasses grow rapidly; the cool and dry winter, when the grass and the land dry up; and the hot and dry late spring and early summer, when strong winds blow great clouds of dust across the land.

Savannas occupy large areas in the South American countries of Venezuela, Colombia, and Brazil, a great belt across

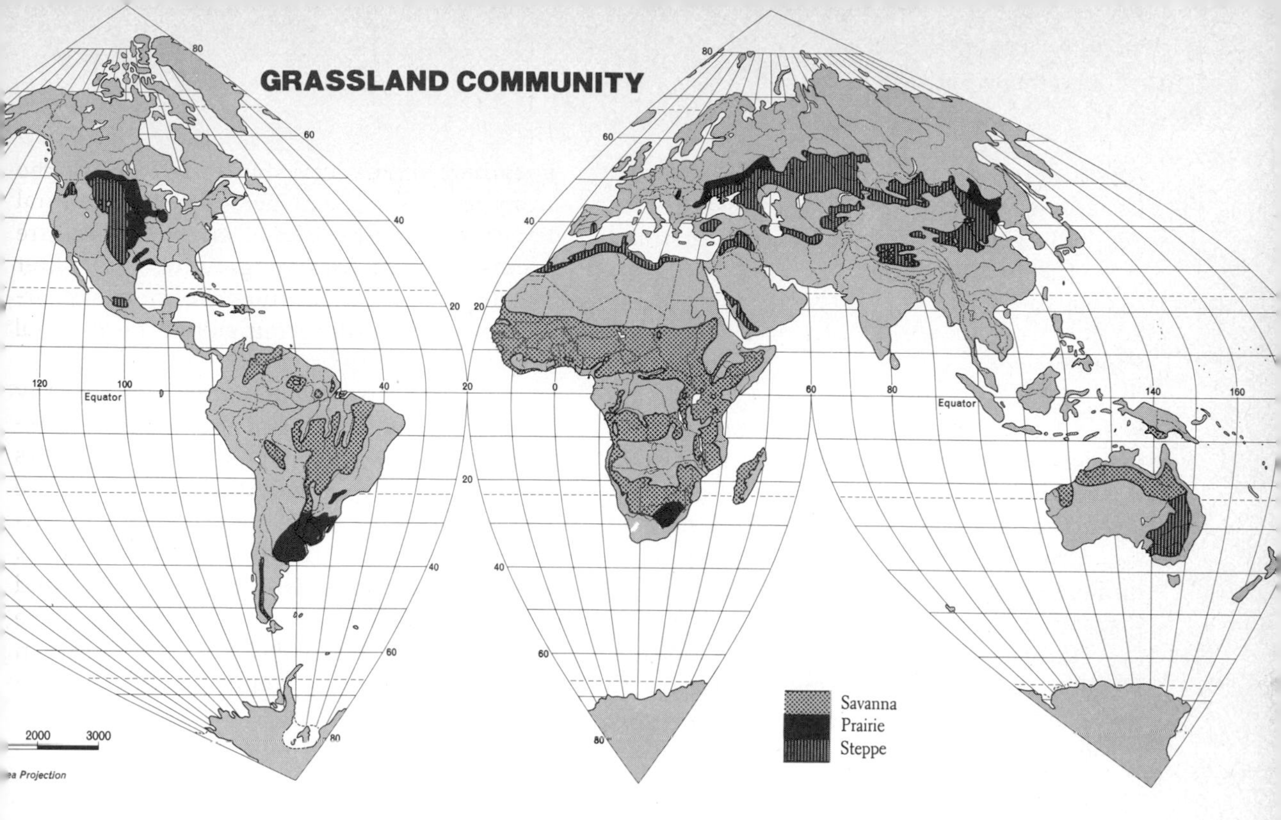

FIGURE 13–16

The *grasslands* of the world cover large portions of all the continents except Europe and Antarctica. Some of the areas that were originally grasslands, especially those occurring as prairies, are now croplands.

FIGURE 13–17

The *scattered tree savanna* of eastern Colombia seen from the foothills of the Andes. Note the *galeria forest* along the stream courses. (Standard Oil Co., N.J.)

Africa south of the Sahara, the plateaus of East Africa, the western side of Madagascar, and a large part of north-central Australia (Figs. 13–16 and 13–17).

Prairie

Most of the world's prairie grass has been destroyed, either by plowing, due to the advance of agriculture from wetter lands, as in the westward spread of settlement in the United States, or by the invasion of domesticated plants, as in the Argentine Pampas (Fig. 13–18).

Prairie is a distinct type of middle latitude vegetation. It is most extensive in the Americas (Fig. 13–16), extending from the eastern edge of the southern Canadian Rocky Mountains into the United States, and from southern Brazil across the Argentine Pampas. In Africa, a small triangle of prairie occurs in eastern Orange Free State and Lesotho, where it occupies an unusual position as a transition from steppe to savanna (Plate V). Its occurrence there results primarily from lower temperatures experienced by this portion of the South African Plateau. In Europe, the Hungarian plain, the Walachian plain, Bessarabia, and a wedge-shaped strip through south-central Russia to the southern end of the Ural Mountains are comparable prairie regions. The only extensive area in Asia is that of the Manchurian plain. Elsewhere on that continent, the transition from steppe to forest is so rapid or so poorly marked that no true prairie regions can be said to exist.

FIGURE 13–18

Grassland areas in the midwestern United States. Left, a cultivated *prairie* with a second-year stand of a grass similar to the original prairie cover, near Mankato, Kansas. Right, a typical *steppe* with its short-grass vegetation, on the high plains of Kansas. (U.S. Department of Agriculture)

Steppe

The term *steppe* is applied to short-grass areas within the middle latitudes. Sometimes the expression *tropical steppe* is used to describe the drier margins of the savanna. Although characterized by relatively short grass, the tropical steppe frequently includes many scattered xerophytic shrubs and even a few spindly, stunted trees, and thus, they differ radically from the middle latitude short-grass areas. In the latter, a carpet of grassy plants extends for miles, with no noticeable interruption by trees or shrubs, except near some wet spot or along a stream (Fig. 13–18).

The worldwide distribution of steppes is shown in Figure 13–16. It should be noted that, in the middle latitudes (Plate V), steppes fringe the edges of the deserts.

Desert Community

Desert Scrub and Barren Desert

In the drier parts of the low latitudes and on into the drier middle latitude areas of the arid climates (BW), only xerophytes have sufficient moisture to form a permanent, though thin and scattered, vegetation cover. Some places are so dry that even precarious footholds of stunted bushes and bunch grass are prevented; in such areas, there are seemingly interminable stretches of sand and rock wastes, with no vestige of plant cover. Actually, there is very little absolutely barren surface, but it is only along the occasional watercourses, or close-grouped about the infrequent springs, that the vegetation is even temporarily heavy. Over most of the land, there are only small patches of short grass or isolated bushes, which are brown and lifeless most of the time. When water becomes available, this growth takes on a bright green coloring that contrasts sharply with the soft browns and yellows or the brilliant reds of the rocks and the sand. Desert vegetation begins where dominant grass cover ends; grass borders a desert area on both its equatorward and poleward sides (Plate V).

Most of the world's desert areas extend inland and poleward from tropical west coasts, and are thus known as *warm deserts* (Fig. 13–19). In all of them, the plant cover is sparse and many plants are xerophytic (Fig. 13–20). The greatest of all deserts is the Sahara, stretching 3,500 miles across the northern part of the African continent, and seldom less than 1,000 miles wide throughout its whole length. From the Red Sea, the desert continues, with occasional interruptions, into the heart of Asia. Other less extensive deserts exist in Africa, Australia, South America, and North America.

Tundra

North of the boreal forest, the tundra, known as the *cold desert*, forms the outer fringe of the continents of the Northern Hemisphere. The tundra is of little significance to man, and, thus, no distinction of its variety has been made on the vegetation map. Near its forest margin, it is more luxuriant than elsewhere, being intermingled with stunted forest specimens, some of which are so small that they can barely be considered trees. Trunks about the size of an

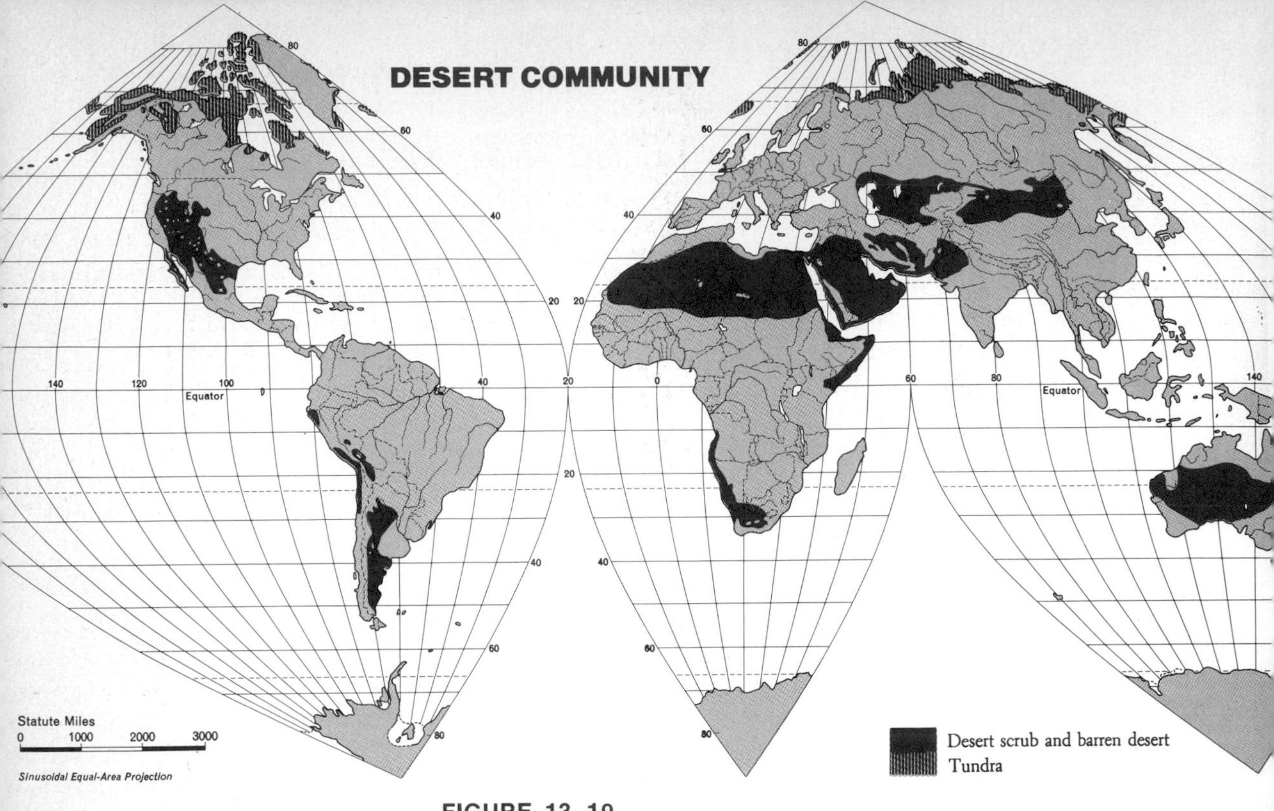

FIGURE 13–19

Deserts cover large portions of all the continents except Europe and Antarctica. *Tundra* is restricted to the highland and poleward portions of the Northern Hemisphere.

FIGURE 13–20

Desert vegetation in the southwestern United States. (Trans World Airlines)

ordinary lead pencil, with other features in proportion, represent many years of exceedingly slow growth. Farther from the forest margin, woody plants and bushes are absent, and the grasses, mosses, lichens, and sedges form an even mat over the whole surface (Fig. 13–21). Still farther poleward, the cover is reduced to isolated patches of these plants, with much bare ground between. In the most extreme situations, all vegetation disappears, and bare rock or ice forms the surface. *Permafrost,* a permanently frozen zone of soil lying a foot or so beneath the topsoil, is common throughout the tundra, and marshy conditions prevail there throughout the short summers.

The larger areas of tundra occur on the Arctic fringe of North America and Eurasia; from there, long fingers stretch southward along highland zones. In Antarctica, tundra is almost entirely lacking, for most of the continent is covered by the icecap.

FIGURE 13–21

Tundra on the Colville River delta in northern Alaska. Note the flowering plants (marsh marigold) that fringe the small pond in the foreground, and the ice still covering the larger pond in the background. (Urban C. Nelson, U.S. Fish and Wildlife Service)

Mountain Vegetation

In mountainous regions, there is a pronounced vertical zoning of vegetation that corresponds to the vertical zoning of mountain climates. The sequence of vegetative types approximates that of the zones running from the equator to the poles. In a middle latitude mountain region, the lower slopes are clothed with deciduous broadleaf vegetation, followed, in succession, by zones of coniferous forest, a tundralike cover known as *alpine meadow,* bare rock, and, near the high peaks, even a zone of permanent snow. In tropical mountains, such as Mount Ruwenzori in east-central Africa, the sequence includes tropical forest, broadleaf evergreen forest similar to that of the middle latitudes, broadleaf deciduous forest, coniferous forest, alpine meadow, and, finally, a total absence of vegetation. The particular zoning depends primarily upon the type of vegetation found at the base of the mountain. In addition, the number of zones in the sequence decreases as one proceeds poleward from the equator.

This zoning, however, does not always occur with diagrammatic simplicity. In mountain climates, details of position and exposure result in a spotty distribution of climatic types, and, thus, similar spottiness is reflected in the details of vegetation. Because of local dryness, the expected pattern of forest zones may be broken by the occurrence of a grassland or a desert vegetation type, or a broadleaf forest may be interrupted by a needleleaf variety. While the general sequence can be stated, local departures from it are frequently so great that the whole pattern appears to be upset. For that reason, the collective expression, mountain vegetation, provides a category that includes all of the types previously discussed.

Although there does seem to be a strong affinity between mountains and forests, the detailed distribution from place to place is not indicated on the world map; thus, the world patterns of mountain vegetation are the same as the world patterns of mountain climates (Plates IV and V).

REVIEW AND DISCUSSION

1. Explain, by providing as many examples as you can, how vegetation may be considered a vital factor in the total life complex of the earth. How does vegetation interact with the total physical environment? What is the relationship between the types of vegetative growth and insolation?
2. What is meant by a bioclimatological approach to the study of vegetation geography? Within the scope of physical geography, why is this the preferred approach? Discuss several ways in which vegetation is of significance to us.
3. The earth's land-vegetation mantle may be divided into four major systems, or, global communities. What are they, and how do they differ? Are they sharply defined regionally? Do they always relate to climatic and soil variations? Draw a profile of the vegetation succession of each of the four communities.
4. What are the four main environmental controls that affect vegetation distribution? Which do you think has the greatest influence? Why?
5. Much, but not all, is understood concerning the influences that individual climatic elements have on vegetation. Discuss the most vital of these influences, and show the importance of each to an understanding of vegetation, and of the environment as a whole. What are transpiration and optimum temperature?
6. Define: perennials, photosynthesis, solar attitude, edaphic controls. How is it that humankind may rank as the most potent of all biotic influences?
7. Is it likely that change is the normal order of things in the realm of vegetation? If so, why? What is the meaning of change in this context? What is plant succession?
8. Explain the contention that the four major plant communities represent vegetative structures of descending orders of size, complexity, and groundcover development—each corresponding to descending orders of moisture availability.
9. Compare and contrast the four major plant communities using the geographical criteria you consider to be most logical and important. Define: evergreens, deciduous trees, shrub, grassland, tundra, bioclimatological.
10. Discuss the characteristics and global distributions of the forest community, placing principal emphasis on its bioclimatological aspects. Compare and contrast tropical forests with middle- and high-latitude forests, particularly in terms of responses to climate.
11. Outward from the regions of tropical wet climate, optimum rainfall conditions within the tropics decline markedly. Discuss these declining conditions and the forest conditions that result from them. Define: jungle, selva, lianas, galeria, monsoon forests. Map the tropical forests of the world. Within what climatic regions are the most dense tropical forests found?
12. In the middle and high latitudes, as elsewhere, moisture availability is important to forest growth. What other climatic factors have a considerable influence on forest growth in these latitudes? Discuss the presence of these factors and the forests that result from them. How do vegetation species in these latitudes differ from those in the tropics?
13. What are broadleaf evergreens? Where winter seasons are severe, what forest type is found? Define: coniferous, edaphically dry, moors, boreal forests, muskeg. Map the middle- and high-latitude forests of the world. Why are extensive boreal forests limited to the Northern Hemisphere?
14. The world's scrubland community occurs, basically, in two general locations. What and where are they, and to what climate types are they related? Discuss

fully the tropical scrubland community and account for its world climatic distribution. Discuss the characteristics and climatic distribution of the Mediterranean scrublands. Define: maquis, chaparral. List six regions in which Mediterranean scrublands are found.

15. Where a scrubland or forest community gives way to a grassland community, what climatic change is evidenced. Into what three basic types can the grassland community be divided? Define each of these types and explain their climatic differences.
16. Compare and contrast savannas with each of the other grassland types in terms of visual appearance and human uses. Map the world's grasslands.
17. Discuss the world distribution and vegetation characteristics of the desert community. Explain the relationship between this distribution and climatic types. What are the warm deserts? Why is tundra included within the desert community? What are cold deserts?
18. Describe mountain vegetation. What is vertical zoning of vegetation? Does it relate to vertical zoning of climate?

SELECTED REFERENCES

Baker, H. G. *Plants and Civilization,* 2nd ed. Belmont, Calif.: Wadsworth, 1970.
Billings, W. D. *Plants and the Ecosystem.* Belmont, Calif.: Wadsworth, 1965.
Cain, S. A. *Foundations of Plant Geography.* New York: Harper, 1944.
Chapman, V. J. *Salt Marshes and Salt Deserts of the World.* New York: Wiley, 1960.
Dansereau, P. M. *Biogeography: An Ecological Perspective.* New York: Ronald, 1957.
DeLaubenfels, D. S. *A Geography of Plants and Animals.* Dubuque, Iowa: Brown, 1970.
Eyre, S. R. *Vegetation and Soils: A World Picture.* Chicago: Aldine, 1963.
Farb, P. *Living Earth.* New York: Harper Colophon, 1968.
Farb, P., and the editors of *Life. Ecology.* New York: Time, Inc., 1963.
———. *The Forest.* New York: Time, Inc., 1961.
Gleason, H. A., and A. Cronquist. *The Natural Geography of Plants.* New York: Columbia University Press, 1964.
Grass: The Yearbook of Agriculture, 1948. Washington, D.C.: U.S. Dept. of Agriculture, 1948.
Jaeger, E. C. *The North American Deserts,* Stanford, Calif.: Stanford University Press, 1957.
Leopold, A. S., and the editors of *Life. The Desert.* New York: Time, Inc., 1961.
Money, D. C. *Climate, Soils and Vegetation.* London: University Tutorial Press, 1965.
Newbegin, M. I. *Plant and Animal Geography.* New York: Dutton, 1948.
Polunin, N. V. *Introduction to Plant Geography and Some Related Sciences.* New York: McGraw-Hill, 1960.
Richards, P. W. *The Tropical Rainforest.* Cambridge: Cambridge University Press, 1952.
Robbins, W. W., T. E. Weir, and C. R. Stocking. *Botany: An Introduction to Plant Science.* New York: Wiley, 1964.
Trees: The Yearbook of Agriculture, 1949. Washington, D.C.: U.S. Dept. of Agriculture, 1949.
Watts, D. *Principles of Biogeography.* New York: McGraw-Hill, 1971.

A well-controlled pattern of irrigated land near Spanish Fork, Utah. (U.S. Department of the Interior)

14

Soils and the Environment

Although relatively inconspicuous in the landscape, soils are nevertheless an essential element of the physical environment. Without them, vegetation could not exist, and there would be no plant or animal resources to sustain life. Even life forms in the seas and oceans are affected by soils, for it is water, flowing over the lands and carrying soil particles and minerals to the water bodies, that supplies much of the food basic to marine life. Though largely indirect, the functions of soils within the environment are as fundamental to humans as those of air and water. Variations in the qualities of soils from one region to another result in differences in an area's habitability, just as do variations in climate, landforms, and vegetation. The way in which people use or misuse the soils frequently determines the success of human settlement in a particular region and its degree of sustained agricultural productivity.

Soil Characteristics: Physical and Chemical

Soil is a loose complex of weathered rock materials, minerals, organic matter, air, and water. In some instances, the mineral element is dominant in the soil, while in others, organic matter (both living and dead) prevails; but regardless of the actual proportions, both elements are always present. Soils may cover some areas to a depth of only one inch, or it may be scores of feet deep; in some places, soil may not be present at all and, instead, unweathered bedrock will be exposed at the surface.

Soil is evolving and changing constantly. Such environmental factors as climate, vegetation, parent material, slope, time, and bacteria contribute to this evolu-

FIGURE 14–3

Characteristic soil structures. Top left, *platy.* Top right, *prismatic.* Bottom left, *blocky.* Bottom right, *spheroidal.* (Roy W. Simonson, USDA)

have lost their granular structure, through excessive plowing and cultivating, become a "soupy mud" when watered heavily and later dry into hard crusts, or clods.

Soil Chemistry

Soil Colloids. Soil colloids are extremely minute gluelike mineral particles that, being ionized (electrically charged), are capable of performing numerous important functions in the soil. They have a profound influence on soil fertility, for they contain plant nutrients that become available to plants in an electrochemical reaction referred to as *base exchange.* In this reaction, a base, which may be ions of calcium, magnesium, or potassium, bound to the colloid compound is freed and becomes available to plants in the form of vital nutrients mixed in a soil solution. Another important function of soil colloids is their ability to serve as a binding agent for soil particles, thus facilitating the soil's ability to retain water. For example, both inorganic and organic colloids are sometimes carried to the lower levels of the soil where, cementing the other subsoil particles, they form a concentrated "hardpan" layer that cannot be penetrated by water.

Soil Acidity or Alkalinity. A soil is considered either acidic, alkaline, or a combination of both, depending on its chemical composition, which is in turn determined largely by the climate. For example, in regions of heavy rainfall, soils are usually acidic; in regions of low rainfall, alkaline soils most frequently develop. In a very significant way, the degree of acidity or alkalinity controls the adaptation of various crops and native vegetation to soils. For example, cranberries usually grow best in moderately to strongly acidic soils; alfalfa and various other legumes ordinarily grow best in slightly alkaline or only weakly acidic soils. The degree of acidity or alkalinity of the soil is expressed in *pH values* that range from 1 to 14: a pH of 7.0 indicates neutrality, while higher values indicate alkalinity, and lower ones indicate acidity.

Soil-Forming Agents and Processes

Major Agents

The evolution of soil is advanced by four major agents: water, air, plants, and animals. Each of these performs certain functions, and the resulting soil reflects their combined action. Soil in one place differs from that in another because the contribution of each agent varies with location.

Water in the Soil. As it soaks into the ground, water becomes either *hygroscopic, capillary,* or *gravitational*. While the effect of hygroscopic water is negligible, capillary and gravitational water are vital agents in soil formation.

Capillary water is the major source of plant moisture. Its amount depends upon the number and size of the spaces between soil particles as well as upon the total amount of water available to the soil. In soils composed of fine particles, water is abundant; in coarse, porous soils it may be almost totally lacking. *Capillarity* is the action by which water is drawn upward across the surface of a crystal, a grain, a tube, and so on. As the soil surface dries out, capillary water from the lower horizons may be drawn upward if the soil spaces are small enough and the grain surfaces sufficiently continuous. In this manner, moisture necessary to plants is made available in periods of protracted drought. Where the movement toward the surface is pronounced and long-continued, it often causes, in the topsoil, a concentration of chemical salts, some of which may be injurious to the plant life. Such accumulations usually occur in arid or semiarid regions, where rainfall is not sufficient to return the soluble salts to lower levels in the soil.

Gravitational water is the major agent of color, texture, and structure changes, both in the topsoil and the subsoil. It is thus responsible for easily observed differences between the parts of any soil profile. Of the major changes produced by gravitational water, two are physical. As the water moves downward through the soil, it tends to pick up minute soil particles (clay and silt particles) and carry them downward out of the topsoil; this process is called *eluviation*. The deposition of these fine particles lower in the profile is referred to as *illuviation*. The result of this process is a coarsening of the texture of the topsoil, particularly in its lower layers. Thus the structure is generally altered, and the size of the spaces between the particles is increased. Consequently, less water is retained as capillary water, for capillary water exists only in fine, netlike spaces.

As gravitational water moves downward toward the water table, it has less ability to carry the load of fine particles, and eventually, the fine particles are deposited in the spaces in the subsoil. The texture and structure of the subsoil are thereby altered: the layer becomes more compact; both hygroscopic and capillary water are increased in amount. This action, in combination with certain chemical changes, may be carried on to such a degree that a very stiff, impervious layer, a hardpan, is formed in a portion of the subsoil. This retards soil evolution and precludes the growth of deep-rooted plants.

The third of the major functions performed by gravitational water is chemical rather than physical. As water moves downward through the topsoil, it tends to dissolve the more soluble of the chemical elements. This process, called *leaching*, may also change both the texture and the structure of the soil. It definitely changes the chemical composition, and thus, it materially alters the value of the topsoil as a source of plant food. The exact nature of the chemical change varies with the reaction of the water. If the water is mixed with large amounts of carbon dioxide from the air or, if it seeps through a heavy layer of humus, it becomes a weak organic acid. Chemical reactions between this acidic water and the soil particles differ from those produced by "purer" water. However, leaching always tends to increase soil acidity.

Chemical changes frequently affect the soil color. When the action is such that it leaves the iron compounds in the topsoil, a reddish or yellowish color results. When the iron is removed, a grayish or whitish color prevails. When the water is strongly affected by the humus through which it passes, it often imparts blackish or brownish coloring to the topsoil. Elements that are leached from the topsoil are usually concentrated in the subsoil. Thus, the passage of water through the soil effects changes in both topsoil and subsoil. If enough change occurs, definite A- and B-horizons are created.

Air in the Soil. The presence of some air in the soil is essential for plants and for soil animals. Life processes cannot go on without an adequate supply of oxygen, and though some plants and a few animals can secure it in ways other than directly from the air, most of them are unable to do so. Their survival depends essentially upon air that is present in the spaces between the soil particles—spaces that are not permanently filled with water. In addition, when the soil air is charged with water vapor or with carbon dioxide given off by the plant or animal life within it, it is capable of bringing about a certain amount of chemical change by solution. The air is thus a direct agent of soil change.

In most soils, there are times when little air is present, particularly during and after heavy precipitation. At those times, gravitational water occupies practically all of the openings, but it soon drains away and is replaced by air.

Organic Matter in the Soil. Both plants and animals live in the soil and are agents of its evolution. When plant roots establish themselves, they draw water and absorb as food many of the chemical elements in the soil. Under the influence of sunlight and carbon dioxide from the air, the elements drawn from the soil are converted into various new compounds known as *organic compounds.* These are the direct foods on which the plants live and from which they develop their structures. The plants, in turn, provide food for animals. When the plants and animals die, both organic and mineral substances are returned to the soil, either at the surface, as in the accumulation of leaves, twigs, branches, trunks, and animal remains in a forest, or in the form of dead root fibers and animals within the soil itself. The processes of decay and associated bacterial action convert the organic matter to humus. At the same time, many of the mineral elements in the decaying structures are released to the soil.

Many small animals, such as the earthworm, feed upon humus and carry it downward into the soil. Large numbers of bacteria feed upon humus, often converting it to other forms, and mix it with the mineral parts of the soil. *Fungi*—plants that are without green color—also feed upon it and, in turn, produce other organic compounds. The remains of all of these plants and animals provide a further source of organic matter when they die. Certain of the bacteria have the ability to take nitrogen directly from the air and transform it so that it can be used by plant roots. The presence of these *nitrogen-fixing bacteria* in large numbers ensures high soil fertility.

The mixing process and the downward movement of altered organic compounds is furthered by the action of water. Some of the compounds are extremely soluble and hence are rapidly leached away; others are quite insoluble and remain to impart black coloring to the soil. Organic matter, either living or dead, makes for greater soil fertility and for ease of soil cultivation; normally, the more humus, the greater the workability, or *friability* (*friable soils* are those that are easily crumbled in the hand), of the soil. In general, where soils are well drained, high humus content implies high fertility, and low humus content implies low fertility.

Soil-Forming Processes

Soils are formed from regolith by the combined activities of the soil-forming agents: no one of them is alone responsible, and the combinations are not everywhere the same. In some places, as in areas of tropical wet climate (Af), water is plentiful, and its functions can be carried on continuously. In other places, as in areas of arid climate (BW), water is infrequently available. In some regions, as in the middle latitude grasslands, organic matter accumulates rapidly; in other regions, as in the deserts, it accumulates slowly. Earthworms are important soil "cultivators" in some regions, but are absent in others. Despite the infinite variety of possible combinations, it may be said that the evolution of soils occurs through three major soil-forming processes: *latosolization, podzolization,* and *calcification.* In addition to these three major ones, various other soil-forming processes exist, the more important of which include: *salinization,* the accumulation of various kinds of salts (many soluble) in the soil; and *gleization,* occurring in poorly drained soils and bringing about the formation of a waterlogged horizon of bluish or greenish color (called the true glei or gley) overlain by a layer of partially decomposed vegetation.

These principal soil-forming processes serve as the basis for classifying the earth's great number of soils into various groups. One process alone, or any of them in combination, may be involved in the evolution of any one soil. The worldwide great soil groups depend upon the balance among these processes.

Latosolization. The *latosolization process* (Latin: *latera* = brick – referring to the red color and clayey texture the process develops in soil) is one in which leaching and eluviation are very highly developed. It occurs only where there are large quantities of moisture at temperatures high enough to allow soil formation to continue throughout the year, and is best developed under forest cover. Because latosolization is continuous, soils in which it is the dominant process are deeper than others. The whole soil profile occasionally exceeds 25 feet in depth. Under the most extreme conditions, as represented in areas of tropical wet climate (Af), bacterial life is plentiful and prevents large accumulations of humus in the upper part of the A-horizon. Organic matter is rapidly converted into a form that is quickly and easily leached away. Most of the soluble minerals are also removed: leaching is so strongly developed that even much of the silica is removed, leaving only iron and aluminum oxides in the A-horizon. Eluviation is likewise at a maximum; the A-horizon becomes extremely coarse-textured and porous, while the B-horizon becomes compact and clayey. One prominent indicator of the process is color: the iron oxides give reddish or yellowish colors to the soil. Latosolization is most common in A and C climates (Plates IV and VI).

Podzolization. The *podzolization process* (Russian: *podzol* = alkaline ashes) is the end result of all the soil-forming agents working under conditions different from those that produce latosolization. Here, acid organic matter accumulates at the surface and decomposes very slowly. The process is basically one of leaching of soils by highly acidic solutions. Bacterial life is meager and soil animals are few in number, thus there is no intensive conversion, or mixing, of the organic elements with the mineral elements of the soil. The A-horizon is leached of its iron and aluminum compounds, and what remains is dominantly silica. It is, thus, gray or even whitish. The B-horizon is compacted by the deposit of leached and eluviated materials from above. Both horizon have relatively shallow depth developments. The process occurs in humid areas with a pronounced cool season and, consequently, does not continue uninterruptedly throughout the year; it is best developed under a coniferous forest cover. The soils formed by podzolization are generally very shallow, with an average depth of only 14 or 15 inches. Podzolization is most common in the mid- and high-latitude areas of C, D, and E climates (Plates IV and VI).

Calcification. The *calcification process* is best developed under grass cover rather than forest cover. Grass feeds on large amounts of calcium brought from the lower parts of the soil by capillary action. When the grass dies, calcium is returned to the upper soil layers. Grass roots and leaves provide heavy mats of organic matter, and bacterial life is plentiful. There is not a sufficient amount of water to cause heavy leaching or eluviation. In contrast to soils in which the processes of latosolization and podzolization dominate, those soils in which calcification is the most important process are mostly nonacidic in their reaction. Soil depths are moderate to shallow. Calcification is most common in B climates (Plates IV and VI).

A brief summary of the more prominent effects exerted on both the A- and B-horizons by each of these three soil-forming processes is presented in Figure 14–4.

EFFECTS ON THE A-HORIZON

LATOSOLIZATION	PODZOLIZATION	CALCIFICATION
1. Calcium and other very soluble elements are lost completely from the entire profile. Most other chemical elements are removed, leaving only iron and aluminum.	1. Most chemical elements except silica are removed by leaching.	1. There is no removal by solution; some emplacement of calcium and magnesium occurs by capillarity from below.
2. Fine particles are removed, producing coarse texture.	2. Fine particles are removed, producing coarse texture.	2. Only the very finest particles are removed, leaving the texture moderately fine.
3. Organic matter is destroyed rapidly by bacteria; there is no accumulation of humus.	3. A moderately large amount of humus is accumulated on the surface, but little is mixed through the horizon.	3. A large amount of humus is accumulated on the surface and is well mixed through the horizon.
4. There is considerable depth development.	4. There is only shallow depth development.	4. There is moderate depth development.

EFFECTS ON THE B-HORIZON

LATOSOLIZATION	PODZOLIZATION	CALCIFICATION
1. There is a concentration of iron and aluminum leached from the A-horizon. There is no calcium or other very soluble element present.	1. There is a concentration of elements leached from the A-horizon.	1. Calcium and magnesium are concentrated in the form of carbonate nodules.
2. There is a concentration of particles eluviated from the A-horizon, producing a heavy clayey texture.	2. There is a concentration of particles eluviated from the A-horizon, producing a clayey texture.	2. The texture is little changed by particles eluviated from the A-horizon.
3. Humus is absent.	3. There is little humus mixed through the horizon.	3. Humus is well mixed through the horizon.
4. There is considerable depth development.	4. There is little depth development.	4. There is little depth development.

FIGURE 14–4
Some major effects of the soil-forming processes on the A- and B-horizons.

Controls of Soil-Forming Processes

The end product in the evolution of any one soil is determined by which of the three processes—latosolization, podzolization, or calcification—dominates, and by the rate at which the process operates. The following five controls determine the dominant process and its rate: *climate, vegetation, parent material, slope,* and *time.* The first two controls are worldwide in their effects; the third and fourth are of extreme importance locally; and the fifth control is important in all soil evolution, whether on a worldwide scale or on a local one.

Climate

Because water and air are such significant agents in the soil-forming processes, climate is the single most important influence on soil development. Usually, a large amount of precipitation means a large amount of soil water. The continued presence of moisture in the soil depends largely upon the distribution of precipitation during the year. Temperature is also a very significant influence, for the effectiveness of water as a soil agent depends upon it.

In a tropical wet climate (Af), with constant high precipitation distributed throughout the year, and with regular high temperatures, there is a large amount of water available for both leaching and eluviation. Not only is the amount large, but it can continue its functions throughout the whole year, for there is no period when temperatures are low enough to freeze the water, and there is no dry season. In addition, air and water temperatures are relatively high, and chemical reactions proceed more rapidly at high temperatures than at low ones. An arid climate provides a sharp contrast to such conditions, for there, the amount of precipitation is extremely limited, and it comes at irregular intervals. Before evaporation has dried out the surface and capillary activity has drawn the water toward the surface again, hardly any of it can sink far into the ground. Leaching and eluviation in such a climate are at a minimum. Similarly, in a polar climate, water is seldom present in large amounts in a liquid state. When water is present in the form of ice, the processes that it carries on under warmer temperatures cannot take place.

Some of the relationships between climate and the processes carried on by water are shown in Figure 14–5.

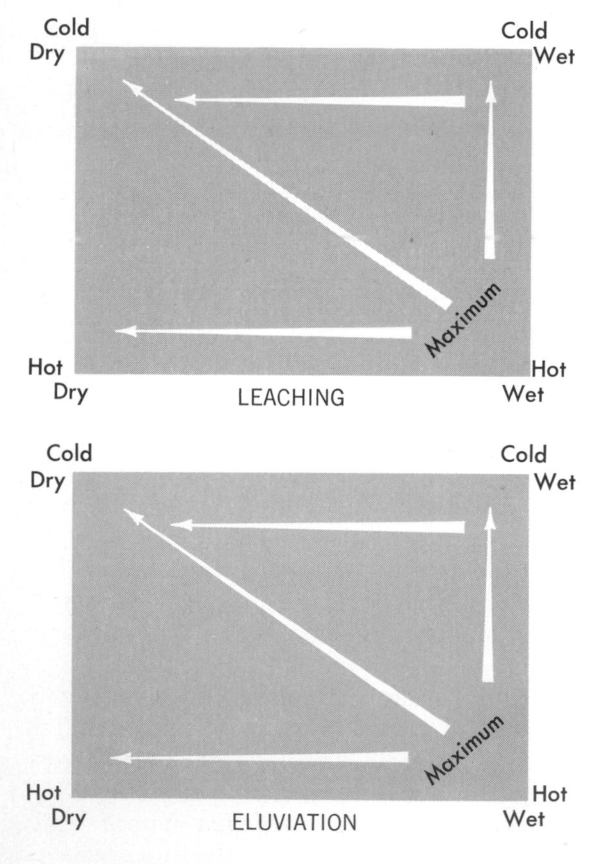

FIGURE 14–5

The relations of the *leaching* and *eluviation* processes to climate. The effectiveness of each process decreases in the directions shown by the arrows. The corners of the rectangles represent climatic extremes.

Vegetation

Soil control by vegetation is exerted largely through the supply of organic matter. The largest amount of humus accumulates under grass cover, and the next largest, under forest cover. The transfer of this humus to the soil is dependent mostly upon living organisms. Some climates favor plentiful soil life, while others discourage it; similarly, some types of vegetation produce organic material that can be readily converted, while others do not.

In forested tropical areas, there is an enormous and continuous production of dead organic matter, but the climate is such that bacterial life is plentiful. This bacteria decomposes plant remains rapidly, reducing them to a form that living plants can use. Hence, with this continuing decomposition, there is no considerable humus accumulation on the soil, nor any great permanent inclusion of organic compounds within it.

In forests of the higher middle latitudes, the cool, moist climate fosters the accumulation of a thick layer of dead organic matter on the surface, but the formation of humus is very slow. It develops in a form that is easily leached out of the A-horizon. Climate, along with the acid reaction of the A-horizon, discourages the presence of many earth animals. Hence, organic matter is not made easily available to plant roots in a form that they can use. Despite this, much of it is actually carried down into the B-horizon.

A definite relationship exists between climate and vegetation on the one hand, and the dominating soil-forming processes on the other. Areas of tropical wet climate, with its accompanying types of forest, are areas where latosolization is at its height; areas of continental climate, with its accompanying types of forest, are areas where podzolization is paramount; and regions along the moister margins of semiarid climates (BS), with their associated grassy cover, are areas where calcification dominates. These relationships are graphically summarized, in part, in Figure 14–6.

Parent Material

The differences among soils are quite evident, especially when fields are being plowed. Some soils are yellowish, sandy, and extremely light; some are sticky, red clay; some are yellowish-brown and rich looking. Often, these differences can be traced to differences in the underlying rock (the parent material from which the soils are derived).

The nature of the parent material contributes greatly to the nature of the soil. Unless the parent material contains certain chemical elements, the soil must, of necessity, lack them. If the products of weathering are fine-grained, then the soil can also be expected to be fine-grained. The parent material provides the specific combination of

FIGURE 14–6

The relations of the soil-forming processes to climate. The effectiveness of each process decreases in the directions shown by the arrows. The corners of the rectangles represent climatic extremes.

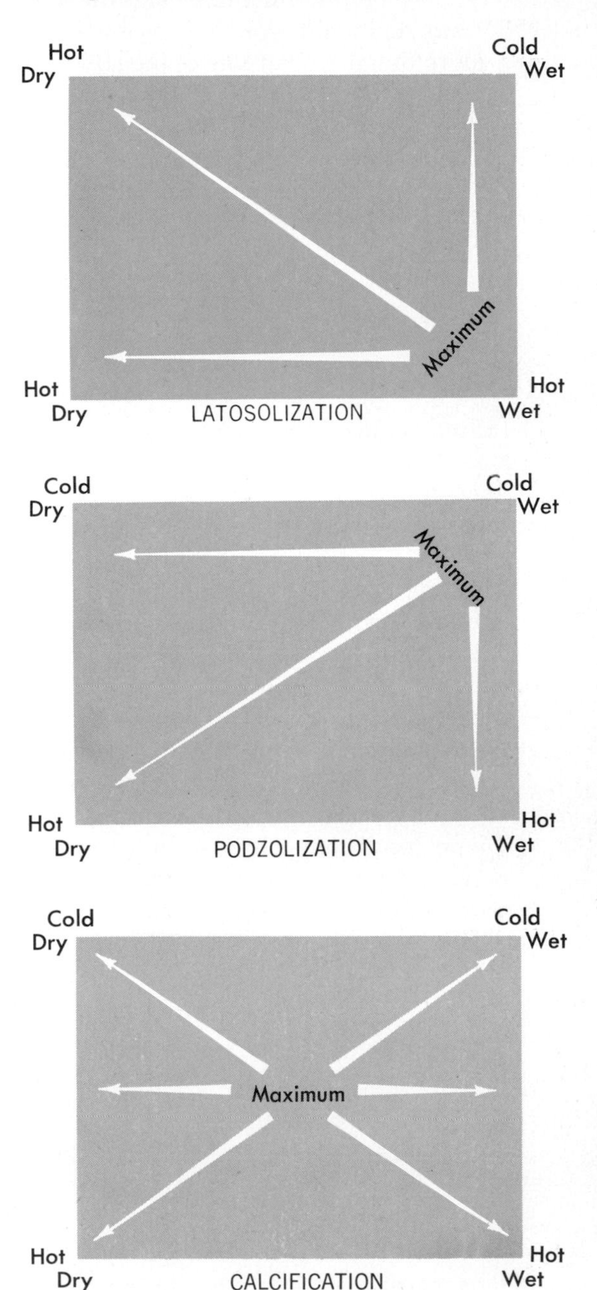

chemical elements in the regolith and thus greatly influences its texture. All of those traits are passed on to the soil.

Differences induced by a soil's parent material may be of tremendous significance locally and regionally, for they provide the finer, more detailed, patterns of the soil, but in a global study, they are of little significance. From that point of view, the controls of climate and vegetation combine to establish certain major climax types toward which all soils are developing, thus making the parent material insignificant.

Slope

Slope, as does the underlying rock, induces certain local and regional variations. On steep slopes, the runoff of water is rapid and large in amount. This runoff is not only an effective erosional agent, but it also diminishes the water available to sink into the ground and carry on the soil-forming processes. Frequently, slope lands are so badly eroded that an A-horizon is lacking altogether. In extreme cases, the whole profile may be lost, and the slope may have no soil at all; it may be covered, instead, by parent material or by coarse rock fragments.

Slope conditions may also induce poor drainage. In poorly drained depressions, the water table is close to the surface and thereby prevents the downward percolation of surface water. Under this condition, the normal processes of leaching and eluviation cannot go on, and soil evolution is stopped. Over wide areas of gently sloping surface, the effect of slope is minimized, and the other controls operate more strongly. Gentle slopes decrease both the rate and amount of runoff, thus reducing erosion and making more water available for participation in the processes of soil formation.

Time

Time is the prime requisite in the production of all soils. Given sufficient time, even the differences induced by the parent material can be overcome. Under each climate and each type of vegetation cover, a considerable variety of soil can develop at different rates and over varying spans of time. For example, on the steppes of European Russia, deep black soils occur across the entire zone of grass that follows the wet margin of the semiarid climate (BS). Under these soils, crystalline rocks are found in some places, and sedimentaries are found in others. This indicates that some of the surface was once covered with glacial debris, while some of it was not. The soil mantle, however, shows similarity throughout. Here, enough time has elapsed so that the controls of climate and vegetation have left their unmistakable imprint.

Classification of Soils

Orders of Soils

The earth's soils can be divided into three major orders: *zonal, intrazonal,* and *azonal.* Zonal soils are more important and more widespread than the soils of the other orders. In well-developed soil profiles, they display characteristics that reflect the dominant influences of climate and vegetation. Intrazonal soils are of a much lesser extent, and their profiles display characteristics that reflect the dominating influences of terrain, parent materials, or time. More widely scattered than intrazonals, azonal soils also show the influences of parent material and terrain; however, owing to extreme youth or to unusual conditions affecting their controls, these soils offer few or no developed profiles.

The Great Soil Groups

The major soil groups, usually called *great soil groups,* of the zonal order have both internal and external features reflecting the combined controls of climate and vegetation. Where they are well developed, they occur over many kinds of parent material, and their profiles are the result of long-term action by soil-forming processes that operate over wide areas. Five principal soil groups are

FIGURE 14–7
The classification of soils. (Based on *Soils and Men*, U.S. Department of Agriculture, *Yearbook of Agriculture,* 1938, U.S.A.)

DOMINANT CLIMATIC ASSOCIATION			DOMINANT VEGETATION ASSOCIATION
	ZONAL ORDER		
	Suborders	*Great soil groups*	
HUMID CLIMATES (Pedalfers)	Latosolic soils of warm, humid tropical and subtropical areas	Latosol Red-yellow podzolic-latosolic	FORESTS
	Light-colored podzolized soils of cool, humid forest areas	Gray-brown podzolic Podzol	
SUBHUMID TO ARID CLIMATES	Dark-colored soils of the subhumid grasslands	Prairie Chernozem Chestnut and brown	GRASSLANDS
(Pedocals)	Light-colored soils of the arid areas	Gray-desert Red-desert	DESERTS
	Tundra, subarctic, polar soils	Tundra	
	INTRAZONAL ORDER		
	Suborders	*Great soil groups*	
	Hydromorphic soils of bogs, marshes, swamps, upland flats	Humic-glei Alpine meadow Bog Planosols	
	Halomorphic soils of poorly-drained arid areas	Saline (solonchak) Alkali (solonetz) Soloth	
	Calcimorphic soils	Rendzina Terra rossa	
	AZONAL ORDER		
	Suborders	*Great soil groups*	
		Lithosols Regosols Alluvials	

recognized: where the process of latosolization is dominant, the resultant soils are known as *latosols;* where podzolization is the principal process, the soils are called *podzols;* where calcification is the prime process, the resultant soils are called *chernozems;* in arid regions, where all soil-forming processes are very slow, the soils are called *desert soils;* and where continued cold persists and all processes are very slow, the soils are called *tundra soils* (or cold desert soils).

Between these extreme forms there are many gradations, certain of which are sufficiently pronounced to warrant recognition as distinct zonal soils. Thus, as one passes through the transitions from areas of latosolic soils to areas of podzolic ones, latosolization becomes less and less significant, and podzolization becomes more and more so. Hence, from the tropical rainforest to the borders of the polar tundra, there exists the following soil progression: *latosol, podzolic-latosolic, podzolic,* and *podzol.*

Similar transitions are observable between humid areas and deserts. In tropical latitudes, as one passes from areas of great moisture to areas of less, the latosols of the tropical rainforest give way to the chestnut soils of the tropical grasslands, which, in turn, are succeeded by the highly calcified soils of the tropical steppes and deserts. In more temperate latitudes, gray-brown podzolic soils dominate the most moist areas, and, as increasingly drier areas are crossed, one encounters dark prairie soils, chernozems, chestnut soils, brown soils, and eventually, the gray or red soils of the deserts. A full list of the great soil groups is presented in Figure 14–7.

In summary, it should be stressed that the zonal great soil groups are the fundamental soils, the *mature soils* toward which all soils are developing. The close relationships that exist between climate, vegetation, and soils are represented schematically in Figure 14–8.

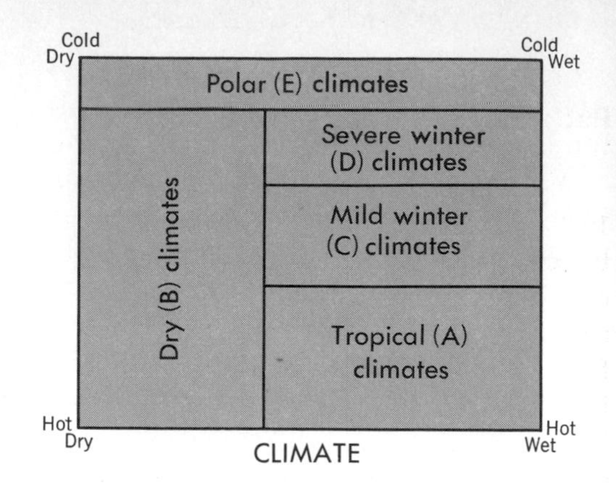

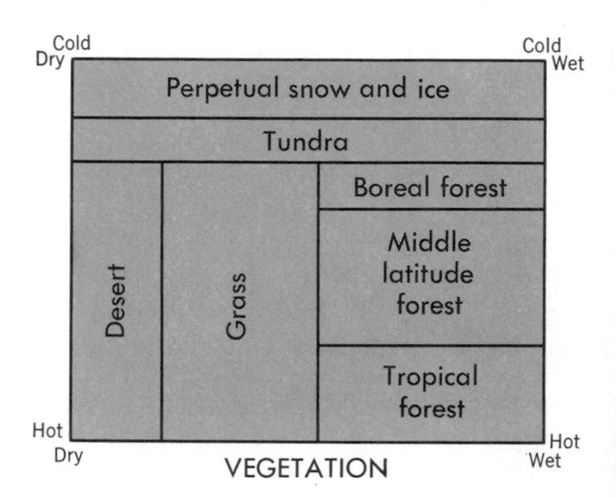

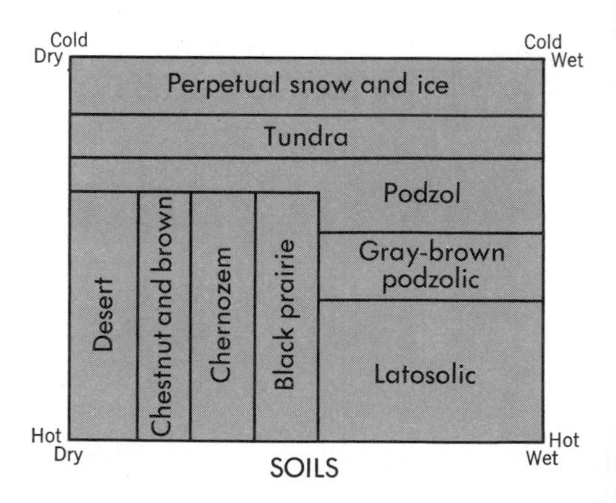

FIGURE 14–8

The distribution of climatic types, natural vegetation associations, and zonal soils. The corners of the rectangles represent climatic extremes. Comparison suggests the interrelationship of these three elements of the physical setting. (After Blumenstock and Thornthwaite, "Climate and the World Pattern," in *Climate and Man*, U.S. Department of Agriculture, 1941.)

Pedalfers and Pedocals

Two major soil groups devised about 1920 recognize this close relationship between climates and soils. One such group consists of the *pedalfers*. In general, pedalfers are soils of the humid regions whose profiles show a concentration of alumina and iron oxide, with no development of a carbonate horizon. The second of the soil groups is the *pedocals*. These tend to be soils of the subhumid to arid regions whose profiles show a horizon of accumulated calcium carbonates.

Subclassifications

In detailed soil study, further division is made within each of the zonal soil groups. Those soils that have profiles similar in their major characteristics and that have been developed locally over the same kind of parent material are considered a *soil series*. Such a series is subdivided into *soil types*, determined largely by the texture of the A-horizon. The types are further differentiated into *phases* that represent features of significance in soil use, such as slope, stoniness, and the like. Applying the foregoing subclassification to a specific soil might yield the following:

Group	Gray-brown podzolic
Series	Hillsdale
Type	Sandy loam
Phase	Gravelly

The soil in this sample classification is referred to as a *Hillsdale sandy loam, gravelly phase*. It is recognized as being within the gray-brown podzolic zonal soil group.

Zonal Great Soil Groups

The great soil groups considered in the following study are organized under climate–vegetation group titles, for purposes of organizational clarity and maximum appreciation of the interrelationships of elements of the physical environment (Fig. 14–9).

The zonal order incorporates most of the great soil groups. Each zonal soil is really a group, or family, of soils, most of which have developed fully mature profiles. The following discussions of the characteristics of zonal soils apply to their mature profiles in particular, although it must be remembered that most soils with immature profiles are evolving, under the dominant controls of climate and vegetation, toward one or another of the several mature profile forms (Fig. 14–10).

Forest Soils

Latosol Soils. Under heavy forest cover in warm to hot, humid climates, latosolization is the major soil-forming process. The latosol soils thereby produced exhibit considerable depth of profile and a reddish to yellowish coloration. Humus or organic

FIGURE 14–9
The zonal great soil groups.

FOREST SOILS	GRASSLAND SOILS	DESERT SOILS
Latosol	Prairie	Gray-desert
Red-yellow podzolic-latosolic	Chernozem	Red-desert
Gray-brown podzolic	Chestnut and brown	
Podzol		Tundra

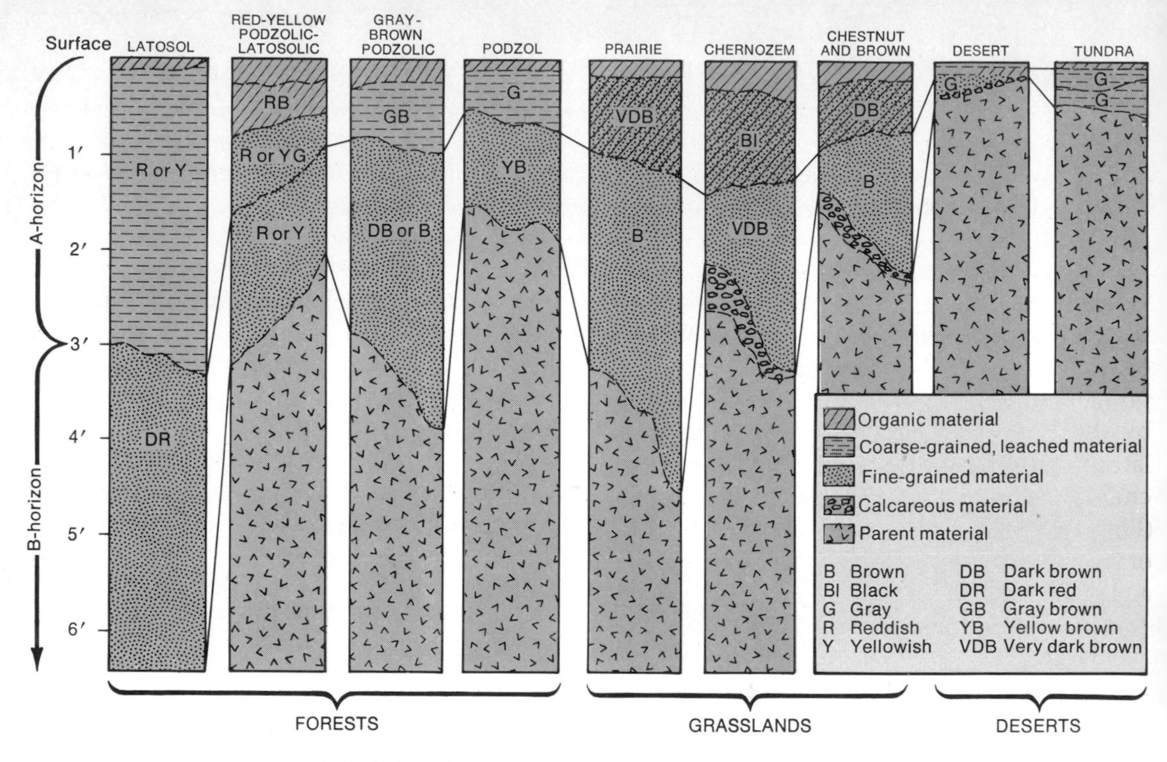

FIGURE 14–10

Comparative schematic profiles of representative zonal soils. The scale at the left applies to all the profiles; the lines between profiles connect the bottoms of the respective A-horizons and the bottoms of the respective B-horizons. Cross-profile brackets and labels below the columns show the dominant vegetation community associations.

material is in small supply or completely absent, due to the rapid destruction of organic matter by bacterial action; their soluble minerals, and even most of their silica, have been removed by leaching (Fig. 14–10), but they have high iron and aluminum contents. In the most completely latosolized forms, the A-horizon is reddish, coarse-grained, and very porous, while the B-horizon is deeper red in color, relatively fine-grained, and often clayey.

The intensity of leaching and the small accumulation of humus tend to make these soils relatively infertile; thus, they do not lend themselves to heavy cropping without heavy fertilization. However, large trees do well in such soils, since their roots are able to penetrate below the leached upper zones to extract plant nutrients from lower levels of the soil.

Principal regions of the latosols are the tropical rainforests, the tropical scrublands, and the transitional zone bordering the tropical grasslands. In the Americas, these soils cover much of Brazil, and extend northward into the coastal portions of southern Mexico and the islands of the Caribbean. In Africa, they form a broad belt across the center of the continent—from Senegal in the west, to Tanzania in the east, and from the southern Sahara to Angola—and they dominate the island of Madagascar. They are also found from southern and central India across southeast Asia and Indonesia to the islands of the southwest Pacific and northern Australia. Climatically, they are associated with Af, Am, and Aw environments (Plates IV, V, VI; Figs. 14–10 and 14–11).

Red-Yellow Podzolic-Latosolic Soils. Developed under warm, humid climatic conditions, and associated with deciduous and coniferous forest covers, the red-yellow podzolic-latosolic soils are produced jointly by the dual processes of podzolization and latosolization. Like the latosols, they exhibit characteristic red and yellow colorations (Fig. 14–10) resulting from a concentration of iron hydroxides, and they commonly have a clayey texture due to their high aluminum content. Also like the latosols, they show strong leaching of their A-horizon, resultant loss of their soluble minerals, and a low humus content due to rapid bacterial action in the prevalent warmth and moisture. Thus, the fertility of these podzolic-latosolic soils tends to be rather low, and they need both fertilizer and careful management to be highly productive agriculturally.

They are found in regions of subtropical humid climates, such as in the southeastern United States, southeastern China, southern Japan, southeastern Brazil, southeastern South Africa, and the southeastern coast of Australia (Plates IV, V, VI; Figs. 14–10 and 14–11).

FIGURE 14–11
The forest soils of the low latitudes have developed under warm to hot, humid climatic conditions.

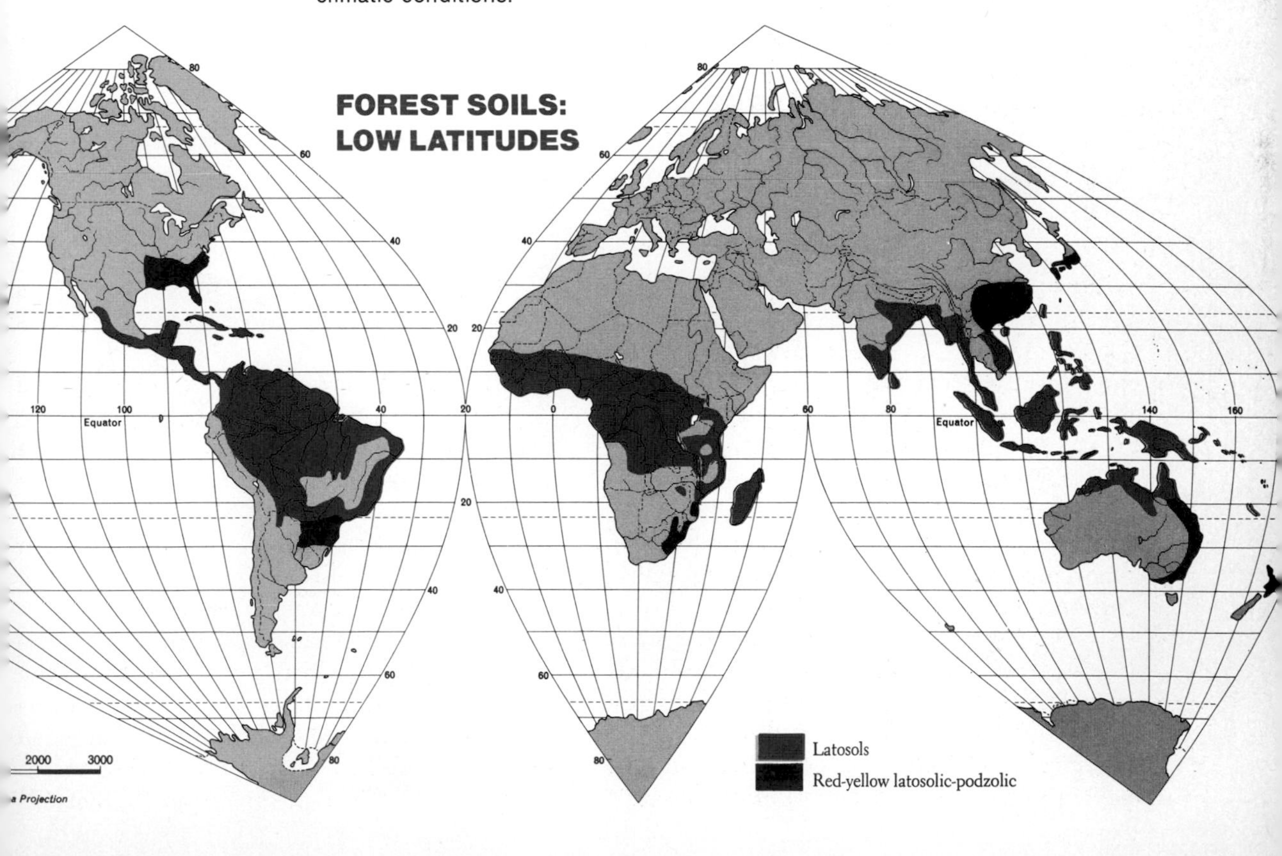

Gray-Brown Podzolic Soils. Poleward from the regions of the red-yellow podzolic-latosolic soils, but still within the humid forested regions, is another zonal soil that shows even more marked features of podzolization. Though thicker than in the latosols, the surface accumulation of humus is still relatively thin (Fig. 14–10) and less leached than the true podzols farther poleward. Below the humus, a grayish-brown, rather heavily leached A-horizon gives way to an illuviated dark-brown, fine-grained B-horizon. The whole profile is shallower than that of the latosols, usually reaching a depth of about 4 feet. Leaching is less prominent in these soils than in those of the tropics and subtropics, and more organic matter is incorporated in them. Consequently, the gray-brown podzolic soils are of medium fertility and are much more retentive of moisture.

Principal areas of gray-brown podzolic soils are in the northeastern United States; in Europe, a wedge of this soil stretches inland from the Atlantic coast to the heart of European Russia; and in Asia, interior and northern China, Korea, Siberia, and northern Japan. It should be noted that some of these are areas of relatively high population density and that they are important agricultural regions. (It should also be noted that such soils are virtually absent from the

FIGURE 14–12

The forest soils of the middle latitudes have developed mainly within the humid forest regions under cool, humid climatic conditions.

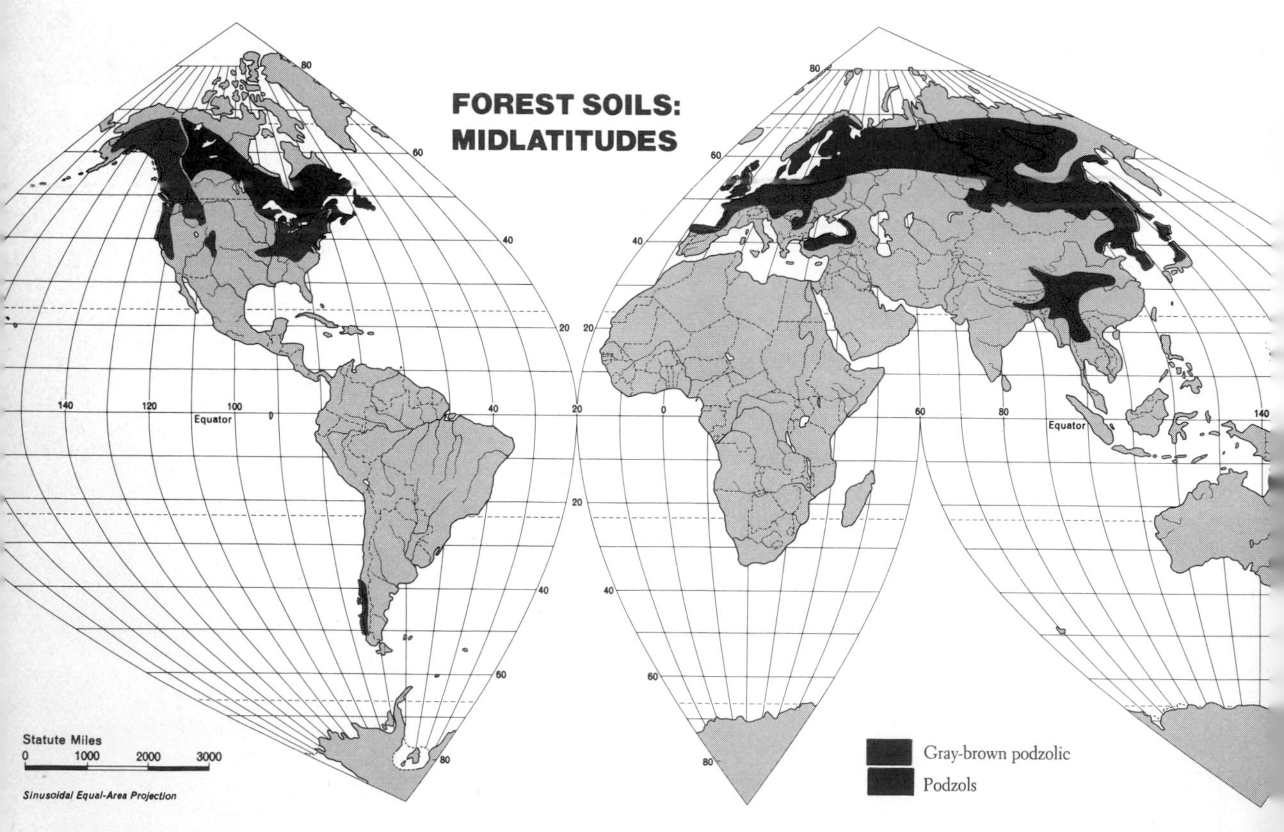

Southern Hemisphere, appearing in Tasmania, New Zealand, and southern Chile only.) That agriculture is still prominent in all of these areas is, in large part, a reflection of careful human management of the soil. These gray-brown podzolic soils are developed mostly under Cfb and Dfb climates under deciduous forest covers (Plates IV, V, VI; Figs. 14–10 and 14–12).

Podzol Soils. These zonal soils develop mainly in coniferous forests in relatively cool, humid climates, and are primarily the result of the process of podzolization. A layer of humus lies upon a coarse-grained, whitish or gray, heavily leached horizon (Fig. 14–10), which, in turn, changes sharply to a very compact yellow-brownish illuviated layer. The whole profile is very shallow, with a total depth of usually less than 2 feet. The natural fertility of podzols is low. Conifer forests do not require the calcium and magnesium bases, so these elements are not returned to the surface of the soil by the leaf and branch humus. Podzol soil A-horizons are usually high in humus and acid content, with their lower portions heavily leached. The B-horizon is a layer of illuviation, with a high colloid base content that often forms a hardpan.

These soils dominate in the northern forested lands of North America and Eurasia. In America, they extend uninterruptedly from the Yukon eastward to the Atlantic coast, where they reach from eastern Maine to central Labrador. In Europe, the podzols stretch from the wetter and cooler sides of the United Kingdom, across Norway, Sweden, Finland, and the northern Soviet Union, to spread in a progressively widening belt across northern Europe and Asia to the Pacific shores. These podzol soils develop largely under Cfc, Dfb, and Dfc climates, and under coniferous forest covers (Plates IV, V, VI; Figs. 14–10 and 14–12).

Grassland Soils

Prairie Soils. Soils of very high fertility develop under tall grassland cover in regions of humid climate. Thick mats of grass roots merge into dark-brown horizons (Fig. 14–10), which grade through lighter browns to light-colored parent materials below. The whole profile varies from 3 to 5 feet in depth. The high percentage of available organic matter and the relatively small amount of leaching maintains considerable plant food. These soils have a rich humus, but have no marked layer of calcium accumulation. Prairie soils are extremely fertile and highly productive.

Chernozem Soils. Also developed under grassland conditions, the chernozems are found on the dry margins of the prairie soils. In these soils, calcification is the principal soil-forming process. They show a thick mat of grass roots and humus at the surface, grading downward into a black A-horizon (Fig. 14–10). With depth, the color shades through the browns to a lighter-colored parent material. At a depth of from 2 to 4 feet, a layer of whitish nodules of calcium usually occurs. Chernozem soils are only slightly leached, have a high organic content, and have large amounts of available calcium in the B-horizon. Consequently, their natural fertility is high. However, they are not well suited to the production of plants that demand large supplies of water, for they develop under conditions found on the drier margins of the humid climates. Nevertheless, the chernozem belts have become the great commercial wheat-producing regions of the world, because of the adaptation of wheat to moisture-poor conditions.

In Plate VI and Figure 14–13, the chernozem and prairie soils are mapped together. There are two belts of these soils in the Americas: one stretches from the Peace

River Country of western Canada southeastward through the Dakotas to Kansas; the other lies in the Pampas of Argentina and southern Uruguay. In Eurasia, these soils extend from Rumania across southern Russia into central Asia, and other belts occur in northern China and Manchuria. A small area of prairie and chernozem soils also occurs in the inland portion of Queensland, in northeastern Australia (Plates IV, V, VI; Figs. 14–10 and 14–13).

Chestnut and Brown Soils. On the dry side of the chernozems, the soils are lighter in color, and the layer of white nodules is closer to the surface (Fig. 14–10). A thin mat of grass roots is underlain by slightly leached horizons that grade through lighter and lighter brown down to the calcium layer. In the chestnut soils this layer may be as deep as 3 feet below the surface, but in the brown soils it is much closer to the surface.

The principal areas of chestnut and brown soils in North America stretch from central Alberta southward along the Rocky Mountain front to central New Mexico, and some are to be found in central Mexico. South America's chestnut and brown soils are located mainly along the eastern edge of the Andes Mountains in Argentina. In Africa a belt of chestnut and brown soil fringes the southern and northwestern margins of

FIGURE 14–13

Grassland soils of high fertility develop under grassland cover in regions ranging from humid to subhumid climates.

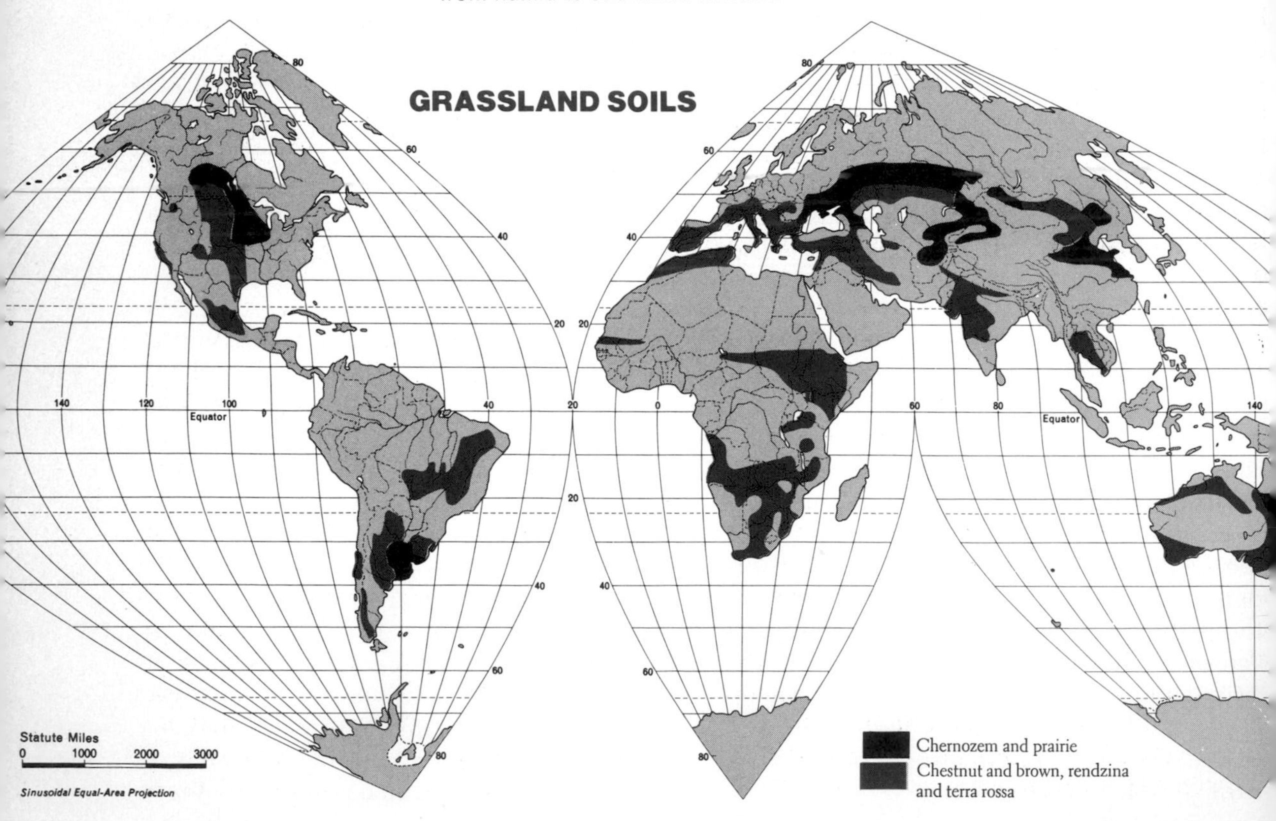

the Sahara and the wet eastern margins of the Kalahari; and in Australia, these soils border that continent's desert heartland. In Asia the chestnut and brown soils form a belt, south of the chernozems, from the Caspian shores to northeastern China, with a second large area in the drier parts of central and northwestern India (Plates IV, V, VI; Fig. 14–13).

Desert Soils

Where desert (both arid and cold) soils exist, they are very shallow, seldom measuring more than a few inches from the top to the bottom of their profile (Fig. 14–10). Slightly altered brownish to gray as well as reddish top materials are underlain by thin zones of calcium nodules. Because of the low degree of precipitation and the high degree of evaporation in arid regions, desert soils are very dry and thus highly susceptible to rising capillary action and to salinization. In those areas of the desert where a slight degree of leaching is possible, these soils are often of relatively high fertility when water is made available to them. The distribution of desert soils corresponds closely to that of arid climates (BW).

Gray-Desert Soils. Gray-desert zonal soils are typical of middle latitude deserts where excessive aridity promotes only sparse plant growth. Consequently, these soils have little humus, and their horizons are shallow and poorly developed, but these exhibit considerable accumulations of crusted, or of more or less cemented, subsurface deposits of calcium carbonate known as *caliche*. Coloration among these desert soils ranges from that of the light grays to that of the weak grayish-browns, which appear even lighter under a limecrust (caliche). These desert soils develop mainly under dry, cold (BWk) climatic conditions under sparse plant covers, and are quite infertile—even when irrigated.

Red-Desert Soils. Red-desert zonal soils are typical of the more severely arid and hotter deserts of the tropics where plant growth is minimal. Humus is at a minimum, if not totally lacking, and horizons are very poorly developed within the coarse-textured soils where mechanical weathering predominates. Colors range from reddish-gray to fairly strong reds, depending upon the accumulations of small amounts of iron oxides. Caliche is common, and fertility is exceedingly low. These soils develop largely under dry, hot (BWh) climatic conditions and minimal plant covers (Plates IV, V, VI; Fig. 14–14).

Tundra Soils. The tundra soils, characterized by a surface mat of dark-brown, partially decayed mosses and lichens, are found poleward from the podzol soils and in some places in high mountain areas (Fig. 14–10). Under this surface mat of vegetation, in favorable locations, occur grayish soil horizons ranging from a few inches to about a foot in thickness. As a result of excessive moisture and a lack of aeration, *glei horizons,* composed of sticky, compact, and often structureless materials of bluish- or olive-gray color may be present. Under these shallow horizons there may be deep permanently frozen soil layers known as permafrost, which range from a few inches to hundreds of feet in thickness. Here, the more common soil-forming processes have had little effect upon the parent material, as it has been broken up largely by mechanical weathering, with some mixing of mineral and organic materials through the alternate actions of freezing and thawing. Where the tundra soils have sufficient depth, the growing season is long enough, and the soil dries

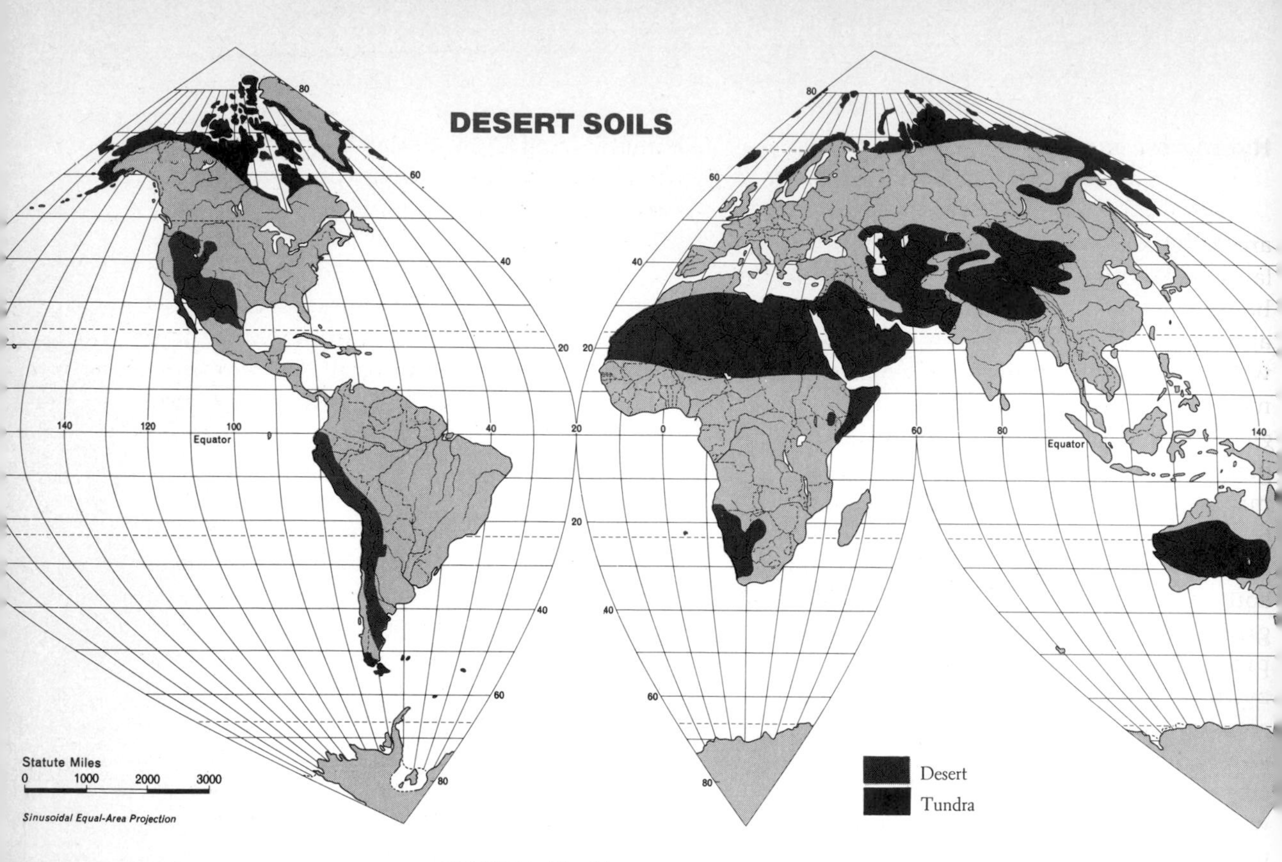

FIGURE 14–14

Desert soils (both arid and cold) develop in considerable variety, but, without irrigation, all are generally shallow and infertile.

sufficiently to allow some small amount of cultivation, natural fertility may have some small importance. These zonal soils (occasionally considered intrazonal because they are so poorly drained) have formed under tundra climatic conditions and under plant growths of mosses and lichens (Plates IV, V, VI; Fig. 14–14).

Intrazonal Great Soil Groups

Within the *intrazonal order,* there are three soil suborders: the *hydromorphs, halomorphs,* and *calcimorphs,* each of which is composed of several great soil groups (Fig. 14–7). The principal features distinguishing intrazonal soils are those characteristics of their moderately developed horizons that reflect the dominating influence of a local soil-forming process controlled by slope (terrain) or by parent materials. Climate and vegetation, dominating the zonal soils, influence the intrazonal soils only slightly.

Hydromorphic Soils

The *hydromorphic* (water-formed) soils are developed on poorly drained humid lowlands and on relatively flat uplands, such as bogs, swamps, marshes, meadows, and alpine meadows. They are typically soils with waterlogged profiles. Four major hydromorphic soil groups are recognized: *humic-glei, alpine meadow, bog,* and *planosol.* Humic-glei soils have dark-brown to black organic-mineral horizons, with moderate depths of 6 to 30 inches that grade into a glei-horizon composed of grayish soils. This intrazonal soil group occurs under swamp-forest or grass-marsh covers in humid climates. Alpine meadow soils have dark-brown horizons of moderate depth that occur under grass covers at higher mountain altitudes, where climatic conditions resemble tundra climates. Bog soils are dark-brown to black, with a rich humus that forms a muck or peaty top layer over a glei-horizon. Planosols are intrazonal soils that have horizons separated sharply from adjacent horizons because of cementation or high clay content, and they are usually associated with fluctuating water tables.

Halomorphic Soils

The second suborder of intrazonal soils, *halomorphic* soils are found in poorly drained arid regions, and are noted for a high salt content. They comprise three great soil groups: the *saline soils* (solonchaks); the *alkali soils* (solonetz); and the leached, or degraded, *soloth soils* (solodi or solodee). These soils are usually found in poorly drained depressions in arid or semiarid regions and have distinct saline horizons formed by the action of rising capillary water brought from saline groundwaters or parent materials rich in salt. Saline soils (solonchaks or salt soils) have a high concentration of soluble salts, and are not excessively alkaline (their pH value is less than 8.5). They are found mostly in arid and semiarid climatic regions, where the process of salinization is most active. In many parts of the world, these soils are of considerable economic importance as a major source of salt. Most of the saline soils are devoid of a vegetation mantle, for they are toxic to most plants, even to salt-tolerant ones, or halophytes. Thus, though they have high productive potential when flushed free of salt, most saline soils are unsuitable for agriculture because they have not been properly irrigated and drained. Alkali soils (solonetz, or metallic salt soils) contain alkali salts, usually sodium carbonate, and have a pH value of 8.5 and higher. They are developed mostly in semiarid or subhumid climatic regions under grass or shrub vegetation. Soloth soils, the third group of halomorphic soils, represent a further stage in desalinization. They have a thin surface layer of brown friable soil underlain by a gray leached horizon that rests upon a brown or dark-brown horizon. Soloth soils are more productive than alkaline soils. They are normally found in semiarid or humid climates under shrubs, grasses, or combinations of grasses and trees (Fig. 14–7).

Calcimorphic Soils

Calcimorphic soils, the third suborder of intrazonal soils, are characterized by the high calcium content they develop due to their lime-rich parent materials. The great soil groups included in this suborder are the *rendzinas* and the *terra rossas*. Rendzina soils are formed from parent materials with a high lime content, and are usually characterized by brown or black friable surface horizons underlain by light-gray or yellowish calcareous material. Terra rossa soils, like the rendzinas, are developed from limestone parent materials and hence have a highly calcareous content. Both of these types of soils are found in humid and semiarid climatic regions under grass or mixtures of grass and forest vegetation. Because of their calcareous composition, the soils in both these great soil groups are often associated with chestnut and brown soils, as is indicated by the grouping of soils that is presented in Plate VI.

Azonal Great Soil Groups

Azonal soils are fairly uniform structurally, and are classified largely on the basis of the origin of their parent materials or the terrain with which they are associated, rather than on the basis of their soil-forming processes. Azonal soil profiles are either very poorly developed or not developed at all, because of the soil's youth, the characteristics of its parent material, or the nature of the terrain. There are three azonal great soil groups: the *lithosol, regosol,* and *alluvial* soils, all of which have incomplete horizons or no horizons at all (Fig. 14–7). The azonal soil profiles that do exist generally have no B-horizons, and the A-horizons are darker colored than the C-horizons because of the accumulation of some humus there. The lithosols all lack horizons, and consist, instead, of a freshly and imperfectly weathered mass of rock fragments, which is largely confined to steep slopes. The regosols are usually composed of dry and loose dune sands, but some are made up of unconsolidated loess. The alluvial soils are primarily layers of such alluvium materials as silt, sand, mud, or other sediments transported by moving waters. They have little or no soil-profile development because of their youth and frequent burial under new sediments; however, in many instances, alluvial soils do possess some accumulations of humus.

People and Soils

Soils, being no more than a thin veneer over the earth's rock crust—often, only inches deep—are delicate and highly susceptible to the actions of nature and of humans. Nature makes continual changes in soils and in the rates at which their natural processes operate: it causes some of these processes to be accelerated, some to be retarded, and still others to become fully inactive. In the search for food and shelter, humans have positive and negative effects upon the agents of soil formation: through them, waters are increased or decreased in quantity, humus is improved or destroyed, and vegetation covers are extended and improved, or eliminated.

Cultural Impacts and Soil Changes

Denudation. When uncultivated land is settled, one of the first actions is generally the removal of much of its native vegetation. For example, among some peoples in the Amazon forest, it is customary to clear the land by burning off the smaller plants and girdling the larger and taller trees, thus causing them to die; more technologically advanced peoples effect the same end through the use of axes and saws. Humans are also indirectly responsible for vegetation removal through the grazing of domestic animals. When too many animals are pastured in a given area, or when they crop the grass too closely in their feeding, as sheep often do, bare soils are exposed. Gradually, the native vegetation is either completely destroyed or substantially replaced by weaker strains.

Whether in forests or in grasslands, the widespread destruction of vegetation removes one of the natural controls of soil formation, and largely ends the formation of humus, a major soil constituent. Thus, such destruction upsets the balance of the total soil environment—and with this comes a change in the soil itself.

Cultivation. When soil is brought under cultivation, both its texture and its structure are modified. Constant working and cropping alters a soil's structure by breaking up the grouping of soil particles at the surface. In consequence, natural processes of soil formation are materially altered.

FIGURE 14–15

A newly planted cornfield that was subjected to heavy rains soon after it was plowed and harrowed. In the absence of plant cover, runoff was rapid and erosion was great. Most of the damage seen here was done in a few days' time. (USDA)

Beyond the mechanical changes that accompany cultivation, there are those that result from the growing of a new vegetative cover. New plants may have different root structures, as, for example, those of a cereal crop and those of a forest. This results in different plant feeding habits; cereals derive most of their food from the soil proper, whereas trees, with deeper roots, obtain a large share of theirs from deeper parent materials. In addition, when the crops are harvested, the greater part of the plant cover is removed. Thus, the chemical elements on which the crops feed are permanently removed, and, unless they are artificially replaced, the soil is gradually impoverished. Continued cultivation of the soils makes them easily susceptible to increased surface erosion by running water (Fig. 14–15). Without plant cover, the amount of runoff is large, and therefore, the influence of erosion is considerable.

Management. When people produce the foodstuffs and resources they require, they may follow certain practices calculated to make the soil more suitable for the plants they wish to grow; this is called *soil management.* For example, an area may be cleared for the purpose of cultivation; after it is cleared, it may become apparent that the soil is too moist for the desired use. In such a case, drainage must be improved, either by ditching or by some other means. In another situation, soil may be so coarse as to have excessive drainage. Then, some procedure must be adopted that will make up for the lack of moisture. In many areas where moisture is deficient either climatically or edaphically, irrigation must be instituted before the soil can be made productive. In short, whatever people do in soil management alters the soil environment—and eventually the soil itself.

Soils for the Future

Soil Improvement. There is a distinction between soil fertility and soil productivity. Fertility implies the presence of adequate amounts of the chemical elements necessary for specific plant growth—both properly balanced with the other elements present and readily available to the plant roots. For example, a soil that is extremely fertile for growing potatoes may be less fertile for growing wheat, since the potato needs certain chemicals in relatively large amounts, while wheat needs others. By proper management, people can alter soil productivity quite rapidly and, in terms of their chosen purposes, can and do improve it. They can supply more water or remove excess water from a soil; they can change its structure by tillage; they can rotate crops so that certain soil minerals are replenished and proper balance insured; and they can add to the humus content by growing cover crops and plowing them under. Similarly, they can change the rate of surface erosion by planting certain crops to hold the soil in place. Thus, by good management, people have improved many soils throughout the world.

Soil Exhaustion. All too commonly, however, people have brought about the destruction of soil, either by exhausting its fertility, by subjecting it to accelerated forces of surface erosion, or by pollution. Soil exhaustion, or loss of fertility, does not occur rapidly when natural forces are unhampered. Although heavy leaching and eluviation remove chemical elements from the topsoil and change the texture and structure of the soil, there is some return of chemical elements from soil organisms and from the plants that mantle the surface. Growing plants take chemicals directly from the soil. When these plants die or lose leaves and branches, the chemicals are, in large part, returned to the topsoil. From there, they are eventually reincorporated into the soil proper by the action of water and soil organisms. In the same way, the soil organisms themselves are part of the cycle.

When people clear land and harvest crops, they indirectly remove parts of the soil. When crops grow, they use certain chemicals just as the native vegetation does; but in harvesting crops, there is almost a complete removal of the plant cover, leaving no plant remains to be reincorporated into the soil. If the process is repeated over long time spans, and if no procedure is established to replenish the elements that are lost, the soil becomes exhausted insofar as those elements are concerned. Although weathering of the underlying rock releases a new supply, the natural replenishment rate is slow. From the human viewpoint, this results in smaller and smaller crop yields until, eventually, it is uneconomic to continue planting and harvesting. In the past, this soil exhaustion problem has frequently been solved by land abandonment rather than by corrective management.

Soil Erosion. The abandonment of exhausted soil leaves an opening for accelerated erosion, especially if the soil is an upland one. Even without human intervention, all upland soils are normally eroded, and their soil material deposited in lowland areas. Youthful streams cutting into an interfluve wear away soil and carry it downstream, where it may be deposited on the floodplain of an old valley or in a delta. This normal process, however, is slow, because, among other things, it is restrained by the vegetation cover. If people use land until the soil is exhausted and then abandon it, if they manage it poorly or simply remove the vegetation (as in lumbering operations) and abandon it, the work of running water and of wind is greatly speeded up. Rills or sheet erosion, developing on bare sloping fields during every rainstorm, wash away parts of the topsoil (Fig. 14–15). Repetition of this soon creates gullies that may be deep enough to segment the land, remove much of the soil, and lower the local water table. The soil is thus lost through human-induced erosion.

People have not yet devised a process capable of speeding up the creation of new soil from the parent materials. The relatively slow process of weathering, followed by the alteration of the regolith by water, air, and living organisms produces soil only after very long periods of time. Once destroyed by humans, this resource is, for all practical purposes, irretrievably lost. By good management, flexibly applied for each soil type, people can do much to conserve soil and to increase its productivity. They probably build up more soils than they destroy, but this positive influence could be even greater if society would have it so.

REVIEW AND DISCUSSION

1. Although relatively inconspicuous in the physical landscape, soils are of great significance to humans and to the earth's entire biotic system. Explain.
2. Define the term *soil* within the context of physical geography. Is finely weathered rock considered a soil? Are variations in soil depth of any significance in the natural functions of soils and in the utilization of soils?

3. Are soils static or evolutionary? What are the environmental factors that contribute to soil processes? In the study of physical geography, why does the study of soils logically follow that of landforms, climate, and natural vegetation?
4. The great soil groups all have common general characteristics, yet they vary in one or more respects. Outline and define several of these characteristics.
5. After sketching a soil profile and labeling the horizons, discuss the development of horizons in soils. What is regolith? A-horizon? humus? How does color provide a clue to soil-type differences? How does organic content control soil color? How do chemical changes affect it?
6. Discuss soil texture, structure, and chemistry. Explain any apparent relationship that these three aspects of soil have with one another or with parent materials and climate.
7. Each of the four major soil-forming agents performs certain functions in the evolution of soils. Identify these agents and discuss the processes and results of each. Define: capillary water, eluviation, leaching, organic compounds.
8. Compare and contrast the three major soil-forming processes: latosolization, podzolization, calcification. What are salinization and gleization? With which climatic types are each of these processes most frequently associated? With what soil color or colors?
9. The controls that determine a soil's dominant formation process and its rate of formation are climate, vegetation, parent material, slope, and time. Explain the role of each of these controls.
10. Justify the need for a global classification of soils that is similar to the global classifications of landforms, climates, and vegetation.
11. The earth's soils can be divided into three major orders. Name and define them, and list their major characteristics. Within which order are the majority of the world's soils found?
12. Describe the five principal great soil groups of the zonal order. Define: pedalfers, pedocals.
13. Discuss the distribution of the zonal great soil groups of the world across the four major vegetation associations. What is the principal distinguishing feature of nearly all zonal soils?
14. Name and describe the major intrazonal great soil groups. What is the principal distinguishing feature of all intrazonal soils?
15. Discuss the azonal great soil groups. Why are azonal soil profiles either very poorly developed or nonexistent?
16. In their efforts to survive, people affect soils both positively and negatively. Approaching the subject from several viewpoints, discuss humanity's past impact upon soils and your estimation of how soils will be affected by humans in the future.

SELECTED REFERENCES

Basile, R. M. *A Geography of Soils.* Dubuque, Iowa: Brown, 1971.

Bauer, L. D. *Soil Physics,* 3rd ed. New York: Wiley, 1956.

Bennett, H. H. *Elements of Soil Conservation,* 2nd ed. New York: McGraw-Hill, 1955.

Buckman, H. O., and N. C. Brady. *The Nature and Property of Soils,* 7th ed. New York: Macmillan, 1969.

Bunting, B. T. *The Geography of Soil.* Chicago: Aldine, 1965.

Donahue, R. *Soils: An Introduction to Soils and Plant Growth,* 2nd ed. Englewood Cliffs, N.J.: Prentice-Hall, 1965.
Eyre, S. R. *Vegetation and Soils: A World Picture.* Chicago: Aldine, 1963.
Farb, P. *Living Earth.* New York: Harper Colophon, 1968.
Gibson, J. S., and J. W. Batten. *Soils—Their Nature, Classes, Distribution and Care.* Montgomery, Ala.: University of Alabama Press, 1970.
Glinka, K. D. *The Great Soil Groups of the World and Their Development.* Ann Arbor, Mich.: Edwards, 1927.
Held, R. B., and M. Clawson. *Soil Conservation in Perspective.* Baltimore: Johns Hopkins Press, 1965.
Hillel, D. *Soil and Water: Physical Principles and Processes.* New York: Academic Press, 1971.
Jenny, H. H. *Factors of Soil Formation.* New York: McGraw-Hill, 1941.
Kellogg, C. E. *The Soils That Support Us.* New York: Macmillan, 1941.
Mickey, K. B. *Man and the Soil.* Chicago: International Harvester Co., 1945.
Miller, C. E., L. M. Turk, and H. D. Foth. *Fundamentals of Soil Science,* 3rd ed. New York: Wiley, 1958.
Mohr, E. C. J., and F. A. Van Buren. *Tropical Soils.* New York: Interscience Publishers, 1954.
Money, D. C. *Climate, Soils and Vegetation.* London: University Tutorial Press, 1965.
Robinson, G. W. *Soils: Their Origin, Constitution, and Classification.* New York: Wiley, 1959.
Soil Classification, A Comprehensive System (7th Approximation). Washington, D.C.: Soil Conservation Service, U.S. Dept. of Agriculture, 1960.
Soils: The Yearbook of Agriculture, 1957. Washington, D.C.: U.S. Dept. of Agriculture, 1957.
Soils and Men: The Yearbook of Agriculture, 1938. Washington, D.C.: U.S. Dept. of Agriculture, 1938.
Stallings, J. H. *Soil Conservation.* Englewood Cliffs, N.J.: Prentice-Hall, 1957.

Citrus groves in southern California's Santa Clara Valley. (Spence Collection, UCLA

15

The Physical Environment: A Type-Study

Humans have lived on the earth for thousands of years. The direction and extent of their progress has been dictated partly by the physical environment, which has sometimes presented some insurmountable obstacles. In recent times, however, humans have become more capable of altering the physical environment to better suit their needs or desires. While some of these alterations have improved the environment, others have created major problems, and still others have been utterly destructive. Today, the environment in many areas has been transformed so greatly that few elements of its original nature are detectable.

Physical geography deals with only one part of the total subject matter of geography, as its basic concern is the study of the earthly elements and forces that operate naturally. However, most of us live in a situation created by the interaction of natural and artificial (cultural) forces. Thus, to ignore our role in the development of the environment by considering only its natural origins, would be unrealistic. The study of physical geography, then, is only the first step toward an understanding of the totality of geography; one to be followed by studies of human, or cultural, geography and, later, political and economic geography. Only thus can the full impact of the interrelations between people and their environments, cultures, institutions, and technologies be truly recognized and appreciated.

These numerous and varied interactions between humans and the environment are often exceedingly complex. In order to illustrate this complexity, as well as some of the basic principles involved in this kind of study, the following discussion will assess, in some detail, a small geographic area in southern California.

Physical Environment

The small area selected for this type-study in physical geography lies about fifty miles up the coast from Los Angeles and about five miles inland from the Pacific, where the area slopes gradually upward from the normally dry-river channel of the Santa Clara River to the low foothills of the Coast Ranges (Fig. 15–1).

Landforms

The landforms in this area can be divided into three distinct regions (Fig. 15–2): a smooth, gently-sloping plain that floors the broad Santa Clara Valley and comprises the lower southeastern half of the area (Fig. 15–3); a steeply rugged hills area that touches the western slopes of the Coast Ranges and comprises the higher northwestern half of the area; and Wheeler Canyon, a deep valley cutting from north to south through the rugged hills area, with a flat floor (a former floodplain) that has recently been incised so that it now features a deep, steep-sided gully (Fig. 15–4).

FIGURE 15–1

A location map of the type-study area showing the area's approximate position along California's Santa Clara River.

Climate

The area is characterized by a Mediterranean climate (Cs): precipitation occurs only in winter (chiefly between November and March), with virtually no rainfall in summer (between June and September). Temperatures throughout the year are mild, but, despite the small size of the area, local variations are very great. In summer, the Santa Clara Valley is cooled daily (from about 11 A.M. until sunset) by a strong sea breeze, and temperatures seldom exceed 75°F., while in the sheltered upper portion of Wheeler Canyon there is no cooling, and temperatures of 100°F. are frequently recorded. On calm winter nights, cold air drains downslope from the hillsides, leaving them relatively warm, and the cold air accumulates at the bottom of both Wheeler Canyon and the Santa Clara Valley, often producing frost. During the summer, the sea breeze brings very moist air into the Santa Clara Valley; thus at night, when the air cools, heavy fog and dew are formed. In contrast, in the upper part of Wheeler

FIGURE 15–3

The broad, smooth floor—composed of alluvium—of the Santa Clara Valley, slopes gently from the hills and the mouth of Wheeler Canyon, at the left, toward the Santa Clara River, at the foot of the distant mountain on the right.

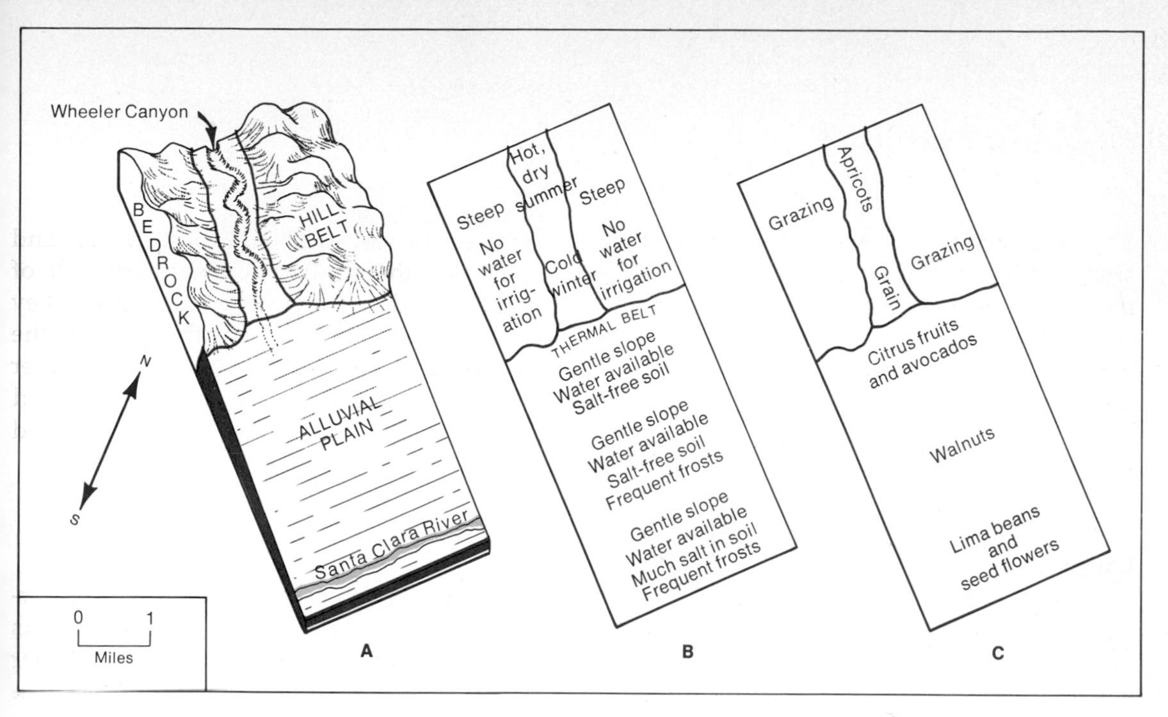

FIGURE 15–2

The physical environment of the type-study area. A. Landforms. B. Physical controls. C. Land use.

FIGURE 15–4
Wheeler Canyon, a narrow valley with a relatively flat floor, is set amidst steep-sided, brush-covered hills.

Canyon, very dry air predominates in summer, both during the day and at night, and dew is almost unknown.

Vegetation

The vegetation, which once covered most of this area, is a low growth (less than 3 feet high) of highly aromatic, drought-resistant shrubs, sometimes called the "coastal chaparral." It is well adjusted to the long, dry summers, and accomplishes most of its growth during the cool, moist winters. Along the creek bed at the bottom of Wheeler Canyon, a *riparian* (riverbank) association of trees (alders, sycamores, bays, oaks, and the like) and bushes (such as willow, mulefat, and poison oak) grow densely, the trees reaching heights of 50 feet. In the hilly areas, marked differences in exposure to the sun cause marked differences in the rate of evaporation: the north-facing slopes, being cooler and more retentive of moisture, have broken patches of oak-walnut-bay forest, with tangled undergrowths of poison oak and other brush; the south-facing slopes, baked dry of their moisture soon after it falls, have only open growths of shrubs, occasionally punctuated by tall, straight stalks of yuccas. Today, the native vegetation in the valleys has been replaced by crops and, on some hillsides, by grasses.

Soils

On the hillslopes, soils are thin; in areas where the bedrock is sandstone, the soils tend to be stony, while in areas underlain by shale, they are stone-free, very clayey, and liable to very serious soil erosion through slumping and landsliding. The floors of both Wheeler Canyon and the Santa Clara Valley are composed of alluvium material recently deposited by sheet floods associated with heavy winter rains. Having been derived from a variety of sources, they are quite rich in most plant nutrients. Their thin vegetative cover, however, provides little organic matter, and much of what is available is destroyed by the long, dry summer—hence, humus and nitrogen are sorely lacking. Certain trace elements, especially copper and manganese, are virtually absent, and there is an oversupply of boron. Since all of the lowland soils are very young (as recent as last winter's rains), there has been no opportunity for a mature soil to develop, and so the soils are noticeably deficient, both in profile and in structure.

Water

In winter and spring, both the Wheeler Canyon stream and the Santa Clara River carry a good flow of surface water; but with the increasing drought of summer, the supply gradually decreases until both are completely dry. However, due to the sand of the stream beds, a seepage that supports the riparian woodland continues all year.

Water also fills the pore spaces of the sand and silt comprising the floor of the Santa Clara Valley, and since that material is hundreds of feet thick, it constitutes a very sizable reservoir of water that can be obtained by wells and pumps.

Sequent Occupance

The ways in which people live in, and make use of, an area at any given time is termed its *occupance*. Over the ages, the same area has often been used by different groups of people in different ways, and the succession of such cultures is called the *sequent occupance* of the area. In the area being discussed here, the succession is rather brief but, nonetheless, it is quite varied.

Indian Occupance

This area was first occupied, through several millenia, by the Chumash Indians who lived by hunting and gathering, thus making very intensive use of the natural products of the land. The animals they hunted—deer, raccoons, badgers, rabbits, squirrels, snakes, and lizards—provided their food, clothing, and related equipment. The Chumash gathered a wide variety of vegetable products as well, such as the acorns of the oak, which they leached of their tannic acid and ground into a meal that served as the mainstay of their diet. They collected the seeds and bulbs of many other plants and found many uses for the yucca: they baked the tender, young stalk of this plant, and the long, tough fibers of its daggerlike leaves were used as twine, woven into nets for trapping or fishing, or

FIGURE 15–5
Beef cattle grazing in the grassy hill country.

fashioned into sandals. One of the greatest problems for these first inhabitants was finding sufficient water during the long summers; thus, villages were usually located close to a reliable water source and a grove of oak trees.

Ranching Period

In 1782, Spanish Franciscan friars established a mission at Ventura, near the mouth of the Santa Clara River, a mission dedicated to the "christianizing and civilizing" of the Indians. Within a few years, all the local Indians were settled at the mission, and their hunting–gathering way of life was a thing of the past.

During the next century, the Wheeler Canyon–Santa Clara Valley area was devoted almost entirely to cattle raising—at first, solely for the hides and tallow of the animals, and, later (after the Gold Rush created a market 400 miles to the north), for beef. Cattle were raised in large herds on vast unfenced ranges and given only minimal care. During this period, the Indians gradually disappeared, being victims of violence and imported diseases to which they had no resistance. The mission eventually disappeared as well and was replaced by Mexican *rancheros* who operated their own ranches; eventually, even they were replaced by Anglo cattle ranchers. Throughout this period, however, the same basic way of life predominated (Fig. 15–5).

Agricultural Period

In the 1880's, the construction of railroads connecting southern California with the rest of the United States opened vast new markets for agricultural products, which resulted in a rapid development of agriculture in the area. At first highly experimental, the agriculture has evolved into what is probably the most highly scientific, intensive, commercialized, and mechanized agriculture in the world today.

Physical Controls of Modern Utilization

The physical environment (Fig. 15–2) exerts certain strong controls over the modern utilization of this area. One important reason for this is the sharp distinction between the hilly lands and the plains (Fig. 15–6). The hilly lands are almost totally devoted to grazing, as they are too steep and too dry for agriculture.

FIGURE 15–6

In most areas, there is a very sudden transition from the flatter lands, used intensively for agriculture, to the hilly areas, used almost exclusively for grazing.

The climate is another major determinant of the agriculture throughout the area. It is too cool in winter and not hot enough in summer to grow such things as bananas, rubber, coconuts, or dates; but within the area's climatic limits, a very wide range of crops is possible. Though precisely which ones are to be grown at a given time is largely determined on the basis of external economic factors, climate usually determines just where a particular crop will be grown. The following are some examples of this climatic determination.

Apricots are raised only in the middle and upper portions of Wheeler Canyon. Although they do not need irrigation—they are able to subsist and bear a crop on the strength of the winter rainfall—since they are raised to be sold as dried fruit they do require the summer heat and dryness (and absence of fog and dew) that is found only in the canyon area.

Lemons and avocados, being highly sensitive to frost, are planted on the higher parts of the sloping valley floor and on the adjacent lower hillslopes. There, the cold air drains away to the valley bottom, creating a warmer zone known as a "thermal belt," where freezing conditions are unknown (Fig. 15–7).

FIGURE 15–7

Avocados (foreground) and lemons (background) grow on the terraced lower hillslopes and on the higher parts of the valley floor—areas from which, on winter nights, cold air drains away, creating a "thermal belt" in which freezing conditions never occur.

FIGURE 15–8
During spring, summer, and fall, vegetables are grown on a large scale on the lower valley bottoms. In these areas, the winters are too cold for growing citrus fruits and avocados.

Oranges, also sensitive to frost, but less so than lemons, are planted in positions downslope from the thermal belt, where temperatures are still much warmer than in the valley bottoms.

Walnuts, being dormant in winter and thus able to stand any amount of cold, are planted in the cold-air enclaves below the oranges. They cannot be grown on the lowest parts of the valley, however, because there, poor drainage in the soil induces an accumulation of salts, especially alkali, to which the walnut is particularly susceptible.

Shallow-rooted plants—crops that are not in the field during the cold periods of winter, and whose roots do not reach down to the levels impregnated with alkali—are grown on the valley flats. Among these crops are lima beans and many kinds of vegetables and seed flowers (Fig. 15–8).

People have also been forced to adjust to the odd seasonality of the area's precipitation. To make use of the floor of Wheeler Canyon, where the soil is good but where water for irrigation is unavailable, barley and oats are sown after the first rains in the fall; the crop grows in winter, and the grain is harvested in late spring (Fig. 15–9).

FIGURE 15–9
In Wheeler Canyon, the winters are too cold for the growing of citrus fruits, and water is not available for irrigation. Therefore, after the first winter rains, grains such as barley and oats are sown on the flat valley floor, and, by late spring, they are ready for harvest.

Human Alteration of the Landscape

In a fervent desire to reap the greatest possible economic return from the area, humans have greatly modified its soil, vegetation, water supply, and its climate. These alterations are manifested in many ways. In order to extend the area of flat land in the thermal belt, terraces have been carved into the lower hillsides (Fig. 15–10). To remove waters impregnated with alkali from the soil of the flatter valley lands, deep ditches have been dug and porous tiles have been laid under the fields. To overcome summer drought and thus be able to grow plants imported from the more humid parts of the world, irrigation water is pumped from wells tapping the underground reserves. When this water proves to be tainted with boron to the degree that it is toxic to plants, water is imported by pipeline from distances often in excess of ten miles (Fig. 15–11). To extend the area of lemon, avocado, and orange production downslope from the thermal belt, the orchards are protected on cold nights by "orchard heaters" burning fuel oil, and by

FIGURE 15–10
Where soils are not too stony, terraces have been constructed on the lower hillsides. These extend the area of the thermal belt, thereby creating more land for the growing of avocados and citrus fruits.

FIGURE 15–11
Often, pipelines are installed on the land surface to carry irrigation water to the fields and orchards. Here, a field of strawberries growing on low ridges that are covered with sheets of plastic (to eliminate weeds, reduce water loss by evaporation, and prevent sand and grit from getting into the berries), is irrigated by water running down the furrows.

FIGURE 15–12

When trees that are easily damaged by frost, as is the avocado, are planted on the valley floor outside the thermal belt, it is necessary to prevent them from freezing. This is usually done either by heating the surrounding air with oil-burners ("orchard heaters") or by stirring it with fans mounted on tall poles ("wind machines").

"wind machines"—motor-driven propellers that blow warm air down from aloft to replace cold air settled near the ground (Fig. 15–12). To protect against limb breakage, fruit droppage, and leaf dehydration by rare but violent winds from the deserts to the east, extensive rows of tall trees are close-planted to act as windbreaks (Fig. 15–13).

The greatest damage to this environment, however, may well be the alteration of the vegetation itself, for in this agricultural area, scarcely a single plant growing is native to it. All of the crop plants have been introduced from other parts of the world: lemons and oranges from southeast Asia and Indonesia; avocados from Central America; lima beans from the Yucatan, by way of Peru; walnuts from Persia; barley from Ethiopia. Even the weeds are largely European in origin, and the trees and shrubs of the home lawns and the windbreaks come from all over the world as well: eucalyptus from Australia, star pines from Norfolk Island in the Coral Sea, oleanders from the Mediterranean, deodars from the Himalayas, and redwoods from the Pacific Northwest.

FIGURE 15–13
Rows of tall eucalyptus trees, interplanted with shorter cypress trees, protect this lemon grove against the damaging effects of high-velocity winds.

Thus, in this one small area, although some activities have been partially curtailed by the physical geographic complex, it has been possible, in part, to overcome those obstacles by decidedly altering the landscape. Such adjustments and alterations vary markedly both with time and with the degree of technological development attained by the people involved.

The hope for humanity's future on this planet as well as the many threats to it lie, to a very large extent, in the handling or mishandling of such basic interactions. Although the earth is large, an alteration in the environment at one place may still have profound effects upon the environment at another far-distant place. This has always been true, but in the past, alterations by humans were only rarely of any appreciable magnitude. Today, however, rapidly advancing technology is producing profound effects upon many aspects of the physical environment: in many places, interactions are being altered, very delicate balances are being disturbed, and worldwide effects are being put in motion. It is absolutely vital, therefore, that careful checks be maintained on all cultural environmental changes, so that the many vital physical forces will neither be seriously disturbed, nor have irreparably detrimental results. Because physical geography studies the worldwide interlinkage of such forces, it can be a key factor in their evaluation and maintenance.

REVIEW AND DISCUSSION

1. The direction and progress of human activities are dictated, to some extent, by the physical environment. Assess this statement, explaining some of the environment's limiting effects upon humans as well as some evidence of their growing ability to alter and control it.
2. Today, in many parts of the world, the natural environment has been transformed so greatly that few elements of its original nature are detectable. Discuss some parts of the world in which such transformations have occurred due mainly to human intervention. In each instance, characterize the transformations as either destructive or constructive, and explain your reasons.
3. Assess the human role in the transformation of the physical environment of the area in southern California studied in this chapter.
4. Discuss briefly how each of the major elements of the physical environment have influenced the utilization of this southern California area. Describe, also, the influences on these elements and on their interrelationships.
5. In your opinion, which of the major elements of the physical environment in this type-study area have exerted the greatest influences on human utilization of it—either positively or negatively.
6. How, and to what extent, have people functioned as a geomorphological agent in this portion of southern California?
7. Show, in some detail, the interrelationships between the areal patterns of the bedrock, landforms, climatic conditions, native vegetation, soils, and natural water supply that existed in this type-study area before the coming of cultural developments.
8. What is sequent occupance? Although human utilization of this area in southern California has been relatively recent, several marked changes in the area's occupance have nevertheless occurred. Discuss these changes, with attention to both the "what and the why" of their development.
9. How does this type-study area now lend itself to an intensively scientific and commercialized agriculture? Discuss the merits and probability of its becoming an urban area, the site of an industrial complex, or a major recreation area.
10. Give examples of the ways in which people have modified the landforms, soils, vegetation, water supply, and even the climate and weather of this type-study area. Had they not effected these modifications, what would the area be like today? Of what human use would it be in its unmodified form? What modifications might and should be effected in this area in the future?

SELECTED REFERENCES

Dickinson, R. E. *Regional Ecology: The Study of Man's Environment.* New York: Wiley, 1970.

Finch, V. C. "Geographical Science and Social Philosophy." *Annals of the Association of American Geographers,* Vol. 29 (March 1939), 1–28.

James, P. E. "Toward a Further Understanding of the Regional Concept." *Annals of the Association of American Geographers,* Vol. 42 (September 1952), 195–222.

James, P. E., and C. F. Jones, eds. *American Geography: Inventory and Prospect.* Syracuse, N.Y.: Syracuse University Press, 1954. (See esp., Derwent Whittlesey, "The Regional Concept and the Regional Method.")

Thomas, W. L., Jr., ed. *Man's Role in Changing the Face of the Earth.* Chicago: University of Chicago Press, 1956.

Part Two

Cultural Geography

As
to grow
respons
ples wh
toward
tural ge
deals w
social o

In
tory, h
resourc
tats. Th
the sm
afoot ir
ties of
with sa
and sh
perishe
gathere
societie
tions o
munica
were ir

FIGURE 1–7

Different approaches to alpine living. Above, terraced rice paddies in the mountains of Japan reflect the pressure of human population on the land. When the food needs of the people exceeded the capacity of the small amount of naturally level land, it became necessary for the Japanese to begin to level the hills for cultivation. And since usable land cannot be wasted, many mountain villages in Japan, including the one shown here, are established at the base of high slopes too steep for terracing. (Courtesy Consulate General of Japan) Below, a mountainside in the European Alps, where, since there is abundant level land nearby, terracing is not necessary and pastoralism is the predominant way of life. (Albert Steiner, Black Star)

something on the basis of past experiences with similar or identical stimuli. Similarly, *environmental perception* involves both the way one senses elements of the natural environment and, more importantly, the way one interprets them.

No two societies perceive any natural environment in exactly the same way. Even when given very similar natural settings, different groups of people tend to create quite different cultural landscapes. Thus, for example, the alpine regions of Europe contrast sharply with physically similar areas of Japan. The European mountain range known as the Alps is an area where seasonal pastoralism, or transhumance, is a common cultural trait. In search of good grazing land, herdsmen move their livestock up and down the mountain slopes, shifting from the lowland valleys in winter to the alpine meadows in summer. The region's permanent settlements are found mainly in the valleys. Halfway around the world, in Japan, there is another such natural alpine environment. Although its steep mountain slopes and humid, midlatitude climates make it very similar to the European Alps, this area is used quite differently by the Japanese, that is, largely as farmland rather than as pasture. The sea-oriented Japanese culture includes little use of domestic animals, except for a few draft purposes. The raising of livestock for food and clothing being of much less importance in Japan than it is in Europe, many of the lower slopes of the Japanese mountains have been intricately terraced for intensive agricultural use (Fig. 1–7).

The contrasting ways in which these two societies have responded to their similar environments are due, in part, to the differences between their environmental perceptions. Individuals in a pastoral society tend to view the elements of their natural environment differently than do individuals in an agricultural society. The particular way one interprets natural environmental elements is based on one's values, needs, and desires—factors generally influenced by one's societal group. Thus the cultural objectives and methods of any given society are important in the shaping of its needs and wants and, ultimately, in the shaping of its perception of the environment as a whole.

A society's objectives and technology determine which of the naturally occurring substances it will use and in what manner it will use them. Only those substances that are actually used by a society are considered its *resources;* thus, just as natural resources vary from one society to another, they also vary within any single society as its culture develops. For example, coal, once considered nothing more than black rock, became a natural resource only after people discovered its value as a source of energy. Similarly, while most people consider land an important natural resource, to the "boat people" of Hong Kong, who spend almost all of their lives on water, it is of little value (Fig. 1–8).

The manner in which it uses its natural

FIGURE 1–8

Homes on the water—sampans and junks in Hong Kong harbor. The way of life they represent is gradually disappearing, however, as many of Hong Kong's offshore dwellers are resettling ashore in the island's new high-rises. (Northwest Orient Airlines)

resources generally reflects the concern a society has for its environment. Resources can be used destructively or constructively, depending upon whether the society perceives the environment as an endlessly exploitable commodity or as a prerequisite for survival to be used judiciously and treated with respect.

There is no one way of living in a given physical setting, but rather as many different ways as there are cultures. Thus the characteristics of a given portion of the earth are determined, in part, by the culture of the group that inhabits it. Much of the individuality of any area is also related to the length of time it has been occupied—either by one group passing through different cultural stages, or by successive groups of different cultures, providing a *sequent occupance* of the area.

REVIEW AND DISCUSSION

1. The growing trend toward world interdependence reflects the fact that modern communication and travel systems allow no human group to remain wholly isolated. Discuss the relationship of this trend to cultural geography. How is it evident within your own cultural group?
2. Through what methods have modern peoples learned about the prehistoric cultural development of human beings?
3. In addition to values and beliefs (*ideology*), all cultures have several other characteristics in common. Discuss those you think are the two most important ones.
4. Some scholars are of the opinion that the prehistoric discovery of the usefulness of fire still represents the greatest contribution to human cultural development. Discuss the merits of this opinion.
5. Compare and contrast invention and diffusion with reference to specific examples. Which phenomenon is more instrumental in bringing about cultural change? Explain.
6. Explain the relationship between diffusion and acculturation. Describe two situations that illustrate the growing influence of the latter around the world.
7. List some of the outstanding differences between Oriental and Western cultures.
8. Discuss some of the probable reasons for the spread of Western culture over a larger portion of the earth than Oriental culture.
9. What is the basis for thinking that a new culture may be developing within the Soviet Union?
10. How does the cultural landscape differ from the natural landscape? What are their relationships to the human habitat?
11. Is the quality of change constant throughout the human habitat? From your own experience, give some examples that support your contention.
12. Describe two geographical areas that have essentially the same physical characteristics but are populated by human groups with very different ways of doing things.
13. Illustrate the meaning of the term sequent occupance by reference to the area in which you live or the one you know best. Explain in detail at least one major occupance "change" in your example.

SELECTED REFERENCES

Barrows, H. H. "Geography as Human Ecology." *Annals of the Association of American Geographers*, Vol. 13 (March 1923), 1–14.

Bresler, J. B., ed. *Human Ecology*. Reading, Mass.: Addison-Wesley, 1966.

Brookfield, H. C. "On the Environment as Perceived." *Progress in Geography, International Reviews of Current Research*, Vol. 1 (1969), 51–80.

Chorley, R. J., ed. *Water, Earth, and Man*. London: Methuen, 1969.

Clark, J. G., and S. Piggott. *Prehistoric Societies*. London: Hutchinson University Library, 1965.

Gould, P. R. *Spatial Diffusion*. Washington, D.C.: Commission on College Geography, Resource Paper No. 4, Association of American Geographers, 1969.

Henry, J. *Culture Against Man*. New York: Random House, 1963.

Ittelson, W. H., and F. P. Kilpatrick. "Experiments in Perception." *Scientific American*, Vol. 185 (August 1951), 50–56.

Jefferson, M. "The Geographical Distribution of Inventiveness." *Geographical Review*, Vol. 19 (October 1929), 649–61.

Kroeber, A. L., and C. Kluckholm. *Culture: A Critical Review of Concepts and Definitions*. New York: Vintage Books, 1952.

Lowenthal, D., ed. *Environmental Perception and Behavior*. Chicago: University of Chicago, Department of Geography, Research Paper No. 109, 1967.

Rodnik, D. *An Introduction to Man and His Development*. New York: Appleton-Century-Crofts, 1966.

Saarinin, T. F. *Perception of Environment*. Washington, D.C.: Commission on College Geography, Resource Paper No. 5, Association of American Geographers, 1969.

Shepard, P. *Man in the Landscape*. New York: Random House, 1967.

Smythies, J. R. "The Problems of Perception." *British Journal for the Philosophy of Science*, Vol. 11 (May 1960), 224–38.

Stoddart, D. R. "Geography and the Ecological Approach: The Ecosystem as a Geographic Principle and Method." *Geography*, Vol. 50 (July 1965), 242–51.

Theodorson, G. A., ed. *Studies in Human Ecology*. Evanston, Ill.: Row, Peterson, 1961.

Wagner, P. *Environments and People*. Englewood Cliffs, N.J.: Prentice-Hall, 1972.

Watt, K. E. F. *Ecology and Resource Management*. New York: McGraw-Hill, 1968.

Whittlesey, D. "Sequent Occupance." *Annals of the Association of American Geographers*, Vol. 19 (September 1929), 162–65.

A Bushman hunting gazelle in southern Africa's Kalahari Deser
(N. R. Farbman, Time-Life Picture Agency

2

Culture and Nature

In trying to understand the present natural environment, many people tend to mentally separate and juxtapose nature and humanity, as though human life could exist apart from the rest of the environmental complex. Among primitive peoples, who lived in close harmony with the natural environment, such a conceptualization would have been most unlikely. But today, less conscious of their dependent role in nature, many human beings think in terms of a struggle between themselves and nature and wonder when they will be able to maintain total control over the natural environment.

Influences of Nature on Human Activity

Environmentalism

A time-honored concept, *environmentalism* is the view that human attitudes and activities are variously influenced by the elements of the natural environment. Reflected today in many biological and social sciences, this concept has been receiving support since the time of Hippocrates (c. 460–377 B.C.), the "father of medicine." Environmentalists have tended to regard certain elements of the natural environment—for example, climate—as key factors in promoting or retarding cultural progress.

In its most extreme form, environmentalism is *environmental determinism*, the contention that the natural environment is the sole determining factor in shaping all human culture. The implication here is that human behavior is dictated by the natural environment and that humans are thus slaves to their surroundings. While the environment is unquestionably an influence on human activities, it seems doubtful that it is the only one; for, as discussed in the previous chapter, human groups inhabiting similar environments often develop quite different ways of life.

Possibilism

In the early twentieth century, in response to what they viewed as environmental determinism's overemphasis on the role of nature in human life patterns, many students of human social development gave their support to a more moderate doctrine known as *possibilism.* Expounded first in Europe and increasingly accepted in recent decades in the United States, possibilism maintains that, in any aspect of culture, nature presents human beings with a number of alternatives from which to choose and thus, in some sense, allows them to determine their own future.

While they agree that nature, at any given time or place, does impose some limitations on human development, the possibilists emphasize that, in the last analysis, human beings must be considered at least partly responsible for formulating their attitudes and activities. From their point of view, the environment is a much less domineering factor than the determinists claim it to be; for it does allow human choice some latitude, even if within rather strict limits. Further, the alternatives it allows are dynamic, in that they vary both from place to place and from time to time. Having its own objectives and ways of doing things, one societal group may discover possibilities within the physical complex that another group has overlooked or avoided.

Influences of Human Activity on Nature

During the relatively brief existence of human beings on this much older planet, they have become a force capable of exerting significant and far-reaching influence on its physical complex. Since they first began to implement the methods for their survival they have never inhabited any portion of that complex without modifying it, intentionally or not, in some measurable way. Even the most primitive peoples, living by hunting and gathering, modified the environment in which they lived, simply by utilizing, either partly or completely, various elements of the biotic life therein.

As human tools and methods have become more sophisticated, the potential for influencing the environment has expanded. In a brief moment of geologic time, modern societies have evolved to the point where they can exert as much influence on their local living spaces—constructive or destructive—as have all the natural forces of the past several million years (Fig. 2–1).

Since it may appear that today, more than ever before, human beings are capable of altering or retaining any environmental condition almost at will, it requires some effort to keep their role in nature in proper perspective. The popular notion is that, through their modern technologies, human beings have become "masters of the world" —a somewhat exaggerated claim, especially when viewed through the shadow of today's world energy crises, food shortages, floods, droughts, and global population explosion. On the other hand, as a broad but ultimately fair and useful generalization, it seems valid to say that, within relatively small geographic areas and short time spans, human influences on an environment tend to super sede those of nature; whereas, within larger areas and longer time spans, natural forces tend to predominate. It must be understood, however, that not even cultural or human forces within small areas can function in a manner completely independent of natural forces. For, in the final analysis, human influences, even within these areas, reside ultimately in the ability of humankind to modify and/or control the many constant processes of nature.

As human activities increase in quantity and complexity, they continue to alter the landscape more sharply and more extensively. All such activities invariably upset delicate balances established in nature over long periods of time, and, when these balances are upset, other natural forces and processes over which humans have little or no control are often set in motion. Most commonly, these forces have existed beforehand but have been held in check by other features

FIGURE 2–1

The building of a dike across Lake IJssel, in the Netherlands. The enclosed area thus created will be drained and eventually used as cropland. (Netherlands National Tourist Office)

of the environment. In any case, there is an ecological goal to which all peoples should aspire: they should attempt to understand the balances in nature well enough to maintain or alter them constructively, that is, for the long-term service of humanity.

Spatial Order and the Regional Concept

Spatial Variation

The cultural traits with which any given group of people identify manifest themselves in various and complex interrelationships with the physical environment. Much geographic inquiry serves primarily as a framework for studying this so-called *spatial variation,* the assumption being that some degree of order can ultimately be discerned by studying the seemingly chaotic associations of cultural and physical phenomena. The science of geography itself has even been broadly defined by some geographers as *the study of spatial variation on the earth's surface.*

A principal tool in the geographer's quest for greater insight into the subject of spatial variation is the *map,* which provides a simplified picture of spatial reality. Unless pertinent data are scaled down to a size suited to map presentation, the distributions and associations of cultural and physical phenomena cannot be fully appreciated.

Interlocking Spatial Concepts

Spatial location is the basic point of reference used in discussions of either physical or cultural variations; therefore, location, or place, is fundamental to most geographic thought. A full geographic understanding of an environmental feature comes only through knowing its precise location, or *site,* as well as its *situation,* or spatial relationship to other features (Fig. 2–2). Such concepts as *scale, pattern, density, distribution,* and *interaction* are also all directly involved with the location of environmental features and with their interrelationship within the human habitat.

The Regional Concept

The *regional concept,* like the map, is a device, or tool, long used by geographers as a means of comprehending the arrangement of spatial variations within the human habitat. A geographic *region* is defined as any portion of the earth's surface that has been delimited, or recognized, for a particular characteristic. Other general terms that are used in a similar manner include *area, belt, zone, realm,* and *landscape.*

There are many different types of geographic regions, and they can be large or small, simple or complex. Regardless of its type and size, however, each region possesses some areally unifying element (or elements). These elements may be either cultural or physical ones or some combination of the two. An island nation, for example, can be considered a region by virtue of both its administrative scope and its shorelines. Most regions, however, do not have such clear-cut physical boundaries. The term *Moslem world,* for example, describes a cultural region distinguished by a single, well-established religion, but it does not represent only contiguous landmasses (Fig. 2–3). A smaller, and somewhat more clearly defined, cultural region is the Mormon "nation," an area centered on Salt Lake City Utah.

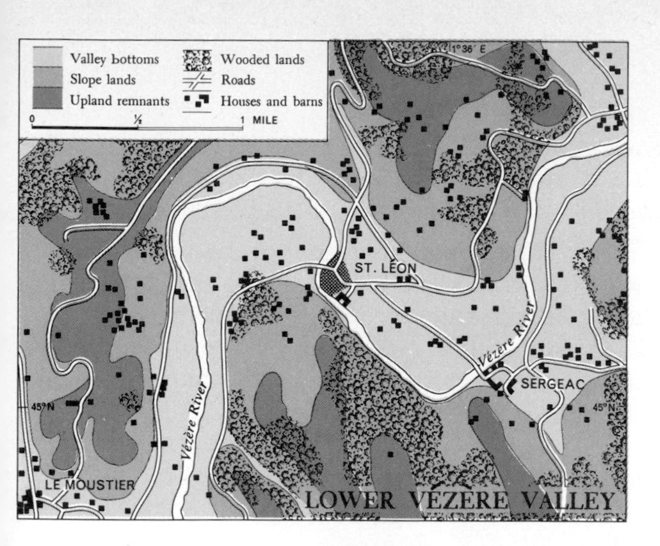

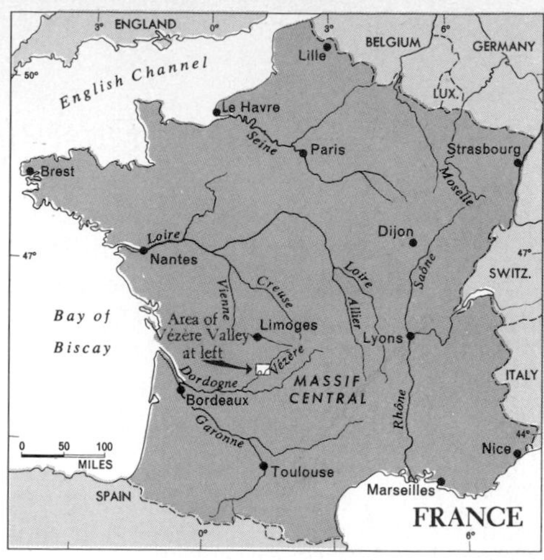

FIGURE 2–2

Site and situation. Left, the *site* of St. Leon: on the right bank of the lower Vézère River, midway between the towns of Sergeac and Le Moustier. Right, the *situation* of St. Léon (and the rest of the lower Vézère Valley): approximately 50 miles above the confluence of the Vézère and Dordogne rivers, on the west side of the Massif Central in southwest central France.

FIGURE 2–3

The Moslem world. Including most of northern Africa and southern Asia, this vast region extends, with some interruptions, from the Atlantic to the Pacific. Several of its countries share not only Islam's strong and long-established religious beliefs but many other cultural features as well.

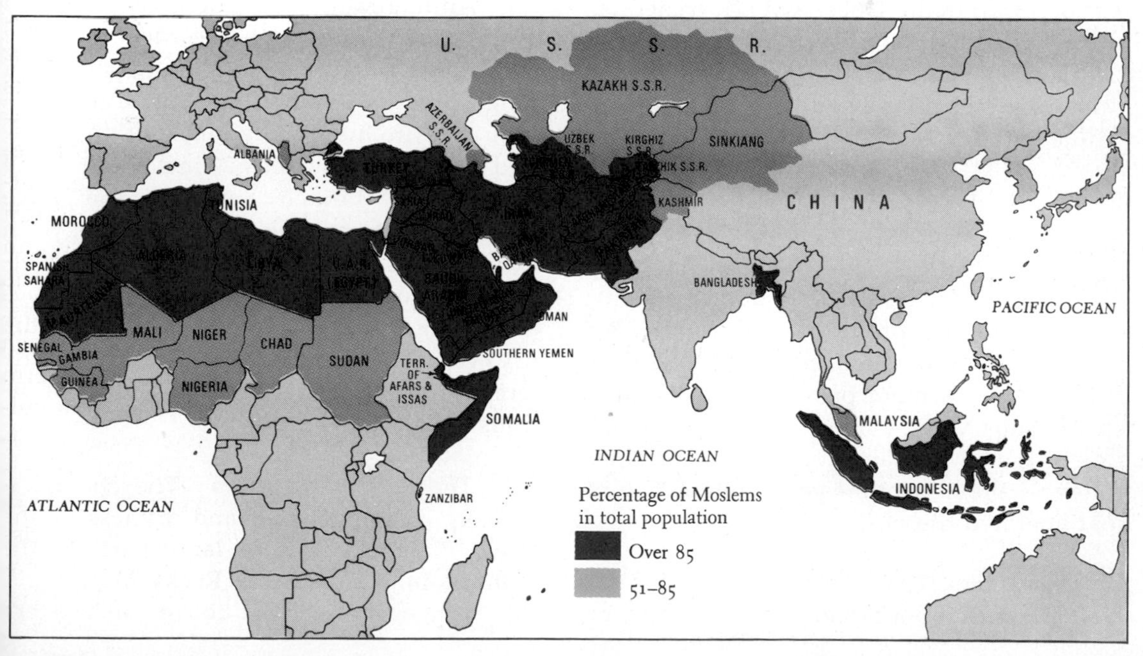

A Regional Study: The Dust Bowl

The following study examines the reciprocal influences of human beings and nature as demonstrated in the 1930s in a section of the Great Plains of the United States. The core of this area included coterminous parts of Colorado, Kansas, New Mexico, Oklahoma, and Texas (Fig. 2–4).

At one time, this vast steppe region supported a short-grass cover that served as a natural protection against extensive soil erosion by holding dry soils in place. However, with the overgrazing and inappropriate agricultural practices that followed the growth in human settlement in the region in the late 1800s, this grass cover completely disappeared, leaving the region's characteristic dry, fine-grained soils, which were no longer held in place, very susceptible to erosion. The erosion process actually began when strong westerly winds—wind being one of the chief agents of erosion—swept over the region during the dry-weather cycle that occurred in the 1930s. By 1934, great dust storms engulfed this dry-farming area and stripped it of most of its fertile topsoil. In this way stable steppeland suddenly became desertlike and barren.

The Original Landscape

Obviously, the effects of human activity in any area cannot be correctly appraised without first understanding the processes of nature at work there prior to human influences. Thus what follows is a brief discussion of the Great Plains region in its original state and a review of the geographical factors that operated therein.

Location. The Great Plains of the United States are a broad, flattish steppe, or short-grass, area sloping eastward from the Rocky Mountains toward the Mississippi River and straddling the 100th meridian, a theoretical line often used to demarcate the dry western portion of the country and the more humid eastern portion. This vast plains region was formed not as the result of one or two isolated geographic factors but rather by the long-term interaction of several such factors.

Precipitation. The Great Plains exhibit a progressive decrease in precipitation from east to west, with an annual average of 40 inches in eastern Kansas and 12 in eastern Colorado. When readings taken over several decades are averaged and plotted, those areas where the average precipitation is 20 inches align roughly with the 100th meridian; but in any particular year they may be areas hundreds of miles east or west of that line (Plate IV).

Annual rainfall variations on the order of 15 inches have been recorded in many parts of the Great Plains over long periods of time. Such variations are of little consequence in humid regions, but in semiarid areas with, say, only 20 inches of rainfall annually, a 15-inch variation can have serious effects on the regional ecology. In these plains, as elsewhere, relatively wet- and dry-weather years tend to run in cycles of varying duration and intensity. It was the dry cycle of 1933–38 that helped to trigger the temporarily disastrous environmental conditions of that period.

The Rocky Mountains. The characteristically limited rainfall and dryness of the Great Plains is due, in large part, to the climatic influence of the Rocky Mountains. This range, rising high above the western edge of the region, serves as a barrier to the natural inflow of moist air from the Pacific Ocean and thus effects a rainshadow leeward over the high plains. As a result, the Gulf of Mexico, the Great Plains' other major source of moist air, becomes the dominant supplier of precipitation over the region. But since the air from the Gulf of Mexico tends to be deflected to the northeast by the Coriolis force, considerable moisture from these waters is supplied to the eastern Great Plains, while relatively little reaches the western portion of the region (Plate III).

Winds. The Great Plains are frequently plagued by a high-velocity westerly wind that descends the leeward side of the Rocky Mountains and moves eastward. Referred to as a *chinook,* or *foehn,* this wind is compressed and greatly warmed during its de-

FIGURE 2–4

The site of the Dust Bowl of the 1930s. (For its situation, see North America's Great Plains region on Plate III.)

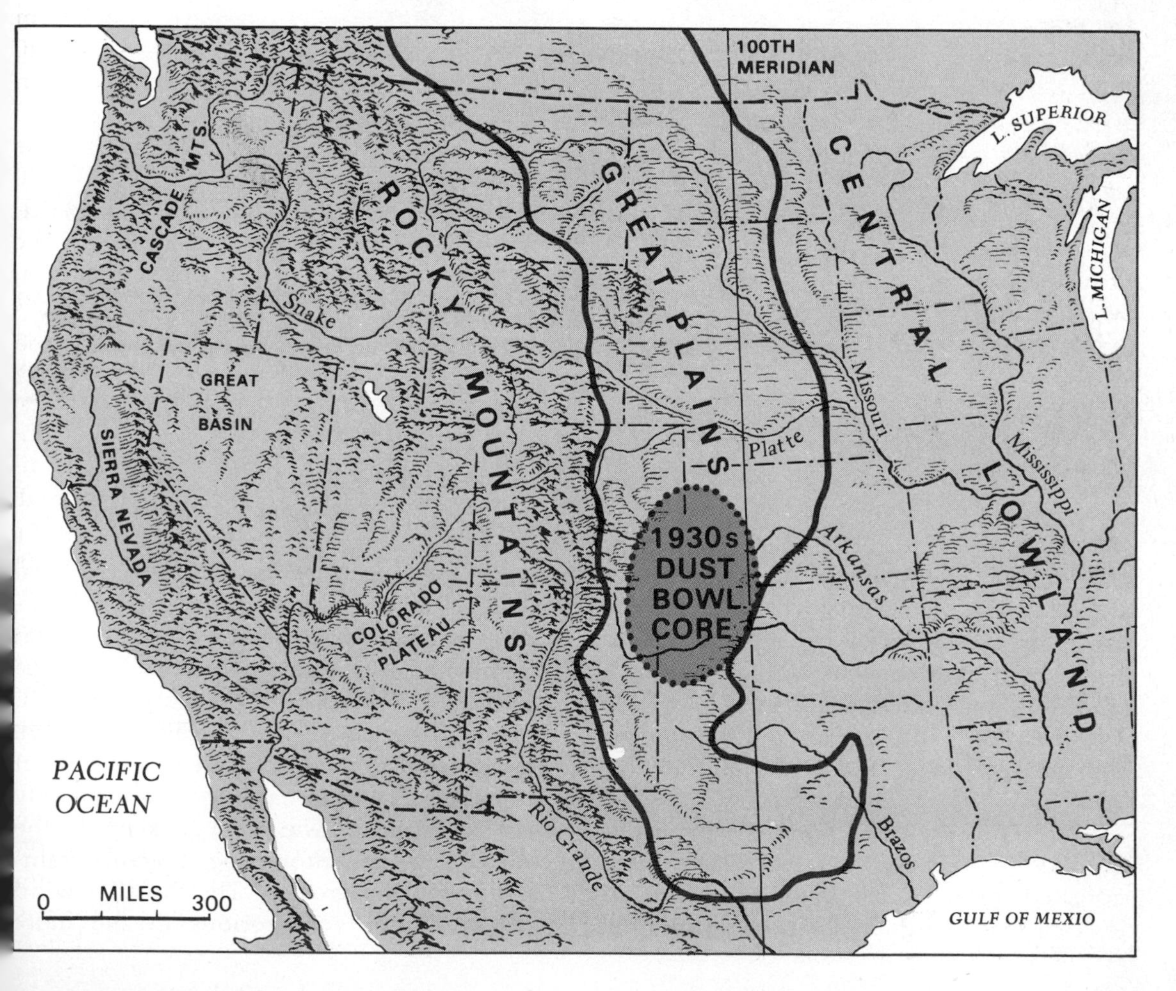

FIGURE 2–5

The flatness of the semiarid land of the Great Plains is illustrated by this area along the old Deadwood Trail, in southwestern South Dakota. Though, since the 1920s, most of the steppe of the Great Plains is usually covered by wheat crops, here one can still see the ruts created by stagecoaches and ox-drawn wagons as they took gold miners to the Black Hills in the 1870s. (South Dakota Department of Highways)

scent; at the same time, its relative humidity is greatly reduced. As it sweeps across the Plains, the chinook may cause surface temperatures to rise as much as 50 degrees in a few hours and, evaporating most available moisture, may cause the area to become almost desiccated.

Vegetation and Soil. Under natural conditions, the soils of the Great Plains were protected by a tough mantle of vegetation—primarily a grass-sod cover of prairie, steppe, and buffalo grasses. Before the local introduction of beef cattle in the 1870s, luxuriant stands of these grasses supported great numbers of native grazing animals, particularly bison, and also helped protect the soil. The importance of the area's protective grasscover cannot be overemphasized: it served as a windbreak, and its relatively deep-root system was vital in keeping the soil in place and protecting it against rainwash and wind erosion. The threat of wind erosion is not very serious in the more humid eastern margins of the Great Plains, but it becomes more serious on the drier western borders (Fig. 2–5).

A Balanced Ecosystem. Before the introduction of farming to the Great Plains, the natural environmental factors in its ecosystem were in delicate balance: all functioned together in a precise system of checks and balances. Thus if even one of these ecological factors was disturbed, it would invariably upset and alter the entire balance.

Influence of Agrarianism

The natural characteristics of the Great Plains region foreshadowed the difficult future it was to face when subjected to extensive human settlement and should have been sufficient as a warning to those attempting to exploit the region to that end. The area's dry-steppe grasslands—especially their vegetation and soil types—indicated that there was a precipitation deficiency and that the land was most suited for limited grazing and, in its eastern portion, for crop farming.

However, most of the people who settled in the region were not well-acquainted with its characteristics nor with the techniques of semiarid cropping necessary to cope with its inherent limitations; in the main, they had migrated from more humid regions in the eastern states. Many of them saw in the region only a chance for easy exploitation and immediate financial return. After 1910, moreover, several new psychological influences and financial encouragements made themselves felt in the western part of the Great Plains (Fig. 2–6). High wheat prices during the war and postwar years, in combination with the development of the tractor, the combine, and other power machinery, resulted in a frenzied westward expansion of the wheat area.

Dry Farming and Grazing

The method of farming most widespread in the region during the 1920s and 30s is known as *dry farming,* which makes use of groundwater stored naturally in the soil. The practices employed to conserve as much of the vital groundwater as possible include: frequent cultivation to eliminate moisture-consuming weeds and to develop and maintain a dust mulch on the surface to break up capillarity; and the alternation of cropping and fallowing, with only one crop planted every second or third year. These methods were used in the region on the assumption that the amount of moisture stored in the soil during the fallow period would be enough to satisfy the moisture needs of the subsequent crop.

Dry farming is fine in theory and, in some parts of the world, it often succeeds in practice; but it was not employed with satisfactory results in the drier parts of the Great Plains, especially during the 1930s. During periods of low or no rainfall, the simple act of turning over the sod with a plow caused the soil to dry out excessively; and, as happens in any dry or semiarid region, the extremely fine-grained soil became very powdery when it dried (Fig. 2–7). Thus, overall, continued cultivation tended to make the soil even more susceptible to drying than it had been before. While the wheat and other crops that were planted had relatively low moisture requirements, they still used more moisture than the natural grasses they replaced. During periods when rainfall was abundant, the wheat thrived; but when precipitation was below its requirement, the wheat began to draw on the moisture contained deep within the soil. In addition, the stubble left when the wheat was harvested was not able to protect the soil against erosion; thus exposed, the soil became more vulnerable than under its natural grasscovers.

Even areas not plowed for wheat experienced major problems, particularly those used for grazing. Because of such factors as higher wartime prices and unusually lush pastures (the result of a wet-cycle period), most of the region's grazing lands had been overstocked. The most serious overgrazing occurred in areas that were naturally sparsely vegetated and arid or semiarid in climate, as, for example, the southwestern fringe of the region. The first livestock to be introduced into the area and raised in significant numbers were cattle; then came sheep and goats. This order represented an increase in the destruction done to the area's grasscover; for while cattle eat only the uppermost part of the grass, sheep and goats often pull it up by the roots. The result was decided deterioration and, in extreme places, even complete destruction, of the original native grasses. Once it was stripped of its grasscover and binding root systems and exposed to constant trampling by livestock and direct sunlight, the soil soon became very loose and dry and thus highly susceptible to erosion by wind and running water.

FIGURE 2–6

The harvesting of a Kansas wheat field after World War I. During the postwar period, increases both in consumer demand and in food prices prompted a rapid expansion of the United States Wheat Belt. (Grant Heilman)

FIGURE 2–7

The drilling of winter wheat on a farm in the semiarid steppe of eastern Colorado in October 1939. Such dry farming is successful today in many parts of the world, including the Great Plains, but it was a dusty and often vain endeavor in the Dust Bowl states during the 1930s. (FSA, Library of Congress)

Dust Bowl Manifestations

Throughout most of the 1920s, a wet-cycle period in the region provided enough rainfall to protect the exposed steppe soils from scattering winds. But in the 1930s, the rainfall average dropped sharply, resulting in severe droughts in 1930, 1934, and 1936. Perhaps in its natural state the dry-steppe region could have survived this severe dry-cycle period with minimal injury to its landscape. However, because of the decades of poorly managed human activities, few areas were immune to the action of wind erosion. Whenever strong chinook winds from the west blew across the region, they shifted the soil's loose surface layers and carried away much of the topsoil (Fig. 2–8).

FIGURE 2–8

A shallow basin in northern Nebraska in 1936, one of the countless watering holes in the Great Plains that succumbed in the 1930s to wind erosion and overuse by livestock. (Monkmeyer Press Photo)

Throughout the early '30s, almost the entire ground surface of the Great Plains region was subject to *deflation* (the removal of surface materials by lofting into the air or rolling along the ground). In the heart of the Dust Bowl—southeastern Colorado, southwestern Kansas, the Oklahoma Strip, and the Texas Panhandle—more than one-third of the land was seriously affected by wind erosion.

Within a short period of time, the entire region was engulfed in severe dust storms, which appeared as great, dark, ominous clouds rolling along the ground and obscuring everything within their reach (Fig. 2–9). From Denver to Dodge City, quantities of good topsoil were carried to considerable distances—sometimes even as far as the eastern coastal states.

The continuous shifting of dry sand, dust, and other soil materials across the countryside often resulted in the formation of sand dunes that covered much of the artificially created landscape. Many fields, farm buildings, machines, and fences were buried; roadways were blocked; and water supplies were clogged (Fig. 2–10). In areas invaded by dunes, the original vegetation cover was seldom reestablished. Where a new cover did develop, it usually represented an ecologically retrogressed form, characterized by fewer and poorer types of perennial grasses, less nutritious annual grasses, weeds, and edaphically useless bushes—in sum, a completely new and less valuable vegetation cover.

Most of the native animal population of the region's rangeland was also either severely disrupted or completely eliminated. For example, when the natural ecology was intact, the area's rodents were kept in check by their natural predators. But with human

FIGURE 2–9

A dust storm approaching Springfield, Colorado, in May 1937. (U.S. Department of Agriculture)

hunting and land clearance and, eventually, the Dust Bowl conditions, many of these predators were eliminated, leaving such animals as the prairie dog, the jack rabbit, and the gopher to increase in number.

Conclusion

The ecological history of the Great Plains region shows that human activity over a relatively short period of time was instrumental in transforming a balanced natural steppe community into an artificially created, desolate, and desertlike environment. Droughts had come and gone in the area before human settlement, and the ecosystem had endured and recovered from them. But when careless human activity was added to natural disasters, a more cataclysmic situation arose.

It may seem, at first, that one area's loss was another area's gain, since the soil blown away from the Great Plains was deposited on lands farther east, making them more fertile. But there was no gain to any area that was capable of compensating for either the Dust Bowl's appalling loss of land and fertility or for the pain and tragedy of families displaced by the drought, dust, and economic catastrophe (Fig. 2–11).

The single positive result of the Dust Bowl experience was that it made the nation and its government aware—perhaps for the first time—of the great danger involved in the thoughtless manipulation of its ecosystem. In the ensuing years, the efforts made to rehabilitate the devastated area—most of them sponsored by the Soil Conservation Service—proved that people can sometimes repair the environmental damage they cause—albeit at great expense. Along with human efforts, nature provided its own ways of

FIGURE 2–10

An abandoned framstead near Guymon, Oklahoma, in October 1937, where wind-blown sand has all but engulfed a dump rake. (U.S. Department of Agriculture)

FIGURE 2–11

Migrants whose trek has been temporarily halted near Stockton, California. They were among the thousands of destitute farm families that left Oklahoma and other drought-ravaged states in the 1930s. (FSA, Library of Congress)

repairing itself. Thus the productivity of the Dust Bowl region was only temporarily interrupted. Beginning in the late 1930s, with the return of more favorable environmental conditions (particularly a more normal precipitation pattern), settlers in the region were gradually able to resume farming and livestock-raising activities. And, with the knowledge gained through experience, along with the assistance of expert agriculturists, these activities were controlled by a far more enlightened form of management. Even today, however, the threat of another "Dust Bowl" still confronts many agrarian developments located in similar semiarid transitional lands, whether in the United States or in the desert fringes of Eurasia, Africa, Australia, or South America (Plates IV, V, and VI).

REVIEW AND DISCUSSION

1. Support or refute the statement that "the natural environment is the sole determining factor in shaping all human culture."
2. Give an illustration, from your own area, of the limitations imposed upon cultural development by climate.
3. Are the limits of the human habitat fixed permanently, or can they be altered? If alteration is possible, suggest how it might be accomplished. If it is impossible, explain why.
4. Is it fair to characterize the relationship between human beings and nature as one based on conflict, or is the relationship more often a harmonious one? Cite at least three societies that have different relationships with their natural environments and explain these differences.
5. Do human beings themselves actually effect changes in the natural environment, or do they simply trigger natural forces that eventually bring about such changes? Explain your opinion.

6. Give examples of three different types of regions. Compare and contrast the Moslem world and the Mormon "nation" as cultural regions.
7. Which of the major geographical factors operating in the Great Plains region in its original condition had the greatest influence on the arrival of Dust Bowl manifestations? How was that influence exerted?
8. Assess the role of human activities in creating the Dust Bowl conditions in the Great Plains. From an ecological point of view, in what way was the area perceived incorrectly? Explain how it was particularly mismanaged agriculturally in the 1920s and '30s.
9. How is this Dust Bowl region perceived today? What steps have been taken to prevent another ecological disaster?
10. Dust Bowl conditions are not unique to the Great Plains area of the United States. Under what circumstances have similar conditions developed in other regions of the world?

SELECTED REFERENCES

Bailey, R. W. "Land Erosion—Normal and Accelerated—in the Semiarid West." *Transactions of the American Geophysical Union* (1941), 240–50.

Bates, M. *Man in Nature.* Englewood Cliffs, N.J.: Prentice-Hall, 1961.

Bennett, H. H. *Soil Conservation.* New York: McGraw-Hill, 1939.

Davis, D. H. *The Earth and Man.* New York: Macmillan, 1950.

Golomb, B., and H. M. Eder. "Landforms Made by Man." *Landscape,* Vol. 14 (Autumn 1964), 4–7.

Heathcote, R. L. "Drought in Australia: A Problem of Perception." *Geographical Review,* Vol. 59 (April 1969), 175–94.

Hewitt, K., and F. K. Hare. *Man and Environment.* Washington, D.C.: Commission on College Geography, Resource Paper No. 20, Association of American Geographers, 1973.

Huntington, E. *Civilization and Climate.* New York: Harper & Bros., 1920.

Jacks, G. V., and R. O. Whyte. *The Rape of the Earth.* London: Faber, 1939.

———. *Vanishing Lands.* New York: Doubleday, Doran, 1939.

Johnson, V. *The Dust Bowl Story.* New York: Farrar Straus, 1947.

Lewthwaite, G. R. "Environmentalism and Determinism: A Search for Clarification." *Annals of the Association of American Geographers,* Vol. 56 (March 1966), 1–23.

MacFadden, H. C. "The Dust Bowl." Unpublished manuscript, 1975.

Meggers, B. T. "Environmental Limitations on the Development of Culture." *American Anthropologist,* Vol. 56 (1954), 801–24.

Meinig, D. W. "The Mormon Culture Region: Strategies and Patterns in the Geography of the American West, 1847–1964." *Annals of the Association of American Geographers,* Vol. 55 (June 1965), 191–220.

Semple, E. C. *Influences of Geographic Environment.* New York: Holt, 1911.

Tatham, G. "Environmentalism and Possibilism." In G. T. Taylor, ed., *Geography in the Twentieth Century.* London: Methuen, 1957.

Thomas, W. L., Jr., ed. *Man's Role in Changing the Face of the Earth.* Chicago: University of Chicago Press, 1956.

Wagner, P. *Environments and People.* Englewood Cliffs, N.J.: Prentice-Hall, 1972.

Whittlesey, D., et al. "The Regional Concept and the Regional Method." In F. E. Dohrs and L. M. Sommers, eds., *Introduction to Geography: Selected Readings.* New York: Crowell, 1967.

Huge ancient monoliths on Easter Island in the South Pacific. (Eugene Gordon)

3

Human Origins and Diffusions

The Beginnings of Human Life

While its earliest beginnings cannot be located precisely, the human race seems to have evolved in the portion of the earth known as the "Old World" (or, more formally, the Eastern Hemisphere) and to have been centered initially in several locations, ranging from eastern Asia to southern Africa. The first year of human existence is, of course, impossible to determine. But it is known with some certainty that as far back as three million years, several humanlike forms were already evolving in portions of eastern and southern Africa. Among these beings were a number of short, humanlike apes of the genus *Australopithecus*. Members of this genus are thought to have walked erectly and thus represent a critical evolutionary development. Inhabitants of savanna areas, these apes, who had binocular vision, once walked on all four feet. But this posture, which limited their viewing angle, caused them to experience great difficulty in seeing predators amid the savanna's clumps of vegetation. Their need for self-preservation led them to discover that they could obtain a much better view by rising erectly on their hind legs while swivelling their heads to survey the countryside. Eventually, they learned to walk upright, and this new posture freed their forelimbs for use in other activities.

Early Tools and Social Units

About a million years ago, certain primate groups began making simple tools, which they used for such purposes as cutting meat. While various primate groups are known to use wood and stone as implements, the intentional manufacture of tools requires intelligence, planning, and foresight and is thus usually considered one of the traits that distinguish humans from all other animals (Fig. 3–1). Most of the tools at-

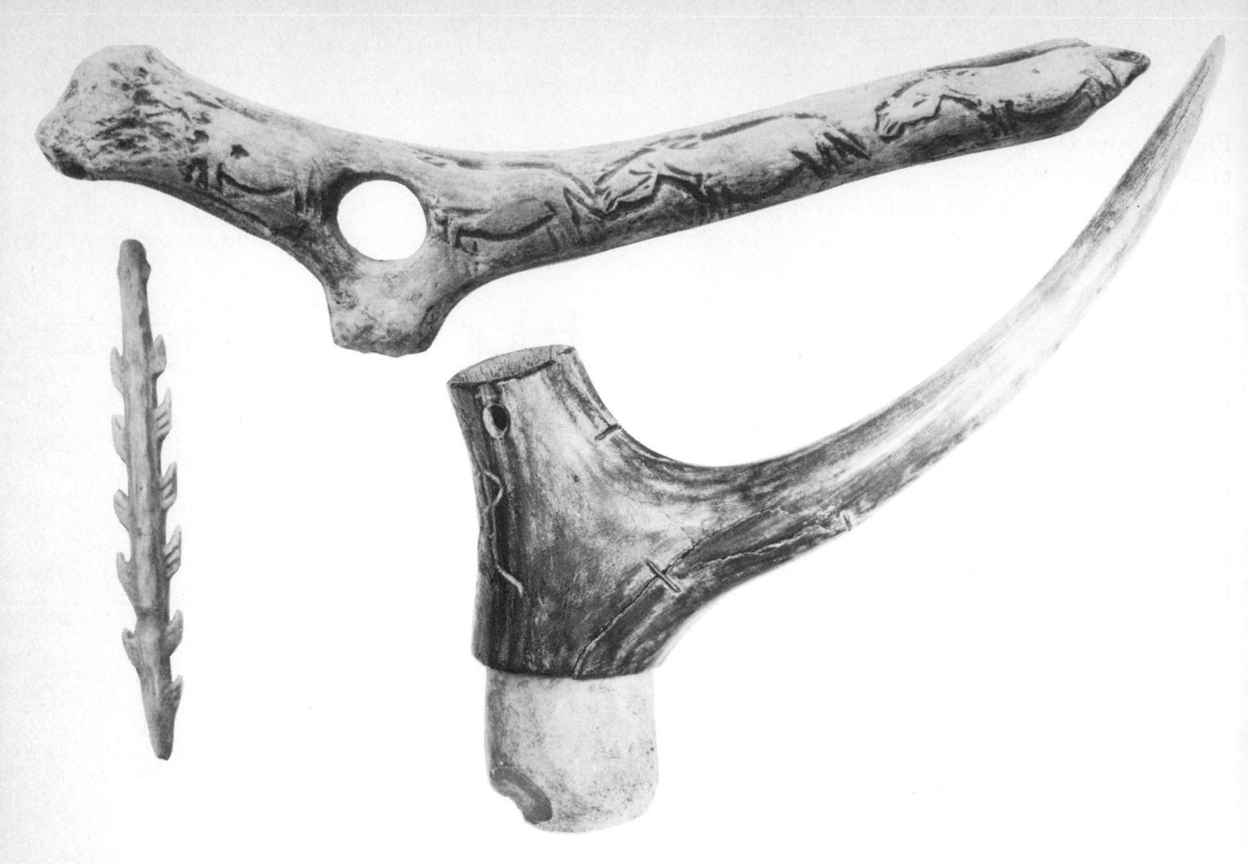

FIGURE 3–1

Some tools made and used by early human inhabitants of Europe. From the Late Paleolithic, or Old Stone Age: top, a *baton de commandement* carved from a reindeer antler and, left, a harpoon. Right, from the Neolithic, or New Stone Age, a pointed axe. (American Museum of Natural History)

tributed to these ancient human types, or *hominids,* were made of stone. Consequently, the first phase of human culture is called the Old Stone Age, or, borrowing from the Greek, the Paleolithic period.

Also about a million years ago, hominids began to form social units and to make sounds that gradually evolved into words and, eventually, into patterned speech. Such applications of intelligence increased in number and complexity as they proved to be invaluable aids to survival. The use of tools, weapons, and clever thinking was the only way the hominids could elude their enemies and obtain sufficient food; they had no sharp claws, no teeth or tusks, nor did they possess superior strength or fleetness of foot.

The Recent Ice Age

The development of the genus *Pithecanthropus,* earliest of the hominids, coincides roughly with the early stages of the Pleistocene Epoch, the geologic age comprising the greater part of the last million years. The most recent ice age, the Pleistocene Epoch, is divided into four periods of large-scale glaciation, during which massive ice sheets covered much of northern Europe and the northern part of North America (Fig. 3–2). Three interglacial periods are also included in this epoch, the climate of at least one of which is thought to have been warmer than the present period. It is quite possible that the radical changes in climate during

Pleistocene times and the accompanying changes in vegetation and animal life through forced adaptation played a significant role in human evolution.

As early as 200 thousand years ago, there were some species that became very much like present humans in both appearance and action; thus, they have been included in the genus *Homo.* One of these visibly human species is known as *Homo sapiens neanderthalensis,* that is, intelligent man from the Neander Valley (a valley near Düsseldorf, Germany, where relics of the species were first discovered in a cave in 1857). Neanderthalers are believed to have evolved about 100 thousand years ago, in western Asia and Europe. They were rather brutish in appearance, with eyes deep-set under overhanging brows, protruding upper jaws, massive but chinless lower jaws, and stocky, sturdy bodies. They were also intelligent and sensitive, as evidenced by the development of a rudimentary religion, the careful burial of their dead, and the demonstrating of a technology (especially in the making of stone tools) far more advanced than that of their predecessors.

The emergence of the Neanderthalers was more-or-less synchronous with the beginning of the last interglacial stage—a

FIGURE 3–2

The areas in which the first human beings developed were widely separated areas of the Eastern Hemisphere, all far removed from the portions of the earth subject to long periods of glaciation.

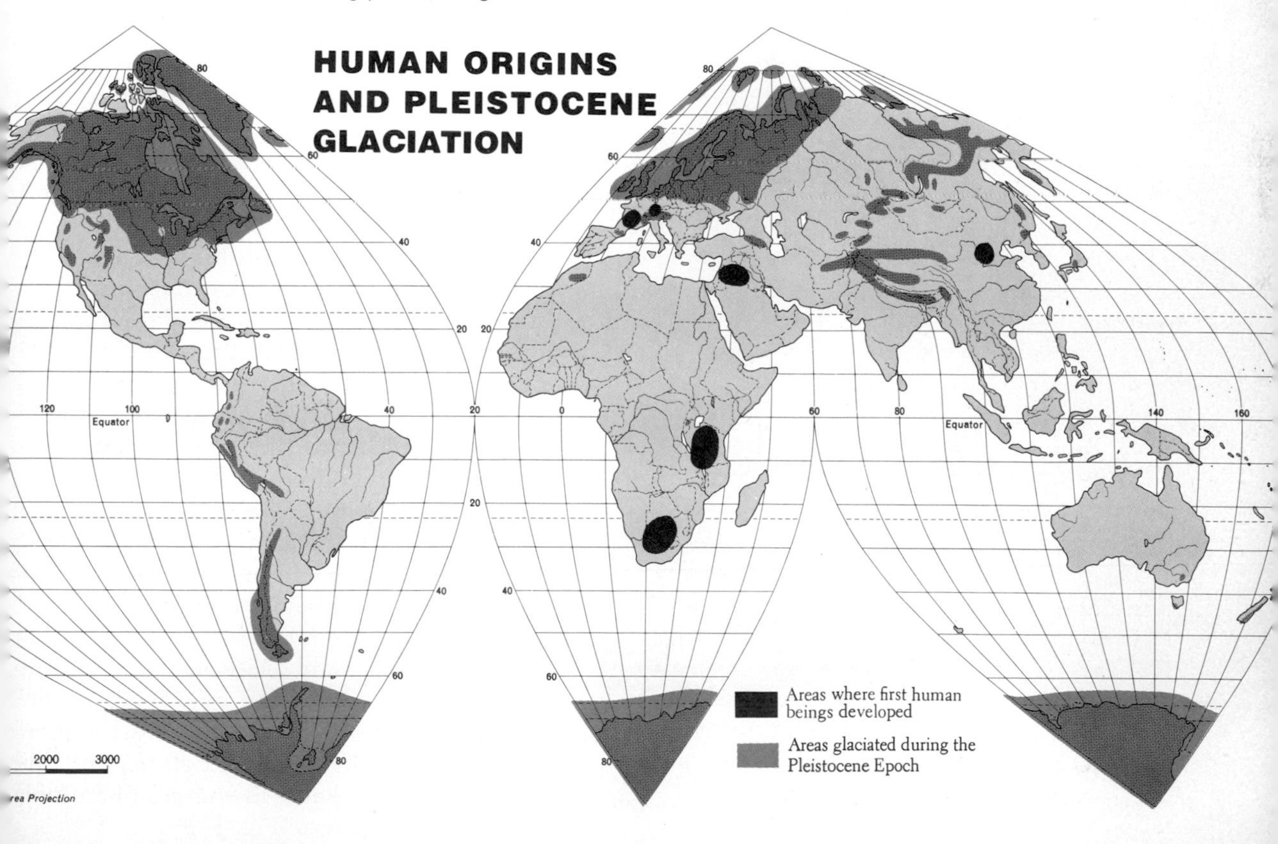

FIGURE 3–3

Neolithic paintings from caves in the Sahara. (Jean-Dominique Lajoux)

period characterized by warmer winters and longer summers than those of the present, and by melting ice sheets, to which was related the northward migration of most plant zones in Europe and Asia. Under these mild conditions the race prospered, developing new technologies and spreading out widely over Europe and Asia. About 70 thousand years ago, the climate began to reverse itself again, bringing lower temperatures and greater snowfall. As these glacial conditions returned to northern Europe and Asia, however, the Neanderthalers, rather than retreating to warmer climates farther south, made use of their newly acquired technological skills and found ways to survive, even in the more northerly regions, which often were in relatively close proximity to the ice front.

The Emergence of *Homo Sapiens*

The present-day human species, *Homo sapiens,* first appeared in parts of Europe and the Middle East during the last glacial stage, about 50 thousand years ago, and gradually replaced or absorbed the last of the Neanderthalers. In several major aspects the new species represented the highest developmental level reached by the genus: first, their brains were larger than those of their ancestors, with frontal lobes, the centers of intelligence and reasoning (nearly nonexistent in earlier species), particularly well-developed; second, the newcomers possessed and applied greater intellectual ability, as demonstrated, especially, by their achieving of ways to communicate over long distances and with permanence; third, they also exhibited greater manual dexterity, which enabled them to produce finely made weapons, tools, and clothing. Among the early achievements of the new species were the development of spear throwing, the carving of designs in ivory and stone, and the painting of remarkable likenesses of animals on the walls of caves (Fig. 3–3).

The Peopling of the Earth

Semitropical Asia and Africa

It has been determined that the initial development of the human race occurred in the tropical and semitropical areas of Asia and Africa, where the climate is characteristically warm at all times of day throughout the year. Sometime later in the race's history, with the utilization of fire, it became possible for humans to move into the colder parts of the world. And 100 thousand years ago, human settlements were spread over most of southern and western Europe, most of southern Asia, and much of Africa. It was not until about 30 thousand years ago, however, that the first people occupied areas in the Americas and Australia.

FIGURE 3–4

The probable location of an ancient Asia–North America land bridge. The similarity of many fossil and living animals and plants found in the two continents contributes to the belief that such a path existed throughout much of the Pleistocene Epoch.

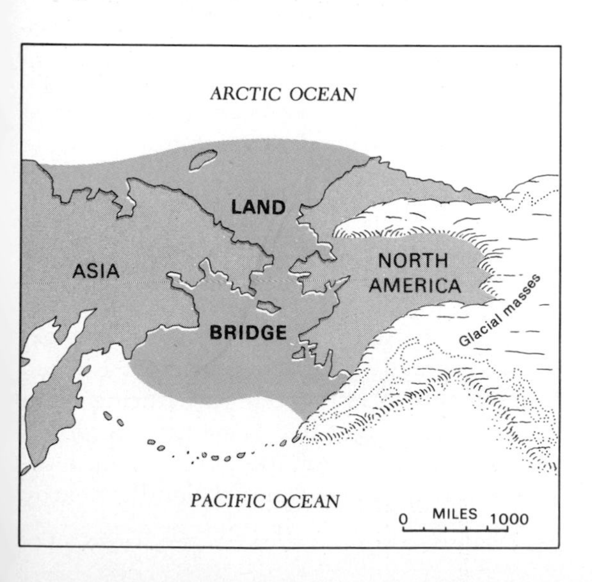

The Americas and the Pacific

At the time of the first peopling of the Americas, Siberia and Alaska were joined by a strip of land that separated the Arctic Ocean and the Bering Sea; this land bridge has since been replaced by the Bering Straits (Fig. 3–4). Thus the first humans to settle in the Americas simply drifted across this land bridge from Asia, without realizing that they had entered another hemisphere (Fig. 3–5). In the same way, the descendants of these

FIGURE 3–5

The probable migration routes of the first human groups to inhabit the Americas. These nomads are thought to have entered the Western Hemisphere from eastern Asia and gradually expanded their hunting ranges southward and eastward.

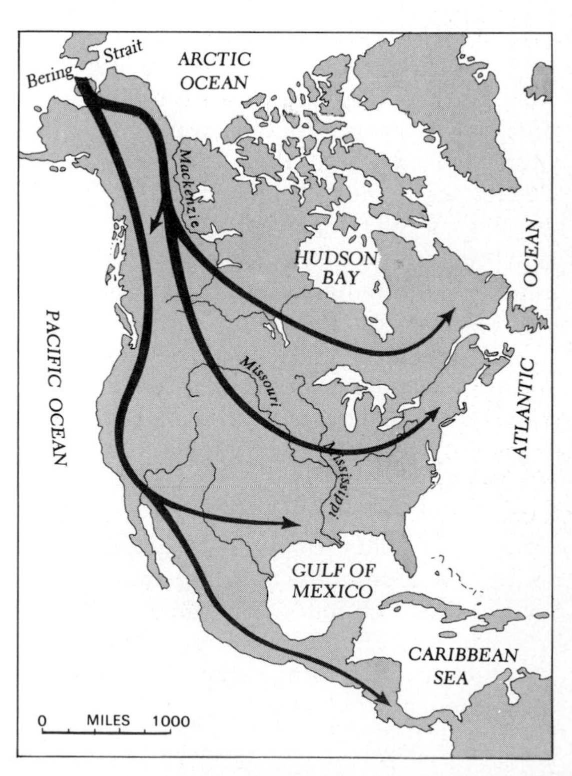

nomads gradually expanded their hunting ranges farther and farther into the new land, and, from time to time, they were followed by new groups from Asia. Eventually, after some thousands of years, the Americas—from Alaska to Tierra del Fuego and from the Pacific to the Atlantic—were populated by many separate groups.

In most cases, then, the settlement of the world was accomplished by people traveling afoot. The habitation of certain remote island areas, however, was accomplished only after the development of relatively sophisticated seagoing vessels. The farthest removed islands of the Pacific were thus among the last parts of the world to be peopled. In large sailing canoes, which were constructed on the outrigger principle and capable of carrying fifty or more people and supplies for a month, Polynesians sailed from island to island, gradually extending the portions of the inhabited world eastward. They appear to have lived in Samoa and Tonga as early as 500 B.C., to have reached Tahiti in about A.D. 200, and Hawaii, in about A.D. 1000.

Early Migrations and Invasions

Relocations and Resettlements

Even after the peopling of virtually all areas of the earth's surface was accomplished, resettlements of many groups continued. Although it has often been said that a person's greatest desire is to have a stable and secure home, wanderlust has often affected whole tribes and nations. This desire, then, must almost certainly be the stronger of the two, as evidenced by the fact that, throughout history, the large-scale relocation of people has been almost continuous in most parts of the world.

There have been some resettlements, of course, that were undertaken out of desperation, motivated by such things as war, oppression, famine, or overpopulation—forces that tend to "push" people from an area. But, more often, there is no extreme propelling force; other fields simply look "greener," offering (or apparently offering) more fertile land, milder climates, or some particular freedom or economic opportunity. Such advantages are the things that "pull" people from one land to another.

Some generalized relocations of people have been the warlike thrusts of highly militarized tribes or nations. Others have been gradual, peaceful migrations of simple farmers and herders. Some have been planned, well-financed colonizations of distant lands. And some have been the separate, unplotted movements of rather poor families. In both large-scale and individual migrations, the pressures of population growth are often found to be a contributive factor.

Routes for Invasion and Travel. Steppe and prairie lands have been the areas most continuously affected by migration, and the ones most commonly occupied by nomadic peoples. Before modern technology, the alternation of plentiful rainfall and drought in such areas resulted in great fluctuations in their life-sustaining capacities. In wet years, the flocks, herds, and human populations expanded greatly. In dry years, they faced hunger and starvation. It was during such periods of drought that some groups would set out to search for food and water elsewhere. The better-watered lands nearby were usually already occupied by sedentary farming peoples. Wanting to secure those lands for themselves, the nomads often violently attacked these less war-ready people, slaughtering, enslaving, or driving them away and occupying their lands, sometimes on a permanent basis. Thrust out of their homes, the displaced farmers displaced others in turn, setting off a chain of events that affected a great number of people over great distances. Such situations occurred repeatedly for over two thousand years, across the area between central Asia and western Europe, and came to a halt only in the last century. These mass movements involving millions of people are sometimes referred to collectively as the *Völkerwanderung* (German for the folk-wandering).

Consider, for example, the invasions by the Huns during the fourth and fifth centuries (Fig. 3–6). In A.D. 370 those nomadic pastoral people, who lived originally in central Asia, far beyond the Caspian Sea, suddenly appeared near the Volga River on the border between Europe and Asia. A great, warlike horde, they overthrew the Alani, a pastoral people settled on the grassy plains between the Volga and the Don rivers, in present-day European Russia. Surging westward, the Huns then quickly overthrew the empire of the Ostrogoths, in the Ukraine between the Don and the Dnestr rivers, and by 376 had defeated the Visigoths and, taking most of Rumania, reached the Danube, the frontier of the Roman Empire, then at its zenith.

For the next fifty years, the Huns were relatively quiet in their new lands, their military involvements consisting only of minor border skirmishes. In 441, however, under their powerful leader Attila, they attacked the Roman Empire itself. By 443 they had defeated the Roman armies in southeastern Europe and had surrounded Constantinople (now Istanbul). In 451, they invaded France, and, in the following year they reached northern Italy, thus pushing right into the heart of the Roman Empire. When Attila died in 453 the contest for leadership between his sons caused fighting within the tribe, and, as a result, its military power sharply declined. By the end of the century the Huns had lost their unity and much of their identity and were being absorbed into the local populations throughout eastern and southern Europe. Once a nomadic dry-lands people living in tents pitched wherever there was grass and water for their herds and flocks, the Huns were now settled as farmers in relatively humid lands and harvested the same

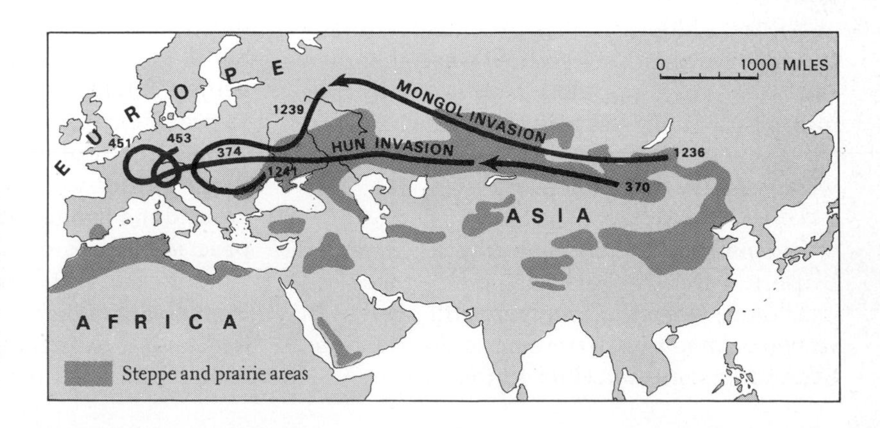

FIGURE 3–6

The route taken by the Huns, during the fourth and fifth centuries, in their advance on Europe. The invasion of central Europe by the Mongols in the thirteenth century took a similar route through the steppe and prairie areas of western Asia.

crops in the same fields generation after generation. In less than a century a whole tribe had been transplanted from semiarid desert borderlands to humid lands a few thousand miles away.

Reports of the ferocity of attacks by the Huns caused many groups to flee their lands even before they were invaded. Their flight started several relocations that were as significant as the movements of the Huns themselves. The Ostrogoths, fleeing the Ukraine, eventually settled in northern Italy. The Visigoths, displaced in Rumania, raided much of the Balkans, ravaged Italy, and took over much of southern France and Spain. The Vandals, originally residents of the area that is now eastern Germany, were also pushed westward. Mounting their own great, warlike drive, they crossed the Rhine and fought their way across France and Spain, eventually settling in North Africa and, in 455, attacking Rome (the term *vandalize* recalls their pillage of the sacred and temporal relics of that city). The Franks, another Germanic tribe, crossed the Rhine and settled in France, giving it its name. The Angles, the Saxons, and the Jutes, all peoples of northern Germany and Denmark, crossed the North Sea and settled in England (German for land of the Angles) and other parts of Britain.

Throughout Europe in the fifth and sixth centuries, tribes pushed from their homelands moved small or great distances and fought bloody battles to settle eventually in some new land. Often they were absorbed into the very population they had conquered, so much so that after several generations their distinctive features and habits were scarcely discernible.

Southern Britain. Southern areas of Britain were first populated 15 thousand years ago, when glaciers were still present in its central and northern regions. At that time, Britain was still attached to the continent of Europe by a land bridge, rather than separated from it by the Straits of Dover and the English Channel. The first people to cross this strip of land into Britain were a short, rather dark people who sustained themselves by hunting and fishing and by gathering wild plants. They lived in small groups, had no domesticated plants or animals (except, perhaps, the dog), and used the very simple tools and weapons of Paleolithic times (the Old Stone Age). Their language was unlike any spoken today.

About five thousand years ago, Neolithic (New Stone Age) farmers and herders began to arrive in Britain from continental Europe, bringing with them flocks of sheep and herds of cattle as well as a good deal of experience at growing wheat and barley. They took over the area's best lands and crowded its previous occupants into the less desirable areas—forested lowlands, rugged mountains, and peninsulas where the soils were thin, the skies overcast, and the temperatures cool even in summer. Lacking effective tools—they made no use of iron—these newcomers had great difficulty in felling trees. Hence, the "best" lands, from their point of view, were those of the *downs*, treeless, chalky uplands, where they could plant their crops most easily. These farmers avoided—in fact, they actually feared—the dense forests that covered the lowlands. This fear was probably reinforced by the presence there of bands of the short, dark people who had been the first group to settle in Britain. British folklore is full of references to the "Little People," "Bogey Men," and "Dark Ones"—each phrase probably an allusion to these primitive people, who long existed around the fringes of the settlements of more progressive groups.

By the fifth century B.C., Celtic tribes from Spain, France, and southern Germany had begun to invade Britain, and they continued to arrive in recurrent waves until the

FIGURE 3–7

Hadrian's Wall, a wall of defense in northern Britain constructed by order of the Roman emperor Hadrian (A.D. 117–138). This reminder of the Roman conquest of Britain extends approximately 73 miles, from Solway Firth to the mouth of the River Tyne. (Reece Winstone)

first century B.C. The people taking part in the earlier of these Celtic invasions were speakers of the Gaelic language; the later groups spoke Welsh. The name of one of the Welsh-speaking groups, Brythons, is the source of the appellations *Briton* and *Britain.*

Technologically, the Celts were relatively advanced, using iron tools and possessing both an agricultural technology and a military strength far superior to that of their predecessors in Britain. With their iron tools, they began clearing the lowland forests and transforming them into agricultural areas. They also drove all the earlier settlers into the mountains and the western peninsulas. Some of these peoples, such as the Picts, eventually banded together and, for several centuries, offered the Celts considerable military resistance.

Although the Romans had conquered Britain by A.D. 43 and held it for about four hundred years, the Roman occupation was only a military and administrative one. It did not involve the mass resettlements characteristic of other invasions of Britain in ancient times (Figs. 3–7 and 3–8).

When the Roman forces withdrew, due to military concerns elsewhere, raids by seafaring peoples of northern Germany and Denmark (Angles, Saxons, and Jutes), which had begun around A.D. 300, developed into full-fledged invasions. The legends associated with King Arthur are based upon the attempts of the various Celtic groups to thwart these invasions. Despite their efforts, however, the Celts were gradually pushed into the mountains and peninsulas to the west, just as they had pushed the Picts a few centuries earlier. By A.D. 500, the Anglo-Saxons had occupied all of the eastern lowlands of Britain; many of the Welsh-speaking Britons of southern England had fled across the Channel into northwestern France, where they occupied and named such regions as Brittany; the remainder of the Celts had retreated into southwestern England, Wales, and southern Scotland; the Scots (an early Celtic group) had been occupying Ireland and southwestern Scotland for some time; and the Picts still held the Highlands of Scotland.

FIGURE 3–8
The network of major roads built in Britannia by the Romans during their long domination of western Europe. Antoninus' Wall, built about A.D. 142, marks the northern extent of the Roman military conquests.

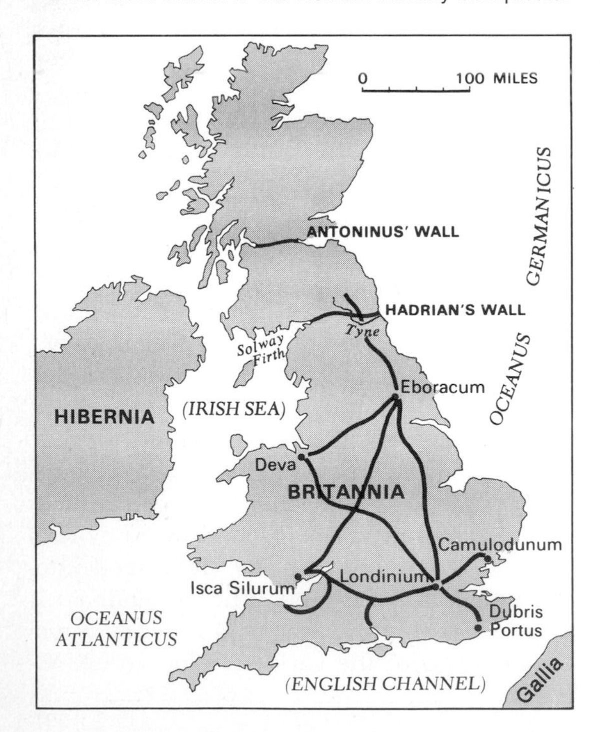

Eventually, the Anglo-Saxons were also faced with an onslaught by invaders from over the sea. Early in the ninth century, the Norsemen, or Vikings, of Norway and Denmark began looting the British coasts and soon thereafter had established a series of bases along the northern shores. By the end of that century, they had pushed their forces well into the interior of the north, either displacing the Celts and Anglo-Saxons living there or coexisting with them.

Another group of Norsemen had conquered part of northwestern France and had settled there in large numbers. They were generally called *Normans,* and the land they settled became known as Normandy. In time, they made many concessions to their new environment, adopting the French language as well as a wide range of French customs.

The last military invasion of Britain took place in 1066, when the Normans attacked and defeated the Anglo-Saxons at Hastings, on the southern coast of England, and took over most of Britain (Fig. 3–9). The Normans had a profound influence on all aspects of the country; they altered its language and introduced a new form of government as well as new customs and styles.

African Migrations and Invasions

Large-scale migrations have not been limited to the Eurasian world alone. For many centuries, for example, Bantu tribesmen had been migrating southward from their main homeland in the northeastern part of Nigeria (Fig. 3–10). Tall, able warriors, equipped with iron-headed spears and tough cowhide shields, certain of these tribes, perhaps, especially, the Zulus, were practically undefeatable; in gaining control of an area, they drove out the weaker tribes, or they killed or enslaved them. The warriors' relatives and possessions always followed close behind the battlefront.

FIGURE 3–9

A scene from the Bayeux Tapestry (eleventh century) depicting the military battle that occurred near Hastings, England, in October 1066. The invading Norman cavalry vanquished the infantry of the Saxon king Harold II. Harold was slain, and William, Duke of Normandy, succeeded him as King of England. (Giraudon)

A Bantu tribe often continued such onslaughts for several years, usually through the will of some powerful and energetic chieftain. But, eventually, it would settle down—at least for a time—and begin clearing fields, raising crops, and building homes and villages. This activity would continue until, suddenly, a new, warlike chieftain would arise and impel the tribe to again surge forward in a wave of conquest. In much the same way, during the centuries before European colonization, various other African tribes moved southward, traveling from areas in present-day Kenya and Tanzania south to the central part of Namibia (South West Africa), Botswana, and the vicinity of East London, in South Africa. In this area, they almost completely replaced the Bushmen and Namas (southern African tribes belonging to the larger group of people known collectively as the *Hottentots*), who had occupied it throughout time immemorial. In the nineteenth century, the Bantu were still pushing southward, but their advance was abruptly halted because of colonization by the Dutch in South Africa and by the Germans in South West Africa. Were it not for

FIGURE 3–10

The diffusion of Bantu-speaking peoples from the first century to 1000. The principal homeland from which these groups moved was in the area that is now northeastern Nigeria; by 1000 they had migrated westward and also conquered most of Africa to the south and east of their original sites.

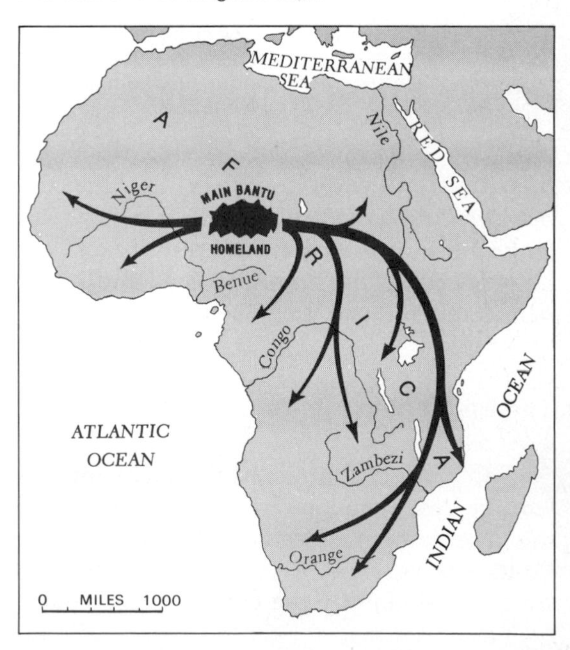

FIGURE 3–11

The siege of Jerusalem in 586 B.C. by the forces of the emperor Nebuchadnezzar of Babylonia. The destruction of this capital city marked the end of the Judean kingdom and the beginning of the Babylonian Captivity, a period in which many Jews were taken to Babylon as captives. (Frank J. Darmstaedter, Photographic Archive of the Jewish Theological Seminary of America, New York)

this outside intervention, it is likely that the Bantu would have reached the Cape of Good Hope early in the twentieth century and totally destroyed the original Nama and Bushman populations.

The Plight of the Jews

In marked contrast to these violent and warlike resettlements are the migrations of the Jews, known collectively as the *Diaspora*. From 586 B.C.—when they were made captives in Babylon by the emperor Nebuchadnezzar—to the present, the Jewish people have had to flee religious persecution and seek out new homes (Fig. 3–11). Such movements have always been domestic, that is, undertaken by an individual family or by groups of closely related families.

The many expulsions of the Jews from their traditional homeland (variously known as Canaan, Palestine, and Israel) and their first resettlements have caused them to be spread widely over the world. Since the first century, Jewish colonies have developed in most of the cities near the Mediterranean Sea and, in the Middle East, as far as Iraq. In each of these areas, Jews have played important roles in trade and commerce, in finance, and as patrons of the arts.

In the fifteenth century, the Spanish Inquisition caused the large Jewish population of Spain to be dispersed over much of the Mediterranean area. They formed settlements in North Africa (Tunisia, Algeria, and Morocco), in southern Europe (southern France, Italy, and Greece), and in the Middle East (mainly in Turkey, Lebanon, and Palestine). These people are known as the *Sephardim,* or Sephardic Jews (Hebrew: *s'pharadh* = Spain), and their vernacular language, Ladino, is a mixture of Hebrew and Spanish.

During the Middle Ages, other Jewish groups migrated to the cities of northern Europe. There, they gradually developed a distinctive language known as *Yiddish* (German: *jüdisch* = Jewish)—a mixture of Hebrew, German, and Slavic words written in Hebrew characters. These migrants are known as the *Ashkenazim* (Hebrew: *Ashk'naz* = Germany), and were centered primarily in Germany, Poland, and Russia.

Greek and Chinese Dispersals

Similar long-term migrations to distant lands include the two-thousand-year movement of Greeks into the towns and cities of the Mediterranean area and northern Africa. Greece, a country with few valuable mineral resources and little arable land, has been faced with overpopulation for much of its history; the main source of relief from the resultant grinding poverty has been migration to other lands.

The traditional flows of Chinese emigrants have come mainly from southern China, especially southern Fukien province and Kwangtung province, which includes the port city of Canton. Since the mid-seventeenth century, some 15 million people have migrated from these largely rural provinces to about 100 different countries throughout the tropical and subtropical world. There are now Chinese populations of over 3 million in Thailand and Hong Kong, over 2 million in Indonesia and Malaysia, and over 1 million in Singapore and Vietnam; sizable Chinese populations also exist in such non-Asian countries as the United States, Cuba, and Australia (Fig. 3–12).

	Number of Chinese	*Percentage*
Asia	15,859,820	96.6
United States and Canada	295,489	1.8
Latin America	148,709	.9
Oceania	52,572	.3
Africa	43,734	.3
Europe	20,586	.1
Total	16,420,910	100.0

FIGURE 3–12

The worldwide distribution of Chinese in lands other than China, as of 1963. Most of the large majority residing in other parts of Asia is located in the southeastern portion of the continent.

The Age of Colonization

The Western world (the part of the earth whose present inhabitants are descendants of people who came from northwestern or Mediterranean Europe) has experienced two long-lasting periods of massive migration. The first was the *Völkerwanderung,* which was discussed earlier in this chapter. The second, which occurred from about 1500 to 1900, will be discussed here as the *Age of Colonization.*

Precolonial European Life

Throughout the Middle Ages (from approximately 500 to 1500), living conditions in western Europe were arduous. An extremely high birth rate was offset by an equally high death rate; life expectancy was short, averaging about thirty years. The high incidence of disease reflected the period's lack of adequate sewage systems, limited medical knowledge (including an ignorance of germ theory), and futile attempts to find ways of preserving food. Infant mortality was particularly high, with a very large percentage of babies dying in their first year.

With little long-distance trading, extensive crop failures often led to famine and widespread malnutrition, which made the population extremely susceptible to disease. During the winters, especially, most people suffered serious dietary imbalances, due to the absence of ways to preserve vitamin-rich foods.

Farming equipment was crude, planting methods were simple and inefficient, and crop yields were correspondingly low. By the end of the fifteenth century, the population of western Europe was very close to the carrying capacity of the land. In fact, had not the "Black Death" (the outbreak of bubonic plague that swept through Europe from 1347 to 1350) greatly reduced the population, widespread famine would probably have occurred (Fig. 3–13). It was thus fortunate that, early in the fifteenth century, the Portuguese and Spanish began the series of explorations that led to the discovery of the full extent of the African continent, a sea route to Asia, and the continents of North America, South America, and Australia.

Migrations to New Lands

The discoveries mentioned above opened up a vast new world to the over-

FIGURE 3–13

Detail of a contemporaneous illuminated manuscript that depicts the burning of infected clothing during the Black Death, the epidemic of bubonic plague that ravaged Europe's urban populations in the fourteenth century. (Bodleian Library, Oxford)

crowded inhabitants of Europe. Asia was already well-populated, highly civilized, and under the rule of strong local governments, and it was thus not open to any large-scale settlement and colonization by Europeans; Australia, southern Africa, and the Americas, however, provided a situation that was quite different. In those lands, population densities were low, and, in most cases, the level of material civilization was equally low; the few capable governments in existence, such as those in Mexico and Peru, could be overthrown far more easily than could those in Asia by European military might (due, especially, to their superior weaponry).

Great numbers of Europeans flocked to these new lands to administer and control, to trade and, especially, to farm. The majority of those who set out were youths; frequently, the journey was made by whole families; and, usually, they all shared the same intention—to make new homes for themselves. The new lands they reached were productive and provided raw materials that could be traded for goods manufactured in the homeland, thus stimulating its economy.

There was, of course, a tendency for settlers to emigrate to the lands controlled by their native countries. Portugal colonized Brazil, and Spain controlled most of the rest of the Americas from Mexico (even California for a time) southward. The French became established in Canada and northern Africa; the Dutch, in New York, the East Indies, and South Africa; and the British, in North America, eastern and southern Africa, and Australia. In each of these places today, the influence of the early European population is still strong, even though many of the areas have gained political independence, some as much as two centuries ago; in each, a significant portion of the population, the language, and many other aspects of local culture are all derived from the European country that first, or most lastingly, exercised control.

Not all of the countries of Europe were involved in the exploration and colonization of the New World. Many portions of the continent had not yet developed a national consciousness and, hence, were not up to the task of developing colonies. Germany and Italy, for example, were still fragmented into numerous independent, or quasi-independent, principalities, duchies, and the like. Eastern European countries in this era appear to have had no aspirations beyond their own local spheres. However, certain individuals and groups from these areas did emigrate to colonies of other nations and were gradually absorbed into their populations.

The gradual movement of individuals and families from Europe, for the purpose of colonizing and developing the newly found areas of Australia, southern Africa, and the Americas is, perhaps, the most significant migration of all time. And, while the term *colonization* has acquired many negative connotations, when this migration is viewed in context, its accomplishments are impressive, and there is much to be said in its favor.

Modern Human Distributions

Despite all the migrations throughout the centuries, the way in which the human population is distributed over the surface of the earth is still very uneven. This is due, mainly, to the uneven distribution of arable land and other natural resources and, increasingly, to the esthetic appeal of certain areas to mobile populations.

FIGURE 3–14

Mountains in the Arctic fringe of the Canadian northwest—cold, barren, and desolate. (Canadian Department of National Defense)

The Empty Areas

Even with the high density of population in many parts of the world and the rapid increase in its total population, some sections of the earth still remain devoid of human inhabitants (Plates II, IV, and VII).

Water Bodies. Obviously, the largest of such areas are the water bodies—the oceans, seas, and lakes. Aside from the people living on boats in such places as Oriental harbors and American marinas, and the fishers, scientists, and sailors on long-voyage vessels, the water bodies of the world are virtually unpeopled.

Arctic Fringes. The extreme cold of winter, the shortness of summer, the impossibility of producing crops by normal agricultural methods, and the lack of available sources of water throughout much of the year have all militated against the development and peopling of the more poleward parts of the earth. But the exploitation of their valuable mineral resources—coal in Spitsbergen, uranium in northern Canada, iron ore in Labrador, and petroleum on the Alaskan North Slope, for example—has resulted in localized human settlements wherein miners and oilers live under comfortable but artificial conditions, their necessities and luxuries imported from afar, often at great cost. Similar situations exist at many of the military bases that various countries have established in the more remote of these places. In Arctic areas, where open water seasonally occurs along the coast, a scattering of primitive hunters, such as the Eskimo, also eke out a bare existence.

Otherwise, the Arctic tends to be quite empty. The interior of Greenland is totally uninhabited, as is the whole of Antarctica. Except in mining areas, population density is very low throughout northern Canada, Alaska, and the northern part of the Soviet Union (Fig. 3–14).

Desert Lands. Owing to their critical shortage of water, deserts are usually occupied only where water is available in quantities sufficient for irrigation; where mining is profitable; and along border areas where, in rainy years, nomadic herders may temporarily graze their animals. Elsewhere, desert areas are either empty or sparsely populated, chiefly by primitive hunting-and-gathering peoples. Such areas include the interior of Australia, the Sahara, the deserts of Saudi Arabia, Iran, and Afghanis-

FIGURE 3–15

Sand dunes in a desert area of Saudi Arabia. The oasis in the background offers the wayfarer some relief from the dry and barren terrain that surrounds it. (Standard Oil Co., N.J.)

tan, large portions of central Asia, the coastal desert of Namibia in southwestern Africa, and, in the United States, parts of New Mexico, Arizona, and Nevada (Fig. 3–15).

In the United States, rapid growth of large urban concentrations on the desert borders and technological advances (in air conditioning and specially designed vehicles, for example) have helped to engender great interest in deserts as areas for both recreation and retirement. Particularly in the southern parts of California, Nevada, and Arizona, desert areas have experienced a rapid growth in population. They are frequented, on a short-term basis (that is, weekends and short vacations), by residents of the nearby metropolitan areas and, on a long-term basis, by vacationers and retirees from the colder parts of the United States.

Mountains and Plateaus. For a number of reasons, high mountains and plateaus are often areas of low population density. Among these reasons are: the diminished supply of oxygen at high altitudes, which necessitates certain physiological adjustments, often difficult for humans; the lower temperatures, which reduce the length of the growing season; and the frequent shortage of flat land suitable for cultivation. The few inhabitants of these areas are likely to be herders and miners. In recent years, selected areas in Norway, the Alps, the Sierra Nevada, and the Rockies have become important ski resorts as well as successful summer retreats.

Included among the empty, or sparsely populated, mountain-and-plateau areas are Tibet, the altiplano of Bolivia, the Andes in southern Chile and Argentina, large portions of Norway, and much of South Island, New Zealand.

Tropical Forests. In the dense forests of tropical-humid and equatorial regions, leached soils, high humidity, and endemic diseases have acted to retard development; these areas remain sparsely populated. This is illustrated by the emptiness of much of the interior of Brazil, parts of central Africa, and most of Borneo and New Guinea.

The High-Density Areas

There are basically two types of areas that today possess dense populations—the agricultural areas in emerging countries, and the highly urbanized and industrialized portions of more advanced nations (Plate VII).

Agricultural Areas. On alluvial plains in tropical and subtropical areas with heavy rainfall, intensive agriculture, supplemented, as needed, by irrigation can usually produce such a high-food value cereal crop as rice; such a crop can be grown in quantities that will support high densities of people. In Java, where the richness of the alluvial soil is amplified by recent volcanic materials, the density of the population dependent upon agriculture alone may exceed 2,000 persons per square mile. Other areas of very high density include the agricultural areas of Japan, China, India, Bangladesh, Korea, Vietnam, Cambodia, Thailand, the Philippines, and the United Arab Republic (Egypt).

However, despite the richness of the soil, the fact that there are so many people dependent upon the land results in a *per capita* yield that is, necessarily, very low; the people live on a very limited diet, and cash income is meager. Any slight reduction in the yields—due to plant disease, inclement weather, or war, for instance—may cause widespread hunger, if not outright famine. Today, some hundreds of millions of people in these overcrowded parts of the world live a precarious existence, the threat of starvation being with them always.

FIGURE 3–16

The city of Tokyo, in central Japan. Like most of the world's high-density urban areas, it is a center of industry and trade, served by a complex of modern transportation systems. (Japanese Information Service)

Urban-Industrial Areas. Due to modern means of transportation, it is possible for large numbers of people who do not produce their own food to live in one area and still be well-supplied with food. Airplanes and trucks make it possible to provide urban areas with raw materials for industry, with great amounts of energy, and with adequate markets for their manufactured goods. Thus it was not until the present technological age that urbanization has been possible on a grand scale.

Large-scale modern urbanization is based upon industry and trade, both of which, in turn, are based upon the availability of several resources; fuel and energy, skilled labor, transportation facilities, and a stable political environment. These have been fully combined on a long-term basis most successfully in the northeastern part of the United States, northwestern Europe, the western part of the Soviet Union, and, recently, central Japan (Fig. 3–16).

In these regions, urban population densities are often very high; but this does not necessarily mean poverty or the threat of starvation. Due to the value of manufactured products, the income from their sales and related trading activities, urban populations are usually fairly affluent. Japan provides an excellent example, for the marked increase in its population density over the past few decades has been accompanied, not by increased poverty, but rather by an obvious rise in the standard of living. This economic change has been caused, in part, by the country's shift from subsistence agriculture to large-scale industrialization. Many of Japan's new workers were formerly farmers dependent upon the yield of small rice fields; as wage earners, they now have cash incomes with which to purchase a wide variety of foods and goods.

The Moderate-Density Areas

The population density found in most areas of the world, however, is more modest, falling somewhere between the extremes discussed above. Generally speaking, densities are fairly low in areas where the economy is based on large-scale forestry, livestock raising, or the cultivation of grains (see Plate VII). For such agricultural regions, it seems that one rule of population density—the more fertile the soil, the denser the population—is countered by a second—the higher the degree of mechanization, the lower the density—and, sometimes, by a third—the longer the history of settlement, the denser the population. Thus India, for example—with fertile soil, little mechanization, and a very long history—has a high population density; while the Corn Belt in the United States (mainly the states of Iowa, Illinois, Indiana)—with fertile soil, a short history of settlement (150 years), and a very high degree of mechanization—is an area of moderate population density.

REVIEW AND DISCUSSION

1. What advantages did the earliest human beings have over lower animals in the struggle for existence?
2. What major evolutionary changes has the human race undergone over the past three million years or so?
3. According to most researchers, in what order were the various parts of the world peopled? What major technological innovation is believed to have allowed humans to move into the colder parts of the world?
4. Why has the large-scale relocation of people been almost continuous in most parts of the world? What are some of the forces that tend to "push" people from an area? What forces "pull" them to another?
5. In what specific ways has the natural environment influenced human migrations?
6. Compare and contrast the migrations of the Jews with the migrations of such other groups as the Huns, Celts, and Polynesians.
7. Which areas of the world were most open to the large-scale colonization and settlement begun by Europeans in the fifteenth century? Explain some of the reasons for their suitability.
8. "Despite all the migrations throughout the centuries, the way in which the human population is distributed over the surface of the earth is still very uneven." Discuss and evaluate the implications of this statement in terms of both the physical and cultural factors.
9. Which parts of the world constitute the "empty areas"? Why are these areas relatively devoid of people? Do you feel that some day large concentrations of people will be living and working in any or most of these areas? Explain.
10. What are the major characteristics of the world's moderate-density areas?

SELECTED REFERENCES

Braidwood, R. J. *Prehistoric Men,* 7th ed. Glenview, Ill.: Scott, Foresman, 1967.

Brodrick, A. H. *Man and His Ancestry.* London: Hutchinson University Library, 1960.

Butzer, K. *Environment and Archeology: An Ecological Approach.* Chicago: Aldine, 1971.

Campbell, B. G. *Human Evolution: An Introduction to Man's Adaptations.* Chicago: Aldine, 1966.

Carter, G. F. *Man and the Land, A Cultural Geography.* New York: Holt, Rinehart and Winston, 1968.

Clark, Sir W. E. Le Gros. *The Antecedents of Man.* Chicago: Quadrangle Books, 1960.

———. *History of the Primates,* 4th ed. Chicago: University of Chicago Press, 1963.

Dobzhansky, T. *Mankind Evolving: The Evolution of the Human Species.* New Haven: Yale University Press, 1962.

East, G. *The Geography Behind History.* London: Nelson, 1940.

Hawkes, J., and Sir L. Woolley. *History of Mankind: Cultural and Scientific Development,* Vol. 1. London: Allen and Unwin, 1963.

Hoebel, E. A. *Anthropology: The Study of Man,* 4th ed. New York: McGraw-Hill, 1972.

Hooton, E. A. *Up From the Ape.* New York: Macmillan, 1960.

Howell, F. C., and the editors of Time-Life Books. *Early Man.* New York: Time, Inc., 1965.

Howells, W. W. "The Distribution of Man." *Scientific American,* Vol. 203 (September 1960), 113–27.

Korn, N., and N. R. Smith, eds. *Human Evolution.* New York: Holt, Rinehart and Winston, 1963.

MacFadden, C. H. *An Atlas of World Review.* New York: Crowell, 1940.

Montagu, A., ed. *Culture and the Evolution of Man.* New York: Oxford University Press, 1970.

Outhwaite, L. *Unrolling the Map.* New York: Reynal & Hitchcock, 1935.

Pfeiffer, J. *The Emergence of Man.* New York: Harper & Row, 1972.

Piggott, S., ed. *The Dawn of Civilization.* New York: McGraw-Hill, 1961.

Simons, E. L. "The Early Relatives of Man." *Scientific American,* Vol. 211 (July 1964), 50–62.

Washburn, S. L., ed. *Social Life of Early Man.* Chicago: Aldine, 1961.

———, ed. *Classification and Human Evolution.* Chicago: Aldine, 1963.

White, L. *Medieval Technology and Social Change.* Oxford: Clarendon, 1962.

Willey, G. R., ed. *Prehistoric Settlement Patterns in the New World.* Publications in Anthropology No. 23. New York: Viking Fund, 1956.

Masai of central East Africa. (Maxwell Coplan, DPI)

4

Race and Cultural Diffusion

The peoples of the world are by no means all alike. Their many differences are both physical and social. Often the average members of two different human groups show obvious differences in such traits as height, build, hair form and color, skin color, and shape of eye and nose. They may also speak differently, worship in different ways, and utilize different clothing, foods, tools, and art forms. The first areas of difference, those involved with physical attributes, are passed on genetically, and thus are considered substantially racial. The others are characteristics acquired after birth, through inadvertent imitation or through deliberate education and training; these are considered cultural.

The early part of this chapter will examine the racial differentiation of humankind. Then it will consider certain highly visible aspects of culture and their spatial distributions. The latter part of the chapter will be devoted to several specific cultures—to their major characteristics, areas and modes of development, and diffusions into other regions of the world.

Racial Groups

A *race* is a segment of humankind whose members possess a set of physical characteristics so distinctive as to differentiate them from the rest of humanity. Most often these distinguishing characteristics relate to physical stature, shapes of facial features, eye and skin color, hair texture and color, and blood type. All are inherited characteristics or traits transmitted genetically from generation to generation. Since humans have been extremely mobile over the millennia, and the intermixing of both greatly contrasting and similar groups has occurred almost everywhere, there are no pure racial groups—nor is it likely that any have ever existed. Descriptions of racial groups are thus generalizations; the races cannot be differentiated by means of sharp lines. They blend.

Classification

Nonetheless, anthropologists endlessly make attempts to delineate the human races, providing nearly as many racial classifications as there are experts. Within this array the best primary classification is one that divides humankind into three groups: Caucasoid, Mongoloid, and Negroid—each having several subdivisions.

The Caucasoid race is characterized by light skin coloring, light eyes (blue, green, gray, brown), light wavy hair, thin lips, and prominent, narrow noses. The Mongoloid race exhibits light-brown to yellowish skin; brown eyes beneath epicanthic eye-folds, the so-called "almond eyes"; flat noses with rather depressed bridge areas; high cheekbones; broad flat faces; and straight, thick, dark hair. The Negroid race is characterized by light brown to very dark brown skin, dark woolly to kinky hair, eyes with dark brown irises and often brownish pigmentation in the sclera, broad and flattish noses, prominent lips, and forward jutting jaws.

Origins and Diffusions

It is commonly assumed today that humankind was once a single homogeneous population and that racial differentiation occurred during the late Pleistocene, when some groups, having drifted away from their common "nursery," became isolated by environmental phenomena. Thus the present Mongoloid group was cut off from the others by the glaciers of the Alpine-Himalayan mountain chain; the present Negroid group became separated by the increasing dessication of interior North Africa; and the present Caucasoid group was also separated. Genetic changes occurred, gradually, in *all* the groups, partly by natural selection and partly by mutation, and the groups grew less and less alike in physical characteristics.

The Negroid and Mongoloid races developed in areas of bright sunshine; the former in the savannas and steppes of sub-Saharan Africa, and the latter in the steppes and deserts of central Asia. Their more heavily pigmented skins and dark eyes probably represent an adaptation to the effects of bright sunlight and glare and excessive amounts of ultra-violet radiation. By contrast, the Caucasoid race (despite the eastern associations of its name) may well have developed in the cooler, moister climates of western Europe, under less sunny skies and often in the partial shade of the forests.

Current Distribution

Today the Negroid race is still most dominant within its original abode, the major part of sub-Saharan Africa (Fig. 4–1). Variant groups within this area include: the Bantu, long-time residents of central Africa and relatively recent arrivals in southern Africa, with dark brown skin, and slightly broad noses and protruding jaws; and the Nilotic Negroes of southern Sudan and adjacent parts of Uganda and Kenya, taller and slimmer, with thinner lips and narrower noses.

For thousands of years, Negroes were abducted in various parts of sub-Saharan Africa and sold as slaves in distant parts of the world. In the United States, from the seventeenth to the mid-nineteeth century, the labor of black slaves was an important factor in the functioning of the plantation agriculture of the South. Proclaimed free in the

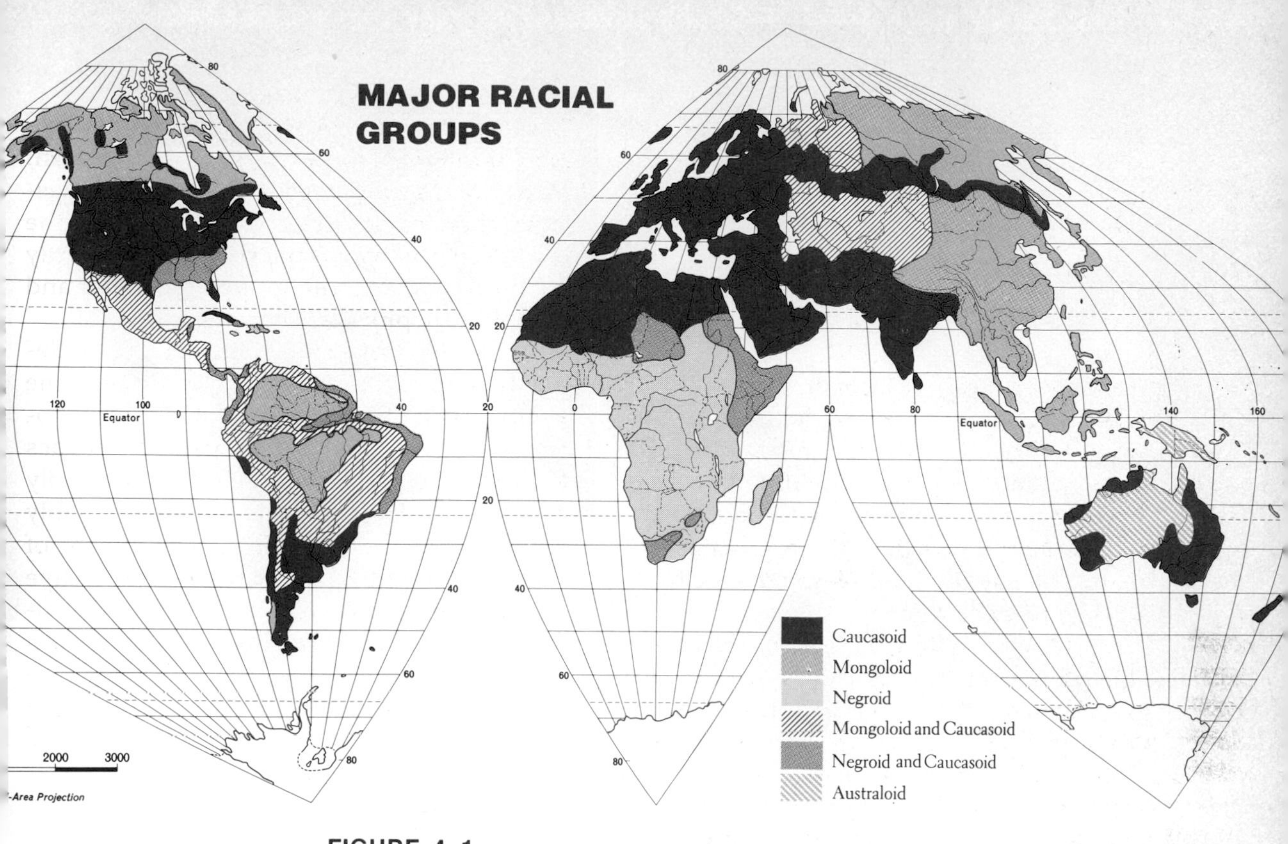

FIGURE 4–1

Caucasoids comprise about 55 percent of the world's people; Mongoloids, 37 percent; Negroids, 7 percent; and Australoids, 1 percent.

1860s, many former slaves remained in the South; but others resettled in the northern states and western territories. And since the 1900s, and especially since the mid-1940s, large numbers of Southern blacks have migrated to the industrial cities of the Northeast and Middle West and to the urban areas of California. The enslavement of Africans is an historical fact elsewhere in the hemisphere as well. Descendants of former slaves comprise significant portions of the populations of coastal Brazil, French Guiana, Surinam and Guyana, the West Indies, and the Caribbean coasts of parts of Central America. Blacks were also transported as slaves to Arabia, where they now form an important part of the population of many of the coastal areas to the east (Fig. 4–2).

The Mongoloid race is well developed in its central Asian nursery and, with only minor variation, over most of the Asian continent north of the Himalayas. The great treks of the *Völkerwanderung* brought Mongoloid groups well into eastern Europe, the latest and most conspicuous of the invading groups being the Magyars, who settled in Hungary, and the Finns and Lapps of Scandinavia. The American Indian, both North and South American stocks, and the Eskimo are usually thought to be Mongoloid peoples who migrated, at different times, across the Bering Strait and along the Aleutian Island

istics; thus, while predominately Nordic, these areas are genetically complex.

The Alpines are a physically stockier group than the Nordics, with straight to wavy brown hair, darker eyes, and broader heads. As their name implies, their area of greatest concentration is in central Europe, with an extension west across France and east toward central Asia. The Mediterranean Caucasoids occupy most of the Mediterranean basin from Portugal to the Middle East, and thence onward to India. They are distinguished from the other branches of the stock by larger noses, darker eyes, skin colors ranging from olive to light brown, and straight to wavy brown or black hair. In India, particularly in the south, the skin coloring is often very dark (Fig. 4–4).

None of these Caucasoid groups is "pure." In the core area occupied by each group the physical characteristics by which it is known may be in the majority, but they become increasingly mixed with the characteristics of other groups as one ventures outward toward the periphery. Thus there are many brown-haired, brown-eyed Scandinavians with light-brown skin—quite in contrast with the features of the typical Nordic person.

Since the beginning of the sixteenth century the Caucasoid race has spread itself widely over the world, partly through the large-scale colonizations undertaken by various European nations, and partly through the overseas migration of great numbers of European citizens on an individual basis. As a consequence, Caucasoids, mainly Nordics, comprise a large proportion of the population of places very distant from Europe—Australia and New Zealand, for example; in the United States and Canada, closer European settlements, they are the basic matrix of the population. Although Nordics predominate among the Caucasoid peoples of the United States, the country's population includes significant numbers of Mediterraneans, chiefly in the urban areas of the northeastern states, and a considerable population of Alpines, in the same areas and in the Middle West. Large numbers of Mediterraneans also live in the eastern and southern parts of Brazil, where they are the major Caucasoids and have been intermingled with Negroids from Africa. Argentina is another South American country occupied largely by Mediterraneans, with scatterings of Alpines and Nordics in the cities. Populations in the rest of Latin America are a blend of Mediterranean Caucasoids and native Mongoloids (American Indians)—a mixture termed *Mestizo*, or "mixed blood."

The eastward expansion of the Soviet Union has sent great numbers of Alpines from the European Soviet far into Asiatic Siberia, where they have settled on the agricultural lands of the steppes and in the cities along the Trans-Siberian Railway. They present a striking contrast to the indigenous inhabitants of that Mongoloid source area.

Unclassified Groups

Many groups of people do not fit into the classification of major races presented here, and thus in most studies are appended to it as unclassified populations. One such group is the Khoisan people of southern Africa; others in the same area include Bushmen and the Namas, or Hottentots. A short people with yellow-gray skin, the Khoisan have many other Mongoloid traits: pronounced development of the epicanthic eyefold, high cheekbones, V-shaped faces with small pointed chins, and a high incidence of the Oriental birth-patch (a highly pigmented patch of skin at the base of the spine, conspicuous at birth but disappearing in early childhood). Khoisan hair, on the other hand, grows in "peppercorns" (tiny, tight, spirals

FIGURE 4–4

Caucasoids of: left, France (Mark Antman); right, India (Alan Keler, EPA); bottom, Saudi Arabia (Standard Oil Co., N.J.)

of hair separated by bare scalp), and the women exhibit pronounced steatopygy (storage of fat in the buttocks); neither of these traits is associated with any Mongoloid group. The Khoisan formerly occupied nearly all of southern Africa, but were first pushed southward by Bantu invaders and then displaced and partly absorbed by recent European settlers. They remain in a relatively pure form in Namibia (South West Africa). Many theories have been advanced as to the origin of these peoples, but none being conclusive, their place in any racial classification is still uncertain.

The place of the Australian aborigines is equally confusing. A very dark-skinned people with Caucasoid facial features, they have been called Australoid, or "Archaic

Caucasoid," by some authorities. Among these are those who account for the aborigines by hypothesizing that a Caucasoid group found its way into Australia and absorbed a small Negroid population that had preceded them—when the first modern Europeans visited Tasmania, the island south of Australia, its inhabitants were dark-skinned peoples.

Languages

Languages are one of the most vital aspects of culture, for they are the means by which ideas and concepts are transmitted from generation to generation and from group to group, the medium through which much of human experience is preserved, transferred, and expanded. Each language is a group of sounds and symbols; it may or may not be written, and may even include body movements or gestures (Fig. 4–5).

While repetition tends to stabilize elements of a language, use also allows a language to be a dynamic entity, changing with the times and conditions. Commonly, too, the implementation of a language varies considerably from region to region and from one social level to another in a given area. Such variations represent *dialects* and usually involve differences in pronunciation, grammar, and choice of words. The English language, for example, adapted liberally throughout the world, is even spoken variously in different parts of Great Britain, where it is represented by such dialects as London Cockney, Yorkshire, and Scottish.

History indicates that spoken languages preceded their written versions, and even today a large number of people speak languages that either have no written form or have acquired one only recently through the

FIGURE 4–5
A Bushman storyteller and his audience. (N. R. Farbman, *Life* Magazine, © Time, Inc.)

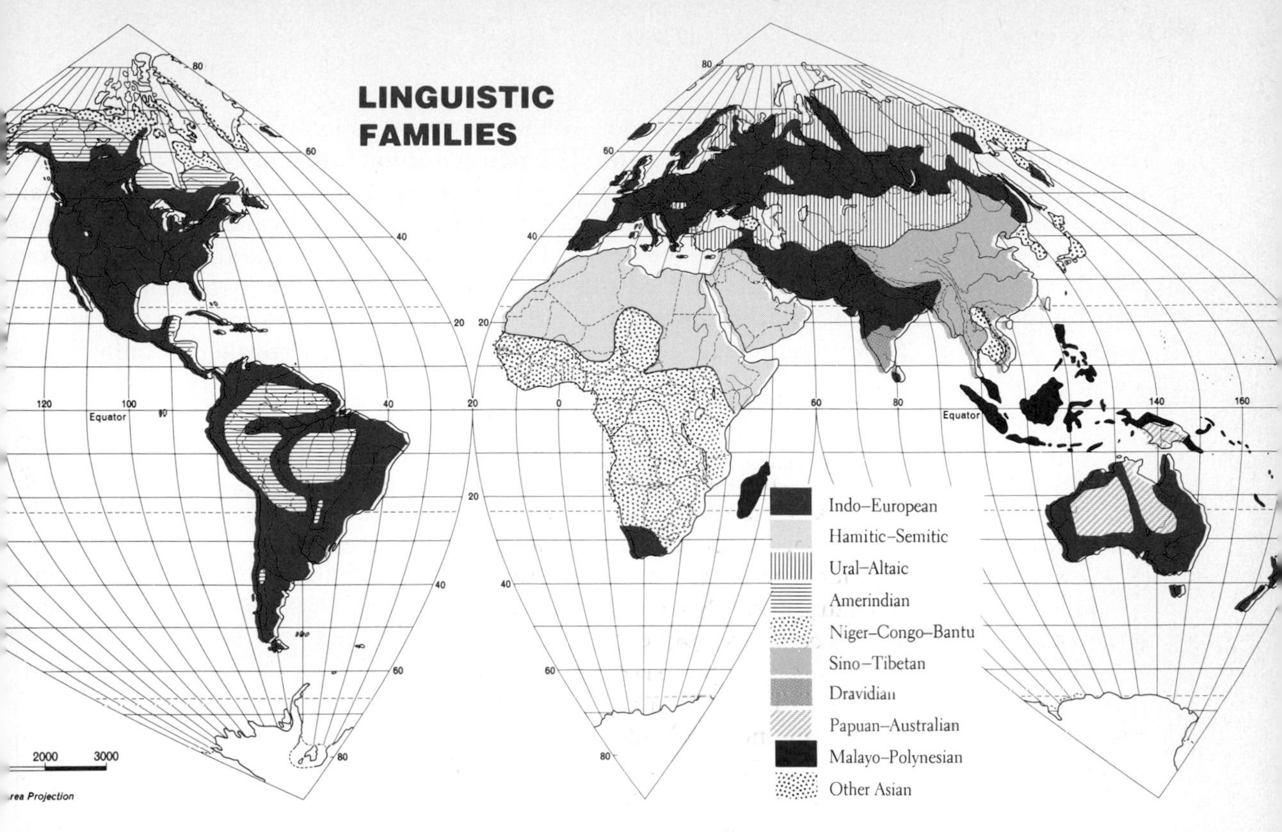

FIGURE 4–6

The Indo-European languages serve about 49 percent of the world's people, and the Sino-Tibetan, about 24 percent; each of the other major linguistic families accounts for less than 5 percent.

efforts of foreign linguists or missionaries. Berber, the language of some ten million Caucasoid inhabitants of North Africa, is written only in an Arabic script, never having generated a written form of its own. No Negroid African group has created a written language, and only among the Aztecs and Mayas of Mexico did there exist any semblance of a written language in pre-Columbian America.

Linguistic Families

Languages change gradually through time, and, in many cases, a dialect diverges so far from the parent tongue as to eventually constitute a separate language. But even when so separated, a language and its dialects can usually be related through a comparison of their vocabularies, grammatical structures, and distinctive sounds. A group of languages so interrelated is called a *linguistic family*.

The several thousand languages used by today's general populations can be variously grouped into linguistic families. For the purposes of this geographical survey, they can be divided conveniently into nine language families—Indo-European, Ural-Altaic, Sino-Tibetan, Malayo-Polynesian, Papuan-Australian, Hamitic-Semitic, Niger–Congo–Bantu, Amerindian, and Dravidian—and a group of miscellaneous Asian languages (Fig. 4–6).

Indo-European Languages. The Indo-European family includes nearly all of the languages of Europe and the Americas, the Soviet Union, Iran, and a good part of the Indian subcontinent—in all, the languages of nearly one-half the world's people. The western region of Europe alone has generated several language groupings; chief among them are the Celtic, Romance, and Germanic language groups (Fig. 4–7).

The Celtic languages, remnants of a language that once prevailed over much of France, Spain, and Britain, are now spoken by only a relatively few people in the westernmost extremities of Great Britain, Ireland, and France. The last speaker of *Cornish*, the

FIGURE 4–7

Major languages of the European nations. Although the staunch advocacy of one language is a common expression of nationalism, attempts to draw language boundaries in exact conformity with national political boundaries are all but futile. Such maps as this on language distribution are necessarily generalizations.

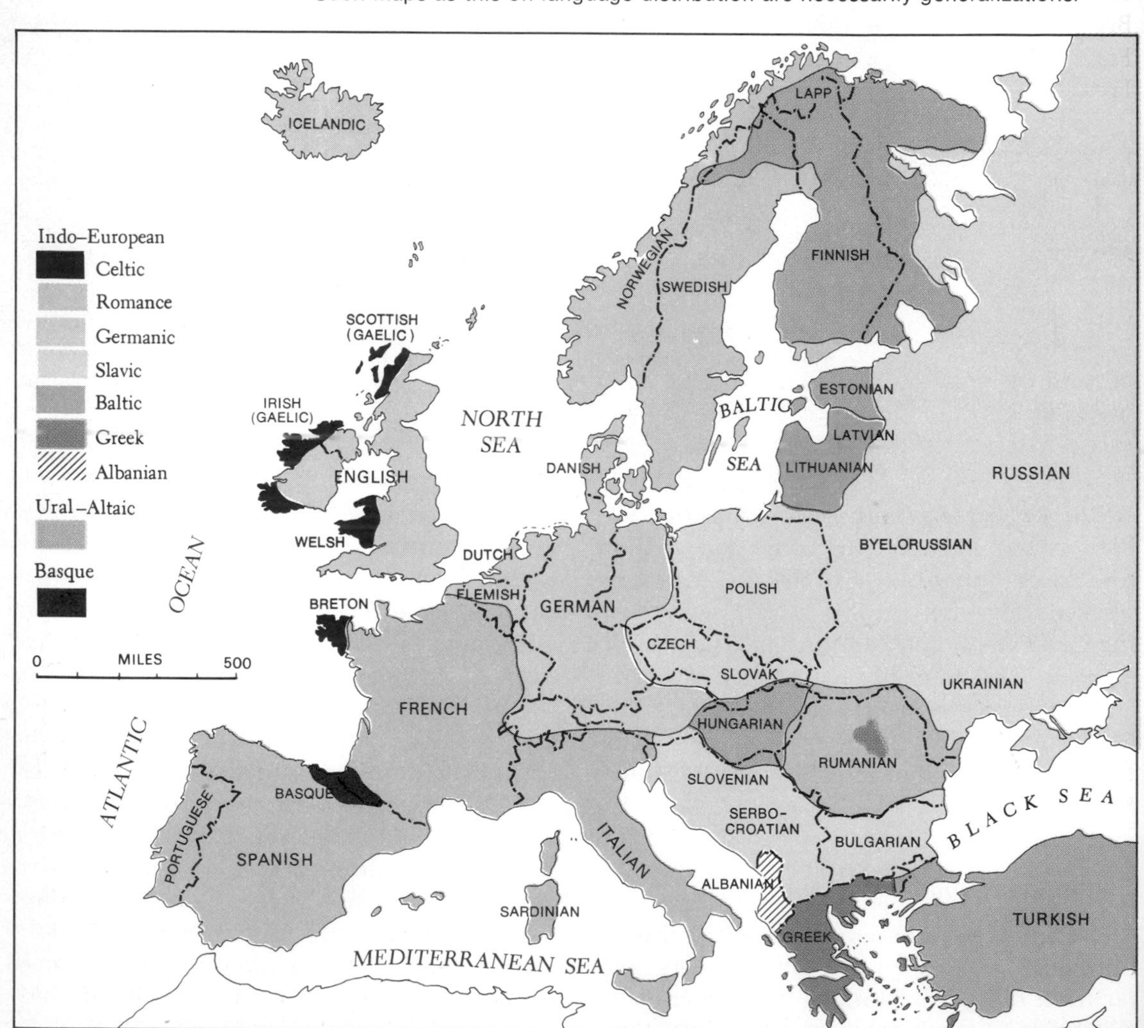

ancient Celtic language of southwestern England, died in 1840; and the last speaker of *Manx,* once the dominant language on the Isle of Man, died early in the present century. *Scottish Gaelic* is limited to northwestern Scotland and the adjacent islands of the Hebrides; and *Erse,* or *Irish Gaelic,* while the official language in the Republic of Ireland, is commonly spoken only in the western part of that country. *Cymric,* or *Welsh,* is spoken chiefly in remote areas of Wales, but it has acquired some prominence recently as part of a Welsh nationalist movement. And in Brittany, the northwesternmost peninsula of France, *Breton* is being gradually replaced by French.

The other major Indo-European language groups are now much more widespread than the Celtic. The Romance group, based upon Latin used two thousand years ago within the Roman Empire (and today as the official language of the Roman Catholic Church), includes Spanish, Portuguese, French, Italian, and Rumanian, the languages of most of southern Europe and virtually all of Latin America. The Germanic languages include German, Swedish, Norwegian, Danish, Dutch, and English, which is the primary language of such former British territories as the United States, Canada, Australia, and New Zealand and the "second language" of much of the world's remaining landmass. Although classified as Germanic, modern English is actually an admixture of Germanic and Romance languages. The English language has not been purely Germanic since A.D. 1066, when, with the Norman conquest of Britain, a large number of French words began to find their way into the vocabulary.

The eastern region of Europe has generated an even larger number of languages than the western region—among them Greek, Albanian, Baltic, and the very large group of Slavic languages. The latter dominate most of eastern Europe, and include such languages as Czech, Slovak, Slovene, Serbo-Croatian, Bulgarian, Macedonian, Polish, Byelorussian, Ukrainian, and Russian. Today the dominant language of the whole Soviet Union, Russian is thus also Asian, having been diffused eastward across central Siberia to the Pacific.

The major southwest Asian extensions of the Indo-European languages are the Iranic and Indic language groups. Included within the Iranic is Persian, a medieval language spoken throughout Iran in a number of local dialects and a large number of minor languages spoken by rather isolated societies in the Caucasus Mountains and in mountainous areas of Iran, Afghanistan, Baluchistan, northern Syria, and eastern Turkey. Most of the Indic languages are obvious derivatives of Sanskrit, the language of northern India in ancient times and very likely the source of *all* the Indo-European languages. In India today, a great number of languages are used—a fact that poses a major problem in the unification and administration of the country. *Hindi* is the official and by far the leading language of India; yet it is the language of only one-third of the population, mainly those who live in the north central states of Uttar Predesh, Madhya Predesh, and Bihar (Fig. 4–8). Its official status is resented by many Indians who speak one of India's twelve other major languages, most of whom have no command of Hindi. Among these tongues are *Urdu,* a Hindi dialect much used in parts of northwest India and in nearby Pakistan, and *Bengali,* the major language of Bangladesh and parts of eastern India.

Ural-Altaic Languages. The Ural-Altaic languages apparently originated in Siberia, but were carried into eastern Europe with the great migrations of the *Völkerwanderung.* Prominent among them is the Finno-Ugric

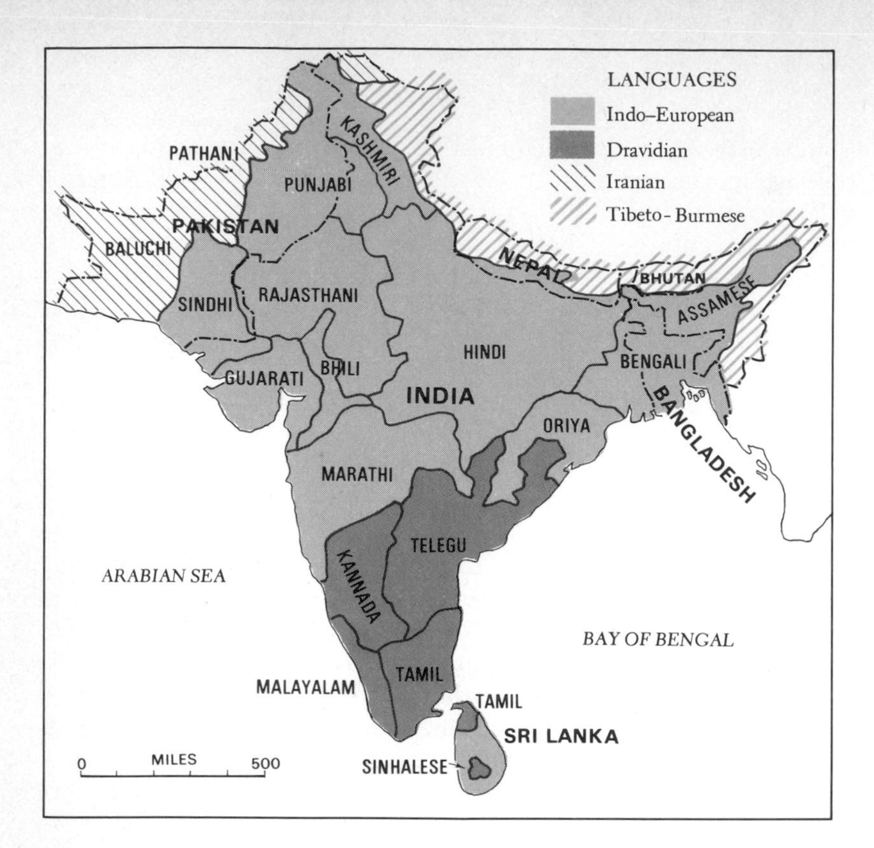

FIGURE 4–8

The languages of India. Since 1947 India has made major changes in its state boundaries in order to have them reflect the nation's pattern of languages.

group, which includes *Magyar,* the Ugrian language carried far into eastern Europe when the Magyars invaded Hungary at the end of the ninth century; and it still exists as a linguistic enclave surrounded by Romance, Germanic, and Slavic languages. Other major languages of the group are *Finnish,* spoken in Finland; its close relative *Karelian,* used in parts of the Soviet Union adjacent to Finland; and *Estonian,* spoken in the area across the Gulf of Finland to the south. The Lapps, reindeer-herding nomads of the mountains of Norway and Sweden, and of the tundras of northern Finland and the northwestern Soviet, also speak a Finno-Ugric language.

Since it is impractical here to describe all the languages of the world, Figure 4–9 gives a general classification of the major languages and their principal countries of use. The world map in Figure 4–6 should also be consulted at this point, for it portrays the spatial arrangement of modern languages.

Bilinguality

Bilinguality, the ability to speak two languages with nearly equal facility, occurs in several types of situations. For example: the language group finding itself under the political, economic, or cultural domination of another may retain its own language for use in private conversation and correspondence while employing the host tongue in broader social contacts and governmental activities. Such is the case in Quebec, a French-speaking province in otherwise English-speaking Canada. So strong is the French feeling and influence that Canada is officially bilingual, with even the postage stamps employing the two languages.

In other cases, two languages share linguistic dominance in the country. Such is the case in South Africa, where English and Afrikaans (a form of Dutch) are both widespread and exist without exclusive terri-

INDO-EUROPEAN

English:	United Kingdom, United States, Canada, New Zealand, Australia, Ireland
Spanish:	Spain, Latin America
Hindi:	North-central India
Russian:	Soviet Union
Bengali:	Bangladesh
Portuguese:	Portugal, Brazil
German:	West Germany, East Germany, Austria, Switzerland
French:	France, Belgium, Switzerland, Canada
Italian:	Italy, Switzerland
Ukrainian:	Ukraine (USSR)
Marathi:	Maharashtra (India)
Polish:	Poland
Punjabi:	Punjab (India)
Rajasthani	Rajasthan (India)
Sinhalese:	South Sri Lanka

HAMITIC-SEMITIC

Arabic:	North Africa, Middle East
Berber:	North Africa

URAL-ALTAIC

Finnish/Komi/Samoyed/Yakut:	Northern Eurasia
Tatar/Kazakh/Khalka/Turkaman:	Central Asia
Hungarian (Magyar):	Hungary, Central Rumania
Turkish:	Turkey

AMERINDIAN

Athabascan/Algonquin:	Northern North America
Arawak/Carib/Quechua/Aymara:	Amazonia

NIGER-CONGO-BANTU

Mandingo/Yoruba/Mossi/Ibo:	Western Africa
Kanuri:	Central Africa
Fang/Bulu/Congo/Swahili/Mbundu/ Makua/Bushman/Hottentot/Zulu:	Southern Africa

SINO-TIBETAN

Chinese Mandarin:	China, Taiwan, Southeast Asia
Cantonese:	South China
Fukienese:	South China
Tibetan:	Tibet

DRAVIDIAN

Tamil:	Tamil Nadu (India), North Sri Lanka
Telegu:	Andhra Praedesh (India)
Kannada:	Mysore (India)

PAPUAN-AUSTRALIAN

Australian:	Papua, Central Australia

MALAYO-POLYNESIAN

Indonesian:	Indonesia
Malagasy:	Malagasy Republic

OTHER ASIAN LANGUAGES

Japanese/Korean/Cambodian/Lao/Eskimo

FIGURE 4–9

Selected languages in the world's major language families and some of their territorial occurrences.

tories. The regional separation of national languages is found in Belgium, for example, where Flemish, a Germanic language closely akin to Dutch, is spoken in the northern half of the country, and Walloon, a French dialect, prevails in the south; it occurs also in Sri Lanka (Ceylon), where Tamil is spoken in the northern third of the country and Sinhalese in the southern two-thirds.

Even if it is too small a portion of the total population to secure official bilinguality, a minority group may still persist in the use of its language in a local area, thus creating an unofficial bilinguality. The Cajuns of Louisiana, for instance, speak a *patois* (dialect) of French that has nearly the same currency as English in their areas and that influences speech in other areas of the state as well.

Two countries are officially trilingual. In Switzerland, German, French, and Italian are used, but in three different regions; and in Namibia, Afrikaans, English, and German are thoroughly intermixed.

Lingua Francas

In some parts of the world, so many different languages are spoken within a relatively small area that traders and others who travel from place to place can seldom become fluent in all of them. Consequently, a *lingua franca,* or common language, sometimes develops; it is usually not an existing language but rather one composed anew of fragments of several tongues. In the southwest Pacific, from southern China to Samoa, Pidgin English is the prevailing language of trade (*pidgin* being a Chinese corruption of the English word *business*); it consists chiefly of English words that are badly mispronounced and arranged in Chinese word order. A similar jargon with a French base is widely used in parts of the southwest Pacific, where it was named *Bêche de mer* after a sea cucumber that is caught on the sea floor, dried in the sun, and exported to China and other countries in great numbers for use in Chinese cooking; it was the brisk trade in *Bêches de mer* between the Pacific isles and China that gave rise to the language.

In the gold-mining areas of South Africa, Bantu workers representing a dozen countries and scores of languages communicate with each other and with their European foremen in *Fanagalo*—a *lingua franca* in which English, Afrikaans, Portuguese, and Bantu words are combined in an essentially Bantu matrix. Similar languages have developed in the Caribbean—among them *Papiemento,* with a vocabulary borrowed from Spanish, French, Dutch, English, Carib Indian, and West African dialects; and in East Africa—most notably *Swahili,* which is composed of Arabic and Bantu words with a Bantu grammar.

Religions

Religion has become one of the most pervasive factors affecting social and political relationships. It can unite people of different languages and regions and it can help to bring about the alienation of neighbors who speak the same language. For some, values and priorities in all areas of life are governed, in varying degrees, by religious beliefs and taboos. And different religions often have very different sanctions on the same human activities. Thus Moslems eat no pork, and Hindus eat no beef; Jews refrain from labor on Saturday, and Christians refrain from labor on Sunday.

It is likely that religions have grown somewhat in response to the needs of human beings to have a place and purpose in the world and to transcend the frustrations and monotony of daily needs—individual and societal. Primitive tribal peoples worshipped many animate as well as inanimate objects: animals, the sun, and numerous images fashioned of stone and wood in both human and animal form. Such *animism* still prevails throughout much of central Africa, parts of Indonesia and Australia, interior Brazil and Peru, and in northern Siberia (Fig. 4–10).

But gradually in most areas, such direct nature-god worship has been transformed into more abstract worship and complex philosophies. The most advanced religious societies have also developed complex hierarchies to formalize the worship, guide the worshippers, and administer the faith's physical property.

Compared with its languages, the major religions of the contemporary world are relatively few in number. Five religions accommodate perhaps three-fifths of the world's four billion people (Fig. 4–11). Three of these five—Christianity, Judaism, and Islam (Moslem)—are *monotheistic* religions, for their doctrine contends that there is only one god; the others—Hinduism and Buddhism—are *polytheistic* religions, for their doctrine supports a belief in more than one god.

FIGURE 4–10

Animistic funeral dancers near a Dogon village in Sangha, Mali. (Eugene Gordon)

Monotheistic Religions

The major monotheistic religions, Christianity, Judaism, and Islam, all originated within the same small Middle East region: in arid and semiarid parts of Palestine, Jordan, and Syria (Christianity and Judaism) or in western Arabia (Islam) (Fig 4–12). In addition to this cradleland, they also share parts of the same religious tracts, such as the Old Testament. However, they still have many differences. For example, the Christians recognize Christ as the long-awaited Messiah, while the Jews recognize Christ as merely one of several prophets and await the true Messiah; the Moslems (Islams) also await the true Messiah but recognize Mohammed as the greatest of the prophets.

Since their place of origin is the arid environment of the Middle East, these three religions all have histories among the pastoral nomads and agrarian-herders of its barren steppe-lands. It is even quite possible that the monotheism of the religions was influenced by the harshness of these lands. They were demanding, and, in order to survive, their human inhabitants may have had need of a philosophy that stressed strength and discipline and provided strong leadership in their own image. Judaism, Christianity, and Islam are all decidedly patriarchal: in each, God is a male, the holiest followers are males, and women play only minor roles—circumstances that reflect the *pastoral nomadic societies* in which these religions developed.

All three monotheistic religions (and Buddhism as well) are traditionally dynamic, having extended their influence far beyond their small bleak cradlelands. Christianity, born in Palestine, has spread over the world and is today the dominant religion in four of the six inhabited continents. Judaism, born in the same locale as Christianity, has taken hold in all the continents, even though its followers have seldom enjoyed the power or favor of rulers. And the followers of Islam have carried their religion from Arabia across North Africa, eastward through the Middle East, and far overseas into maritime southeastern Asia.

Christianity. With the preachings of the apostle Paul in Greece and Rome, Christianity passed out of its cradleland in Asia and into the lands of Europe. Christianity

FIGURE 4–11

The three major monotheistic religions comprise approximately 41.5 percent of the human population; and the two major polytheistic religions account for approximately 19 percent.

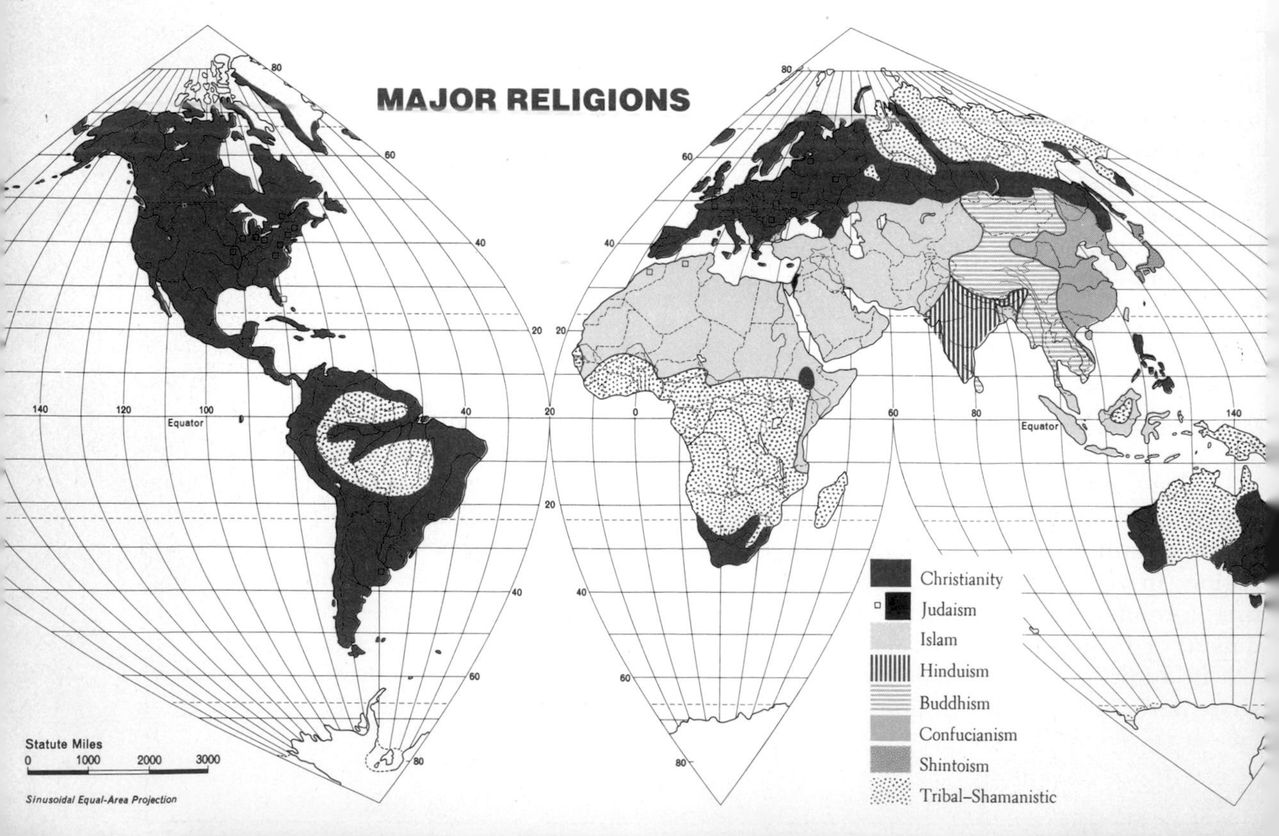

grew with Europe, and vice versa, and Europe has long since become the modern fountainhead of Christianity's three major subdivisions: Catholicism, Protestantism, and Eastern Orthodoxy (Fig. 4–13). *Catholicism* is dominant throughout most of western and central continental Europe, from beyond the English Channel to the Mediterranean to the Baltic Sea—in ten countries and adjoining country border areas. Catholicism also prevails in all the European-settled regions of Latin America and throughout French-speaking Canada (especially Quebec); and it is well represented in the northeastern, midwestern, and southwestern parts of the United States. As it has been for centuries, the Catholic Church is presided over by a Pope, who lives in Vatican City, an independent territory within Italy's capital, Rome (Fig. 4–14).

In the early sixteenth century the German monk Martin Luther posted 95 theses attacking the established papal authority of the Roman Catholic Church. Luther protested that true Christianity lay not in a hierarchy headed by a Pope but in the humble layman's belief and direct communion with God. In the 1530s a strong German Lutheran Church was established; the Church of England had previously broken with the papacy, and many other dissident sects followed Luther's. Today, *Protestantism* is the dominant religion in central West Germany, East Germany, Denmark, the United Kingdom, and Scandinavia; and it has spread overseas to dominate the United States and most of

FIGURE 4–12
Cradlelands of the world's major religions.

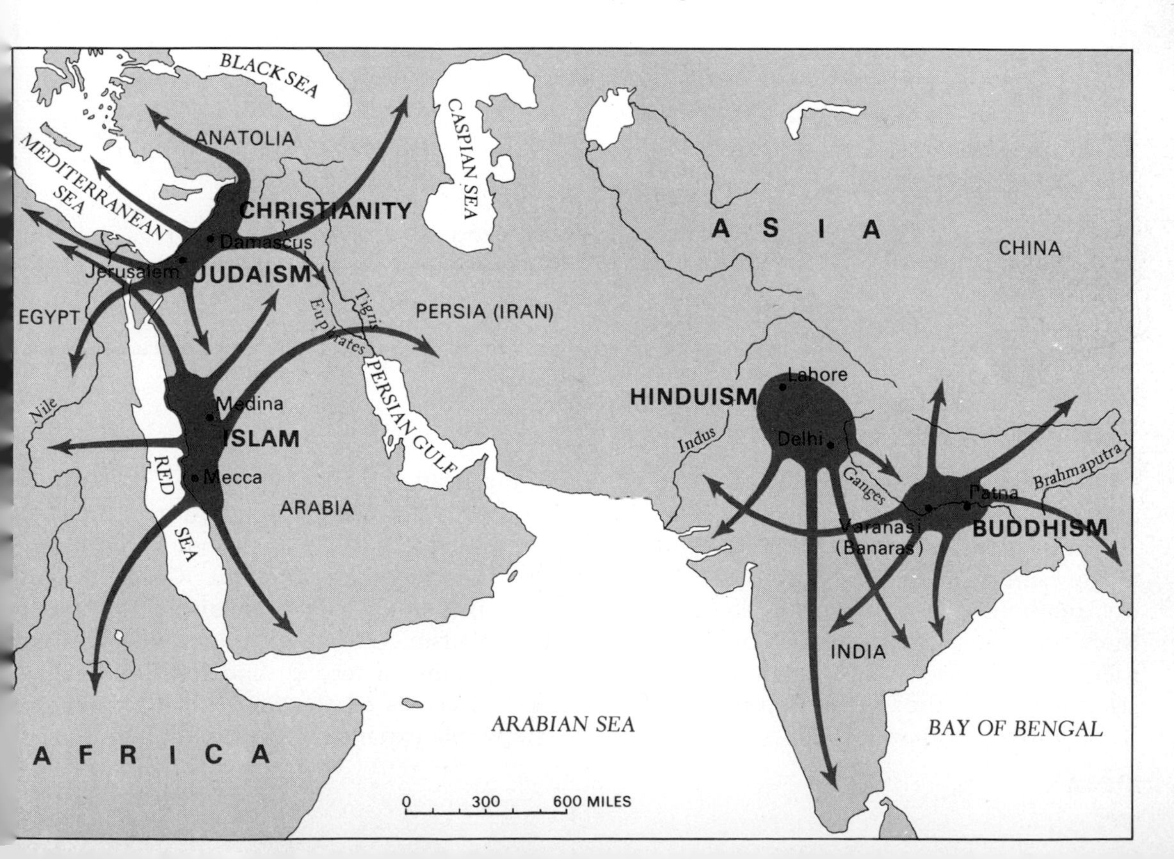

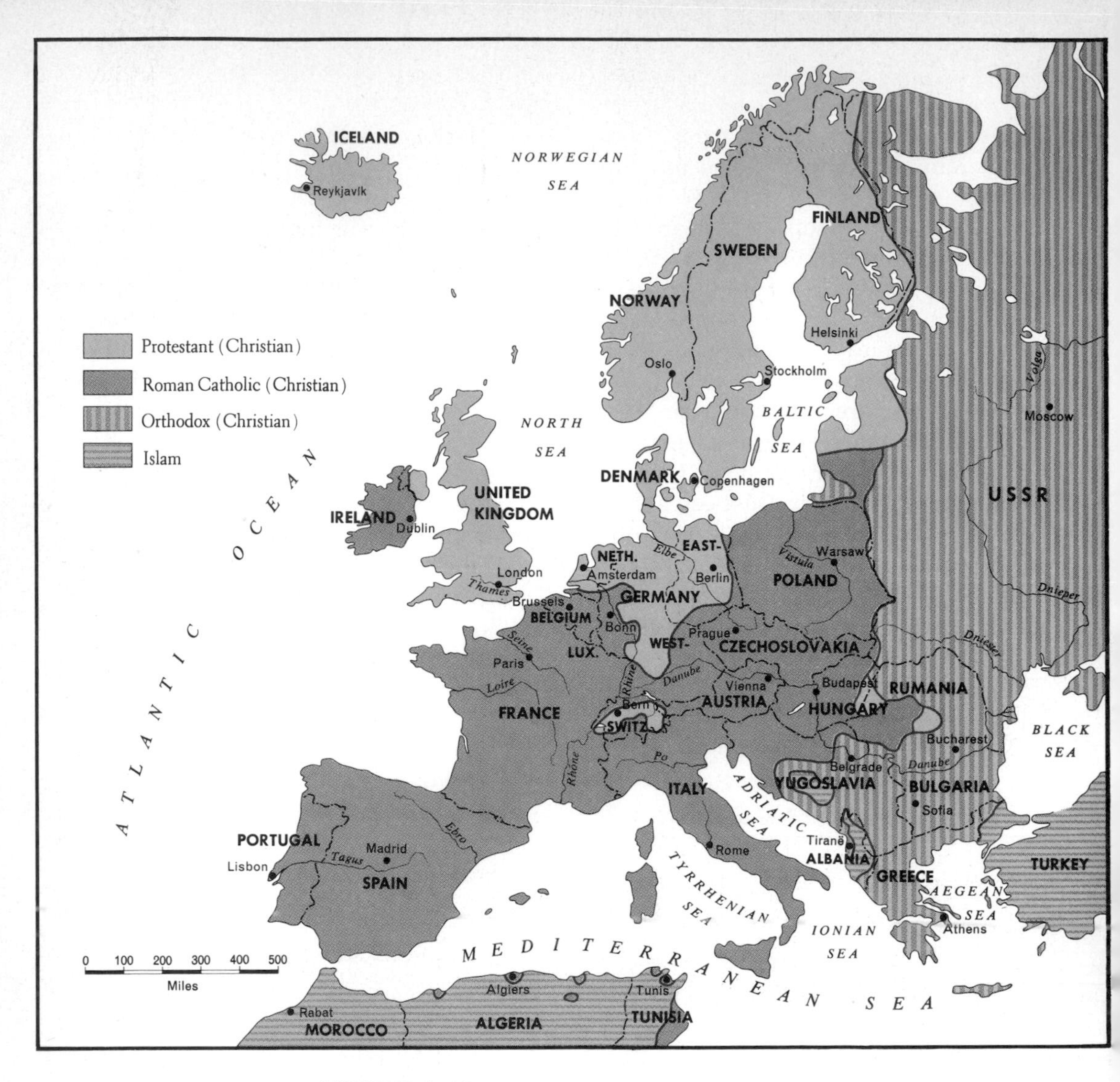

FIGURE 4–13

The dominant religious affiliations of European countries. (Modified from Karl Wenschow, *Weltatlas*.)

Canada, as well as to South Africa, Australia, and New Zealand.

The third largest branch of Christianity is the *Eastern Orthodox* Church. The Orthodox religion began in the Holy Land before there were any Christians in Rome, and its followers consider Jesus its founder. Constantine shifted the Roman Empire throne to Constantinople in A.D. 330; and thereafter Rome, headed by the Pope, and Constantinople, headed by its Patriarch, competed for converts. Seven centuries later they excommunicated each other, and the Great Schism began; it still persists. Today, while Roman Catholicism continues to dominate western Europe, the Eastern Orthodox religions have spread across all the Balkan countries, and the whole expanse of the Soviet Union.

FIGURE 4–14
The celebration of a Catholic mass. (*Ebony* Magazine)

Judaism. While the fundamental precepts of Judaism have not changed for thousands of years, new and differing views on the value of ritual and tradition give the religion its three subdivisions—Orthodox, Conservative, and Reform. Orthodox Judaism strictly observes all traditional Jewish laws and traditions; Conservative Judaism believes that the concept of revelation is open to "reasonably varied" interpretations; and Reform Judaism believes that each new generation has the right to accept or reject the traditions it inherits. Today, Judaism is well established in the new State of Israel and in many urban centers of the Middle East, North Africa, Europe, and the Americas (Fig. 4–15).

Islam. Islam is the youngest of the major religions, having emerged in the desert lands of western Arabia over 500 years after the death of Jesus. Its founder, Mohammed, explained himself as the best-informed of several prophets (among them Jesus), as a

FIGURE 4–15
The passing of the Torah, or written Jewish law, through a congregation. (Israeli Ministry of Tourism)

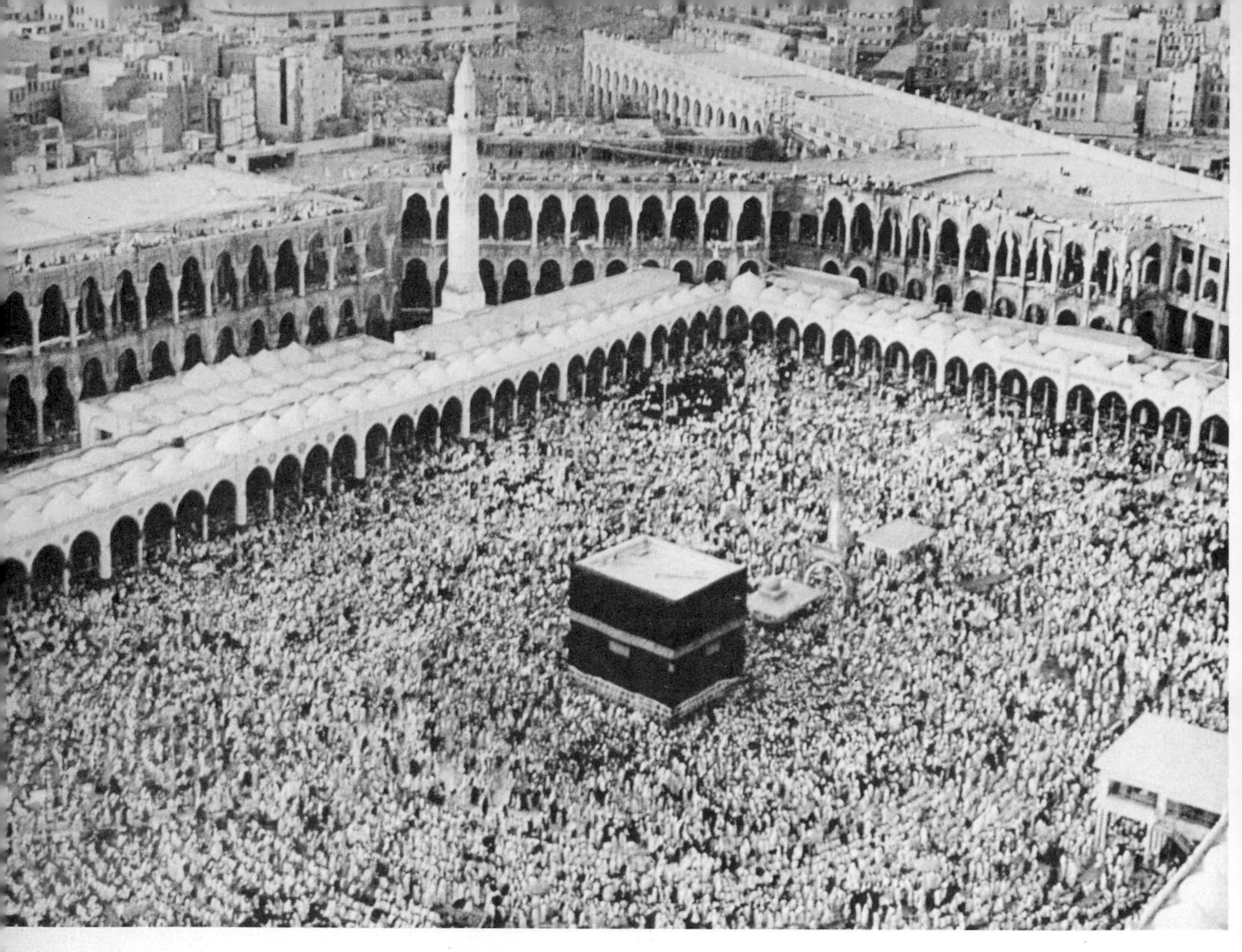

FIGURE 4–16

The Holy Mosque at Mecca, with thousands of devout Moslems performing the Tawaf, or circling ceremony, around the Ka'bah, the most sacred of all Islamic shrines. (Saudi Arabian Consulate)

common man through whose lips God spoke to the believers. Within a century after Mohammed's death (in A.D. 632), the Islamic influence extended from Gibraltar to the Himalayas. Today it spreads across northern Africa and the Middle East to South Asia's Pakistan and Bangladesh and has significant acceptance in northern India and the Malayo-Indonesian archipelago: from Morocco to the Malacca Straits. The teachings of Mohammed, reflecting aspects of Judaism and Christianity, comprise the Koran, the sacred text of the world's Moslems; Mecca, his home, is their holy city (Fig. 4–16).

Polytheistic Religions

Hinduism. Hinduism, the oldest contemporary religion and the dominant one in India, emerged in the broad fertile plains of the Ganges Valley of northern India, near the now sacred Ganges River (see Fig. 4–12). It has no known founder, no hierarchy, and no rigid moral code. A polytheistic religion, Hinduism embraces a great number of gods and goddesses and many lesser deities; they occur in human form and as animals, and others are resplendent combinations of human and beast or humans with many limbs (Fig. 4–17). The best known of the more recent Hindu leaders include Shankara (ninth century A.D.) and Ramanuja (eleventh century).

During its long and loosely structured history, Hinduism has suffered the rise of many splinter faiths. In the sixth century B.C., the sect known as *Buddhism* was founded in protest against the established Hindu caste system and adherence to ancient Hindu laws and doctrines. Other derivatives of Hinduism include *Jainism* (also sixth century B.C.) and *Sikhism* (sixteenth century A.D.). As it has grown stronger through the centuries, Hinduism has absorbed most of its splinter faiths and their reforms. However, it has not been a dynamic religion in other respects; for example, it has not spread to neighboring or distant lands and societies. By far the majority of the followers of Hinduism are located within a single national and cultural unit, India. Even within Asia other Hindu territories are few in number; and like the small island of Bali, east of Java, such areas exist in a vast sea of Islam.

Buddhism. When disagreement with the practices of Hinduism led to the founding of Buddhism in the sixth century B.C., the Hindus responded by appointing the leader of the new movement, philosopher Gautama Buddha, to the Hindu pantheon and by accepting many of the Buddhist protests. Consequently, Buddhism languished for a time as a movement without a cause. But Gautama Buddha continued his teachings, and, after his death, in the early fifth century B.C., they spread rapidly across India and were accepted by a majority of the Indian people, most of them formerly Hindus. Later, a reformed and rejuvenated Hinduism absorbed Indian Buddhism—and thus the Buddhist life-cycle in India came full circle.

But by then Buddhism, unlike Hinduism, had spread extensively within lands beyond India. In the third century B.C. Buddhist missionaries carried the faith to Sri Lanka, where it became the dominant religion; and, in the same way, it reached China by the first century A.D., Korea by the fourth century, Japan and Burma by the sixth century, and Tibet by the seventh century (Figs. 4–18 and 4–19).

FIGURE 4–17

A statue of Siva, the god representing destruction in the great Hindu triad of gods. Brahma, worshipped as the Creator, and Vishnu, the Preserver or Renewer, complete the triad, which is known as the Trimurti. (Leslie Holzer)

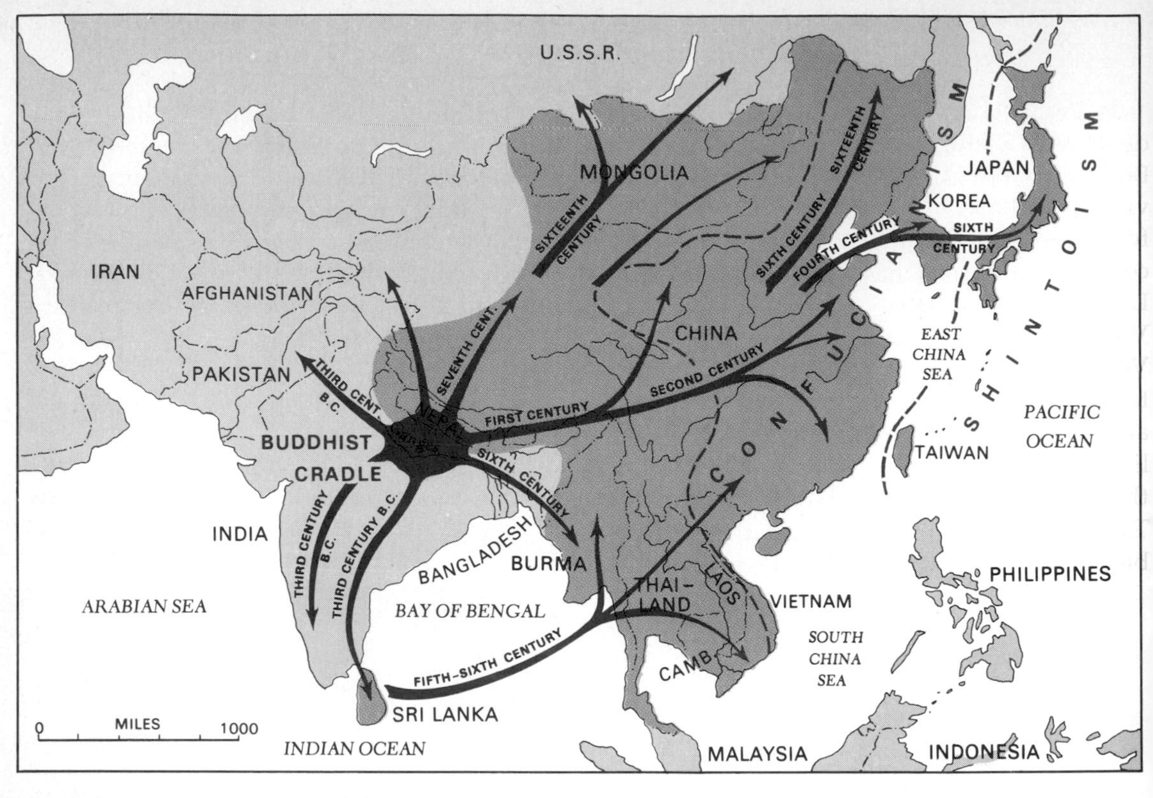

FIGURE 4–18

The early diffusion of Buddhism—from India throughout several countries of southern and eastern Asia.

FIGURE 4–19

The altar of a Buddhist temple in Chieng Mai, Thailand. Its dominant statue is a large representation of Gautama Buddha. (Steve Hershey)

Two schools of Buddhist doctrine have developed through the millennia. Hinayana Buddhism, the "Lesser Vehicle," exalts individual austerity and has its contemporary following mainly in Asia's southeastern countries—Sri Lanka, Burma, Thailand, Laos, Cambodia, North Vietnam, and South Vietnam. Mahayana Buddhism, the "Greater Vehicle," stands for individual salvation by faith and has its largest following in eastern and central Asia—in China, Taiwan, Mongolia, Tibet, Korea, and Japan. In several of these countries, local philosophies, such as Taoism, Confucianism, and Shintoism, blend with Buddhism, giving this religion a slightly different tone in different parts of the Eastern world.

Secularism

In many societies today, there is a drift away from organized religion on the part of a sizable segment of the population. Usually, this noninvolvement is on a purely personal basis, often reflecting the preoccupation of the individual with material concerns. However, in the Soviet Union and in China, Marxism and Communist ideology have made the minimization of religion a matter of national policy.

Early Culture Hearths

Many of the cultures the world has known have been very short-lived and have influenced only a handful of people. Others have been more long-lasting but equally localized; those that persist today are often described as "curious, unique, and exotic" by travel brochures offering expensive "journeys into yesterday." Some cultures, on the other hand, have been widespread and highly influential; and, by enduring through centuries and even millennia, have become the basis of modern civilizations. Each of these cultures originated in a particular region and, while it may have since spread widely over the surface of the earth, is still associated primarily with that source region, or *culture hearth.* The Old World comprises several such culture hearths, most notably: Mesopotamia, Egypt, Palestine, Crete and Greece, the Indus Valley, and North China. Within the New World there are two of comparable significance: Middle America and the Andes (Fig. 4–20). These areas have certain features in common: each has long been either a focus of trading or a crossroads of trade routes; each is, or lies close to, an area of early plant and animal domestication; and each possesses a climate that presents a certain challenge (a pronounced alternation of wet and dry seasons, a need for irrigation, a need to store food for use in a cool season), which in turn fosters planning for the future and organized group activity.

Old World Hearths

Mesopotamia. Mesopotamia, "the land between the rivers," is situated between the Tigris and the Euphrates rivers, in modern Iraq. In ancient times in the adjacent hilly lands to the northeast, a number of plants were domesticated, among them: wild grasses that evolved into wheat, barley, oats, and rye; legumes that were developed into peas, beans, and lentils; several kinds of nuts and such fruits as apples, peaches, pears, plums, cherries, and grapes; and a number of

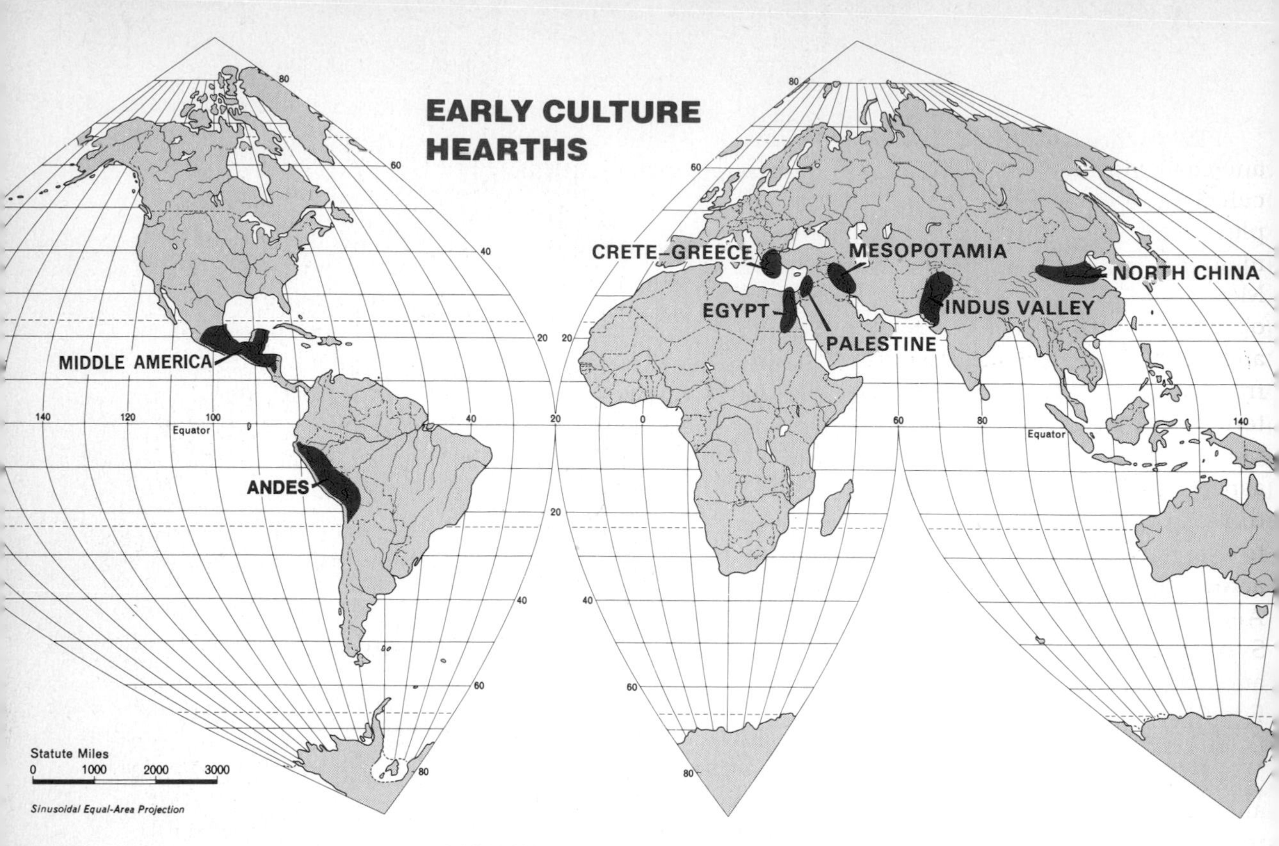

FIGURE 4–20

Each of the eight areas considered among the world's early culture hearths was, by nature, a center of agriculture or trade.

root crops. The animals domesticated in the region include cattle, sheep, goats, and pigs.

With a reliable food source thus under their direct control, some human groups in Mesopotamia were able to give up their nomadic hunting existences and become sedentary farmers, living in a fixed abode and tending their crops and herds. Thus as early as 10,000 years ago, many of the area's people lived in agricultural villages.

These people were not isolated in their land. From time to time, barbaric, militarily powerful peoples attacked and conquered Mesopotamia. At other times, victorious Mesopotamian armies brought home large groups of conquered peoples, to serve as slaves. And continuously over the centuries, outside groups migrated into the area, attracted by its food supply. Each of these groups—conquering armies, captured slaves, and peaceful migrants—adopted the Mesopotamian culture and added to it technique, language, and philosophy from their own.

Out of all this a civilization was eventually forged—the first great civilization in the world. The first written language probably evolved in Mesopotamia—a *cuneiform* script with wedge-shaped or arrowhead characters, usually engraved into clay tablets (Fig. 4–21). There the first multistoried buildings were constructed, and communities with public buildings and organized street patterns first appeared. This civilization persisted through three millennia, in spite of invasion and conquest, linguistic and ethnic change, and economic transformation.

From this early culture hearth, the most ancient center of civilization, innumerable cultural traits were passed on to other peoples. This transfer was accomplished in many ways. The conquest of other lands by Mesopotamian armies, the migration of Mesopotamian civilians into other countries, and, above all, the movements of traders from country to country spread both the material culture and the ideologies of Mesopotamia to near and distant places. By 3000 B.C., aspects of Mesopotamian culture had been diffused to the southeast into southern Iran, to the northeast into Central Asia, westward across the Anatolian Plateau of Turkey to the Aegean Sea and on to Greece, by way of Syria and Lebanon to the Mediterranean, and southwestward across Palestine into Egypt.

FIGURE 4–21

A Babylonian boundary stone (thirteenth to tenth century B.C.) that bears examples of *cuneiform* writing. (© British Museum)

Egypt. In contrast to Mesopotamia, ancient Egypt was somewhat protected from invading armies by its vast surrounding reaches of desert. It developed a strong and highly ritualistic culture that resisted all change for literally thousands of years. One element of this culture was a written language composed of *hieroglyphics,* or pictorial symbols. These were carved on the rocky walls of buildings, monuments, and tombs or were written on papyrus, a form of paper produced from a plant that grows abundantly along the Nile (Fig. 4–22).

Today, ancient Egypt is best known through its massive architectural constructions and statuary: the Sphinx and the pyramids, the temple of Karnak, the tombs of the Valley of the Kings, and the massive carvings of Abu Simbel. Because of the enduring nature of the rock from which they were made and the aridity of the climate, such structures have been beautifully preserved. They are not representative of most construction in ancient Egypt, however. The dwellings of ordinary people were usually made of a hardened mud much like the adobe used by Indians of the southwestern United States; these structures have succumbed to wind and rain. What is known of the life of the people has been learned from the paintings, stone carvings, and hieroglyphics that graced their places of worship and the homes of their leaders, and from the objects buried with them in tombs.

Ancient Egypt was visited occasionally by outsiders, mainly traders, who carried elements of its culture back to their homelands. The influence of this civilization is

FIGURE 4–22

Hieroglyphics, or pictorial writing, on pillars of an ancient temple at Karnak in southern Egypt. (Courtesy TWA)

thus detectable in contemporaneous and subsequent cultures of Palestine, the Middle East, Crete and Greece, and Rome.

Palestine. In many regards—physical geography, ethnic groups, historical developments—the Palestinian culture hearth was more complex than either the Mesopotamian or the Egyptian. Palestine, known also as Israel, the Land of Canaan, and the Holy Land, comprises a narrow strip connecting Mesopotamia and Egypt; it is hemmed in on the west by the Mediterranean Sea and on the east by the Arabian deserts. Through it, over thousands of years, has passed an endless procession of traders, nomads, armies, and whole nations. Its inhabitants have been conquered by several groups of outsiders, each affecting the area's landscape, genetics, customs, and ideologies.

Unlike Mesopotamia and Egypt, Palestine is not, and never has been, a fertile land easily capable of large-scale agriculture. Its great importance in ancient times reflected its position on the land bridge between Asia and Africa, not natural wealth or fertility. The principal "port-land" of the Middle East, it was a place for the cross-fertilization of cultures, a land of contrasts (maritime versus land-bound, sedentary versus nomadic, agricultural versus pastoral, conservative versus liberal) and endless change. Consequently, the culture it produced became the very antithesis of the unchanging Egyptian culture.

Thus it is not surprising that it was in Palestine that the world seems to have had its first town (as distinct from village), and that that town was founded at least in part on trade. By 6850 B.C., Jericho, located in the Rift Valley near the northern end of the Dead

Sea, was a flourishing settlement, existing partly on intensive agriculture and partly on trade in locally produced salt, sulphur, and asphalt.

In the following millennia, a regional settlement pattern evolved based upon a number of fortified towns and ports, each developing on a spot with natural protection (such as a hilltop), with defensive walls of stone, and houses of mud. While the walls persisted, defying invaders and the elements alike, the mud houses melted down in the rains or were destroyed in the invasions; at some later date, new houses (also of mud) were constructed on the ruins of the old, only to be eventually destroyed themselves. Thus layer accumulated upon layer (in most cases at least a dozen of them), each representing a different period of peace. Today, such sites can be clearly seen, even from a distance, as conspicuous mounds of earth rising well above their surroundings and disclosing, upon close inspection, shards of pottery, fragments of bone, and an occasional bead. This is a *tell,* an artificial hill where there was once, for a thousand years or more, a thriving community.

All the ancient empires of the Middle East exerted influence upon Palestine. Mesopotamia and Egypt both controlled it at times. So did the Hittites, a people of the Anatolian Plateau, whose capital was near the site of Ankara, in modern Turkey; the Assyrians of Kurdistan, with their capital at Ninevah across the Tigris in modern Iraq; the Babylonians, whose chief city was on the Euphrates south of Baghdad; and the Persians, from present-day Iran, who controlled all the area from India to Ethiopia, then the greater part of the "known world." And in addition to direct military and administrative control, these and other nations exerted influence upon Palestine through trade and commerce.

Many commodities were produced in Palestine for shipment to markets near and far—wines and figs, and olives that provided food and oil for cooking, cosmetics, and lamps. But most of the products traded there were produced elsewhere and used elsewhere. Goods brought from other lands, by camel caravan and by sailing vessel, were sorted, reworked, and sold in Palestine, then shipped away to still other destinations. Ports were developed by various groups, but especially by the Phoenicians, who carried on commerce with all parts of the known world. From the south, from Ethiopia, Sudan, the Red Sea and adjacent parts of Arabia, and from Egypt, they brought slaves, pearls, ivory, gold, gum arabic, frankincense, myrrh, perfumes, and cotton cloth. From the east, from Babylonia by camel caravan and even from India to Babylonia by sea, came handicrafts in metal, leather, and cloth, as well as pearls and exotic spices. Afoot from the north came herds of cattle and horses, flocks of sheep and goats, and caravans transporting wines, wool and woolen cloth, and garments. From the shores of the Mediterranean came a rich purple dye made from a type of snail, as well as wine and olive oil, and copper ore from Cyprus and timber from the mountains of Lebanon, Cyprus, Crete, and Greece. And from far away Cornwall, the southwesternmost extremity of England, by way of the Atlantic and Straits of Gibraltar, came tin ore that was mixed with copper to produce bronze, the wonder metal of the pre–Iron Age.

Many cultures developed in this area, ran their course, and disappeared. Many languages were spoken, and many gods were worshipped. The Judaic religion and the Christian religion both had their origins in this area of the world.

Crete and Greece. Because of their geographical proximity, and because their developments represent an historical continuum, it is logical to consider Crete and Greece together in this discussion. An amphibious landscape dominates each of these areas: long, rugged promontories separate deep, sheltered bays, and the sea is dotted with islands; on a clear day, even in the midst of the sea, land is always in sight. This setting inspired its early inhabitants to maritime interests: there was always a new land beckoning; always a cove to run to for safety, with a sandy strand at its head on which to beach a boat; always warm navigable waters; and always fish to be sought as food. With three continents contributing to its growth, this was a place, too, for the development of a new civilization.

At its peak, from 2000 to 1400 B.C., Cretan (or Minoan) civilization was on a level with the greatest periods of ancient Egypt or Mesopotamia. Its craftsmen produced exquisite pottery, glazed earthenware beads, finely wrought gold jewelry, and polished stone vases of flawless white marble. Architects and builders created great open palaces that could receive the sea breeze on hot summer days; these structures were equipped with running water and sewers and decorated with wall paintings whose elegant colors and flowing lines portrayed graceful animals and children amidst fields of flowers. There were no fortifications; the towns and the palaces were open and unprotected, perhaps because of the relative isolation of the island and the strong naval power of the government. The city associated with the chief palace, Knossos, probably included as many as 100,000 persons—a very large number considering the small size of the island of Crete, its even smaller area of arable land, and the difficulties of transporting food for so many people by sailing vessel and primitive land conveyances. The island's system of writing (still undeciphered) was both inscribed on clay tablets and written in ink on clay and on papyrus.

Cretan civilization came to an abrupt end about 1400 B.C., when the island of Thera (also called Santorin), about 75 miles north of Crete, literally exploded in a vast volcanic eruption. Knossos was instantly destroyed, very likely by the resultant tidal wave and fall of ash.

Well before catastrophe overwhelmed Crete, a strong culture had developed on the mainland of southern Greece, at Mycenae in the area called the Peloponnesus. It was a military culture and thus a marked contrast to the peaceful, artistic Minoan civilization. Its cities were hilltop fortresses; its people the aggressive warriors immortalized centuries later by Homer in his two epic poems, the *Iliad* and the *Odyssey*. At its peak, this culture held sway over all of Greece and the Aegean; but like many another civilization, it eventually mellowed, weakened, and was overthrown by an invading force.

The invaders were the Dorians, nomads from the Balkans who swept southward into Greece about 1100 B.C. They were acquainted with methods of smelting iron, and their weapons of that metal were far superior to the bronze ones of the Mycenaeans. Their attack initiated a period of chaos that lasted for three hundred years, during which time Greece came to be divided into many small, autonomous city-states. Some were Dorian, having been taken over by the invaders; others remained Mycenaean, having either resisted the invaders or worked out a shaky peace with them. In addition, there were Ionian settlements, colonies founded on the Aegean islands by former Mycenaeans.

These city-states would very likely have remained isolated and independent rather indefinitely had not a common enemy appeared. In 480 B.C. a large Persian army attacked the Greek mainland and occupied Athens. Faced with annihilation, some of the city-states unified to combat the Persians and, in 479 B.C., defeated them. Following this unification, Greek (or Hellenic) civilization achieved its Golden Age.

During the brief fifty years of their Golden Age, the Greeks attained one of the highest levels of culture the world has experienced. A new Athens was built on the ruins of the old. The Acropolis, long the citadel above the city, became the religious center; and it remains an architectural model to this day (Fig. 4–23). Knowledge and learning were revered, and the philosopher ranked in importance with the priest, the governor, and the financier. Poetry, literature, drama, sculpture, logic, and science flourished; and democracy, which had its beginnings in Greece, continued—still greatly qualified, however, by the denial of the franchise to slaves and to women.

Seeking to control trade and ease the overpopulation of their homeland, the Greeks established colonies in all parts of the Mediterranean, from Spain to Egypt and the far reaches of the Black Sea. From them, they imported grain and timber and ores; to them, they exported wine, olive oil, and pottery, jewelry, and carvings. And most important of all, they exported their laws, philosophy, and ideals. In this way, Greek culture became the foundation of Roman civilization, and later of European civilization, and eventually all the Western world.

Indus Valley. Pakistan, the country that comprises the Indus Valley, is a fairly new political entity, but its cultural roots lie deep in antiquity. The recent discoveries of the ancient cities of Mohenjo-Daro, in the lower valley, and Harappa, in the upper valley, suggest that, between about 5000 and 1500 B.C., a highly developed civilization

FIGURE 4–23

The Acropolis, Athens, Greece. This flat-topped hill, once a fortified retreat, bears the remains of temples built in the fifth century B.C. Its most prominent structure, the Parthenon, has inspired such relatively recent constructions as the Lincoln Memorial. (Courtesy TWA)

flourished in this arid but once fertile northwestern part of the Indian subcontinent; this was a pre-Aryan culture, created by a people whose name is not now known. Mohenjo-Daro lies buried 40 feet below the sandy plains of the Sind Desert, and attempts to excavate it were first made over 30 years ago. They have now been halted, by order of the Pakistan government, because groundwaters rising in the diggings threaten the mud bricks of the buried structures; it is feared that, if the excavation continues before the water recedes, parts of Mohenjo-Daro may be lost forever. The portions of the city uncovered thus far include wide aligned streets with elaborate drainage systems and multistory buildings with spacious courtyards.

The people of this ancient civilization apparently worked in bronze and utilized an alphabet, which has not yet been deciphered but seems to have influenced many of the languages of Asia and Europe. Sumerian seals of terra cotta found in the excavations of Mohenjo-Daro and Harappa are strong evidence that the Indus Valley people traded with peoples of Mesopotamia. Exactly why the civilization of the cities came to an end about 1500 B.C. is not known, but the decline and fall of the civilization may have been prompted by the invasions of Aryan nomads from southwest Asia—ancestors of present-day Hindus. The Aryans did adopt much of the religion of the Indus people, a religion that is thought to be the basis for modern Hinduism.

North China. Around 1500 B.C.—about the time of the decline of the Indus Valley civilization—another civilization arose in the valley of the Yellow River (Huang Ho), in North China. Like the Mesopotamian and Indus peoples before them, the North China people built large cities, with palaces of wood; one such site has been excavated near Anyang, in northern Honan province. These people also cast fine bronze vessels and inscribed, on bits of bone, the most ancient examples of Chinese writing that have yet been found. This early North China civilization diffused widely, influencing areas from Central Asia to Japan and the Vietnamese shores far to the south. It is perhaps most important to note that North China is the only major cradleland that has a clear cultural continuity through several millennia to the present.

New World Hearths

One of the most remarkable aspects of the ancient American civilizations was their development completely apart from the civilizations of the Old World. Early inhabitants of the Americas apparently did not suspect that the rest of the world existed, and there seems to have been no knowledge of their existence on the part of Europeans and Asians.

Middle America. It is very likely that the peoples of Mexico began growing a few vegetable crops to supplement their hunting and gathering as early as 7000 or 8000 B.C. By 1000 B.C. large ceremonial buildings had been constructed in the area, and by 300 B.C. large cities were developing. Under the Toltecs, the city of Teotihuacán, in the Valley of Mexico, where Mexico City now stands, was the principal scene of such cultural activities as crop-raising and the building of stone pyramids. By 1519, when the Spanish conquistador Cortes arrived in the Gulf of Mexico, Aztecs had supplanted the Toltecs and, controlling much of central Mexico, ruled a formidable empire.

The Indians in northern Guatemala and Yucatán also developed a strong civilization—the Mayan, which reached its peak about A.D. 700. The Maya grew crops, built limestone temples, and created fine paintings and pottery. They also developed a complex calendar, a system of hieroglyphic writing, and several advanced mathematical concepts.

Andes. Excavations along the Andean coasts of northern Peru have revealed evidence that village societies existed there perhaps as early as 2500 B.C., or before. The most enduring of these was the Incan, whose ancient peoples probably lived mainly by hunting and gathering and only secondarily cultivated such crops as squash, beans, potatoes, and maize. Cities developed in the area, but probably no earlier than A.D. 800; and road networks and suspension bridges no earlier than A.D. 900. The early Incas also cast vessels of bronze, gold, and silver, and their weavers made fine textiles of cotton and alpaca wool. Ancient stone walls built, without mortar, of huge fitted blocks still stand within the lands of the Incan Empire as one of humanity's early great structural achievements (Fig. 4–24). The empire itself remained intact until the sixteenth century, when it was destroyed in the Spanish conquest of the Americas.

FIGURE 4–24

Machu Picchu, the "lost city" of the Incas, an ancient stronghold that sprawls along a ridge 2,000 feet high in the Andes Mountains. Its gabled granite-block structures—including palaces, temples, barracks, and houses—cover more than 100 acres and testify to the ingenuity and hard work of their builders. (Eugene Gordon)

Modern Culture Realms

A *culture realm* is a region occupied by a group of similar, interrelated cultures, these usually sharing a multitude of elements, particularly language and religion. But no two authorities will ever agree *in toto* on the criteria to be applied in differentiating culture realms or on their boundaries. Furthermore, any possible classification can become invalid as time passes, even on the basis of its own criteria.

The modern world is usually divided into six major culture realms: Occidental (European), Islamic, Indic, East Asian, Southeast Asian, and African (Fig. 4–25). Each of these can be subdivided and is named for its core area, the region in which its set of distinguishing features first occurred; some contemporary cultures have expanded into areas very far removed from their original locales.

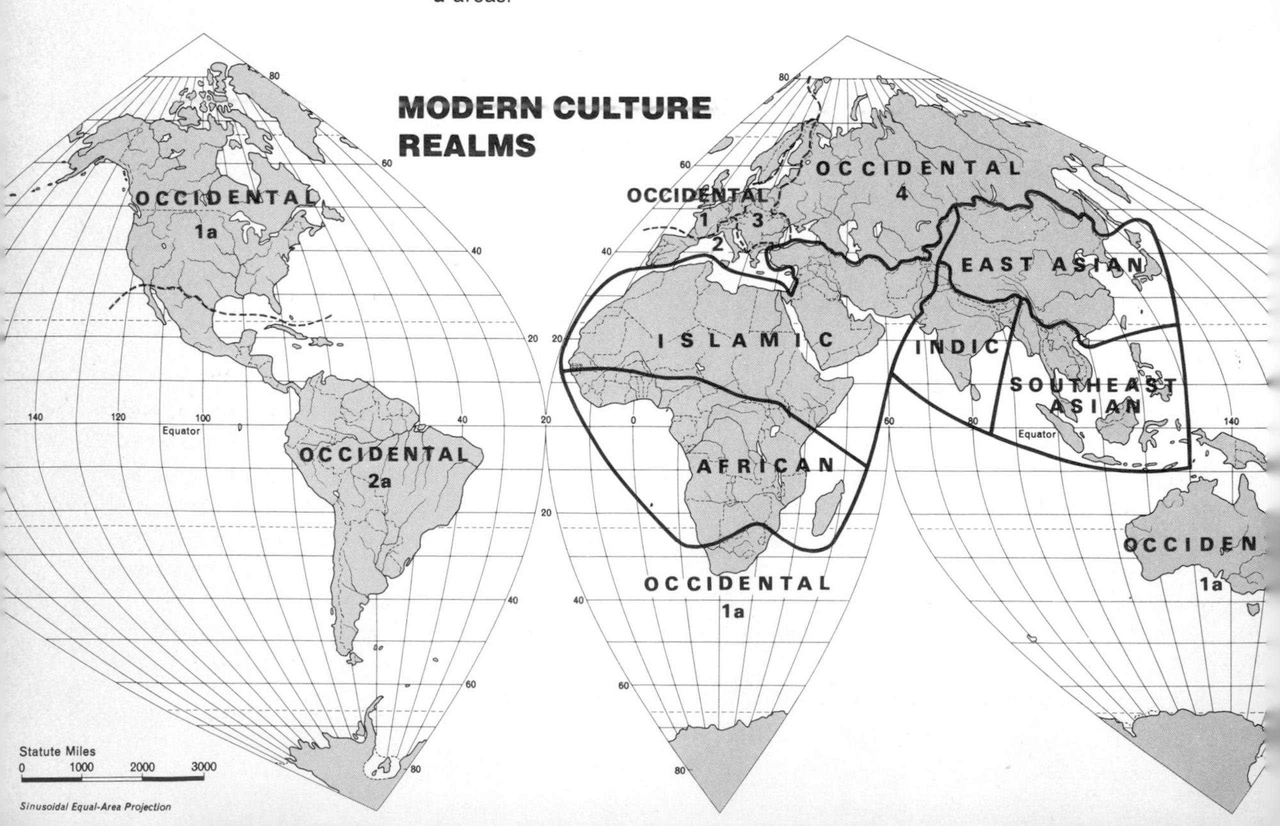

FIGURE 4–25
The most widespread of the six major culture realms is the Occidental, which has several subrealms: 1) the Northwest European, 2) the Mediterranean, 3) the Central European, and 4) the Russian. Two of these subrealms have shaped the present culture of large territories beyond Europe; such territories are indicated on the map as **a** areas.

FIGURE 4–26
An evening street scene in Brussels, Belgium. (© Lynda Gordon)

Occidental Culture Realm

The Occidental (European) realm can be broken down into four subrealms: Northwest European, Mediterranean, Central European, and Russian. The first two of these comprise not only a portion of the European continent but overseas appendages.

Northwest European Subrealm. The homeland of the Northwest European subrealm includes the British Isles, the Benelux countries, West Germany, Switzerland, France, and Scandinavia. And, beginning in the seventeenth century, steady emigrations from this area have resulted in the development of large overseas outliers in the United States, Canada, South Africa, Australia, and New Zealand.

The Northwest European subrealm is characterized by adherence to Protestant and Roman Catholic Christianity and by the use of Germanic and Romance languages. Its governments are, for the most part, democratic and stable, and they possess an international outlook and exert great influence upon the international community. The individual cultures within the realm have long written histories, and their present populations are, overall, highly literate and well educated and have substantial access to the fine arts. These cultures have achieved the world's highest degrees of technological knowledge and thus high levels of industrialization and very high *per capita* incomes. Their diets are relatively high in protein, placing strong emphasis on meats and dairy products, and they utilize wheat more than any other grain and large quantities of vegetables and fruit—thus providing the world's most balanced diets and its healthiest populations (Fig. 4–26).

Mediterranean Subrealm. The Mediterranean subrealm includes Portugal, Spain, Italy, Crete-Greece, the island of Cyprus, and Israel—and its influence has spread overseas to most of Latin America. This Occidental subrealm is characterized by adherence to the Roman Catholic and Eastern Orthodox churches and by the use of Romance and Greek languages. Its countries' governments are at times relatively unstable, being prone to coups d'état, revolutions, dictatorships, and military juntas; and their present interest in international affairs is, for the most part, only regional. The subrealm has an extremely long written history, and maintains a fairly high rate of literacy. Most technological developments are known here, but there is still widespread dependence on manual labor—especially in agriculture and the many cottage industries. On a world scale, *per capita* income is moderately high in the more progressive areas, such as northern Italy, but extremely low in many rural agrarian areas, including large parts of Spain and Greece. In certain circles in the urban areas, there has been a large-scale development of the performing arts, art, literature, and music, but a great portion of the rural population lives with access to only the simplest arts and crafts. Most diets within the region are low in meat and dairy products and high in starches, especially wheat; seafood is plentiful, and there is much use of the olive, the grape, and spices.

Central European Subrealm. The Central European subrealm consists of the Balkan-Slavic cultures that lie between Germany and the Soviet and north of Greece. While many individuals from these cultures have migrated to other parts of the world, Balkan-Slavic peoples have never constituted a dominant population in any country beyond Europe.

The cultures of this realm are characterized by their adherence to the Roman Catholic and Eastern Orthodox churches and by a use of Slavic languages—although a Romance tongue predominates in Rumania and a Ugrian tongue in Hungary. Their governments, long monarchial and retaining many elements of feudalism until well into this century, have now come largely within the Communist sphere of influence; political involvement is largely regional in nature. These cultures have long written histories and, in recent times, fairly high literacy rates. With modest records of technological development, they have achieved a fair degree of industrialization and moderate *per capita* incomes. While the arts are highly evolved in this subrealm, a very large portion of the population lives in areas far removed from any artistic center and, therefore, encounters only rustic arts and crafts. The typical human diet in the subrealm is high in starches derived from potatoes, maize, wheat, rye, oats, and barley, but also contains moderately high amounts of proteins, from meat and dairy products, and root crops and leafy vegetables.

Russian Subrealm. The Russian subrealm once included only the area of the present Soviet west of the Urals. However, it gradually extended eastward during the eighteenth and nineteenth centuries and, since the 1930s, has vigorously encompassed all of Soviet Asia. In some areas, this expansion has been through essentially original settlement, but in other areas, the Russian culture has been superimposed over indigenous cultures—sometimes spontaneously, sometimes by governmental edict.

Modern Russian, or Soviet, culture is officially atheistic, although strong feeling toward traditional Eastern Orthodox Christianity still remains in some circles; its several languages are all Slavic. Soviet government is strongly socialist and pervades every facet of its citizens' public and private lives; the central government of the Soviet Union is also heavily involved in international affairs on a worldwide scale. The subrealm has a long written history, and its literacy rate has been pushed to very high levels during the last generation. Modern technology is widespread, both in agriculture and in industry, and, though difficult to measure in a Communist system, *per capita* income is now fairly high. The general population's involvement in the arts has also been increased tremendously in the last two generations, and thus a very sizable portion of the working class is exposed to most contemporary arts. Human diet throughout the subrealm is dominated by starchy foods, especially potatoes, and is relatively low in meat and dairy products; wheat and rye are the major bread grains, and roots the major vegetables.

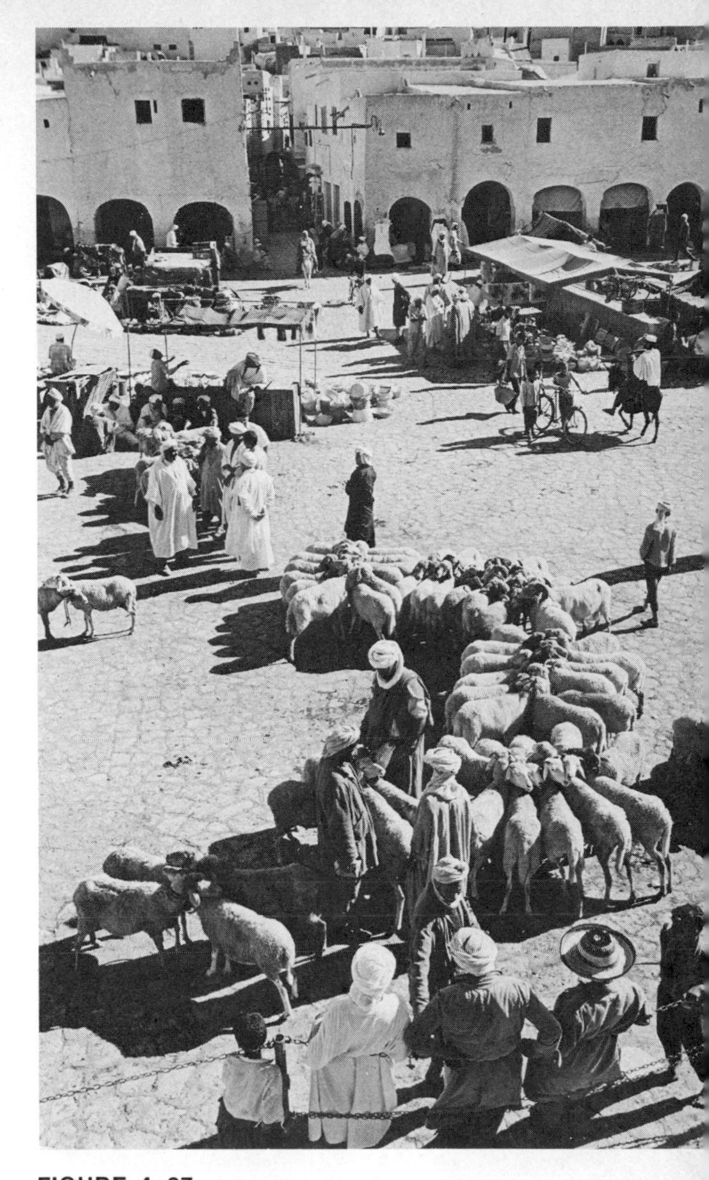

FIGURE 4–27

Ghardaia, an oasis market town in north-central Algeria. (Klaus D. Francke from Peter Arnold)

Islamic Culture Realm

The Islamic culture realm is associated primarily with the arid lands of the Old World, for its areas extend from Morocco across northern Africa to Turkey, the Arabian countries, Iran, and Pakistan (Fig. 4–27). Once several outlier regions in central Asia were also included, but the extension of Soviet influence since the 1930s has essentially removed them from the Islamic realm. Thus the realm's present subdivisions are: its Arabian core region; a Berber-European area in northwestern Africa; Negroid areas in western Mali and the Sudan south of the Sahara; Turkey; Iran; and Pakistan.

The *Arabian subrealm* adheres strongly to the Islamic, or Moslem, religion and speaks Arabic, a Semitic language. Political attitudes are relatively naive and simplistic, and the outlook of many is extremely provincial; only in regard to marketing petroleum and resisting an Israeli presence in the Middle East is there substantial interest in international affairs. These concerns are also the only ones that generate any semblance of unity among the subrealm's otherwise feuding nations and sheikdoms, despite much vocalization in support of Arab unification.

Although it has a very long written history, and in ancient times was a seat of philosophy and literature, the subrealm now possesses a very low literacy rate. Technological research, mechanization, and industrialization are virtually absent, except where introduced by foreigners in pursuit of the area's abundant petroleum. Economic wealth is poorly distributed, and there is no sizable middle class to bridge the gap between the poverty-stricken, illiterate majority and the extremely wealthy aristocracy. Poorly balanced, the subrealm's typical human diet consists mainly of rice, wheat, and dates, with small amounts of milk and milk products and little meat. Alcohol and pork are forbidden by Islamic religion.

FIGURE 4–28

Silver Street, in the heart of Old Delhi, India. (© Screen Traveler from Gendreau)

Indic Culture Realm

The Indic culture realm is roughly coincident with the countries of India, Bangladesh, and Sri Lanka. Its many religions are polytheistic, highly spiritual and mystic, and deeply concerned with the passage of the soul through a series of reincarnations. The use of many languages and dialects in this realm poses great problems in its development of unified nations. The Indian subcontinent long consisted of many separate, and largely autonomous, states. Only since 1947, when it achieved independence from British rule, has India been completely unified under a fairly stable central government. The country's international involvement is still usually minimal, partly because of its many internal problems.

India and the rest of the Indic realm have a very long written history—certainly longer than most European cultures—but are nevertheless beset by widespread illiteracy. The realm has also achieved very little technological development and mechanization and only a relatively minor degree of industrialization; *per capita* income is extremely low, and few people can pursue an interest in the arts or leisure activities (Fig. 4–28). Indic populations are segregated by birth into several socioeconomic strata, or castes, from which there is little escape. The bulk of the people are impoverished and receive inadequate medical attention. Their diet, obtained usually at the minimum subsistence level, consists mainly of starchy foodgrains—rice, pulses, chick peas, lentils, millet, and wheat; and oils—peanut, coconut, sesame, and linseed. It is largely lacking in meats and in dairy products, in part because of religious restrictions. Most Indic dishes are heavily seasoned with a wide range of spices, and their exotic flavors have gained them worldwide acceptance.

East Asian Culture Realm

China, Japan, Mongolia, North and South Korea, and Taiwan comprise the East Asian culture realm, which thus includes about one billion people, or one-quarter of the world's human beings. None of these countries has been under the direct control of a non-Asian power, but the realm has been marked in modern times by turbulent politics and ideological conflict. Warfare and political unrest were hallmarks of the rise and fall of Japan as a major Asian force a generation ago, and since then the ascendance of Communist China has had similar effects within East Asia and beyond. Despite their similarities in race, religion, language, and life styles, the countries in this realm have often been far from harmonious neighbors.

For centuries, while Western societies were experimenting with new ideologies and new technologies, China was the "sleeping dragon" of the East. Since 1949 it has been undergoing a socioeconomic awakening under a Communist government; the country still has undeveloped natural resources that can one day bring it tremendous wealth (Fig. 4–29).

FIGURE 4–29

A commercial street in Peking, China. (© Georg Gerster, Rapho/Photo Researchers)

Abundant arable land is not one of them, however; most of the land suitable for food production—10 to 15 percent of China's total and located mainly in the eastern third of the country—is already in use. Thus, many observers question whether China can long feed its growing society, a population expected to include 1.5 billion persons by the end of the century. In both agrarian and industrial pursuits the Chinese are already highly capable, but many substantial technological advancements will have to be made in the near future if so large a population is to continue to experience economic growth.

Japan and Korea have both derived much of their cultural heritage from historical China. Traditional Japanese life styles bear distinct imprints of borrowings from China—often carried by means of the Korean land bridge. Examples of this cultural diffusion include Japan's acceptance of both the Buddhist religion, in the sixth century by way of China and Korea, and China's Confucian philosophy of family relationships. And, perhaps even more important, the Japanese borrowed and adapted the basic characters of Chinese in developing a written form of their own language. It should be remembered, however, that these borrowings were made in the distant past and that, long before their first contacts with modern Western cultures, the Japanese had developed a distinctive society of their own—like, yet unlike, the Chinese.

Southeast Asian Culture Realm

Stretching nearly 4,000 miles—from Burma through Indochina and across the vast archipelagos of the Philippines and Indonesia to the shores of New Guinea—and including several extremely diverse peoples, the Southeast Asian culture realm is a truly unified region in name only. Different parts of this realm are almost as different from one another culturally as they are from the surrounding cultural realms. What cultural unity they possess is largely a matter of common indebtedness over the millennia to their mainland neighbors. Most of Southeast Asia's languages, religions, political systems, art, and music had their origins in India and China; and much of the remainder is European or American. Each of the countries in the Southeast Asian realm—excepting Thailand, that is—was for long periods part of one or another of the several European overseas empires of the sixteenth to twentieth

FIGURE 4–30

Stalled traffic in Sumatra, Indonesia. (© Georg Gerster, Rapho/Photo Researchers)

FIGURE 4–31
The village of Torke in Ethiopia. (United Nations)

centuries, and many Americanisms were introduced when the region became a principal Asian battleground of the Second World War. Thus in the Southeast Asian realm East truly meets West, producing much chaos but also infinite richness and variety (Fig. 4–30).

African Culture Realm

The coasts of central Africa became known to Europeans in the fifteenth and sixteenth centuries, but explorations of the interiors did not make much progress until near the close of the eighteenth century. Thus the development of Western social systems in the African cultural realm is fairly recent. For some African peoples the change from tribal isolation to European colonialism to sociopolitical independence occurred within a human lifetime. The vast territory of central Africa was not unknown to foreigners before the European arrivals, however. A considerable Moslem influence preceded the Europeans. The most developed areas of the realm when intense European colonization began, about 1885, were probably areas of West Africa along the Niger River, areas within the tract of savanna bounded by the Sahara to the north and the forest lands to the south. Today European colonial power has dissipated, and independence prevails over most of the African realm. However the disposition of sociopolitical powers inherent to newly independent states will probably continue to generate serious conflicts and problems throughout the area for some time to come.

The African culture realm constitutes one of the world's major underdeveloped regions; agriculture is not extensively mechanized, and the most successful industrial operations have been those undertaken on a rather small scale (Fig. 4–31). Like virtually all parts of the modern world, the area is also undergoing a major cultural revolution. Its cultural change is characterized by a new acceptance of foreign ways and products. Chief among these is the pursuit of socioeconomic stability through planned agro-industrialization.

REVIEW AND DISCUSSION

1. What is the meaning of the term *race* in cultural geography? What type of features are commonly cited as distinguishing the races? Evaluate the claim that there are probably no "pure" racial groups.
2. A relatively simple racial classification is widely used today in scientific study. Characterize each of its three major divisions, and indicate some of the probable reasons for their major differences.
3. Discuss the origins and present world distributions of the three major races; include references to pertinent geographical patterns, involving, for example, continents, nations, latitudes, or climates. Where would one find "classic" examples of each race in large numbers?
4. Name and characterize the nine major linguistic families. Summarize the global distribution of each family.
5. Discuss the geographic distribution of the Indo-European language family. Trace the diffusion of three or four of its principal languages.
6. Describe several of today's "classic" bilingual societies; include one example for each of the following continents: North America, South America, Europe, Africa, and Asia. What do these examples show about the relationship between bilinguality and race or bilinguality and culture?
7. Discuss the three major monotheistic religions collectively and singly with respect to their origins, tenets, and regional and national patterns.
8. Discuss the two major polytheistic religions collectively and singly with respect to their origins, tenets, and regional and national patterns.
9. Incorporating as much recent data as possible, discuss the positions of religion within the Soviet Union and China. Do you see any strong relationships between religion and socioeconomic development? Explain.
10. Describe the physical environments of each of the major culture hearths. What features do these environments have in common?
11. Compare and contrast the Old World's four "river-valley" civilizations, with particular attention to development, diffusion, and worldwide influences. How did these river-valley civilizations differ, if they did, from the two non-river-valley civilizations of the Old World?
12. Compare and contrast the two principal New World civilizations with the two non-river-valley civilizations of the Old World, with particular attention to their development, diffusion, and worldwide influences.
13. What is a *culture realm?* How does it differ from a *culture hearth?* In what ways can culture realms be considered extremely dynamic, and of what importance is this dynamism in the use of the culture realm as a research and study tool in cultural geography?
14. What are the distinguishing features of each of the modern world's six major culture realms?
15. The Occidental (European) culture realm is quite extensive. Justify its claim to each of its four European subrealms. Summarize also the characteristics that make each of these four subrealms distinctive.

SELECTED REFERENCES

Benson, P. H. *Religion in Contemporary Culture.* New York: Harper & Row, 1960.
Black, M. *The Labyrinth of Language.* New York: Mentor, 1968.
Chiari, J. *Religion and Modern Society.* London: Jenkins, 1964.
Coon, C. S. *The Origin of Races.* New York: Knopf, 1962.

Coulborn, R. *The Origin of Civilized Societies.* Princeton, N.J.: Princeton University Press, 1959.
Fellows, D. K. *A Mosaic of America's Ethnic Minorities.* New York: Wiley, 1972.
Fisher, C. A. *Southeast Asia: A Social, Economic and Political Geography.* London: Methuen, 1966.
Fisher, W. B. *The Middle East,* 6th ed. London: Methuen, 1971.
Fishman, J. A. *Socio-Linguistics: A Brief Introduction.* Rowley, Mass.: Newbury House, 1971.
Hawkes, J., and L. Woolley. *History of Mankind: Cultural and Scientific Development,* Vol. 1. London: Allen and Unwin, 1963.
Hoebel, E. A. *Anthropology: The Study of Man,* 4th ed. New York: McGraw-Hill, 1972.
Hymes, D. *Language in Culture and Society.* New York: Harper & Row, 1964.
James, P. E. *Latin America,* 4th ed. New York: Odyssey Press, 1969.
Lenski, G. *The Religious Factor: Sociological Study of Religion's Impact on Politics, Economics, Family Life.* New York: Doubleday, 1961.
Lerner, M. *America as a Civilization.* New York: Simon & Schuster, 1957.
MacFadden, C. H., H. M. Kendall, and G. F. Deasy, *Atlas of World Affairs.* New York: Crowell, 1946.
Murdock, G. P. *Africa: Its Peoples and Their Cultural History.* New York: McGraw-Hill, 1959.
Nash, M. *Primitive and Peasant Economic Systems.* San Francisco: Chandler, 1966.
Noss, J. B. *Man's Religions,* 3rd ed. New York: Macmillan, 1963.
Oppenheim, A. L. *Ancient Mesopotamia.* Chicago: University of Chicago Press, 1964.
Pei, M. A. *The Story of Language.* New York: Lippincott, 1965.
Sapir, E., and D. Mandelbaum, eds. *Selected Writings in Language, Culture and Personality.* Berkeley: University of California Press, 1951.
Simoons, F. S. *Eat Not This Flesh.* Madison: University of Wisconsin Press, 1961.
Sopher, D. E. *Geography of Religions.* Englewood Cliffs, N.J.: Prentice-Hall, 1967.
Srinivas, M. N., and A. Beteile. "The Untouchables of India." *Scientific American,* Vol. 213 (December 1965), 13–17.
Welch, G. *Africa Before They Came.* New York: Morrow, 1965.
Willey, G. R. *Prehistoric Settlement Patterns in the New World.* New York: Viking, 1956.

Homage to a chieftain in Uganda. (George Rodger/Magnum Photos)

5

Societies and Political Order

Political order is basic to cultural development; for without the rule of law and some governmental structure there can be no civilization, only chaos and anarchy.

It is evident that since early in their existence, and long before recorded history, humans have often surrendered much of their individual freedom to the will of society. In fear of numerous animals and environmental hardships, they have usually chosen to live in organized groups, with recognized leaders and rules of behavior. While they do limit the freedom of the individual, such rules have always been regarded as essential to the survival of the group. Their adoption marks the actual beginnings of government, government being no more than the machinery through which the rules of a society are enforced. And, once established, the subjection of the individual to political authority continued to increase in scope and complexity.

Human political groups have taken a variety of forms, of which the *family* is generally believed to be the oldest and most enduring. As families came into contact with one another, they often coalesced to form *clans,* more extensive units that were headed by a chieftain or group of elders. A still broader unit was the *tribe,* also headed by a chieftain or council. In most instances the cohesive element in these early political units was a common ancestry, language, or religion, coupled with geographic proximity and/or an awareness of the advantages to be gained through cooperation and centralized leadership.

It was not until the early nomadic groups settled into more sedentary ways of life that the ownership of definite territory became a strong determinant of human political units. With this new-found attachment to land, small agricultural communities began to appear in those regions of the world having warm climates, adequate water supplies, and fertile soils. Such geographical conditions were found, for example, in the valley areas of the Tigris-Euphrates, Yangtze, Huang, Nile, Ganges, and Indus rivers. The agricultural communities located in these areas and others like them became the earliest examples of the *city-state,* that is, an independent political entity made up of a city and its surrounding tributary region. This kind of central organization allowed some of the area's inhabitants to accumulate wealth and to expand their lands and power through trade and commerce. In short, in such areas control of territory came to equal political and economic power.

The State: Territory and People

The concept that there exists a bond between a group of people and the territory they occupy is a basic one. And, along with the other major cohesive elements of early political units, it is also a basis for contemporary political organization. Today, the entire land surface of the earth is sharply divided into politically organized territorial units referred to as *countries, states,* or *nations.* The people inhabiting each of these land units are also frequently classified in terms of political organization; that is, an American is distinguished from a German, a Frenchman from a Luxembourger, a Thai from a Burmese. The life style followed by each—including clothes, foods, currency, laws, and manner of speech—is generally a reflection of the particular country of which that individual is a citizen.

The State and the Nation

Although considered synonyms, the terms *state* and *nation* should be used with some degree of differentiation. While *state* properly connotes political unity, *nation* properly connotes cultural unity—in effect, then, each word represents a distinct and essentially different type of social bond.

The State. Formally, the modern state is comprised of specific territory and of a body of people who, while often economically and sometimes culturally varied, are bound together politically on many significant levels. Statehood is achieved only after the people occupying a definite territory have been united under a single, *sovereign* government, that is, one officially possessing supreme authority over all of the territory's domestic and foreign affairs. In practice, of course, the degree of authority a government has over the affairs of its territory varies with the relative strength or weakness of the state in the world community.

The Nation. A nation, or nationality, on the other hand, is a body of people who may be only bound together culturally. Several nations may be represented in a single state. Belgium, for example, includes two distinct national groups, the Flemings and the Walloons. Or, a nation may extend beyond the territorial limits of a particular state. For example, the nation of the Kurds is divided politically by several states, including Iran, Iraq, and Turkey (Figs. 5–1 and 5–2). The people of a state owe their allegiance to its sovereign power; the people of a nation can find unity in their sense of "belonging together," even though they may be separated by political boundaries. In short, the modern state is a political community; whereas the nation is a population group characterized by cultural ties, most frequently formed around language, race, religion, place of origin, and/or historical experiences and traditions.

FIGURE 5–1

A Kurdish tribe approaching the mouth of the Prince Ali Gorge in Iraq. The Kurds are a pastoral, tent-dwelling people. (United Nations)

The National State

Ideally, a state should be a reflection of a nation; that is, it should be composed of a single population group, which, for various reasons, regards itself as constituting a unified political community. In this sense, the prime objective of "nation-building" is the achievement of the point at which the membership of the state coincides with that of the nation. The state would then be the political expression of the nation, the instrument through which the nation could protect and assert itself. Such a political unit would properly be called a *nation-state.*

As suggested previously, the nation-state concept is an ideal; it is not demonstrated by the majority of the world's political units. Even a cursory examination of current political organization worldwide shows that most states are multinational.

In most cases, therefore, the state

FIGURE 5–2

The core of the Kurdish homeland, the lower slopes of the Taurus Mountains of eastern Turkey and the Zagros Mountains of northwestern Iran. The Kurds dwell also in the interiors of this area's contiguous states—Iran, Iraq, the Soviet Union, Syria, and Turkey.

requires a more exact loyalty than that given the nation. Lucky is the state that has a homogeneous population and no "nationality problems." And even luckier is the one that, usually with the help of a responsive and efficient political system, has been able to unify its people despite their national differences.

When a state develops a nationality of its own, its population is said to have a "concept of nation"; this means that a significant majority of the citizens of the state are aware of their government, prefer it over others, and identify themselves with its policies. The members of such a political unit place loyalty to the group as a whole above any internal political loyalties.

Nationalism. The positive feelings for the modern national state expressed by its citizens are usually described as *nationalism,* a somewhat nebulous term that has been defined as follows:

> At its most elementary level, nationalism appears as a conversion and enlargement of the concept of kinship: the individual identifies himself with, and therefore gives willing support to, the political organization and region which he feels belongs to him, and to which he belongs, because the people of all that region and those operating the organization are kin to him. Not that they are kin in the literal sense of common biological origin, but in the sense more meaningful to him that they are like him (Richard Hartshorne, *American Geography Inventory and Prospect* [Syracuse: Association of American Geographers and Syracuse University Press, 1954], p. 193).

Europe Sets the Pace. The rise of the modern national state and of nationalism, which accompanied it, are relatively recent phenomena, first appearing as major political forces in Europe. There, they evolved from the feudal system that developed and flourished in the agricultural life of the Middle Ages. Characterized by extreme territorial fragmentation as well as a near absence of any contact, even between neighboring communities, feudalism divided land into sections owned and ruled either by members of the nobility or the clergy. These feudal landowners, or lords, allowed others to live on their lands, subject to their control, and for a share in the fruits of their labors. Although they may have sworn allegiance to a king, the lords ruled their estates independently, usually from within the security of fortified castles (Fig. 5–3). The king, a landlord himself, collected taxes from the other lords and led them in battle against foreign adversaries, but he was not the direct authority in the lives of most of his subjects. A tenant's first allegiance was to the landlord, and it was the landlord and the laws of the estate that most influenced a tenant's economic and personal life.

Such a system could not cope with the growth of commerce. The expansion of trade required a single, portable currency and guaranteed rights-of-way. It also produced a commercial class, whose members found it profitable to live close together in towns, and whose landlords were paid only in money, not fidelity. As trade increased within a realm and eventually extended beyond its borders, and as people became less tied to the land, the need for a centralized authority and source of identity became greater.

The national state filled that need. Based on cultural ties, it gave a more mobile population an expanded geographical identity. And since it drew on a much wider base for its financial support and directly controlled its army, the nonfeudal state surpassed feudalism in maintaining internal order and repelling foreign invasion. The basic change, of course, was the centralization of domestic power, which had long been fought by the feudal lords. With this change the state assumed responsibility for a

FIGURE 5–3
A scene from a fifteenth-century French manuscript depicting farm laborers on the grounds of a feudal estate. (Giraudon)

wide range of important human concerns, including the establishment of uniform laws and the administration of justice. Most of the new states were absolute monarchies, but these gradually gave way to parliamentary governments and to the extension of the franchise to more and more members of the society (Fig. 5–4).

Among the first national states to emerge in Europe were those of France, England, Spain, Switzerland, the Netherlands, and the Soviet Union. Among the last were Italy and Germany, both of which remained divided into small principalities and dukedoms until very late in the nineteenth century, at which time they achieved unification under the leadership of centralized monarchies. In present-day Europe there are still some reminders of feudalism existing in the form (if not the administration) of such "ministates" as Andorra, Liechtenstein, Monaco, and San Marino (Fig. 5–5). The political unit that predominates is the national state, which, since its European beginnings hundreds of years ago, has spread to nearly every part of the world.

Threats to National Unity

It is virtually impossible for any modern nation-state, even those in Europe, to remain totally unified at all times. At one time or another internal frictions are bound to develop; some may be resolved peacefully, others may erupt into riot, or even civil war. Whatever the results may be, this is never-

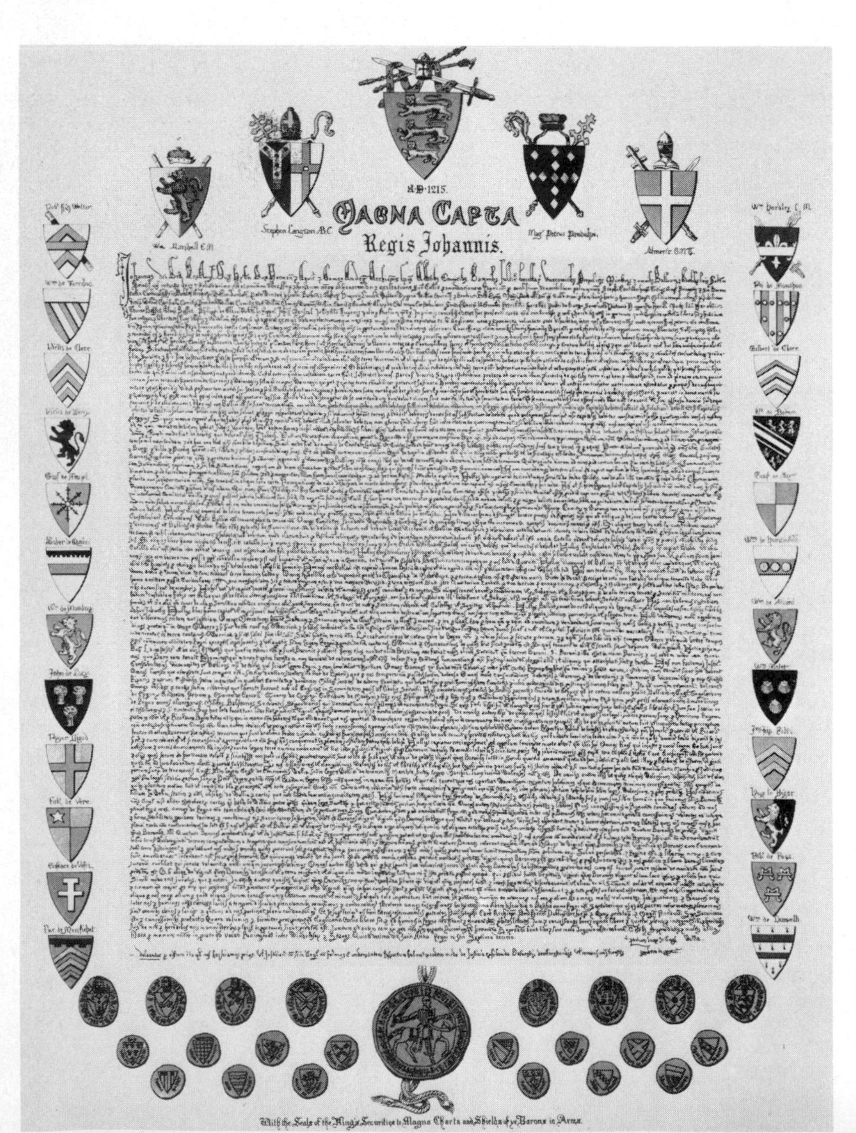

FIGURE 5–4

Magna Carta, the vital document to which King John reluctantly affixed his seal in 1215. Essentially a guaranteed "bill of rights" for England's nobles, it limited the legal powers of the king and is regarded as a landmark in the long development of the British "constitution." The words are Latin. (Bettmann Archive) theless true for the newest states as well as for the oldest ones.

Communalism. When individual citizens of a country identify themselves with, and give their primary loyalty to, a racial or cultural group rather than to the state, they are said to have a *communal outlook.* This so-called *communalism,* based primarily on religion, was instrumental in 1947 in the division of the Indian subcontinent into Moslem-dominated Pakistan and Hindu-dominated India. In those countries that are comprised of several nationalities, or subnations — such as Switzerland, the United Kingdom, Belgium, Yugoslavia, Canada, India, and Sri Lanka (Ceylon), to mention but a few — communalism is often the most obvious potential threat to national unity.

Tribalism. In general, tribal groups do not strongly identify with the states in which they live. They possess their own sets of laws and customs and are usually located in remote areas of their states, far removed from population centers. As a result, the state that contains tribal groups cannot be totally united until it has penetrated their areas and gained the strong support of, or control over, the powerful, rural tribal chieftains.

In several parts of Africa, tribal participation in the state is unusually strong. Some African governments are dominated by a single, large tribe, with varying participation by one or more smaller tribes. Dissatisfaction on the part of one or another of the less-powerful tribes is occasionally the cause of much internal strife. Such was the case in Nigeria in the 1960s, when a rebellion by the Biafran minority resulted in a major civil war and their own near-annihilation.

Economic Protectionism. Disharmony seemingly based on cultural differences may actually be, more fundamentally, a confrontation between different economic groups. Those in lower-income groups may join together on the basis of their cultural ties to seek greater governmental power and improvements in their economic condition. The communal problem in Northern Ireland, for example, resulting from the friction between the Catholic minority and Protestant majority, has definite underlying socioeconomic implications.

Regionalism. When sharp cultural and/or economic disparities exist between different areas of a state, they may foster serious regional disunity. A region that feels it is un-

FIGURE 5–5

San Marino, the world's oldest and smallest republic. The hilltop castle shown here was built in the thirteenth century, and it served as a defensible retreat during invasions of the agricultural lands below. (Bob and Ira Spring) fairly treated by the government may demand greater local autonomy. If that demand is not satisfied, then a movement for total independence from the state may ensue.

Separatist movements of varying magnitudes can be found today throughout the world. Significant ones include those within the French Catholic community of the Province of Quebec, the Basque region of Spain, and remote parts of Western Australia.

The State: Organization and Power

Every state reflects a different approach to the exercise of sovereign power. Each has a different influence on the way its people live and the way its natural and cultural resources are utilized. The power of a state depends, to a large extent, on how well it is able to organize and manage its territory.

Territorial Administration

As discussed earlier, there are forces within the boundaries of all states that are capable of threatening national unity. A state's ability to deal with these forces, especially those that develop regionally, depends largely on its system of territorial administration. With few exceptions, the territory of most national states is divided, for administrative purposes, into smaller units. For example, the states of the United States, India, or Mexico, the provinces of Spain, and the republics of the Soviet Union represent the highest order of civil division within their respective national states. Although all such administrative subdivisions within a single state are totally subordinate to its government, some may be given special status.

In order to maintain as much national unity and power as possible, civil administrative divisions are delegated some degree of local autonomy; the amount depends on the internal conditions of the state and on its type of government. At one extreme, a state can have a *unitary* form of government, that is, a form in which authority is highly centralized; at the other, it can have a *federal* form of government, in which authority is highly decentralized.

Whatever the state's official form of government, adjustments toward greater centralization or greater decentralization of power are frequently made during its lifetime in response to changing problems. In a crisis, such as war or natural disaster, the national government may very quickly take over a far more extensive range of powers; in a national emergency, the central government of India, for example, has the constitutional power to take over complete control of any of its states and put it under presidential rule. On the other hand, during periods of relative stability, the national government may give more power to its subdivisions, thus implying that regionalism, though probably not eliminated, no longer threatens the state with disunity. In most instances the unity and power of a state depend on the achievement of a proper balance of authority between the national government and its civil divisions.

Unitary States. Civil divisions with very little authority are characteristic of states that have a unitary system of government; such a system is best suited to populations that are homogeneous and cohesive. However, even a state that is essentially centralized may allow a significant degree of local autonomy. The government of the United Kingdom, for example, though primarily unitary, grants significant local autonomy to many of its civil divisions, notably Scotland, Wales, and Northern Ireland. China, the largest unitary state in the world, has five major "autonomous regions," including Inner Mongolia, Kwangsi, Sinkiang, Ningsia, and Tibet. None of these once-rebellious areas was granted autonomous status, however, until it was deemed to be securely under Chinese control.

Federal States. Civil divisions with the greatest degree of authority are normally found in those states with federal systems of government, such as Brazil, Switzerland, Canada, Australia, and the United States. However, just as some unitary states have made adjustments toward greater decentralization, some federally organized states actually function more like centralized, or unitary, ones, as is illustrated by the governments of South Africa, Yugoslavia, and the Soviet Union.

In some states, the administrative subdivisions each represent a different national group. For example, India under Nehru reorganized its administrative divisions to correspond to its major national linguistic groups. Only time will tell whether this will help to unify or divide the Indian people in their attempt to build a national state.

Socioeconomic Organization

Typically, only a small percentage of its total territorial area furnishes most of the important socioeconomic elements that allow a state to function. This vital percentage, the *ecumene*, generally includes the country's densest communication and transportation networks, its chief cities and largest aggregates of people, the major concentrations of natural resources and industry, and the best agricultural land. For example, the vast open spaces of Siberia contribute relatively little to the maintenance of the Soviet Union, which depends more strongly on the well-developed region around Moscow and other parts of Soviet Europe.

The areas that make up the ecumene may vary considerably in their contributions to the nation. Those that are the most effective are commonly referred to as *core areas* (Fig. 5–6). They are usually the nuclei from which the state grew, and, frequently, the national capital is located in or near one of them (Fig. 5–7). As time passes, however, it is not unusual for core areas to develop in other parts of the state. Thus, for example, in addition to the core area around Sydney, Australia also contains core areas near Melbourne and Perth. Each state, within its own territorial confines, has environmental characteristics that differentiate it from other states. Some, due to their location or size, are endowed with an abundance of inherent wealth; others are not as fortunate. However, no matter how potentially rich a state may be, its actual wealth and power depend on how well its territory is organized and managed.

Today, the national government of each state has the chief responsibility in formulating and executing a national development policy and in determining national priorities. In short, the national government can, and usually does, set the stage for progress or decline in its territory. To this extent, the national government and its leaders bear supreme responsibility for the quality of life within the territory.

Whether a state is young or old, it may

be quite skillful, or rather inept, at the fine art of governing and at setting developmental policies. Of course, the choice between the available alternatives depends upon each state and its particular ideology.

Those governments that actually own or control various economic means of production, such as industrial plants, transportation facilities, banks, and postal services (communist governments control all economic activities), tend to have a much more direct influence on the direction and speed of national development. Thus, for example, China has surpassed India in governing and in the setting (and implementation) of developmental policies.

In order to strengthen its power and to ensure the well-being of its people, most states pursue national policies that will strengthen relations with areas beyond their own territorial limits. Such is the stimulant that has led to international trade and, in a bygone era, to the worldwide spread of European colonization.

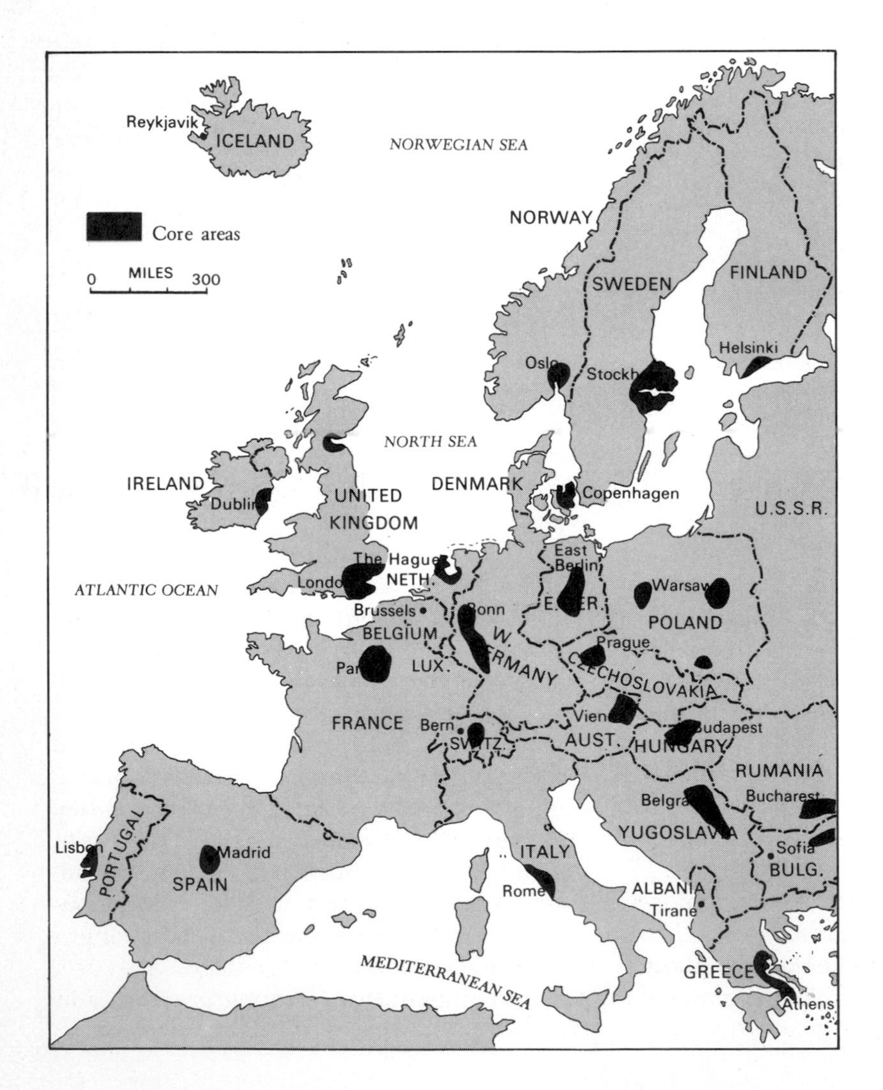

FIGURE 5–6

Core areas of Europe. On this continent, as elsewhere, an economic center is often the political heart of the nation.

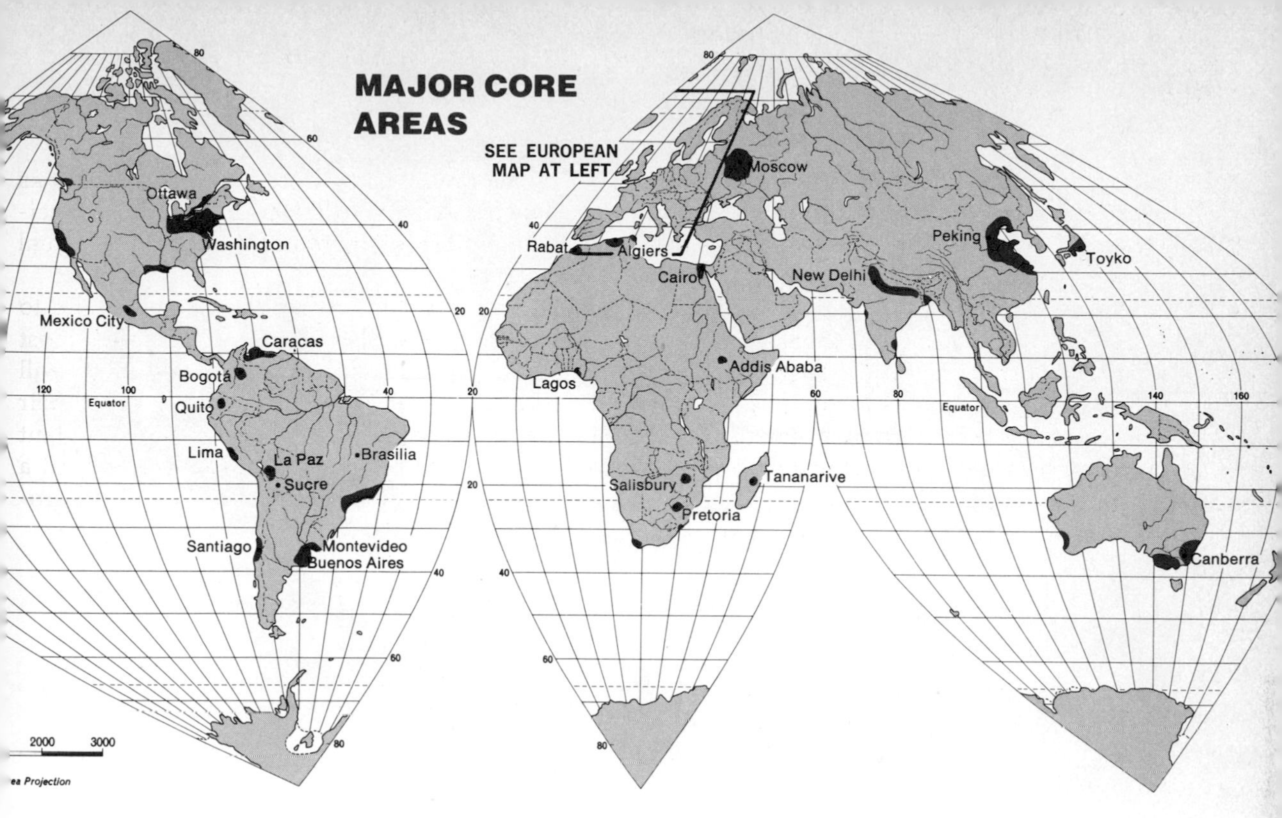

FIGURE 5–7

Few countries of the world have extensive or multiple core areas. In the vast majority of states, the capital city is essentially the only core area.

International Relations

The State System

At present the international community is composed mostly of about 150 independent states or nations, with a small number of other political units of subordinate rank making up the balance (Plate II). The primary goal of each of these is the self-preservation of the unit as a whole along with the attainment of the best possible standard of living for its people. Since, to date, there is an absence of any active or genuine form of world government, each state feels that it must work separately to jealously guard its own national interest. The operation of this territorial methodology is called a *state system,* and it can include relations based on conflict as well as those based on cooperation.

Colonialism

The state, it has been theorized, is like any organism in nature, in that it goes through stages of growth and decline. And, just as the strength of an organism depends on the sustenance it obtains, so the strength of a state depends on territory. A state's strength and ability to grow is thus often in direct proportion to its ability to expand.

FIGURE 5–8
Ruins of the Roman settlement Leptis Magna, on the southern shore of the Mediterranean, in Libya. (Rev. Raymond V. Schoder, S.J.)

Colonialism, an expression of the expansionist tendencies characteristic of many states, involves the control taken by a state of areas and peoples other than its own. Apparently a cyclic phenomenon, it is by no means a new one. Examples of extensive colonization, or "empire-building," occurring long before the most recent wave of European colonialism are numerous. At its zenith, the empire built by the ancient Romans extended far beyond their homeland and subjugated peoples in Europe, Asia, and Africa (Figs. 5–8 and 5–9). In the Americas, the Aztec Empire also grew by colonial acquisition, and its expansion was not halted until the empire was defeated by the Spaniards in the early 1500s (Fig. 5–10). Today, many independent states, some of which were themselves part of colonial empires within the very recent past, can be said to follow colonialist policies. India and the United States are but two of many examples.

European Colonialism

A unique characteristic of the European period of colonialism, which began in the fifteenth century, was its unprecedented scale. This wave of territorial expansion affected every continent of the world and lasted, with significant strength, for over five hundred years.

This growth was prompted, at least initially, by economic interests; and one of the major influences on the newly industrializing European states was their acceptance of the mercantilist philosophy that the prosperity of a state depends on its ability to sell as much as possible and buy as little as possible. Since opportunities elsewhere in Europe were somewhat limited, each major European state was eager to obtain overseas possessions that would serve as exclusive markets for its expanding economy and provide a duty-free supply of raw materials.

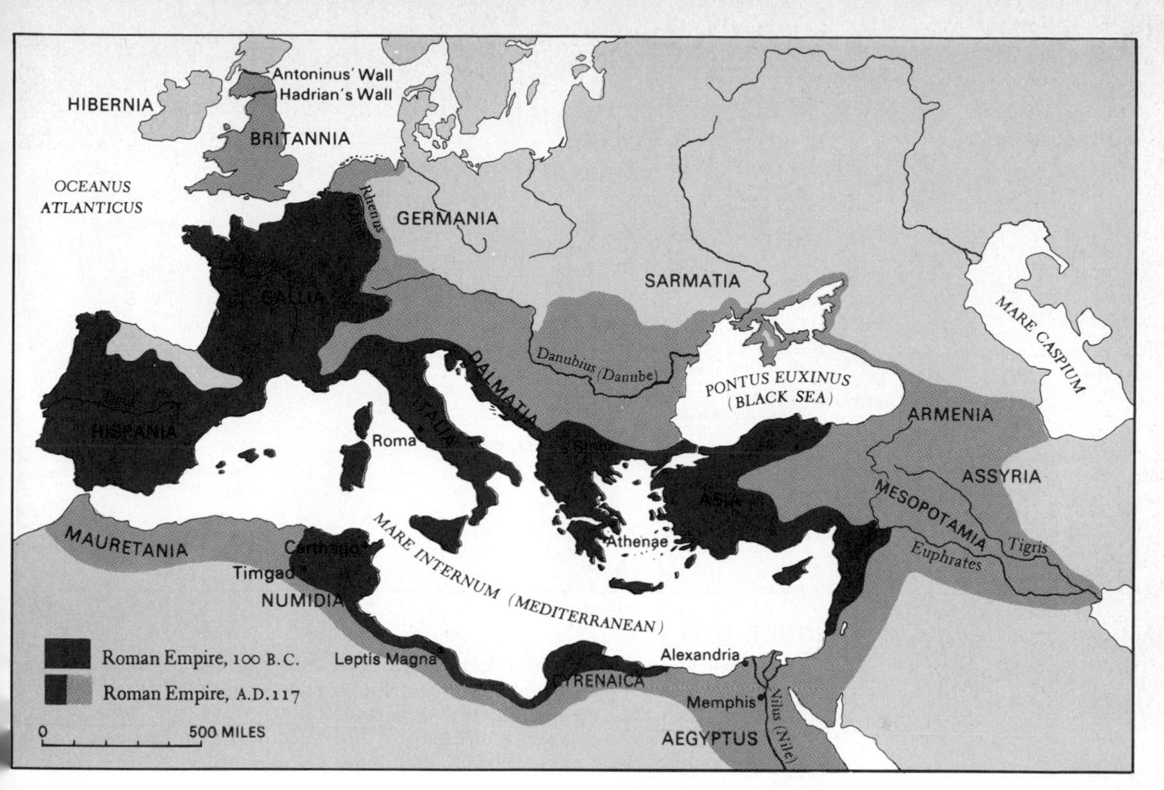

FIGURE 5–9

The Roman Empire. As early as 100 B.C. the authority of Rome extended far beyond Italia; and by A.D. 117, when the empire reached its zenith, Rome controlled the major part of the world then known to Europeans.

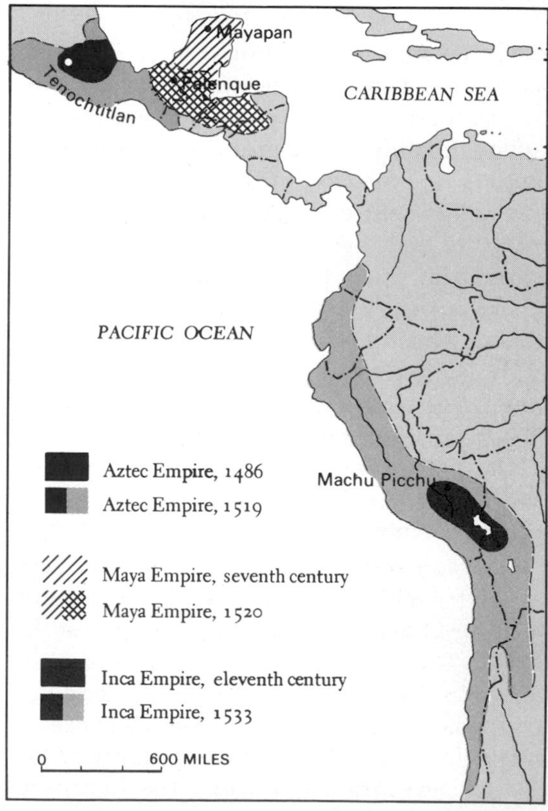

FIGURE 5–10

Empires of Middle and South America in the early 1500s. Each of these empires—Aztec, Maya, and Inca—had grown extensively beyond its original territory by the time the Spanish fleets arrived at their respective shores.

Economic penetration through trade was usually followed by political penetration, and this occurred until almost the entire non-European world was divided among the European powers. The dependencies that resulted represent a wide range of political entities with varying degrees of sovereignty, including such forms as colonies, dominions, condominiums, mandates, trusteeships, protectorates, and spheres of influence.

Usually one thinks of a dependent area

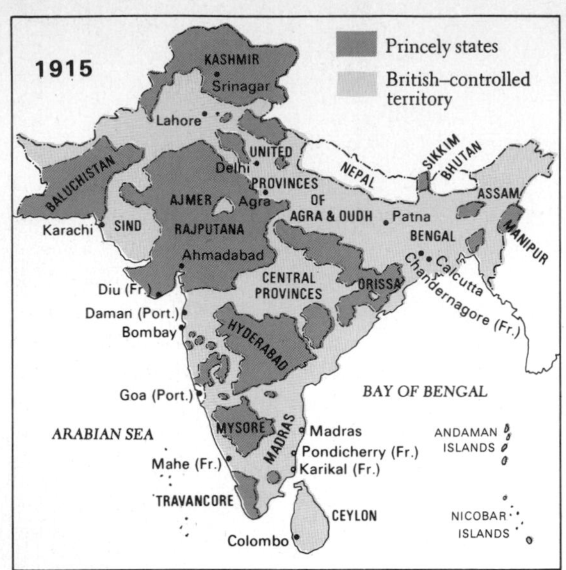

FIGURE 5–11

The nationalization of the Indian subcontinent. Left, throughout the British occupation the subcontinent comprised over 600 princely states. Right, when British rule came to an end in 1947, these princely states were unified into two independent nations, India and Pakistan (West and East). In 1971 East Pakistan became the independent state of Bangladesh, and in 1975 India absorbed Sikkim.

as a noncontiguous territory remote from the *metropole,* or mother country. But many instances of national expansionism have involved not only overseas areas but areas contiguous to the expanding state as well. Several present-day states have dependent areas located even within their national boundaries; such areas—among them the Yukon Territory in Canada, Amapa in Brazil, and the Northern Territory in Australia—often have limited self-governing powers, and they are generally administered directly from the national capital.

Decolonization and New States

Colonialism has been the primary means by which Western European patterns of culture have been diffused throughout the world. The political consequences of this, both good and bad, have been tremendous.

Colonialism, in the traditional sense, is today almost totally nonexistent. Most of the colonial territory acquired by the European powers during the Age of Colonization now comprises independent states; only a handful of colonial dependencies remain. Since 1776, when the United States proclaimed its independence from Great Britain, colonial ties throughout the world have been breaking.

The Spread of Ideologies. In some ways, the mere fact of European colonialism helped set the stage for its decline. For with the Europeans came political unification, including even the joining together of many areas, once noninteracting or at odds, into colonial units. This promoted the development of common interests among some of these previously separate areas, which sometimes led, in turn, to a new and wider sense of political community and purpose (Fig. 5–11). When combined with the spread of such ideas as political equality, democracy, progress, and nationalism—the mainsprings of Western culture—these coalescences made it all but impossible for the colonial powers to refuse demands for self-determination. Still, until about World War I, independence came to European colonies sporadically and slowly.

It was not until World War II that this process was greatly accelerated and intensified. Over 50 percent of the independent states that exist today have come into being since that period; it may be said that the actual beginning of this tremendous increase in the number of sovereign states was Lebanon's declaration of independence, in 1941. Between that year and the present, only a generation after the emergence of the Lebanese nation, over eighty new states have emerged—mostly in Africa, the Middle East, and Asia (Fig. 5–12). In 1960 alone, seventeen new states emerged—all in Africa.

In the meantime, such states as China, Iran, and the United Arab Republic—which, though sovereign in name, were long subject to degrees of control by various foreign powers—strongly asserted their independence; nationalism and the desire for self-determination were finally coming of age in the non-Western world. The process begun such a short time ago continues today, but it does so at a much reduced rate.

Nation-State Building. All nation-state building, in the modern sense, involves not only the creation of a unified central government with effective administrative control over a precisely defined area but also the development of a truly national and relatively uniform culture; it is not easy and cannot be accomplished overnight. Most states grow as a dominant group, or core area, imposes its control and culture over an even broader, and usually more culturally diverse, area. It was through this process that most of the European states developed.

The Global Pattern of Independence. Today's preponderance of sovereign, independent states, tightly packed worldwide, is historically unprecedented; and it has necessitated a total reordering of world politics.

As in the past, there is little reason to believe that the present pattern of state boundaries will remain constant. New states are constantly being born, and existent ones are frequently altered; as a result, inventories of the world's independent states and international boundaries are constantly in need of revision. In 1965 Singapore seceded from the newly formed Federation of Malaysia, and became a completely independent state; in 1971 Bangladesh, formerly East Pakistan, won its independence; in 1975 a separate Turkish-Cypriot state was proclaimed in the Turkish-occupied northern sector of the island of Cyprus.

In addition to those formed as a result of political division, there are new states that

FIGURE 5–12

A Ghanaian woman celebrating the election of Kwame Nkrumah as the first president of the republic. (Camera Press from Photo Trends)

Global Sovereignty

As land dwellers, human groups throughout history have had as a primary concern the establishment of sovereignty over the dry lands of the earth, areas that constitute slightly less than 30 percent of its surface. The modern state, the expression of this desire for territorial sovereignty, is distinguished by boundary lines that, while clearly marked and recognized on maps, are not detectable on the landscape itself, except when they coincide with such natural features as rivers and mountain ranges. Thus, where none of these exist, states establish artificial boundary markers and designate border check points. Of all the dividing lines created for contemporary states, however, none equals the scale of the ancient Great Wall of China (Figs. 5–15 and 5–16) or the walls built centuries ago in Great Britain by Roman occupation forces (see Fig. 3–7).

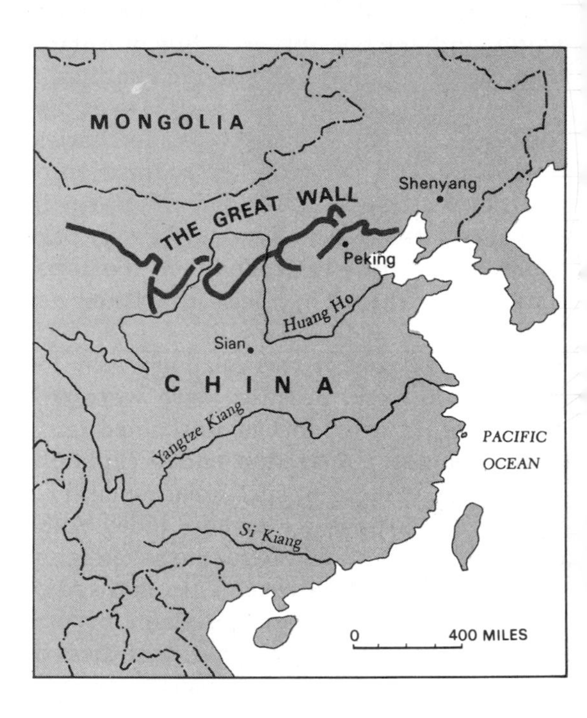

FIGURE 5–15

The situation of the Great Wall, a portion of which is shown in the photograph at left.

FIGURE 5–16

The Great Wall of China. Replacing a wall created in the third century B.C. as a protection against nomadic invaders, the present wall was begun about five hundred years ago, during the Ming dynasty. It is about 20 feet high and 20 feet wide and extends across northern China from Peking to the Kansu Corridor, a distance of approximately 1,500 miles. (Brown Brothers)

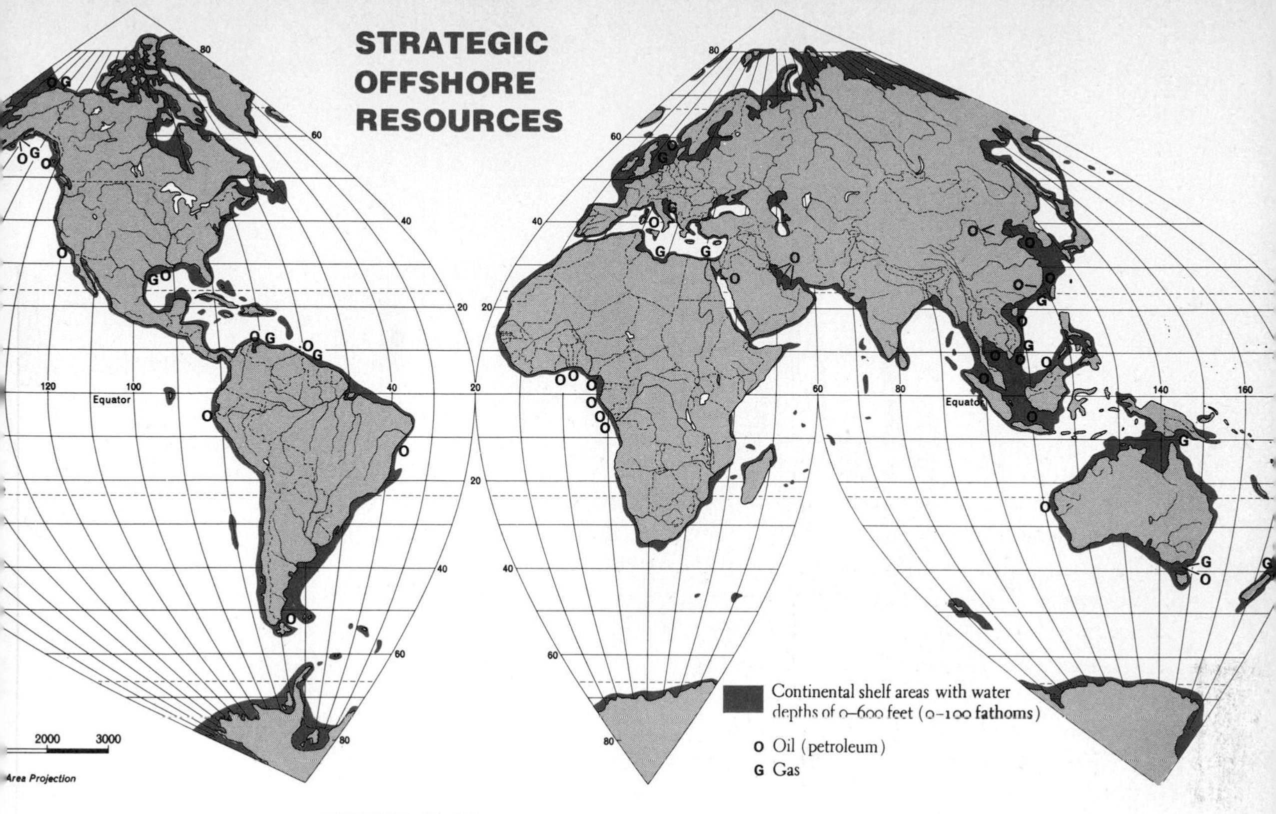

FIGURE 5–17
In addition to being an increasingly vital source of oil and gas, the world's continental shelves are becoming more and more important as sources of salt, sulphur, magnesium, and other minerals.

The New Frontier

Now that humans have extended their domain over practically all the early land frontiers, they are taking an ever-greater interest in the only surface frontier that remains—the oceans, especially in the continental shelf areas of the world. Today, the bulk of the world's fish catch continues to come from shelf waters, such as the fishing banks off the coasts of New England and Newfoundland and those in the North Sea. Nearly 20 percent of the world's petroleum and natural gas now comes from the shallow waters of the continental shelves (Fig. 5–17). The current bonanza in the exploitation of these vital submarine resources is taking place in shelf areas, such as those in the North Sea, the Persian Gulf, the United States Gulf Coast, and those off the coasts of California, Alaska, Venezuela, western Africa, Southeast Asia, and China (Fig. 5–18). This activity, along with the increased size and use of seagoing oil tankers, has helped to intensify the desire of coastal states to protect their shorelines against accidental pollution.

While about one-fifth of the countries of the world are landlocked, a vast majority of them have territorial waters; sovereignty over these areas gives them not only greater national security but, equally important, the right to exploit sought-after marine and submarine resources. To this end, the rapid development of oceanic technology is now a high priority of many states.

FIGURE 5–18

An oil rig off the coast of England. Such rigs are a mixed blessing; for while they provide access to sorely needed energy reserves, they also pose the threat of serious environmental pollution through spillage. (Mobil Oil Corp.)

Conflict Over Breadth of the Territorial Sea

Although all states accept, and understand the purpose of, the concept of the territorial sea, they have yet to formulate any international agreement as to a uniform breadth for this space. This situation, combined with the fact that boundary lines are much more difficult to regulate on water than on land, has been the cause of much confusion; it has also greatly increased the possibility of armed conflict between countries over the control and ownership of ocean waters and ocean-bottom areas. Such rivalries over disputed waters reflect, not only the age-old struggle for commercial and naval control of the high seas, but also the active competition for the mineral resources and fisheries of shelf areas. A case in point is the "Cod War" of the mid-1970s, in which Iceland claimed that British trawlers were illegally using its coastal waters, over which it claims exclusive fishing rights.

It may be observed today that the breadth of the territorial seas of various states ranges from a few claims of over twelve miles (protested but still effective), to several claims of twelve, nine, six, four, and three miles. Most of the major maritime nations, including the United States and Great Britain, claim a three-mile limit.

Three miles has been the extent generally preferred by those states upholding the freedom of the seas, a concept accepted by—or imposed upon—the world in the nineteenth century by the major maritime nations, to whom order on the seas was essential. But this convention has since given way to the current state of chaos. The present impetus for the extension of sovereignty over sea space has developed primarily within the last few decades, especially among developing countries with comparatively few land resources.

In order to appreciate more fully the position of numerous states seeking to maintain traditional limits and the freedom of the seas concept against expanding claims of offshore sovereignty, one must understand some of the consequences of adopting even a twelve-mile territorial limit on a global basis. For example, numerous strategic straits—such as Bab-el-Mandeb (at the mouth of the Red Sea), Gibraltar, Dover, and Malacca—and narrow channels located beside continental margins or between islands in archipelagoes would have to be reclassified from high seas to territorial seas; and, as a result, it is possible that they would no longer be open to all ships. If "high-seas corridors" were strictly maintained through such waterways, then the problems of a new limit could be minimal; but if strategic straits and channels were totally nationalized, many difficult questions would be raised. Would custom give all shipping a right of "free transit," or would only the right of "innocent passage" be honored? Of course, each coastal state would be free to determine what constitutes "innocent passage," since, by international law, it is defined as that which is *not prejudicial to the peace, good order, or security of the coastal state.*

An associated problem concerns archipelagic states, such as Indonesia and the Philippines, that want to extend their dominions to include the waters between even their outermost islands. Allowing all archipelagoes to legally claim such intervening waters would reduce significantly the total area of the high seas and thus further restrict freedom of access to the traffic lanes and treasures of the seas.

Future Expansion of the Territorial Sea

The trend since 1958, when the first Law of the Sea Conference, sponsored by the United Nations, was convened, has been to extend territorial waters. At the 1958 conference in Geneva, most of the participating states were in favor of a three-mile limit. Today over half of the member states have claimed limits of six to twelve miles, and about a dozen have asserted control over more than twelve miles. A few states, including Chile, Ecuador, and Peru, have actually claimed exclusive fishing rights in the waters up to 200 miles off their shores.

The law of the sea, like all international law in a state system, is created by consent, through formal agreements, or through custom. In light of this, it seems rather unlikely that any future international tribunal will decide that twelve-mile claims are illegal, *per se.*

Recent claims to additional territorial waters are, no doubt, harbingers of things to come. As long as there is serious concern over national security and substantial economic advantage to be gained from greater limits, there is little possibility of worldwide agreement on the width of the territorial sea. And, if an international settlement of the matter is not soon reached and enforced, the resultant scope and havoc of "sea-grabbing" may overshadow the colonial "land-grabbing" of past centuries.

FIGURE 5–19

Cities of the Hanseatic League. Although most of its member cities were German, this medieval association, which flourished during the fourteenth century, influenced trade in cities as distant as London and Novgorod and Bergen and Krakow.

International Organizations

The evolution of political society has progressed to the present-day's voluntary grouping of states into multistate organizations. The best-known state groupings of the past are, perhaps, the Greek city-states, the Hanseatic League of northern Europe, and a number of other similar trading leagues of the late Middle Ages (Fig. 5–19). Twentieth-century multistate organizations are distinguished from these mainly by their number and scale.

In the modern world, any sovereign, independent nation can be in great peril if it relies only on itself; but, on the other hand, the creation of global union, or world government, is still a utopian dream. What is plain, however, is that between complete national isolation and total interaction among the earth's peoples, an important middle ground is developing. Sovereign nations are attempting to work together in large groups for some purposes, while maintaining their rights to independent action for others.

Global Organizations

An encouraging sign in the promotion of global cooperation has been the relative success of the United Nations. Since its inception in 1945, this organization has increased its membership from about 50 to over 140 states. Formed to replace the League of Nations, which had been relatively weak and ineffectual from its inception in 1919, the United Nations continues the purpose of the

League in several ways. Like its predecessor, it has as its primary purpose the prevention of war and political conflict and the establishment of international peace and security. In the pursuit of this goal, the United Nations has served as an international forum, where political disputes can be aired and negotiated. Although its record in this regard has not always been successful, the United Nations, most observers agree, has been, perhaps by its mere presence, of inestimable value to the world.

Beyond its role in political conflicts, the United Nations has achieved its importance largely through its cultural and socioeconomic work. Together, the several subsidiary agencies of the United Nations have been, perhaps, the most energetic international advocates of cultural, economic, and social progress among the peoples of the world. On the premise that the development of international as well as regional economic and social cooperation is a major prerequisite for the ultimate solution of political conflicts, these agencies have attempted to improve every aspect of human welfare—among these, health, education, and economic development (Fig. 5–20). Within the realm of cultural

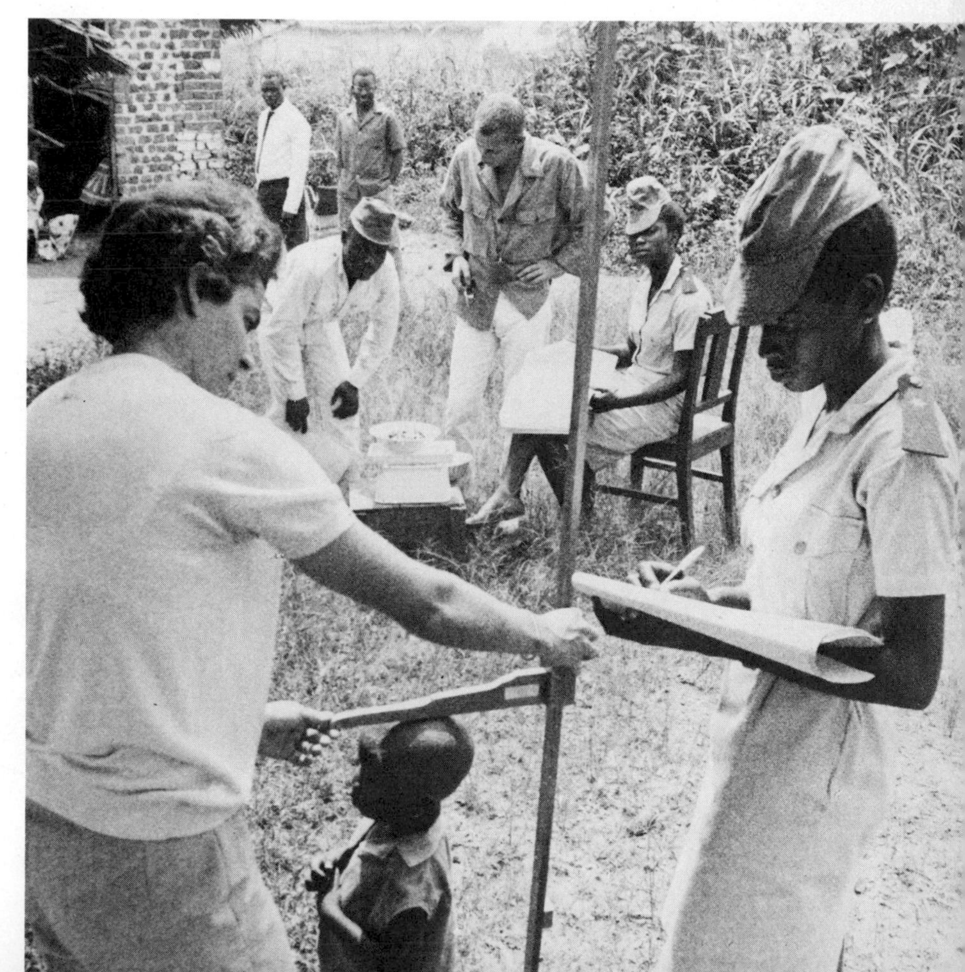

FIGURE 5–20

The measuring of a child's height during a nutrition survey conducted in Gabon by the United Nations Food and Agricultural Organization. (United Nations)

and economic development, four regional Economic Commissions established by the United Nations encompass nearly the entire world: ECAP, for Asia and the Pacific, 1947; ECE, for Europe, 1947; ECLA, for Latin America, 1948; and ECA, for Africa, 1958. Each commission serves as an important instrument for multistate cooperation within its own region.

Regional Organizations

Regional cooperation is a relatively recent phenomenon in international relations. Mainly in Europe and Asia, but in Africa and the Americas as well, the thrust of regional integration has accelerated rapidly since the end of the Second World War. If continued, it may provide a valid substitute for colonialism, which is now socially outmoded and uneconomical, and new paths toward the equalization of human rights and opportunities among all the peoples of the world.

Regional integration is a phenomenon of both the developed and the developing worlds; it can be seen in operation in both the Communist and non-Communist worlds and across all the continents as well. The most sophisticated and successful of the new regional integrations is the European Eco-

FIGURE 5–21

The memberships of two modern trade associations—ASEAN and the Common Market.

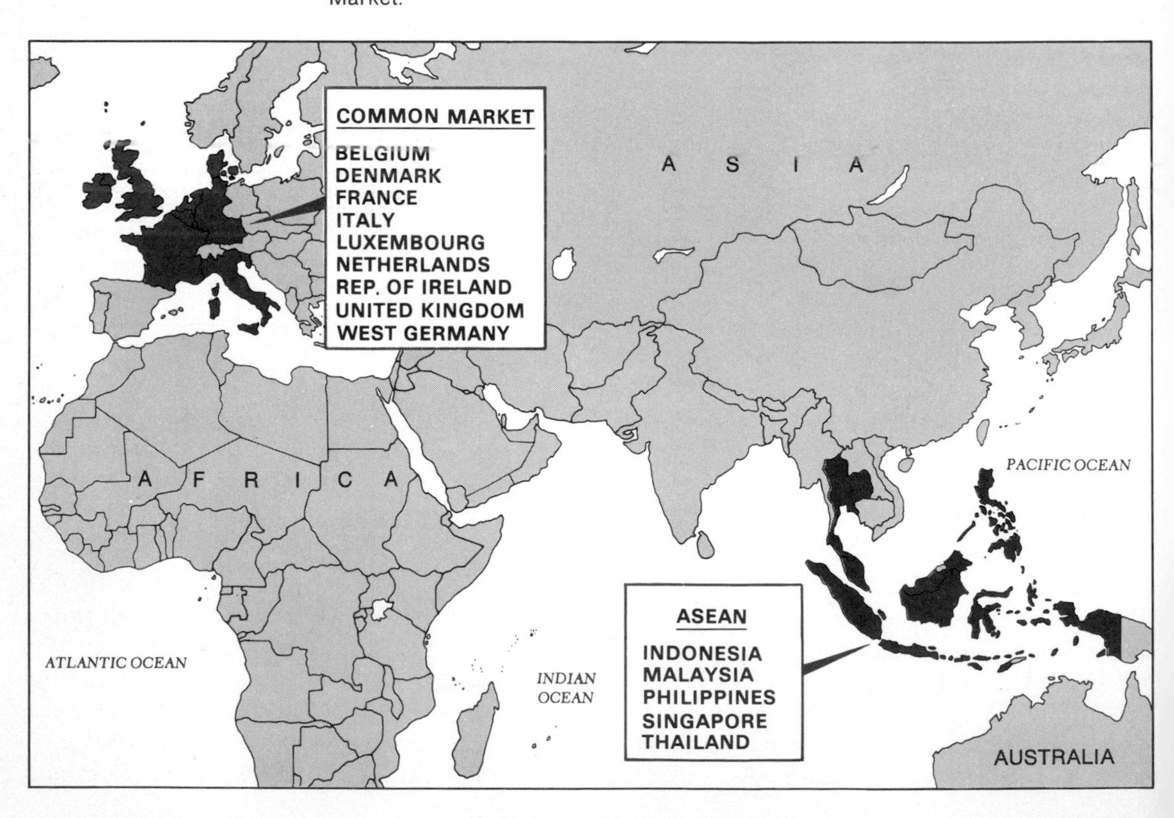

nomic Community (the Common Market), which helped to raise Europe from the ashes of the last world war to its present socioeconomic heights. Another, quite different in structure and original purpose, is the Organization of Petroleum Exporting Countries (OPEC), which in the mid-1970s proved the theory of "strength in unity" to a dismayed world energy market.

Along with economic organizations, regional integration may also be represented by military, political, or cultural bodies, or by organizations that often serve as a combination of two or more of these. For instance, most collective defense associations—such as NATO, CENTO, SEATO, and the Warsaw Pact—whose *raison d'être* tends to be rather short-lived, usually supplement their central purpose with projects in various cultural and economic spheres. Undoubtedly, the most promising regional organizations for the future are the cultural-economic groupings; the two best models of these are probably Europe's Common Market (the European Economic Community) and Asia's ASEAN (the Association of Southeast Asian Nations) (Fig. 5–21).

It should be kept in mind though, that all the various groupings, so far organized since World War II, have at one point or another lacked one or more of the essential elements that would mark any one of them as an unqualified success. But several of them have become qualified successes—which is all that one can say for any such grouping, even the European Common Market.

Comparative Regional Cooperation Studies: The Common Market and ASEAN

Western Europe and the Common Market

After World War II, Western Europe lay in ruin, and Europeans realized that in order to rebuild their societal and economic structures, they would have to mount a cooperative effort. Europe's postwar recovery program was very successful, and perhaps the most important aspect of this new cooperative vision was that it helped pave the way for broader economic unification.

There were many cooperative institutions established to help in the economic unification of Western Europe after World War II. Among these was the Organization for European Cooperation (OEC), which was designed to administer the aid received from the United States through the Marshall Plan. The European Payments Union (EPU) also helped to administer foreign aid and was also instrumental in achieving a tighter economic integration among the European nations. It encouraged the governments of each Western European country to think of its separate financial accounts in relation to the whole of Western Europe rather than as separate and unrelated problems.

The European Coal and Steel Community (ECSC), formed in 1952, served as the first true step toward the eventual economic unification of Western Europe. Its primary goal was to promote the production of coal and steel and the free movement of these between the member countries.

In 1958, encouraged by the success of the European Coal and Steel Community, Belgium, France, Italy, Luxembourg, the Netherlands, and West Germany formed the European Economic Community (EEC), more commonly known as the *Common Market.*

Before the formation of the Common

Market, trade among these six countries was greatly restricted, because of high tariffs, import quotas, exchange controls, and other customs regulations. Thus, to insure strong socioeconomic coordination, the European Economic Community was designed as both a customs union and a common market. Within a customs union, the member countries eliminate tariffs and other restrictions to trade on products exchanged between them, while adopting common tariffs on products exchanged with nonmembers. Relaxing the trade regulations among the six countries increased the market for their respective products and allowed for greater mobility of capital and labor throughout their territories. Only through the creation of close, cultural and economic ties between the six could the final goal of complete political cohesion among them be accomplished.

The formation of the Common Market launched its six member countries on a bold adventure, designed to stimulate socioeconomic growth, ensure governmental maturity, and provide lasting mutual prosperity. The door was left open for other countries to join the organization, as full members and/or as associates. And in 1973, when Denmark, Great Britain, and the Republic of Ireland joined, as full members, the Common Market became a union of nine and a stronger stabilizing force in the economic and political worlds.

Southeast Asia and ASEAN

Southeast Asia is a region of separate and very distinct nations. For hundreds of years, wide cultural and political diversities (and disparate levels of economic development) have, quite naturally, prevented feelings of unity among them, while discouraging the interdependencies that foster regional cooperation.

After the Second World War, the newly independent states of Southeast Asia stood isolated from one another; although, geographically, they were neighbors, few of them were accustomed to sharing their problems or goals. Their lines of communication led outward from Asia to the capitals of Europe, as did, to a very large extent, their thinking and interests. Externally at least, intraregional contacts and communications had remained undeveloped under colonial rule; and intraregional trade and commerce had been largely discouraged.

Although very close ties have not existed among the nations of Southeast Asia for centuries, there are forces at work today helping to bring about a sense of regional involvement and identity. And, there does seem to be a growing feeling of regional consciousness.

Among the most outstanding regionalist forces are: first, a growing awareness, on the part of Southeast Asian leaders, that their countries share many major problems, largely in the area of socioeconomic development; second, the increase in intraregional communications (brought about, in part, by regional airline services, telecommunications systems, and increasing diplomatic exchanges); third, the growing trend toward institutionalized cooperative ventures in the world, including the Asia Pacific region; and fourth, the realization by Asian leaders that in the postcolonial era their nations must provide for their own political and economic security and shield their region from the threat of outside political intervention or internal socioeconomic decay.

The important idea to keep in mind though, is that all four of these very prominent factors have helped to facilitate and encourage the involvement, by Asian leaders, in the affairs of their neighbors. This has led to the development of a sense of regional consciousness which has led further to an interest by Southeast Asian leaders in establishing institutions of regional cooperation for their mutual benefit.

The most important step so far taken toward effective regional cooperation has been the successful formation of the Association of Southeast Asian Nations (ASEAN). In 1967 the ministers of five Southeast Asian nations met in Bangkok to form this new organization, a successor to the then dormant Association of Southeast Asia (ASA). ASEAN's five charter members included Malaysia, the Philippines, and Thailand—the three original ASA members—as well as Indonesia and Singapore.

Officially, ASEAN's chief aims are those of socioeconomic development and regional strengthening; but the organization's activities also seem directed toward independence from all outside influences. The economic and social projects undertaken by ASEAN since 1967 include a host of cooperative ventures in agricultural technology and production, industry, transportation, trade, tourism, and the development of educational facilities and student-faculty exchanges. Because of the present low degrees of economic development in each of the member nations, the goal of ASEAN is not simply to integrate the economies of its member-nations in the manner of the European Common Market but also to provide a means by which they can assist one another in social, scientific, and administrative matters. ASEAN has also been important as an unofficial forum, where political disputes between member countries (such as the dispute between Malaysia and the Philippines over Sabah) can be arbitrated.

Even in the face of steady progress, it must be recognized that Southeast Asia's economic climb will be a long and frustrating one. The increase in economic development and trade within the region since the conception of ASEAN has been relatively small, in terms of its goals; many of the nations within the region still tend to be dependent upon trade with areas outside of it. (And this outside trade is often mutually competitive and barely economical, as, for the most part, the nations of Southeast Asia sell raw materials on the world market in competition with one another.) Nevertheless, ASEAN must be considered a viable experiment in regional cooperation. Its new Secretariat, built in Jakarta, symbolizes both ASEAN's intention to remain and grow and Indonesia's leadership within the organization. It is possible that by the late 1970s ASEAN may be as unified as the Common Market, perhaps even enjoying some direct association with its European counterpart.

REVIEW AND DISCUSSION

1. Would civilization, as we have come to know it, be possible without law and political order? Discuss your position on this question and support it with examples.
2. Compare and contrast the concepts of the state and the nation in terms of their "bonds of unity." Evaluate your findings in terms of the concept of the nation-state.
3. What is nationalism? Could it exist if there were no national states? Explain.
4. Of what significance is the fact that national states and nationalism first appeared as major political forces in Europe?
5. No national state is totally immune from internal dissension. With the aid of appropriate examples, identify and discuss some of the major threats to national unity.

6. Compare and contrast the unitary state and the federal state; cite examples of each. How does the ecumene contribute to the power of a state?
7. How does a state system function?
8. What distinctions can be made between colonialism and imperialism? In what places has one occurred in the absence of the other?
9. In your opinion, was European colonialism a positive influence or a negative influence on the world in terms of diffusion and acculturation?
10. How has the state system been affected by decolonization and the recent increases in new states? Have world tensions increased in intensity, or have they just assumed new forms? Explain.
11. Do only new states follow irredentist policies? Explain. When a state makes a territorial claim against another state is that claim always based on historical and cultural grounds? Discuss.
12. What is a buffer state, and what is its significance? Cite two examples of such states.
13. The importance of the territorial sea has been increasing in recent decades. Why? What, if any, related changes should be made in international maritime law?
14. In what areas other than politics and military alliances have global and regional state organizations made substantive contributions? Discuss with the aid of specific examples.
15. Compare and contrast the Common Market and ASEAN in terms of their goals, programs, and achievements.

SELECTED REFERENCES

Alexander, L. M. *World Political Patterns.* Chicago: Rand McNally, 1957.

Boggs, S. W. "National Claims in Adjacent Seas." *Geographical Review,* Vol. 41 (April 1951), 185–209.

Bowett, D. W. *The Law of the Sea.* London: Manchester University Press, 1967.

Bruel, E. *International Straits,* Vols. 1 and 2. London: Sweet and Maxwell, 1947.

Burghardt, A. F. "The Bases of Territorial Claims." *Geographical Review,* Vol. 63 (April 1973), 225–45.

Cobban, A. *National Self-Determination.* Chicago: University of Chicago Press, 1944.

De Blij, H. J. *Systematic Political Geography,* 2nd, ed. New York: Wiley, 1973.

East, W. G., and A. Moodie, eds. *The Changing World: Studies in Political Geography.* New York: World Book, 1956.

Emerson, R. *From Empire to Nation.* Boston: Beacon Press, 1960.

Fawcett, J. E. S. "How Free Are the Seas?" *International Affairs,* Vol. 49 (January 1973), 14–22.

———. "The Development of International Law." *International Affairs,* Special Issue (November 1970), 127–37.

Fisher, C. A. "The Malaysian Federation, Indonesia and the Philippines: A Study in Political Geography." *Geographical Journal,* Vol. 129 (September 1963), 311–28.

Frank, T. *Roman Imperialism.* New York: Macmillan, 1914.

Fried, M. *The Evolution of Political Society.* New York: Random House, 1967.

Ginsburg, N. S. "The Political Dimension: Regionalism and Extra-Regional Relations in Southeast Asia." *Focus,* Vol. 23 (December 1972), 1–8.

Gordon, B. K. *The Dimensions of Conflict in Southeast Asia.* Englewood Cliffs, N.J.: Prentice-Hall, 1966.

Gottmann, J. "Geography and International Relations." *World Politics,* Vol. 3 (January 1951), 153–73.

Hallstein, W. *United Europe: Challenge and Opportunity.* Cambridge, Mass.: Harvard University Press, 1962.

James, P. E., and C. F. Jones, eds. *American Geography: Inventory and Prospect.* Syracuse, N.Y.: Syracuse University Press, 1954. (See esp., Richard Hartshorne, "Political Geography.")

Kohn, H. *The Idea of Nationalism.* New York: Macmillan, 1944.

Kristof, L. K. D. "The Nature of Frontiers and Boundaries." *Annals of the Association of American Geographers,* Vol. 49 (March 1959), 269–82.

Lipset, S. M. *Political Man: The Social Bases of Politics.* New York: Doubleday, 1959.

MacFadden, H. C. "Toward Regional Cooperation Among Asian Nations." Unpublished manuscript, 1974.

______. "Changing Pattern of Sovereignty Over the Straits of Malacca and Singapore." Masters Research Paper (Thesis), University of California Los Angeles, 1973.

Mayfield, R. C. "A Geographic Study of the Kashmir Issue." *Geographical Review,* Vol. 45 (April 1955), 181–96.

Pearcy, G. E. "Geographical Aspects of the Law of the Sea." *Annals of the Association of American Geographers,* Vol. 49 (March 1959), 1–23.

Pollard, V. K. "A.S.A. and A. S. E. A. N. 1961–1967: Southeast Asian Regionalism." *Asian Survey,* Vol. 10 (March 1970), 244–55.

Pye, L. W. *Aspects of Political Development.* Boston: Little, Brown, 1966.

Regional Co-operation in Asia. Annual meeting of Directors of Development Training and Research Institutes, Tokyo, 10–14 March 1969. Development Centre of the Organisation for Economic Co-operation and Development.

Roy, S. J. *The Theory of Sovereignty.* Calcutta, 1923.

Shanks, M., and J. Lambert. *The Common Market Today—And Tomorrow.* New York: Praeger, 1962.

Silvert, K. H. *Expectant Peoples: Nationalism and Development.* New York: Random House, 1964.

Soja, E. W. *The Political Organization of Space.* Washington, D.C.: Commission on College Geography, Resource Paper No. 8, Association of American Geographers, 1971.

Thompson, J. E. *The Rise and Fall of Maya Civilization.* Norman: University of Oklahoma Press, 1956.

United Nations, Sea-Bed Committee. "Archipelagic Principles as Proposed by the Delegations of Fiji, Indonesia, Mauritius and the Philippines." Document No. A/AC.138.SC.11/L.15, 14 March 1973.

Vaillant, G. C. *The Aztecs of Mexico.* New York: Doubleday, 1950.

Whittlesey, D. *The Earth and the State.* New York: Holt, 1939.

FIGURE 6–4

The Harvest, a Theban tomb painting (about 1415 B.C.) that depicts the reaping, measuring, and recording of a wheat crop. The *registers,* or rows, of the painting are read in sequence from bottom left to top right. (Photograph by Egyptian Expedition, The Metropolitan Museum of Art)

Major productive sectors	*World labor force (in percentages)*
Agriculture	51.0
Grazing	1.0
Forestry	.5
Hunting–fishing	.5
Mining–drilling	1.0
Industry	20.0
Services	26.0
Total	100.0

FIGURE 6–5

The differentiation of the world labor force into major productive sectors.

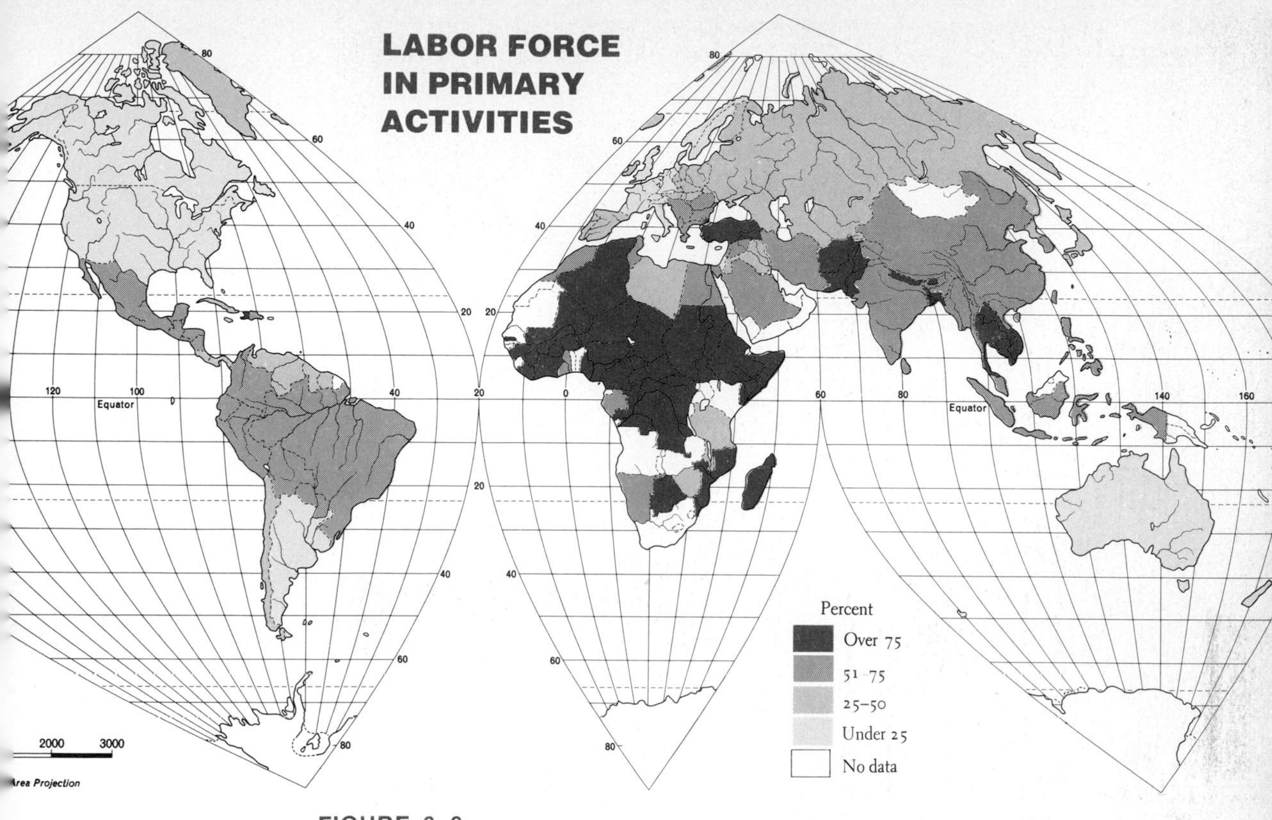

FIGURE 6–6
Well over 50 percent of the total world labor force is employed in primary activities. On a country-to-country basis, the percentage usually correlates inversely with socioeconomic advancement.

Foods for Survival

Human Nutritional Needs

Never before has there been so much food for so many, and so many with so little (Fig. 6–7). One of the major questions facing humanity today is whether, and how, the earth can provide food for the human populations of tomorrow. If present population trends continue, today's population of 4 billion will increase to one of 7 billion by the year 2000—nearly doubling in a single generation. This projection is staggering for several reasons, not the least of which is the fact that food supplies, too, will have to double by the year 2000, in order to just maintain present consumption levels (Fig. 6–8). If allowances are made for the much-needed raising of dietary standards for all peoples to acceptable nutritional levels, the same population projection demands that food output be increased threefold by the year 2000.

To begin to comprehend the problems of feeding today's masses and tomorrow's even greater human numbers, we must understand a little of the basic nutritional needs of all people, needs that ignore ancestry, location, or custom. If their traditional foods are available in sufficient quantity, and with

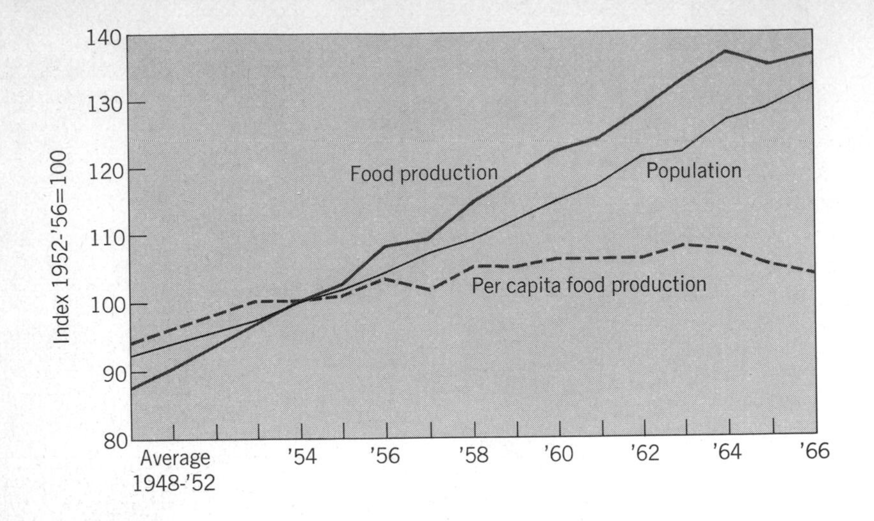

FIGURE 6–7

There have been substantial increases in world food production in recent decades; however, nearly equal rises in world population have kept the *per capita* increases in food production relatively small.

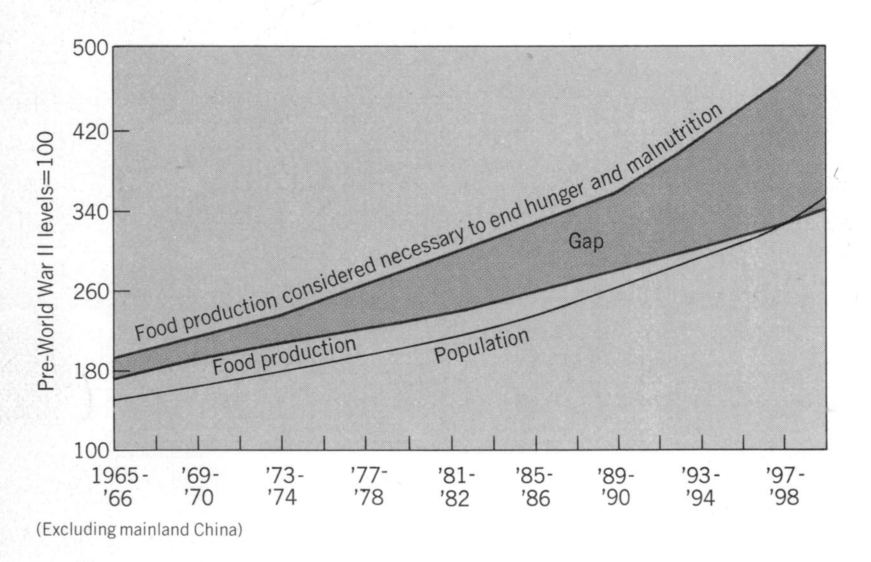

FIGURE 6–8

Projected increases in world population, probable food production, and necessary food production to the year 2000 show a widening gap between foods needed and foods in supply. (© 1966–67 by the New York Times Company. Reprinted by permission.)

reasonable regularity, most people will have a food intake that is fairly adequate. The undernourishment and malnutrition suffered by most peoples of the world stems mainly from the insufficient availability of some, or all, of their traditional foods, a lack of purchasing power, or ignorance in the selection and preparation of the foods available. Even a global abundance of foods equitably distributed, however, would not immediately solve all the world's food problems.

Foods in Supply

Most human nutritional needs can be satisfied by a few simple foods, most of which are produced at relatively little expense in developed and developing countries alike. Those produced worldwide include the true staple foods of humankind: the cereal grains—rice, wheat, and maize (corn)—and fruits and vegetables. The others, considered staples by the affluent and luxuries by the less affluent, are largely such animal-source foods as meats, milk, and dairy products. Cereal grains and produce—mostly carbohydrates—account for most of the food intake of the world's population, particularly among the 80 percent that live in poverty. Most human diets, then, supply largely carbohydrates; they are usually lacking in proteins and fats, as well as in many of the vitamins and minerals most easily obtained from meats and dairy products.

Needed Foods

Estimates of actual world food production and consumption, and of the quantities of food needed to achieve a higher standard of living in this and future generations, are frequently difficult to assess and interpret, because of the many different methodologies available. The most widely accepted index is the United Nations Food and Agricultural Organization's listings of "standard daily *per capita* caloric requirements," which estimate human caloric food needs by age, body size, sex, work activity, and environmental differences. On the basis of these FAO estimates, it is projected that the ideal minimum *per capita* average daily caloric need is 2,300 calories; and that a *per capita* daily food intake of less than 1,700 calories constitutes starvation. The FAO also estimates that, at present market levels, the *per capita* world average daily caloric production is about 2,400 calories. On the surface, these figures appear to suggest that there is a food sufficiency, and even a food surplus, throughout the world. But, when "spoilage" (averaging 10 to 25 percent in most foods) and failures or inequities in distribution to consumers are considered, the reality of the gap between caloric *need* and caloric *availability* becomes all too evident (Fig. 6–9).

FIGURE 6–9

A food dole in India during a local drought. (Henri Cartier Bresson/Magnum)

Estimates of the *per capita* average intake of protein are probably better guides to the quality of the human diet than are estimates of caloric consumption. Malnutrition is largely a matter of protein deficiency. Where animal source foods are essentially lacking, a wise and liberal consumption of many kinds of cereal grains and produce (usually requiring a knowledge of nutrition) is necessary if adequate amounts of the nutrient are to be obtained (Fig. 6–10).

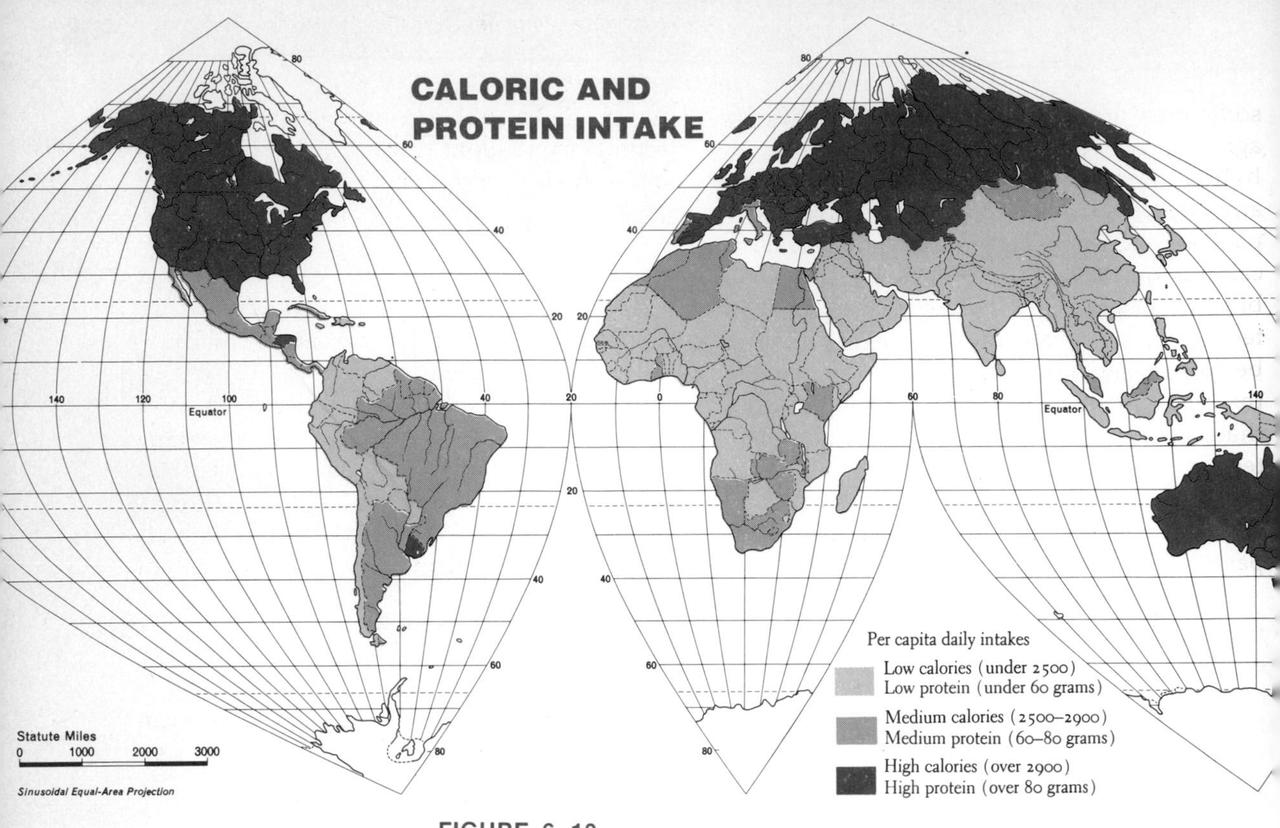

FIGURE 6–10

Human nutritional levels vary considerably throughout the world. And since different foods have different values to the body, *per capita* protein intake is a better guide to world nutrition than either caloric consumption or total absorption by weight.

Environmental Potential and Cultural Development

Spatial Fragmentation

The magnitude and variety of agrarian development have always been limited by the environment. The present, somewhat uneven global distribution of agricultural production can thus be explained, at least in part, as being a consequence of the continuing need for adjustment of cultural development to environmental potential (Fig. 6–11). This situation should be considered only a temporary condition of a changing world; for it is all but certain that the distribution of agrarian centers will change in the future as it has in the past. The more difficult questions are how much it will change, and in what ways; at present, these can be answered only with rough predictions.

The development of a new technology in food production has progressed to levels unimagined even a generation ago. But can this pace continue? Will there be scientific and technological developments to revolutionize agriculture in the future, or has a stasis been reached? Assuming that human inventiveness knows no bounds, as the saying goes, it is reasonable to expect that as new ideas continue to evolve, there will be some capable of affecting humankind's agrarian pursuits directly. Such things as hybrid corn, "miracle," or high-yield, rice and wheat, and *hydroponics* (the growing of plants in solutions and without the mechanical support of soil) may well be only harbingers of an accelerated pace in agrarian technological development, a pace that may be able to keep food supplies in balance with the growth in population. However, with all that can reasonably be expected of future technology, there will probably always be areas of the world where crops cannot be grown profitably and where most, if not all, pastoral pursuits cannot be practiced. The map in Figure 6–11 gives a generalized view of the present global spatial potential for crop and animal production.

Spatial Controls

The distribution of agricultural activity is determined by a number of controls—some environmental and some cultural—operating interdependently. The predominant environmental controls are climate, landforms, drainage, soils, and vegetation. The most influential cultural controls are cultural inheritance and social attitudes, plant

FIGURE 6–11
Human habitation is strongly influenced by local environmental conditions, especially those that have a direct effect on agricultural production.

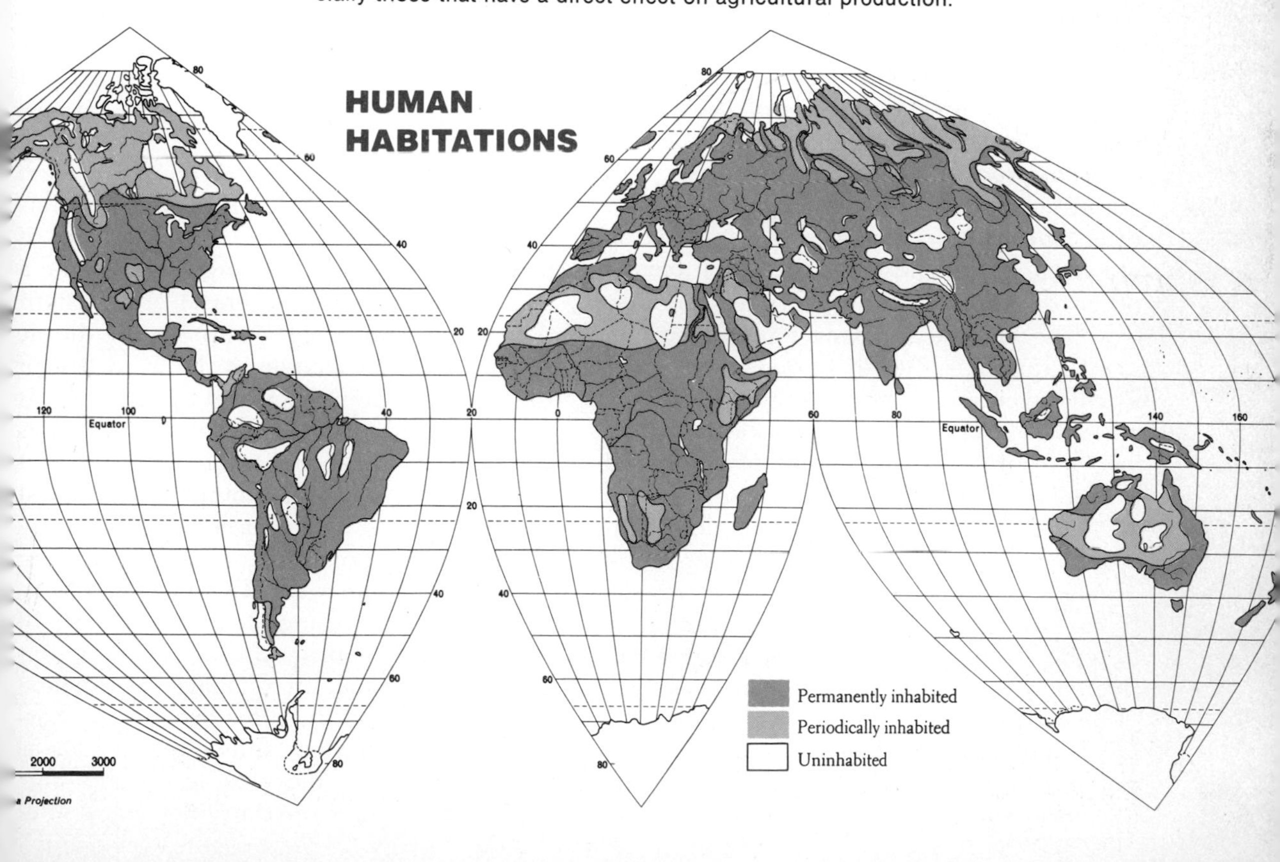

and animal selection, capital and labor, socioeconomic status, government, transportation and markets, and competing activities.

In general, the influence of environmental factors, considered both singly and as a group, declines as cultural factors advance in sophistication and complexity. For example, as a rule, environmental factors tend to exert more influence on simple agrarian societies than on more technologically advanced agrarian societies.

Since these environmental and cultural factors do not operate singly, it is somewhat artificial to assess them singly. However, their interactions are so complex, and their effects on agriculture sometimes so indirect, that a multiple-factor assessment tends to be suitable only in specific type studies.

Environmental Controls Over Agriculture

Climatic Controls

Each plant species has specific environmental requirements for and limitations to successful growth. Some plants are very tolerant, that is, capable of adapting to many varied conditions; while others have very restrictive environmental requirements. Notwithstanding the importance of landforms, the most potent environmental control over agrarian production is climate. Every plant, domesticated or wild, thrives best under some specific combination of temperature, moisture, wind, and light; each will tolerate other combinations for only limited periods.

The climatic element most critical in plant growth is that of temperature. Poleward, lower and more varying temperatures as well as shorter growing seasons progressively inhibit the production of crops and the growth of all wild plant life as well; while equatorward, higher and more uniform temperatures coupled with year-round growing seasons impose far fewer restrictions. Some crops, such as maize, can adapt to wide temperature ranges, and so are cultivated at several different latitudes. But other crops have a much smaller tolerance area. Commercial cotton, for example, has a poleward limit that lies approximately along the mean summer isotherm of 77° F., about 35° North latitude in the United States and 40° North latitude in China; and such food crops as sugar cane and bananas do not thrive in areas very far beyond the tropical regions. Temperature factors directly control the range of food crop and nonfood crop production, and thus there is strong competition by nonfood crops, such as cotton and other fibers, for the world's favored agricultural lands.

The situation is more complex than it appears at first glance, however. For, although maize can grow in a wide range of temperatures—thus allowing it to become a staple crop in southern, central, and western Africa, in much of South and Middle America and, of course, in the Corn Belt of the United States—to mature as a grain, it needs a lengthy period of hot summer days and warm summer nights. Hence it does not grow well in cool-summer, west-coast marine climates; and, therefore, it is virtually absent among the crops of Europe growing north of Italy and west of Hungary. In Minnesota, Wisconsin, and Michigan, all north of the Corn Belt, short and cool summers preclude the production of maize as ripened grain; thus, in these states the ears are harvested green, ground, and stored in silos as winter fodder for dairy cattle. In both this western portion of the Dairy Belt and that of the Corn Belt, maize is produced for animal feed, but, in each region, the crop is harvested, processed, and fed to the animals differently.

Within certain limits, humans are able to encourage plant growth by altering natural temperature conditions. The growing areas of several crops highly sensitive to freezing, for example, have been extended by the artificial raising of minimum temperatures in the fields at critical times. In southern Italy, citrus fruits are protected against frost by the building of roofs over entire orchards; each roof is made of boards spaced far enough apart to admit sunlight but close enough to reflect downward a considerable amount of ground radiation, which might otherwise be lost into space at night. In California's lemon and orange groves, cold lower layers of air are dispelled by currents of warm air blown earthward by motor-driven propellers set atop high towers; and, especially at night, air temperatures are raised by ground-level oil-burning heaters.

Moisture availability is another climatic element that serves to set limits on plant growth. Every crop and wild plant has a "dry limit"—some also have a "wet limit"—and an optimum moisture requirement. Crops are almost totally nonexistent in the world's vast desert areas, except where assisted by irrigation, and, in semiarid lands, their growth is seriously restricted (Fig. 6–12). Most major crops are produced universally in a pattern that closely mirrors moisture availability. An exception, wheat is commonly produced in semiarid to moist subtropical and midlatitude regions—but seldom in regions where heavy rainfalls occur during harvest periods, or within the tropics.

Landform Controls

Both major and minor features of the earth's crust exert considerable control over agrarian activities. The influence of landforms operates both indirectly, through the climatic complex which, in turn, affects vegetation and soil, and directly, through land-use adaptability. Most of the world's agricultural lands are concentrated within broad regions that are comprised mostly of plains and undissected plateaus. Wherever the terrain is rough, as in mountain areas, cultivation is markedly limited, if not altogether precluded. Even in hilly regions and in areas having dissected plains and plateaus, cropping is often seriously restricted by the limited use of machines possible in such areas and the resultant prohibitive costs (Plates II and III).

Agrarian societies forced from flat lands into rougher terrains have survived. But, in general, mountain areas are notably lacking in agrarian development, with small mountain valleys accounting for most of their crop production. Adverse climatic conditions, due largely to higher elevations, and the problems of soil scarcity and short growing seasons all contribute to the lack of extensive agrarian development in the world's mountain regions—and, consequently, to the scarcity of ancillary cultural features, such as roads, housing, market centers, and irrigation systems.

Under some circumstances, however, even the most rugged mountain regions may become agricultural areas. If population densities are high in the flat lands, if the cost of labor is low and the food supply even lower, and if there is no market for labor other than the direct production of food at a subsistence level, great effort may be made to extend arable land to the mountain areas. This often necessitates the construction of *hillside terraces,* sometimes so narrow that only a single row of trees or a few rows of wheat or rice can grow on them. These terraces may form seemingly endless tiers up a mountainside, and individual trees or vines may be planted wherever the landform offers a pocket of soil big enough to support a plant. California and

FIGURE 6–12
The effect of irrigation on Quincy Valley in Washington. Above, an aerial view of the valley as it appeared in 1952—an unproductive, semiarid landscape. Below, the same view in 1957, when five years of irrigation under the Columbia Basin Project had transformed the valley into a rich agricultural region. (Bureau of Reclamation)

Illinois require that a field be big enough and flat enough for a tractor or combine to maneuver on. But in Greece, Crete, and Spain, for example, grape vines and walnut, carob, and olive trees are grown singly on steep, rocky hillsides; they are later harvested manually and the crop is carried away in baskets or sacks on the pickers' heads or backs. Similarly, in Indonesia and the Philippines, rice is grown on terraces located on the steepest of hillsides (Fig. 6–13). Perhaps the most extreme examples of how planters overcome nature in this manner are to be found in Scotland and Ireland; there, to permit the raising of potatoes, hillside terraces hewed in rock have been filled with soils concocted of peat from nearby bogs, sand composed largely of shell fragments from the beaches, and seaweed gathered along the shores. Al-

FIGURE 6–13

Terraced agricultural lands. Above, olive groves in the province of Alicante, Spain. (Heleva Kalda/Monkmeyer) Below, rice paddies in Luzon Island, Philippines. (Ewing Galloway, N.Y.)

though certain landforms act as deterrents to agriculture and other forms of human activity, when the pressure is sufficiently great, human beings can often overcome many terrestrial disadvantages and make use of seemingly unusable areas.

Drainage Controls

In many parts of the world, poor drainage conditions almost completely preclude human attempts to develop agriculture over extensive areas. Such conditions range from low wet spots in river floodplains to extensive coastal marshes too low to drain and reclaim for productive uses. Other regions are too frequently subject to damaging flooding (*natural drainage*) to sustain regular cropping; still, people often gamble on them, and some suffer periodic devastation as a result—roads, crops, dwellings, irrigation systems, and a host of other cultural developments being destroyed in a few hours.

The decision as to whether or not certain areas in a country should be drained or left unchanged commonly has depended upon both the socioeconomic value of the crops that could be raised on such land and the population pressure present in adjoining dry areas. The Netherlands a heavily populated nation, hard pressed for agricultural land, undertook the draining of its coastal marshes centuries ago; today, much of its most productive land lies well below sea level, protected by dikes against the incursion of the sea. In central California, the chance to extend the area in which to grow such high-value crops as asparagus and rice has made the draining and diking of delta lands at the confluence of the Sacramento and San Joaquin rivers worthwhile. On the other hand, little serious thought has been given to draining and improving such areas as the vast coastal swamps of eastern Borneo,

FIGURE 6–14

A herd of goats in Big Bend National Park, Texas. Sharp-hoofed and voracious, such animals prod and bite the grass cover close to its roots and thus can do great injury to the soil of their grazing area. (National Park Service)

areas where population density is very low and huge areas suitable for agriculture are still unused.

Soil Controls

Agrarian development also depends, of course, on soil, which cannot be divorced from the elements of climate and vegetation in terms of its role as an environmental control. However, soil can be quickly destroyed or gradually improved independently of these other factors. Overpopulation, resulting in overcropping and overgrazing, along with poor cultivation practices can do much to destroy a good soil. And, once started, this destruction is difficult to halt and the soil, difficult to redeem (Fig. 6–14). Generally, the more economically advanced a country is and the more recent its agrarian practices, the more deliberate is the handling and maintenance of its soils.

Almost any soil can be rendered more suitable for the raising of a specific crop, through the use of appropriate fertilizers. And, it may well be that a better understanding of plant fertilizers and their proper application will play a large part as steps toward solving the ever-increasing world food crisis.

Vegetation Controls

Native vegetation also has a share of control over the world's agrarian activities, mostly inasmuch as it serves as an impediment to human efforts. Until the late 1800s, for example, the tough sod cover of the world's vast grasslands largely discouraged their cultivation; only after the availability of machinery like the steel plow were such regions, including the prairies and steppes of the United States, efficiently utilized for crop production. And, even with modern technology, the equally vast forest regions of the world are still forbidding obstacles to such human endeavors as crop tillage and animal husbandry.

Nevertheless, all crops are but domesticated versions of wild plants, bred to human tastes and, in many cases, dispersed widely into new environments and cultures. Thus potatoes, tobacco, maize, cacao, and rubber were domesticated in the New World and spread to the Old World; and, similarly, wheat, sugar cane, cotton, coffee, tea, oranges, and many lesser crops were domesticated in the Old and then introduced into the New World.

Cultural Controls Over Agriculture

The nature and global distribution of agrarian development have always been influenced, in large measure, by a variety of cultural controls, at times in conjunction with environmental ones and at other times in direct opposition to them. For example, cultural factors are at least partly responsible for the fact that there have always been areas of the world in which crops well-suited to the local environment have not been produced; and, as an even more frequent result of such factors, there have been areas in which ill-suited crops have been produced. These situations have usually developed through the misapplication of cultural controls due to an inadequate understanding of them, or through the misapplication of fully understood controls due to strong pressures from socioeconomic, political, traditional, and/or religious segments of the community.

Cultural Inheritance

Personal life styles and social attitudes passed down from generation to generation exert considerable influence over not only the type of agrarian activities any given human group adopts but, indeed, whether it will adopt one at all. For, in addition to those cultures that exist where no agrarian activities are possible, such as the Eskimo culture of the Arctic fringelands, there are others that, given their environmental conditions, could have developed very satisfactory agrarian life styles, but did not. Among these latter groups are the hunting-and-fishing peoples to be found in parts of tropical and midlatitude coastal lands, for example, the boat people of Hong Kong. The cultural inheritance of these groups includes no knowledge of or experience in cultivation or pastoralism, and the life styles followed by many of them are rather precarious.

However, in the majority of cultures by far, at least traces of agriculture have been present. Even most early cultural groups had knowledge of the rudiments of plant production and animal raising and some understanding of climatic conditions and soils as well. As these early peoples migrated, they took their biotic technologies with them, packing their valuable seeds and tools on the backs of domesticated animals. Thus they extended their agrarian ways to new areas, benefiting both themselves and the people with whom they came in contact. The cereal grains were probably the first foods to be shared with distant peoples in this fashion. Rice production, for example, probably spread from its source area in southeastern Asia until, today, it is practiced throughout Asia, wherever natural conditions are promising. Cultural groups in other parts of the world have also become aware of the food value of rice and adopted its production. Thus, ironically, commercial rice-production in the southern United States, which is characterized by the use of modern production techniques and by high per acre yields, now serves as a model for the traditional rice-producing peoples of the world to study and emulate.

But, despite their dispersion, most staple crops are still more closely identified with the cultures of their source areas; for instance, the most distinctive "rice cultures" are those in Southeast Asia, and the primary "maize culture" is that in Mexico.

Plant and Animal Selection

Throughout the first millennia of agrarian activity, most successful human attempts to control the development of plants and animals were probably either accidental or the results of trial and error. In today's world, selection and breeding have advanced to new levels, with specialists able to develop plants and animals suited not only to particular environments but to particular cultural needs as well. As an example, scientific efforts have resulted in the creation of new types of wheat that are more resistant to drought conditions, better adapted to shorter growing seasons, less susceptible to the diseases common to humid areas and, perhaps of greatest importance, capable of much higher yields (Fig. 6–15). Such agricultural feats would hardly have been possible even a century ago.

The possibility of selecting and breeding crops with genetic resistances to diseases and pests has attracted plant breeders for half a century or more but, as yet, only limited beginnings have been made. Success in this area would not only release farmers from worry and costly disease and pest controls, but also spare consumers many of the scarcities and price escalations that result from crop failures.

Capital and Labor

To bring land under cultivation and insure its productivity, capital (money) is usually necessary. In some instances, the amount of capital needed is small; in others, it is very large. The establishment of a farm in the southern portion of Canada's northern forest region requires comparatively little capital: land costs there are relatively low; the acreage needed for such a farm is small; many of the materials needed for fences, barns, homes, and fuel can be obtained from the forest by the farmers themselves. In addition, access to hunting-and-fishing in the area can augment the family's food supply, while an abundance of wild hays can be cut to help feed the farm animals.

In contrast to this situation is the large amount of capital needed to set up a wheat ranch in North Dakota: here, the price of land is high; large tracts of land are necessary for successful commercial production; several kinds of expensive machinery are needed; fertilizers must be purchased; and wages must be paid. Also, reserve funds must be available to "ride out," not only possible crop failures resulting from drought, storms, abnormally heavy precipitation, plant diseases, and occasional locust plagues, but also those times when the market price for wheat is low. Even more capital is needed to establish a tropical plantation, especially one in a new area: land must be purchased, or leased; vegetation must be cleared away and continual work must be done to prevent its return; planting and cultivation costs are high; and fertilizers must be purchased. In addition, several years must elapse before there will be any return on the investment. For example, it takes anywhere from seven to nine years to bring a rubber plantation into production, during which time expenditures are heavy and income is nil.

Conditions in some parts of the world are such that land can be made productive only through huge investments of capital; and, often, government assistance as well as private capital may be necessary. This is particularly true in the development of desert and steppe areas, which require the building of such elaborate water-supply systems as reservoirs, main canals, lateral canals, small ditches, and flood-control works; and additional capital is required to meet the water costs that continue after these systems have been completed. If the agricultural use of these lands depends on the extraction of groundwater, then drilling, pumping, and maintenance costs will be involved. Heavy capital outlays are also necessary to develop swamp or marshlands for agricultural use: expensive drainage ditches must be dug and

FIGURE 6–15

Wheat—the staff of life for half the world's people. Left, the traditional wheat of Pakistan; right, MEXIPAK 65, an improved, or "miracle," wheat. (Ford Foundation).

FIGURE 6–16

Different agricultural labor requirements. Above, a family farm in an Amish area of Pennsylvania that is worked by the owner and horse-drawn equipment. (Grant Heilman) Below, a California vegetable farm on which harvesting involves a hired labor force and complex machinery. (Blackwelder Corp.)

maintained; drain pipes must be laid in the fields; and, very often, pumping must be carried on. Thus, in one way or another, the availability of capital plays a vital part in the shaping of the world pattern of agriculture.

Some crops and some systems of agriculture have high labor requirements; others have low ones (Fig. 6–16). In any case, labor is always involved in agriculture to some degree; and, since labor must be paid for, except where a family can itself furnish all the labor needed, it is closely related to capital. On the Canadian farm mentioned above, the cash outlay for labor is practically nil. But on the Dakota wheat ranch, despite the wide use of machinery, labor costs are appreciable. And on the tropical plantation, they are still higher; for, even if individual wages are low, the great number of laborers required makes the total cost very high.

The problem of labor, of course, cannot always be solved by the availability of capital. In some agricultural regions, there is a severe shortage or a complete deficiency in the supply of labor; and often, the problem is compounded by the inability to find laborers elsewhere or, if found, to get them to go to the areas where they are needed. These conditions present one of the major obstacles to the production of rubber in such remote areas as the Amazon Basin. In contrast, in industrialized nations, particularly the United States, urban life has proved so attractive that a great influx of manpower into the cities has occurred. This has drained farms of much of their labor force, and, since there has been no compensating flow from the cities, it has thus created another, different kind of labor problem.

Socioeconomic Development

Two-thirds of the world's population lives in so-called developing countries, and the other third, in relatively developed ones (Fig. 6–17). The agrarian sector in most of the developing countries dominates the economy, which is based largely on subsistence farming and only sparingly on technology and industrialization. Such lack of development does not denote an absence of industrialization alone, however. In most cases, the agrarian sectors are themselves inefficient, often due, in part, to social institutions that make the adoption of new procedures slow and difficult.

Rapid agrarian advancement within this large segment of the world is obviously greatly handicapped wherever technology cannot be extensively and skillfully employed. Most of these countries are poor, not in terms of land, climate, or individual abilities, but in terms of organized application of modern techniques. India, Indonesia, and Zaire are only some of the agrarian countries whose populations live at low economic levels on land of considerable unused potential. On the other hand, Japan, New Zealand, and Denmark offer examples of people living at much higher economic levels on land of relatively little natural potential.

Government and Politics

Under any and all forms of government, some physical and cultural controls over agriculture may have less than their usual importance. If agricultural activities take the form and location dictated by the government, wheat may be grown in a region that is better-suited for maize, cotton may occupy land better-suited for soybeans, and these and other food crops may be raised in regions more easily adapted to grazing or forestry. Such misapplications are often the result of a government's attempts to promote self-sufficiency for its people, regardless of cost, or to ensure a harvest large enough to accommodate sales to other nations. Strong governmental control of this sort affects the agriculture of only a few nations, but among these are two of the major powers in world agriculture, the Soviet Union and China.

Control of agriculture within national boundaries is by no means limited to totalitarian states; nontotalitarian countries may also establish direct or indirect agricultural controls. At times of overproduction, a government may pay its independent farmers not to plant particular crops and may deny them certain privileges if they fail to cooperate. In other instances, a collapsing market may be reinforced by government price-setting or by large-scale government buying.

DEVELOPED (ABOVE WORLD AVERAGE GDI)

Africa	*Americas*	*Asia*	*Europe*	*Oceania*
South Africa	Argentina	Israel	Austria	Australia
	Canada	Japan	Belgium	New Zealand
	Chile	Soviet Union	Czechoslovakia	
	Cuba		Denmark	
	Mexico		Finland	
	Puerto Rico		France	
	United States		Germany, E.	
	Uruguay		Germany, W.	
	Venezuela		Hungary	
			Ireland	
			Italy	
			Netherlands	
			Norway	
			Poland	
			Sweden	
			Switzerland	
			United Kingdom	

DEVELOPING (BELOW WORLD AVERAGE GDI)

Africa	*Americas*	*Asia*		*Europe*
Algeria	Bolivia	Afghanistan	Syria	Albania
Cameroon	Brazil	Bangladesh	Taiwan	Bulgaria
Egypt (U.A.R)	British West Indies	Burma	Thailand	Greece
Ethiopia	Colombia	Cambodia	Turkey	Portugal
Ghana	Costa Rica	China	Vietnam	Rumania
Liberia	Dominican Republic	India	Yemen	Spain
Libya	Ecuador	Indonesia		Yugoslavia
Malagasy	El Salvador	Iran		
Morocco	Guatemala	Iraq		
Mozambique	Haiti	Jordan		
Nigeria	Honduras	Korea		
Rhodesia	Nicaragua	Laos		
Sierra Leone	Paraguay	Lebanon		
Sudan	Peru	Malaysia		
Tanzania		Nepal		
Tunisia		Pakistan		
Uganda		Philippines		
Zaire		Saudi Arabia		
Zambia		Sri Lanka		

FIGURE 6–17

A categorization of developed and developing nations on the basis of *per capita* gross domestic income (GDI).

For example, during the Great Depression, when the world coffee market dropped to a very low level, the Brazilian government bought huge amounts of coffee from its citizens in an attempt at providing price support. Though much of the coffee was stored in warehouses, thousands of pounds of it were wasted, being dumped into the sea or burned. These same agricultural controls have also been used at various times in the United States. Agriculture may face other types of governmental controls as well. In many parts of the world, vast areas are set aside to serve as national forests, parks, and game preserves. Within them, all private uses of land, including agriculture, are stringently regulated or prohibited.

The quality and packaging of those exported agricultural products that cross international boundaries are often subject to strict governmental regulations; and the importation of crops is sometimes prohibited from certain areas but allowed from others. Many of these rules represent attempts to ensure the consumption of high-grade, disease-free products; but, sometimes, such regulations are imposed for purely political reasons.

Tariff laws also affect the movement of farm products. The extremely high prices set by many of these laws are established in order to prevent imports; others are made just high enough to protect domestic farmers from the competition of foreign products that can be sold cheaply due to their low production costs. The United States tariff on foreign sugar, for example, though not high enough to prevent imports, does enable domestic raisers of sugar cane and, especially, sugar beets, to meet foreign competition. This law is very effective because, without it, there probably would be no commercial sugar-cane or sugar-beet production in the continental United States.

Transportation and Markets

The problems of handling, moving, and exchanging agricultural products are extremely varied. If products are consumed by their producers, transporting them is usually fairly simple: in China, most of the rice produced is consumed within sight of the field where it grows; in Ireland, most of the potatoes grown are consumed on the farms that produce them. When commercial production of goods is involved, however, rapid and efficient means of transportation are of paramount importance. If a product such as sugar cane is not moved quickly from the fields to the mills, the juice in the stalks will sour; if bananas are not picked and shipped while green, they will be overripe when they reach distant ports. In short, the problem of transportation is particularly critical when any perishable product is involved. Where the proper type of transportation is not available, certain kinds of agricultural production cannot succeed, no matter how favorable all other physical and cultural factors may be.

Wheat, once harvested, is practically imperishable; yet, even the handling of this durable crop demands special transport equipment and storage facilities. The harvested wheat must be hauled to the railroad, where it is stored and kept dry in large elevators until it can be transferred to railroad cars and then moved to milling centers for processing or to ports for export (Fig. 6–18). Canadian wheat from southern Manitoba, for example, is transported first by rail to Fort William, on Lake Superior, then by lake boat to such ports as Kingston, on Lake Ontario, or Montreal, on the St. Lawrence River, and then across the Atlantic to a large wheat-receiving port, such as Liverpool, from which it may be carried to Asia or Africa.

Obviously, there is no point in producing commodities for the purpose of moving them over long distances unless there is a market for them at their destination. The price, which is extremely variable, which any product can command on the market is the final determinant of much farm production. The southwestern United States could be used as a supplier of huge quantities of natural rubber (from the guayule plant, a desert shrub), if rubber could command a selling price high enough to cover the extraordinary costs. Bananas might be produced in Greenland, under artificial conditions, if they could be sold at the high prices such an operation would require. When prices are high, even distant and marginal areas might be able to get their goods to market and sell them at a profit. When prices are exceptionally low, even regions favored by optimum natural conditions and proximity to markets may have trouble showing a return on their investment. Thus, a region may produce a cer-

FIGURE 6–18

Crop transportation and storage. Left, a newly cut cargo of Philippine sugar cane as it arrives at a refinery from fields nearby. (Philippine Travel and Tourist Association) Right, wheat elevators near a long-haul railway in Colorado. (Grant Heilman)

tain crop in some years but not in others, or farmers may withhold their goods from market in order to save additional, unredeemable costs or in hope of increased prices. Throughout the world, the production and distribution of agricultural goods is more closely determined by market values than by consumer need. Thus local scarcities, even in the area in which a crop is produced, may continue as long as markets elsewhere provide a greater monetary return.

Competing Activities

There are many parts of the world where certain crops are not grown, in spite of their being admirably suited to such production. That these lands are not under cultivation is usually because they have been given over to more valuable uses. In metropolitan Los Angeles, the once-numerous thriving and profitable walnut and orange groves, as well as vegetable and dairy farms, have been displaced by "urban sprawl"—by homes, factories, and freeways. As profitable as it once was on these lands, agriculture could not compete with urbanization. An interesting result has been the encouragement of agriculture in new areas removed from the spreading metropolis. Much the same process has occurred in many other metropolitan areas. As large cities have grown larger, and as nearby towns have become their satellites, land that once held farms has become the site of apartment houses, shopping centers, and factories; and the farms have had to move on (Fig. 6–19).

Some displacements of agriculture are more abrupt: in north-central California, for example, during the late 1800s, huge dredges in search of gold chewed up many farms and left, in their stead, only debris-piled wastelands. Since the early 1900s, ranches in Texas and Oklahoma have given way to "forests" of oil derricks. In this same period, parts of New England and the mid-Atlantic states have seen their farmlands become resort and recreation areas, and other sections of the country have had power shovels and mining scars take the place of barns and planted fields. Though it does not represent an abuse of the environment, many mountain valleys and basins have also been lost to agriculture through their inclusion in national forests and parks. Thus, in one manner or another, competing activities, some regrettable and some commendable, continue to curtail or displace agricultural activities in many areas.

FIGURE 6–19

A recently built shopping center in Lancaster, Pennsylvania, on a site formerly occupied by small farms. (Grant Heilman)

Von Thünen's Location Theory

Because of the multitude of cultural and physical variables, there is no simple way to explain the spatial pattern of agricultural activities in the real world. The job is made easier, however, when approached by means of theoretical models, in which the number of variables can be controlled.

One of the first theoretical models for agricultural land-use was developed by Johann Heinrich von Thünen (1783–1850) and published in 1826 in his work *The Isolated State*. In the formulation and development of this theory, von Thünen drew heavily on his experience as a gentleman farmer on the North German Plain as well as on his knowledge of works by such prominent contemporary economic theorists as Adam Smith and David Ricardo.

The general intent of von Thünen's work, especially from a geographic point of view, was to demonstrate the influence of various cultural and physical factors on the spatial arrangement of different types of agricultural activities. For example, he was especially interested in determining the role of agricultural prices and distance from the marketplace in forming the rural land-use pattern in a free-enterprise system.

In order to do this, he developed a series of models; the first was the simplest, and the others, progressively complex modifications of the original. Each of the models represented an "isolated state," a totally self-contained, self-sufficient country with no outside connections of any sort—in a sense, each was a country in a test tube, to which he could add various elements as desired. His first model, or isolated state, for example, included the smallest number of elements: a single marketplace with a rural tributary area, or *hinterland*, surrounding it. In order to more closely approximate reality, this arrangement was complicated in subsequent models by the addition of multiple markets, a navigable river, tariffs, and so on.

About his initial model, von Thünen made these assumptions:

(1) There is a single urban market centrally located on a flat plain with uniform environmental characteristics (topography, climate, soils, fertility, and so on), and this market and its hinterland, solely dependent on each other, freely exchange all goods and services.
(2) The state's economic system is free and competitive, and it encourages the entrepreneur to maximize profits as permitted by the process of supply and demand.
(3) For any given product, all land is equal in terms of on-site production costs and productivity per unit.
(4) Finally, there is one mode of transportation (the horse-drawn wagon) throughout the plain, providing equal accessibility to the marketplace from all parts of the hinterland.

The first major variable in the model is the cost the farmers incur in getting their products to market. These transport costs are directly proportional not only to their distance from the marketplace but also to the weight, bulk, and perishability of the products involved. The second major variable is the price of each agricultural product, which is determined in the marketplace by the law of supply and demand.

Thus, the spatial pattern of agriculture in this von Thünen model depends primarily on the interaction between transport costs and market price. It is through the mechanism of *economic rent* that these variables exert their influence. The term *economic rent* refers to payments required for the use of scarce, nonreproducible resources. In the context of von Thünen's model, it is the amount paid by a farmer for a unit of land, a resource that is scarce and in fixed supply. Since the allocation of land around the marketplace is based on competitive bidding, the cost of this nonreproducible resource is determined solely by the existing demand for it.

The more profit a farmer can expect to make per unit of land, the higher his bidding power (or manageable economic rent) and, thus, his chance of occupying land close to the central market. Note that a farmer's profit is measured not in terms of return per unit of weight (pound, kilogram, and so on), but rather in terms of return per unit of land (acre, hectare, and so on). Since, due to the farmer's transportation costs, the margin of profit per acre for a particular product decreases with increasing distance from the market, there is a distance from the market beyond which a product would bring no profit at all. Thus, to be profitable, all products must be produced on the market side of their critical distance. However, since only the most valuable product can provide the amounts commanded by the most desirable areas in competitive bidding, this product will be the only one adequate in the near-market areas. Farms in adjacent areas, with higher transportation costs, will be at a disadvantage if they produce the same product; their chances of showing a profit are better if they avoid competing with near-market areas and produce the next most profitable product. Farms even farther from the market will find it advisable to avoid both the most profitable product and the runner-up or any product produced in an area that is closer to the market than they are.

The result is the formation of a series of distinct circular zones around the market, each with its own agricultural products (Fig. 6–20). The farmers whose products yield the higher per acre profits are found on the more expensive land closer to the market, and those whose per acre profits are lower occupy the less expensive land farther from the market.

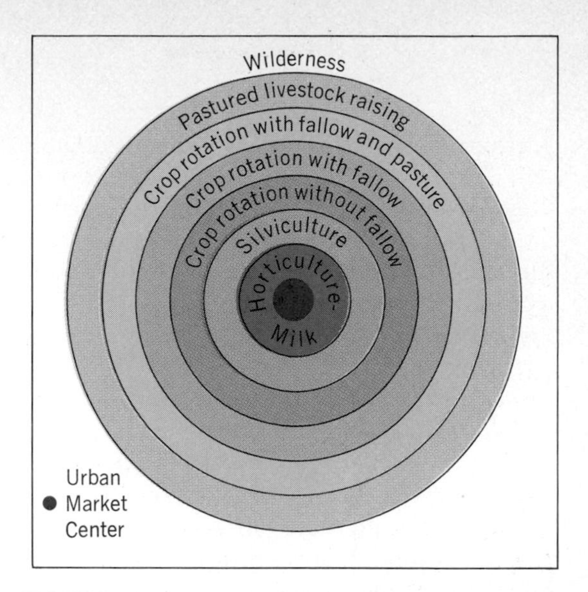

FIGURE 6–20

The agricultural zones that develop around von Thünen's urban market center.

On the basis of von Thünen's assumptions about transportation costs, the maximum distance from the market at which a given product can be produced profitably tends to vary inversely with the difficulty involved in transporting it. In von Thünen's time the restrictions of heavy, bulky, and perishable products were particularly acute.

On the land closest to the market center, farmers maximized their profits by selling milk and such horticultural items as fruits and vegetables. The market demand for these highly perishable products (due, mainly, to poor techniques of food preservation and slow means of transport) determined the extent of this first zone. Inhabitants of zone two specialized in silviculture. This seemingly odd use of such expensive land actually made good sense during von Thünen's time, because forest products were the main source of both fuel and building materials.

Furthermore, the rapid decrease in economic rent (profit) due to the increasing cost of transportation meant that these bulky commodities could not be produced too far from the market. Zones three, four, and five all featured commercial crop and livestock farming, with the intensity of cultivation decreasing outward from the market. Zone three had a one-field system of intensive crop rotation, and its marketable items were grains, potatoes, and any surplus livestock products. Zone four had a two-field system that allowed it to alternate sowing and fallow periods without halting production; its most profitable products were rye, livestock, and butter and cheese. Zone five had a less intensive, three-field rotation system wherein each farm was equally divided between cropland, fallowland, and pasture; this zone offered the same marketable products as zone four, but in smaller quantities. Zone six was devoted to pastured livestock-raising—an *extensive*, or low-labor, form of land-use; its commercial products included livestock, which could be walked to market, and such processed dairy foods as butter and cheese, which, less bulky and perishable than the products of zone one, could also be transported economically. Beyond the last zone was a wilderness area, considered unprofitable for any commercial agriculture.

Following this basic and very simplistic model, von Thünen extended his theoretical analysis through models with variables more complex than those attending a single centrally located market. Although none of his models corresponded to any real-life situation exactly, the basic locational principles he developed have been the foundation for many subsequent land-use theories and are still of value today in the spatial analysis of various cultural activities.

REVIEW AND DISCUSSION

1. What is the significance, in terms of cultural change, of findings suggesting that the early development of many advanced agricultural practices and techniques occurred somewhat simultaneously on opposite sides of the earth?
2. One of the world's major problems today is supplying the food needs of its growing population. Discuss and evaluate this problem, in light of the projected population increases by the year 2000. What are some of the major causes of undernourishment and malnutrition in different parts of the world?
3. FAO estimates place the world *per capita* daily caloric production at about 2,400 calories and the minimum *per capita* daily caloric need at about 2,300 calories. On the basis of these figures, there would appear to be a world *per capita* caloric sufficiency, but in fact there is a deficiency. Discuss.
4. Which probably serves as a better guide to the "quality" of the human diet, estimates of protein consumption or estimates of caloric consumption? Explain.
5. "In general, the influence of environmental factors, considered both singly and as a group, declines as cultural factors advance in sophistication and complexity." Discuss and evaluate this contention. In what ways is it valid and/or invalid with respect to the Dust Bowl area of the United States?
6. Of all the environmental controls over agrarian production, which exerts the most influence? Why? Provide two examples that illustrate how human beings have begun to overcome this environmental control as well as others.
7. Of all the cultural controls over agrarian production, which is the most potent? Does this cultural control exert equal influence in all parts of the world where it is operative? Explain with appropriate examples.

8. What was von Thünen attempting to demonstrate by means of his theoretical models?
9. What cultural controls are operative in von Thünen's location-theory model?
10. Explain the pattern of agricultural land-use in the area around von Thünen's central marketplace. Why does this pattern develop? To what extent would it be likely to occur in the real world?

SELECTED REFERENCES

Braidwood, R. J. "The Agricultural Revolution." *Scientific American*, Vol. 203 (September 1960), 131–48.

Chisholm, M. *Rural Settlement and Land Use: An Essay in Location.* London: Hutchinson University Library, 1962.

Clark, C. *Population Growth and Land Use.* London: Macmillan, 1968.

Clark, C., and M. Haswell. *The Economics of Subsistence Agriculture.* London: St. Martin's, 1964.

Cutler, H. "Food Sources in the New World." *Agricultural History*, Vol. 28 (April 1954), 43–49.

Ehrlich, P. R., and A. H. Ehrlich. *Population, Resources, Environment.* San Francisco: Freeman, 1970.

Flannery, K. F. "Ecology of Early Food Production in Mesopotamia." *Science*, Vol. 147 (12 March 1965), 1247–55.

Grotewold, A. "Von Thünen in Retrospect." *Economic Geography*, Vol. 35 (October 1959), 346–55.

Hall, P., ed. *Von Thünen's Isolated State.* Oxford: Pergamon Press, 1966.

Harris, D. R. "New Light on Plant Domestication and the Origins of Agriculture." *Geographical Review*, Vol. 57 (January 1967), 90–107.

Higbee, E. *American Agriculture.* New York: Wiley, 1958.

———. *Farms and Farmers in an Urban Age.* New York: Twentieth Century Fund, 1963.

Holden, C. "Food and Nutrition." *Science*, Vol. 184 (3 May 1974), 548–50.

Isaac, E. *Geography of Domestication.* Englewood Cliffs, N.J.: Prentice-Hall, 1970.

Kramer, S. N., and the editors of Time-Life Books. *Cradle of Civilization.* New York: Time, Inc., 1967.

MacFadden, C. H. "Food and Man in Ceylon." *Papers of the Michigan Academy of Science, Arts, and Letters*, Vol. 38 (1953), 323–29.

Malthus, T. R. "An Essay on Population," Vol. 1. New York: Dutton, Everyman's Library, 1933.

Peet, J. R. "The Present Pertinence of von Thünen Theory." *Annals of the Association of American Geographers*, Vol. 57 (December 1967), 810–11.

Sauer, C. O. *Agricultural Origins and Dispersals.* New York: American Geographical Society, 1952.

———. "Sedentary and Mobile Bents in Early Societies." *Social Life of Early Man.* Publications in Anthropology No. 31. Viking Fund, 1961.

Spencer, J. E. *Shifting Cultivation in Southeast Asia.* University of California Publications in Geography No. 19 (1966).

Staley, E. *Future of Underdeveloped Countries.* New York: Praeger, 1961.

Von Thünen, J. H. *Der Isolierte Staat in Beziehung auf Land-wirth-schaft und Nationalökonomie*, 3 vols. Hamburg, Germany: Friedrich Perthers, 1826.

"The World Food Problem." *Report of the Panel on the World Food Supply*. Vols. 1, 2, and 3. Washington, D.C.: U.S. Government Printing Office, 1967.

Al Qatif, a Saudi Arabian coastal town surrounded by irrigated date gardens.
(Standard Oil Co., N.J.)

7

Agrarian Societies: Changes and Diffusions

For perhaps three million years human beings lived by hunting and gathering; only about ten thousand years ago did there develop on this planet more complex forms of agrarian living. The types of agricultural societies now in existence thus represent the latest stage in a relatively short period of human responses to natural and cultural agrarian controls. Their functional diversity is, with certainty, even more recent, developing, for the most part, within the last two hundred years and owing its extent mainly to spatial variations in the incorporation of technology and scientific knowledge. For the purposes of this text, therefore, contemporary agricultural peoples are divided into *preindustrial agrarian societies* and *industrial agrarian societies* (Fig. 7–1).

Preindustrial agrarian societies:
- *Subsistence*
  - Nomadic hunting and gathering
  - Nomadic herding
  - Shifting cultivation
  - Sedentary subsistence farming
    - Rice-dominant, or sawah, regions
    - Non-rice-dominant regions

Industrial agrarian societies:
- *Commercial*
  - Crop and livestock farming
  - Dry-grain ranching
  - Tropical plantations

FIGURE 7–1

The major types of agrarian societies in existence throughout the world.

Preindustrial Agrarian Societies: Subsistence

Nomadic Hunting and Gathering

Nomadic hunting-and-gathering people still exist over some sparsely populated but extensive regions of the world (Fig. 7–2). Like most nomads before them, they occupy some of the earth's least favorable lands, those rejected by the inhabitants of permanent settlements because of extreme weather conditions or rugged terrain (Fig. 7–3).

Theirs are largely the fringe areas of the inhabited world, seemingly barren lands with little apparent potential: the poleward margins of the Americas and Eurasia; scattered interior areas of Middle America, South America, and Africa; and the highland portions of Southeast Asia. Anthropologists have long since learned that geographic isolation is an important safeguard to the continuing existence and life style of these hunting-and-gathering societies. Most of these groups have dwindled drastically in the last hundred years, and today they comprise no more than a quarter of a million people, the majority of whom are found in Africa. Notable among them are the Bushmen of the Kalahari Desert, in southern Africa, the Negritos of the Philippines, and a number of Indian tribes of South America.

Nomadic Herding

With the earliest domestication of animals (which may even predate the first plant domestications), many nomadic peoples turned from hunting as their principal livelihood to herding. Hours once spent stalking and attacking game were now devoted to tending groups of relatively pacific animals —to moving them between areas of pasture

FIGURE 7–2

The regional development of predominant livelihoods has been influenced not only by local environmental conditions but by technological levels, cultural practices, and political policies as well.

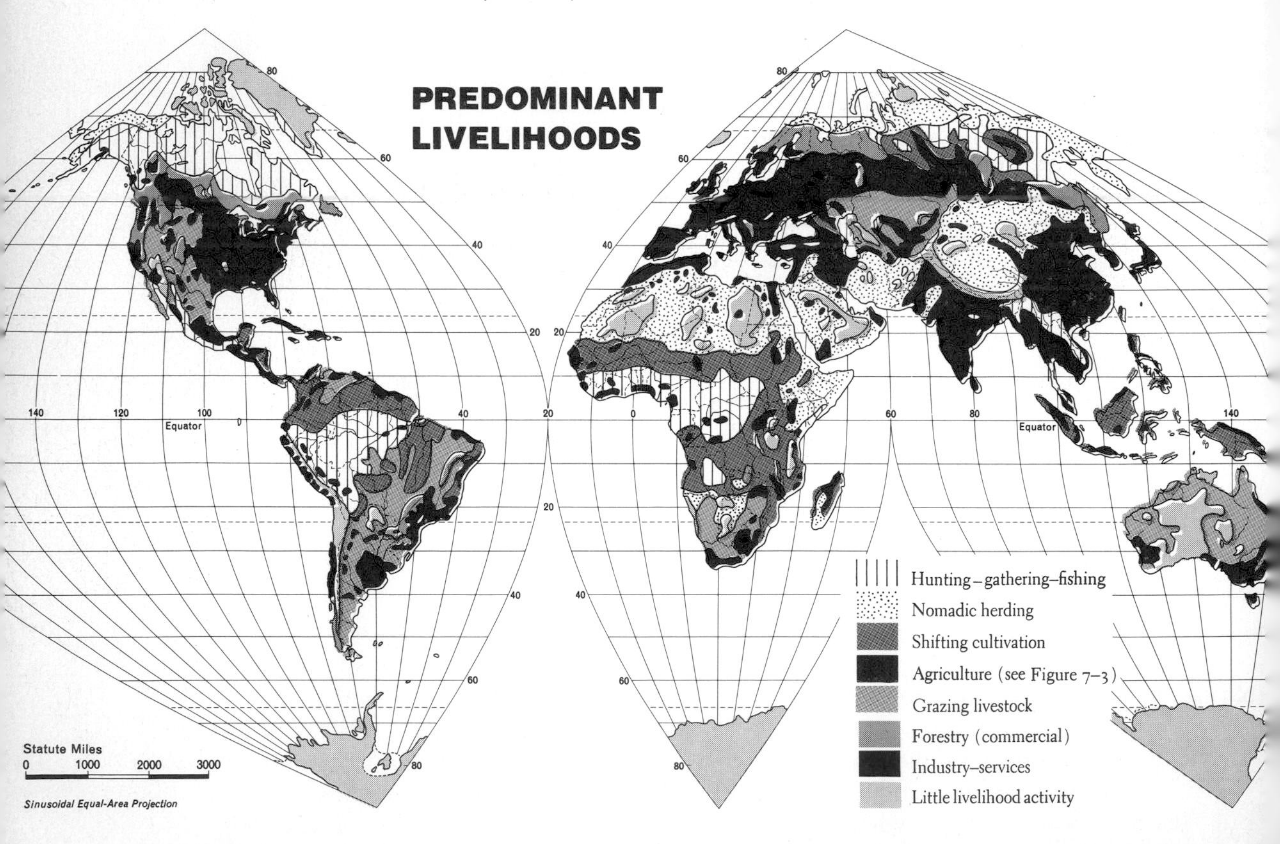

and water and protecting them from predators and the elements. By their labors, the herders gained fairly dependable supplies of milk and, with selective slaughtering, blood, meat, clothing, and building materials.

The contemporary herder can generally support more people than the hunter-gatherer, but faces, almost equally, the disadvantages of forced group transience. Although not daily at the mercy of their surroundings for food, nomadic-herding societies, like hunting-gathering peoples, can be quickly devastated by almost any natural disaster. Without adequate shelters and artificial support systems, their herds can be wiped out by even a mild drought or flood. And, since they reckon wealth in terms of animals owned, and seldom receive cash, nomadic herding societies rarely have any accumulated purchasing power with which to replace a flock or herd (Fig. 7–4).

Present-day nomadic herders are located largely in the arid and semiarid regions of the Eastern world, where their areas form a broad belt across North Africa, the Middle East, and much of central Asia, an east-west distance of well over eight thousand miles. The domestic animals characteristic of the nomadic-herding societies in these regions are cattle, sheep, goats, camels, and horses.

FIGURE 7–3

Only about 10 percent of the world's land area is used for crop production—a very low percentage given the average yield per acre and the number of people to be fed. Moreover, some of the lowest yields occur in the most densely populated areas.

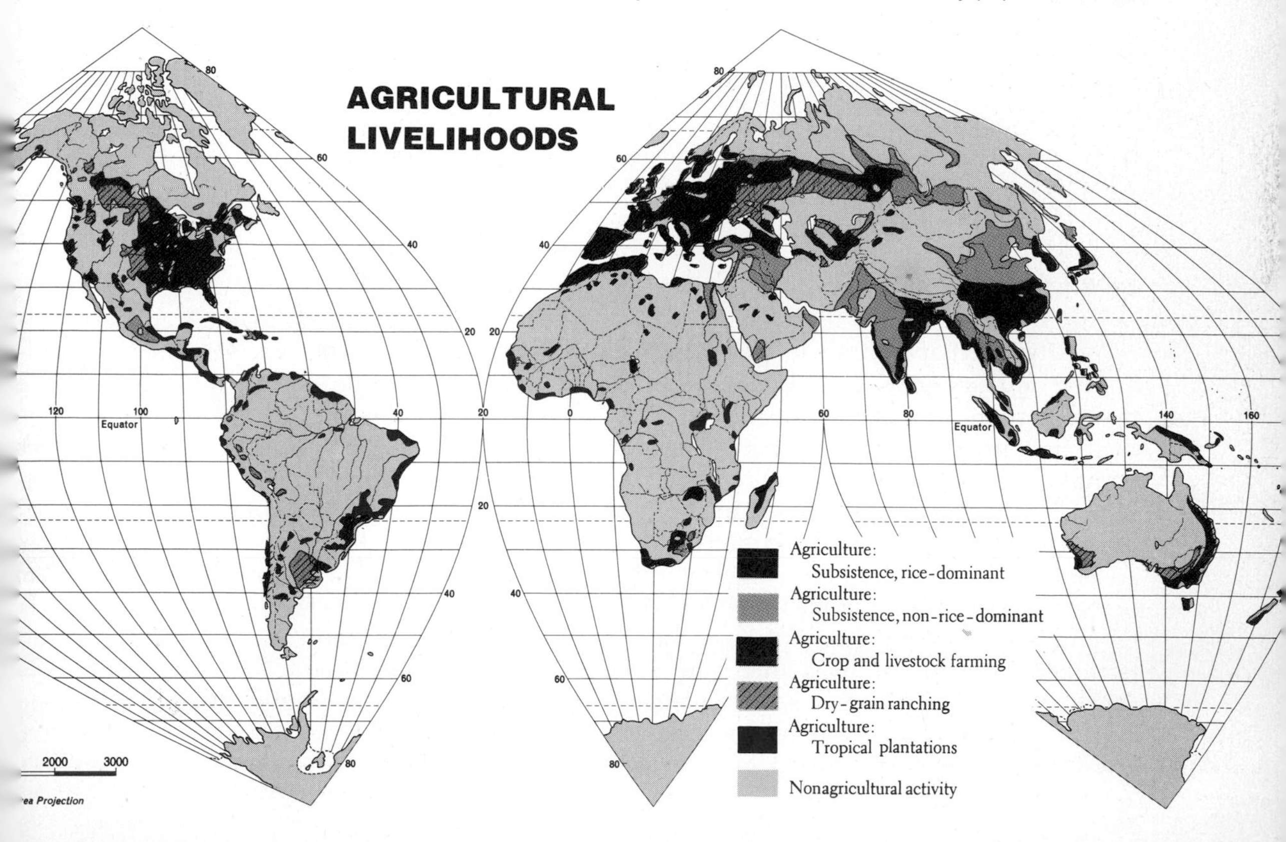

FIGURE 7–4
Malian nomads in Upper Volta's Christina Wells. These nomads migrated from Mali in search of water, but found only more drought and starvation, hardships widespread in northwestern Africa in the late 1960s and early '70s. The woman pounds scarce bits of grain for her family, but the trees and a steer have already succumbed to the drought. (United Nations)

Shifting Cultivation

After perhaps three million years as hunters and gatherers, and a brief time as nomadic herders, human beings advanced to their first involvement in rudimentary cultivation, by tending a few plants in a succession of small forest clearings. The trees in these clearings were usually either felled or girdled and left standing for firing, the ashes of the small trees being mixed with forest litter and spread over the dark soils as natural fertilizer. The ground was then dug with sticks or hoes, and plants or seeds—usually a variety of beans, yams, taro, cassava, melons, and other hardy horticultural plants—were inserted. When the fertility of a clearing declined and weeds overpowered the crops, a new area was cleared (Fig. 7–5).

With this shifting cultivation there developed, for the first time in human history, the possibility of semipermanent settlement and attachment to the land. And the need to traverse vast areas in pursuit of game, or to move continuously in search of pasture for herds and flocks, was thus greatly reduced.

Shifting cultivation became widespread over many parts of the world, particularly within tropical-humid regions of Africa, the Middle East, Middle America, northeastern China, and southern and southeastern Asia. It is still the prevailing form of agriculture in many parts of the world and is known, variously, as *milpa,* in Latin America; *fang,* in Africa; *ladang, taung-ya,* and *chena* in parts of southern and southeastern Asia; and the Old English term *swidden* is sometimes used by Western research scholars to mean a "burned clearing." Today, there are probably fifty million people practicing shifting cultivation; more than half of them are located in central Africa and Indonesia (Fig. 7–6).

Sedentary Subsistence Farming

Humanity's first major step into agriculture, through the practice of shifting cultivation, was followed, in the Middle Stone Age, by the development of *sedentary subsistence farming,* the working of a specific piece of land on a long-term basis to satisfy personal food needs. This prehistoric form of agriculture persists in many contemporary societies and affects about half the people of the world; it is still the major livelihood in the tropical and midlatitude lands of southern and eastern Asia. This way of life produces all or most of the available foods and other essential materials used by almost two billion people; but seldom does it provide

FIGURE 7–5

The practice of slash-and-burn agriculture in the Amazon Basin, Brazil. (Loren McIntyre, Woodfin Camp)

FIGURE 7–6

Much of the world's land area is ill-suited to human occupance based upon agriculture. Many societies thus eke out an existence within environments that are too cold, too dry, too hot and wet, and/or too high and steep for easy cultivation.

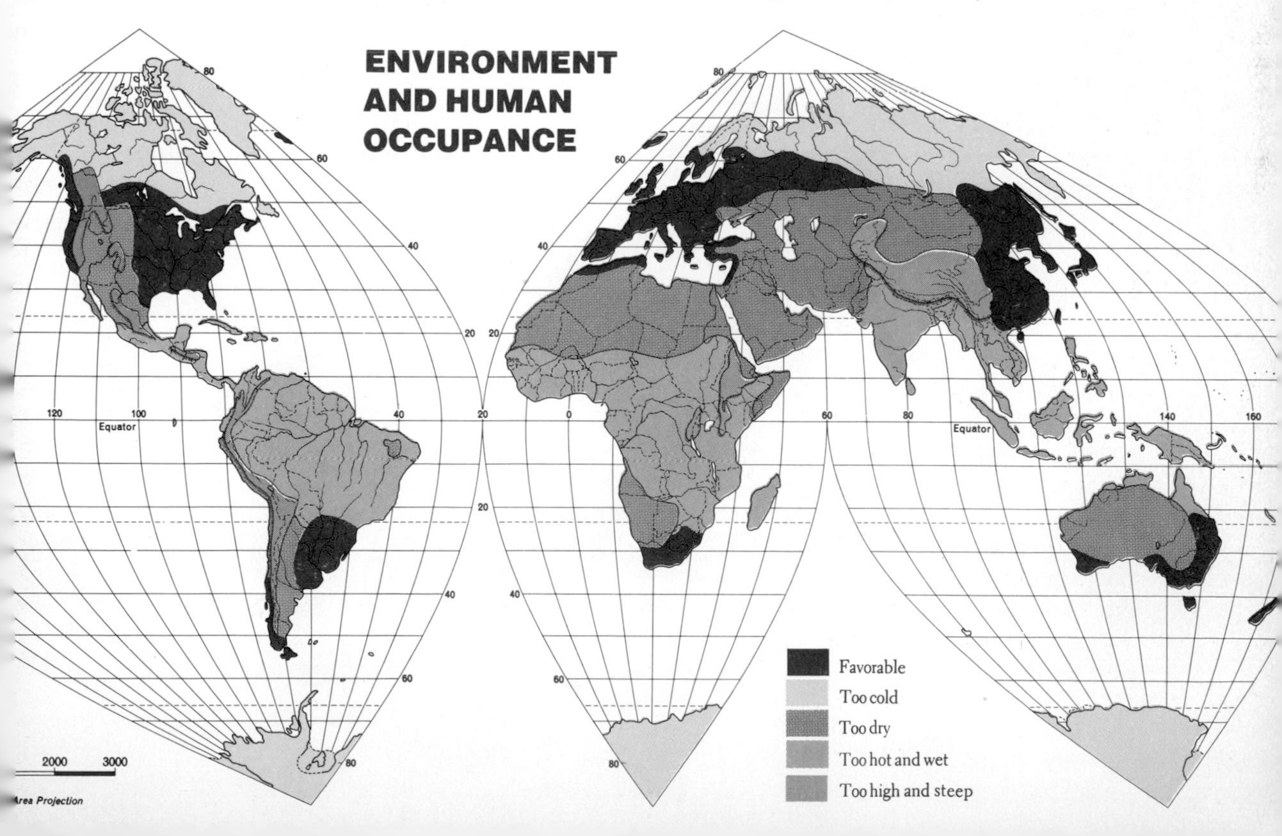

surpluses to be stored or sold. Subsistence agriculture is also *intensive*, that is, it uses high amounts of human labor per acre.

The sedentary subsistence-agriculture societies of today differ very little from those of ancient times. Typically, their farm plots are only one or two acres in extent and are often fragmented into several widely separated miniplots; these are usually devoted to the growing of different crops, in order to provide the family with some semblance of dietary balance, and intensively worked by family members, with the help of few tools, few or no draft animals, and practically no mechanical power (Fig. 7–7).

The dominant crops of most sedentary subsistence-agriculture societies—rice, in wet areas, and wheat, millet, and leguminous plants, in drier areas—are all foods high in starch and low in proteins and fats. In some areas these starchy foods are accompanied, in the human diet, by leafy vegetables, usually in fair variety but small quantity, by bananas and other fruits, and by small and infrequent additions of meats, poultry, and diluted dairy products.

FIGURE 7–7

Near New Delhi, India, a small subsistence farm tended by family members with the aid of a pair of bullocks and some simple tools. (Marc and Evelyne Bernheim, Woodfin Camp)

Rice-Dominant, or Sawah, Regions. Rice, the staple food of much of southern and eastern Asia, so strongly dominates life and fortune from India to Japan that this entire area is sometimes called the "rice crescent," or the *sawah,* or wet-field rice, culture (Fig. 7–8). A major crop in world agriculture for several thousand years, rice today feeds more than one-third of the world's people. Its dominance throughout southern and eastern Asia is due primarily to the region's terrain and climate (Plates III and IV; Fig. 6–13). Because they generally thrive under irrigation, rice fields are largely free from cyclical harvest failures and produce continuously high yields over long periods; their crop is relatively simple to produce and store and makes a highly adaptable and easily prepared basic food.

The vast Asian rice crescent includes the countries of Pakistan, India, Sri Lanka, and Bangladesh—in South Asia; Burma, Thailand, Cambodia, Malaysia, Singapore, Vietnam, Laos, the Philippines, and Indonesia—in Southeast Asia; and Taiwan, China, North Korea, South Korea, and Japan—in East Asia (Plates II and III). Most rice production in this region is confined to the flatter coastal lands and flat-bottomed river valleys. In some areas, however, spectacular yields have been developed from hillside terraces, the oldest and most spectacular of which are those on Luzon Island, in the Philippines; others can be found in Sri Lanka, Japan, southern China, and Java.

In only a few regions—among them, Burma's Irrawaddy Delta and Thailand's peninsula—is rainfall usually sufficient for

FIGURE 7–8

Most of the world's rice production occurs in eastern and southern Asia, with China and India accounting for over 55 percent of the annual total. Most of the world's rice trade takes place within the same areas.

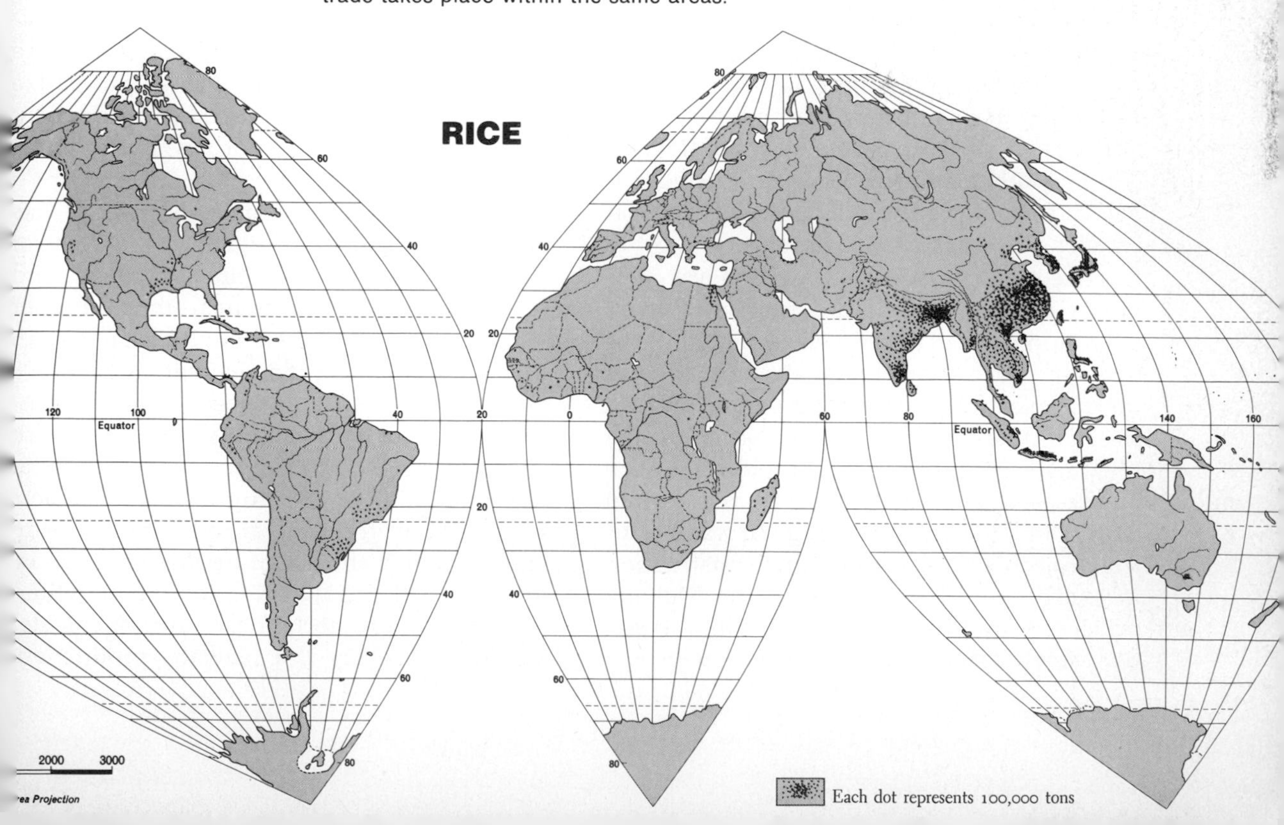

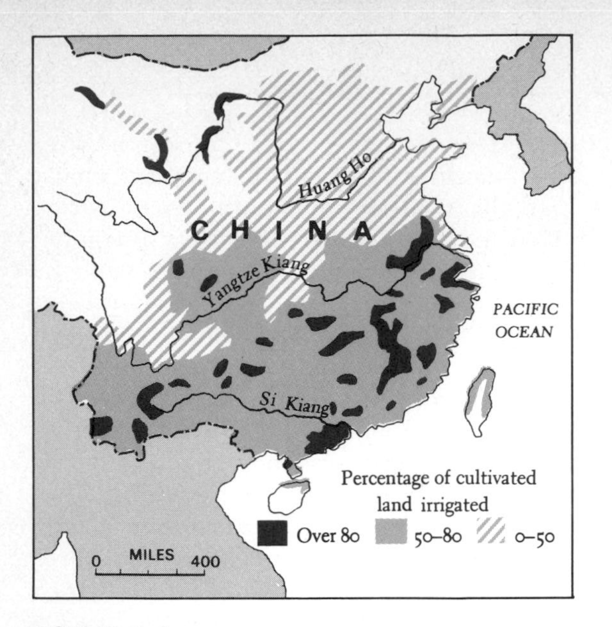

FIGURE 7–9

The extent of irrigation in China, the most intensively irrigated portion of the vast rice crescent. (Adapted from R. R. C. de Crespigny. *China: The Land and Its People*, St. Martin's Press, New York, 1971.)

rice production; in most rice-producing areas, some type of irrigation system has had to be developed and must be used every planting season. In the most intensively irrigated portion of the rice crescent, central and southern China, over 50 percent of all cultivated land is irrigated, with the percentage surpassing 80 in some sections (Fig. 7–9). Many other countries have developed irrigation as intensively as China, though over smaller areas relatively. Throughout most of the rice crescent, two rice harvests are commonly achieved in any one year; but where irrigation is particularly well regulated, three crops a year are not unusual. Illustrative of the complex irrigation systems necessary for rice production in much of Asia is the Mekong River Development Project. Like the Tennessee Valley Authority, in the United States, and most of the other great water-control projects completed or under construction in southern and eastern Asia, the Mekong project is multipurpose, combining irrigation, flood control, and power production.

The production of rice in Asia is an arduous task, one in which human labor is used almost exclusively. Following the first monsoons the soils are turned, by hand with a mattock or by bullock-drawn plows, and then flooded and worked into a thin sludge. Then, young rice plants are transplanted from nursery beds to the newly prepared fields (in Asia, a job done almost exclusively by women); finally, the plants are tended, weeded, and harvested (undertakings involving mixed labor). Most harvesting is done by hand with a sickle or knife, each head of rice being cut from its stalk separately to ensure minimum loss of the precious grains.

The breeding and raising of animals for human consumption is relatively minimal throughout most of the sawah system; however, many animals are used in the fields for draft purposes (but even they are provided little feed or pasture and thus must forage on the roadsides and irrigation embankments). As long as the traditional underemployment in sawah areas prevails—and few other occupational outlets seem in the offing—there will be no real incentive to further mechanize any major phase of Asian rice production. Humans will continue to carry the brunt of the workload (Fig. 7–10).

Where extensive mechanization has been applied to rice production—in areas of southern Europe and the southern and southwestern United States—much higher yields per acre have resulted. Rice harvests in California, for example, average about 4,500 pounds per acre; while those in Asian countries average 1,600 pounds per acre, with China at 2,100 pounds and Japan leading at 3,800 pounds. Japan's high yields per acre result largely from advanced techniques in seed selection, the use of quality commercial fertilizer, plentiful irrigation, and widespread use of mechanical equipment designed especially for sawah production

China's relatively high productivity is due largely to extensive use of organic fertilizers, including much "night soil" (human waste), and fairly advanced techniques of seed selection and irrigation; some successful harvests have also been attributed to the extended use of deep plowing and close planting. India's generally low efficiency levels in rice production, lower than either Japan's or China's, are attributable, in part, to less favorable climatic conditions, less well developed irrigation, and rather decided deficiencies in seed selection, fertilization, and pest control. However, the introduction of some Japanese agricultural methods into India has already shown some beneficial results.

IR-8, the so-called miracle rice—a high-yielding hybrid-grain plant developed in recent years at the International Rice Research Institute in the Philippines—has been cultivated experimentally in several Asian countries with good to mediocre results—varying usually with environment, expertise of farmers, informed use of commercial fertilizers, proper irrigation, and proper weed and pest controls. If most of these variables can be controlled, IR-8 seems to promise yields of two to six times those of traditional rice strains. Thus it may be the key to sustained increases in Asian rice production and thereby a stimulus to the region's socioeconomic progress (Fig. 7–11).

Non-Rice-Dominant Regions. Although adjacent to the sawah lands, most of the dry interior areas of India, Pakistan, and China north of the Yangtze River, as well as inland portions of Southeast Asia, differ from them considerably in terms of agriculture. This disparity reflects marked differences in moisture availability and a perhaps related regional contrast in individual wealth. Since their rainfall is deficient and their people cannot afford the high costs of irrigation, these interior areas of Asia must produce

FIGURE 7–10

Women planting rice in Vietnam. (Air France)

FIGURE 7–11

The threshing of IR-8, or "miracle" rice, at Manila's International Rice Research Institute. (The Rockefeller Foundation)

FIGURE 7–12

Terraced loessal fields in the highlands of northern interior China. (Marc Riboud from Magnum Photos)

food grains other than rice; their fields are planted with wheat, kaoliang, millets, and sorghum and, secondarily, with such crops as grams (chickpeas), peanuts, cotton, and sugar cane.

In the interior areas of India and Pakistan, *per capita* productivity is among the lowest in the world; there are too many people, on too little land, with too little irrigation and agricultural technology. In addition, in some places an antiquated land-tenure system severely hampers both the incentive and the efficiency of the people. Recently, land-reform movements, some meaningful governmental extension services and financial support, and improved seed grains and fertilizer supplies have helped to increase *per capita* production in some of these areas for the first time in generations. Much more is needed, however, to cope with the minimal needs of the burgeoning population and with the often extreme weather. Many of Asia's hot and dry interior lands, especially those in India, are heavily dependent on the rains of the monsoons; when the monsoons fluctuate, in either arrival time or amount of rainfall, crops in these areas suffer and frequently fail. With little or no food in reserve, the subsequent food shortages are often severe.

In China, where the landscape grades rather suddenly from the rice fields south of the Yangtze River to the winter wheat and kaoliang fields north of that river, agriculture is not quite as precarious as it is in Pakistan or India. The northern interior areas of China are mainly flat expanses of fertile and less fertile soils, with rainfall less frequent and certain than that in the area to the south, and many periods of alternating drought and devastating flood. However, in spite of its dryness and poor irrigation systems, northern China is a highly productive region, providing much of the country's major domestic food grains, wheat and kaoliang, and a wide range of secondary crops, including barley, peas, soybeans, peanuts, maize, fruits, vegetables, sweet potatoes, and tobacco. Particularly in the highlands to the northwest, this productivity is helped by broad deposits of rich eolian silt, or loess. Several hundreds of feet thick in places, these deposits can support good agricultural production despite marginal rainfall and little irrigation. Many parts of these loessal highlands have been terraced in order to check the erosion of their valuable soils (Fig. 7–12). The lands of China north of the Yangtze could, of course, produce even more food and fiber if sufficient irrigation were available. Thus, building an adequate irrigation system is a task the Chinese have undertaken with great vigor in the past two decades; and the millions of laborers at work in this cause, though equipped with hand tools and few machines, have apparently made surprising progress (Fig. 7–13). If widespread famines are to be avoided, this kind of development must soon be undertaken on a large scale throughout Asia.

FIGURE 7–13

Dam building on the Huang Ho. To improve irrigation and flood control near the river, the Chinese have completed many such arduous construction projects in the past two decades. (© Henri Cartier-Bresson and Magnum Photos)

Industrial Agrarian Societies: Commercial

Many of the world's agrarian societies have been transformed, through the application of modern industrial technology, from various stages of crop- and livestock-raising, into industrial agrarian societies. These changed societies have learned how to utilize industrial planning, organization, and marketing techniques in the production and distribution of raw materials. In contrast to the subsistence farming described earlier, the average farm in an industrial agrarian society produces goods primarily for sale rather than solely to meet the nutritional needs of its occupants.

Crop and Livestock Farming

Typical of the industrial agrarian society is the practice of combining large-scale crop production and livestock raising, or *mixed farming*. Although different farms may raise different sorts of crops and livestock, most commercial mixed-farming operations follow similar practices of crop rotation (Fig. 7–14). Such practices are essential to high-productivity farming, for they maintain and build natural soil fertility by varying, with each crop, the net chemical output of the land.

Alternating fields so that there are always some left fallow, in order for their soil to regain its strength, is probably the oldest form of crop rotation, but it is usually considered too wasteful to be used in commercial farming. More viable is the practice of rebuilding exhausted soils by following a soil-depleting crop with the planting of soil-enriching legumes. Leguminous crops allow a soil to replenish its chemical supply, mainly by providing it with nitrogen. They also serve as fodder for livestock and allow a farm to raise more animals with little increase in its feed costs. These animals, in turn, supply manure (which can be used in soil maintenance), food, and, if they are marketable, cash income. Such coordination of crop and livestock production is essential to maximum utilization of agricultural land.

The crops generally included in crop-rotation systems in Europe, the United States, and the Soviet Union comprise three groups: row-tilled plants, such as maize, potatoes, beans, and sugar beets; close-drilled grains, such as wheat, oats, barley, and rye; and deep-sod-rooted hays, such as alfalfa, clover, and lespedeza. Typically, a rotation cycle can successfully include the raising of these crops in three, four, or even five stages. For example, a four-stage rotation cycle might begin with the planting of wheat or barley, as the cash crop; followed by alfalfa, for hay and pasture and as the means of adding nitrogen to the soil; followed by sugar beets or potatoes, to clean and cultivate the soil; and, lastly, by maize, for livestock feed and further soil cultivation (Fig. 7–15). Whatever the number of stages or the specific crops, a sound rotation system is one in which crops are mixed so as to provide not the highest possible immediate profits but the most profitable and most serviceable yields consistent with long-term, high soil-fertility.

In Europe, the average commercial crop and livestock farm consists of about 50 to 100 acres; the typical farm in the United States is a little larger, about 80 to 240 acres; and, in the Soviet Union, collective farms in the Ukraine, each housing hundreds of families, extend over several thousand acres (Fig. 7–16). The working of rented land, or *tenant farming,* is quite common in present-day commercial agriculture; and, among farmers

FIGURE 7–14

A large mixed crop and livestock farm in Pennsylvania. (Grant Heilman)

FIGURE 7–15

Maize production is heavy in North and South America, Europe, Asia, and Africa, with the United States accounting for half the world total. The world trade in maize is meager in comparison to the world trade in rice or wheat.

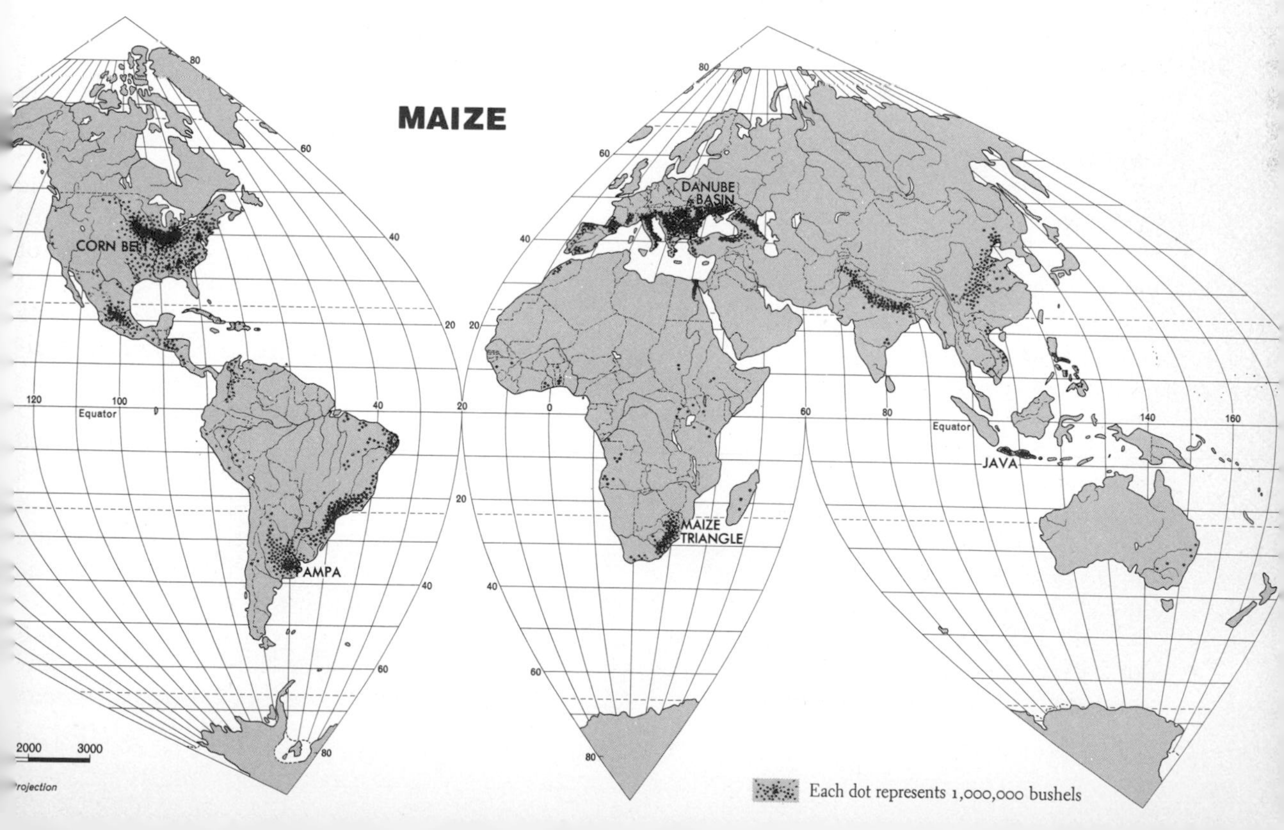

in Europe, it is even preferred. Not as widely practiced in North America, tenant farming in the United States today usually serves as a hopeful step toward ultimate ownership of the land. Europe, because of its smaller farms and the general strain on its economy following two world wars, trails considerably behind the United States and the Soviet Union in terms of agricultural mechanization but not much in type and variety.

Commercial crop and livestock farming dominates a broad region of Europe, from the Mediterranean to the North and Baltic seas, and extends across the Soviet Union, in a narrowing wedge, to beyond the Ural Mountains. In the United States, it is most evident in the eastern half of the country, from the Gulf and Atlantic plains northward to the Great Lakes and westward to beyond the Mississippi River.

The Corn Belt (the small middle region of the United States stretching from central Ohio to central Nebraska) comprises the world's most highly concentrated examples of large-scale commercial crop and livestock farming (Fig. 7–17). Much of the huge quantities of feed grains this region produces is converted into meat and fats by its large populations of commercial beef cattle and swine (Fig. 7–18).

Commercial agriculture in midlatitude regions of the southern continents is similar in form to that in Europe and North America, but it is much smaller in scale and less developed. Agricultural regions of South America, Africa, and Australia have relatively low per-

FIGURE 7–16

A collective farm in the Ukraine. In addition to tending the commune's fields, the member families raise fruits and vegetables in small private plots behind their homes. (Sovfoto)

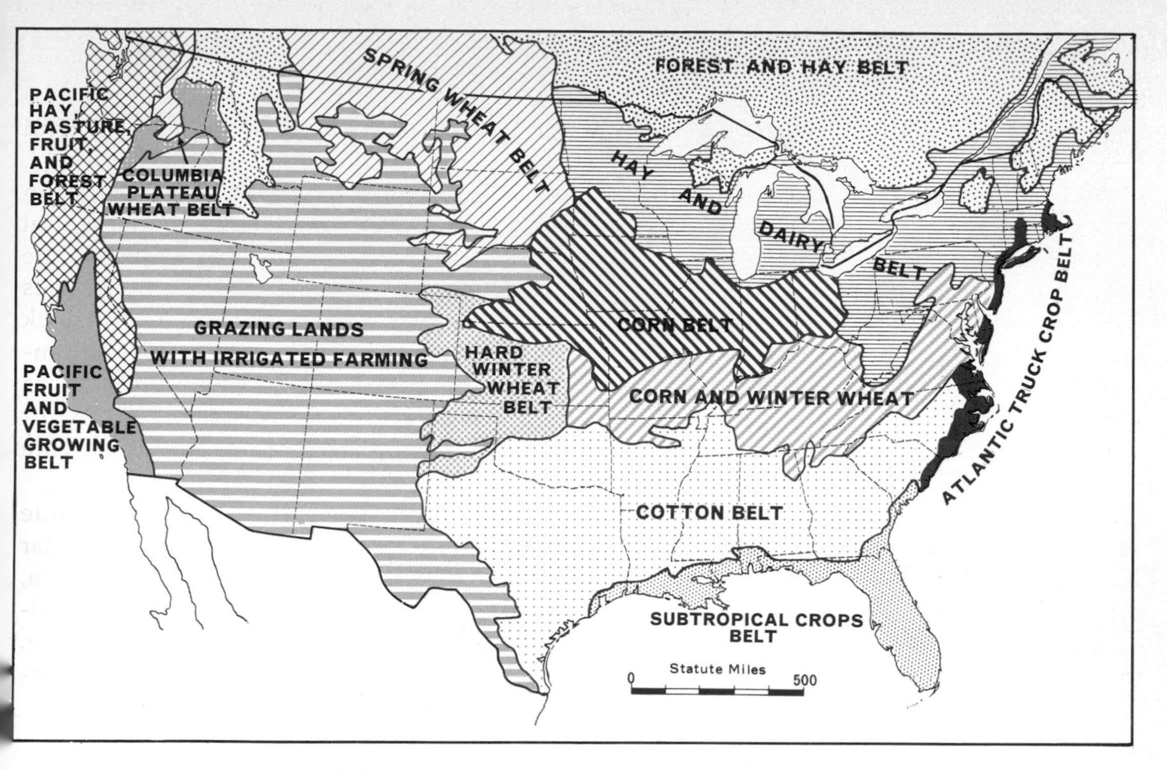

FIGURE 7–17

Agricultural regions of the coterminous United States and southern Canada. The array of physical conditions in this vast area allows it to produce a wide variety of products throughout the year. Even in the winter months, fruits and vegetables are being harvested in southern California and along the Gulf Coast.

FIGURE 7–18

Livestock production in the coterminous United States. A. Beef cattle, which, though raised in most sections of the country, are most numerous in the western part of the Corn Belt and elsewhere in the Great Plains. B. Swine, which are also major animal products of the Corn Belt.

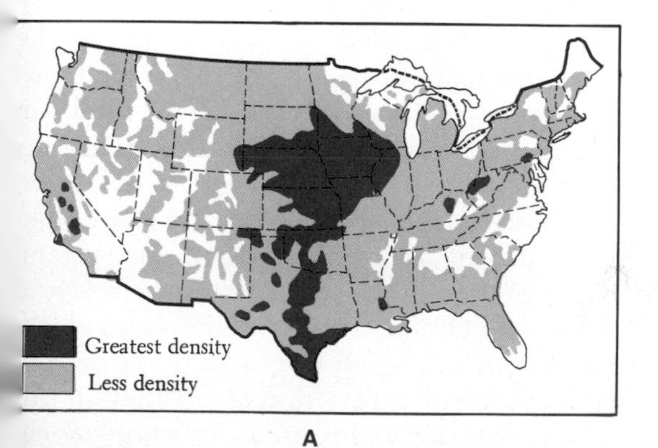

A

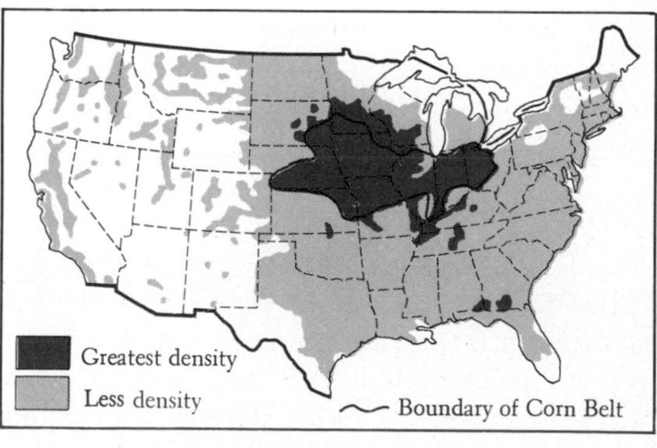

B

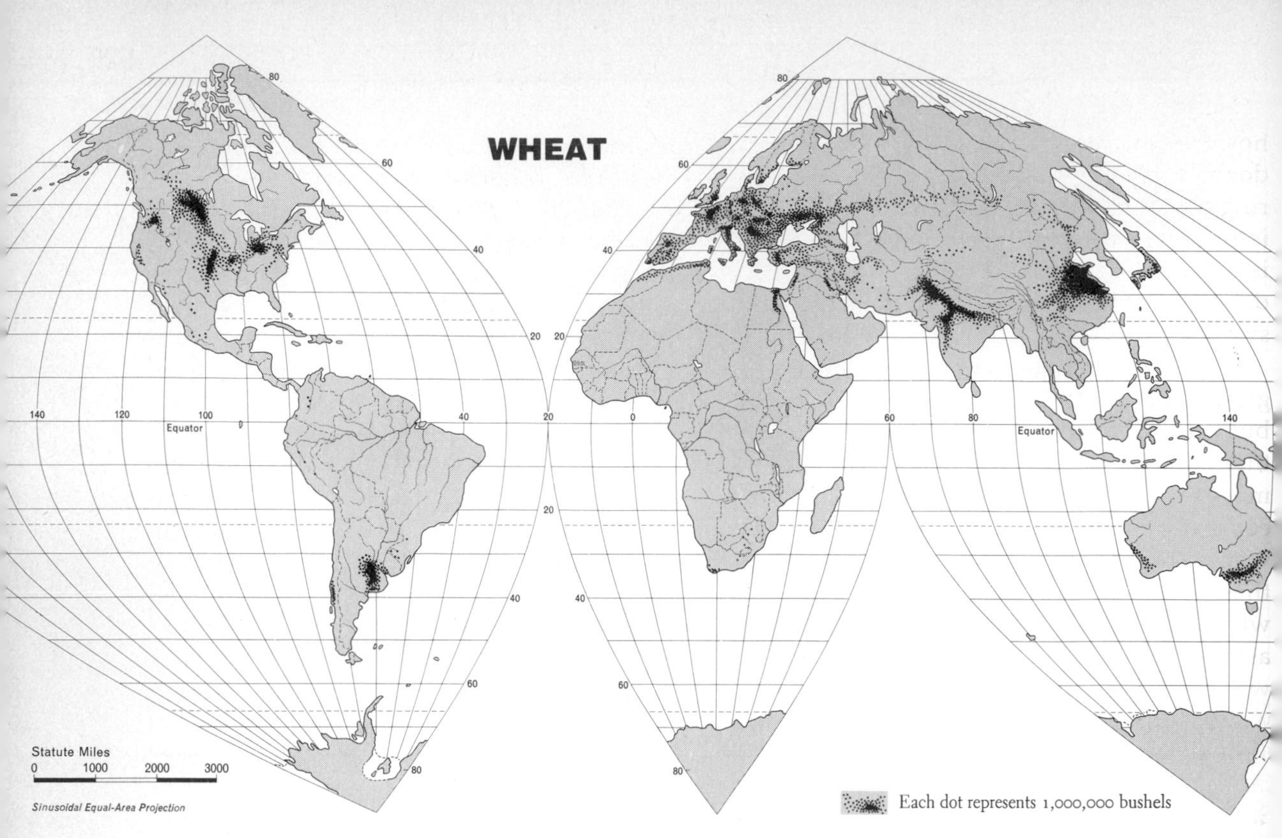

FIGURE 7–19

Wheat production takes place in all the continents, particularly in the midlatitudes of the Northern Hemisphere. Trade in wheat is also nearly worldwide.

centages of their total land devoted to the growing of crops, as much more of it is given over to use as semipermanent pastureland. This illustrates the fact that large-scale crop and livestock farming has not yet become widespread beyond Europe, North America, and the Soviet Union, where it has been a necessary way of meeting the expanding needs of these heavily industrialized societies.

Dry-Grain Ranching

The large-scale agricultural system developed most recently is *dry-grain ranching,* in which a farm produces a single dry grain to the virtual exclusion of other crops. As a provider of human foods, wheat-dominated monocultures of this sort constitute a system that is second only to that of sawah agriculture (Fig. 7–19). For, today, the dry grains are as essential to the diet of one half of the world as rice is to the other half.

With dry-grain ranching, many of yesterday's frontier grassland-prairies—grazing lands at best, vast wastelands if the sparse summer rains failed—are today among the world's most specialized commercial food-producing regions, spilling their bounty of wheat and lesser crops into practically every country of the world. Commercial dry-grain ranches usually comprise more acreage than is typical of any other agricultural system. Many of these ranches in North America and the Soviet Union are a thousand acres in extent, and, in these areas, one comprising several thousand acres is not considered unusual; per-acre yields on a dry-grain ranch however, are low, with returns of under a dozen bushels (720 pounds) per acre occurring frequently; harvests are uncertain; labor is minimal; mechanization is high; tenancy is negligible; and population is probably less dense than in any other major agricultural system (Fig. 7–20).

Wheat, the chief product of most dry-grain ranches, is raised in many countries, but is consumed at so high a rate that few of them can produce as much as they need. To meet their resultant deficiencies, then, most countries must buy wheat from the few wheat-surplus countries that exist. Among these few are the United States and Canada, which devote large portions of the lands astride their common border to commercial wheat production; thus, a highly productive wheat region extends across the plains of the United States, from Texas to the Dakotas, and far into those of Canada, in the provinces of Saskatchewan and Alberta.

The Soviet Union's large commercial wheat region extends 2,000 miles in length, from the southern plains of European Russia into Asia, beyond the Ural Mountains. The European part of this region, that which lies west of the Volga River, includes the soil-rich Ukraine, the "breadbasket of Russia," and is thus the most productive agricultural area of the Soviet Union.

In the Southern Hemisphere, the most productive commercial wheat areas are the *pampa* of Argentina and the southern part of Australia (Fig. 7–21). Although these regions account for relatively little of the world's total production (3 percent each), both Argentina and Australia are important for their large wheat exports, which often total as much as half their annual production. The world's largest exporter of wheat is the United States, providing 40 percent of the total wheat export, and Canada, whose exports total about half that amount, is second; France and the Soviet Union are the third and fourth largest, and Australia and Argentina follow. Australia exports its wheat principally to southern Asia, largely to India, but some goes to its Commonwealth partner, the United Kingdom, a country that over many decades has had only about a 30 percent self-sufficiency in staple foodstuffs.

IGURE 7–20

he harvesting of wheat, by utomatic combines, on a oviet State Farm near the lack Sea. (Sovfoto)

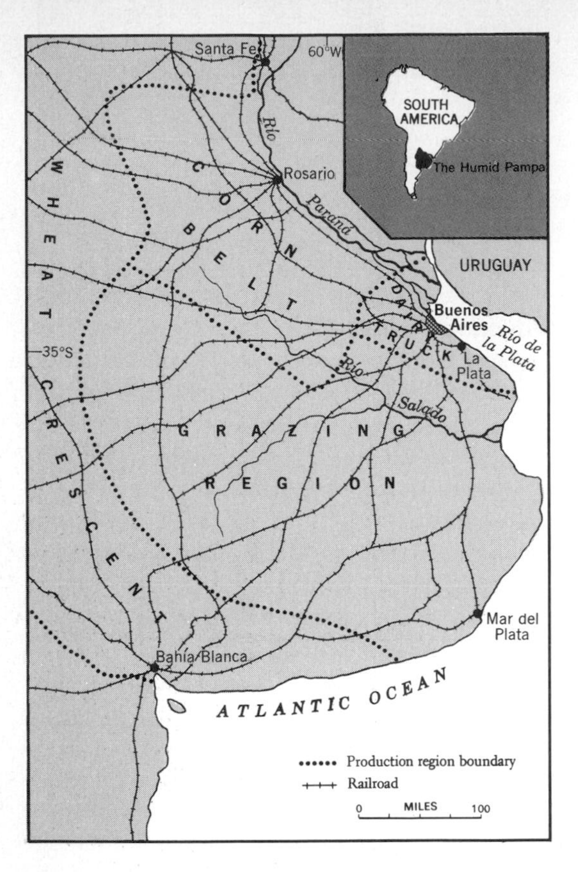

FIGURE 7–21

The humid *pampa* (grassland) of Argentina. Its fertile, level lands and mild-humid climates allow it to be the country's major food-producing area. Particularly productive is the semiarid western wheat crescent.

Tropical Plantations

In the eighteenth and nineteenth centuries, European colonialism led to the development of new agricultural societies across much of the world's humid tropical forest lands. Highly specialized, extremely commercial, and technologically advanced, these settlements, or *plantations,* still have an economic importance to the world beyond that normally attending the limited amount of land they occupy, as well as a substantial political importance to their host countries.

The islands in the Gulf of Guinea were probably the first sites of such plantations, but it was the sugar plantations of tropical northeastern Brazil and, later, those of the Caribbean Islands that firmly established this new type of agricultural society. Beginning in the late eighteenth century, colonialism brought the plantation system into Asia as well (Fig. 7–22). India and Ceylon (now Sri Lanka), Southeast Asia, and the East Indies were all introduced to the "great experiment" (Fig. 7–23). Unlike most agricultural societies, whose development has been gradual and evolutionary, the commercial tropical plantation was conceived and planned and

FIGURE 7–22

Tropical plantations. Left, cacao groves in the Caribbean. (Hershey Foods Corporation) Right, a tea plantation in Sri Lanka's central highlands. (United Nations)

then imposed, rather suddenly, for the satisfaction of personal or national coffers and international markets.

The original tropical plantations ranged in size from corporate estates of thousands of acres to small units of a few acres; most were owned and managed by foreigners, and some employed a foreign work force as well. Today, many of the small holdings are locally owned and operated, and a considerable number of the estates have been taken over by local corporate groups or local governments. There has always been a high degree of specialization on commercial tropical plantations, and, usually, each produces one crop exclusively; this is most commonly sugar cane, pineapples, citrus fruits, cacao, bananas, coffee, tea, coconuts, spices, or rubber. Often the specialization is so complete that the plantation does not produce even basic foodstuffs for its labor force and must acquire these at high cost from distant markets.

It is not surprising that, in the course of its development, the commercial tropical plantation, unique and in some respects unnatural, has had to face innumerable natural and cultural obstacles, not all of which have been negotiated to the satisfaction of its host countries. Yet, despite the great difficulties that remain, the disappearance, or even decline, of this type of agricultural society in the near future seems very unlikely. This is because socioeconomically, it is much too important to all concerned, especially to those host countries attempting to establish a sound economy under a newly independent government.

FIGURE 7–23

Aspects of agriculture in Sri Lanka. About two-thirds of the nation's agricultural lands are in tea, rubber, and coconuts; the relatively small acreage devoted to vegetables and livestock gives the country only about a 60-percent self-sufficiency in foods.

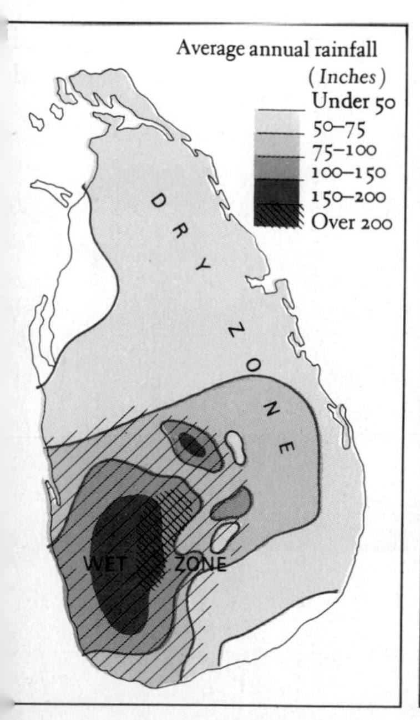

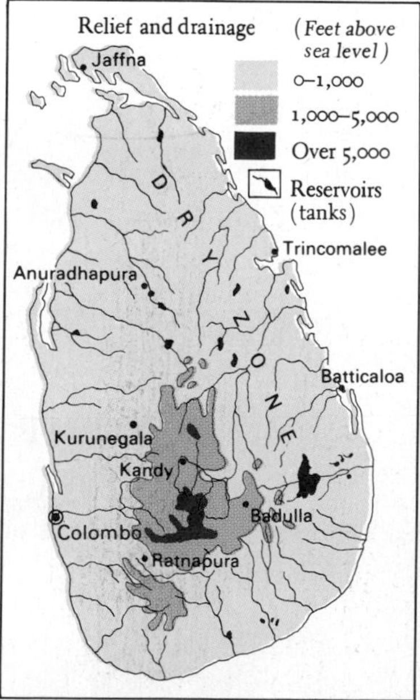

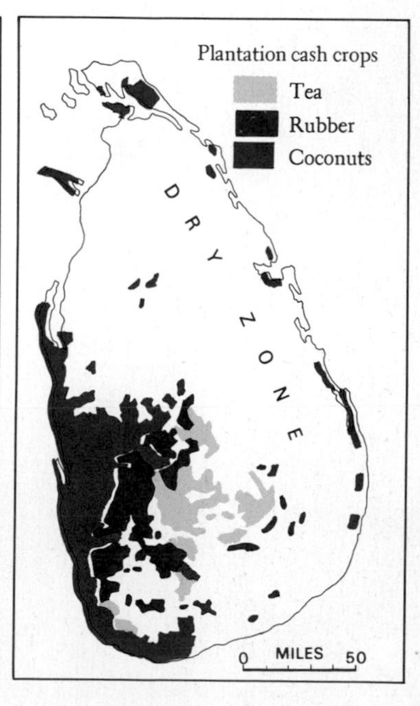

Trends in the Agrarian Infrastructure

Agrarian Reform and Productivity

Changes in rural agrarian institutions and concepts, usually considered together as *agrarian reform,* are certain to take place in the near future, in both the developed and developing countries of the world (Fig. 7–24). In many areas of Asia, Africa, and Latin America, outdated systems of land tenure still slow down the adoption of new and improved production methods, making it next to impossible for farmers to achieve relatively decent living standards. Often without the security of proprietorship over their land, some farmers have little incentive to improve their holdings through the use of expensive irrigation or drainage systems, fertilizers, or storage facilities. Excessive fragmentation of land holdings, through antiquated inheritance practices, also frequently prohibits efficient agriculture. A dozen miniplots scattered over a several-mile area are too distant to be profitably fertilized and cultivated; underuse and misuse of land and labor are the obvious results.

However, large tracts of land can also be a handicap to greater agricultural efficiency, for the successful utilization of such holdings requires an expertise beyond that developed

FIGURE 7–24

The pitch for a cooperative rice mill at Abakaliki, in eastern Nigeria. Agrarian reform has been underway in Nigeria for several decades; and agricultural cooperatives, in existence there since the 1930s, are today important factors in the country's socioeconomic development. (United Nations)

on the small farm, as well as huge investments of capital. They are also particularly subject to absentee landlordism, which, by dividing ownership and management, has exacted a heavy toll on effective land utilization. And, in some of the more densely populated countries, especially in Asia and Africa, the ownership of large tracts of land has also led to *rack-renting*, an ancient system in which tenant farmers are charged rents equal, or nearly equal, to the full annual returns from the land. To save their families from the squalor of overcrowded cities, where they will find high costs and little chance for employment, farmers working under this system live year after year in debt and make it a burdensome legacy to their children.

If properly undertaken, the redistribution of the world's agricultural lands, in terms of both ownership and use, could not only directly cause the living standards of many farmers to be improved but could also indirectly effect the advancement of whole countries toward a condition of food sufficiency and, thereby, proper nutrition for their populations. Such agrarian reform will be slow and painful and, of course, prone to the evils of self-interest; but, being a key to the further spread of agricultural technology and greater motivation on the part of the farmer, it seems a prerequisite to solving the world food crisis and to maximizing the earth's agricultural productivity.

The Agricultural Labor Force

Authoritative estimates place the size of the world's labor force at approximately 40 percent of the total world population, or an estimated 1.5 billion people. Included in this number, supposedly, are all the men, women, and children engaged in productive employment, whether full- or part-time, salaried or unpaid. In considering such estimates, it should be remembered that interpretations of "productive employment" as well as demographic reporting systems vary dramatically from country to country. While, in such countries as the United States and Canada, much detailed information on population characteristics is both collected and published, there are some others that do not even conduct a regular general census.

To increase the accuracy of international employment studies, the United Nations Statistical Commission has devised a system that combines dozens of diverse occupations into several general groupings, which in turn are divided into three broad sectors (designated as *primary, secondary,* and *tertiary*) of societal activity (Fig. 7–25). Based on estimates made by the United Nations International Labor Organization, primary activities (mainly agriculture) generally employ about 54 percent of the total world labor force; secondary activities (mainly industry) account for nearly 20 percent of the total; and tertiary activities (varying with each particular country), represent the remaining 26 percent of the labor force.

Certainly among the most revealing facts to emerge from the United Nations' data on world employment is the absolute

FIGURE 7–25

The general tripart division of basic types of labor.

Primary societal activities:
agriculture, grazing, forestry, hunting–fishing–gathering, mining and drilling
Secondary societal activities:
industry, construction, energy production, etc.
Tertiary societal activities:
services, transportation, commerce, communications, banking, education, etc.

predominance of the agricultural labor force, which comprises 51 percent of the world total (see Fig. 6–5). And even when modern agricultural techniques offer an alternative, throughout much of the world heavy expenditures of human labor are still required to raise food. In many countries of Asia and Africa the percentage of the labor force in agriculture rises well above the average, reaching as high as 70 percent in India and 85 percent in Thailand, for example. These percentages are particularly startling when compared with the 6 percent of the labor force engaged in agriculture in the United States.

Generally, the percentage of its people engaged in such primary activities as agriculture declines as a country develops economically; one reason for this is the simple fact that labor from agriculture will be siphoned off to fill jobs in new areas of the economy. Since national economic development can be arduous, especially where illiteracy or opposed customs are entrenched, it will probably take decades for some countries to experience a substantial decline in their high levels of agricultural employment.

REVIEW AND DISCUSSION

1. Enumerate some of the extensive regions in which nomadic hunting-and-gathering peoples live. What physical or situational characteristics do these regions have in common that explain their use by such groups? What cultural or political characteristics do these regions share that explain their type of occupance?
2. Sedentary subsistence farming, which began some ten thousand years ago, persists in many contemporary societies. Assess this major form of agriculture with respect to geographical distribution and population patterns.
3. Discuss the basic characteristics of the rice-dominant, or sawah, agricultural society that dominates the lives of most Asians. How would the adoption of modern technologies reduce the weaknesses and enhance the strengths of this system? What role do animals play in this form of agriculture? Can their role be increased, and how would such an increase affect the whole society? What role can "miracle" grains play in Asia's future?
4. In matters important in cultural geography, compare and contrast the non-rice-dominant societies of southern and eastern Asia with their neighboring rice-dominant, or sawah, societies.
5. What factors have influenced the transformation of many agrarian societies into industrial agrarian societies? What are the sociocultural advantages of such a change? How would the transformation of all agrarian societies into industrial agrarian societies affect the welfare of humankind?
6. Commercial crop and livestock farming dominates a broad region of Europe and much of the Soviet Union and the United States. Is it a natural consequence of the physical environments and cultures involved, or does it seem to be an imported and ill-suited system in one or all of these areas?
7. In what regions and in what ways did European colonialism promote the development of plantations? What were some of the major cultural impacts of this "imposed" agricultural economy in Asia and Africa? Why have many of the Asian and African countries that became independent in this century retained and even expanded their plantation economies under their new governments?

8. What type or types of agriculture are practiced in the region in which you live? What are the main physical and cultural factors that have influenced their development?
9. How do you think the world pattern of agrarian societies might change if there were heavy restrictions on the movement of agricultural products in international trade?
10. In light of its many surplus agricultural productions in recent decades, why is the United States concerned with increasing its agricultural productions? How does its agricultural labor force compare with those of other countries?

SELECTED REFERENCES

Anderson, E. *Plants, Man, and Life.* Boston: Little, Brown, 1952.

Brown, J. R., and S. Lin. *Land Reform in Developing Countries.* International Seminar, 1967. Phoenix, Ariz.: The Lincoln Foundation, 1967.

Cantor, L. M. *A World Geography of Irrigation.* New York: Praeger, 1970.

Chang, Jen-hu. "The Agricultural Potential of the Humid Tropics." *Geographical Review,* Vol. 58 (July 1968), 333–61.

Chao, Kang. *Agricultural Production in Communist China.* Madison: University of Wisconsin Press, 1970.

Chen, Cheng-siang. *Taiwan: Economic and Social Geography.* Taipei: Fu-min Geographical Institute, 1963.

Gregor, H. F. *Geography of Agriculture.* Englewood Cliffs, N.J.: Prentice-Hall, 1970.

Griffin, P., R. Chatham, A. Singh, and W. White. *Culture, Resource and Economic Activity.* Boston: Allyn & Bacon, 1971.

Hayama, Y., and V. Rutlan. *Agricultural Development: An International Perspective.* Baltimore: Johns Hopkins Press, 1971.

Highsmith, R. M. "Irrigated Lands of the World." *Geographical Review,* Vol. 55 (July 1965), 382–89.

MacFadden, C. H. "Mechanized Rice Production in California." *Il Riso* (Milan, Italy), Vol. 14 (December 1965), 325–29.

Nath, V. "The Growth of Indian Agriculture." *Geographical Review,* Vol. 59 (July 1969), 348–72.

Nulty, L. *The Green Revolution in Pakistan.* New York: Praeger, 1972.

Schultz, T. W. *Transforming Traditional Agriculture.* New Haven, Conn.: Yale University Press, 1964.

Staley, E. *Future of Underdeveloped Countries.* New York: Praeger, 1961.

Ucko, P. J., et al. "Swidden Systems and Settlement." *Man, Settlement and Urbanism.* London: Gerald Duckworth, 1972.

Udo, R. K. "Sixty Years of Plantation Agriculture in Nigeria: 1902–1962." *Economic Geography,* Vol. 41 (October 1965), 356–68.

Sunlit transmission towers along the road from Karangi to Karachi, in Pakistan
(United Nations

8

Societal Pacemakers: Energy

The Nature of Energy

The constant increase in human dependency on environmental energy sources has made them important pacemakers in societal advancement. These energy sources and their innovative uses have been the key to some of humanity's highest attainments to date, as well as to its dreams of creating a better world in which to live. Gradually, humans have learned to control and utilize many energy forms far greater than their own limited muscular capability. Ever since early cave dwellers first experimented with fire, humanity's quest for ways to survive, and to improve its physical well-being, has been tied, in both concept and action, to the effective harnessing of one, or several, of nature's many energy forms.

But, one might ask, what *is* energy? For, though human beings are always surrounded by it, some may have great difficulty in defining energy; it is easier to explain what energy *does*. To most of the world's people, energy is that which gives them the added strength and extra hands with which to fulfill their needs and wants. In a very direct way, energy controls not only all the world's societal systems but also all the activities that take place within them—from such large-scale activities as agriculture, industry, mining, transportation, communications, and national defense down to such smaller-scale activities as the provision of food supplies, automobiles, and home appliances for the satisfaction of individual daily needs.

The only significant source of the huge and constant flow of energy on the earth is the sun, which, as it speeds through space, emits tremendous amounts of energy, a small fraction of which is intercepted by the earth as *solar radiation*. Of the minute fraction of solar radiation obtained by the earth, an even smaller fraction is synthesized and stored by the earth's vegetation. It is this stored chemical energy that serves as the major energy source for the earth's animal life, including humans. When plants and animals die, much of this energy is buried with their remains under other earth materials. Thus subjected to long burial amid the earth's internal heat and pressure, the remains of ancient plants and animals have been transformed into *fossil fuels,* that is, coal, petroleum, and natural gas. These forms are the principal sources of stored energy upon which modern humanity depends.

Energy Thresholds and Cultural Evolution

The Use of Fire

The ability to produce and control fire was certainly the greatest triumph won by early humans in their struggle toward achieving mastery of the earth's major energy sources. The use of fire enabled these primitive peoples to expand their hunting-and-gathering activities, both in variety and extent. For example, many more kinds and parts of plants and animals, some indigestible by humans when raw, could now be made edible by cooking; and, perhaps, most important, human settlements could now be extended into cool, and even cold, climates. In these and other ways, this single cultural achievement allowed human beings to gradually increase their numbers and to establish their dominance over lesser primate forms.

The First Agricultural Revolution

About ten thousand years ago, as discussed earlier, more effective utilization of human energy, through the domestication of plants and animals, led to the first *agricultural revolution.* No longer forced to rely solely on hunting and gathering, human beings progressed beyond such practices and began to develop an entirely new behavioral pattern, one that was based on the raising of plants and animals. With this major step into agriculture, they crossed their first important energy-use threshold. However, for many millennia, human life remained a monotony of labor, as all work was still done by the muscle power of humans and beasts. Only gradually were the new and improved tools developed that made it possible for agriculture to require less human labor.

The Use of Water Power

After the first agricultural revolution, nearly eight millennia passed before human beings, approaching their second major energy-use threshold, learned that some of their tasks could be performed by simple mechanisms operated by the natural forces of water. The first of these was the *water wheel,* a huge wheel turned by the weight of falling water or by the momentum of flowing streams. This primitive device supplied the energy necessary to do such things as raise water from wells and cisterns or grind grain into flour. In order to take advantage of the water wheel's help in these and other essential tasks, people began to establish their settlements on the banks of streams (at waterfall sites) from which substantial water power could be obtained (Fig. 8–1).

But, for reasons one can only postulate, it took another millennium—or more than thirty generations—for humans to advance to a point at which they could make any more complicated applications of the water wheel. Moving water, whose energy was harnessed essentially by means of the water wheel, continued to serve as humanity's dominant source of energy until the time of the Industrial Revolution; it was widely available, easy to control and utilize, and essentially inexhaustible.

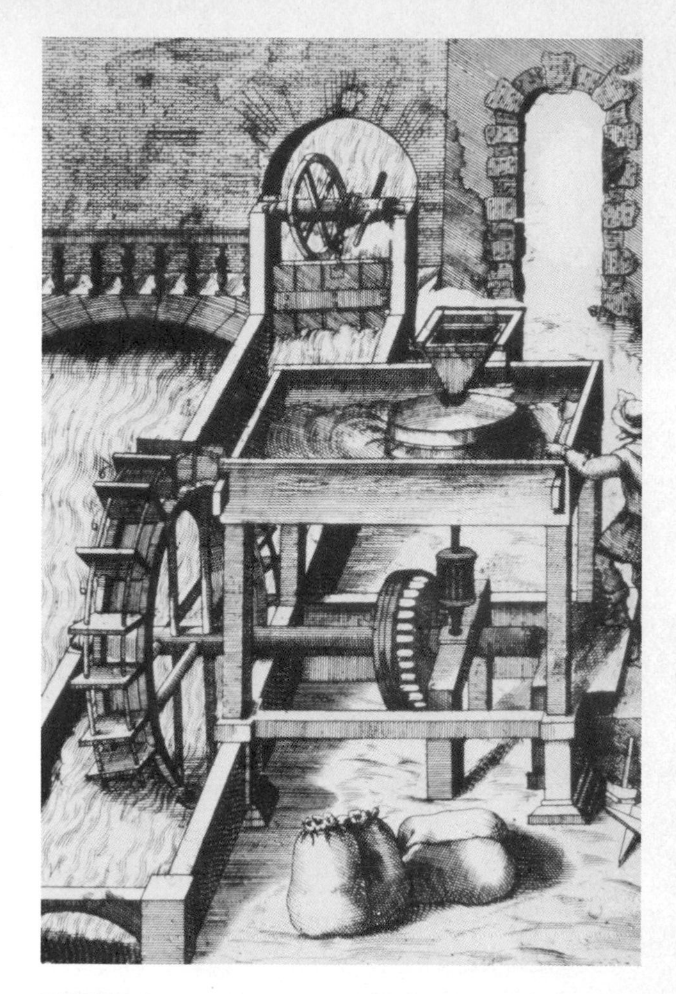

FIGURE 8–1

An early water wheel being used in the grinding of grain into flour.

Energy Diversification

The peoples of Europe reached a third energy-use threshold by about the tenth century, when they began to use the energy supplied by water wheels for a greater range of purposes. The mechanisms themselves had been improved, and they were now harnessed to new or more effective hoisting and pumping equipment and to various new specialized devices. In these combinations, they became effective in a multitude of basic tasks performed previously only through muscle power; these included the sawing of wood, the hammering of metals, the pressing of fruits, and the processing of textiles.

Two other important mechanical developments also occurred in this era. The first of these was the successful adaptation of the *cam,* a conveyor of motion that soon came to be a feature both of machines used to perform rather precise operations, and of heavy-duty machines used in such operations as stone crushing (Fig. 8–2). The second development was the adaptation of the *crank,* a mechanism that converts circular motion to linear motion, as in the operation of a pump cylinder or a piston.

The Use of Wind Power

By the fifteenth century, some European societies had attained a fourth energy-use threshold by improving on the concept of the *windmill,* which they borrowed from older Eastern civilizations. This enabled them to make wind another major source of energy, a development that was of particular importance in areas remote from dependable sources of flowing water. This ability to exploit several energy resources allowed medieval Europe to achieve more rapid and more extensive economic advances than were made during the same period in other parts of the world. European countries began to surge economically at home and to reach out to other continents with the first hints of overseas colonialism in modern times. In a few other relatively advanced regions of the

FIGURE 8–2

An early cammed axle in combination with a water wheel (beyond the wall). As the wheel turned the axle, the axle's *cams*, or knobs, raised the hammers (at left), thus converting the wheel's rotary motion into vertical motion.

world, much of the same advanced technology was introduced and accepted—but without much interest or vision for development. Nowhere else was the developmental equipment as complete, or the urge to master the forces and gifts of nature so great, as they were within European societies. The results were that Europe advanced socioculturally and socioeconomically relatively rapidly, while other societies with less cultural drive remained as they had been for ages, some even being content to retrograde through inaction. But today those same "other societies" are rapidly reversing their views and their actions.

Coal: The Industrial Revolution

In the sixteenth and seventeenth centuries many new machines were invented for use in Europe's rapidly expanding textile and engineering industries, but the capacity of all such machinery remained subject to the availability of harnessible wind and water power. More convenient and more controllable energy sources were urgently needed. The many operable but highly complex and inefficient steam engines developed during these two centuries represent the long-time European attempt to make coal one of these new, more practical sources of industrial energy. Success did not come until 1769, when James Watt (1736–1819), a Scottish engineer and inventor, patented a radical redesign of a coal-burning steam engine (or "fire engine") invented in England by Thomas Newcomen half a century earlier. Far more efficient and economical than any of its predecessors, Watt's steam engine was put on the market in 1776 and immediately became industry's most effective energy-converting machine (Fig. 8–3).

This new "fire engine" and the improved versions that followed brought about tremendous increases in the industrial output and wealth of all the European societies that adopted them. In England, they were major factors in the developing industrial revolution, which eventually affected not only the country's economic structure but many of its social systems as well. Yet, although European colonialism was already spreading around the world, the new coal-based technology and its resultant industrial development did not always follow the colonial flags.

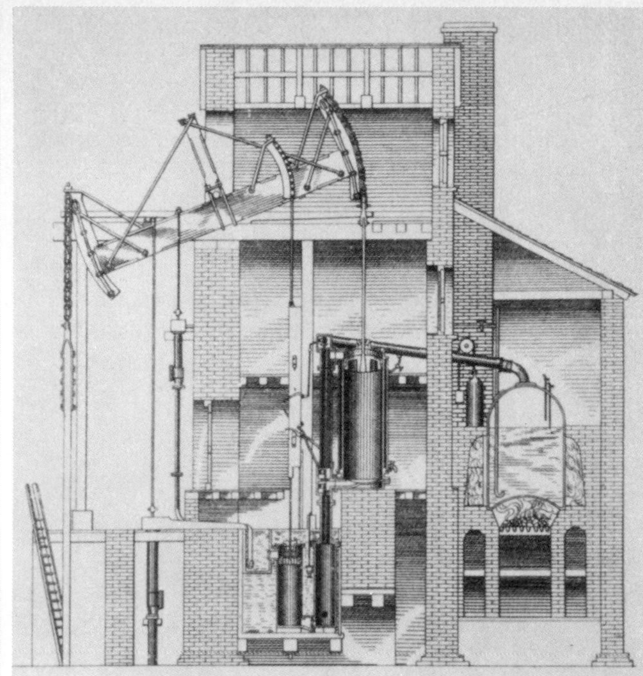

FIGURE 8–3

Eighteenth-century steam engines. Left, Newcomen's engine, which used steam to move a vertical piston that in turn activated a horizontal "rocker-beam." (Smithsonian Institution) Right, Watt's more efficient steam engine; it used steam both to move the piston upward and to return it to its original position and required only one-fourth the coal used by Newcomen's engine in comparable tasks. (Brown Brothers)

Petroleum: The Transport Revolution

Although they were used in raw forms in the Eastern world over three thousand years ago and in the Western world at least a thousand years ago, it was not until very late in the nineteenth century that petroleum and natural gas became of major significance to human beings as energy sources. The widespread use of *petroleum,* a flammable liquid found in portions of the upper layer of the earth, followed the production of the first internal-combustion engines, most of which operated on the refined petroleum product known as *gasoline.* Together, the internal-combustion engine and its portable fuel made possible the first commercial means of motorized transportation, and thus they played a vital part in one of the most noticeable and far-reaching changes in human living patterns.

Electrical Power

Without the use of electricity, the worldwide, and even regionwide, distribution of the vast energy available from coal, petroleum, and natural gas would seem an insurmountable undertaking. It is fortunate for modern humankind that the basics of electrical technology have been largely understood since the late nineteenth century and that their implementation has progressed rapidly. Today the conversion of fossil fuels, wind, water, and nuclear fuels into *electrical energy,* which can be transmitted readily in great quantities by means of high-voltage grids, has become a common practice throughout much of the world, in most of the underdeveloped countries as well as in the highly industrialized ones. Among humanity's energy-use achievements to date, the mastery of electrical-energy creation and long-distance transmission undoubtedly represents one of the major thresholds.

The Use of Nuclear Energy

Since the mid-twentieth century, when human beings discovered the basic principles of *nuclear fission* and *nuclear fusion*, a tremendous new source of energy has been available to the world. Like so much of the other technologies born of war, the ability to penetrate the nucleus of an atom has proven to be highly adaptable to peaceful purposes. For example, there are thermoelectric plants in which the intense heat produced during nuclear fission is converted into electricity. Such applications of nuclear energy are still relatively few in number, but they give promise of being at least a partial answer to the world's expanding energy requirements.

Modern Energy Bases and Imbalances

The Changing World Energy Bases

During the last few decades, much of the world has been rapidly converting from the use of coal to the use of petroleum and natural gas, both of which have been inexpensive and convenient sources for everything from gasoline and other fuel oils to neon signs and plastics. And, due in part to the ease and versatility of these two energy sources, human beings have consumed more of the earth's energy resources in this brief period than they did in all their previous history; moreover, this sort of acceleration is expected to continue, with total energy consumption doubling at least every generation. And the result? All known petroleum deposits are being depleted much more rapidly than most of the world's other energy resources. Oil and gas supplies are shrinking rapidly, costs are rising alarmingly, and availabilities are being seriously threatened.

Although most people give little thought to their daily use of the earth's energy resources, much of humanity now depends, directly or indirectly, upon energy consumption for its survival. In many societies of the world, the supplies of human food, clothing, and shelter evolve largely through the utilization of energy and the other natural resources in increasingly complex systems. Uses of energy resources, in particular, are so interwoven into the fabric of the modern industrial society that its socioeconomic foundations suffer markedly when denied their accustomed flow of energy. And if this denial were prolonged, many people in such a society would find it difficult to feed themselves at even subsistence levels. Nuclear science and necessity may soon provide the world with more consistent flows of energy, but, at present, high-energy-consumption societies remain almost totally dependent on unevenly distributed fossil fuels, water, and nuclear fuels. Indeed, *per capita* consumption of these basic energies is perhaps the truest index of the socioeconomic level of a contemporary society.

World Energy Imbalances

Wood, dung, peat, charcoal, and other biotic materials were used by human beings as domestic and commercial fuels for thousands of years; but during the period of the Industrial Revolution, coal became the "energy king," and for half a dozen generations, its use continued to exceed that of all other energy sources combined. Until well into the present century, when petroleum and gas became the new "miracle" fuels, coal supplied much more than one-half of the total energy consumption of most countries of the world; it powered the factories, drove the ships, and largely supplanted firewood in the heating of homes and public buildings. Thus, coal became one of the principal prerequisites of national economic development; countries with little or no coal could not, and did not, become industrialized.

But the world's energy preferences have changed dramatically in recent decades, with coal now accounting for only a little more than 35 percent of total energy consumption and petroleum and gas consumption rising to about 40 percent and 20 percent respectively. (Waterpower accounts for another 2 percent, as hydroelectricity, and other sources supply the balance.) In the United States, the energy shift has been even more pronounced than elsewhere. Between 1940—when coal supplied 45 percent of the energy consumed in the nation—and the mid-1970s, the United States shifted to an energy-consumption pattern in which both oil and gas played larger roles than did coal, which then represented only 20 percent of national energy consumption (Fig. 8–4).

Currently, two nations produce and consume about one-half of all the energy resources used by human beings; the United States consumes about 30 percent and the Soviet Union, about 20 percent (Fig. 8–5). The more than 140 other nations of the world, representing nearly 90 percent of its population, divide the other half among them, with 20 percent consumed by western Europe. This means that only a relatively small percentage of the world's energy production and consumption takes place in all of Asia, for example, where more than half the world's people live. These facts serve to illustrate the obvious and great energy imbalance that presently exists among the world's human societies.

	United States		Europe	
	1940	1975	1940	1975
Coal	45*	20	80	45
Petroleum	35	40	10	40
Natural Gas	15	35	1	6
Hydroelectric	3	4	8	8
Other energy sources	2	1	1	1

* In percentages.

FIGURE 8–4

The major energy sources in the United States and Europe in 1940 and 1975. Between these years, both these high-energy-consuming regions underwent decided shifts in their energy bases, relying less and less on coal and more and more on petroleum and gas.

Energy Crisis–or Evolution?

With the energy consumptions of new and old industrial countries alike continuing to spiral upward at almost unbelievable rates, reserves of petroleum and other fossil fuels are already in jeopardy (Fig. 8–6). Although this situation had been predicted by some observers for decades, most political representatives did not even begin to sound warnings of overuse until the mid-1970s. These belated indications of the rapid dwindling of energy supplies have finally awakened world concern to the possibility of a severe energy shortage. Some of the present sense of crisis has, no doubt, been created to foster particular political or monetary concerns, but most of the alarm reflects the hard fact of decreasing energy supplies and rising energy demands.

FIGURE 8–5

Electric lighting in use over the eastern United States, as photographed by a meteorological satellite 450 miles above the earth. The largest and brightest spots coincide with the locations of major metropolitan areas. (Westinghouse Electric Corp.)

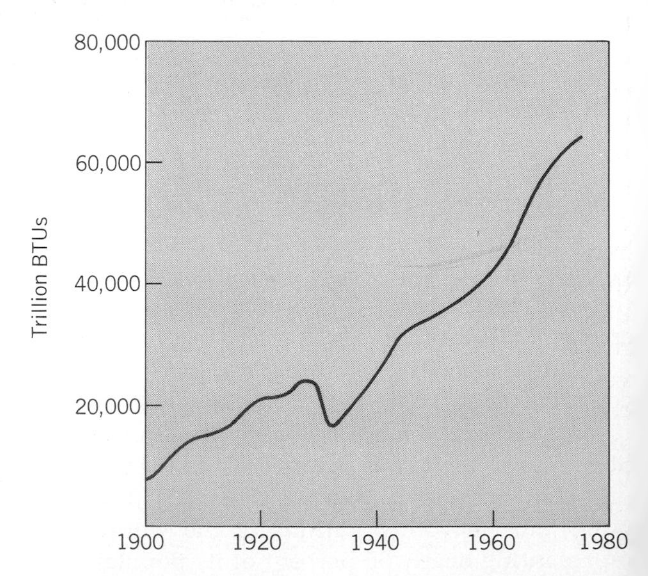

FIGURE 8–6

Energy production in the United States since 1900—by means of both fossil fuels and hydroelectric sources.

Future Energy Alternatives

Traditional energy resources are rapidly declining, creating an international energy crisis—as exploding populations and skyrocketing consumption rates overtax finite supplies. Energy consumptions have doubled in recent decades and are expected to double again within the next twenty-five years. In response to the decline in traditional energy sources, new *energy alternatives* are needed for the future—entirely different sources to sustain rapidly expanding populations and their complex technological societies. Some of the most promising energy developments for the future are solar energy, geothermal energy, tidal power, and nuclear energy. But what are the likely benefits and problems of such new energy sources for future societies?

Solar Energy

Among the energy alternatives inhabitants of the earth may pursue in the future, certainly the most abundant and most readily available is *solar energy*, or solar radiation, which each day provides the planet with energy equal to one-half of *all* its remaining fossil-fuel reserves. A great hope of human beings, through all eras, has been to harness the vast energies of the sun. Historical records reveal many notable attempts at this and some successes—although one suspects that fact has sometimes been mixed with fiction. For example, using a line of shields as reflectors, the Greek mathematican and inventor Archimedes is supposed to have concentrated enough solar energy, or heat, to set afire the sails of a Roman fleet as it besieged the Sicilian city of Syracuse, in 212 B.C.; Antoine Lavoisier, a French chemist, is said to have used concentrated solar energy, or heat, to melt iron in 1774; and Swedish-American engineer and inventor John Ericsson is credited with creating, in the late 1860s, the first steam engine powered by direct solar energy. Modern research on the uses of solar energy began in the 1930s, when American solar scientist Charles Greeley Abbot invented a steam engine that operated on solar energy concentrated by means of parabolic, or bowl-shaped, mirrors.

Many such *solar steam engines* have since been used in small-scale electric generation and water pumping, particularly within the developing countries of the tropics; but no engine with a solar boiler has yet been able to compete with the relatively cheap combustion engine. *Solar cookers*, or stoves with parabolic reflectors, have been used in everyday food preparation for the past century, again most successfully within tropical countries, where, however, high cost prevents their widespread adoption (Fig. 8–7). *Solar distillation*, which uses flat rather than parabolic collectors, has been employed worldwide for many centuries in salt production; but its first large-scale commercial application probably occurred during the nineteenth century in the desalting of water for Chilean nitrate mines. Today, solar distillation is still widely used in commercial projects, such as the operation of salt pans that precipitate considerable salt annually from sea water. With the growing recognition of the energy crisis, *solar heating* for industry and homes has also attracted considerable attention in recent years and, within this generation, may become the first widespread application of solar energy (Fig. 8–8).

Solar furnaces are designed for the high-temperature-heating needs of many industries and, utilizing a huge parabolic mirror and several flat mirrors, can usually achieve

FIGURE 8–7

The testing of a *solar cooker* at the National Physics Laboratory, New Delhi, India. In the pressure cooker shown here the boiling time of water by means of this solar device is about 10 minutes; in a saucepan, about 30 minutes. (United Nations)

temperatures well above 1,000°F. Felix Trombe, a French scientist who has built a number of solar furnaces at Mont Louis in the French Pyrenees, in 1970 completed the world's largest solar furnace (near Odeillo); it develops more than 1,000 kilowatts of power, reaches temperatures over 6,000°F., and can melt through a steel plate three-eighths of an inch thick in less than 60 seconds (Fig. 8–9). Trombe's huge furnace has a central parabolic mirror that extends 148 feet high, and this awesome centerpiece receives and concentrates solar rays directed to it by sixty-three flat mirrors set higher up the mountainside. Each panel mirror is itself about 18 feet square, and turns slowly with the moving sun.

Solar batteries are fairly recent devices that can convert rays from the sun directly into electricity. They represent an application of the photovoltaic method announced by Bell Telephone Laboratories in 1954. By 1958, solar batteries had become a feature of space exploration equipment.

FIGURE 8–8

An industrial-scale, flat-mirror solar energy system, at Valley Forge, Pennsylvania. (Time-Life Picture Agency)

Solar energy research has been greatly accelerated in the United States since the general recognition of the energy crisis. For example, through a special program of the National Science Foundation, the monies available to support the work of solar researchers were increased from practically zero in the 1960s to $13 million in 1974.

Geothermal Energy

In response to the energy crisis, increasing attention has been focused in the last decade on the potential of geothermal resources in the generation of electric power. As the term implies, *geothermal energy* is the "natural heat" of the earth. The source of this heat is the earth's highly plastic interior.

As a result of the decay of radioactive materials and such forces as pressure and friction, the temperature at the molten center of the earth reaches 1,000°C. (1,832°F.). This core heat migrates outward to the earth's crust in a constant flow that becomes less intense as it approaches the surface. Thus, the heat of the earth's crust normally increases 30°C. (54°F.) with each additional kilometer toward the center, or 48°C. (89°F.) with each radial mile. Existing as exceptions to this rule are countless "heat pockets," or *geothermal reservoirs*, places at which unusual geologic conditions have created pressures and heat much more intense than is typical at such depths. These pockets are frequently associated with areas of past volcanic activity and tend to occur at fairly shallow depths—that is, within a few thousand feet of the surface (Fig. 8–10). Where subsurface waters, or groundwaters, accumulate and come into contact with the interior's superheated rock materials, geothermal hot water and steam are formed. These forms of geothermal energy are most apparent on the earth's surface when they are vented as geysers and active volcanoes. In the absence of such natural venting, subsurface steam, superheated

FIGURE 8–9

The huge parabolic mirror of the solar furnace at Odeillo, France. (© Sipa-Press, Liaison Agency)

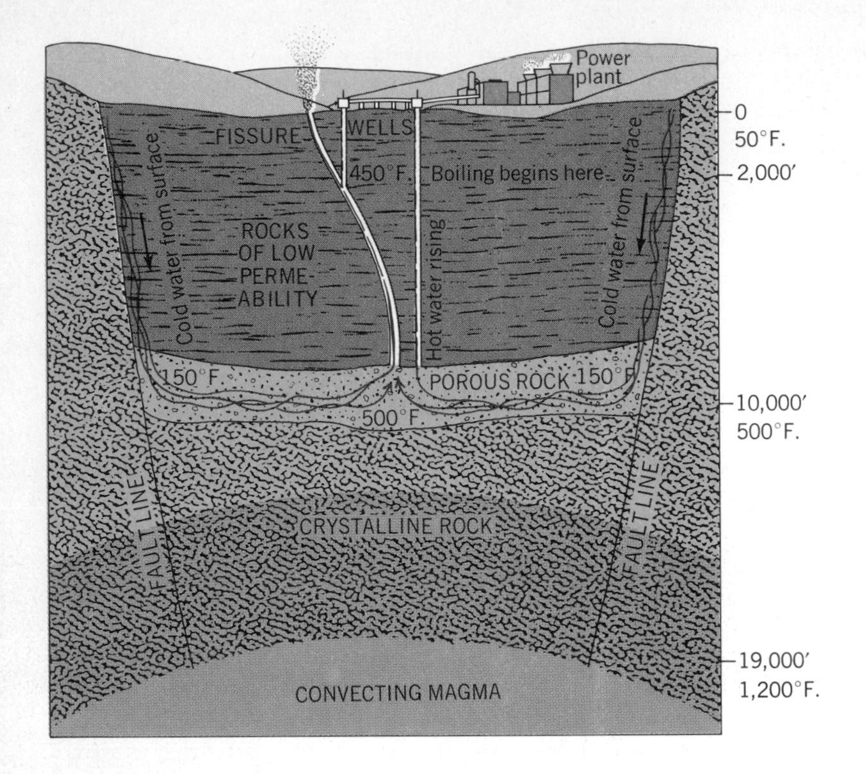

FIGURE 8–10

A diagrammatic cross-section of a natural hot springs and geyser area. Surface water seeps down through the faults to deep layers of porous rock, which are heated, through convection, by the still deeper magna. Wells driven into the geyser, or fissure, or into the deep porous rock can tap their hot water and steam for utilization by electric power plants on the surface.

water, and dry heat can be tapped by means of artificial wells. Geothermal energy thus tapped may be used for many purposes, including the operation of conventional thermoelectric power plants and space heating.

Studies by the United Nations suggest that the total energy contained in subsurface dry-steam, wet-steam-and-water, and dry-hot-rock materials is practically unlimited; or, in other words, that the geothermal energy constantly "stored" in the upper 25,000 feet of the earth's crust is more energy than all the probable human populations of the earth could ever use. The question is no longer *whether* geothermal energy is present and available in sufficient quantities to be utilized on a large scale in many parts of the world, but in what manner it can be utilized most efficiently and with the least environmental disturbance.

The commercial use of geothermal energy is already a long-established practice in some countries of the world. The first facility using geothermal steam to generate electric power went into operation at Larderello, in the Italian province of Tuscany, in 1904 (Fig. 8–11). New Zealand has had a system of geothermal wet-steam electric power plants since the early 1950s, and Iceland, the Soviet Union, and Hungary use geothermal wet-steam-and-water for home heating and for some industrial purposes; most of the homes and buildings of the town of Hveragerdi, Iceland—and 45,000 homes throughout the country—are heated from these sources. Two cities of the United States, Boise, Idaho, and Klamath Falls, Oregon, have used geothermal energy for heating buildings since 1890 and 1930 respectively. Japan has successfully used geothermal steam for generating electric power for several years, and the Mexican government has recently completed a 75,000 kilowatt geothermal electric power plant at

FIGURE 8–11

The geothermal electric power plant at Larderello, Italy. The pipes in the foreground carry steam from nearby wells to the plant, where it is used to turn turbines. The large spool-shaped condensers cool the volcanic steam after it has been used. (Jack Birns, Time-Life Picture Agency, © Time, Inc.)

Cerro Prieto, which is located near the Mexico–California border.

By the mid-1970s, the countries of Italy, Iceland, New Zealand, the United States, the Soviet Union, Japan, and Mexico were collectively generating some 1,000 megawatts of geothermal steam electric power—at costs averaging one-third to one-half less than those of other energy alternatives.

Commercial geothermal interests in the United States have been centered in the Geysers area, north of San Francisco. Private companies interested in using geothermal dry steam for electric power generation have been drilling there since 1960 (Fig. 8–12). The present total capacity of the Geysers field is about 200 megawatts, but by 1980 the area's geothermal steam should be generating several thousand megawatts of electric power—at costs less than those of comparable fossil or nuclear operations. A second geothermal area, one rapidly gaining attention since the early 1970s, is the Imperial Valley in southern California; its wet steam and water are of interest to almost all of the country's major oil companies. The total estimated power-generating capacity of the Imperial Valley's geothermal energy is 30,000 megawatts. The Mono Lake area of California's eastern Sierra Nevada is also commonly thought to have geothermal power potential; and, while less frequently mentioned in this context, an area near the Gulf of Mexico covering 150,000 square miles of Texas and Louisiana may offer even greater promise. Several other promising areas of geothermal energy exist in the western United States, but many of them—included in federal lands, which were not available for development until 1970—have not yet been leased.

Of the three distinct types of geothermal energy—dry steam, wet steam and water, and dry hot rocks—dry steam is the type preferred for power generation, because it is environmentally clean and easy to use; however, it is the least-available type worldwide. Dry-steam fields are restricted mainly to northern California (the Geysers fields), Japan, and Italy. Wet-steam-and-water fields are found in New Zealand, Mexico, Hungary, Iceland, the Soviet Union, and southern California. Geothermal energy in the form of wet steam and water is twenty times more abundant than dry steam, and its high temperatures allow it to be used directly to heat buildings or to generate electric power. However, only the third type of geothermal energy, dry hot rocks, seems to be nearly lim-

FIGURE 8–12

Geysers north of San Francisco, California. (Dow Chemical Co.)

itless and universal. The temperatures of this type of geothermal energy are usually lower than those of the two other types, for dry hot rocks reflect the normal temperature levels of the earth itself, rather than some irregularity of nature. Yet, since the heat of the earth increases at the rate of 30°C. (54°F.) per kilometer of depth below the surface, rocks even 5 kilometers deep would suffice in a home-heating system, and those 9 kilometers deep would be sufficient in the operation of an electric power plant.

The most logical and feasible way of tapping the hot-rock heat of the earth's crust is by "drilling and hydrofracturing," a technique somewhat similar to oil drilling and recovery in pressureless fields. One well is drilled, perhaps to a depth of 15,000 to 20,000 feet. Cool water is then pumped to the bottom of the well under very high pressure; there, as it heats, the water fractures the surrounding rock. After this stage of fracturing is completed, a second well is drilled to "tap" the top levels of the fractured zone—and thus create a circulation system, or "hydro-loop." Water is then continuously pumped into the first well, heated within the fractured zone, and then forced to rise to the surface through the second "tap" well. This entire procedure involves only relatively inexpensive present-day technologies.

The environmental impacts of geothermal-energy recovery vary considerably. Iceland's geothermal waters are "fresh" enough to drink; but California's geothermal steam and water is high in salts and minerals. However, most geothermal steam and water impurities can be desalted, demineralized, and degassed quite economically. But, at the same time, all geothermal-energy-recovery technology should be carefully monitored for possible earthquake effects. Geothermal energy, particularly in the form of dry hot rocks, has great potential and, if handled with reasonable care, will count importantly among future energy alternatives.

Tidal Power

The harnessing of *tidal power* has been a human practice for centuries. While some early European settlers in the New World dammed streams and rivers and built traditional water wheels, others found an energy source in the rhythmic ebb and flow of the tides. In 1734, a complex tidal water wheel that generated some 50 horsepower was put into operation at Slades Mill in Chelsea, Massachusetts, where it was used in the grinding of spices. Another eighteenth-century tidal mill was located in Rhode Island and was equipped with water wheels that weighed 20 tons and measured 26 feet wide. Numerous other tidal mills were constructed in the eighteenth and nineteenth centuries, especially on Passamaquoddy Bay, between Maine and Canada on the Bay of Fundy—the only area of North America that is today considered a suitable site for the large-scale generation of electric power by means of tidal energy.

The ultimate demise of these and other eighteenth- and nineteenth-century tidal power experiments must be attributed to the competition provided in electric power production by cheaper energy sources, such as fossil fuels and water power. Early tidal plants were small and could not meet the needs of the new large industries for high volume and low costs. This situation remains essentially true today, having been accentuated by the competition of nuclear fuels.

Few of the world's coastal areas offer the tidal extremes and locational conditions necessary for large-scale electric power generation. The Bay of Fundy, the only one of these areas located within North America, includes nine commercially exploitable tidal power sites (Fig. 8–13). Two other areas considered highly suitable for electric power development are the Severn River estuary in southwestern England and the Rance River estuary in northwestern France. Other possi-

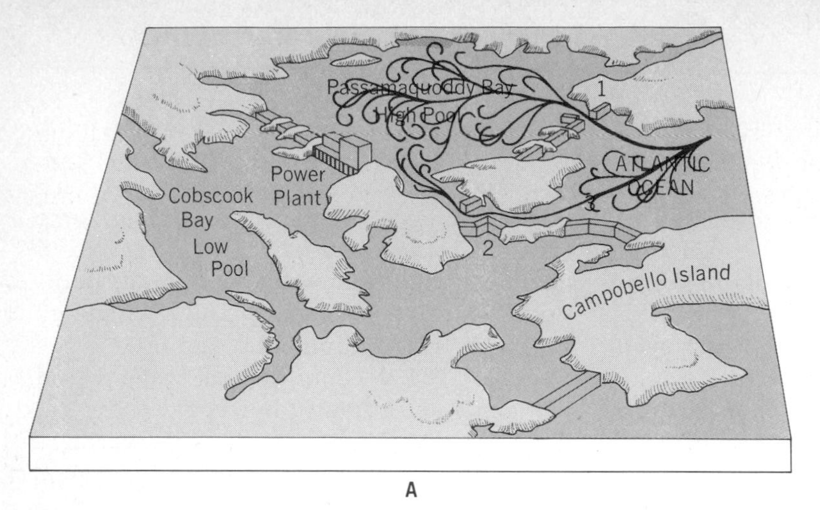

A

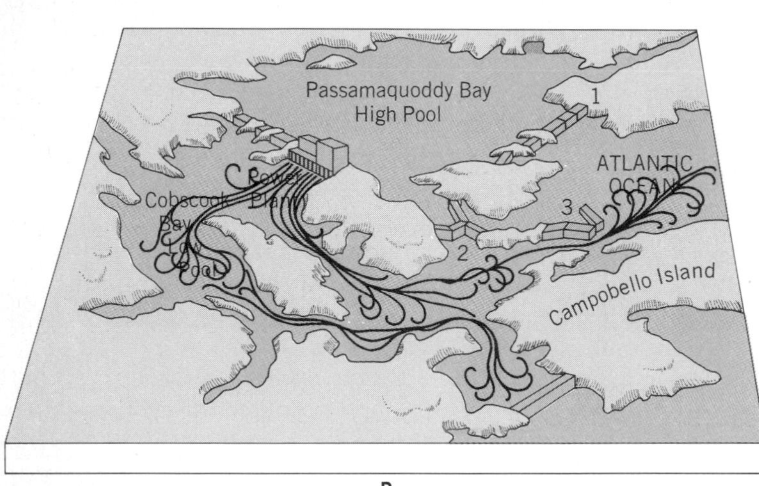

B

FIGURE 8–13

The expected effect of a tidal power project proposed for Passamaquoddy Bay. A. Water is trapped in Passamaquoddy Bay, the high pool, by the opening of gates at points 1 and 2 to the rising tide; the gates are closed just before the outgoing tide begins. B. Water in Cobscook Bay, the low pool, flows through the gate at point 3 during low tide, creating a maximum difference between the water levels of the two pools. Water from the high pool can then flow downward into the low pool, in the process passing through generators in the power plant.

ble areas, all of which require further study to determine their suitability, are in France, the Soviet Union, and the Cook Inlet area of Alaska.

France pioneered in the development of tidal electric power on an industrial scale by completing, in 1965, a unique dam and 240,000-kilowatt generating plant that arcs across the tidal estuary of the Rance River in Brittany. The Soviet Union later built an experimental 400-kilowatt tidal electric power plant near the White Sea in the northwestern part of the country. Soviet engineers say that the project's objectives were reached and that further study into the uses of tidal energy is now warranted.

There are definite environmental disadvantages inherent in building such gigantic dams across any river estuary or inlet. The marine population is disrupted and sometimes permanently damaged, and the landscape is often marred by the facility and its related activities. But these same effects occur around a fossil-fuel or nuclear plant, and perhaps to a greater degree. Therefore, as high-consumption rates deplete traditional energy supplies, tidal power will no doubt be developed wherever practicable—including Passamaquoddy Bay in Canada, which made a "false start" in this direction in the 1930s, and comparable sites in England, France, and the Soviet Union. However, since it is restricted in its occurrences and, for practical purposes, unportable, tidal power cannot be regarded as one of the future's principal energy alternatives.

Nuclear Energy

Humanity is now on the threshold of the nuclear age, which offers much more promise of cheap and abundant power than any other period in history. Within a generation or two, nuclear energy will be the lifeblood of all human societies.

Nuclear Fission and Fusion. There are two ways of producing nuclear energy, by fission and by fusion. All conventional nuclear reactors burn "cores" of natural uranium fuels, whose atoms split, or *fission,* and release vast amounts of energy, or heat. Today's nuclear electric power plants and nuclear submarines are all powered by energy produced by the fission process. In natural uranium the isotope uranium 235 is fissionable—when the nucleus of a uranium 235 isotope is struck by a neutron (one of the particles of an atom), the atom splits into two, releasing still more neutrons; these in turn strike other nuclei and prompt a *chain reaction* in which quantities of heat are emitted. This heat can be transferred to a liquid coolant (usually water) surrounding the fissionable "core" and, through a system of mechanical heat exchangers, used to raise the steam necessary to turn the turbine generators often found in electric power plants and submarines (Figs. 8–14 and 8–15).

Nuclear *fusion* takes place when two lightweight atoms unite, or fuse, to form a heavier nucleus. This occurs only under tremendous heat and pressure and releases immense quantities of energy. The sun is itself a gigantic nuclear fusion device, for it constantly fuses ions (atoms) of light hydrogen to create the heavier atoms of helium. During this process, mass is "lost" and changed into energy—the energy released becomes heat. When science understands how to control this fusion, which it has already employed in an "uncontrolled" form in the hydrogen bomb, humanity will have virtually unlimited power at its command.

Major Uses of Nuclear Energy. Controlled nuclear energy (fission) is already playing a constructive role in the generation of electricity. Nuclear energy is now the fastest-growing means of electric power production in the United States and Canada, as well as in most of Europe and the Soviet Union; and many other countries of the world have already constructed experimental and commercial nuclear electric power plants. Within another decade, most countries, including many developing ones, will have soundly established industrial complexes that can use nuclear energy to achieve major health and welfare benefits for their people.

But the nuclear revolution will by no means involve only the productions of vast quantities of electric power. Other uses of nuclear energy, including the production of fresh water from sea water and other brackish waters, will be extremely important worldwide. Developed and developing countries alike have an increasing need for cheaper and more plentiful water supplies. By 1990, nuclear energy should be sufficiently cheap—and available—to be used in most agro-industrial and municipal desalinization operations, which are particularly necessary in the semiarid and drought-prone regions of the world. In some areas this and other uses of radioactive substances in agriculture, industry, and medicine will contribute as much to future living standards as will nuclear electric power production.

Some Nuclear Problems. Although it offers great benefits in many areas of society,

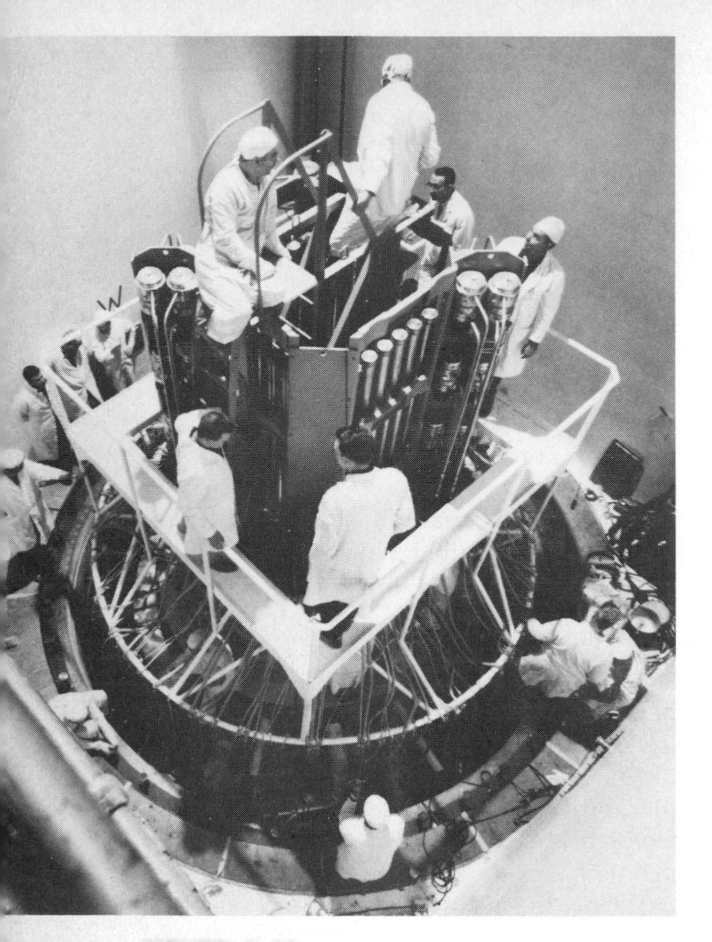

the use of nuclear energy is also fraught with more potentially dangerous problems than is the use of traditional fossil fuels. If they are to live safely with the benefits of nuclear energy, human beings must first solve the problems it presents; its damage cannot be undone. Foremost among these problems is the need to control nuclear weapons production, excessive radiation, and nuclear wastes.

The awesome and lethal power of *nuclear weapons* is attested to by the mass destruction wrought in the Japanese cities of Hiroshima and Nagasaki, by the detonation of two relatively low intensity nuclear bombs in August 1945. Some 78,000 people in Hiroshima and 27,000 in Nagasaki did not survive the blast, and thousands more were harmed by radiation. Many of today's nuclear weapons are fifty times more powerful than the bombs dropped on Japan, with destructive powers equivalent to one million tons of TNT (Fig. 8–16).

FIGURE 8–14

The construction of the reactor core at the Shippingport nuclear electric power plant, near Pittsburgh, Pennsylvania. Completed, the core is now shielded by thick layers of stainless steel, water, and concrete. (Fritz Goro, Time-Life Picture Agency, © Time, Inc.)

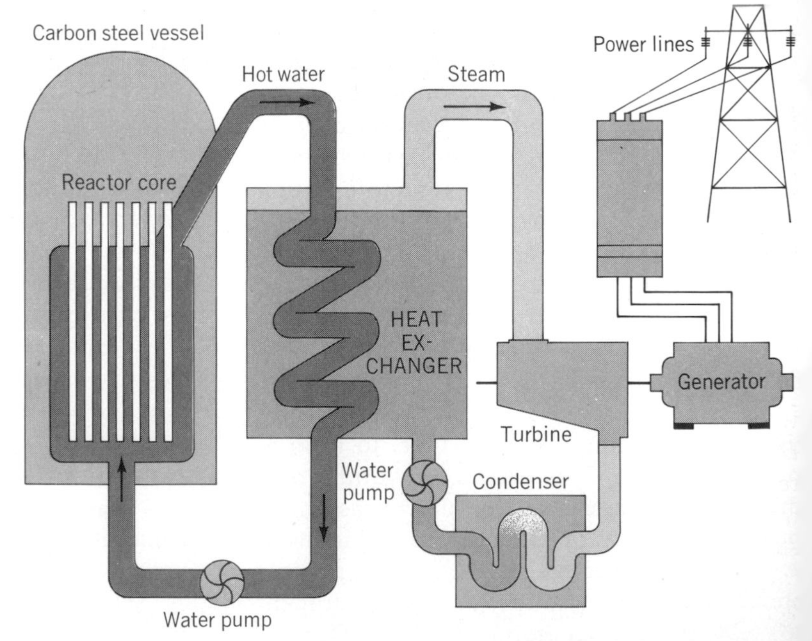

FIGURE 8–15

Electricity through atomic energy. The radioactive *reactor core* heats water, which then transfers its heat to a second water system in a *heat exchanger;* the steam that is thus formed within the exchanger is used to drive a turbine, which operates a generator that feeds electricity to the power lines.

FIGURE 8–16

The mushroom cloud that arose over Nagasaki, Japan, following the August 1945 nuclear bombing. The death and devastation at ground level were much more enduring and horrific. (USAF)

Excessive radiation may occur from both wartime and peaceful uses of nuclear energy. Even in peacetime the accidental escape, or release by a saboteur, of radioactive particles from nuclear power plants or nuclear bomb stockpiles is an ever-present possibility. Although to a much reduced degree, the dangers of radioactivity extend as well to more common occurrences, such as the operation of x-ray machines, micro-wave ovens, and color television sets.

Nuclear, or *radioactive, waste* must be disposed of in a safe manner so its radioactivity will not contaminate any part of the environment. Disposal by burying in the

FIGURE 8–17

A mechanical arm manipulating a capsule in the encapsulation and underground storage of radioactive wastes at a government facility built for their disposal, in Hanford, Washington. The operator of the remote controls is protected from radioactive contamination by the glass panel, which is 3 feet in thickness. (Atlantic Richfield Co.)

ground or dumping in sealed containers into the open ocean have both been tried with fair success and as yet with no major adverse consequences (Fig. 8–17). But such waste-disposal problems will become more difficult as more nuclear electric power plants are put into operation in the decades ahead. As yet there is no fail-safe solution.

Today's Nuclear Countries. Since the early 1950s, more than twenty-five countries of varying sizes and socioeconomic levels have conducted nuclear research and built nuclear electric power plants for their expanding energy needs. This group includes the world's leading agro-industrial nations, the United States, the Soviet Union, Japan, West Germany, the United Kingdom, France, Italy, and Canada; several smaller industrial and industrializing European countries, including Belgium, the Netherlands, Luxembourg, East Germany, Denmark, Norway, Sweden, Switzerland, Spain, and Ireland; the several industrializing countries of India, Pakistan, China, South Korea, and Israel in Asia; Brazil and Argentina in Latin America; and Australia (Fig. 8–18).

The United States achieved its first production of electric power from nuclear energy at Arco, Idaho, in 1951; and its first small-scale use of nuclear energy for commercial power production occurred four years later. The nation's first full-scale nuclear electric power plant, the second in the

FIGURE 8–18

Some of the world's proliferating nuclear power plants. Top left, the experimental superheat reactor at Pleasanton, California. (General Electric) Top right, Sweden's first commercial nuclear power plant, in Oskarshamn, Sweden. (Swedish Information Service) Center right, Wylfa, Britain's most powerful nuclear power plant, at Anglesey, North Wales. (Wide World) Opposite, the Peach Bottom nuclear electric power plant, in San Jose, California. (General Electric) world, did not begin full operations until December 1957, at Shippingport, near Pittsburgh. Ushering in a new era of nuclear transportation, the United States Navy launched the first nuclear-powered submarine, the *Nautilus,* in 1954, and a nuclear-powered merchant ship, the *Savannah,* in 1959. Today, there are several nuclear submarines, and dozens of nuclear electric power plants dot the United States from coast to coast (Fig. 8–19). The government's Energy Research and Development Administration (formerly the Atomic Energy Commission) predicts that nuclear electric power plants will be generating about 30 percent of the nation's electricity by 1980 and about 60 percent by the year 2000.

The United Kingdom established the world's first full-scale nuclear electric power plant, at Calder Hall, on the west coast of England, in October 1956. The production of nuclear electric power has progressed rapidly within the United Kingdom in the last two decades, the country's need for cheap power being greater than that of comparable nations. (Electrical production from coal costs twice as much per unit in Britain as it does in the United States.) Today, many nuclear electric power plants have been constructed in Britain, mainly along the western coast-

FIGURE 8–19

United States nuclear electric power plants: 22 operable, with 9,131,800 total kilowatt capacity; 55 being built, with 46,605,000 total kilowatt capacity; and 49 planned, with about 48,524,000 total kilowatt capacity. (After: U.S. Atomic Energy Commission, September 1971)

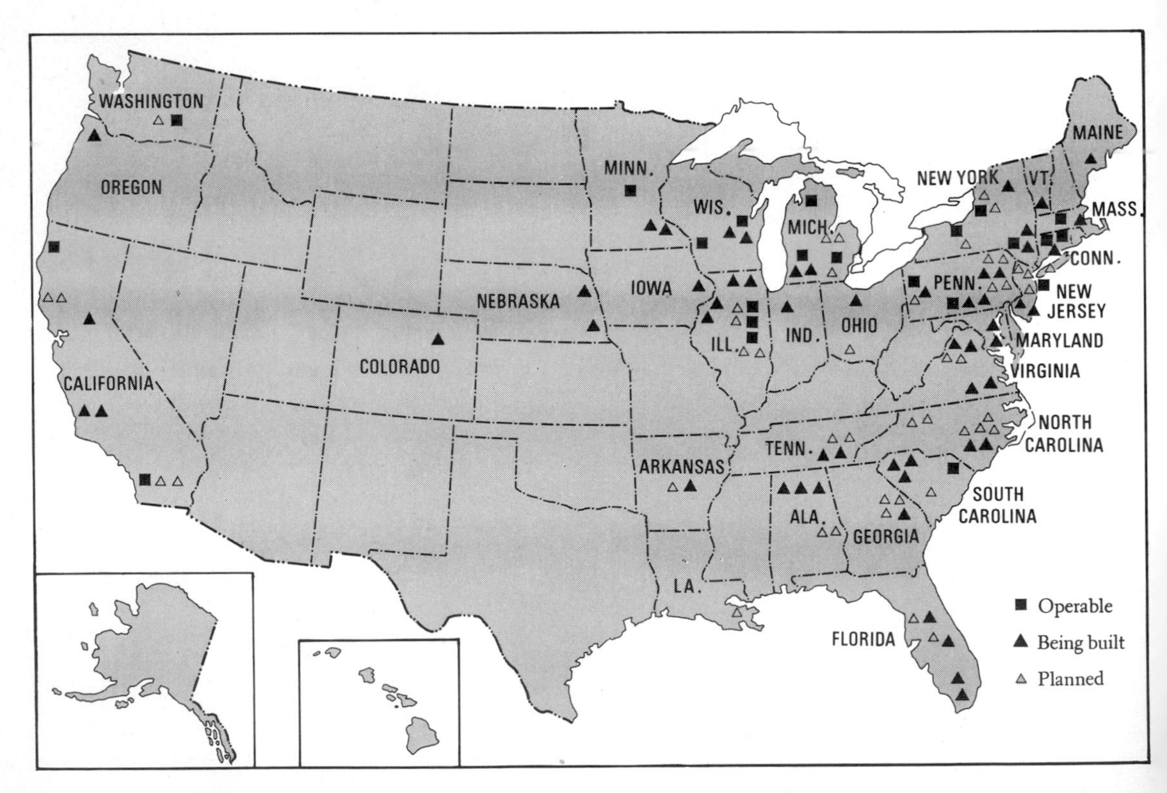

line; they include a large new experimental "breeder" reactor (which produces power and converts natural poor-grade uranium into additional fissionable fuel), at Dounreay in northern Scotland.

The Soviet Union undertook intensive research into nuclear energy shortly after World War II and, in the same period, built some of the world's largest hydroelectric plants to take care of its short-term energy demands. The vast area of the country makes nuclear plants very attractive, for they do not require long hauls of such relatively bulky materials as coal and oil. The Soviet Union built its first small-scale nuclear electric power plant in the 1950s, at Obninsk, and several large-scale plants in the 1960s; one of its most modern large-scale plants, an installation capable of generating 1,000 megawatts of electricity, is located on the Don River, at Novovoronezhskiy.

In Canada, basic nuclear research and power production are both featured in the national energy program. Canada's first nuclear electric power plant was completed in 1962, near Chalk River, Ontario, on the Ottawa River. And the first 2,000-megawatt unit of a new nuclear electric power complex has been completed recently at Pickering–Port Hope, near Toronto. Canada's reactors use natural (low-grade) uranium as fuel, and *heavy water* as a cooling agent.

In Europe, many aspects of nuclear-energy use are being approached regionally. Several countries of western Europe—France, West Germany, Italy, Belgium, the Netherlands, and Luxembourg—joined together in 1957 to form the European Atomic Energy Community (Euratom); in 1973 the group was extended to include the United Kingdom, Denmark, and Ireland. Together, Euratom members are planning an electric power base that will insure economic prosperity and promote regional self-sufficiency; thus nuclear energy figures prominently in their plans. Six European countries—France, Italy, Norway, Sweden, Denmark, and Switzerland—each have established their own independent nuclear research programs as well.

REVIEW AND DISCUSSION

1. Discuss the several energy thresholds humankind has reached since its first controlled use of fire. Include references to specific energy sources and their impact on human activities.
2. If world population continues its present trend and doubles by the year 2000, massive increases in energy consumption will soon occur. What do you think the next decade holds in store with regard to energy types, consumption controls, regional availabilities, and technologies?
3. Can the current excesses in energy utilization be reduced effectively on a purely local basis, or must any correction be attempted worldwide in order to be effective? Why? How would a change in the energy base help to correct the problem? What have been some of the reasons for and the consequences of such changes in the past?
4. Compare and contrast the major forms of energy available today in terms of their convenience and control by humans.
5. Although several future *energy alternatives* are evident, the expectation is that for some time to come humans will remain largely dependent on the traditional and rapidly dwindling fossil fuels—coal, petroleum, and natural gas. Discuss the reasons for this continuance, especially as it relates to *culture lag*.

6. One of the many drawbacks to solar energy's becoming a major source of large-scale power in the immediate future is the fact that the arrival of sunlight on the earth is interrupted in many locales by thick cloud cover. Discuss some of the implications of this problem, including some possible solutions.
7. What is geothermal energy? Name and describe the three major types. In what areas of the world does each type predominate? Which type has the greatest potential for human use in environmental, technological, and economic terms?
8. What are some of the major disadvantages of tidal power as a means of solving the present energy crisis?
9. Explain, in nontechnical terms, the difference between nuclear fission and nuclear fusion.
10. Nuclear power has been described as humanity's best hope for plentiful "clean" energy. Present several arguments that support or oppose this point of view.

SELECTED REFERENCES

Baldwin, C., and E. McNair. "California's Geothermal Resources." *Report to the 1967 California Legislature*. Sacramento: Joint Legislative Committee on Tidelands, 1967.

Barnaby, F. *Man and the Atom: The Uses of Nuclear Energy*. New York: Funk & Wagnalls, 1971.

Brinkworth, B. J. *Solar Energy for Man*. New York: Wiley, 1972.

Clark, W. *Energy for Survival: The Alternative to Extinction*. Garden City, N.Y.: Doubleday Anchor, 1975.

Daniels, F. *Direct Use of the Sun's Energy*. New Haven, Conn.: Yale University Press, 1964.

Ehrlich, P. R., and A. H. Ehrlich. *Population, Resources, Environment*. San Francisco: Freeman, 1970.

Engler, R. *The Politics of Oil*. Chicago: University of Chicago Press, 1969.

Garvey, G. *Energy, Ecology, Economy*. New York: Norton, 1972.

Glaser, P. E. "Power from the Sun: Its Future." *Science*, Vol. 162 (November 1968), 857–61.

Gregory, D. "The Hydrogen Economy." *Scientific American*, Vol. 228 (January 1973), 13–21.

Hammond, A. "Geothermal Energy: An Emerging Major Resource." *Science*, Vol. 177 (September 1972), 978–80.

Hammond A., W. Metz, and T. Maugh. *Energy and the Future*. Washington, D.C.: American Association for the Advancement of Science, 1973.

Hull, O. *A Geography of Production*. London: Macmillan, 1968.

Inglis, D. R. *Nuclear Energy: Its Physics and Its Social Challenge*. Reading, Mass.: Addison-Wesley, 1973.

Ion, D. C. *The Significance of World Petroleum Reserves*. Proc. 1B: 25–36. Seventh World Petroleum Congress, Mexico City, April 1967.

Landsberg, H., L. Fischman, and J. Fisher. *Resources in America's Future*. Baltimore: Johns Hopkins Press, 1963.

Lincoln, G. A. "Energy Conservation." *Science*, Vol. 180 (13 April 1973), 155–62.

Manners, G. *The Geography of Energy*. Chicago: Aldine, 1967.

Mischke, G. "The Search for Fusion Power." *Naval Research Reviews*, Vol. 24 (April 1971), 1–16.

National Academy of Sciences—National Research Council. *Resources and Man.* San Francisco: Freeman, 1969.
Odum, H. T. *Environment, Power, and Society.* New York: Wiley—Interscience Press, 1971.
Oort, A. "The Energy Cycle of the Earth." *Scientific American,* Vol. 223 (September 1970), 54–63.
Rocks, L., and R. Runyun. *The Energy Crisis.* New York: Crown, 1972.
Schurr, S., and B. Netschert. *Energy in the American Economy: 1850–1975.* Baltimore: Johns Hopkins Press, 1960.
Seaborg, G. T., and W. R. Corliss. *Man and Atom.* New York: Dutton, 1971.
Skinner, B. J. *Earth Resources (Energy).* Englewood Cliffs, N.J.: Prentice-Hall, 1969.
Thirring, H. *Energy for Man: Windmills to Nuclear Energy.* Bloomington: Indiana University Press, 1958.
Twentieth Century Fund. *Europe's Needs and Resources.* London: Macmillan, 1961.
United Nations. "Natural Resources." *Science and Technology for Development,* Vol. 2, 1963.
Vennard, E. *The Electric Power Business.* New York: McGraw-Hill, 1962.

A fertilizer plant on Mindanao Island, Philippines. (United Nations)

9

Modern Societies and Agro-Industry

The Evolution of Modern Societies

The modern factory system, characterized by expensive machinery and wage labor, originated in England during the last decades of the eighteenth century. It had an immediate impact upon the culture of that country and has since given rise to revolutionary cultural changes throughout much of the world. The introduction of this system into a given society is thus aptly referred to as the beginning of its own industrial revolution. In most instances, this revolution has meant more jobs, better food supplies, more consumer goods, better housing, and more leisure time. Unfortunately, however, it has also given rise to unsightly factory districts, pollution, alienation, and a sharp decline in craftsmanship (Fig. 9–1).

Today, even after several generations of intense technological development, the industrialization of the world is still incomplete. For, while England's Industrial Revolution has spread to many societies throughout the world and brought remarkable rises in their standards of living, even during periods of population growth, many other societies have remained virtually untouched, in any direct way, by the industrial process and must therefore endure without its benefits.

Early Industrialization

The first instances of industrialization were an array of machines, some old and some new, and the development of new sources of energy to run them. Most of the early advances were related to the production of such goods as textiles, ironwork, and pottery, all goods for which there was already an established and expanding demand.

The production of cotton textiles first gained prominence in England in about 1600, but, at that time, such products provided no serious challenge to the popularity of domestic woolens and cotton prints imported from India. Textile production pro-

FIGURE 9–1
A massive and less than picturesque industrial complex in the upper Ruhr Valley, near Dortmund, West Germany. (United Nations)

gressed rapidly during the early 1700s, when a ban was placed on the importation of India's cottons, which had become a threat to the English wool industry. The first of the major inventions in cotton manufacturing was not long in coming. The "flying shuttle," as it was called, appeared in 1733, and, replacing the hand-thrown shuttle, it soon increased the capacity and speed of English looms considerably. This created an immediate need for improvement in the traditional spinning process, and many tried their hand at finding a suitable solution to the problem (Fig. 9–2).

In the 1760s and 1770s, several of these attempts produced inventions that soon revolutionized the spinning process; chief among these was Hargreaves' spinning jenny, Arkwright's spinning water frame, and Crompton's spinning mule (Fig. 9–3). By 1785, Cartwright's power loom had also appeared, but its widespread use was delayed somewhat by violent opposition from hand-loom weavers, who intimidated fellow workers and committed acts of sabotage, including the burning of several textile mills.

In iron production, as in textiles, a series of technological developments carried England and the other early industrial countries through several rapid stages of advancement, each stage bringing both higher production levels and greater social problems. Perhaps the most significant of these developments occurred in 1709, when one Abra

FIGURE 9–2

The simple sort of spinning wheel in use in the early 1700s. (Bettmann Archive)

FIGURE 9–3

Improved mechanical spinners. Above, James Hargreaves' spinning jenny, invented in 1770. The model shown here was built by the U.S. Patent Office for the Chicago Exposition of 1893. (Smithsonian Institution) Below, Samuel Crompton's spinning mule, which combined drawing rollers and Hargreaves' stretching device. (Bettmann Archive)

ham Darby succeeded in processing coal into coked cinders that could perform as well in smelting and was without the toxic sulfur content of unprocessed coal. The many other technological advances in the iron industry included air pumps, steam hammers, and lathes for making heavy tools.

As has been mentioned in early chapters, the major medieval sources of mechanical power, the windmill and the water wheel, could not meet the new demand for power in the dawn of the industrial age. And throughout the seventeenth century and most of the eighteenth, attempts to perfect a steam engine also failed to provide an adequate response. A successful version of the steam engine, though patented by James Watt in 1769, was not marketed until 1776. But once available, Watt's steam engine was soon being used widely in cotton mills, steel mills, and pottery works throughout Europe. The perfection of the steam engine had removed a stubborn impediment to the expanded use of heavy equipment in factories of all sorts and in many mining operations as well (see

drous developments in the technology of transport were to take place: the spread of the railroad, almost worldwide, between 1825 and 1900; the addition of the now ubiquitous truck, after 1900; the one-day-to-anywhere airfreighter, after 1950; the ocean-going supertanker, in the 1960s; the mechanized container ships, after 1970 (Fig. 9–10).

However, the nuclei of today's major industrial regions and many of the lesser ones were already established by the early nineteenth century, and all were associated with tidewater or navigable lakes or rivers. During the last century and a half, railroads and highways have filled these regions with a network of ancillary land routes, but they have not shifted the major industrial regions to any noticeable degree.

Capital

Capital availability has traditionally been rated as an important factor in the location of industry. For even if sufficient financial resources are available for initial installations, expanding production and marketing costs can soon outstrip original plant investments. Tremendous amounts of capital are usually needed to buy large inventories of raw materials and fuel, finance new machinery and technology, pay wages and salaries, maintain plant and transportation systems, develop and expand markets, and pursue research and planning. Countries or regions in which financial institutions offer industry abundant capital, or credit, attract both new and relocating enterprises. Societies that do not provide liberal financial assistance stand little chance of developing strong industrial centers. Both Spain and India, very different in many other aspects of industrialization, are among the nations that face this problem; the United States and Switzerland are two of those that do not. A complete classification of the world's nations on the basis of capital availability reinforces the implication of these examples: that a country's industrial success does not depend on land area, as does most of its success in agriculture.

FIGURE 9–10

Controversial but efficient, the containerization equipment shown here can move a large cargo from ship to truck far more swiftly than is possible through manual operations. Containerization can thus help to decrease per-unit shipping costs and secure the transportation of high-value goods. (Seatrain)

Industrial Location Theory

It is obvious from the preceding discussion that the problem of where to locate a factory is a complex and difficult one. Thus, it is often helpful to represent locational principles in terms of theoretical models. In studies of industrial location, the model very often used is one developed early in this century by Alfred Weber. Despite some criticism over the years, this theory has served as the basis for many subsequent locational and industrial theories and still provides a valuable simplification of an often complex topic.

The Weber Theory

Alfred Weber, a German economist, was the first to publish a general theory on the location of industrial activities. Published as the *Theory of the Location of Industries,* in 1909, it was intended to be applicable to all types of industry under all types of economic, political, and social systems. Even though his major area of concern was the distribution of secondary activities, the basic principles Weber developed are applicable to other economic locational problems as well. The major criterion Weber considers operative in the selection of a site is *least cost.* He makes clear that his method of analysis is not intended to predict or explain actual industrial locations but merely to provide a basis for evaluating the numerous variables that influence them.

Weber begins his theory with a number of assumptions:

(1) There is a single isolated political state with a homogeneous culture (including technological development) and uniform physical characteristics.
(2) The products from each producing unit in the country are sold at a single market.
(3) The number and combination of natural resources (raw materials and fuels) as well as their distribution varies from case to case.
(4) The country has a uniform mode of transportation (the railway), with transportation costs increasing in direct proportion to distance and weight. Natural resources and finished products have equal transportation costs.
(5) The distribution of the labor force and wage rates is restricted to certain locations. Whereas at each of these locations the labor supply is unlimited, the prevailing wage rate is fixed and differs from place to place.

Within the context of these assumptions, industrial site selection in Weber's analysis is based on two major variables. The first and more important is transportation costs.

Transportation Costs. In Weber's analysis, the problem is simply to determine the point at which a factory will incur the least transport cost in (1) procuring its necessary raw materials/fuels and (2) delivering its finished product to market. In order to solve this problem it is necessary to know (1) the relative weights (weight loss) of the raw materials/fuels and the finished product involved and (2) the distance over which they have to be transported. (Since it is assumed that the natural resources and the finished product incur equal transportation costs, it would not be necessary to calculate the weight loss of the finished product.) The next step, then, is to establish weight and distance criteria for the natural resources.

In regard to weight, the natural resources can be categorized as either *pure* (that is, they lose no weight in processing) or *gross* (that is, they lose some or all of their weight in processing); all fuels, for example, suffer a loss of almost all their weight in processing. In regard to distance, the natural resources can be either *ubiquitous* (that is, they are available nearly everywhere) or *localized* (that is, they are available only at specific locations).

To demonstrate the influence of least transport cost on the location of industry, Weber uses a number of cases as examples.

Let us begin with the simplest case, which involves only one natural resource and the market. (1) If the raw material is ubiquitous and either pure or gross, then the factory would tend to be located close to or at the market in order to avoid paying any transportation costs. (2) If the raw material is localized and pure, then the factory can be located either at the market, at the raw material source, or anywhere in between these two points; the transportation costs in each instance would be the same. (3) If the raw material is localized and gross, then the factory would tend to be located at the source of the raw material, for there would be no sense in paying to transport waste material closer to the market.

This case helps to demonstrate that, as a general rule, localized rather than ubiquitous natural resources tend to force the location of factories away from the market. In addition, the same basic principles used for determining industrial location under simple conditions can also be used in cases where more than one natural resource is involved.

In cases involving more than one raw material, no two of which lie on the same direct line to the market, Weber makes use of various locational figures. For example, in a situation with two natural resources and one market, he uses a locational *triangle,* each corner of which represents one of the three factors involved. If, in such a situation, both of the localized resources are pure, then chances are high that the factory will be located at the market. If, on the other hand, either one or both of the resources are gross, then the location of the plant will be at the point in the locational triangle that achieves a balance, or equalization, in the transportation costs of the two resources; such a site would not necessarily be at the market.

In a situation involving three raw materials and a market, Weber uses a locational *polygon.* For still more complicated cases, where four or more resources are involved, he employs a mechanical device known as *Varigon's frame.* In this device the relative influence of the various factors—resources and the market—is represented by appropriately scaled weights that are suspended off the edge of the frame's circular top by wires over pulleys. The point at which the connected wires are in balance on the top of the circular disk is the theoretical point of least transport cost and therefore of the factory location.

Labor Costs. Whereas in the first part of Weber's analysis the ideal location for a factory is based solely on least transport cost, in the second part the ideal location is the point at which the combination of transport and labor costs is the lowest.

In order to illustrate the method Weber uses to determine this point, let us use a simple example involving one localized gross raw material and a market. In the diagramming of such a situation, shown in Figure 9–11, each concentric circle, or *isotim,* that is drawn around the individual market and raw material areas represents one unit of transport cost per ton of product. In this particular instance the least-transport-cost location for a factory would probably be at the raw mate-

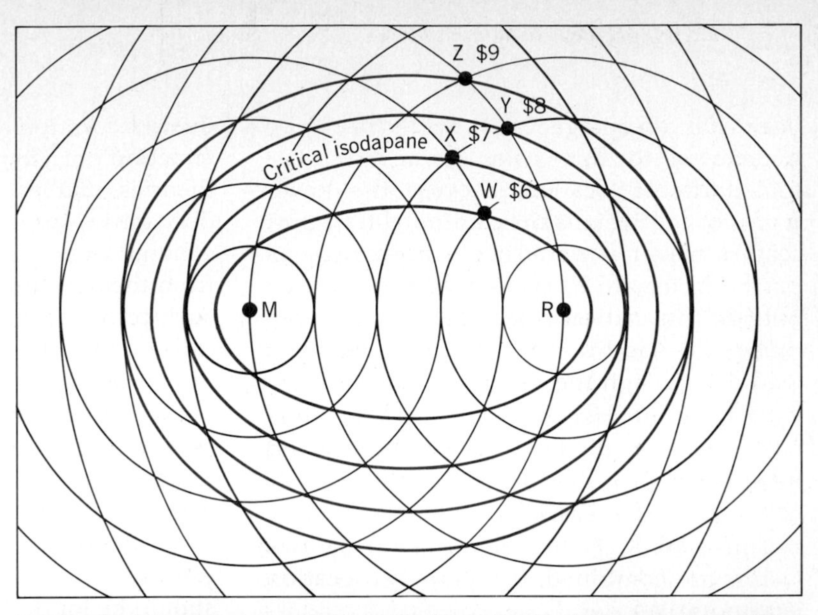

	R	W	X	Y	Z
Transportation cost	$ 5	$ 6	$ 7	$ 8	$ 9
Labor cost	$ 8	$ 6	$ 6	$ 6	$ 6
Total cost	$13	$12	$13	$14	$15

FIGURE 9–11

A diagrammatic representation of Weber's industrial location theory.

rial source (R), in which case each ton of product sent to market would incur 5 units of transport cost.

If, however, the factory were located at point W, the total transport cost to market, based on the total number of isotims involved, would be 6 units per ton of product (two concentric circles from the raw material source plus four concentric circles from the market). The solid red line, which point W is located on, is called an *isodapane* and is used by Weber to connect points of equal transportation cost—in this case, all points of 6 units. In similar fashion, all points along the isodapanes on which points X, Y, and Z are located would incur total transport costs of 7, 8, and 9 units respectively.

The isodapane is of most value when other variables—such as labor costs, taxes, rent, and subsidies—are combined with the basic variable of transport costs in determining the optimum least-cost location for industry. When such variables are involved it is possible that some point other than R would have the greatest advantage. In all cases involving multiple variables, an isodapane can be used to determine how great an advantage must exist at an alternate location in terms of labor costs in order to offset the location's disadvantage in terms of transport costs.

For example, if we know that at point W the cost of labor per ton of product is $2 less than at point R, we first find the isodapane (point X) along which transportation costs are $2 greater than at the point where they are the lowest (point R). Weber calls this the *critical isodapane* because its higher transport cost would exactly equal the relatively lower labor cost at point W, thereby establishing a balance between total costs at points X and R. Since point W lies inside this critical isodapane, it is advantageous for the factory to be transferred to this point from the least-transport-cost location, because the lower labor cost outweighs the additional transport cost. However, it would not be advantageous for the factory to be transferred to any point outside this critical isodapane (Y or Z), because the additional transportation cost would outweigh the savings in labor cost.

To summarize, if the sole factor is transport cost, then the least-cost location would be at point R; if the cost of labor is the sole factor, then the least-cost location would be at points W, X, Y, or Z; and when these two factors are combined, the least-cost location is at point W.

Other Locational Principles

In the real world, where there are no perfect conditions, decisions regarding industrial location are based on a complex interaction of dynamic variables, not on theories. But as a guide in making these decisions, planners often employ various locational principles that are derived, in part, from theory. The following is a sampling of such principles: (1) industries should be located at the point of least production cost; (2) industries should be dispersed interregionally in order to encourage regional self-sufficiency; (3) industries should be located so as to reduce the socioeconomic differences between town and country; (4) industrial location should be based to some extent on defense needs; (5) industries should be located so as to improve the quality of the environment.

At present, the distribution of most industry does not reflect such principles. If they can be more effectively applied in the future, not only the industrial regions but all the world should benefit.

REVIEW AND DISCUSSION

1. Compare the industrial developments in western Europe during the eighteenth century with the present-day industrial development of most societies in the Third World.
2. How did the dramatic technological changes wrought in the early stages of western Europe's industrial revolution enrich the lives of many Europeans? What hardships did they bring?
3. Which seven countries could correctly be called "industrial societies" at the turn of the twentieth century? What national features account for their early industrialization?
4. With respect to socioindustrial development, what are some of the disadvantages faced by nations that have only recently become independent?
5. The beginnings of modern societies have usually been marked by population shifts away from agrarian activities and toward urban commerce and industry. Discuss and evaluate some basic social consequences of these population shifts, both to groups and to individuals.
6. Most modern societies recognize that socioeconomic stability can best be achieved through the planned balance of primary and secondary livelihoods. Evaluate this view in terms of your own culture and locale, and then in terms of some culture and locale not your own.
7. Enumerate and explain some major advantages offered by secondary economic activities (especially industry) but not by primary activities (especially agriculture).

8. With respect to efficiency and productivity, compare and contrast the labor and technology used in agriculture with the labor and technology of industry. How could a better balance be established between them throughout the world, and what would be the socioeconomic gains?
9. Why and how and to what degree are markets increasing in importance as locational factors in world industry?
10. Identify and evaluate the advantages and the disadvantages in using the Weber theory for the purpose of analyzing industrial location.

SELECTED REFERENCES

Alexandersson, G. *A Geography of Manufacturing.* Englewood Cliffs, N.J.: Prentice-Hall, 1967.

Ashton, T. S. *The Industrial Revolution, 1760–1830.* London: Oxford University Press, 1968.

Birnie, A. *An Economic History of Europe.* London: Methuen, 1966.

Burton, I., and R. W. Kates. *Readings in Resource Management and Conservation.* Chicago: University of Chicago Press, 1965.

Froehlich, W., ed. *Land Tenure, Industrialization, and Social Stability . . . in Asia.* Milwaukee: Marquette University Press, 1961.

Greenhut, M. L. *Plant Location in Theory and in Practice.* Chapel Hill: University of North Carolina Press, 1956.

Hammond, J., and B. Hammond. *The Rise of Modern History.* London: Methuen, 1966.

Hull, O. *A Geography of Production.* London: Macmillan, 1968.

Isard, W. *Location and Space Economy.* Cambridge: Massachusetts Institute of Technology Press, 1956.

Lösch, A. *The Economics of Location,* trans. by W. H. Woglom and W. F. Stolper. New Haven: Yale University Press, 1954.

Mantoux, P. *The Industrial Revolution in the Eighteenth Century.* London: Methuen, 1966.

Miller, E. W. *Geography of Industrial Location.* Dubuque, Iowa: Brown, 1970.

Mountjoy, A. B. *Industrialization and Underdeveloped Countries.* Chicago: Aldine, 1967.

Mumford, L. *Technics and Human Development.* The Myth of the Machine, Vol. I. New York: Harcourt, Brace and World, 1967.

O'Brien, R., and the editors of *Life. Machines.* New York: Time, Inc., 1964.

Rosovsky, H., ed. *Industrialization in Two Systems.* New York: Wiley, 1966.

Smith, D. M. "A Theoretical Framework for Geographical Studies of Industrial Location." *Economic Geography,* Vol. 42 (April 1966), 95–113.

Staley, E. *The Future of Underdeveloped Countries.* New York: Praeger, 1961.

Trewartha, G. T. *A Geography of Population.* New York: Wiley, 1969.

United Nations. "Industry." *Science and Technology for Development,* Vol. 4, 1963.

———. "National Resources." *Science and Technology for Development,* Vol. 2, 1963.

Weber, A. *Theory of the Location of Industries,* trans. by C. J. Friedrich. Chicago: University of Chicago Press, 1929.

Industrial Regions of the Soviet Union

Industrial developments in czarist Russia were comparatively minor, but the Soviet Union, built from the remains of the old Russian Empire following the 1917 revolution, has made rapid industrial progress. It now stands second only to the United States in total industrial production. Complete government control of agriculture, mineral and energy resources, transportation, and labor has had one main objective—to strengthen Soviet industry, whatever the cost.

Most Soviet industrial developments are concentrated in four major regions, widely separated by vast agricultural expanses. Three regions are in Europe: the Moscow region, the oldest; the Donbas region, the largest; and the Ural region, the most nearly self-sufficient. The Kuzbas region, the newest, is in Asia (Fig. 10–10). Important factors that have contributed to the concentration and development of industry in these regions are the local availability of such source materials as minerals, textiles, and foodstuffs; such energy supplies as coal, petroleum, natural gas, and hydroelectricity; markets that are rapidly expanding and locally oriented by virtue of limited Soviet transportation; and a communist economy dedicated to the pursuit of regional and national self-sufficiency.

FIGURE 10–10

Industrialization in the Soviet Union. The four major industrial regions of the nation—Moscow, Donbas, Urals, and Kuzbas—are located in its European and west-central Siberian areas, in close association with rich mineral and energy concentrations.

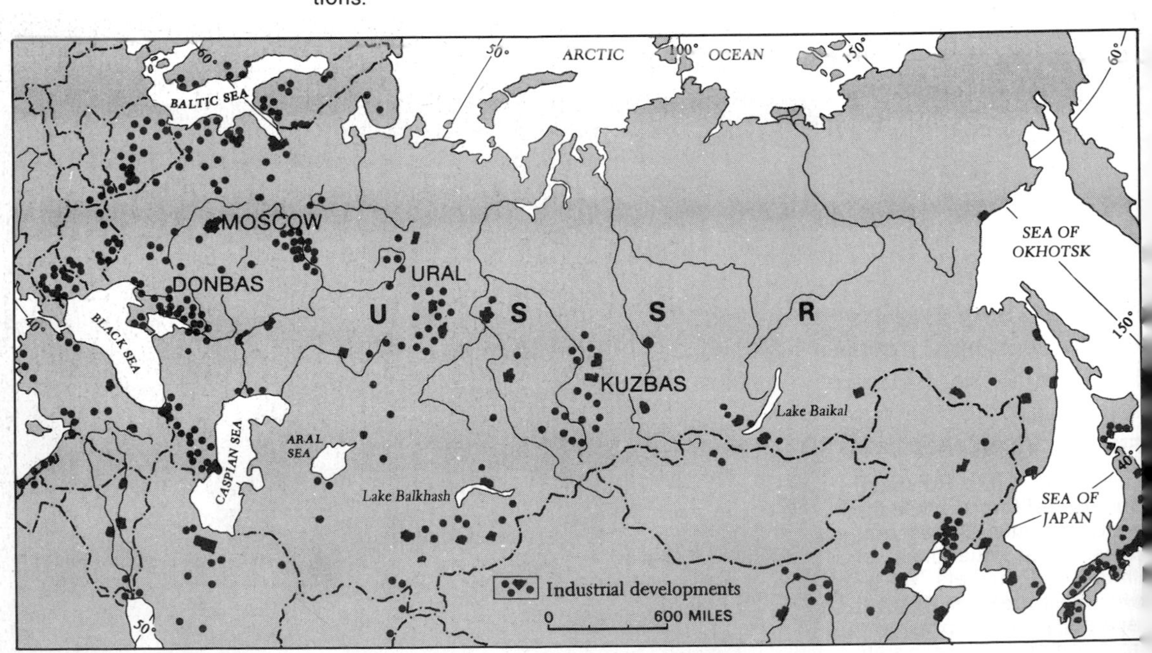

Moscow Region

The Moscow region was the oldest and largest in czarist Russia and had developed, like many others in the world, during the late nineteenth century, primarily because Moscow was the original nucleus of the old Empire. Industry was, and still is, varied, but the emphasis is on light industry and consumer goods to supply the traditionally large capital-city market with such products as textiles, clothing, and foods, and in recent years with various luxury items (Fig. 10–11). The region is not well favored with minerals and energy resources, but the surrounding plains are well developed agriculturally.

Donbas Region

The Donbas region is the principal heavy-industry region of the Soviet Union. It stretches across the eastern Ukraine, near the famous coal fields of the Donets Basin. The rich iron ore deposits near Krivoi Rog to the west are easily available, as is manganese in the same general region. In addition to its very large iron and steel production, the Donbas region supplies other important heavy-industry products, such as locomotives, tractors, industrial chemicals, cement, and armaments. There is a railroad shuttle system between the Krivoi Rog iron ore fields and the Donets coal fields that allows a heavy exchange of coal and iron and thus the development of iron and steel production in both areas. The Donbas industrial region is surrounded by the broad expanse of the Ukrainian Plain—the Soviet breadbasket—where the nation's agriculture still is most heavily concentrated; together the Donbas and the Plain represent an excellent agro-industrial development.

Ural Region

During czarist days the Ural region produced about one-fifth of the Russian Empire's iron and steel, as against two-thirds produced in the Donbas region. But the strategic interior location of the Urals has prompted recent Soviet planners to push the country's industrial development there, as well as even farther eastward into the central Asian regions. These two regions, the Urals and central Asia, now combine to produce half the Soviet Union's iron and steel, representing a significant planned eastward shift in major heavy industry.

The Ural region stretches 500 miles along the southern half of the Ural Mountain range and includes such heavily industri-

FIGURE 10–11

Fur processing near Lake Baikal, for the Moscow region. The wearing of animal furs and skins remains one of the few primitive clothing customs widely pursued in modern societies. (Harbrace by Jacques Jangoux)

important in clothing the nation's millions. Here also, particularly in the Ahmadabad area, has developed one of India's largest concentrations of cane-sugar refining.

Southern India Region. The Southern India region includes the tip of the subcontinent northward to a line from Madras to Mangalore. Here are concentrated major resources of iron ore, very large amounts of hydroelectric power, and large supplies of raw cotton, wool, and coir (coconut fiber). Upon these resources are based major productions of iron and steel, textiles, machinery, airplanes, electronic equipment, railway equipment, and technical instruments, scattered among such industrial cities as Madras, Bangalore, Mysore, and Madura. Variety rather than specialization is the keynote of industrial development in southern India.

China

China is the most recent of the major societies to undertake industrialization. Few industries developed in China during the nineteenth century, when industry was taking hold in western Europe, the United States, Russia, and Japan, or even after the turn of the present century, when modern industry came to India. China's long delay was due to many factors; the most significant of these were the country's age-old, weak, and chaotic political structure; foreign influences; internal strife; recurring famines; poor transportation facilities; and general socio-economic weaknesses. All these factors combined in fervent resistance to foreign ideas and innovations of any kind, whether from the West or the East. But, since 1949, when it came under the totalitarian control of communism, China has been attempting to industrialize "quickly."

China's traditional industrial resources are adequate but not abundant and not always conveniently located for efficient development and utilization. However, such handicaps are often manipulated politically when a totalitarian government is set on promoting a particular sector of its economy; industrial efficiency can be bought, to a degree, at the price of human freedom. China has substantial coal supplies, located largely in the northern provinces of Shansi and Shensi and in Manchuria, but also scattered in nearly every province of the country. The country's moderate supplies of iron ore are found in the central Wuhan region and in Manchuria, and during the early 1970s, relatively large reserves of petroleum and natural gas were found to exist in coastal northern China. The distribution of these resources leaves southern China relatively poorly endowed for modern heavy industrialization, but China's considerable hydroelectric potential has been long established over many regions, especially the Huang Ho Valley in the north, eastern Manchuria, and the yet undeveloped fringes of the Tibetan plateau.

During the last two decades, modern industry has made significant inroads into China's age-old cottage industries; today China's production of most manufactures depends almost entirely upon a "modern" factory system. Cotton textiles have become almost totally factory-oriented; hand-weaving has almost disappeared in China. Iron and steel, machinery, cement, railway equipment, foods, ceramics, rubber goods, electronic equipment, sugar, and ships are the country's major industrial items—all of which are emphasized in national production quotas.

Northeastern China Region. Central Manchuria and the Liaotung peninsula in northeastern China encompass the country's largest heavy-industry region (see Fig. 10–13). The greatest industrial concentration is

FIGURE 10–15

The modern steel complex at Anshan, in northeastern China. Like most other working people in China, the labor force in this center commutes by means of bicycles and buses. (Eastfoto)

in the southern area, where appreciable resources of coal and much of the country's iron ore occur in close proximity. Iron and steel plants have been developed at Anshan (built by the Japanese during their occupation of Manchuria) and at Penki; there are shale-oil reduction plants at Fushun, and auto-truck plants at Changchun; and the largest railroad shops, shipyards, and cement plants in China are located in this region (Fig. 10–15).

Northern China Region. The plains of northern China and the Huang Ho Valley contain a considerable amount of the country's industrial development. Here is found the bulk of China's coal reserves; within the general area of Peking, Tientsin, and Tangshan are China's second-largest coal mines and various associated industries. This region also includes cement plants and cotton textile and flour mills. Eastward, in the Shantung peninsula, the important industries are iron and steel, cotton textile and flour milling, and foodstuff production.

Central China Region. The Yangtze Valley, extending from the great port city of

Shanghai 1000 miles inland to Chungking, constitutes China's most extensive and most diversified industrial region. The lower Yangtze, with such cities as Shanghai, Soochow, and Nanking, is heavily industrialized. Shanghai, one of the world's largest cities, is also China's largest industrial city, with virtually every type of light industry. The middle Yangtze Valley includes the Central Lakes district, which is well supplied with coal and iron and accounts for much iron and steel production and much ancillary heavy industry near Wuhan. The upper Yangtze Valley, with Chungking at its center, accounts for a varied industrial production that ranges from steel and chemicals to cotton and silk textiles.

Southern China Region. On the delta of the Si River, with Canton as the hub, is concentrated considerable light industrial production, including textiles, cement, rubber goods, and electronic equipment, and serving the large markets of southern China and recently some foreign markets. Because of their large and diversified light industry, Hong Kong, though still British territory, and Macao, still Portuguese territory, can both be included in this southern China industrial region.

Industrial Societies of the Southern Continents

South America

South American industry is conspicuously underdeveloped. One of South America's principal industrial developments is the Río de la Plata region. Important centers there are Buenos Aires, La Plata, and Rosario, all in Argentina, and Montevideo, Uruguay. The Río de la Plata region produces a variety of staple foods, textiles, light machinery, cement, meat, and meat products. In Chile, small industrial areas are found in the central Santiago–Valparaiso region, the southern Concepción–Valdivia region, and

FIGURE 10–16

A tractor-assembling plant in Córdoba, Argentina. This and many other large plants in South America were established by foreign interests, mainly European or American, and produce products of foreign design. (United Nations)

the northern Antofagasta region. These centers produce a wide variety of commodities, including foods, clothing, wines, and some metals products. The only other industrial region of comparable significance is that centering on Rio de Janeiro and São Paulo in Brazil, with productions of some iron and steel, textiles, clothing, machinery, cement, chemicals, and processed foods, as well as the assembling of automobiles (Fig. 10–16).

Africa

Africa is primarily a seller of source materials and a buyer of industrial products; foreign rather than local factories supply most of Africa's industrial needs. Small industrial centers are found around Cape Town and Johannesburg in South Africa, and there are other scattered centers such as those at Port Elizabeth, East London, and Durban, along the southeastern coast of the country. Manufactures consist of clothing, processed foods, wines, household goods, some iron and steel, and automobile assembly. Compared to Africa as a whole, South Africa has made fair industrial progress in recent years.

Egypt has also surpassed most other African nations industrially. When it became a republic in 1953, most of its large-scale industry was of recent development. Even in the 1960s the country's industrial achievements exceeded expectations—production doubled, or nearly doubled, in many basic industries, including textiles, cement, iron and steel, fertilizers, foods, and oil refining. Today Egyptian industry also produces tires, refrigerators, radios, automobiles, and trucks and buses. Most industrial development is concentrated in the Delta, near Alexandria and Cairo, with some of the more recent works strung along the Nile Valley to the Aswan Dam.

Together, South Africa and Egypt account for one-half of Africa's industry. Most of the remainder is provided by Ghana, Morocco, Zaire, and Algeria.

Australia

Many of the industrial products used by the people of Australia are not produced within the country and must be purchased from the United Kingdom, continental Europe, the United States, and Japan. Australia has plentiful supplies of coal and fairly large iron ore deposits, with its largest reserves in the centers near Sydney, and smaller ones near Melbourne and Brisbane. However, it is not well endowed with most of the other necessities of industry. Even so, locally important industrial centers based upon the limited source materials available have developed in southeastern Australia around Sydney, Melbourne, and Brisbane.

Australian industry is diversified, but it is mainly light industry. Of chief importance to other parts of the world is Australia's processing of butter, cheese, and condensed milk, as well as meat packing and the processing of raw wool. Products for local consumption include textiles, clothing, flour, sugar, brewery products, metal articles, and some iron and steel.

Southern Continents in Perspective

The industrial regions of the southern continents are small in number and extent compared with those of the northern continents, but it should be recognized that even a limited industrial region may be extremely important to the country directly concerned. Australia is an excellent example. The country's total industrial productions are very small parts of the world total, but they are vital to the national welfare. The importance of industry in Australia can be measured, for example, by the fact that a higher proportion of the labor force in Australia is employed in industry than is true of the labor force in the United States. Furthermore, most of Australia's people live in large modern cities and, as in so many countries, large cities and industry go together.

REVIEW AND DISCUSSION

1. Why is it that, until relatively recently, the major industrial societies have been found within the Western world? Why and to what degree has the distribution of industrial societies changed in the last three decades?
2. Why is industrial production probably more vital to the United Kingdom than to any other major industrial society today?
3. Identify, as precisely as you can, the environmental and socioeconomic factors involved in the creation and perpetuation of the industrial regions of the United States.
4. Compare and contrast the four major industrial regions of the Soviet Union.
5. What are some of the major reasons for the great success of Japan's rather recent efforts to industrialize?
6. What accounts for the fact that several important industrial societies have emerged in regions that by nature lack practically every basic raw material needed for industry? Cite appropriate examples.
7. Both India and China have tremendous population problems and each is struggling to achieve an agro-industrial balance. Compare and contrast these two countries in terms of industrial development. Which one seems to have the greater industrial potential? Why?
8. Identify and characterize the major industrial societies of the southern continents. In terms of international trade, how do they differ from more highly developed societies of the northern continents?
9. Compare and contrast the industrial development of such diverse countries as the United States, the Soviet Union, Japan, China, Egypt, Ghana, and Australia; mention such factors as areal distribution, degree of self-sufficiency, diversification, and growth rate.
10. Using specific examples, discuss the differing effects of culture—that is, negative or positive—on the development of industry in different regions or nations of the world.

SELECTED REFERENCES

Alexandersson, G. *A Geography of Manufacturing*. Englewood Cliffs, N.J.: Prentice-Hall, 1967.

———. *Industrial Structure of American Cities*. Lincoln: University of Nebraska Press, 1956.

Chen, Cheng-siang. *Taiwan*. Taipei: Fu-min Geographical Institute, 1963.

Estall, R. C. *New England: A Study in Industrial Adjustment*. New York: Praeger, 1966.

Fleming, D. K. "Coastal Steelworks in the Common Market Countries." *Geographical Review,* Vol. 57 (January 1967), 48–72.
Li, Choh-ming, ed. *Industrial Development in Communist China.* New York: Praeger, 1964.
Mayer, H. M. "Centrex Industrial Park: An Organized Industrial District." *Focus on Geographic Activity.* New York: McGraw-Hill, 1964.
Miller, E. W. *A Geography of Industrial Location.* Dubuque, Iowa: Brown, 1970.
Nakayama, Ichiro. *Industrialization of Japan.* Tokyo: Center for East Asian Cultural Studies, 1963.
Pounds, N. J. G. *Geography of Iron and Steel.* London: Hutchinson University Library, 1959.
Pryde, P. R. "The Areal Deconcentration of the Soviet Cotton-Textile Industry." *Geographical Review,* Vol. 58 (October 1968), 575–92.
Rodgers, H. B. "The Changing Geography of the Lancashire Cotton Industry." *Economic Geography,* Vol. 38 (October 1962), 299–314.
Rogers, E. M. *Diffusion and Innovations.* New York: Free Press of Glencoe, 1962.
Sharer, C. J. "The Philadelphia Iron and Steel District: Relations to Seaways." *Economic Geography,* Vol. 39 (October 1963), 363–67.
United Nations. "World of Opportunity." *Science and Technology for Development,* Vol. 1, 1963.

An Indonesian village on the island of Bali. (Pan American)

11

Settlement Types and Contrasts

Major Contrasts in Settlement

Mobile Versus Sedentary

Among the human population in some parts of the world, there are still large kinship groups that move from one place to another without permanent abode.

These people are nomads, and, in most instances, their way of life is dependent on herds of grazing animals (usually horses, sheep, or goats) whose keep means a constant search for areas of plentiful grass (Fig. 11–1). Though a nomadic group may attach itself to a specific location long enough to build a complex of semipermanent shelters, the time soon comes when local sources of grass for the animals are exhausted and it is necessary to move to a new area.

Nomads who rely heavily on pasturage are found in such diverse areas as the desert fringes (see Chapter 3) and the Arctic tundra. In other areas, particularly the tropical forests, there are several groups that live entirely upon what they can secure through hunting and fishing, or through the collecting of edible plant products (Fig. 11–2). Such groups are usually not as mobile as nomadic groups that maintain herds and must search for pasturage. Among the least mobile of them are the Eskimos of the Arctic fringelands, who, anticipating the seasonal migrations of their prey, usually move their settlements to locations known to them and to which they have frequently returned.

Unlike such groups, however, most of the world's peoples are tied to specific places on the surface of the earth. These *sedentary*, or fixed, populations provide relatively long-lasting patterns and forms of settlement that can be observed and compared as essential elements in a thorough understanding of the cultural phase of the human habitat.

FIGURE 11–1
Settlements of nomadic peoples, such as this Mongolian yurt, are usually easy to construct and move. (American Museum of Natural History)

FIGURE 11–2
A temporary home constructed of giant palms in the forests of Venezuela. (© Jacques Jangoux)

Dispersed Versus Agglomerated

The types of settlements devised by sedentary populations vary widely. At one extreme is the isolated dwelling that houses a single individual or family; at the other, a group of dwellings that stand side-by-side and house hundreds of people. In some of the world's large cities, where skyscrapers pile one occupant's space on top of another's, the total vertical distance of housing can be greater than the city's horizontal extent.

When individual dwellings are scattered over the land, the settlement pattern is said to be *dispersed.* When dwellings are packed close together, the settlement pattern is described as *agglomerated.* Complete dispersion is understood to mean an arrangement of single dwellings (and their associated buildings) that suggests no grouping or clustering. Complete agglomeration, on the other hand, implies a close massing of dwellings and other buildings with no isolated dwellings between clusters. Possible, too, is the mix, or combination, in which agglomerated settlements are spotted throughout a general background of dispersed settlement.

Being highly gregarious, humans tend to live in groups rather than alone, but not necessarily in a fixed place. There is ample evidence to show that humankind existed first as wandering tribes and that it established permanent settlements only after it had achieved some mastery of agriculture. Reflecting the isolation and internal cohesiveness of these tribes, the first sedentary populations assumed an agglomerated settlement pattern. And in many parts of the world, this pattern has persisted. Dispersion, which came relatively late in the history of human settlement, characterizes relatively few political regions of the world; however, those regions are extensive.

Rural Versus Urban

Any distinction between rural and urban is a relative one. There is no sharp line that divides rural folk from urbanites. But one of the most reliable differentiations is that a *rural* population is one predominantly engaged in securing life's basic necessities directly from the land, while an *urban* population is one that has little or no involvement in the initial production of foods or materials used for shelters and clothing; rather, most urban dwellers are engaged in secondary and tertiary activities, such as transporting, manufacturing, buying or selling raw or manufactured materials, educating, and managing affairs of state. The other criteria commonly used to distinguish between urban and rural communities are much more problematic.

Distinctions between rural settlements and urban ones are often made on a statistical basis, through the use of population figures. However, this method can become somewhat confusing, especially when applied in comparisons of different countries. One finds that there is very little agreement in the international community as to what constitutes the minimum population of an urban community. For example, in Denmark any settlement with a population over 250 is considered urban; but in Canada and Chile the minimum figure is 1,000, and Panama classifies as urban only those places with no less than 1,500 inhabitants. A number of countries—Argentina, Austria, Czechoslovakia, West Germany, and Portugal, for example—require at least 2,000 inhabitants, while, in general, the United States, Mexico, and Venezuela list as urban only settlements of 2,500 persons or more. Still more restrictive are Belgium, India, and Japan, which classify as urban only those settlements with a population above 5,000. Switzerland and Spain exceed even this requirement, however, by making 10,000 their demarkation between rural and urban. Last in this escalating scale is the United Nations, which sets a minimum population requirement of 20,000 when it lists urban communities.

Some classifications of populations as urban or rural are also based on their spatial arrangement. Obviously, rural populations are more often arranged in dispersed patterns than are urban populations. Yet, rural folk need not necessarily be scattered over the land; in fact, there are more agglomerated rural populations throughout the world than there are dispersed ones. And with urban population almost entirely agglomerated, it is estimated that fully three-quarters of the world's population lives in agglomerations of one type or another.

With today's huge populations in such centers as London, New York, and Tokyo, the world's total urban population is about equal to its total rural population. And most of the urban settlements that do exist are found in the Western world, specifically Europe and areas colonized by European peoples. The populations of the Eastern world, where over half of the earth's inhabitants live, are almost entirely rural agglomerations.

Micro Versus Macro

Various terms are currently used throughout the world to suggest a general categorization of settlements, based primarily on their size and on the number and complexity of functions performed. Although useful here and in similar discussions of fundamentals, this classification is not suitable for all settlement analysis.

Within this classification, one of the simplest types of settlement is the individual *farmstead,* which is usually composed of a single dwelling for humans and a barn or shed where domesticated animals are kept and equipment is stored. Another very sim-

FIGURE 11–3

Examples of settlement types usually considered rural. Top left, an isolated farmstead in the wheatlands of South Dakota. (Rotkin, PFI) Top right, a roadside near Glendale, California. (Harry W. Rinehart) Center, a hamlet in Pennsylvania. (Pennsylvania Dept. of Highways) Bottom, a village in Monroe County, Wisconsin. (U.S. Dept. of Agriculture)

ple type of settlement is the *roadside,* usually a lone commercial establishment—such as a gasoline station, drive-in eating place, small general store, motel or inn—located beside a highway (Fig. 11–3). Larger than a roadside is the *hamlet,* a settlement made up of several buildings that serve residential, commercial, and social purposes. Typically somewhat larger and more diversified is the *village,* which generally includes schools and churches and other public facilities. In general, all these settlements—farmstead, roadside, hamlet, and village—are considered rural.

Next in this settlement classification is the *town,* which offers more services than the village but fewer than can be found in the average *city;* the essential distinction between a town and a city is a matter of structural complexity, the city being the more complex settlement. *Metropolis,* a term loosely used to refer to any very large city, implies the presence of *satellite cities,* settlements that have become tied to, or nearly engulfed by, a huge urban nucleus; London, Paris, and Los Angeles are among the examples. The term *conurbation* is used to describe a large area of unbroken urban development created by the growth and coalescence of several neighboring cities; although still separate settlements by statute, these cities function so interdependently that they seem to be a single urban unit. Next in the progression of settlements is the *megalopolis,* or giant conurbation, examples of which include the stretch of metropolises that extends from Boston to Washington, D.C., along the northeastern seaboard of the United States, and a comparable urban belt from Tokyo to Osaka in Japan.

Rural Settlement

Even today, in this machine age, dependence upon primary production from the land is fundamental to human existence. Direct use of the land is still the way of life for the majority of human beings; in other words, most of the world's populations are rural. Their spatial arrangement on the land is the matrix in which urban populations are set.

Agglomerated Rural Settlement

A rural agglomeration is a group of relatively closely built structures surrounded by fields, pastures, or woodland areas. Human dwellings constitute much of the cluster, but there are also storehouses for agricultural produce and tools, and shelters for livestock. There may also be, in more advanced societies, a church, a school, and a town hall. Practically the entire population is concerned with cultivating the surrounding land or with raising animals—in living directly from the land. To augment their living, some of the inhabitants may engage in selling surplus commodities. For example, someone in the settlement may offer various foods for sale and possibly add to that small tools or other farm supplies. Professional people—such as doctors, ministers, and teachers—may reside in the agglomeration. But most of the life of the settlement is tied to the land, and, except as the settlement contributes its agricultural surpluses to the outside world, it is essentially self-contained. The rural settlement is first and foremost the dwelling place of an agricultural society.

Rural settlements in different parts of the world vary widely in appearance and structure; they often share no common characteristic except function (Fig. 11–4). A rural agglomeration is always the dwelling-place of a group engaged in direct use of the land. Beyond that, the number, size, appearance, and arrangement of buildings correspond to the traditions and habits of the particular group of people and to the nature of the land from which they derive their living.

For example, the average village in many parts of Japan consists of a cluster of bamboo-and-thatch dwellings and sheds, so tightly pressed together as to allow only narrow passageways between adjacent buildings. The whole settlement occupies a bit of ground only slightly above the level of the surrounding irrigated rice paddies. Seldom are there roads leading to and from the village. Narrow pathways along the tops of low dikes separating the rice paddies are usually the only ground routes between one village and the next (Fig. 11–5).

The inhabitants of a village leave it early in the day to work their paddies and return again at night. A farmer's land is seldom in one continuous piece, but rather it is in several fragments in scattered locations around the settlement. The daily life of the inhabitants, often members of only a couple of large family groups, is a pulsing out onto the land and back into the village. Both the scattered nature of the landholdings and the intensity of the agriculture demand constant attention. The village is the focal point from

FIGURE 11–4

La Calera, 40 miles north of Bogotá, Colombia. It is typical of South American villages in that it features a church and a public square. (Standard Oil Co., N.J.)

FIGURE 11–5

A settlement on Japan's Izu Peninsula that is representative of most Oriental rural agglomerations. Its bamboo-and-thatch dwellings and their garden plots are surrounded by diked rice paddies. (Bob and Ira Spring)

which a small portion of the human habitat functions.

The hamlets or villages in many parts of Europe are in considerable contrast to those of the Orient. In southern France, for example, villages like Turenne are common. Turenne crowns the top of a low hill with fields spreading down the slopes and away from the settlement. The village is roughly rectangular in shape, with a chateau at one end and a church at the other. Forming the sides of the rectangle are houses built side-by-side. There is thus enclosed a public *plaza*, or "square," with exits on either side of the church and chateau. Immediately back of the houses on the hill slopes are small kitchen gardens. Farther down the slopes are larger fields and, beyond them, pastures and patches of woodland (Fig. 11–6).

There is the same daily movement out from the village to the fields and back, but it does not have the regularity of that of the Oriental village: agriculture in France is less intensive and demands less attention. In addition, the village is not so completely a family settlement. Finally, the European village is linked by roads to the transportation net of the larger area; it is not nearly as self-contained as the Oriental village.

In certain other parts of Europe—in Hungary, for example—rural agglomerations of several thousand people are not uncommon. These villages cover considerable land areas, with each dwelling and its farm buildings set on a small plot. The farm buildings are distinctly clustered, arranged along streets to form collected settlements set in open cultivated or pastured land.

FIGURE 11–6

The hilltop village of Eze, in southern France. In the manner of many European settlements, its houses and terraced fields surround a medieval castle built on the village's highest ground. (American Museum of Natural History)

In the eastern seaboard regions of the United States, particularly in New England and the South, agglomerated rural settlement was the early mode. In colonial New England, farmhouses were built close together, usually fronting on a *common.* The common was originally a pasture area collectively owned and used by all the inhabitants of the village. Where it still remains today, it has become a park. In the South, the establishment of plantations that required large labor forces led to the importation of slaves and to another kind of rural agglomeration, the compact slave quarters.

These few examples suggest that there is great variety in the detail of rural agglomerated settlement throughout the world and within small portions of it as well. The unifying characteristic is the tie of the people to the land.

Dispersed Rural Settlement

Smaller than the hamlet or village is the isolated farmstead, a settlement unit usually composed of a family dwelling and a few barns and sheds. Fields, pastures, and other land from which the family draws its living stretch out from the central unit, uninterrupted by other buildings. The distance between farmsteads may be slight, perhaps a quarter of a mile, or great, as much as several miles (Fig. 11–7). There is none of the compactness of the agglomeration, nor is there the daily flow of large numbers of people to surrounding fields. Land used intensively generally lies close to the farmstead itself. Extensively cultivated lands, pastures, and wooded areas are farther away. Each farm serves as a unit in itself, not as a fragment of a larger integrated hamlet or village.

World Distribution

The farm village characterizes the greater part of the Orient. Except in a few relatively small areas (such as Hokkaido, the northern island of the Japanese group), isolated farmsteads are rare. Throughout the southern islands of Japan, the densely peopled plains of China, Southeast Asia, and India, closely set, compact hamlets and villages are sprinkled thickly through the intensively cultivated flatter lands.

In parts of the Middle East and Africa—the lowland of the Tigris and Euphrates rivers and the valley of the Nile, for example—rural agglomeration persists almost exclusively. In other parts of Africa—such as the Guinea lands, the Congo Basin, or parts of the East African Plateau—the agglomerated pattern of rural settlement is repeated.

The hamlet or village is the characteristic settlement unit throughout most of southern and eastern Europe. Even into the western part of that continent, it continues as a prominent form, but with the dispersed pattern becoming more and more prevalent farther northwestward. Yet even in northwestern Europe, the incidence of dispersed rural population fails to mask the continent's basic agglomerated pattern.

As indicated, colonial settlements in the New World were of the agglomerated sort. Throughout parts of South and Middle America, agglomeration in rural areas is still the rule; but in the United States and Canada it is now virtually absent, having been replaced in most areas by dispersion. In the United States, especially, the farm village is essentially a relict form; in only a few cases, as in some of the Mormon settlements of Utah, does it represent a functioning unit.

It is in the lands settled or colonized in this millennium by peoples of northwestern Europe that the dispersed pattern of rural settlement is most characteristic. Thus, Canada, like the United States, is an area of scattered rather than clustered rural population; so too are the greater parts of Australia, New Zealand, and areas of European colonization in Latin America and south and east Africa. The habit of agglomeration has been lost in these areas.

FIGURE 11–7

A dispersed rural settlement pattern composed of farmsteads, near Madison, Wisconsin. (Rotkin, PFI)

Urban Settlement: Origin and Diffusion

Urban settlements are as old as civilization itself; the word *civilization* is, in fact, derived from the Latin word *civilis*, meaning a citizen of a politically organized society, such as a city-state. Indeed, one of the major features of a civilized society is thought to be the ability to support an urban way of life. And many of the characteristics commonly attributed to a civilized people are similar to those considered typical of urbanites; in fact, they are practically synonymous.

In most analyses, civilized society is described as a group of people who have developed some form of organized religion, a system of writing, and the fundamentals of science. Furthermore, their social structure is advanced enough to allow a rather extensive division of labor. This means that certain specialized groups, such as farmers, herders, and fishers, are responsible for providing the food, while miners, craftspeople, merchants, clerks, soldiers, and others provide services and manufactured goods. Such institutions as government and religion are managed by kings, princes, priests, and other officials.

In order to support such a highly developed division of labor, a society would have to possess a number of attributes more complex than those that characterize a so-called primitive culture. Thus, complexity is an essential distinction between a civilized society and a primitive society.

Similarly the establishment and growth of an urban settlement requires not only technological advances and a favorable physical environment but a workable societal organization as well. An urban society is characteristically organized into classes of specialists, one of which is a ruling elite. In the simplest urban settlements it is the responsibility of this group to see to it that the food needed by the settlement's inhabitants is obtained from source areas and properly distributed or stored. In more complex settlements the ruling group is also concerned with the planning, construction, and maintenance of large-scale public projects, including municipal buildings, travel routes, and water supply systems.

Early Settlements

Mesopotamia. Diverse evidence indicates that human beings spent several millennia in sedentary communities before the development of the first truly urban settlements. These are believed to have appeared around 3500 B.C., mainly in an area comprising the fertile valleys of the Tigris and Euphrates rivers. Throughout ancient times this southernmost part of Mesopotamia (then known as Sumer), and the area to the north (then Akkad), spawned many urban communities, some of the most prominent of which were Babylon, Eridu, Lagash, Ur, and Kish (Fig. 11–8). The birth of these first urban centers marked the beginnings of the *urban revolution*, a development that is as important to the cultural growth of humankind as the agricultural revolution some 5,000 years before it and even the industrial-scientific revolution some 5,000 years after it.

Certain functional characteristics were common to all these early urban settlements. In each settlement, for example, the power of the ruler was based heavily on religious dogma, and the socioeconomic base reflected a limited number of specialized services. Most ancient urban dwellers were either government and military officials, clergy, craftspeople, merchants, or farmers; and each of these groups tended to be found in a different section of the settlement. The average total population of these early urban centers, though difficult to estimate, undoubtedly exceeded that of the typical Neolithic village.

One of the most important developments associated with the early spread of urbanism was the invention of writing, which enabled cities to keep records, chronicle historical events, and codify laws. Only with the help of the written word could a social organization on a large and rather complex scale be developed and maintained.

The Nile Valley and Other Areas. Mesopotamia was not the only ancient culture hearth to generate urban settlements. Several Egyptian villages in the valley of the Nile developed into urban communities by about 3000 B.C., and, within five hundred years, urban centers could also be found in the Indus Valley of present-day Pakistan. Among these were the highly organized and wealthy cities of Mohenjo-Daro and Harappa, which were in their day among the leading metropolises of the world. By about 1500 B.C. urbanization had begun as well in China, principally along the Huang Ho River. In the Western Hemisphere, villages with urban characteristics did not develop until about 500 B.C.; the first examples were located in Middle America.

There is some question as to whether the idea of the city spread around the world or was arrived at independently in each area of early urban development. Since the evidence thus far uncovered shows a considerable time lag in the occurrence of urban centers in different places, it would seem that diffusion, rather than invention, exerted the stronger influence in the growth of urbanization. The earliest known urban centers of the New World may represent exceptions to this

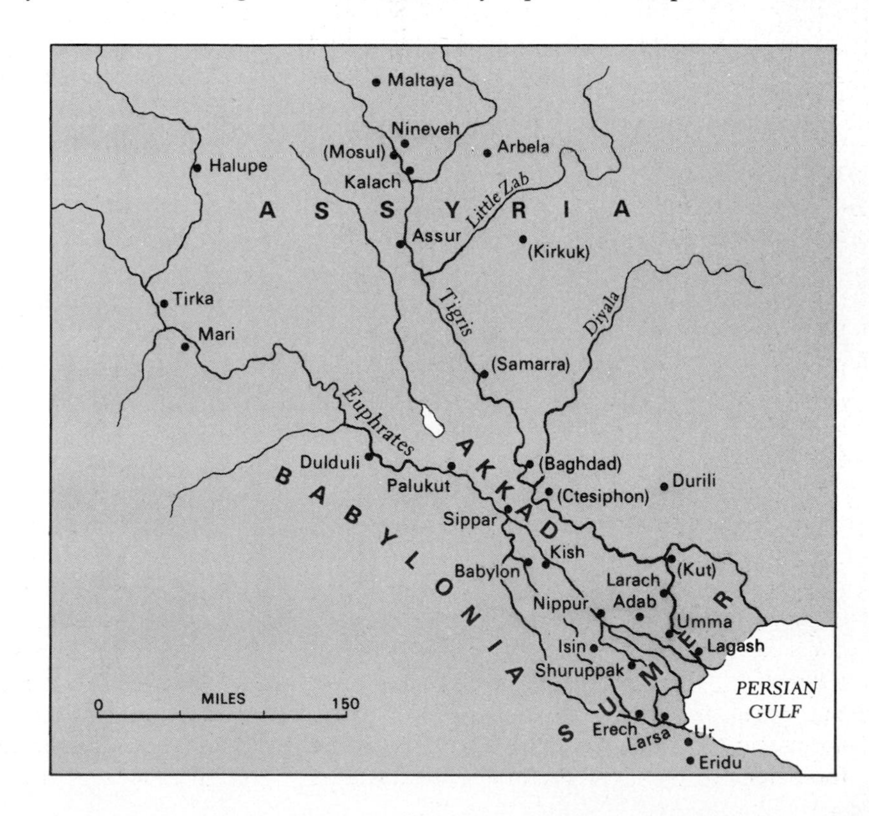

FIGURE 11–8

Mesopotamia's ancient sites and cities. Many of these early urban centers were small agricultural villages long before they developed into cities.

rule, however; for it is very likely that such Middle American urban settlements as the Mayans' Tikal and the Aztec capital Tenochtitlán (now Mexico City) evolved independently of almost any Old World influences.

The Influence of Empire. As the early centers of urbanization grew and became more powerful they became less immediately involved in agriculture and more dependent on the surrounding lands for nutritional support. Control of these lands—and, sometimes, more distant ones as well—was thus essential, and often prompted military invasion and conquest. By conscripting many of their former enemies and taxing or confiscating their holdings, some of the independent urban centers, or city-states, became powerful enough to establish empires and, by blending many of the cultural traits of their colonial territories, even generated new civilizations.

The spread of urbanization around the Mediterranean and into Europe, for example, was closely associated with the development and spread of the Greek, Phoenician, and Roman empires. In order to secure their colonies, the seats of power found it necessary to establish numerous outposts that could serve as administrative, military, and trading centers. And, not surprisingly in the light of their extreme dependency, there seems to be a close correlation between the rise and fall of an empire and the life span of the urban settlements it established or reinforced. The Roman Empire provides a particularly good example of this relationship between imperialism and urbanization. Areas around the Mediterranean that were once part of the Roman Empire are today filled with "dead" cities, settlements that prospered while the empire was strong, withered when it fell, and never recovered their imperial vitality; among these ruined cities are Stobi in southern Yugoslavia and Timgad in Algeria. On the other hand, most cities of the Roman Empire did not die, but many—including Rome, the capital—did experience temporary declines in population.

Medieval European Settlements

Urbanization in Europe reached another important stage during the Middle Ages, the thousand years following the fall of the Roman Empire. European city life, which had flourished at the height of the empire, ebbed sharply after the fall of Rome, in 476. Intercity travel was greatly restricted by the ensuing fragmentation of power, and trade between cities nearly ceased. As their economic base dwindled, the existing cities lost many of their inhabitants to the countryside; and the development of new cities stalled. Dependent on surrounding communities for most of their provisions, cities were particularly vulnerable in the general state of confusion and disorder that gripped much of Europe in the centuries after the imperial collapse: the Vikings were attacking from the north and by the mid-800s were raiding the Mediterranean; various nomadic groups, including the Huns, were invading from the east; and the Arabs were penetrating from the south; moreover, the Europeans were at war with one another.

Thus began what is commonly referred to as the Dark Ages—a period that lasted roughly from A.D. 400 to 1000. During this period European civilization suffered severe setbacks; it was not until the end of the Crusades, in the thirteenth century, that the continent began to emerge from its decline. While Europe was submerged in the Dark Ages, however, most other parts of the world were not; indeed, during this era many non-European urban communities, such as Constantinople (Istanbul), Cairo, Mecca, Baghdad, Canton, Peking, the Mayan city of Tikal in Guatemala, and other Middle American urban centers, were growing and, in some cases, flourishing.

But in spite of the disorder that prevailed during the Dark Ages, most European cities survived, some resorting to massive walls and moats as protection against attack (Fig. 11–9). The population of these urban centers was thus often limited to the number of people who could be accommodated within the confines of the city walls and fed by the agricultural lands within the immediate vicinity. If a city was not walled, then, when under attack, its inhabitants usually took refuge in the area's most fortified and strategically placed castle. Thus, for defense and, generally, for economic reasons as well, the castle of the local noble served as the nucleus of many medieval settlements; and from the ninth to the fifteenth centuries feudalism reigned.

Not until the end of the Dark Ages, with the rudimentary beginnings of modern industry and commerce, did Europe resume the development of an urban way of life. An increase in trade brought about a gradual return of the peasant to the city in search of jobs and, eventually, led to the development of craft guilds. These were the predecessors of the modern trade union, and, in several areas of Europe, they became the means by which the newly created middle class gained

FIGURE 11–9

Dubrovnik, Yugoslavia, a medieval (probably sixth century) walled city located on the Dalmation coast. Although under Venetian control for more than a century (1205–1358) as a result of the crusades, Dubrovnik has, for the most part, enjoyed full autonomy as a city-state, reaching its height as a trading center in the fifteenth and sixteenth centuries. It has also been a model of urban development and planning. Even in medieval times, zoning laws controlled its building construction, sanitary facilities, and water supplies. A public medical service existed in Dubrovnik as early as 1301; its first pharmacy was opened in 1317; and by 1432 the city had its own orphanage. (Harvey Stein)

political power. Cities became strongholds of the traders, whose power was eventually used to end the control of the feudal nobility.

The gradual revival of trade during the twelfth and thirteenth centuries also prompted great trade fairs, which in turn encouraged the growth of both the host cities and the contributing urban settlements. With this rebirth of intercity cooperation, such cities as Genoa and Milan, in northern Italy, were revived as trading centers, and Cologne, London, Danzig, Amsterdam, and Paris also grew in size and stature. Their close ties allowed the cities to maintain political independence from the declining feudal principalities and dukedoms that surrounded them. However, this autonomy was rather short-lived, as the emergence of the nation-state brought centralized power and authority.

Industrial Cities

The industrial revolution had a profound and lasting influence on the nature of urban settlements. Throughout the medieval and early colonial periods, European cities grew and changed at a relatively slow pace; but in the late decades of the eighteenth century the beginnings of large-scale manufacturing generated a completely new and revolutionary type of urban settlement, the industrial city.

Since the industrial revolution had its origins in England, it is not surprising that England was also the home of the first industrial cities. Among these were Manchester, Sheffield, and Stoke-on-Trent, important early manufacturing centers for cotton textiles, iron and steel, and pottery, respectively. The growth of industrial cities in other parts of the world during the eighteenth and early nineteenth centuries quite naturally coincided with the slow and somewhat limited spread of industrialization in this period. And although the industrial revolution was responsible for creating a number of new cities, many of those that came to be industrial centers by the mid-nineteenth century had been major commercial centers in ancient or medieval times. In Europe, these transformed commercial cities included the capitals London, Berlin, and Moscow as well as such secondary cities as Milan, Hamburg, and Poznan.

The advent of the industrial city was largely a consequence of the widespread application of new sources of energy, particularly coal-generated steam power. Up to this time the primary source of power, other than wind and water, had been animate rather than inanimate—specifically, human and animal muscle. The growing importance of coal in such areas as the Ruhr Valley, Appalachia, and the Donets, for instance, was in turn responsible for the development of such cities as Essen, Wheeling, and Donetsk, respectively.

The industrial revolution altered the established social patterns of these cities. Less-than-attractive factories were now often located in or near the heart of town, and within walking distances of them stood the cheaply built and cramped row houses and tenements in which many of the factory workers lived. Frequently the wealthy classes moved from the center of the city to outlying sections in order to avoid the crowding and unpleasant conditions associated with inner-city life. This tendency represented a reversal of the preindustrial residential pattern. But although people of different socioeconomic backgrounds usually congregated in different neighborhoods, as indeed they had done in preindustrial cities, industrialization brought urban populations increased opportunities for social contact and exchange, especially with the development of factory centers and transportation systems. The re-

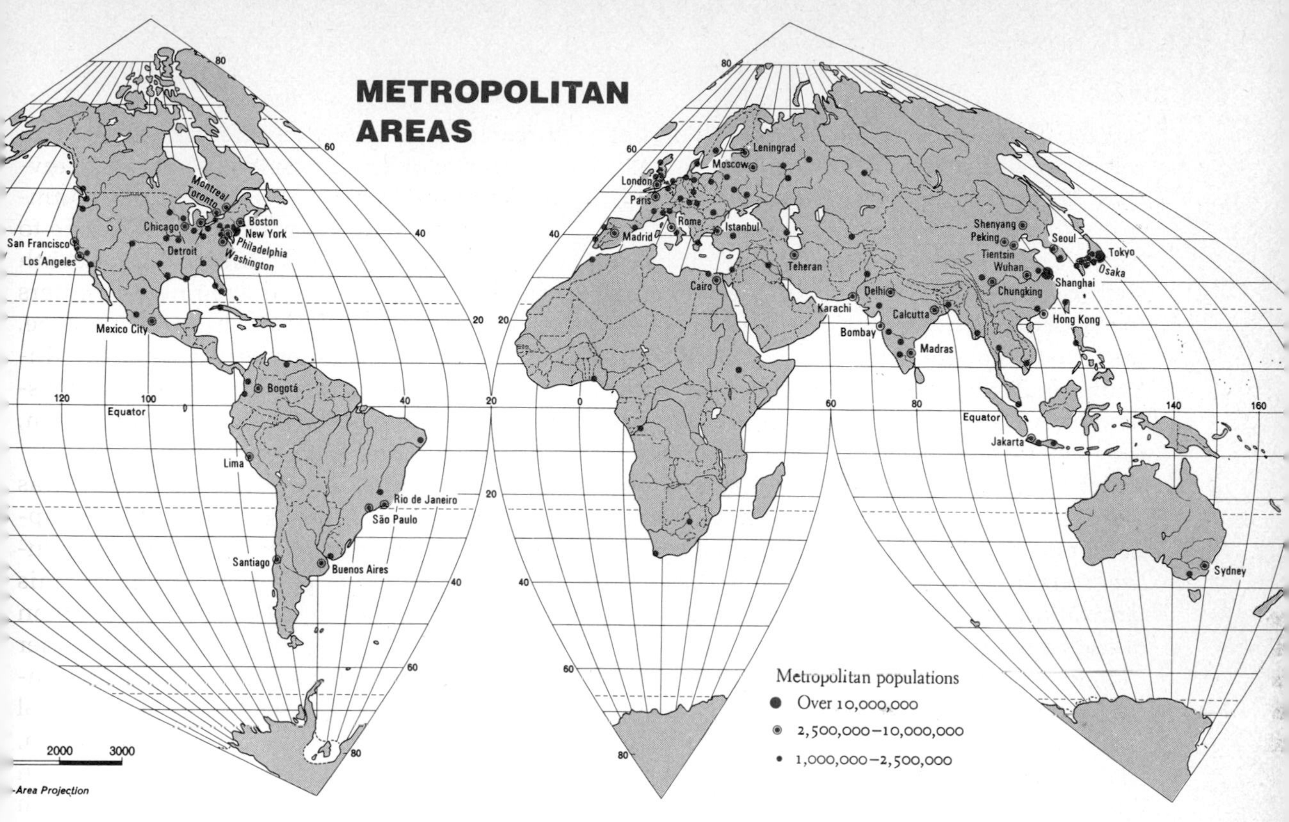

FIGURE 11–10
In this distribution of selected metropolitan areas exceeding one million persons, only those urban centers with more than 2.5 million inhabitants have been named.

sult was a considerably more fluid social structure within most industrial centers than had existed previously.

Industrialization also brought with it marked increases in urban populations; the factories needed workers, and improvements in farming technologies, such as the mechanical reaper and the cotton gin, had reduced the number of jobs available in agriculture. Further, improved methods of food preservation, water supply, and local transportation allowed for the comfortable maintenance of a greater concentration of people. The result was an ever-increasing migration of people from rural to urban centers.

But as early industrial cities grew in size, there was a corresponding growth in the number and complexity of urban problems. Urban dwellers, especially those in the inner city, were constantly troubled by overcrowding, poor sanitation, crime, and pollution—thus, they suffered widespread disease and high death rates.

World Distribution

Despite the long and constant growth of urban settlements, they remain relatively limited in their distribution over the world (Fig. 11–10). The large city is not a feature of areas with low population densities. There are, however, exceptions to this generalization. Madrid, for example, with a population of over three million, stands conspicuously on Spain's thinly populated Meseta (central plateau), and Tehran rises abruptly out of almost empty surrounding lands. But, overall, the thinly populated areas of the world are nearly devoid of great cities.

The large urban center is also not a feature of the great lowland areas of the low latitudes. The relatively few exceptions include such cities as Bombay, Calcutta, Singapore, Djakarta, Canton, Hong Kong, and Manila in the Asiatic sphere, and Rio de Janeiro, São Paulo, and Mexico City in Latin America. And elsewhere in the low latitudes, some of the regions' (and the world's) highest population densities exist in areas without cities. Thus, since cities can exist amid low populations and high populations can exist where there are no cities, it is apparent that urbanization is not a necessary concommitant of high population density (Plate VII).

Examination of the distribution of cities reveals no strict tie to climate. And though mountains usually preclude the growth of large cities, landforms are seldom major determinants in the distribution of urban settlements; nor can native vegetation or soil be demonstrated to be the outstanding control. Of all the physical elements of the environment, minerals and water appear to have had the greatest influence on the growth of new cities. However, while "mining towns" do exist, not all mineral areas contain or lie close to an urban settlement. One is forced to conclude, therefore, that the world distribution of cities cannot be accounted for primarily in terms of physical setting.

Considering the history of urbanization, one comes to realize that the world distribution of cities more sharply reflects the spread of a particular culture than it does the occurrence of any physical element. In modern times, urban living has become more and more an earmark of what we call Western civilization. As the culture of the West spread over the various continents, urban settlement spread with it. Indeed, much of the urban development since the Middle Ages has been the result of foreign enterprise or domination. Certainly, most of the large-scale urban developments in Asia, Africa, and the Americas took place after these lands were colonized or brought within the European sphere of influence; the coastal areas of these continents, in particular, are peppered with such important colonial cities as Quebec, Boston, Rio de Janeiro, Dakar, Cape Town, Calcutta, Singapore, and Manila. Instead of developing slowly from rural settlements, as did the cities of early history in lands near the Mediterranean, urban settlements in these colonized areas were often created almost "ready-made," by adventuresome or fleeing European city dwellers.

In those colonized areas where Western, or, more specifically, European, culture has become dominant—in the United States and Australia, for example—urban settlements now involve more of the population than they do in the regions of their origin. Where Western culture has come only in the form of government or commercial activity, as in India or China, Westernized urban settlements have likewise appeared, but they rest as unassimilated forms on a base of rural settlement types.

Characteristics of Modern Urban Settlements

Site and Situation

Site. The exact position of an agglomeration on the land is considered to be its *site*. For example, the original site of the city of Singapore (established in 1819) included an area of only about 3 square miles (now the most congested central city area) on the southern tip of the island of Singapore, which itself has a total land area of only about 227 square miles (Fig. 11–11). The present site of urban Singapore includes about 50 square miles. However, some people wish to do away with the old distinction between "Singapore City" and "Singapore country" and consider the whole island an urban complex.

Situation. To understand any agglomeration, it is necessary to appreciate its relationship to the area in which it is located, its regional position or *situation*. This does not mean the precise location or site, which is indicated by latitude and longitude, but rather the relationship of the agglomeration to the region it serves. Fortunately, Singapore has geography on its side: it stands less than a mile off the southernmost tip of the Malay Peninsula at the funnel point of the Straits of Singapore and Malacca. Thus, strategically located at the major crossroads of the Eastern world, Singapore could hardly have helped but become a great center of commerce and the strategic hub of power during the long span of European colonialism in Asia. A product of the mercantilist era, Singapore was transformed under British rule from a bit of desolate swamp into a great commercial and strategic center. Today, Singapore is the rubber market for the Eastern world and the *entrepôt* (distribution center) and trading crossroads for all Southeast Asia. And with the growing importance of this region, Singapore may soon become the world's third major port.

Change Through Time. Both site and situation change with the passage of time. Increase in population in any agglomeration brings with it the expansion of its site over a larger area, as has been suggested in the example of Singapore. Loss of population may cause contraction of the site.

With respect to situation, change is perhaps less obvious, but it is nevertheless real. For example, if a new highway is built

FIGURE 11–11
The site and situation of Singapore, one of the major port cities in Asia. Left, the site of Singapore, on the southern coast of Singapore Island. Right, the situation of Singapore, within Southeast Asia.

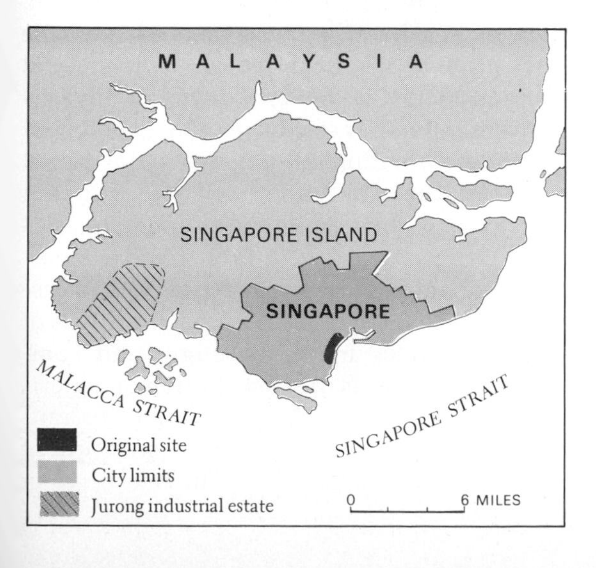

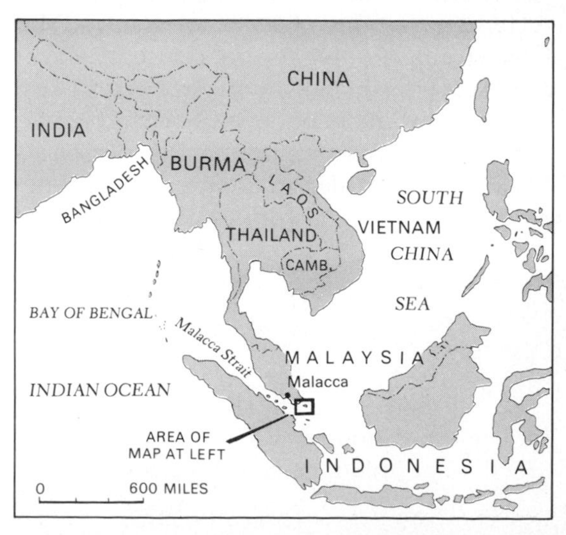

that bypasses an old agricultural market town, the flow of goods to and from the town is thereby diverted, and it may lose its significance as a regional commercial center; the town's situation has been changed.

The situation of an area may also be altered by a regional shift from agriculture to manufacturing; or, new urban centers may arise to replace or compete with an older one. Such was the fate of the port city of Malacca, which lies 100 miles north of Singapore on the Malay side of the Malacca Strait (see Fig. 11–11). With the decline of the sea-based empire of Sri Vijaya and its capital city near Palembang, in the thirteenth century, the way was open for the rise of a new Asian entrepôt. Initially a small fishing village, Malacca began to grow as more and more ships stopped there for provisions on their way through the Straits of Malacca and Singapore; and it gradually evolved into a port of call. Before long, trade became the mainstay of the city. And within a hundred years, Malacca had grown from a trading port to the center of an expanding empire—the strongest state in Southeast Asia. In the fifteenth century it controlled the commerce of the Straits; by the sixteenth century its archipelagic empire encompassed most of the Malay Peninsula and the island of Sumatra, including portions of both sides of the Malacca Straits. Not for several centuries was it to be replaced, by Singapore, as the principal entrepôt in Southeast Asia. The shift to another city occurred in part because Malacca's harbor was not deep enough to accommodate the very large ships of the nineteenth century.

Major Urban Functions

The basic consideration in classifying an agglomeration as urban should be the function, or functions, that it performs; for it is from this factor, not size, that the essential difference between rural and urban agglomeration derives. But what are the distinctive functions of an urban settlement?

When the names of certain cities are mentioned, specific functions immediately come to mind. For example, Washington, D.C., and New Delhi, being the capital cities of the United States and India, respectively, are readily associated with governing; Detroit, with the production of automobiles; Hong Kong, with shipping and commerce; and Jerusalem, Mecca, and Varanasi (Banaras), with religion. But while each of these cities has been largely associated with one particular function, the existence of nearly all large cities—including those mentioned above—is based on a multitude of functions.

Several different classifications of urban functions have been devised. In general, however, the activities that urban settlements perform may be divided into seven major functions.

The first of these functions is *manufacturing,* which, ordinarily, involves the making or altering of goods for human use and may be performed on a variety of scales, ranging from the complexity of heavy industry to the relative simplicity of handicraft. The next most common function is probably *commerce,* which involves the exchange, either through sale or purchase, of goods and services. Third is *transportation,* the physical movement or transfer of people, goods, and services. This function has two major facets. One is the collection, or the bringing together, of large amounts of goods (or services)—for example, the ores used in a particular manufacturing process may be necessary in such large quantities that the yields of several sources must be sought and combined; the other is the distribution of materials received at a focal point—a port or railhead, for instance—in bulk to subsequent destinations in smaller lots. The fourth urban function is *administration,* or government, a necessary feature of social life that, as discussed in Chapter 5, occurs at several broad levels of complexity—the local, state, national, and international. Also necessary is the function of *defense,* which is most graphically illustrated in fortress towns or military garrisons. The *cultural* function, so-called, is a rather expedient grouping of activities in religion, art, and education. Lastly, the function of *recreation* is growing in importance throughout the world. Health and tourist centers perform the function of recreation.

Broadly defined, then, the major urban functions are manufacturing, commerce, transportation, administration, defense, culture, and recreation. They stand in marked contrast to rural functions. When the population of any agglomeration, regardless of actual numbers, is primarily engaged in cultivating the land or in raising animals, the agglomeration can be described as rural; when the population is in greater part engaged in one or more of the urban functions, the agglomeration is generally described as urban.

Internal Form and Structure

Size. Urban agglomerations cannot be reliably distinguished from rural agglomerations in terms of size, whether it is measured in numbers of people or areal extent. To be sure, rural agglomerations are usually smaller than urban ones, and large centers are usually of more importance than small ones. Yet the small number of buildings and residents in a grain-collecting center in the wheatlands of the midwestern United States may have a regional significance far beyond that of a settlement twice as large in, say, the more heavily populated Northeast.

Nevertheless, for convenience, size is often used to distinguish between rural and urban agglomerations, as in the official census of the United States. But such an arbitrary criterion may lead to a misunderstanding of the significance, or even the nature, of any area's settlements. Only broadly, and then only as a contributing element, should size be a consideration in the differentiation of urban and rural settlements.

Street Patterns. The skeleton of any urban agglomeration is the arrangement of its streets, which may involve some geometrical plan or be completely irregular. Often the natural features of a site conspire to favor certain lines of movement between buildings over certain others and thus to determine the settlement's configuration. But if nature imposes no control, the pedestrian pattern of a settlement develops entirely in accord with the desires or needs of its inhabitants.

In urban agglomerations that are old and had their origins in simple groupings of rural people, the patterns of streets may be quite *irregular.* In many instances the alignment of houses and other buildings was not a matter of great importance to early inhabitants. When such settlements grew larger and found it necessary to have formally designated streets, many irregular routes were recognized officially and allowed to remain. Thus, for example, much of the arrangement of streets in lower Manhattan or in the heart of Paris reflects routes that developed spontaneously, before the city's configuration was subjected to planning.

Sometimes the arrangement of city streets has been determined, directly or indirectly, by government, as when, for the most part, the urban settlements of Spanish America were laid out with a central plaza and streets parallel to its sides. This *rectangular,* or checkerboard, pattern, with streets crossing one another at right angles, is one of the simplest city designs and also one of the oldest; it was the layout of cities founded in ancient times, including Kahun in Egypt and Harrapa in Pakistan. In the central and western United States—areas settled mainly after 1850 and under government control—an attempt to have main roads cross one another at right angles is apparent, and, as is generally true in agglomerations that originated at a crossroads, most additional streets are laid out parallel to main roads—and thus, in this case, at right angles to one another.

Increases in the importance and size of an agglomeration create greater need for speed and ease of access to its center from the periphery. Rectangular arrangements are surpassed in this regard by the *radial* pattern, whereby streets lead from points on the periphery directly to a common center, much as the spokes of a wheel connect the rim to the hub. Cross streets then tend to assume a weblike arrangement. The skeleton of streets in Washington, D.C., for example, is a complex combination of radial and rectangular patterns.

Beyond these three basic patterns—irregular, rectangular, and radial—there are many others that have grown out of the vagaries of individual choice or of real estate

FIGURE 11–12

A westward aerial view of mid-Manhattan showing the uniform rectangular pattern of the area's eastern half. The Empire State Building, on Thirty-fourth Street, is just left of the center. (New York Department of Commerce)

skyscraper, the greater its contribution to the prestige of its city. At present the tallest skyscraper in the world is Chicago's 110-story Sears Tower; it is 104 feet higher than its leading competition—the twin towers of New York's World Trade Center—and 204 feet higher than the Empire State Building.

Only recently has some of the technology used in the construction of such multistoried buildings been made available. Not until the late 1800s, for example, did the electric elevator come into widespread commercial use; and it was even later that the cagelike steel frame used in skyscraper construction became practical and safe. The first skyscraper, the 10-story Home Life Insurance Building in Chicago, was built by William Le Baron Jenney in 1884. Up to that time the maximum height of buildings was severely limited, to about 5 or 6 stories—that being the maximum number the average human could climb several times a day and the maximum number that could be built soundly and attractively without a steel frame.

The high cost of the little ground space available has been the main incentive for building vertically rather than horizontally in congested urban centers. This vertical utilization of space, or the "piling up at the center," is rapidly becoming a characteristic of urban profiles around the world. In Hong Kong, for example, the 52-story Connaught Center represents only one of the many highrises recently completed in Asia. It seems, in fact, that the time is rapidly approaching when the larger urban centers of the world—New York, San Francisco, Chicago, Paris, Berlin, Moscow, Tokyo, São Paulo, Hong Kong, and Cairo—will no longer be distinguishable from one another simply by a glance at their profiles.

Functional Areas. As described earlier, urban agglomerations perform various func-

tions for the groups
the area that surroun
functions involves c
buildings as well a
defined quarters, or
hand, there is usually
tween one section of
and another; the form
settlement is compara

In contrast to thi
cated on a coast and
have a definite docki
district that is quite c
residential district or
Similarly, a district of
on the commercial func
agglomeration, contras
scene created by a grou
rying on the function o
one of these more or l
performing a distinct fu
functional area.

Taken together, th
an urban agglomeration
cernible pattern. In larg
tions this pattern can
down into three distinct
business district (CBD), th
suburban fringe (Fig. 11
broadly defined sections
character and its own a
tional areas.

It is common, in th
least, to refer to the centr
or "downtown" section,
city and, in many instanc
for its existence. Tradition
the urban settlement has
over both to the high co
special services as large
hotels, specialty shops, an
and to the employment o
people—most of whom co
from the suburbs. It is us
settlement that is all but

development. These are irregular in a sense, but usually symmetrically so. Circular drives, curving and recurving lines, figure-8's, cul-de-sacs, and dead-end inner courts reached from the four sides of a large rectangular block are some of the familiar forms to be encountered in today's developing urban agglomerations.

All the patterns may occur in combination as well as individually, especially in larger urban areas. As one proceeds uptown, the irregular streets of lower Manhattan give way to streets that form a well-crystallized, rectangular pattern, with only one notable exception, Broadway (Fig. 11–12). The repeating radial patterns of Washington, D.C., are imposed upon a rectangular base. The radial pattern about the Place de l'Etoile in Paris links tenuously to the irregular framework farther into the old city, to the circular boulevards that belt the city, and to the great cross avenues of modern construction that tear across the old irregular pattern (Fig. 11–13). Similarly, where peculiarities of site restrict full development of any one pattern, local modifications are introduced. For example, the city of Syracuse, New York, has a street pattern that is basically rectangular;

FIGURE 11–13

The Place de l'Étoile (Star), in Paris, where the Arc de Triomphe stands in a circle at which twelve radial avenues meet, including (upper left) the broad, tree-lined Champs Élysées (Rotkin, PFI)

FIGURE 11–14

San Francisco, California, a seaport city wh
among those of large urban centers. (Transa

however, much of the city's site is characterized by low but steep hills that prevent an unrestricted rectangular pattern.

Whatever the pattern or combination of patterns formed by the streets of a city, the character of the agglomeration is deeply affected by it. The alignment of houses and other buildings is determined by the position of the streets, and thus the agglomeration acquires a fixed quality. Any planned additions to or detractions from the established pattern, both measures of growth, must in some way be articulated with the existing pattern. Despite changes of detail, the agglomeration as a whole retains a distinctive appearance, determined largely by its pattern of streets.

Profile, or Skyline. Something of the nature of any agglomeration is told by its *profile,* or skyline. In small rural settlements, the buildings are usually low in height and

often widely
above low sh
may overshad
spire of a chu
the crowns of
the profile o
usually rath
jagged.

A very
large urban ag
more highly
The outer edge
typical of rural
is approached,
more massivel
of a low, wide
to be straight o
rically rounded
inate rather th

The skysc
pyramidlike ur
be a measure o

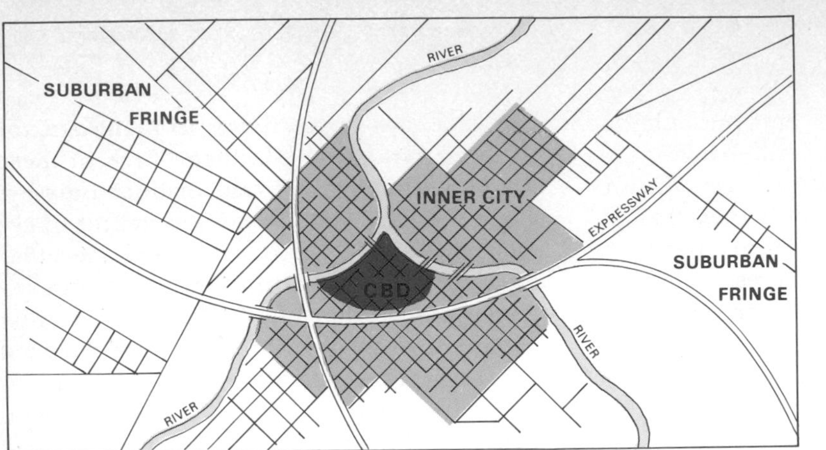

FIGURE 11–15

A hypothetical metropolitan area, or urban center. Its three main sections are interdependent: the *central business district* (CBD), at the core of the area, is also located near the intersection of its major through roads; the *inner city* surrounds the CBD and forms a buffer between it and the third and outermost section, the *suburban fringe.* In reality, each type of section varies in size; the suburbs also tend to be more discontinuous or fragmented.

family residences, although large apartment buildings are often found here in appropriately zoned areas. Similarly, there is generally a manufacturing zone close by. In most urban centers around the world, the central business district is the most accessible section of the metropolitan area, for transportation and communication networks focus on it. In New York City, for example, the CBD is primarily the Wall Street area and the area between 34th and 59th streets; in Chicago, it is based on the Loop at the intersection of State and Madison streets and includes an area five blocks wide and seven blocks long; in London, it includes the area from Saint Paul's Cathedral to the Strand.

Within older settled parts of the Western world, distinctions between the specialized areas of any agglomeration are less clearly marked than in newer settlements. The history of the growth of an urban agglomeration often reveals why its various functions are performed in scattered rather than in collected zones. In European settlements, for example, residence is often maintained close to or in the same building as retail stores; such "shophouses" are very common in Asian settlements as well. The function of commerce may thus be so scattered as to leave no impression of a "downtown" section. Moreover, in Oriental settlements the same building may contain a workshop, serve as the residence of the workers, and, also, house persons engaged in intensive agriculture on a small plot of land nearby.

The arrangement of the functional areas of a city is the result of the interplay of several forces. At an early stage, the intersection of two highways, which may have given rise to the settlement in the first place, may be an attractive site for a store. Here people from the vicinity come to purchase needed supplies or to dispose of surplus agricultural items. In time, a second store may be established because there is a good commercial opportunity. A third enterprise, to be followed by many others, may come next. An inn may be built near the crossroads. Gradually, there appears a nucleus within which all the buildings are used for commerce; they are occupied during the day by the people engaged in the performance of that function but who live elsewhere. A distinct functional area develops—a *commercial core,* or central business district in this instance.

Sometimes, factories are attracted to a commercial core by the presence of a large number of people, the family members of those active in commerce representing a potentially large and readily available labor supply. Wealth accumulated in commercial activity might be available for investment in industry.

Sometimes, however, factories have to be built on the outskirts of a settlement. If past growth of the commercial core has made land values too high and the presence of residences about the commercial core makes adjacent land unavailable, factories are forced to the outskirts. Gradually, under such conditions, there may develop an industrial belt encircling the commercial and residential centers.

The growth of any agglomeration brings with it continual, far-reaching adjustments of its functional areas. If a settlement has developed a commercial core surrounded by residences and that, in turn, is encircled by an industrial ring, further growth in population requires expansion of the residential area. Often the only possibility lies in jumping beyond the industrial ring; soon new residential areas appear on the fringe, sometimes—usually for the workers—close to the factories, and sometimes well away from the factory belt—most often for those who can afford to choose their own site. Here and there, small neighborhood communities may develop, perhaps separated physically from the original nucleus but economically still very much a part of it.

As these newer residential areas spring up, there is usually a decline in the old residential belt located near the center, or commercial core, of the settlement. To such a middle section, or zone, the name *inner zone* or *inner city* is frequently given. Its decline may be accompanied by an outward spread of the commercial core and an inward growth of the industrial belt. Some of its old residences, relics of a more youthful period in the settlement's history, may become rooming houses, gradually deteriorating and eventually being removed to make way for other buildings.

The spread of an urban agglomeration is the result of many factors. Modern transportation, for example, has made it possible for more and more urban folk to live at far greater distances from their work than did most workers in the past. Intangible incentives, such as the desire for space, and tangible incentives, such as lower tax rates, have contributed to the continued migration of urban people away from the urban nucleus. Where this movement has occurred, a zone of interpenetration of urban and rural settlement, the *rural-urban,* or *suburban,* fringe, has come into being.

As its growth and spread continue, distances within the agglomeration, especially between the residential districts and the commercial core, become so great that new scattered nuclei of commerce begin to appear in the outer residential districts. These are the occasional "corner stores," and the "neighborhood shopping centers," so characteristic of growing metropolitan areas in the United States and many other countries. Their appearance begins a trend toward the decentralization of the agglomeration's retail trade, which will tend to adversely affect merchants in the inner city and central business district; they will feel, with increasing intensity, the pinch of competition from the more modern, multipurpose suburban shopping centers.

Similarly, as the agglomeration spreads, details of site, such as steepness of slope or the alignment of a lake shore or a river, may limit the growth of one or another of its functional areas. Chronic traffic congestion may necessitate cutting new streets or widening old ones. This in turn may bring about the rearrangement of functional areas, thus altering the character of the entire urban unit. As its physical and cultural elements continue to interact, the city continues to grow and change.

Centers of Influence and Change

From their first appearance, human agglomerations have been the nuclei of cultural ferment and change. Indeed, they have been both the cause and effect of the innovations in human culture. Usually the larger the agglomeration and the stronger its organizational bonds and purposes, the greater the scope and intensity of its influence. A single well-knit family can strongly influence change in a small settlement; and, allied with its neighbors, it can affect far broader regions.

Thus, throughout history, the heavily populated cities of the world have been the fountainheads of humanity's important social, cultural, and political innovations. Modern-day Paris, for example, exerts a far greater cultural influence, both locally and internationally, than do all the smaller settlements of France combined.

Without such population agglomerations of varying magnitudes, there would be no nations as we know them, nor any large-scale transportation and communication systems, except perhaps on a local, extremely fragmented basis. There would be no international cooperation and competition, and therefore little opportunity for cultural, political, or scientific exchange; indeed, the prime mechanism for the creation and diffusion of cultural, political, and educational institutions would be almost totally lacking. Without urban settlement, people would be living pretty much as they did fifty thousand years ago.

Urban agglomerations have been the places where scientists, philosophers, poets, and traders have met, argued, and often merged their divergent talents and views. The resulting changes have not always been praiseworthy, but there have apparently been more good ones than bad, because civilizations have, on the whole, tended to more than balance their cyclic declines. Certain cities, in fact, have come to be associated with specific societal changes. Moscow and Leningrad established communism as a modern political reality, and Peking adapted that economic system to the particular needs and way of life of China. London may well lay claim to the creation and diffusion, beginning in the late 1500s, of worldwide colonialism. And, during the late Middle Ages, several northern European cities—among them Cologne, Lübeck, Danzig, and Riga—organized the first large-scale commercial-trading structure, the Hanseatic League (see Fig. 5–19).

Christaller's Central Place Theory

In 1933 the German geographer Walter Christaller published a theoretical work titled *Central Places in Southern Germany*, in which he attempts to explain the distribution, size, and number of settlements that might develop in a given region. Although based on the conditions existing in southern Germany several decades ago, many of the principles set forth in this work are of great value in understanding the spatial relationships of settlements in any region or period. Since the study rests on the proposition that a settlement is centrally located in the region it serves and that, on a group basis, smaller settlements will arrange themselves around a larger central place, it is commonly referred to as Christaller's *central place theory*.

The Model Region

In presenting his theory Christaller provides a model based on the following assumptions:

1. There is an evenly distributed number of rural consumers residing on a flat, uniform, and limitless plain.
2. Each of these rural consumers has the same wealth, income, and desires; and thus purchasing power, or propensity to consume (demand), is distributed evenly over the plain.
3. A transportation system serves all the plain uniformly, providing equal accessibility to all central places of the same type.
4. The single function of each settlement is to provide the rural population with goods and services. There are no mining towns or industrial centers, for example.
5. Producers of goods and services cannot be located adjacent to each customer.
6. Producers should receive a profit from the sale of their goods or services.
7. Each good or service has the same market range regardless of where it is produced.
8. A maximum number of consumer demands should be satisfied from a minimum number of central places.

Assuming these circumstances, Christaller proposed that an even distribution of settlements would develop on the plain, supplying the rural consumers with goods and services they could not provide for themselves—such as processed food, clothing, entertainment, formal education, legal aid, and medical services. The area served by each settlement is referred to as its *tributary area,* or *complementary region.*

Theoretically, since the price of goods and services increases with the consumer's distance from the point of production, the ideal tributary area is circular, with the settlement as its focus—in the manner of von Thünen's model (see Chapter 6); this configuration allows consumers in a maximum number of directions to be supplied at a minimum cost. Each centrally located settlement, or central place, is assumed to have a complete monopoly over its tributary area.

But what size tributary area is required in order to support a particular good or service? Obviously each good or service has a specific *threshold,* or minimum level, of demand, at which it can be sold at a profit. Since propensity to consume is evenly distributed, this threshold can be calculated in terms of numbers of customers. If the number of actual customers is below this minimum level, then the product will not be commercially viable. The territory encompassing a commodity's necessary number of customers is known as its *market range.* In order for a good or service to remain commercially viable, its market range must at least equal its demand threshold. So, when competition for customers exists between central places, the size of their tributary areas is determined by both the market range and the threshold level.

While some goods have a *low* demand threshold and can tolerate a short market range (low-order goods), others have a much *higher* minimum level of demand and, consequently, require a much greater market range (high-order goods). Bread, for example, is a low-order good, whereas the services provided by a specialized lawyer are high-order.

If only goods or services of the same order are produced, and if the entire population cannot be served from a single center, then the central places that develop will be evenly spaced over the entire plain, each one

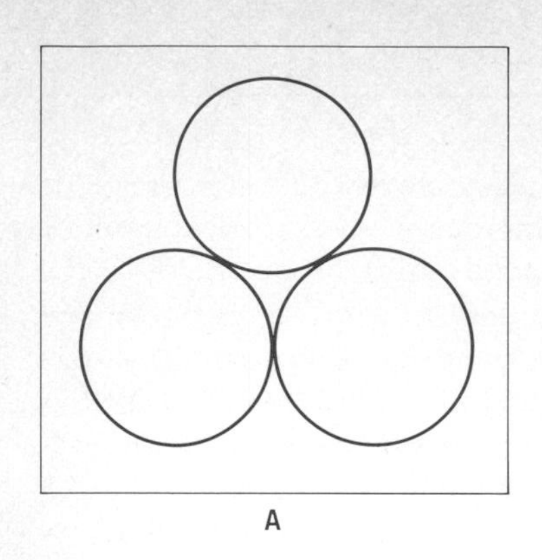

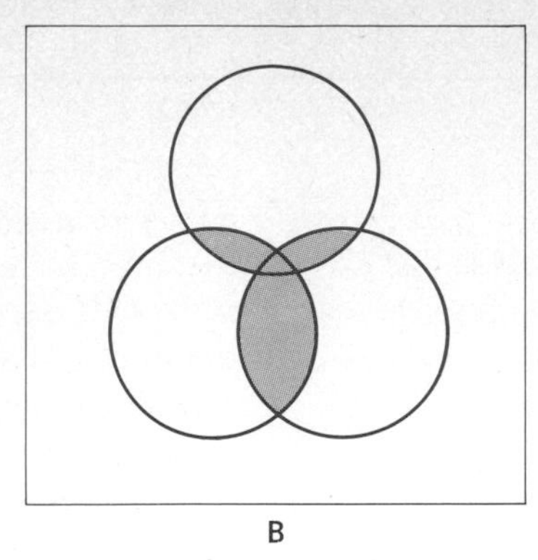

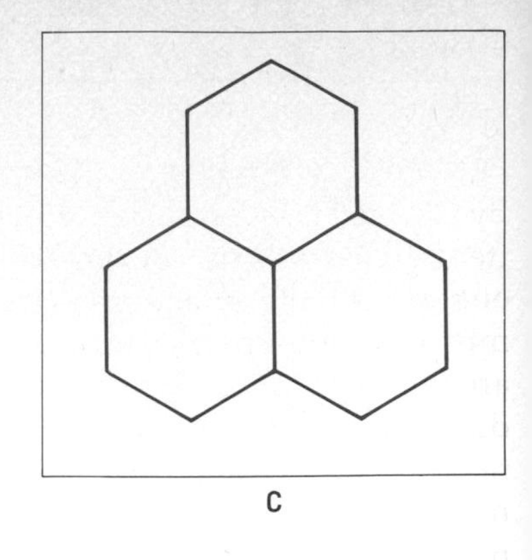

FIGURE 11–16

Some possible configurations of tributary areas. A. Tangent circles, which leave some areas unserved. B. Overlapping circles, which create competition. C. Hexagons, which leave no area unserved and create no competition.

serving a tributary area of identical size with an equal number of customers. However, circular tributary areas cannot exist in such a situation, because—if the circles just touch—there would be areas where people are residing (by assumption) but not being served, or—if the circles overlap—there would be areas of competition. Christaller's solution was to have each tributary area assume the shape of a hexagon—the geometric form that permits the most efficient coverage of the entire plain (Fig. 11–16).

When goods and services of various orders are involved, the high-order goods, those that require the largest tributary areas, are supplied from only a few of the central places. The low-order goods are supplied from a greater number of central places, each serving a relatively small tributary area. All production centers, however, regardless of their size, remain evenly spaced over the plain.

The Hierarchy of Central Places

On the basis of this differentiation of goods and services, Christaller formulated a hierarchy of central places that may range in size from as small a unit as the hamlet through the village, town, city, and up to the large metropolis, and even higher if so desired (Fig. 11–17). In his arrangement the production centers that provide commodities of the lowest order are the smallest in size and the most numerous; these centers are rather closely spaced hamlets. The next largest centers are the villages, providing items demanded so infrequently that the tributary area of no single hamlet can support them. Each village will be located on the site of a former hamlet and include among its goods and services all those provided by a hamlet. Its tributary area will be bounded by six surrounding hamlets and will be larger than the tributary area of any one of them. Consequently, there will be fewer villages on the plain than there are hamlets. Following the same procedure, the third-order center, the town, will have its tributary area delimited by six villages; and the fourth-order center, the city, will be bounded by six towns. This progression, which can be used to establish a hierarchy of any number of orders, assumes that *each center at a given hierarchical level produces its own distinct goods and services as well as those of any center at a lower level.*

One of the main principles illustrated by Christaller's central place model is that an item's cost to the consumer—in time and energy as well as money—is usually determined in large part by distance. Consumers are willing to incur the cost of traveling long distances in order to obtain infrequently needed, expensive items; but they will not make the same sacrifice for frequently needed, cheaper items: consumers expect to find a grocery store in their local area but not the services of a stockbroker. In response to this variation in consumer travel, Christaller suggests, there will evolve in any given region a hierarchy of service centers in which size and availability of goods and services are directly related.

And indeed, in most regions of the world there does seem to exist an approximation of Christaller's hierarchy of different-sized settlements. However, it is rare that a settlement provides all the commodities available in centers at lower levels. And further, the number and distributional pattern of modern settlements never more than roughly coincide with Christaller's model.

The ideal symmetrical distribution of central places is prevented by many factors.

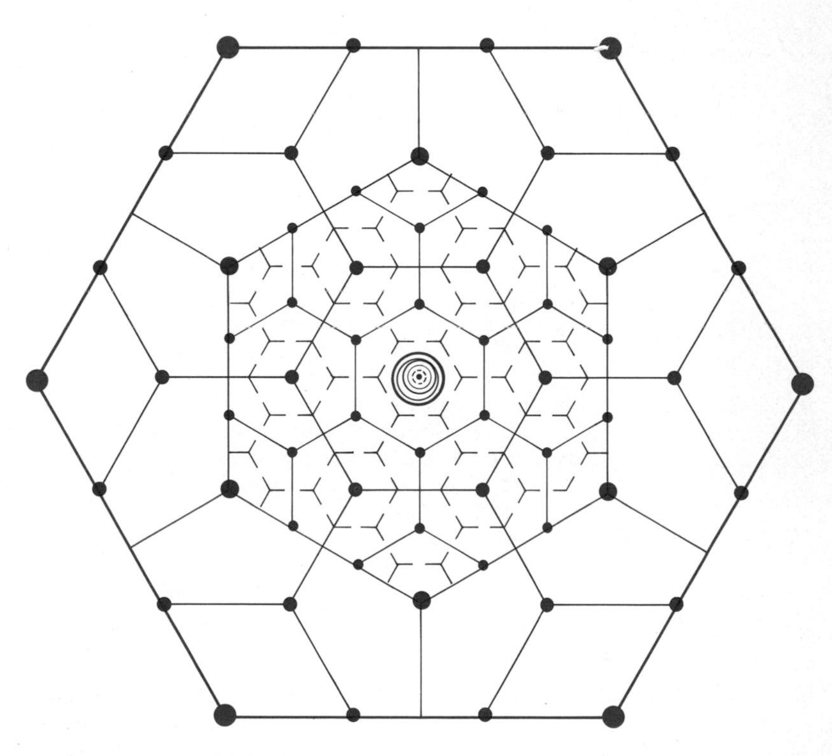

FIGURE 11–17

A diagrammatic representation of Christaller's hierarchy of central places. (It should be understood that hamlets and villages exist over the entire tributary area of the metropolis.)

Settlement	*Limit of tributary area*
Hamlet	– – –
Village	——
Town	——
City	——
 Metropolis	——

One of these is the clustering of settlements around some natural feature, such as a water body or mineral deposit. Another is the tendency for settlements to develop along transport routes, which are unevenly distributed over the land because of local relief features and the uneven distribution of demand; in many areas the result has been a linear settlement pattern. Moreover, within the last few decades, vastly improved modes of transportation have increased the mobility of people and cargo to such an extent that a settlement no longer has a monopolistic hold over its surrounding area. Today's tributary areas are often international, and changes in the network of central places in one country can promote a considerable reorientation of trade areas within another country. In the real world, central places are dynamic entities; it is not uncommon to see the prosperous settlement of an earlier technological age die or decline as new ones emerge. Thus, although certain elements of Christaller's central place theory are demonstrated in the real world, variation from the ideal centrality of settlements appears to be the rule.

REVIEW AND DISCUSSION

1. Why and in what ways do the types of settlements created by a mobile population differ from those created by a sedentary population? Explain the differences between the various settlement types of sedentary populations.
2. Is there any precise means of distinguishing between a dispersed settlement and an agglomerated one? between a rural settlement and an urban one? Explain.
3. List several examples of each of the following: hamlet, village, town, city, and metropolis. Be prepared to explain your classifications. Are their distinctions clear-cut?
4. Trace the early development and diffusion of urban centers throughout the world. Does there appear to be a strong correlation between imperialism and urbanization? Explain.
5. Discuss and evaluate the impact of the industrializing period on the growth and development of urban centers after the Middle Ages.
6. Differentiate between and explain the geographic importance of *site* and *situation*. Give specific examples of each term.
7. Discuss and evaluate the seven major urban functions. Name several urban agglomerations that perform functions other than these.
8. Describe and evaluate the internal form and structure (size, street patterns, and functional areas) of the population agglomeration you know best.
9. As centers of influence and change, in what fields have cities been pioneers? Cite several examples, preferably from both historical and modern times.
10. To what extent is the hierarchy of central places in Christaller's model region a mirror of reality? Support your evaluation with specific examples.

SELECTED REFERENCES

Ahmad, N. "The Pattern of Rural Settlement in East Pakistan." *Geographical Review*, Vol. 46 (July 1956), 388–98.

Boal, F. W. "Technology and Urban Form." *The Journal of Geography*, Vol. 67 (April 1968), 229–36.

Briggs, A. *Victorian Cities.* New York: Harper Colophon, 1965.
Brush, J. E. "The Hierarchy of Central Places in Southwestern Wisconsin." *Geographical Review,* Vol. 43 (July 1953), 380–402.
Brush, J. E., and H. E. Bracey. "Rural Service Centers in Southwestern Wisconsin and Southern England." *Geographical Review,* Vol. 45 (October 1955), 559–69.
Chinitz, B. "Contrasts in Agglomeration: New York and Pittsburgh." *American Economic Review,* Vol. 51 (May 1961), 279–89.
Christaller, W. *Central Places in Southern Germany,* trans. by Carlisle W. Baskin. Englewood Cliffs, N.J.: Prentice-Hall, 1966.
Demangeon, A. "The Origins and Causes of Settlement Types." in P. L. Wagner and M. W. Mikesell, eds., *Readings in Cultural Geography.* Chicago, 1962, 506–16.
Dwyer, D. J., ed. *The City as the Centre of Change in Asia.* Hong Kong: University of Hong Kong Press, 1971.
Getis, A., and J. Getis. "Christaller's Central Place Theory." *Journal of Geography,* Vol. 65 (May 1966), 220–26.
Gottmann, J. "Why the Skyscraper?" *Geographical Review,* Vol. 56 (April 1966), 190–212.
Green, F. H. W. "Community of Interest Areas: Notes on the Hierarchy of Central Places and Their Hinterlands." *Economic Geography,* Vol. 34 (July 1958), 210–26.
Johnson, J. H. *Urban Geography: An Introductory Analysis.* Oxford and New York: Pergamon Press, 1967.
Lampe, F. A., and O. C. Schaefer, Jr. "Land Use Patterns in the City." *The Journal of Geography,* Vol. 68 (May 1969), 301–06.
Lapidus, I. M. *Middle Eastern Cities: A Symposium on Ancient, Islamic, and Contemporary Middle Eastern Urbanism.* Berkeley and Los Angeles: University of California Press, 1969.
Lewis, O. *Village Life in Northern India.* Urbana: University of Illinois Press, 1958.
Mumford, L. *The City in History: Its Origins, Its Transformations, and Its Prospects.* New York: Harcourt, Brace and World, 1961.
Murphey, R. "New Capitals of Asia." *Economic Development and Cultural Change,* Vol. 5 (April 1957), 216–43.
Pirenne, H. *Medieval Cities.* New York: Doubleday, 1925.
Robson, B. T. *Urban Analysis: A Study in City Structure.* Cambridge, England, 1969.
Sjoberg, G. "The Origin and Evolution of Cities." *Scientific American,* Vol. 213 (September 1965), 54–63.
——. *The Preindustrial City.* New York: Free Press, 1960.
Thomas, E. N. "Toward an Expanded Central-Place Model." *Geographical Review,* Vol. 51 (July 1961), 400–11.
Ullman, E. L. "A Theory of Location for Cities." *American Journal of Sociology,* Vol. 46 (May 1941), 853–64.

Disorder at the base of New York's World Trade Center during its construction. (© Arthur Tress, Magnum)

12

Challenge of Urban Growth

Urbanization

The twentieth century, more than any other period in human history, may truly be called the urban age. Though cities as such are not new, the existence of innumerable societies in which more people can be classified as urban than as rural is a unique feature of modern times, a new and revolutionary stage in the long development of cultural institutions. As discussed earlier, cities have been a part of the cultural scene for more than 5,000 years; however, throughout most of that span, they have been relatively small and few in number, their combined populations at any given time accounting for only a small portion of the human total.

In recent centuries, by contrast, not only has the number of urban agglomerations been increasing markedly but, often massive and highly complex, they have been claiming an ever-higher proportion of the world's total population. Until 1800, less than two centuries ago, no more than 2 percent of the world's population lived in urban settlements of 20,000 persons or more. By 1900, this figure had increased nearly fivefold, to 9 percent. And projections indicate that, by the year 2000, urban dwellers will account for well over 50 percent of the total world population, as compared with about 40 percent today. (Of course, such percentages are even higher when based on a definition of urban that takes in settlements of less than 20,000 persons; 20,000 is an arbitrary United Nations standard and, in view of much world practice, a rather high one.)

According to recent studies by the Bureau of the Census, for example, about 75 percent of the people now inhabiting the United States live within the country's urbanized areas (Fig. 12–1). And, also as measured by the bureau, nearly every state in the nation has a population that is at least 50 percent urban, California being the leader with a population that is 90.9 percent urban (Fig. 12–2). Moreover, much the same can be said

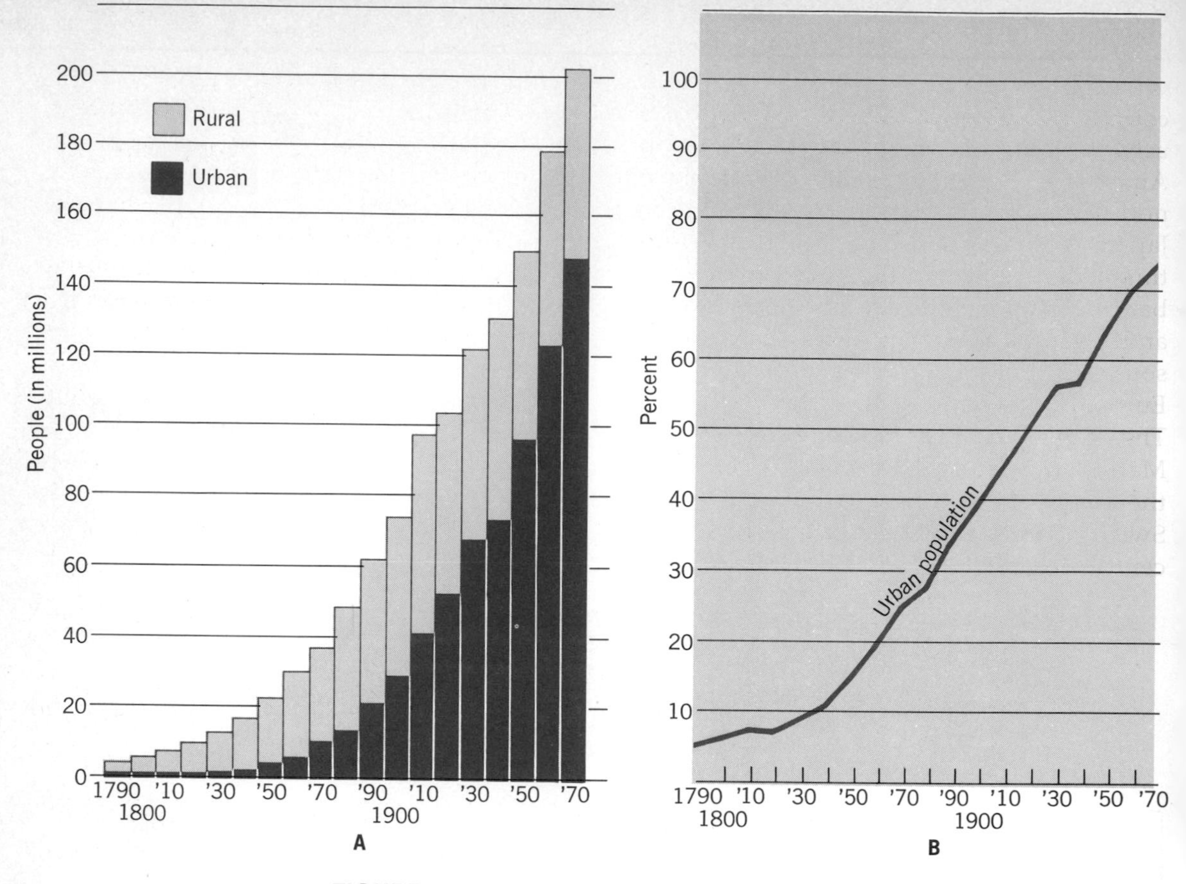

FIGURE 12–1

United States population from 1790 to 1970. A. The rural–urban division. B. The urban population percentage. The country's urbanization has been swift; between 1790 and 1970 the total population grew by almost 200 million, and the urban population rose nearly 70 percent.

of other parts of the world; many societies of western and northern Europe, for example—including Great Britain, the Netherlands, Denmark, and Sweden—as well as several Asian and Southern Hemisphere nations equal, or even exceed, the urban population percentage of the United States.

The trend toward urbanization involves much more, however, than a mere regrouping of people or an expression of the increasing importance of nonagricultural occupations. The high degree of physical and psychological contact related to modern city living has prompted entirely new life styles and philosophies and a general re-evaluation of many concepts traditionally associated with agrarian societies.

World Patterns

As Figure 12–3 shows, a wide belt of countries characterized by a *low* degree of urbanization stretches across Africa, southern Europe, the Middle East, and on across southern and eastern Asia (excluding Japan); there is also a small outlier of low urbanization in Central America. In contrast, the countries characterized by a *high* degree of urbanization comprise most of North America and South America, and most of northern Europe, as well as the Soviet Union, Japan, Australia, and New Zealand. Within this latter category, the most highly urbanized regions include the northeastern and western coasts of the United States, southern South America, northwestern Europe, Japan, and southeastern Australia. The small, anomalous countries of San Marino and Macao have over 90 percent of their population living in urban settlements; Sweden, Israel, and Australia are over 80 percent urbanized; Canada, the United States, Venezuela, the Netherlands, Chile, Bahrain, and Japan are in the over-70-percent group; and Mexico, Colombia, Greece, Lebanon, and the Soviet Union are among those in the 50-to-70 percent category (Fig. 12–4).

Some of the highly urbanized countries of the world have one city in their urban hierarchy that greatly overshadows all others in terms of population size. But, in general, such *primate* cities tend to be found mainly in the developing countries, especially those with a very recent history of political control by a foreign nation. Most highly urbanized countries include several major urban centers.

FIGURE 12–2

United States urban population percentages. Approximately three out of every four Americans (73.5 percent) live in urban areas, which the government defines as settlements with over 2,500 inhabitants. (United States Department of Commerce, Bureau of the Census, 1970)

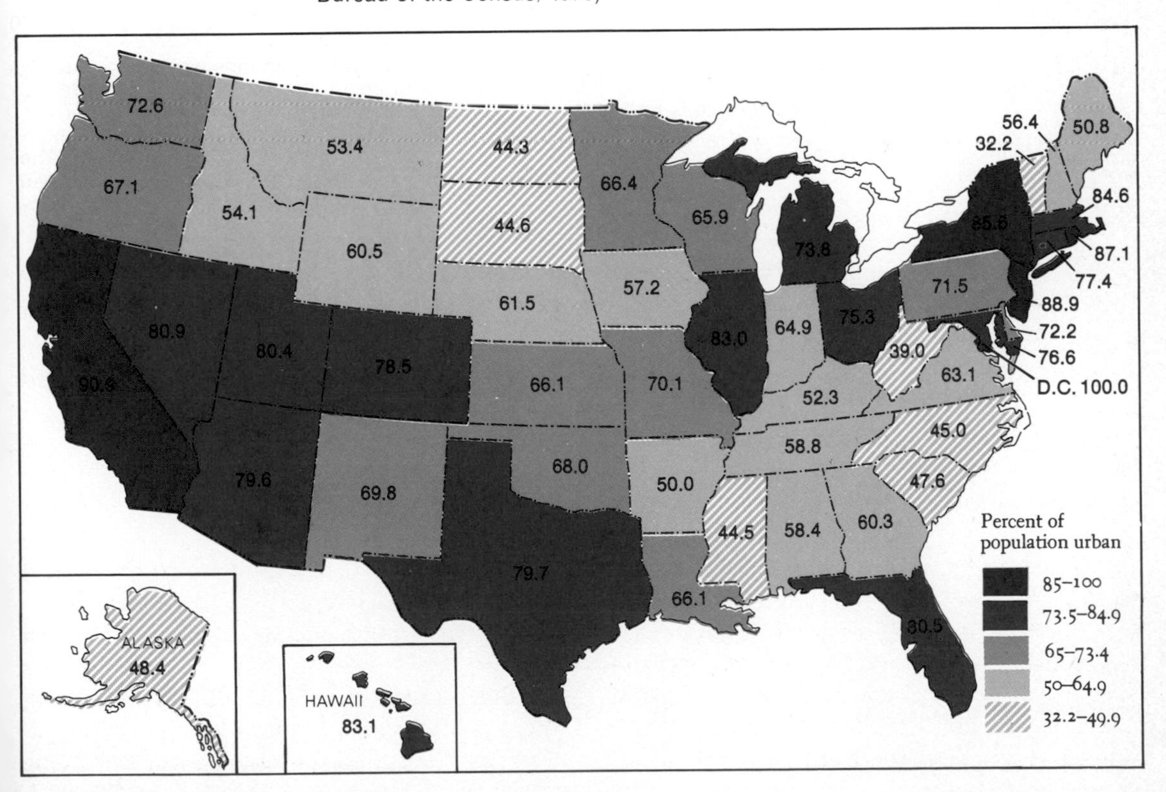

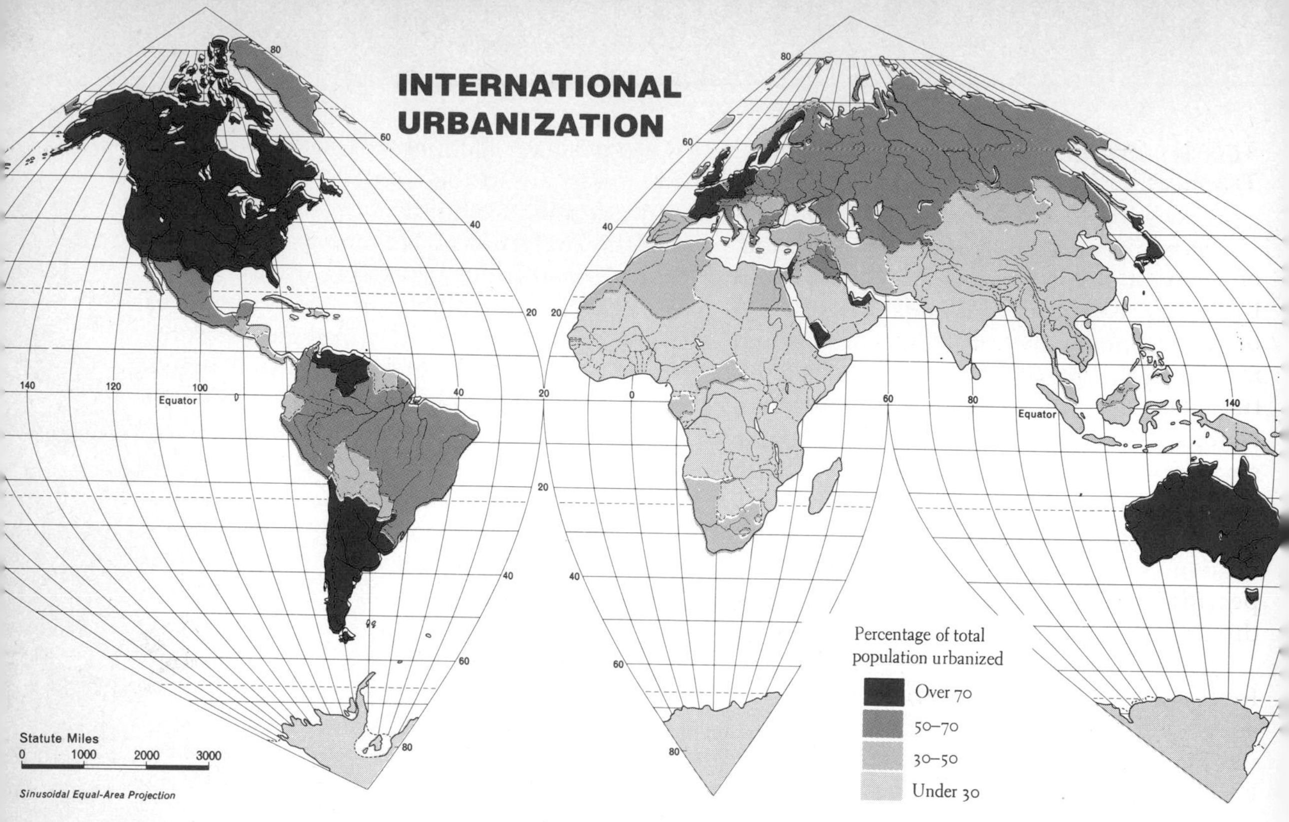

FIGURE 12–3

The average degree of urbanization among the world's nations was approximately 35 percent in the early 1960s; it increased to about 40 percent in the early 1970s, and will probably be well over 50 percent by the year 2000.

FIGURE 12–4

The urban–rural population percentages of selected countries, representing a wide population range.

Country	*Total population (in millions)*	*Urban percentage*	*Rural percentage*
San Marino	0.02	92	8
Australia	13	86	14
Israel	3	84	16
Great Britain	54	76	24
Canada	22	76	24
Venezuela	10	76	24
United States	203	74	26
Japan	104	72	28
Mexico	53	60	40
Soviet Union	247	58	42
South Africa	23	48	52
Yugoslavia	21	39	61
India	563	20	80
China	800(?)	13	87
Nepal	12	4	96

The Socioeconomic Transformation

Close examination of the world pattern of urbanization reveals that degree of urbanization often correlates directly with level of socioeconomic development. Generally the more developed countries have large portions of their populations residing in urban centers, and the less-developed ones have small urban population percentages. In light of this, urbanization may be seen as a collateral process experienced by a country as it undergoes socioeconomic development, especially from an agrarian economy to an industrial economy.

Accordingly, Great Britain, home of the Industrial Revolution, is usually considered as having created one of the first modern urban societies; likewise, intensive urbanization in most of the rest of today's developed societies has occurred only since the mid-nineteenth century and in almost no society of eastern Asia or Latin America until after World War II. Indeed, only a few countries in any region have progressed to the economic level typical of a highly urbanized society. A wide gulf exists between these few and the rest of the world. In Great Britain, for instance, a highly developed country, almost everyone is familiar with cities and the urban way of life, for about 76 percent of the country's population either resides or works in an urban center; but in Laos, a comparatively undeveloped country, not only is the urban population percentage much smaller but the distinctive features of city living are all but unknown to most of the country's rural population. The range of familiarity with cities is wide, and most populations of the world fall between these two extremes.

A Finite Process

The idea that *urbanization* and the *growth of cities* are synonymous is a common misconception. Though the processes they represent have been complementary factors in the creation of urban societies, a distinction should be made between the two terms. In this discussion of cities, *urbanization* is used solely to suggest the percentage of urban dwellers in a given total population. This percentage usually increases with time but may decrease if, for instance, the rural population grows faster than the urban population, either because of a higher birth rate or urban to rural migration. Urbanization, then, should be thought of as a finite process, with a definite beginning and end. Once a population is 100 percent urban, its urbanization process has reached its limit; it cannot go beyond this point.

In contrast, *urban growth,* the establishment of cities and their increases in population, has no fixed upper limit; on the contrary, it can and probably will continue throughout human existence. Even in a country that had become totally urbanized, as none yet has, the growth of its cities could still continue indefinitely if the support of foreign agriculture could be maintained.

The Growth of Cities

In almost every part of the contemporary world, urban centers are on the increase, embracing more and more of the earth's population and areal extent. Most of this increase has involved the expansion of existing urban areas rather than the creation of totally new ones, and the rate of growth has, of course, varied from country to country as well as from one city to another within the same country.

Population

Precise data on many urban populations around the world are somewhat limited; much of the statistical material is outdated or poorly compiled, and data on some areas do not exist at all. In general, however, it may be estimated that a little under two billion of the world's total four billion people now live in urban centers of over 20,000 persons. The United Nations World Health Organization estimates that the proportion of urban dwellers will jump to about 60 percent by the year 2000, when the total world population is expected to be close to seven billion. Thus urban population growth will continue to occur on a scale unknown in the previous evolution of human culture, and what it will mean to the future of the planet is one of the great uncertainties of our times. Among the few predictable aspects of this surge is its dependence on the traditional urban population sources, mainly foreign immigration, domestic rural migration, and natural urban population increase.

Foreign Immigration. Some of history's greatest increases in urban populations occurred during the nineteenth century, when the steamship and other innovations in long-distance transportation improved the economics of passenger travel and made it possible for great numbers of people to seek a new life in a foreign land. When foreign immigrants came to a new country they tended to settle in the most accessible urban centers, where they were more likely to find employment and speakers of their native language. As a result, many United States cities, especially such ports as New York and San Francisco, became the destination of thousands of newly arriving immigrants, most of whom were natives of Europe and Asia. In this same period, immigrants from Europe, specifically from Great Britain, were also in large part responsible for the development of many of the then fledgling colonial port cities, such as Sydney, Melbourne, and Cape Town. The cities of Singapore and Hong Kong, colonial products as well, are composed almost entirely of foreign immigrants, primarily southern Chinese.

In modern times, foreign immigration continues to influence urban growth. For instance, in recent years it has had a strong impact on the urban centers of Canada. Statistics show that about half the people who enter Canada each year under its liberal immigration laws settle in the country's three major urban centers—Montreal, Toronto, and Vancouver. These cities, as well as many others located in countries with liberal immigration policies, are finding the influx increasingly difficult to absorb. Some governments are responding by severely limiting the number of new immigrants allowed to enter their country each year.

Rural to Urban Migration. A factor that has been even more important in urban growth than the immigration of foreign nationals is the intranational migration of people from rural areas to cities. This redistribution of population has been the greatest single cause of the rapid urbanization around the world during the last 100 years or more. The tremendous rush to the cities begun in the late eighteenth century has resulted in the concentration of huge numbers of people in centers not properly prepared to meet their problems. But the migration from rural areas continues, even to cities where there are shortages of such necessities as jobs, housing, and health facilities. This trend reflects the dissatisfaction of rural dwellers with the stagnant economic and cultural life of the village and the lure of vast opportunities, whether real or imagined, to be had in the city. As long as such opportunities exist for some, there seems to be little that can be done to keep thousands of rural dwellers from moving to urban centers in search of better working and living conditions.

There is, on the other hand, little or no evidence that great numbers of urban dwellers are moving to rural areas, except in those few instances where people have been forced back to the countryside by government decree. Even those migrants who come to the city in search of temporary employment are tending to stay for longer periods of time, and more are bringing their families. Furthermore, whereas rural to urban migration once tended to progress in stages — from farm to small town and then to city — adherence to this "law of intervening opportunity" appears to be declining, for more and more migrants are coming directly to the city.

Natural Increase from Within. Urban growth the world over, but most particularly in the developed countries, is becoming increasingly attributable to natural population increase within urban areas themselves. In the expansion of many cities around the world, this population source is now fully as important as domestic migration. Its new prominence results not only from improved health techniques that prolong the human life span but from the relatively high birth rates that exist among many urban populations.

Spatiality and Social Complexity

The spread of urban living since 1900, coupled with unprecedented population increases, has given the world many huge urban centers, more than a few of which are much larger and more complex than the largest city a century ago. London, New York, Tokyo, and Shanghai are only some of the many cities that now have larger populations than many countries. Because of their rising populations, such large cities have "exploded" into the countryside, spreading themselves over greater and greater areas (Fig. 12–5). The degree to which they have grown is somewhat difficult to comprehend in its totality, for increase in size has been attended by even greater increases in complexity. Size increases arithmetically; complexity increases geometrically. Large cities of the modern world represent marked qualitative changes over the past and thus are not merely enlarged versions of traditional cities but entirely new and different forms of human settlement. The comparatively simple way of life typical of urbanites in the past has been replaced by a much more complex way of life that is metropolitan in scope.

Standard Metropolitan Statistical Areas

A metropolitan area can be defined in a variety of ways, but the definition that has become rather widely accepted and used is the one established by the United States Bureau of the Census in its inventories of large urban agglomerations. A basic unit of such studies is one the bureau calls its *Standard Metropolitan Statistical Area* (SMSA) and defines as

> a county or group of contiguous counties which contains at least one city of 50,000 inhabitants or more, or twin cities with a combined population of at least 50,000. In addition to the county or counties containing such a city or cities, contiguous counties are included in an SMSA if, according to certain criteria, they are socially and economically integrated with the central county.

The population living in SMSAs is designated as the metropolitan population. The population is subdivided as "inside central city or cities" and "outside central city or cities." The population living outside SMSAs constitutes the nonmetropolitan population (U.S. Bureau of the Census, 1970. For a complete listing of the criteria used to define such SMSAs, refer to Bureau of the Budget Publication, Standard Metropolitan Statistical Areas, 1967, U.S. Government Printing Office, Washington, D.C. 20402).

There are a total of 243 SMSAs in the United States, ranging in size from New York, with 11,571,899 inhabitants, to Meriden, Connecticut, with 55,959 inhabitants (Fig. 12–6). And, as depicted in Figure 12–7 on a state-by-state basis, about 68.6 percent of the country's total population resides in SMSAs. Most SMSAs have larger populations living outside their central cities than within them; New York is an outstanding exception. The contiguous counties commonly known as the New York City metro-

FIGURE 12–5

A view of Houston, Texas, which has recently mushroomed into one of the largest urban centers of the United States. (Houston Chamber of Commerce)

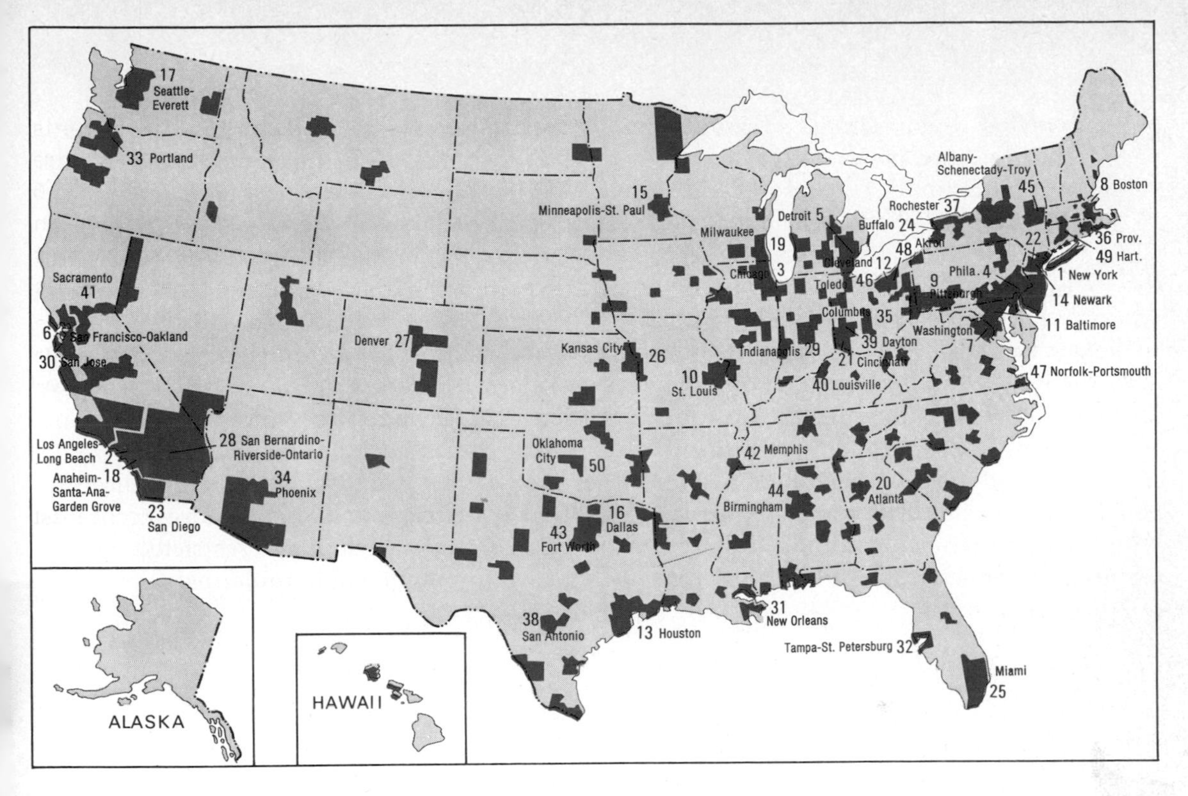

FIGURE 12–6

The Standard Metropolitan Statistical Areas (SMSAs) of the United States. There are 243 in total; the top 50, as ranked by population, are named on the map (space allowing) and listed, with their 1970 populations, in the table below.

1	New York, N.Y.	11,571,899	26	Kansas City, Mo.–Kans.	1,253,916
2	Los Angeles–Long Beach, Calif.	7,032,075	27	Denver, Colo.	1,227,529
3	Chicago, Ill.	6,978,947	28	San Bernardino–Riverside–Ontario, Calif.	1,143,146
4	Philadelphia, Pa.–N.J.	4,817,914	29	Indianapolis, Ind.	1,109,882
5	Detroit, Mich.	4,199,931	30	San Jose, Calif.	1,064,714
6	San Francisco–Oakland, Calif.	3,109,519	31	New Orleans, La.	1,045,809
7	Washington, D.C.–Md.–Va.	2,861,123	32	Tampa–St. Petersburg, Fla.	1,012,594
8	Boston, Mass.	2,753,700	33	Portland, Oreg.–Wash.	1,009,129
9	Pittsburgh, Pa.	2,401,245	34	Phoenix, Ariz.	967,522
10	St. Louis, Mo.–Ill.	2,363,017	35	Columbus, Ohio	916,228
11	Baltimore, Md.	2,070,670	36	Providence–Pawtucket–Warwick, R.I.–Mass.	910,781
12	Cleveland, Ohio	2,064,194	37	Rochester, N.Y.	882,667
13	Houston, Tex.	1,985,031	38	San Antonio, Tex.	864,014
14	Newark, N.J.	1,856,556	39	Dayton, Ohio	850,266
15	Minneapolis–St. Paul, Minn.	1,813,647	40	Louisville, Ky.–Ind.	826,553
16	Dallas, Tex.	1,555,950	41	Sacramento, Calif.	800,592
17	Seattle–Everett, Wash.	1,421,869	42	Memphis, Tenn.–Ark.	770,120
18	Anaheim–Santa Ana–Garden Grove, Calif.	1,420,386	43	Fort Worth, Tex.	762,086
19	Milwaukee, Wis.	1,403,688	44	Birmingham, Ala.	739,274
20	Atlanta, Ga.	1,390,164	45	Albany–Schenectady–Troy, N.Y.	721,910
21	Cincinnati, Ohio–Ky.–Ind.	1,384,851	46	Toledo, Ohio–Mich.	692,571
22	Paterson–Clifton–Passaic, N.J.	1,358,794	47	Norfolk–Portsmouth, Va.	680,600
23	San Diego, Calif.	1,357,854	48	Akron, Ohio	679,239
24	Buffalo, N.Y.	1,349,211	49	Hartford, Conn.	663,891
25	Miami, Fla.	1,267,792	50	Oklahoma City, Okla.	640,889

politan area represent so large a "population pack" that they are no longer considered a single SMSA within the criteria of the Bureau of the Census. The Newark and Paterson–Clifton–Passaic areas, though contiguous to the New York SMSA, do not maintain an economic integration with New York City sufficient to warrant their inclusion in that SMSA. In 1970 the high-population belt, or megalopolis, extending from Boston to Washington, D.C., included one-fifth of the country's total population (or 37.8 million people) and 43 contiguous SMSAs—all within 450 miles north to south and 150 miles east to west. Seven of these 43 SMSAs have a population in excess of 1 million: New York, Philadelphia, Washington, Boston, Baltimore, Newark, and the Paterson–Clifton–Passaic area of New Jersey (see Fig. 8–5).

A system of SMSAs based on counties—in all areas, that is, except New England, where town and city lines are used—is somewhat arbitrary, to be sure, but in many instances it offers definite statistical convenience. Most important, since more and more urban areas include more than one city, it reflects the real distribution and concentration of urban populations more accurately than would a relatively restricted system based on municipal boundaries.

FIGURE 12–7

State populations in Standard Metropolitan Statistical Areas. Approximately seven out of every ten Americans (68.6 percent) live in SMSAs.

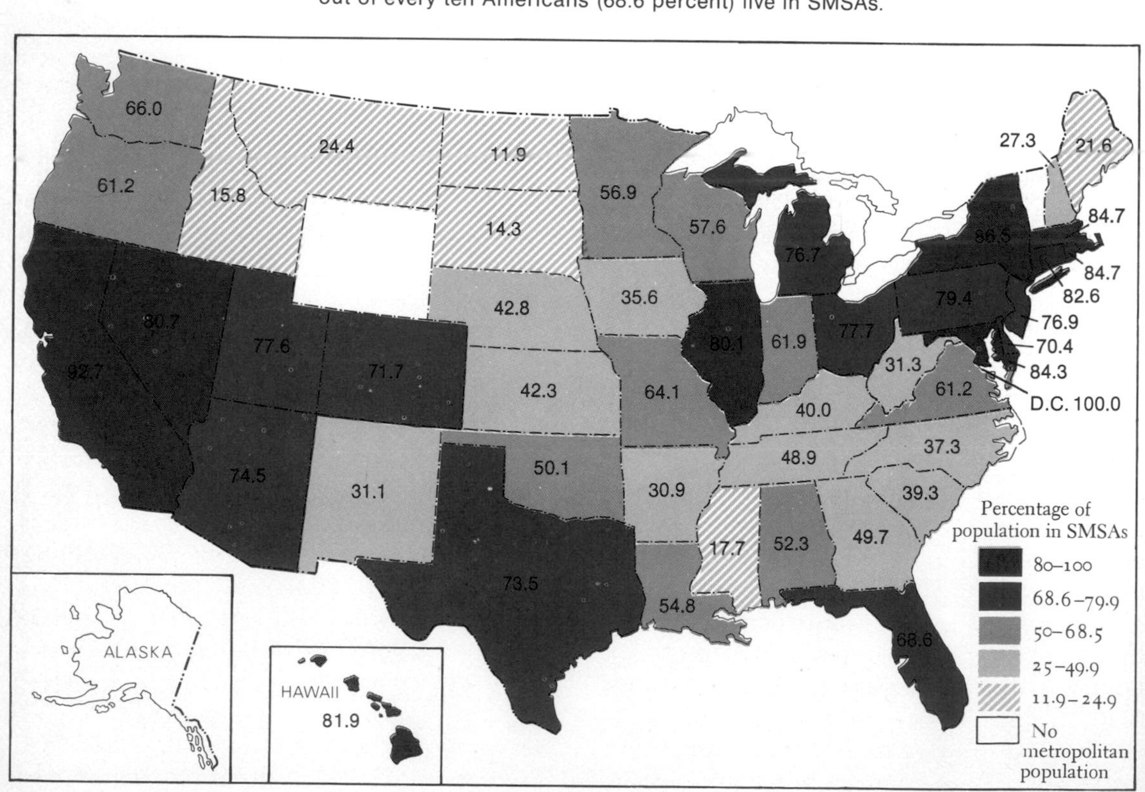

FIGURE 12–8
Commuting by subway in New York City during rush hour. (Sam Falk from Monkmeyer)

Emerging Patterns of Metropolitan Life

The Transportation Tangle

Commuting. That which determines a settlement's form and cohesion is almost exclusively the product of its spatial linkage systems, and the constant movement of people within such systems has become a predominant characteristic of life in contemporary metropolitan areas. In a bygone age most people lived within a short walking distance of their places of employment, but today a very high percentage of urban dwellers have to commute to their places of work (Fig. 12–8).

Among this high number of commuters the majority travel each workday from residences in outlying areas to a central business district; some journey to businesses or factories located in more remote areas; and a few actually go from one metropolitan area to another. In many cases the distances involved are as much as 100 miles each way—although, since expressways and high-speed trains and planes are often employed by commuters, daily commuting distances have come to be measured more in terms of travel time than miles. Commuting has become a necessary evil imposed upon hundreds of thousands of urbanites, and those who must spend an hour or more traveling to work can amass years of unproductive effort before they reach retirement age.

In certain metropolitan centers, among them New York, Paris, and Moscow, commuter trains, subways, and other means of public transit figure prominantly in helping people get to and from work; in Los Angeles, and other sprawling cities, and especially in the developing areas of the world where the automobile is just becoming commonly

FIGURE 12–9
A phalanx of automobiles surrounding a suburban shopping center in Maryland. (The National Observer)

available, the private car is the primary means of local transportation (Fig. 12–9). In still other centers, such as New Delhi and Saigon, a significant number of commuters rely on motorscooters and bicycles.

So much movement by so many people and vehicles has created huge transportation problems for metropolitan areas. The mere task of getting people from one place to another has become a herculean undertaking. The number one culprit, especially in regard to commuting, seems to be the automobile. It is not only expensive to operate and often inefficiently used but also constitutes one of the primary sources of urban air pollution.

During almost any period of the day—but especially within the morning and evening rush hours—automobile traffic in metropolitan areas can become so congested that it slows to bullock-cart speed, if not to a standstill; what vehicular movement occurs is often no faster than walking (Fig. 12–10). Airplane traffic around many large cities is also problematic, for, even in good weather, planes must frequently circle a major airport for an hour or more before landing space becomes available. The result is delay and aggravation, and the surfeit of cars, trucks, buses, trains, and airplanes in such relatively small areas imposes a severe strain not only on the nerves but the pocketbook as well. Obviously there is an urgent need for the development of more rapid and more efficient means of transporting people and goods between and within metropolitan areas.

Some Traffic Solutions. The solution to the problem of traffic congestion most frequently put forth is the building of more and bigger expressways. And such construction has been employed successfully in some instances; however, with the constant increase in the number of vehicles in use, the improved conditions it brings never seem to last—particularly in regard to intracity traffic. Furthermore, the proponents of such solutions rarely take into consideration their effects on other aspects of the urban system. Inevitably such construction involves the destruction of historic landmarks and other distinctive buildings, the displacing of people and businesses, and the waste of valuable acreage (Fig. 12–11). The cost in money and human comfort can be enormous, and there is rarely an opportunity for adequate repayment. Increasingly expensive to construct and maintain, expressways and parking facilities often deplete public funds still further by causing at least a temporary decrease in the tax revenues from the land they occupy.

And, of course, if remedial construction is undertaken, it should be based on an accurate analysis of the reasons for the traffic congestion. A major portion of intracity traffic—sometimes as much as one-fourth—consists of vehicles that are only passing through the central business district on their way to other places. To combat this element of the traffic problem, bypasses and loops have been built around some cities, thereby allowing vehicles to skirt the central business district if it

FIGURE 12–10

A traffic jam in Paris. (M. H. Kellicutt, EPA)

FIGURE 12–11

The striated pattern of vehicular facilities—expressways, roads, and parking lots—in a section of Los Angeles. Single-purpose and often vacant, such highway and parking areas represent a relatively unproductive use of urban land. (California Division of Highways)

is not their destination. However, even after circular expressways, or "inner loops," as they are commonly called, have been built and much of this transient traffic is diverted from the city, the problem of controlling the central city's remaining vehicular traffic still requires a solution.

As a step toward this solution, many people argue for more extensive and more efficient means of public transportation. Their contention is that car owners might be persuaded to use trains, subways, and buses on trips to the city if these facilities offered faster and less expensive travel than the automobile can provide. Acting on this principle and following the lead of several other cities, Washington, D.C., has built a rapid transit system based on the "park-and-ride" concept: motorists from the suburbs are encouraged to leave their cars at parking lots just outside the city and continue their trip to the central business district on public transportation. Various other cities around the world are attempting to create or modernize rapid transit systems of their own. Such schemes, however, are usually shortchanged, and greater emphasis is still being placed on the automobile. For example, Hong Kong has recently completed a four-lane submarine tunnel for automobile traffic between Hong Kong Island and Kowloon. Even though this expensive project was intended, in part, to relieve the pressure on the cross-harbor ferries, it is not having the success imagined, because overall traffic increases already tax its capacity.

Future technology may offer more comprehensive solutions to the traffic problem. Many promising new rapid transit systems are already on the drawing board. However, total solutions are not likely to be achieved through the use of new machinery alone. Much depends on proper planning and the coordination of transportation systems with all other aspects of life in metropolitan areas.

Urban Renewal

As urban centers age, their structures and facilities tend to become run-down or outdated and as a result need to be renewed. Most *urban renewal* projects, to date, have focused on the inner portions of their respective metropolitan areas. Even though these sections usually occupy less than 5 percent of the total land area of the agglomeration, they frequently contain as much as 30 percent of its total population. They are also usually the areas with the greatest needs, having been neglected during the years in which rapid growth and modernization were occurring in the suburbs.

Generally speaking, it is in the older parts of the inner city that one finds substandard housing and schools and other structures in a state of disrepair; the number of slums increases; public utilities are insufficient; narrow streets cannot handle the dense traffic; and there are few open spaces or recreational facilities. With rural people continuing to pour into cities, the pressures on urban housing are intensifying; this is especially true in the developing countries. Here millions of people, recently rural but now knocking on the door of urbanization, live in shanty towns consisting of crude shacks and other dwellings that barely provide shelter (Fig. 12–12). In the more advanced nations, most people are somewhat better housed. But even in these countries, too high a percentage of the housing is poorly constructed, run-down, or unsanitary; the poor and disadvantaged end up living in these dehumanizing environments.

FIGURE 12–12

A hillside slum, or *favella*, in Rio de Janeiro, Brazil—one that overlooks portions of the city's fashionable Copacabana beachfront section. (Paul Conklin from Monkmeyer)

Long overdue, effective work on many urban problems has at last been started worldwide. The job involves both the private and public sectors, and governments all over the world have initiated a variety of programs to rid their countries of substandard housing, replace unsatisfactory streets, and update commercial areas and industrial facilities—all in an attempt to create more modern and livable communities (Fig. 12–13). In the United States, for example, it was not until the Federal Housing Act of 1949 was passed that urban renewal was conducted on a firm financial base. Only then were large sums of federal money available for the renewing of old urban areas, and only with such sums can the job be approached on the scale necessary for success. In the last three decades, the degree of government involvement in urban renewal has varied from total responsibility, as in Cuba and the Soviet Union, to the complicated inducements to the private sector by federal, state, and local government in the United States. But whether communist, socialist, or capitalist, countries are making a growing commitment to meet the need.

In some cases the areas in need of renewal are totally chaotic in design and obsolete in construction, necessitating a complete overhaul. In other instances the problems are much less severe and thus require less extreme measures. The different types of urban renewal can be divided into three broad cat-

FIGURE 12–13

In the distance, the construction of a union members' housing project on the Lower West Side of Manhattan. (Magnum)

egories: conservation, redevelopment, and rehabilitation.

Conservation. Conservation projects generally necessitate no major changes in land use. The areas involved are usually not yet blighted, and the work required may involve nothing more than local clean-up campaigns and stricter enforcement of zoning laws and building and health codes. The objective is to encourage people to take pride in their communities through the proper upkeep and maintenance of their houses, stores, and other buildings and facilities. By putting the emphasis on slum prevention, this type of urban renewal project generally requires relatively little expenditure of public funds, and it can go a long way in strengthening community spirit.

Redevelopment. Redevelopment is a much more extensive type of urban renewal. It involves already blighted areas and usually necessitates, therefore, the total redesign of very large tracts of land and even whole neighborhoods. Some parts of a metropolitan area may be completely torn down and then redesigned and rebuilt to accommodate new land-use patterns.

When entire slum areas are cleared away, they are frequently replaced by more profitable office buildings, cultural facilities, condominiums, or apartment complexes. Many of the high-rise apartment complexes are designed to be self-sufficient and include a variety of stores, restaurants, and facilities for entertainment and recreation—all intended to reduce the dehumanizing qualities frequently associated with this kind of residence. Main Place in the business district of Dallas, Lake Meadows in Chicago, and the Civic Center in downtown Los Angeles are all examples of redevelopment projects designed to redevelop blighted areas and bring them into the twentieth century (Fig. 12–14).

Although urban renewal in this form is a step in the right direction, some people question whether its benefits outweigh its disadvantages. Redevelopment all too frequently involves the elimination of historic landmarks, the dislocation of a great many people and shops, and very high costs. When a slum is to be torn down and its residents relocated, there is frequently not enough low-income housing available nearby to accommodate them; even promises that they will be accepted into public housing are usually very slowly fulfilled. Furthermore, the simple fact of being forced to move and live in a different and unfamiliar environment tends to put a severe physical and psychological strain on people, especially the elderly. Inevitably, too, along with poor housing, redevelopment eliminates the jobs of many area residents.

Social critics of urban redevelopment thus claim that it has done relatively little to help the poor and even less to correct patterns of segregation and discrimination. Further, they argue that those who have benefited the most have been the big land developers.

Rehabilitation. Another approach to urban renewal is the rehabilitation of aging buildings—a procedure that precludes the permanent relocation of their tenants. In slum areas earmarked for rehabilitation rather than redevelopment, the basic structure of most of the buildings is solid enough to remain and can absorb the installation of such improvements as new electrical wiring, fast elevators, modern plumbing, and the like. Only occasionally is there the need to tear down a structure, and then an attempt is made to replace it with a new structure that serves the same purpose; for example, a small corner grocery store is replaced by a modern supermarket. Such attempts at renovation have been carried out in various parts of the

FIGURE 12–14

Urban renewal in downtown Los Angeles. Scores of delapidated buildings in this area have been demolished in recent years in order to make way for new constructions (mainly additional public facilities on the scale of the music center, which includes the flattop in the center of the photograph) and for improved roads and highways. (PFI, © Charles Rotkin)

United States: Wooster Square in New Haven and Society Hill in Philadelphia are but two of many examples. Although such rehabilitation projects tend to cause only minimal disruption to a neighborhood and sometimes greatly improve community spirit, their overall rate of success has been lower than expected.

Whichever of these three broad types of urban renewal is used the overriding goal is to increase the comfort and efficiency of the environment for its residents as well as others. Many municipal governments hope that by reversing the decline of the central city areas, they will attract affluent people and businesses back to these areas from the suburbs. Already a significant number of suburbanites, encouraged in part by the rising cost of suburban living, are resettling in the central city. This flight to the city is particularly popular among young business and professional people; many buy town houses in such areas as Washington's Georgetown or Cincinnati's Mount Adams area and then make improvements to suit their own tastes and pocketbook. Some come so they can live close to their work; others want to be able to draw readily upon the city's cultural resources; and others simply enjoy the stimulation of the urban environment. Thus, the appeal of the city persists; after all, it is the cradle of civilization—however uncivilized it may seem.

Most urban renewal programs have been qualified successes. Many people feel that the returns would be much greater if the approach were more comprehensive—for example, metropolitan rather than local. Focusing haphazardly on different sections of a city, rather than planning the revitalization of the whole metropolitan area at once, has frequently meant a lack of coordination in execution and a less than efficient product. Part of the reason for the continuance of the piecemeal approach is, of course, the sheer size of most metropolitan areas; but a more fundamental problem is the fact that most city governments have fragmented jurisdictional authority and as a result there is an urgent need for a comprehensive metropolitan government structure. It is to this need that people must address themselves in the future.

Metropolitan Management

In every country of the world, the task of urban management is becoming more and more complex. City governments are finding it increasingly difficult to operate adequate urban renewal programs; police, fire, and sanitation services; educational institutions; health and welfare programs; public recreational facilities; and the host of other basic necessities they are expected to provide. Rising numbers of people, inadequate funds, fragmentation and lack of authority, and conflicting areas of jurisdiction are among the major causes of the current crises in urban administration.

The ever-increasing growth of urban populations means that local governments must provide more and more services—and usually at higher and higher costs per unit. The problem is intensified by the fact that, although they lose tax income as affluent residents move to the suburbs, municipal governments must still provide such services as roads and parking facilities for a growing number of commuters and visitors, as well as increased welfare benefits for the low-income people who continue to settle in the cities. To aggravate the problem even further, many of the cities with the greatest needs are also the ones that have the highest proportions of tax-exempt property.

Local governments attempt to increase their tax base in various ways but to little avail. In desperation they have been turning with increasing frequency to federal and state governments for financial support. But, despite federal and state aid, local governments continue to face a serious financial crisis.

Hierarchical Delegation of Authority. The political authority of local governments is delegated from a higher level of government, and the degree of authority granted varies from country to country. Most cities in continental Europe, though subject to the regulations of regional governments, exercise considerable freedom in local matters. In the Orient, state and national governments have tended to retain more control over local governmental units. And in communist countries the centralization of authority is almost total.

In the United States, the federal system predominates. City governments, as is sometimes forgotten, are legally created by the states. Since no reference to local government is made in the United States Constitution, the states have assumed the right to establish and empower whatever forms of local administration suit them. The city's right to adopt its own *charter*, or constitution, and its degree of *home rule*, or autonomy, are defined by the state legislature. Until recently, most states retained fairly tight control over their cities. Until 1962, for instance, state law gave the governor of Massachusetts the right to appoint the police chief of the city of Boston. More recent revisions of the law in such states as Connecticut, Florida, Illinois, and Michigan have all increased significantly the powers of local government. Yet, most state executives and legislatures still retain broad authority over city policy matters, including the nature and rate of local taxation. Undoubtedly, such supervision has its advantages; but it also tends to put local officials, especially mayors, in a kind of legal straitjacket—they have tremendous responsibility but little power.

Fragmented Jurisdiction. Most metropolitan areas are managed by a multitude of local governmental units. In Chicago's Cook County alone, for instance, there are over 1,000 separate and independent governing units. This multiplicity of authority appears to be a characteristic of metropolitan growth. The various units within a metropolitan area may include city, county, and township governments, and, in certain instances, school districts and other special districts and agencies as well. Each one of these local governmental units tends to be concerned only with its own problems and attempts to solve them, without regard for the welfare of the general area. Moreover, neighboring localities are usually in fierce competition for such assets as recreational space, industry, and residents able to pay taxes.

None of these administrative units has authority over the entire metropolitan area. Consequently, in most cases there is no single governmental body to deal comprehensively with areawide problems, such as control of water supplies, zoning regulations, pollution, sanitation, transportation, or civil defense. This lack of coordination puts a severe handicap on the orderly development of metropolitan regions as a whole.

Of course, some attempts have been made at coordinated metropolitan planning and administration, but generally on a very restricted basis. There are special administrative districts that provide for fire protection, garbage collection, sewage disposal, soil conservation, and park management. In addition, special intergovernmental "authorities" control the construction and operation of such facilities as harbors and airports; notable among them is the New York Port Authority, a tristate agency that oversees the New York metropolitan area's bridges, tunnels, and transportation terminals.

Metropolitan Government. Many argue that a far more promising way to handle the problems of large urban areas is to institute metropolitan governments. Crossing present political boundaries where necessary, each of these governments would be responsible for an entire metropolitan area's police protection, water supply, transportation system, civil defense, pollution control, and other services that seem to be most efficiently administered by an areawide authority. A few metropolitan governments have already been established—in Toronto; Nashville; and Dade County (Miami), Florida. So far, reaction to most of these experiments has been mixed. While some form of centralization is no doubt desirable, the metropolitan government is hardly a panacea. No single administrative unit can hope to deal effectively with the diverse problems of the hundreds of local communities that comprise the largest metropolitan areas, such as New York, Los Angeles, London, Tokyo, Shanghai, and Buenos Aires. In some ways it could even help to further intensify and aggravate existing problems. In many of these centers, indeed, the proposals most often heard are for decentralization of the already cumbersome and impersonal municipal bureaucracies. The ideal, probably, would be a balance between the two.

Planned Urban Land Use

The use of land in an urban environment does not have to proceed haphazardly, with little thought to the needs of the people living and working there. Instead, urban growth and development can be accomplished in a sane, orderly, and humane manner through proper planning and supervision.

Although only recently recognized as a distinct discipline, *urban planning,* in at least a rudimentary form, is believed to be as old as the urban revolution itself (Fig. 12–15). For the most part, however, the history of urbanism and urban growth has been devoid

FIGURE 12–15

Among the excavated ruins of Mohenjo-Daro, the Great Bath, a centrally located, brick construction that was served by the 4,500-year-old city's elaborate drainage system, one not equalled until Roman times. (© Frances Mortimer from Rapho/Photo Researchers)

of planning in the modern sense of the word. Until quite recently, the majority of urban planning was limited to such matters as the design of street patterns and housing blocks.

Most urban growth has thus occurred on a hit-or-miss basis. With this unplanned development have come increasingly polluted, crowded, and, in the opinions of some, ugly and difficult environments in which people must live and work.

Planning for Today's Needs. In response to this situation there has been a major rebirth in urban planning within the last several decades. Concern for the quality of urban life is finally coming of age. People are now beginning to realize that, if civilization is to survive, conservation principles must be applied not only to the natural environment but to the artificial environment as well. Nearly all the world's major cities have some kind of urban planning program, and in some of the larger centers—such as Paris, Tokyo, Rio de Janeiro, and Toronto—there are permanent and well-staffed urban planning departments.

In the United States the job is normally done by a city planning commission, which, in most cases, functions as an advisor to the local governing body. The membership of such a commission usually includes municipal officials and concerned citizens appointed by the mayor or the city manager. The product of the commission's deliberations is frequently a master plan, or a series of maps, graphs, and reports, suggesting procedures for the orderly development of the community. Since planning on this scale involves countless variables, such plans are constantly undergoing revision.

Planning Procedure. Before a master plan can be judged adequate to meet the needs of a community, certain well-defined objectives must be established. These objectives should be arrived at only after a thorough inventory has been made of the community's existing structures, other assets, and socioeconomic patterns. Realistic planning for the effective use of any area should also be based upon a complete inventory of regional assets and possibilities.

Furthermore, planning must be an ongoing process; the various environmental factors, both physical and cultural, are continually changing. Thus the effective master plan for land use is flexible and allows for needed alterations without substantial loss. Indeed, many of the current difficulties associated with various areas have resulted from perpetuation of outdated master plans.

Methods of Implementation. Once a plan that suits its objectives has been devised and accepted, a city government must decide which of the many legal methods at its disposal will best effect the changes the plan recommends. One of the government's chief tools in this regard is its right to establish zoning regulations, statutes that limit the ways in which land can be used and the type, height, and size of the structures built in a given area (Fig. 12–16). Zoning has its limitations, but it has been reasonably successful in helping to avoid undesirable mixes of such facilities as factories, stores, junkyards, and residences.

Local governments also have the right of *eminent domain,* or the authority to purchase at a "fair" price private property they wish to devote to such public facilities as roads, schools, municipal administration buildings, parks, and public housing. An *easement,* or the right to develop a piece of land without actually having to own it, may also be used by local authorities.

Urban Space. There are a great many ways in which to add beauty, recreational opportunities, and a sense of open

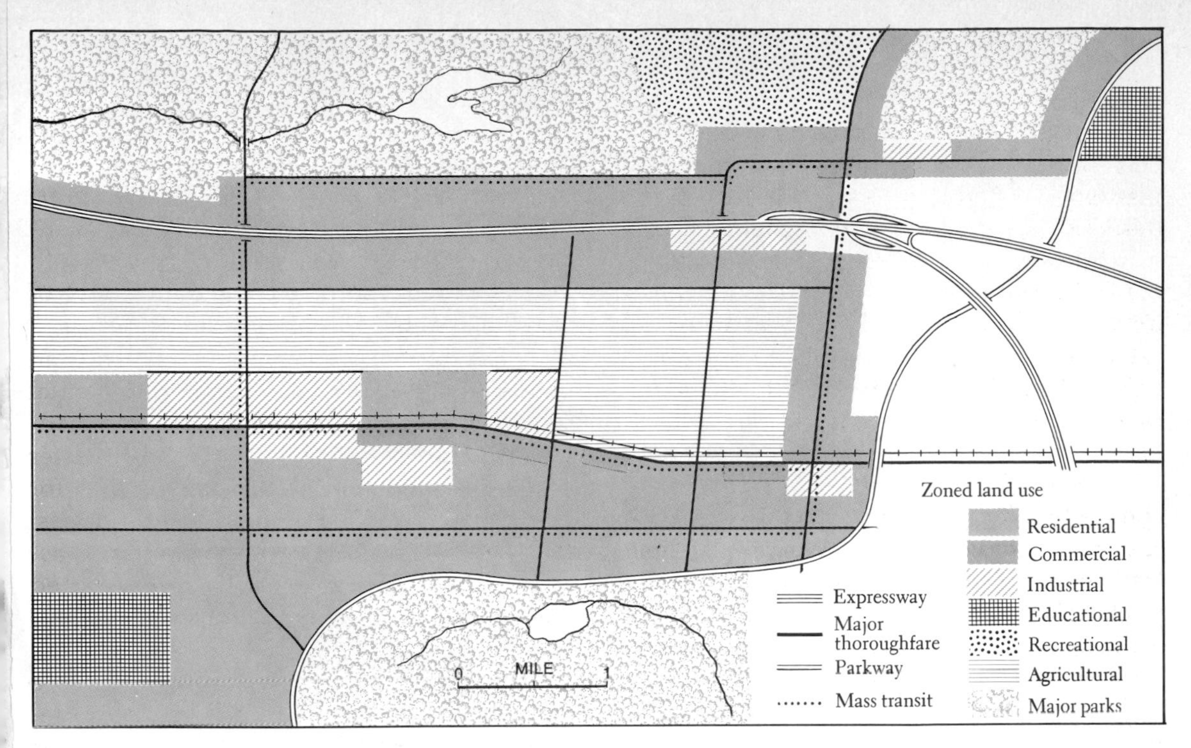

FIGURE 12–16

A simplified zoning map depicting a portion of a hypothetical urban community.

space to densely crowded sections of metropolitan areas. Several cities have successfully retained relatively large areas of land and used them for parks: New York's Central and Prospect parks, Los Angeles's Griffith Park, London's Hyde and Regent parks, and Moscow's Gorki Park of Culture and Rest are examples. However, parks need not be large to be effective. Many small and confined areas have recently been transformed into attractive meeting and resting places known as vest-pocket parks.

In the last few decades, large cities have also encouraged the proliferation of malls and plazas furnished with water fountains, pools, greenery, and sculpture (Fig. 12–17). Since open space at ground level is often in short supply, many cities have also promoted the development of rooftops as gardens and recreation, dining, or shopping facilities. Still another approach to improving the aesthetics of a city is the designation of greenbelts—that is, wide tracts of land zoned for limited development. Devoid of factories and residences, a series of greenbelts can provide a metropolitan area with a refreshing contrast to its otherwise gray and drab landscape. London and Vienna are among the cities that make use of greenbelts quite successfully.

The construction of public buildings in clusters known as civic centers has been a popular trend among urban planners within recent decades; however, the basic idea is centuries old. It represents an attempt to make public offices and auditoriums aesthetically pleasing and convenient to a maximum number of people, as well as cheaper to construct and maintain. Washington, D.C., Cleveland, Denver, and San Francisco have

and urban unmanageability. Further, each new community was to be surrounded by a greenbelt, a permanent reserve of open country used only for agricultural or recreational purposes. It was hoped that, by limiting the area in which construction was permitted, the indiscriminate use of valuable crop land would be avoided and the natural beauty of the countryside would be preserved.

It was from these early English models that the modern version of the new town evolved. Usually part of a comprehensive regional plan, the typical post–World War II new town is located some distance outside an established urban center; however, there are some instances in which new towns are located within metropolitan areas and function as satellite towns or suburbs. By providing needed housing and jobs, modern new towns have been able to handle some of the problems associated with the overflow of population from major urban centers. Many new towns have been quite successful in at-

FIGURE 12–21

Stevenage, England, one of the many new towns in Great Britain. (*The Times*, London, from Pictorial Parade)

tracting industry, business, and commerce—thereby further increasing job opportunities and strengthening the position of the towns as growth points for the surrounding region.

New Town Design. Goals other than decentralization and land conservation have also been given high priority in the design of new towns. For, to be lastingly successful, a new town should be able to accommodate growth and change without undue disruption. Thus, at least in the planning of these settlements, care is taken to balance total population with the level of community employment opportunities and social services. Optimally, each new town is a compact, self-contained entity in which the need for long-distance commuting is minimized or even entirely eliminated and residents are provided with schools; medical centers; markets; police, fire, and ambulance service; parks and playgrounds; and community centers. Planners also seek to enhance the aesthetic qualities of such environ-

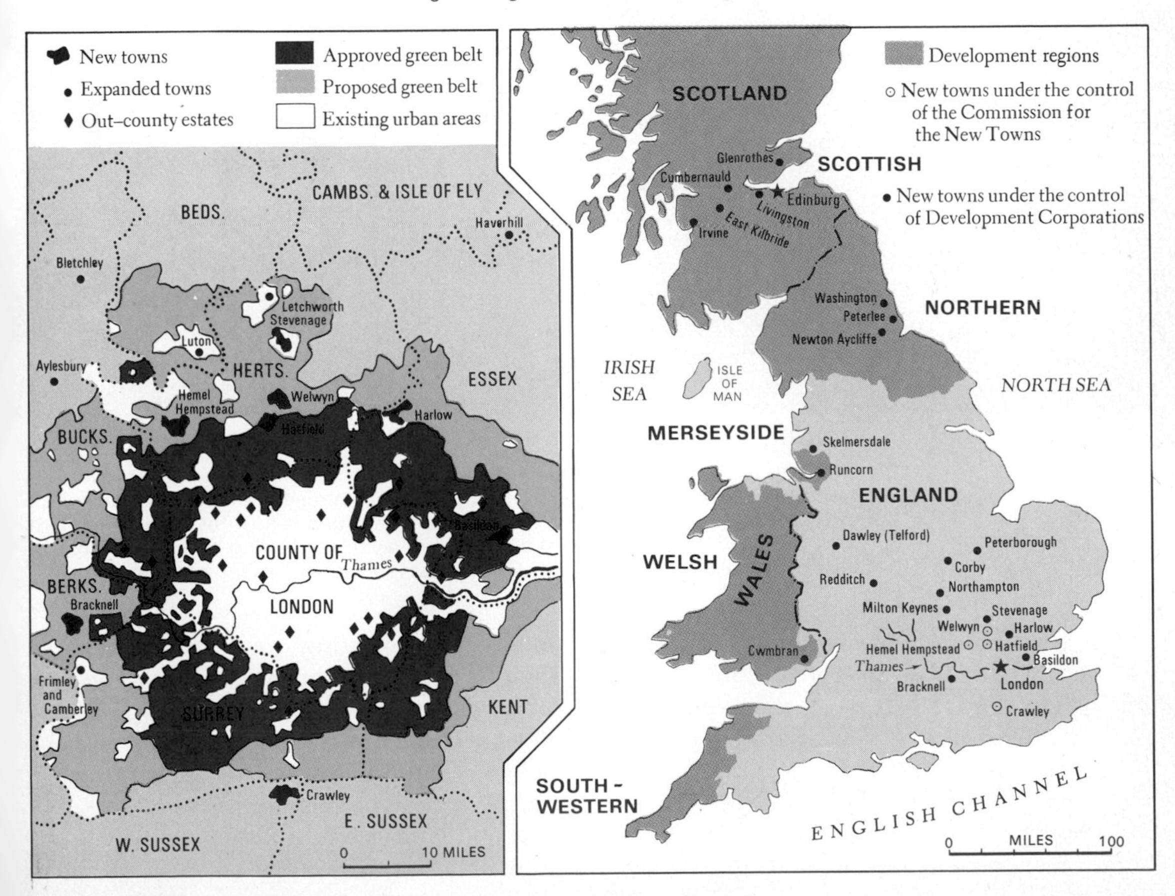

FIGURE 12–22

The new town movement in Great Britain. Left, new and expanding towns in the London region. Right, new towns in England, Wales, and Scotland.

FIGURE 12–23

Southwest over Roosevelt Island, which is located in New York's East River between the boroughs of Manhattan and Queens. Occupied until recently mainly by a few old farmhouses, hospitals, and a prison, the island is now the site of a modern housing development that seeks to combine apartment living and easy access to open land and recreational facilities. (Roosevelt Island Development Corp.)

ments—for instance, by avoiding the eyesore of rooftop antennas and utility poles. Perhaps most important, most town planners make a conscious effort to harmonize permanent artificial features with the natural environment.

So far, planners have only been able to come close to achieving some of their goals; none of the existing new towns is ideal. Some critics claim, for instance, that the new communities look too much alike and that they are lacking in color, character, and social facilities; others point out that there is usually an insufficient number and range of jobs and thus a widespread need to commute. Obviously, much more imaginative work must be done by planners and civic leaders before the ideal community can become a reality.

Future Urban Design

To think of the vast possibilities that lie ahead in urban design is a most challenging pursuit (Fig. 12–23). If history is a proper guide, the prospects for the world's urban landscape appear to be centered on a pattern of rise and decline; some urban centers will decline and perhaps even die (become defunct), while others will continue to grow. Further, while some future urban developments may be totally new, most urban growth will occur around sites already established as urban centers.

Given the population explosion, a continuing growth in both the population and the areal extent of most metropolises is inevitable. The United Nations World Health Organization predicts that, by the year 2000, the world's total urban population will nearly triple in size, from the present estimated total of about 1.5 billion to over 4 billion. In this same period, the number of urban dwellers in the United States, about 150 million in 1970, is expected to nearly double, to somewhere around 250 million. A significant percentage of this additional urban population is expected to live in the larger urban centers. This will mean that, for many newcomers to already overcrowded cities, the move will be literally upwards; for ever-increasing population will bring about the construction of more and taller skyscrapers, even in areas where they now are prohibited. This utilization of vertical space will no doubt bring about many societal changes; for one, much of the home-to-office commuting done in the next century may require only an elevator, with many future skyscrapers including both offices and apartments as well as shopping and health facilities.

However, trends in the United States since 1970 indicate that most of the expected increase in the nation's population will concentrate not in the cities but in the suburbs. As a result, many predict that the most likely basic urban form of the future will be horizontal, with a rather poorly defined nucleus, suggesting the present configuration of Los Angeles. Most people, it seems, prefer to live in detached houses that have a bit of open land associated with them.

As cities and their suburbs continue to expand they will eventually coalesce, forming replicas of the megalopolises that already exist in several parts of the world; these current ones will also expand. It is very possible, for example, that by the year 2000 the megalopolis along the northeastern seaboard of the United States will grow to include such midwestern cities as Chicago and Detroit. Thus, for better or for worse, the megalopolis will probably become the dominant unit of urban development for generations to come —not only in the United States but throughout the world.

REVIEW AND DISCUSSION

1. Discuss and evaluate the contention that there is a correlation between degree of urbanization and level of socioeconomic development.
2. Explain, in general terms, the difference between *urbanization* and the *growth of cities*. Evaluate the geographic significance of these two concepts.

3. Within your own frame of reference, what are the principal causes and consequences of the phenomenal growth of cities in recent decades?
4. What is a Standard Metropolitan Statistical Area? What is its value in urban analysis?
5. Identify and evaluate, in as much detail as possible, the three major types of urban renewal.
6. Describe, in as much detail as possible, the structure and operation of the city government under which you live or by which you are most often affected.
7. What seems to be the reasons for the recent rebirth of urban planning? What problems were created by the trial-and-error approach to urban development commonly used in the past?
8. Discuss and evaluate: some of the legal methods by which city governments can implement their master developmental plans, and some of the ways in which beauty, recreational opportunities, and a sense of openness can be added to metropolitan areas.
9. Explain, in general terms, the basic assumptions inherent in the *new town* concept. Discuss the origin and diffusion of new towns on a worldwide basis and evaluate their impact and record of success.
10. What do you envision will be the major changes in urban design during the next several decades? Through what means are these changes likely to occur?

SELECTED REFERENCES

Adams, R. McC. *The Evolution of Urban Society*. Chicago: Aldine, 1966.

Babcock, R. *The Zoning Game*. Madison: University of Wisconsin Press, 1966.

Blumenfeld, H., and P. D. Spreiregen, eds. *The Modern Metropolis: Its Origins, Growth, Characteristics, and Planning*. Cambridge: Massachusetts Institute of Technology Press, 1967.

Borchert, J. R. "American Metropolitan Evolution." *Geographical Review*, Vol. 57 (July 1967), 301–32.

Chapin, F. S., Jr. *Urban Land Use Planning*. Urbana: University of Illinois Press, 1965.

Down, A. *Urban Problems and Prospects*. Chicago: Markham, 1970.

Eldredge, H. W. *Taming Megalopolis*, 2 vols. New York: Doubleday, 1967.

Gans, H. J. *People and Plans*. New York: Basic Books, 1968.

Goodman, W. I., and E. C. Freund. *Principles and Practice of Urban Planning*. Chicago: International City Managers' Association, 1968.

Gottmann, J. *Megalopolis: The Urbanized Northeastern Seaboard of the United States*. Cambridge: Massachusetts Institute of Technology Press, 1964.

Howard, E. *Garden Cities of Tomorrow*. Cambridge: Massachusetts Institute of Technology Press, 1965.

International Urban Research. *The World's Metropolitan Areas*. Berkeley: University of California Press, 1959.

Linsky, A. S. "Some Generalizations Concerning Primate Cities." *Annals of the Association of American Geographers*, Vol. 55 (September 1965), 506–13.

McGee, T. G. *The Southeast Asian City: A Social Geography of the Primate Cities of Southeast Asia*. New York: Praeger, 1967.

Meyer, J. R., J. F. Kain, and M. Wohl. *The Urban Transportation Problem*. Cambridge, Mass.: Harvard University Press, 1965.

Michelson, W. *Man and His Urban Environment: A Sociological Approach*. Reading, Mass.: Addison-Wesley, 1970.

Milgram, G. *The City Expands: A Study of the Conversion of Land from Rural to Urban Use, Philadelphia, 1945–62*. Philadelphia: Institute for Environmental Studies, University of Pennsylvania, 1967.

Mumford, L. *The Culture of Cities*. New York: Harcourt Brace Jovanovich, 1970.

Muth, R. *Cities and Housing*. Chicago: University of Chicago Press, 1969.

Natoli, S. "Zoning and the Development of Urban Land Use Patterns." *Economic Geography*, Vol. 47 (April 1971), 171–84.

Osborn, F. J., and A. Whittick. *The New Towns: The Answer to Megalopolis*. Cambridge: Massachusetts Institute of Technology Press, 1969.

Owen, W. *The Metropolitan Transportation Problem*. Garden City, N.Y.: Doubleday, 1966.

Perloff, H. "New Towns Intown." *Journal of the American Institute of Planners*, Vol. 32 (May 1966), 155–61.

Putnam, R. G., F. J. Taylor, and P. G. Kettle, eds. *A Geography of Urban Places*. Toronto: Methuen, 1970.

Schaffer, F. *The New Town Story*. London: MacGibbon and Kee, 1970.

Simmons, J. W. "Changing Residence in the City: A Review of Intra-Urban Mobility." *Geographical Review*, Vol. 58 (October 1968), 622–51.

Stouffer, S. A. "Intervening Opportunities and Competing Migrants." *Journal of Regional Science*, Vol. 2 (Spring 1960), 1–26.

Strole, L. "Urbanization and Mental Health: Some Formulations." *American Scientist*, Vol. 60 (September–October 1972), 576–83.

Timms, D. W. G. *The Urban Mosaic*. Cambridge, England: Cambridge University Press, 1971.

Vernon, R. *Metropolis 1985*. Cambridge, Mass.: Harvard University Press, 1960.

Wilson, J. Q., ed. *Urban Renewal: The Record and the Controversy*. Cambridge: Massachusetts Institute of Technology Press, 1966.

Sikhs fleeing West Pakistan after the partition of British India.
(Margaret Bourke-White/Time-Life Picture Agency, © Time, Inc.)

13

Population Trends and Spatial Linkages

Since their beginnings, human beings have been very much a part of the natural environment. At first they accepted the bounty of nature as they found it, changing the environment little in satisfying their need for food and physical security. Later, as a means to improved food supplies, they learned to domesticate plants and various animals, and thus were able to make more extensive use of plant and animal products for clothing and shelter. As human groups learned more effective ways of adapting nature to their simple but expanding needs, their own numbers began to increase—sometimes exceeding the largess of the immediate environment. Many groups so taxed their surroundings that it became necessary for some members, or even all, to seek new living sites. Repeated through the millennia, this pressure for more abundant living space eventually brought about the spread of human beings to most of the present habitable world.

It is estimated that, for three million years or more, the number of humans on earth grew very slowly, reaching no more than a quarter billion by the first century A.D. (Fig. 13–1). During the next dozen or more centuries, until about 1650, the number of humans doubled, to approximately one-half billion; by 1830, less than two centuries later, the total had again doubled, giving the world its first billion human population. The population spiral continued to gather momentum, and, by 1925, the human total reached two

FIGURE 13–1

Estimated millennial world populations, from 4000 B.C. to A.D. 2000.

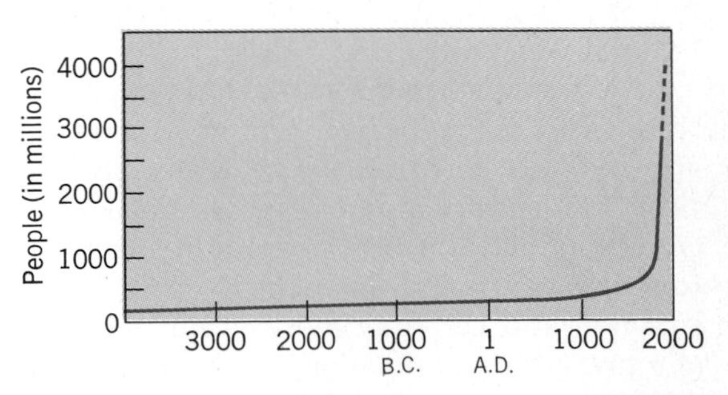

billion; by 1960, three billion; and, by 1975, four. Should this population explosion continue, there will be five billion human beings on the earth by 1985—and close to seven billion by the year 2000, less than a generation from now (Fig. 13–2).

FIGURE 13–2

The accelerating pace of world population increase. Humankind's first 1 billion population required close to 3 million years of human reproduction; the latest 1 billion was added in less than 15 years.

Time periods	*World population*	*Time lapses*
3,000,000 B.C.–A.D. 1	Under 250 million	3 million years
A.D. 1–1650	500 million	1,650 years
1650–1830	1 billion	180 years
1830–1925	2 billion	95 years
1925–1960	3 billion	35 years
1960–1975	4 billion	15 years
1975–1985	5 billion	10 years
1985–2000	7 billion	15 years

Population Distribution and Density

The spatial distribution of population over the earth, perhaps the most significant of the major cultural patterns, is, and has always been, extremely uneven. The entire Antarctic continent, as well as many large, isolated islands scattered throughout the oceans and seas, is still virtually uninhabited, while, for instance, the small tropical island of Java supports a population of over 60 million people and Singapore over 2 million—giving these areas statistical densities of 1,000 and 9,000 per square mile, respectively (Plate VII).

The spatial irregularity of human population can be effectively represented by means of a *dot distribution map,* which gives a clear visual impression of gross areal distribution; the right-hand map in Figure 13–3 contains such a graphic device. But a more common and certainly more useful method of representing population distributions and densities is by means of an *isopleth map,* through which one can indicate with greater precision the number of people living on a given unit of land. This is a quantitative approach—showing the ratio of total population to total unit area—and is known as the *arithmetic density,* or *man–land ratio* (Plate VII).

Both of these graphic devices have inherent limitations. The dot map details spatial distribution, but it makes few quantitative distinctions and therefore shows little in the way of man–land ratios. The density map, on the other hand, shows man–land ratios but suggests only generally the actual geographic, or areal, distribution of charted populations. Thus, where possible, the two types of map should be used together to obtain an adequate representation of distribution and density, concepts closely interrelated in all population studies.

Gross Distributions

On either a distribution or a density map of the world's present population, several major regional differences are discernible. First, the vast majority of the world's people is concentrated in the Northern Hemisphere, this owing to the incidence of birth as well as to migration; less than 10 percent of today's population lives in areas south of the equator. Second, the East is more heavily populated than the West; more than 85 percent of humanity lives in the Old World. Still another distributional pattern is discernible in the concentration of population along coastal lowlands and fertile river valleys inland; at least two-thirds of the world's people lives within 300 miles of the sea. The effect of these disparities is that about 80 percent of all the world's people is found within less than one-third of its land area.

Four areas of the world stand out as major population centers: eastern Asia, southern Asia, western Europe, and eastern North America (Fig. 13–4). And these major population centers share certain environmental conditions: moderately level terrain with low elevations, coastal and riverine plains, relatively high to moderate rainfalls, and temperature regimes of considerable range (see Fig. 13–3). Further, they are all major resource production centers.

In marked contrast to these four population centers, many other regions of the world are nearly, or even completely, empty of human beings. They include the vast arctic fringes—where a total human population of only a few thousand people is scattered widely over a landscape five times the areal

FIGURE 13–3

The terrain and population patterns of Japan. Dense populations crowd the small and scattered plains areas, which are favorable to human occupance; whereas, the mountains and hilly areas are sparsely populated.

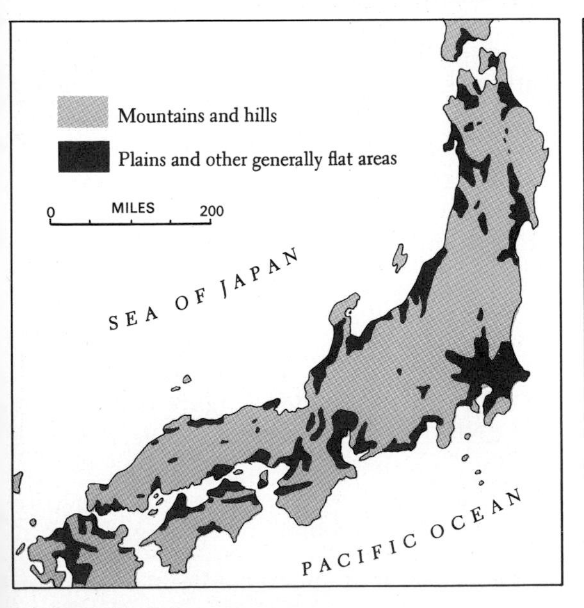

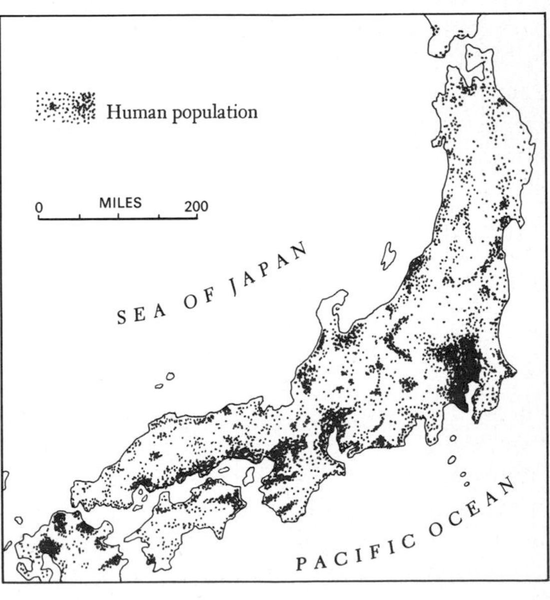

FIGURE 13–4

The beach at Coney Island, New York, on a summer's day. This forest of people is temporary, but it is a reminder that individuals in a megalopolis experience overcrowding on a daily basis—in housing, traffic, public transportation, and recreation. (Ben Ross)

extent of India, which contains over 600 million—and the vast continental interiors of South America, northern Africa, central Asia, northern North America, and central Australia.

National Arithmetic Densities

Population density is usually expressed as a ratio derived by dividing the number of people in a given area by the number of square miles in that area. In 1975 the average world population density was about 70 per square mile, the earth's 57.3 million square miles of land containing nearly 4 billion people. However, since the arrangement of people over the earth is far from uniform, this ratio has very limited uses. National density ratios are much more revealing, and they range from fewer than one to several thousand persons per square mile. For example, French Guiana averages 1.1 persons per square mile; the United States about 59; China about 200; Czechoslovakia about 300; Trinidad about 500; Belgium about 800; Bermuda about 2500; Hong Kong over 10,000; Monaco over 40,000; and Macao over 41,000. Even this sort of national figure is not without distortion, however, for it reflects all the land within the areal unit, inhabited as well as uninhabited areas, and therefore tells little about how the population is arranged within the nation. Thus, as an example, if one computes Hong Kong's population density in terms of its total area, the ratio obtained is slightly over 10,000 persons per square mile —although, in fact, the Mong Kok district within Hong Kong's city of Kowloon itself has a population density of over 400,000 persons per square mile (Fig. 13–5). Despite such inherent handicaps, the concept of arithmetic density is one of the best means of amplifying general population patterns.

Other Densities

In detailed studies of small areas, measures of population other than arithmetic density may be of greater service. In attempts to understand the effects of overpopulation, for instance, it is frequently useful to know how many people there are in a given area per unit of arable land; the arithmetic expression of this relationship is sometimes known as an area's *nutritional density*. Other frequently used population ratios include *agricultural density*, the number of agriculturally occupied people per unit of arable land; *rural density*, the number of people carrying on rural activities in addition to agriculture and living in rural settlements per unit area; and, its companion, *urban density*.

The major difficulty in these more specific types of population ratios, especially in international regional comparisons, is the inadequacy of the contributing data. While fairly accurate general population figures are available for most local areas, at least in the more advanced countries, a woeful lack of data on arable land and individual economic activities exists almost universally. It is therefore usually impossible to establish valid local population ratios on these subjects, and hence most international comparisons deal only with arithmetic densities.

Age Structure

Age structure, or breakdown by age group, is one of the most basic and meaningful ways of describing a population. Data on the size of different age groups have multiple uses in assessing population resources and planning socioeconomic development, for the age make-up of a society largely determines its domestic priorities. A predominantly youthful society has economic and psychological needs that differ from those of a predominantly old-age society, and both sets of needs will be at some variance with those of a predominantly middle-age culture. The populations of most societies are mixtures of the old, the middle-aged, and the young; but these mixtures are seldom even or unchanging through a long period of time.

FIGURE 13–5

A narrow side-street market in Kowloon, Hong Kong. (Paul Conklin from Monkmeyer)

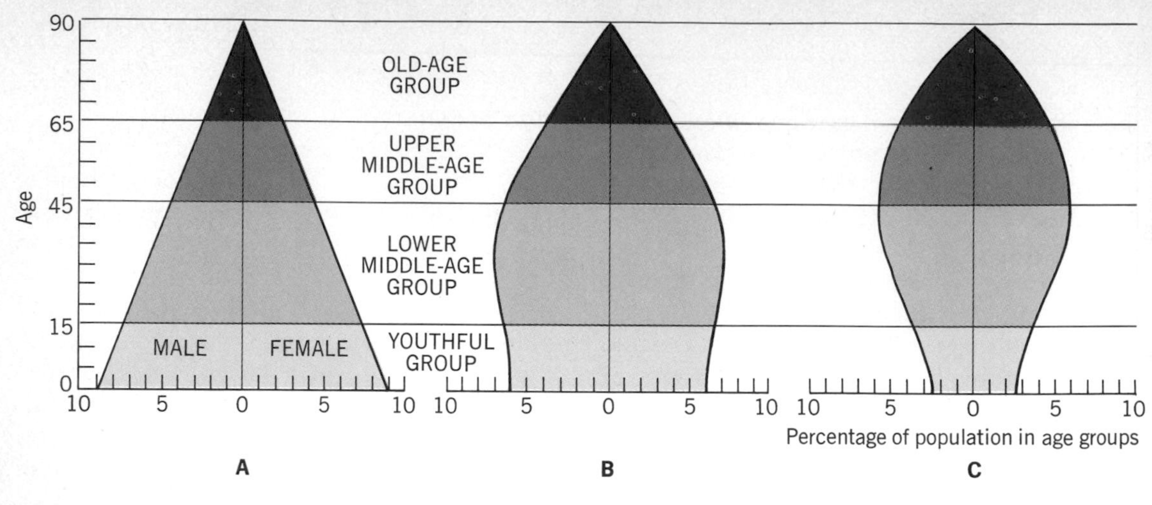

FIGURE 13–6

Some age structure models. A. Youthful population. B. Middle-aged population. C. Old-aged population.

Even barring major migrations and such catastrophes as war and famine, the age composition of a society changes with its birth and death rates; of the two, birth rate is usually the more influential factor.

The division of a population into age groups in any demographic study is usually quite arbitrary. An adequate selection of age boundaries has not been standardized for such purposes; however, some broad age ranges do seem to prevail in most international studies: *young* is normally applied to members of the population below the age of 16; *old* to all persons over 65 years of age; and *middle-aged* to all those between 16 and 65 years (Figs. 13–6 and 13–7).

In countries where crude birth and death rates are high, the youthful group is apt to be the largest of the three, as it is in most underdeveloped societies. Worldwide, the youthful group accounts for perhaps as much as three-fourths of the total population. It is generally an economically unproductive segment, and its housing, food, and educational needs represent a major societal burden—sometimes an overwhelming one.

The oldest group, which comprises a larger percentage of the population in developed societies than it does elsewhere, also requires expenditures that exceed its economic productivity. The middle-age group is the most economically productive of these broad groups, and it is largely responsible for the support of all three.

Population Change

The three principal components of population change in modern societies—mortality, fertility, and migration—tend to be variably influenced by the same factors. Most high national death rates, for example, are fostered by such factors as poverty, illiteracy, low national health standards, and fatalistic religious attitudes, while, to an almost equal extent, these same factors contribute to high national birth rates.

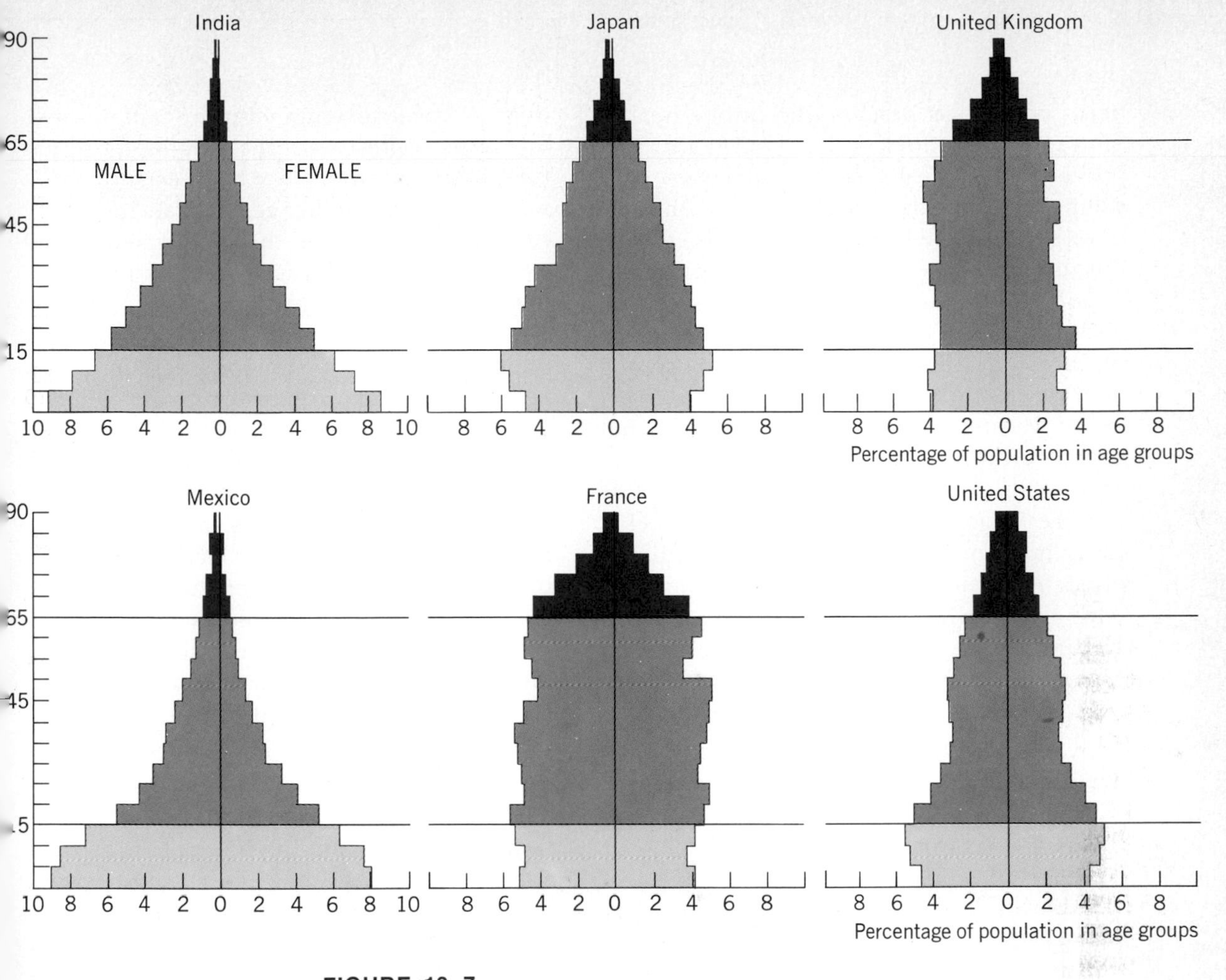

FIGURE 13–7

The age profiles for some selected countries showing their male-female population divisions in the present decade.

Mortality

Throughout most of history the death, or mortality, rates among human groups exceeded those known today in even the least advanced societies. It is estimated that the average human life span in Western societies nearly doubled between prehistoric times and the end of the Middle Ages, after which it remained almost constant until the nineteenth century. During the last century and a half, and especially in recent decades, the death rate in the Western world has declined by 50 percent or more. The first of these spectacular increases in the human life span, from 18 to 35 years, is traceable to a host of socioeconomic changes largely associated with the political consolidation of new territories into powerful states. The more recent doubling, from 35 to 75 years, reflects mainly technological advancements affecting agrarian activities and advancements in the fields of medical practice and public health. Medical innovations and the spread of improved public health services have been so extensive that marked declines in death rates have been effected in many societies even in the absence of the usual contributing rises in socioeconomic well-being (Fig. 13–8). The declining death rate in Asia in the twentieth century is due not so much to the continent's own recent industrialization as to discoveries of Western science and medicine that have been diffused into Asia (Fig. 13–9).

Technology and the Death Rate. Humankind's contest with lesser forms of life, a substantial cause of high human mortality through the millennia, has been largely won through technological advancement: from the turn to agrarian activities in lieu of the hunt to the gradual dominance of science over disease-carrying varmint and insects. But in its contest with the parsimony of nature, humankind has not been as successful, for hunger still challenges human ingenuity throughout much of the world. Improved storage and distribution systems have virtually ended the persistent danger of local famines; but such disasters still occur, and when they do, political considerations and xenophobia often prevent or greatly impede their relief.

There are other, active hindrances to the decline in death rates as well. For countering the victories over disease and famine, humans have applied their technology to the deliberate extermination of their own kind. Indeed, armed conflict seems to be an expected and increasingly large-scale feature of human life, rather than a declining remnant of a barbaric past. Deaths in the First World

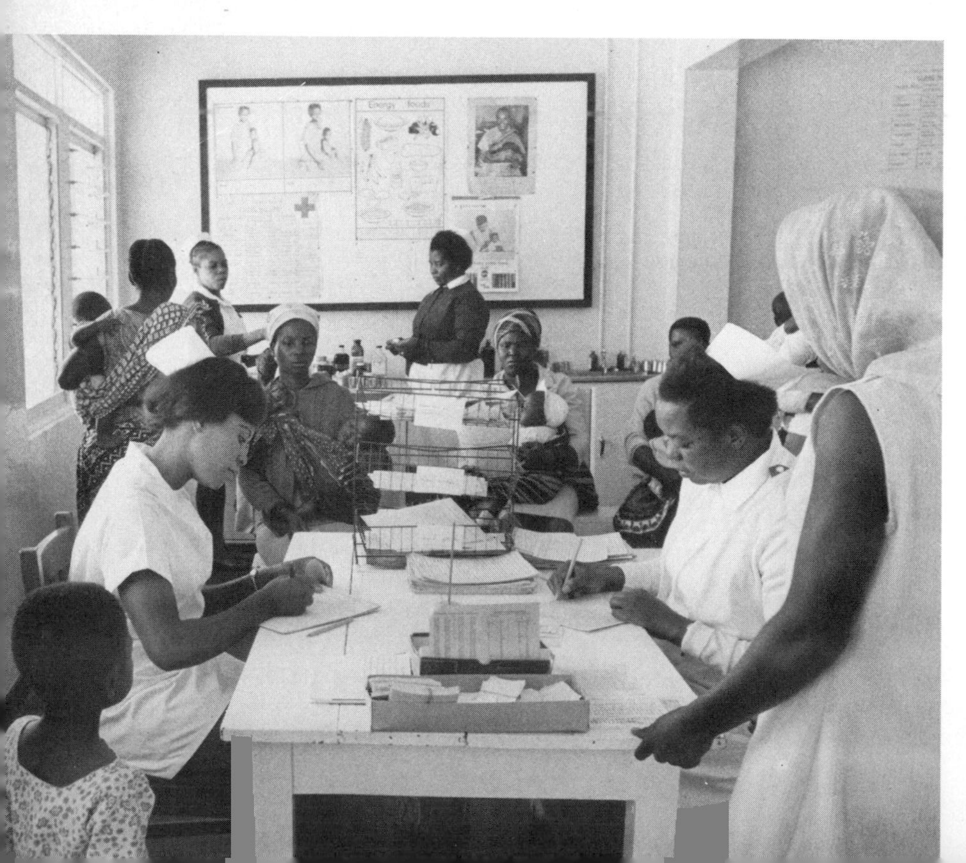

FIGURE 13–8

A clinic for the families of copper miners in Chinoala, Zambia. (Marc and Evelyn Bernheim from Woodfin Camp and Associates)

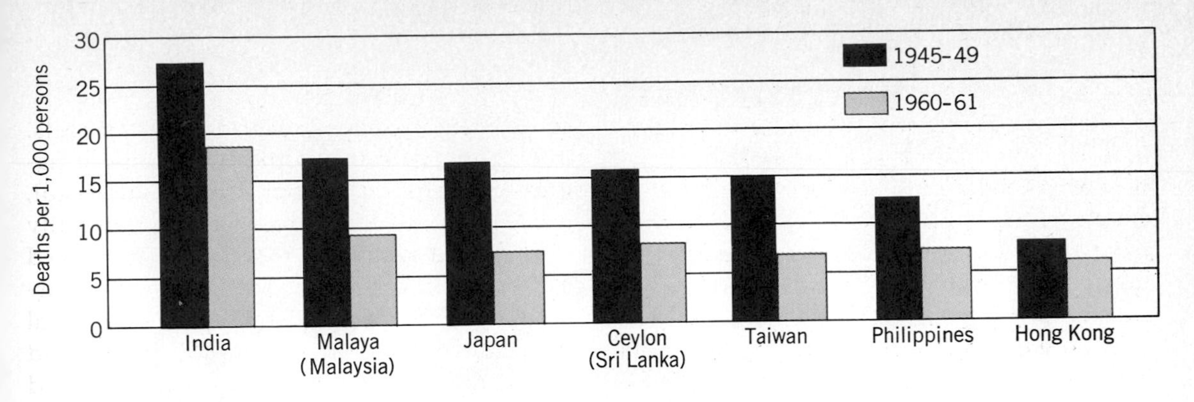

FIGURE 13–9

Death rates for several Asian countries in the periods 1945 to 1949 and 1960 to 1961.

War totalled nearly ten million, and the Second World War produced five times that many. If the present knowledge of nuclear science were applied in another world war, the death toll would possibly include most of the human population.

Measures of Mortality. The *crude death rate,* or the number of deaths within a population unit per year, is the most commonly used measure of mortality and the one most readily available on a worldwide basis. However, since it does not consider the age or sex of the dead, it tends to understate international differences in mortality; for instance, in general, the more advanced societies have a larger portion of their population in the higher age groups, where the probability of death is greatest. A more sophisticated measure of mortality is the *age-specific death rate,* which calculates the number of deaths within each age group, making international comparisons more meaningful. It is difficult to employ either measure of mortality comprehensively on a country-by-country basis, however, since death records for many parts of the world are either nonexistent or very unreliable—those societies with the higher mortality generally having the poorer recording of vital statistics.

Mortality Comparisons. On the continents of Asia and Africa, although reliable, objective measures of population have been made in only a very few countries, death rates have apparently declined considerably since mid-century; but they are still quite high (see Fig. 13–9). The world death rate is estimated at about 15 persons per 1,000—or about half the world birth rate. In the most underdeveloped countries of Asia, Africa, and Latin America, the death rate exceeds this average, ranging from 20 to 30 persons per 1,000. Most of the heavily industrialized countries have death rates below the world average (Fig. 13–10).

Fertility

Estimates made by the United Nations suggest that the world average birth rate has not declined significantly since the mid-twentieth century—while the world average death rate has decreased by a third in the same period. The index such organizations and independent researchers most often use for measuring the reproductive level of a society is its *crude birth rate,* or the number of live births within each unit of population per

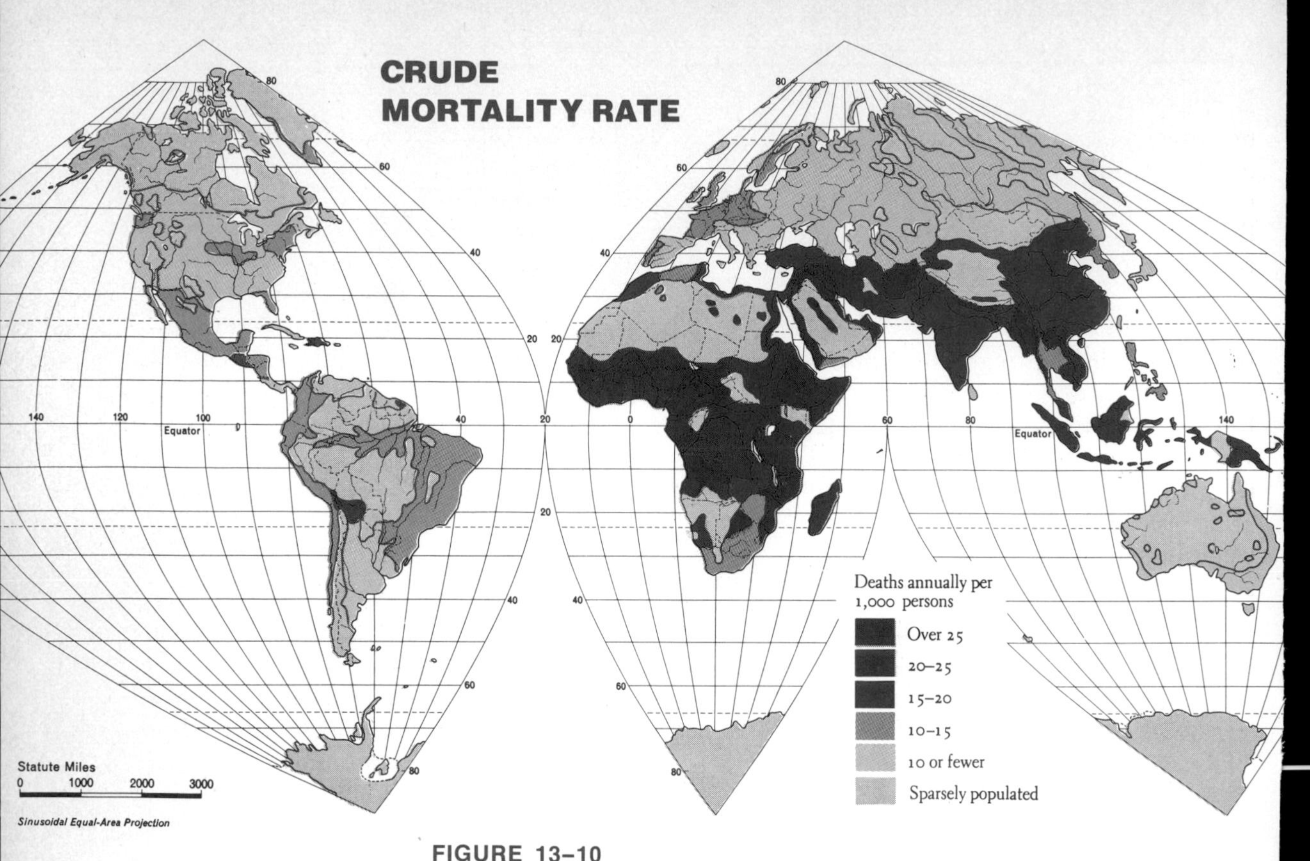

FIGURE 13–10

The world average annual mortality rate is currently about 15 per 1,000 persons.

year. Of course, this index is most easily prepared when the subject country is one in which births are normally registered; but, even supported by reliable data, it has some significant limitations, for, like the crude death rate, it does not consider the population's age distribution and sex ratios.

Determinants of Fertility. If birth rates were determined only by biological factors, the average number of live human births would probably range between 10 and 15 per couple. But since socioeconomic factors tend to intervene in many societies, the number of live births averages no more than 5 or 6 per couple. One increasingly influential social determinant of fertility is the development of medical birth control, but setting the climate for its acceptance, and thus of greater overall importance, are a host of socioeconomic influences of which potential parents are often largely unaware. Among these is the decreasing need for large numbers of offspring as farm helpers and a protection against hardship in one's old age, due primarily to the spread of industrialism and relative economic prosperity. A shift to urban living, with its distractions and spatial restrictions, has also contributed to the slow but persistent reduction in childbearing in many societies. Operating more specifically than such worldwide trends, religious sanctions, marriage practices, levels of education, age and sex ratios, and political and psychological considerations are also important general fertility controls. Political intervention, for example, can be seen to have affected the fertility of major countries both positively and negatively in the present era: the government of Nazi Germany promoted high birth rates in the pre–Second World War period; the government of postwar Japan encouraged low birth rates.

Fertility Trends. The present crude birth rates of the world's countries range from lows of about 15 per 1,000 persons to highs of around 60 per 1,000, with 30 per 1,000 the accepted but arbitrary division between low and high (Fig. 13–11). Future birth rates are difficult to predict, since birth trends are basically products of changing socioeconomic factors. However, the United Nations estimates a "possible drop" in the crude birth rate of most underdeveloped societies, from the current 40 or more per 1,000 to about 35 per 1,000 population by the 1980s and even 28 per 1,000 by the 1990s.

But fertility decreases are not necessarily permanent. In the 1920s and '30s Canada and the United States showed declining birth rates, with the latter reaching its all-time low rate of about 17 per 1,000 in the late 1930s, the receding years of the Great Depression. The marked rise in United States births following the Second World War produced high national rates of 25 per 1,000 in both 1950 and 1957; but birth rates closer to

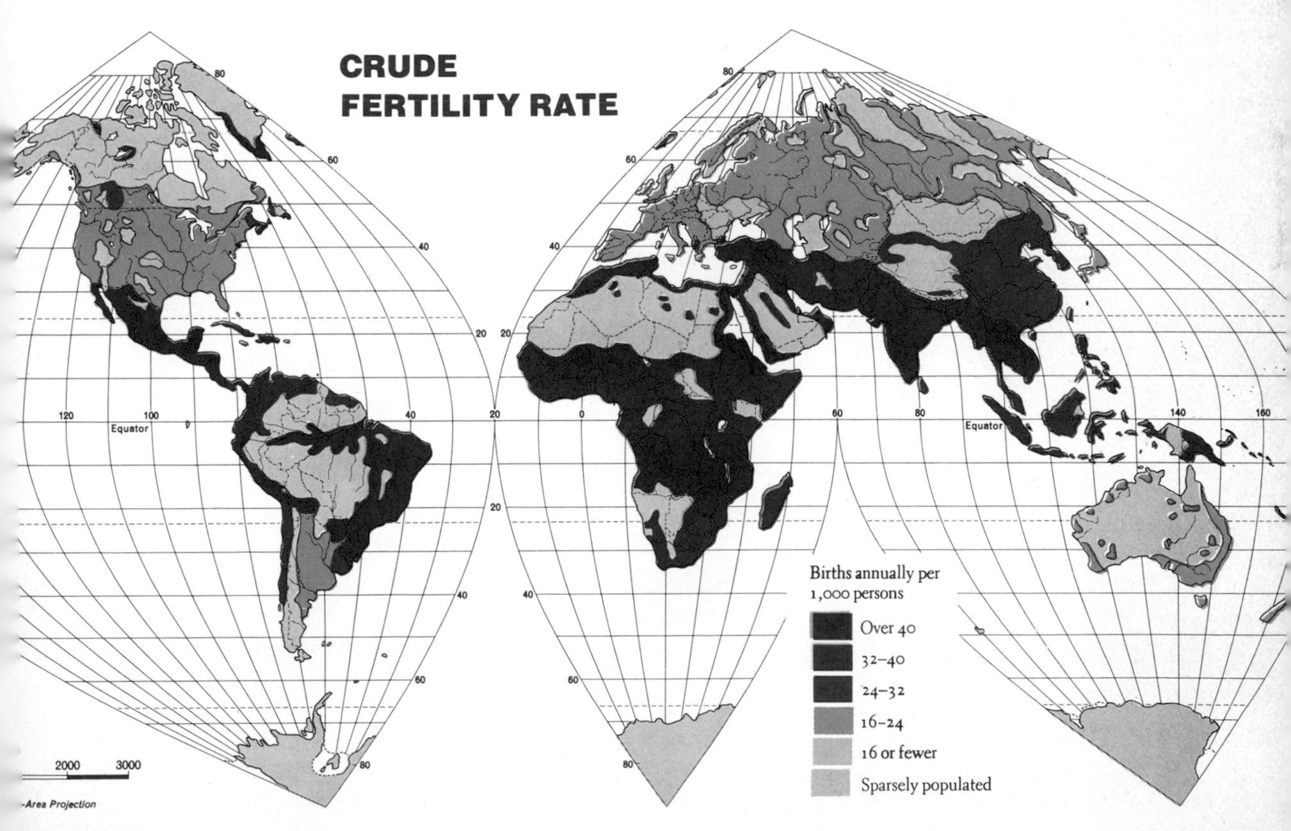

FIGURE 13–11
The world annual fertility rate is currently about 30 per 1,000 persons—that is, about twice the annual mortality rate.

those of the Depression years have characterized the nation in subsequent years. Somewhat similar postwar ebbs have been typical of Canada, many western European countries, and Australia as well.

The Demographic Transition

The succession of changes in mortality and fertility rates that theoretically accompanies the economic modernization of a given society is referred to as its *demographic transition*. Endorsed by many population geographers and cultural geographers, the contention that such changes will occur is based upon the well-documented history of Western countries during the last two centuries.

In evolving from agrarian societies into strong urban-industrial ones, these and most other modern societies have concurrently moved (usually in four stages) from high mortality and fertility to relatively low mortality and fertility (Fig. 13–12). And in this transition, nearly all have experienced intermediate stages in which death rates markedly below birth rates resulted in relatively large increases in total population and, of course, changes in age distribution.

The time consumed by such demographic transitions varies, involving one-half century to several centuries. For western European societies, as an example, the first stage occurred before 1750, years in which a combination of high mortality and high fertility rates resulted in low-growth populations. The second stage began with the Industrial Revolution, when, as the new societal economies expanded and basic health services became more readily avail-

FIGURE 13–12

A *demographic transition* model, involving major sequential changes. A very large *natural increase* (growth) occurs in a population when its mortality (death) rate declines substantially while its fertility (birth) rate does not, as in stage 2; and natural increase is smaller when death and birth rates approach a common level, as in stages 1, 3, and 4.

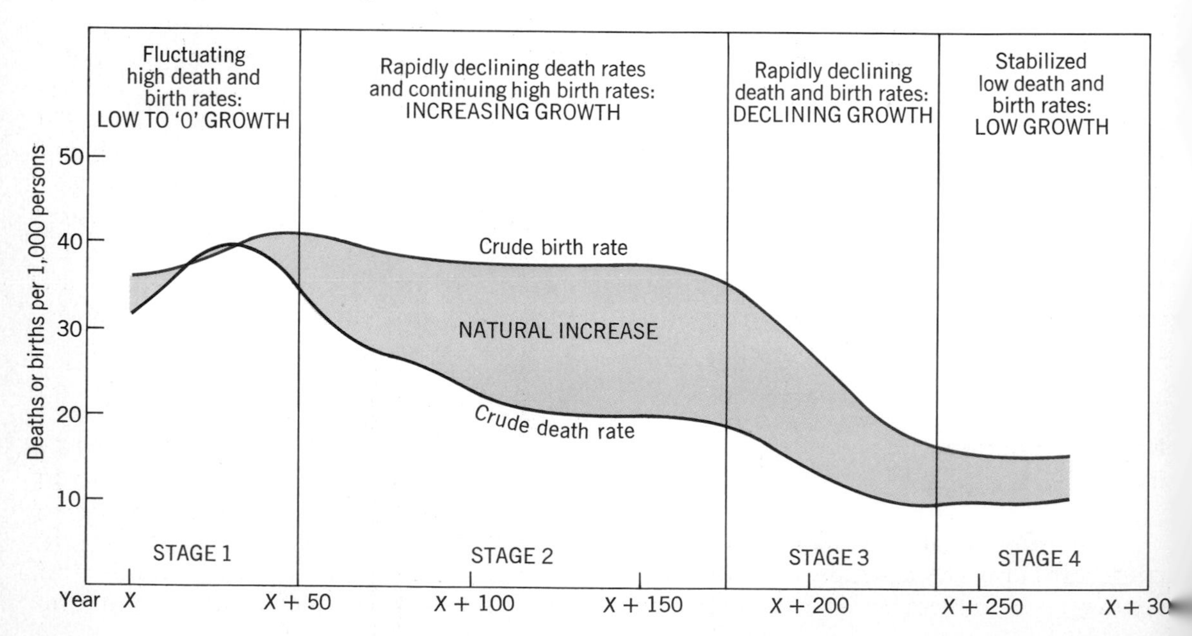

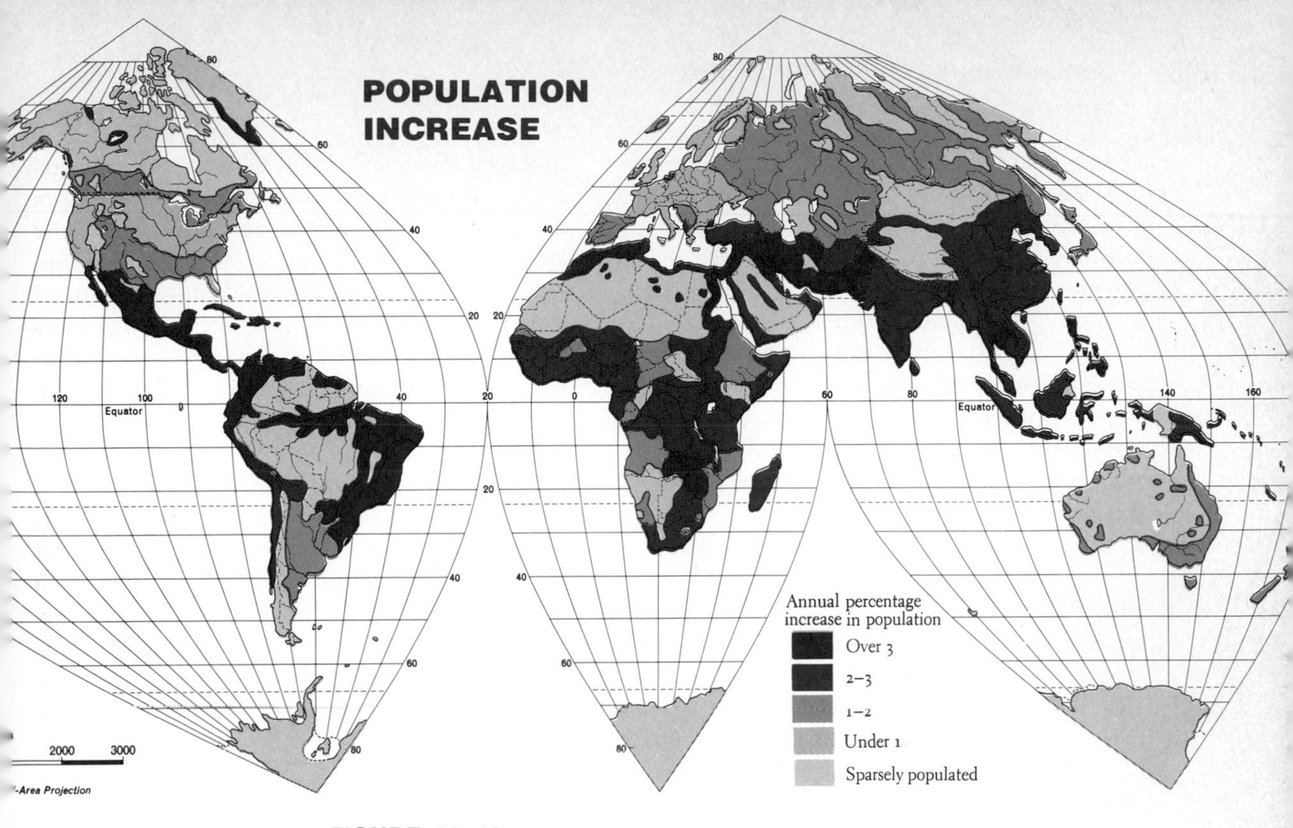

FIGURE 13–13

Recent rapid increases in human populations lie at the base of many of modern society's major problems, particularly food and energy shortages and urban sprawl.

able, mortality rates declined and fertility rates maintained their high levels—resulting in major growths in population. The last quarter of the nineteenth century marked the beginning of stage three. Its declines in total growth rates reflect the first acceptances of modern birth control (now known as *family planning*) and a new view of large families as economic handicaps. The demographic transition of these European societies reached stage four toward the mid-twentieth century, when both mortality and fertility rates stabilized at new low-growth levels.

Natural Population Increase

The numerical difference between the crude birth rate of a society and its crude death rate constitutes its rate of *natural population increase;* this figure will differ from its total rate of increase if there have been net gains or losses through migration. With over 120 million live births and over 50 million deaths annually, the world currently experiences an annual natural increase of about 70 million persons. This represents a growth rate of about 20 per 1,000 persons, or—as demographers usually express it—2 percent, annually. If the world rate of natural increase remains constant at 2 percent, then the human population of the earth will double in about 35 years.

The countries with the highest average annual rates of natural increase—3.0 percent and above—are located in Latin America (Fig. 13–13). Most African countries, with an estimated average of 2.3, and Asian countries, with a 2.2 rate of natural increase, are well below the Latin American levels and only slightly above the world average. The lowest average annual rates of natural increase are found in the heavily industrialized countries of Europe. A half dozen European countries consistently maintain rates of 0.5 percent or lower, while only three such countries have established rates of over 1.0 percent—giving the continent a very low natural growth rate of 0.7 percent. The two largest heavily industrialized societies, the United States and the Soviet Union, and urbanized Australia maintain average annual rates of about 1.1 percent.

Migration

Technological advances in transport and communication linkages in recent times have greatly expanded human capabilities for international migration. The first trips by humans across the Bering land bridge and down the western shores of the Americas undoubtedly represented generations of family and tribal wanderings, with only a few of the strongest members surviving each leg of the journey. In contrast, contemporary human migrations from northern Asia to America, accomplished by ship or by airplane, are matters of but a few days. Today, in technical terms at least, the population of any one continent could be relocated to any other in less time than it once took to travel between New York and California.

But such huge shifts do not occur. Migration has seldom assumed an importance in world population distribution even remotely comparable to that of natural increase; indeed, with more and more countries establishing legal immigration barriers, the relative importance of migration in international demography is probably on the wane. However, the present population of the world is so large that the involvement of even a very small percentage in international migration can constitute a sizable population change. Further, if a society's numerical gains or losses through migration are concentrated within a particular population segment (income level, age range, sex, or professional group)—as they seldom are through natural processes—the socioeconomic effects may be discernible for generations and may even alter the course of the society permanently.

European Colonial Emigrations. Among the most extensive international migrations of modern times is the migration of Europeans to overseas colonial territories during the early sixteenth to mid-twentieth centuries. Over this long period probably 70 million Europeans took up residence abroad, and at least 50 million of them resettled permanently—about 35 million making their new homes in the United States.

This great international migration did not occur as a continuously large and steady stream, however. Until the early nineteenth century, the numbers involved were relatively small, with less than two million immigrants to the Americas throughout the previous three centuries. The more massive migrations from Europe began about 1830 and, by the beginning of the First World War, included more than a million people per year. At first the principal countries of origin were those in northwestern Europe; but by the early twentieth century, most of the new immigrants came from Europe's Mediterranean and Slavic countries.

United States Immigrations. The long period of heavy immigration to the United States, the late nineteenth and early twentieth centuries, reflected the extensive European emigrations (Fig. 13–14). Before 1890 the major countries of origin were Great Britain and Germany, with smaller numbers of immigrants coming from the Scandinavian countries, Italy, Russia, and the Austro-

Hungarian Empire; by the end of the nineteenth century, 70 percent of the immigrants to the United States were natives of Mediterranean and Slavic Europe (Fig. 13–15).

Many less recent Americans became alarmed at the rising number and new ethnic make-up of the European immigrants, and they clamored for some sort of restrictive measure—presumably, something similar to the 1882 act that had curbed Chinese immigration to the west coast states. Thus in 1921 and 1924, the United States Congress enacted immigration laws designed "to maintain the cultural and racial homogeneity of the United States by the admission of immigrants in proportions corresponding to the composition in the present population." In addition, the worldwide Depression of the 1930s reduced all immigration into the country, and, war continuing the reduction, the average annual levels sunk to little more than a quarter million by the middle of the next decade.

For much of this century, therefore, immigration to the United States has not reflected European "population pressures," as it once did. The extension of a selective immigration policy in the 1920s established a set of annual quotas that favored Great Britain, Ireland, and Germany—who never used more than half their allotments. In 1966 the United States Congress abolished the quota system, but it also reduced the maximum number of immigrants and attempted to bar those who would take jobs sought by Americans; the growing prosperity in western Europe has also affected immigration. Thus, in the years following the quota system's abolition, nationality ratios among new immigrants changed substantially; by the late 1960s, over 40 percent of all immigrants to the United States were from Latin America, mainly Mexico and the Central American countries.

Some Post–World War II Migrations. Migration in response to political turmoil did not come to an end with the signing of peace

FIGURE 13–14

The origins of most immigrants to the United States between 1831 and 1910. During this period, over 90 percent of the new settlers in the United States were from northwestern and southern Europe.

Origin	*U.S. Immigrants*
Europe	25,500,000
Canada	1,200,000
Latin America	300,000
Asia and the Pacific	700,000
Total	27,700,000

FIGURE 13–15

European emigrants traveling steerage on the *SS Pennland* in the late 1800s as it headed for the United States. (Byron Collection, Museum of the City of New York)

FIGURE 13–16

Migration in the Middle East. Top, Jewish immigrants arriving in Israel soon after the establishment of the new state. (Consulate General of Israel) Bottom, children of a Palestinian refugee camp bringing home their family's rations. The camp shown is near Amman, Jordan; many similar encampments exist elsewhere in Jordan and in the adjoining nations. (Katrina Thomas from Photo Researchers)

treaties ending World War II. Indeed, as independence came to European colonies in Asia and Africa in the postwar years, one of the results was often the uprooting and forced migration of millions of people from these areas.

The partition of British India, in 1947, into Moslem Pakistan and Hindu India, set in motion a vast, two-way migration. The total number of people uprooted has been estimated at about 17 million, representing, in almost equal numbers, Moslems who fled

India for Pakistan and Hindus and Sikhs who fled Pakistan for India. Another major postwar migration began in 1948, with the establishment of the State of Israel. In subsequent years over 600,000 Jews migrated to the new state from Europe, North Africa, the United States, and other parts of the Middle East. Unfortunately, nearly an equal number of Arabs were uprooted as a result; and many who fled their ancestral lands found that their hardships continued, and sometimes multiplied, elsewhere (Fig. 13–16). Almost simultaneously, Communist victories in China sent millions of Chinese into permanent exile in Taiwan, Hong Kong, and other Southeast Asian centers; these precursors of the 1949 takeover also dislodged the country's large European population, many of whom had come to China as refugees during the First World War.

Unlike European refugees of the post–World War II years, Asian migrants of the same era remain largely unsettled, with little being done to alleviate their misery. The human casualties of Asia's "forced" migrations are still visible in the streets and alleys of such great cities as Calcutta and Bombay, Bangkok and Saigon, and Hong Kong and Taipei.

The Rush to the Cities. One of the most striking population changes of modern times is the rural exodus in search of urban opportunities. If demographic predictions are correct, urban populations will account for over 50 percent of the earth's total population by the year 2000, which is above the present ratio. A universal trend, this rush to the cities is particularly strong in newly developing societies, where, unfortunately, there are few resources to deal with the resultant array of social, economic, and political problems. In Colombia, for example, the rural sector's rate of growth is about 1.0 percent, while the urban sector's is 5.0 percent. And such disparities exist in many other Latin American and African societies. And Israel, in southwest Asia, has one of the highest urban population ratios in the world: with nearly 85 percent of all Israelis living in towns and cities, and with an even higher percentage within the Jewish population. Tel Aviv alone accounts for about 30 percent of Israel's total population.

Twentieth-Century Spatial Linkages

The present world *circulation,* or transportation–communication, network has evolved bit by bit throughout the long course of human history. Its configuration reflects humankind's successes and failures in overcoming the distances and barriers of nature, and thus the progress of the species from primitive isolationism to complex group associations and interrelationships.

The spatial linkages of early human groups were few and irregular, perhaps radiating from each group's crude shelters to its hunting grounds or water sources, and limited by the distances the average adult could travel afoot in the hours of daylight. With the domestication of large animals, human spatial ranges increased—pathways becoming established trails, and hunting and fishing sites developing into temporary rest stations. Local road networks grew and joined those of neighboring peoples, and water bodies provided paths to more distant sites. Always the intent and result were the extension of each group's "known world."

Today, the socioeconomic level of a society determines the nature of its circulation network and the intensity of that network's use. To the subsistence sawah farmer of southern Asia, superhighways and airports are of little significance; southern Asia's circulation network consists largely of paths and cart tracks leading from village to field to town. They are not heavily or speedily traveled by people or goods. By contrast, in such industrial societies as western Europe and the United States, superhighways and a web of air, water, and surface feeders are essential. Along these routes flow the products upon which the livelihoods of an industrial people are based.

The spatial arrangement of any circulation system is subject to strong influences from both physical and cultural elements of the environment. Since the basic function of any route is to link together two points in space, and since the shortest distance between two points is a straight line, it follows that the ideal route is a straight line. However, this ideal rarely occurs, for many obstacles force deviations from its course. Some of these obstacles are physical, such as terrain slope, elevation, and surface condition; others are cultural, such as political boundaries, zoning limitations, and property lines.

FIGURE 13–17

The principal roads of France, a network centered on Paris.

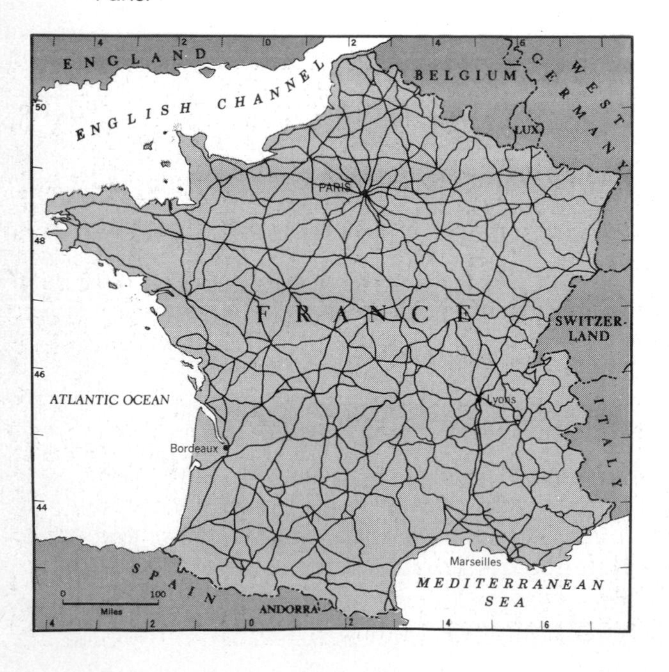

Road Systems

In early times, roads were developed and operated primarily on a local basis. Perhaps the earliest major departure from this tendency was the extensive road system created by the ancient Romans. Under the Roman emperors, surfaced roads were built throughout much of southern and western Europe, all leading eventually to Rome itself. Main trunk lines linked Rome with all parts of its empire, and especially with strategic points on the frontier; the primary purpose of this system was to facilitate troop movements. With the decline of the Roman Empire, this road system contributed to its defeat, for it gave invading armies easy access to lands formerly under strong Roman control and, later, to Rome itself. In modern times, both France and Germany have built national road systems that, like the Roman, center on the capital city and fan out to strategic points on the frontier (Fig. 13–17). And in both countries, in times of war, the road system serving invading armies all too well.

During the Middle Ages the impetus for roadbuilding was lost, and existing road systems declined. It was not until the end of the fifteenth century, and then only in France, that organized road systems again became national interests. Further, it was not until the age of the automobile that such systems showed major improvements over the Roman models. As it answered the need for

FIGURE 13–18

Contrasting road systems. Left, a primitive road in central Africa. (United Nations) Right, heavy traffic on the Harbor Freeway, one of the many freeways that help to make sprawling Los Angeles a functioning unit. (Southern California APCD)

high-speed circulation between centers of industry and commerce, the automobile created a demand for new and better road systems; these were provided throughout much of the United States and Europe by the mid–twentieth century, and they have since been a feature of national development plans in nearly all parts of the world.

Today, among the world's small to medium-sized countries, the highest road density occurs in Japan, which is followed closely in this respect by the United Kingdom, Denmark, France, Germany, and Belgium—the latter all highly industrialized countries of western Europe, the most densely road-crossed multinational region. Among large countries, the United States has the highest road density, but some sizable areas of northeast China and peninsular India have fairly dense road systems as well. India's rather extensive road system, acquired largely under British colonial rule, has benefited from good road maintenance and bridge-building programs since the country obtained independence in 1947. Today, India has over 200,000 miles of good hard-surfaced roads.

The road systems of most other countries and regions suffer by comparison with those mentioned above. South America, Africa, eastern Europe, and Southeast Asia are not characterized by elaborate road systems. Thus for the most part, the density and quality of a society's road system seems to vary directly with its level of economic development (Figs. 13–18 and 13–19).

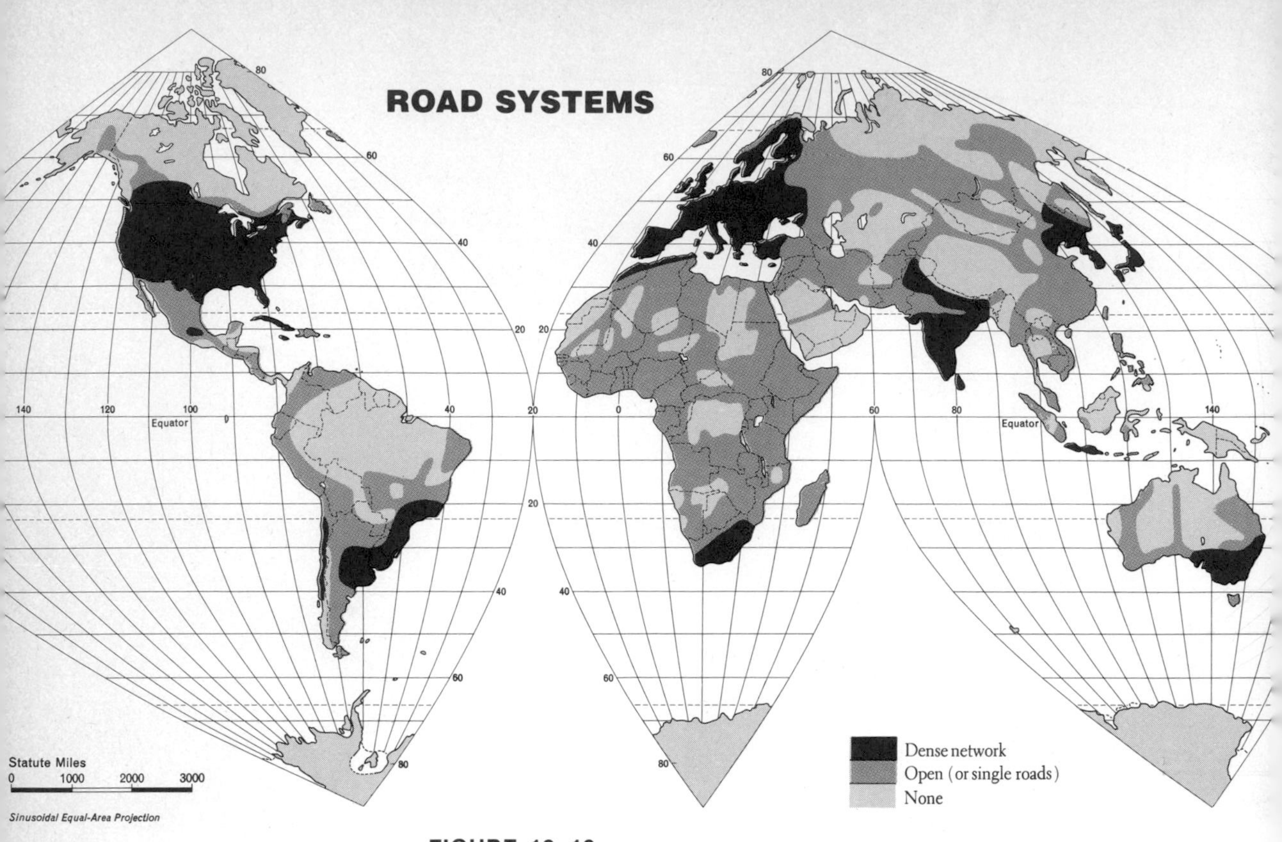

FIGURE 13–19

Dense road networks, which occur over relatively little of the earth's landmass, are found, for the most part, in heavily industrialized nations and in some of their former colonies.

Rail Systems

While most of the world's societies have road systems of some kind, railways are almost exclusively features of industrial societies and some of their client states (Fig. 13–20). The dense rail systems of the eastern United States and northwestern Europe mirror the heavy concentrations of industry in these regions (Fig. 13–21). Most other societies of the world are but lightly served by railroads. Tremendous expanses in Asia are little touched by rail traffic. Some loosely connected railroads extend across Soviet Europe, but the Trans-Siberian is the lone railway across the vastness of Soviet Siberia to the Pacific coast. And elsewhere in Asia, as in Africa, Latin America, and Australia, vast stretches of moderate to densely populated lands have little rail service.

While rail travel has a rather short history—the world's first railway was put into operation in England in 1825—railroads have been so vital to urban society that it is difficult to understand how major cities once existed without them. In the development of the United States, they have influenced the location of major population centers, the outcome of a civil war, and reams of folklore and fantasy. Nevertheless today, at least in both the United States and Europe, other modes of transportation have taken over much of the local and transcontinental traffic of the railways. Automobiles, trucks, buses, and airplanes have largely displaced railroads in the moving of passengers and freight.

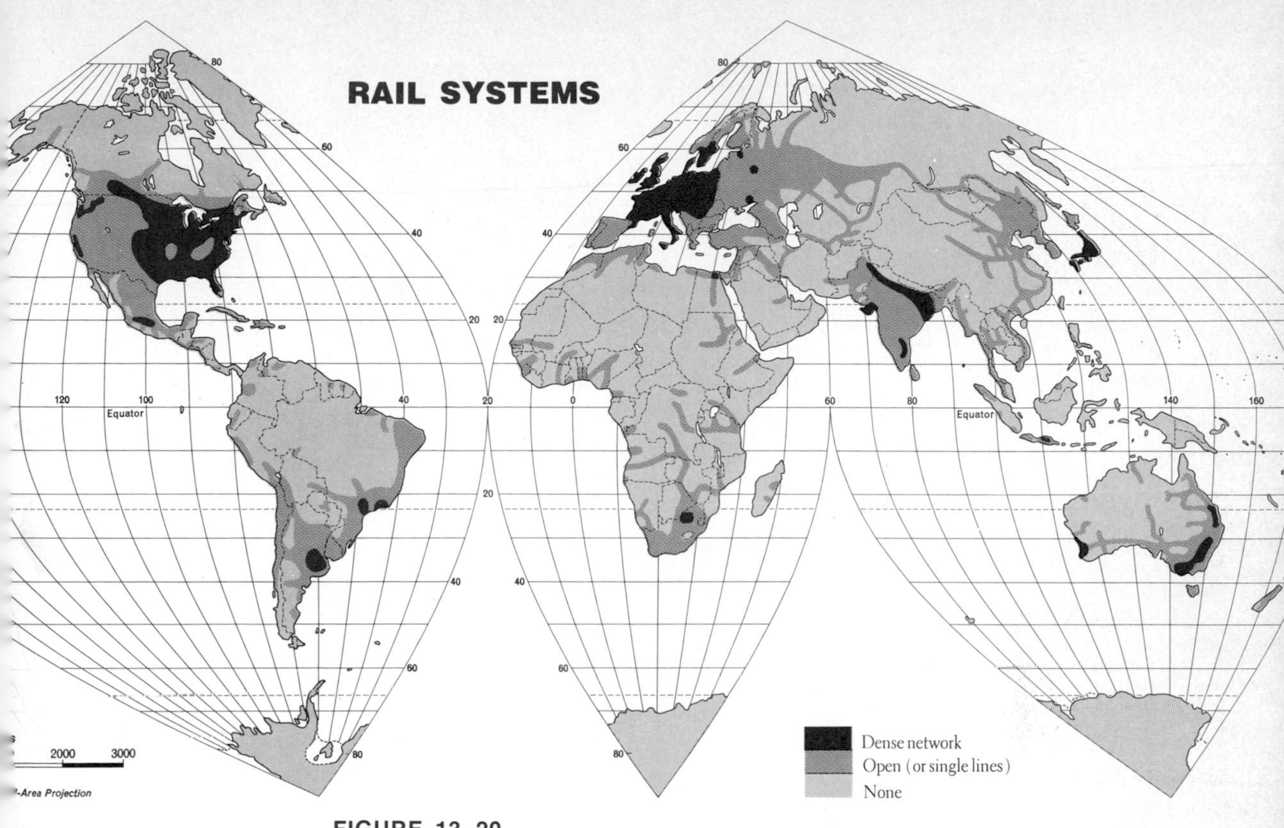

FIGURE 13–20

Dense rail networks are features principally of Western Europe and the eastern half of the United States, the world's most highly industrialized areas.

FIGURE 13–21

An automated classification yard of the Louisville and Nashville Railroad. (Association of American Railroads)

Other Circulation Systems

Existing along with the world's road and rail systems are many other means of societal circulation, most obviously water (Fig. 13–22) and air vehicular routes, but also power lines and pipelines and such communication tools as telephonic and telegraphic lines, radio and television networks, and radar and satellite-tracking stations. Like dense rail and road networks, these communication systems are found largely in urban-industrial societies; yet they too are stretching out to touch practically every individual in every society of the world (Fig. 13–23).

Transport and Western Culture

That the many fragments of world populations and societies have come to be more closely interlinked as a result of the development and spread of Western culture is a fact of record. During an early period of history, the people of a small south European tribe

FIGURE 13–22

Barge traffic near Bonn on the Rhine River, one of the world's busiest water routes. (Lufthansa Archiv)

FIGURE 13–23

The use of closed-circuit educational television in Niger, where experiments in television teaching have been made since 1964. (United Nations)

gradually expanded throughout the Italian peninsula, as well as eastward and southward into the Middle East and Africa and northwestward across Europe beyond the English Channel. They took all this vast territory under their political control and, welding it together with a farflung system of roads and waterways, formed the Roman Empire.

Eventually this vast empire declined as a political society, but its descendant societies later brought even more European peoples into the dynamic and expanding sphere of Western culture. These societies underwent radical economic growth and expansion and, in time, became highly industrialized. Their ever-increasing needs for goods and markets required their contact with more and more of the world's people. The linkages between industrial cores of European societies and their client regions were at first roadways and waterways and later the railways, airways, and early long-distance communication networks.

Imperfect though this intricate circulation system still is today, its transportation and communication networks have profoundly affected humankind's concepts of the physical environment and the human role therein. For most individuals and societies today, the world is no longer a simple collection of small scattered societies and activities, separated one from another by xenophobic barriers. Rather, it is a complex mosaic of societies, interrelated and interfunctioning by means of a vast matrix of transportation and communication linkage systems.

REVIEW AND DISCUSSION

1. What caused the pressures for more abundant living space that eventually brought about the diffusion of human beings beyond their original centers of origin?
2. Cite some specific "empty" areas of the world and explain why they remain "empty." Distinguish between the following types of density: nutritional, agricultural, rural, and urban.
3. The three principal components of population change in modern societies—mortality, fertility, and migration—tend to be variably influenced by the same cultural factors. Substantiate this statement with facts and examples.
4. Discuss the extent and consequences of the death-rate decline in the Western world during the last 150 years. In what ways is technology related to this decline?
5. Discuss the effects of trends in crude birth rates and death rates during the twentieth century.
6. What is the meaning of the term *demographic transition?* Discuss its application to several contrasting societies.
7. What is the meaning of *natural population increase?* What are the uses and limitations of this measurement?
8. How does migration compare with natural population increases as an influence on world population?
9. Discuss some of the socioeconomic consequences of the major post–World War II migrations. How did some of these migrations set the stage for many of today's social and political problems?
10. Contrast the national and international rail system linkages in developed societies with those of developing societies. What is the role of rail transportation in modern socioeconomic development? What major changes will probably take place in the rail system of the United States by the end of the present century?

SELECTED REFERENCES

Barbour, K., and R. Prothero, eds. *Essays on African Population.* New York: Praeger, 1962.

Becht, J. *A Geography of Transportation and Business Logistics.* Dubuque, Iowa: Brown, 1970.

Borah, W. "America as a Model: The Demographic Impact of European Expansion upon the Non-European World." *Actas y Memorias, XXXV Congreso Internacional de Americanistas.* Vol. 3 (1964), 379–87.

Carr-Saunders, A. M. *World Population: Past Growth and Present Trends.* Oxford: Clarendon, 1966.
Cipolla, C. M. *The Economic History of World Population.* Harmondsworth, England: Penguin, 1962.
Clarke, J. *Population Geography.* Oxford: Pergamon, 1965.
Cook, R. "Israel: Land of Promise and Perplexities." *Population Bulletin,* Vol. 21 (November 1965), 101–34.
Cootner, P. "The Role of the Railroad in United States Economic Growth." *Journal of Economic History,* Vol. 23 (December 1963), 477–521.
Demko, G., H. Rose, and G. Schnell. *Population Geography: A Reader.* New York: McGraw-Hill, 1970.
Drake, M., ed. *Population in Industrialization.* London: Methuen, 1969.
Durand, J. "The Modern Expansion of World Population." *Population Problems, Proceedings of the American Philosophical Society,* Vol. 3 (June 1967), 133–93.
Ehrlich, P. R., and J. P. Holdren. "Impact of Population Growth." *Science,* Vol. 171 (March 1971), 1212–17.
Eliot Hurst, M. E. *Transportation Geography: Comments and Readings,* Part 1. New York: McGraw-Hill, 1973.
Farris, M., and P. McElhiney, eds. *Modern Transportation: Selected Readings.* Boston: Houghton Mifflin, 1967.
Kindleberger, C. "Mass Migration, Then and Now." *Foreign Affairs,* Vol. 43 (July 1965), 647–58.
Kosinski, L. *The Population of Europe: A Geographic Perspective.* London: Geographical Publications, Longman, 1970.
Learmonth, A. "Medical Geography in India and Pakistan." *Geographical Journal,* Vol. 127 (March 1961), 10–26.
Murphey, R. "The Population of China: An Historical and Contemporary Analysis." *Geography of Population,* 1970 Yearbook of the National Council for Geographical Education, Palo Alto, Calif., 117–34.
O'Dell, A., and P. Richards. *Railways and Geography,* 2nd ed. London: Hutchinson University Library, 1971.
O'Flaherty, C. *Highways.* London: Arnold, 1967.
Peterson, W. *Population,* 2nd ed. New York: Macmillan, 1969.
Schechtman, J. B. *Postwar Population Transfers in Europe, 1945–1955.* Philadelphia: University of Pennsylvania Press, 1962.
Sealy, K. *The Geography of Air Transport,* 3rd ed. London: Hutchinson University Library, 1966.
Stamp, L. D. *The Geography of Life and Death.* London: Collins, 1964.
Taaffe, E., et al. "Transport Expansion in Underdeveloped Countries: A Comparative Analysis." *Geographical Review,* Vol. 53 (October 1963), 503–29.
Taaffe, E., and H. Gauthier. *Geography of Transportation.* Englewood Cliffs, N.J.: Prentice-Hall, 1972.
Taft, D., and R. Robbins. *Internation Migrations: The Immigrant in the Modern World.* New York: Ronald Press, 1955.
Thomlinson, R. *Population Dynamics.* New York: Random House, 1965.
Trewartha, G. *A Geography of Population: World Patterns.* New York: Wiley, 1969.
Wolpert, J. "Migration as an Adjustment to Environmental Stress." *The Journal of Social Issues,* Vol. 22 (October 1966), 92–102.
Wrigley, E. *Population and History.* New York: McGraw-Hill, 1969.
Zelinsky, W. "Toward a Geography of the Aged." *Geographical Review,* Vol. 56 (July 1966), 445–47.
Zelinsky, W., L. Kosinski, and R. Prothero, eds. *Geography and a Crowding World.* New York: Oxford University Press, 1970.

The effluent takeoff of a jet airplane. (Harbrace Photo)

14

Emerging Environmental Realities

As burgeoning societies exert unprecedented pressure on their environments and as increasing numbers of people seek more living space, greater food supplies, and ever more uses of energy resources, great injury has frequently been done to the earth's air, waters, and land. Today, some people still boast of "man's conquest of nature," seemingly unmindful of the feat's high cost. But among a growing number of people there is a deepening awareness that, if the conquest continues recklessly, the cost will be more than the species or the environment can long stand. Most societies today are thus attempting to achieve a better understanding of nature and carrying out or planning a more careful utilization of their environments.

Movements to check environmental damage, in existence for the past two hundred years, met with some success in Europe and the United States prior to World War I and experienced a brief revival after the war. But most early environmental movements were overshadowed by other concerns during the years of economic depression, World War II, and the cold war that followed. During those demanding decades, however, world environmental problems grew more serious, more complex, and practically impossible for societies to continue to ignore.

Thus the Western world's present-day environmental movements, for the most part creations of the late 1960s, have heritages that date back to the early industrial revolution. Those who today urge Americans and Europeans to "clean up the environment" while there is still time echo the voices of such concerned figures of the past as American naturalists John James Audubon (1785–1851) and John Burroughs (1837–1921), Theodore Roosevelt (1858–1919), German biologist Ernst Heinrich Haeckel (1834–1919), American forester Gifford Pinchot (1863–1946), and Sir Arthur George Tansley (1871–1955), a British ecologist (Fig. 14–1).

In too many instances today, environmental problems have progressed far beyond being local nuisances. Many have already become progressive maladies that threaten the socioeconomic strengths of whole societies. As countries with an urban-industrial base and the attendant environmental pollu-

FIGURE 14–1

A cartoonist's depiction of Theodore Roosevelt the conservationist. (Bettmann Archive)

tion and waste problems become increasingly numerous, many environments are being turned into cores of pollution. Such environmental decline, if unabated, may soon create international "disaster areas," seriously threatening the health and welfare of millions of people.

Today, *environmental pollution* has taken on a broad meaning; for in general the term now incorporates all the ways in which humans pollute, damage, and endanger the environments in which they live and work. Everyone on earth who eats, drinks, and works contributes a small share to today's worldwide environmental pollution. In the production-exchange-consumption complex in which most people are involved, contaminants of various types are dumped into the environment. The day-to-day production of food, clothing, shelter, vehicles, energy, roads, books, and the host of other necessities and luxuries involves the concurrent production of unusable and unwanted by-products that become pollutants and solid wastes. To this there is but one practical alternative. Some of today's abundant technologies must be redirected toward the development of pollution-free methods of producing the luxuries and necessities of life.

Air Pollution

Human Impacts

Although contaminated air has been a problem in most industrial centers since the early nineteenth century, only within the past two decades have people begun to view air pollution as perhaps the most serious threat to the environment. Most prevalent where population is concentrated and societal activities highly developed, air pollution has many harmful effects on human health and welfare, some temporary and others long-range. Many manufactured gases and minute airborne wastes are debilitating to human beings, often seriously damaging their respiratory systems. Airborne wastes, or *particulates,* more than most other pollutants, collect in the lungs and cause or contribute to such respiratory diseases as asthma and bronchitis. Research indicates that particulates may also be contributory factors in many cases of cancer, pneumonia, and emphysema.

Although usually a debilitator rather than an immediate killer, air pollution has sometimes become so intense in specific locations that only a few hours or days demonstrated its lethal potential. In 1930, for example, in the Meuse Valley of Belgium, many people were stricken with respiratory illnesses during a few-hour period of extreme air pollution, and 63 pollution-related deaths were recorded. Some years later, in 1948, a similar period of intense air pollution struck the town of Donora, in the Monongahela Valley of western Pennsylvania, resulting in 17 deaths (Fig. 14–2). And in 1952, during two weeks of severe air pollution in the area, London experienced approximately 4,000 "excess deaths."

During the nearly half century since those tragedies, air pollution problems have received increasing attention among scientists, the general public, and politicians of many countries. Much has been learned and put into practice in the area of pollution abatement during the past two decades, but there is still much work to be done before we can control air pollution. For example, explicit evaluations of the origins and effects of the pollutants in smog are still not yet available, although this atmospheric condition is an old and ever-present health hazard in many areas of the world.

Local meteorological conditions of any sort are almost entirely beyond the control of human beings and, despite radar and satellites, remain largely unpredictable even from day to day. But just as naturalists and environmental planners consider watersheds, or river basins, operational units, it may well be possible to study, plan, and act upon the inherent nature of large atmospheric regions, or "air sheds."

FIGURE 14–2

A smog-shrouded intersection in the central business district of Donora, Pennsylvania, on October 30, 1948. (Wide World Photos)

Thermal Inversion

Precipitation and wind tend to disperse air pollutants and remove them from their source areas. At times, however, weather conditions are such that air pollutants are concentrated over a region instead of being cleared away. One such major weather condition, one that occurs almost anywhere and in many areas repeatedly, is known as a *thermal inversion.* During a thermal inversion, a layer of warm air settles over a layer of cool air that lies nearer, and perhaps in contact with, the ground. The warm upper layer acts as a natural cap and stalls, or minimizes, air currents within the lower layer of air; in so doing it prevents the usual dispersal of air pollutants by means of rising air currents. When, by a continuous build-up, stagnated pollutants are concentrated at or near ground level, the condition is called *smog.* If the air pollution is severe, a thermal inversion that lasts a few days or even a few hours will have noticeable harmful effects upon the comfort of even the healthiest human beings (Fig. 14–3).

Los Angeles County, where *smog* has been a household word since World War II, is topographically and meteorologically ideal for thermal inversions, for it occupies a large basin that is open to the sea and encircled landward by a chain of hills and mountains. Warm air from the interior deserts can slip over the encircling mountains and settle over low, cool air from the Pacific. As the thermal inversion keeps the area's abundant automobile and industrial pollutants from escaping upward, light prevailing breezes from the ocean may exert just enough pressure to keep them from drifting out to sea. And so

FIGURE 14–3

The layering of air over a hypothetical metropolis. A. Under normal conditions. B. During a thermal inversion.

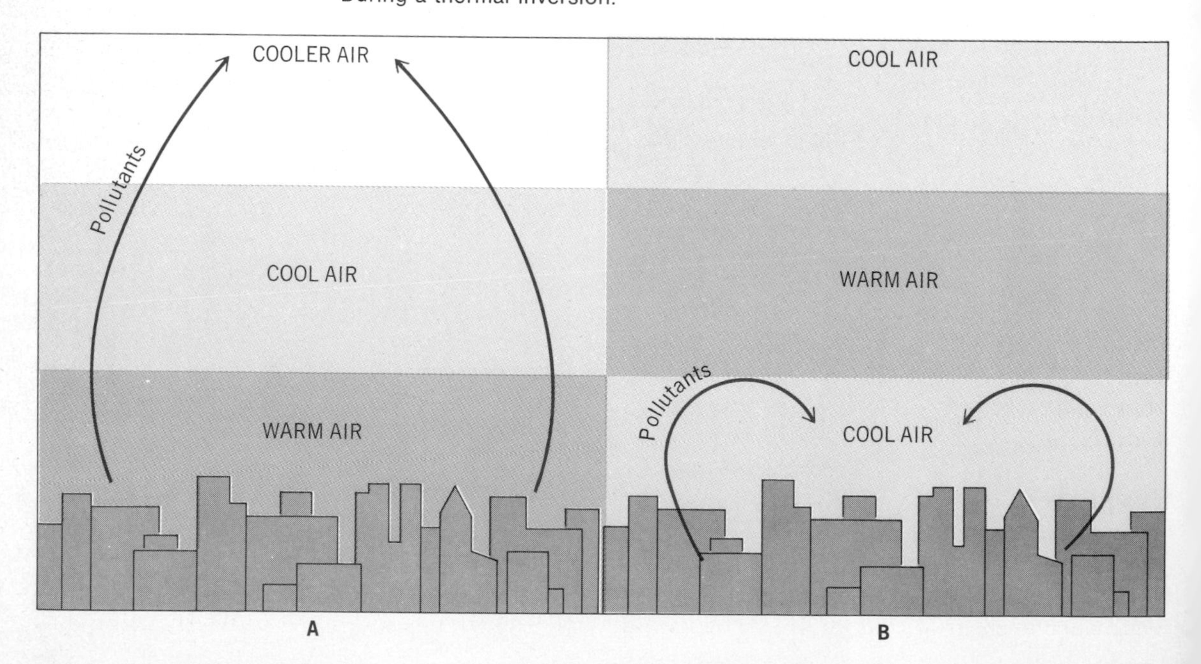

FIGURE 14–4

Downtown Los Angeles in 1956. Above, on a clear day with no inversion layer. Below, on a smoggy day with an inversion layer at about the 300-foot level. (Southern California Air Pollution Control District)

they remain trapped within the basin until a change of weather brings either sea breezes strong enough to carry them over the mountains, or land breezes strong enough to carry them out to sea (Fig. 14–4). During the stagnation, a series of chemical reactions brought on or accelerated by solar radiation, as provided by the area's abundant sunshine, changes some of the air pollutants into *photochemical smog*. Composed mainly of hydrocarbons and nitrogen oxides, pollutants derived mainly from automobiles, this type of smog often smarts the eyes of millions of Los Angeles County residents and damages many of the basin's numerous crops and its natural vegetation.

Major Air Pollutants

The air people breathe, every minute of every day, is their most continuously vital environmental resource. Nevertheless, they pollute that very same resource by using it as a waste disposal. Backed by tradition, many still assume that the air can remain beyond societal regulation, free for all to use and misuse without peril. Land and waterways have been at least partially regulated and protected since early in human history. Air, the most vital environmental resource, is the last to come under societal protection.

There have always been natural emissions, from volcanoes and the like, that temporarily overloaded, or polluted, the atmosphere, but the more harmful air pollutants have usually come from human activities. Since the first human control of fire, technological progress and social development have steadily increased the levels of harmful materials in the air.

From Transport Sources. When the components of air pollution in the United States are estimated by weight, those from transportation devices—automobiles, airplanes, trucks and buses, and trains—comprise the most dominant category, accounting for some 42 percent of the total. Such motorized vehicles spew into the atmosphere vast quantities of *carbon monoxide* and lesser quantities of *hydrocarbons, sulfur oxides, particulates, nitrogen oxides,* and *tetraethyl leads*—automobiles are the primary offenders (Fig. 14–5).

Of these air pollutants, carbon monoxide is probably the most toxic to human beings. It can cause sudden death when it reaches 1,000 parts per million (ppm) in the air, and even levels of carbon monoxide below 50 ppm are thought to be damaging to many people. Nevertheless, in many traffic-congested cities the level of carbon monoxide in the air must be as high as 100 ppm, 200 ppm, or even 400 ppm before a pollution alert is declared.

Only major modifications of the engines and the fuels used in automobiles will produce meaningful improvements in

FIGURE 14–5

Air-pollution emissions in the United States in 1968. A, Their types. B. Their sources. (Both in percentages by weight)

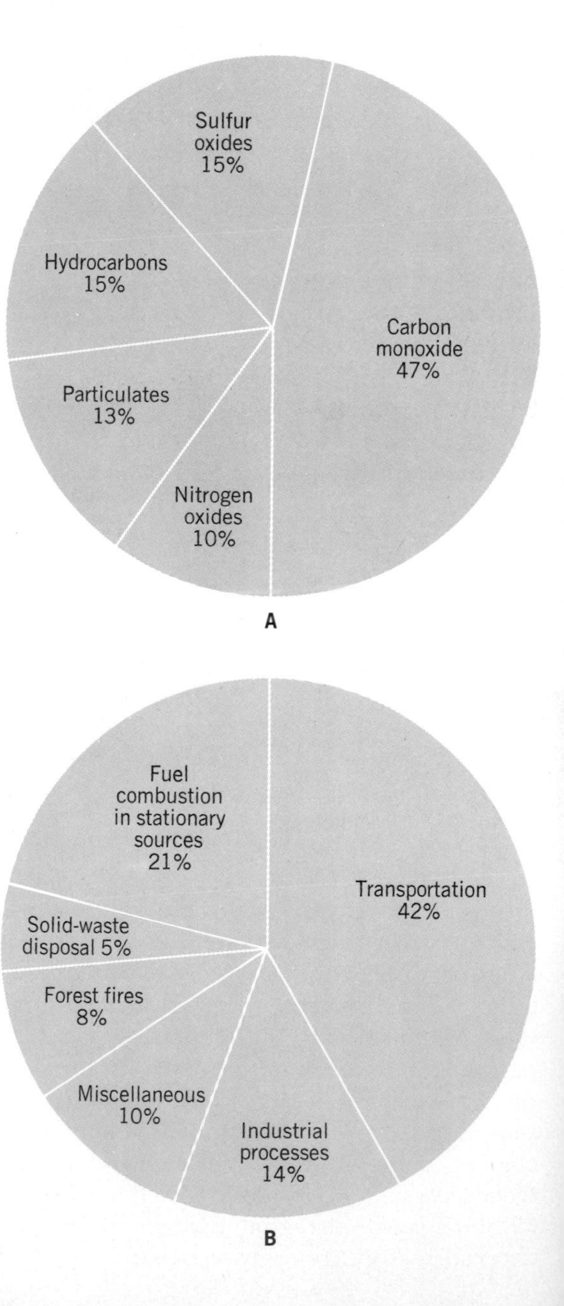

the quality of the atmosphere over most large metropolitan areas. *Catalytic exhaust converters* and similar emission control devices have been developed to help automobiles meet the present and impending government restrictions on harmful vehicular emissions. However, the transition to nearly pollution-free gasoline engines, requiring the cooperation of industry, government, and consumers, will probably proceed at a very slow pace.

The present alternatives to the gasoline-fueled internal-combustion engine are very promising in theory but much less so in operation. The electric car, for example, has a limited driving range between rechargings, and these rechargings take several hours. Moreover, the electrical energy required must be generated somewhere. A switch to the electric car may represent only a shifting of pollutants to someone else's backyard—say, from Los Angeles to Black Mesa, where local low-quality coals would fuel the large electric power plants necessary to charge the metropolis's cars.

With only minor changes most of today's gasoline engines could operate on natural gas. But the threatened condition of natural-gas supplies since the mid-1970s makes this alternative to the gasoline engine somewhat impractical. Among the other alternatives are the steam, gas-turbine, and diesel engines. The diesel already powers most of the world's trucks and buses, but it remains a less than ideal substitute for the gasoline engine in the family automobile because of its greater emissions of nitrogen oxides.

From Stationary Sources: Power Plants. Thermoelectric power plants (Fig. 14–6), domestic and institutional space-heating systems, community incinerators, and mining

FIGURE 14–6

Air pollution from power plants, air conditioners, and space-heating units—in a typical North American city. (Elliot Erwitt, © 1966 Magnum)

operations are commonly grouped together as *nonindustrial stationary sources* of air pollution. They all involve the "burning" of coal, oil, and gas to create steam to operate turbines—a contrast to the "explosion" of fuels in internal-combustion engines. As a consequence, the operation of most motor vehicles and the running of these stationary systems are almost the reverse of each other in terms of pollutants. Whereas vehicular emissions are high in carbon monoxide and hydrocarbons, the stationary systems emit very little of these toxins; but while motor vehicles emit low sulfur oxides, the stationary sources produce emissions in which the level of sulfur oxides is high. As Middle East oil policies stimulate greater use of coal-fueled steam in thermoelectric power plants, it is projected that the percentage of sulfur oxides in the emissions from these plants will increase (Fig. 14–7).

From Industrial Sources. The world's developed societies and many of the developing ones have been so involved in industrial planning and growth that they have ignored the social and environmental problems created by their economic progress. Frequently the industries that are most vital to their economic infrastructures also contribute most of the gases and particulates that pollute their environments. According to the United States Public Health Service, the principal industrial polluters are the manufacturers of iron and steel and other metals; refined petroleum; petrochemicals, including fertilizers, plastics, and pesticides; and organic chemicals, such as synthetic rubber, pulp, and paper (Fig. 14–8).

Industry in the United States accounts for about 20 percent of the country's atmospheric sulfur oxides and one-third of its airborne particulates, but only about 10 percent

FIGURE 14–7

Estimated electrical generation in the United States through major energy sources, from 1965 to 2000.

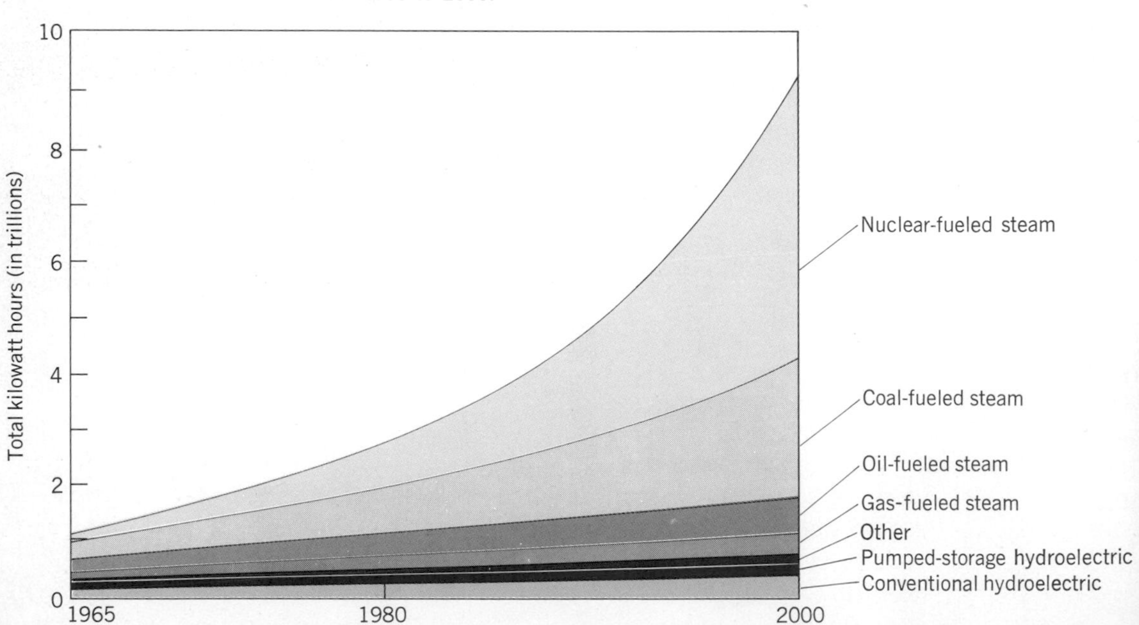

FIGURE 14–8

Industrial air pollution. Above, a chemical plant in Mexico City. (United Nations) Below, lumber mills east of Salt Lake City, Utah. (Grant Heilman)

of the environment's carbon monoxide and hydrocarbons. Thus, while it often receives most of the blame and publicity for the nation's air pollution, industry actually contributes only about 14 percent of the total air pollutants; as suggested earlier, over 42 percent comes from transportation devices and 21 percent from such stationary sources as thermoelectric power plants.

The Atmosphere's Self-Purification

There are natural elements or processes that tend to reduce the levels of most atmospheric pollutants. Precipitation washes large quantities of gaseous pollutants and particulates from the atmosphere into the soils and water of the earth. And the oxygen of the air itself combines with many pollutants, creating compounds that are more readily removed from the air. *Sulfuric acid* droplets, which probably contributed heavily to the London smog tragedy of 1952, and the photochemical smog of many west coast areas are but stages in the atmosphere's own natural process of self-purification. But nature's "scrubbing" processes can no longer keep pace with humankind's environmental polluting.

United States Air Pollution Controls

The United States government's first legislative involvement in meaningful air pollution control did not occur until 1955, when Congress appropriated $5 million for air pollution research and assistance to state and local air-pollution-control programs. And it was not until the passage of the Air Quality Act, in 1967, that further substantial federal action was taken against air pollution. With the passage of this act, the United States government assumed a large part of the responsibility for the improvement and maintenance of the nation's air (Fig. 14–9). Although the act encourages action at state and local levels, federal intervention is assured if lower-level agencies fail to function effectively.

As a step toward fulfilling its responsibility for the atmosphere, the federal government has demanded that all new automobiles produced domestically comply with specific limitations on emissions of carbon monoxide, hydrocarbons, and nitrogen oxides. While these limitations represent a 90 percent improvement over the norm in the recent past, additional restraints on automobiles and their use will probably be instituted by local government in some cities. Those that have exceptionally congested traffic conditions—Los Angeles, New York, Chicago, Philadelphia, and others—may well find it necessary to initiate minimum passenger requirements and limit access to particular streets and highways.

Some Global Aspects of Air Pollution

The relationships between air pollution and urban growth are recognizable today in most countries of the world—in both the highly developed industrial societies and in the newly industrializing ones. Urban populations and industrial activities are increasing the atmospheric pollution in so many metropolitan centers and to such an extent that nature's self-purification processes cannot cope with the problem. And since they are not "scrubbed" or converted by the overtaxed natural processes, urban air pollutants are scattered and deposited over larger areas by the winds and precipitation, eventually

1. Pittsburgh	11. Washington	21. Salt Lake City	31. Portland, Oreg.
2. Detroit	12. Phoenix	22. Toledo	32. Seattle
3. Chicago	13. Atlanta	23. Denver	33. Fresno
4. Springfield, Mass.	14. Philadelphia	24. San Diego	34. San Francisco
5. Los Angeles	15. Milwaukee	25. Cincinnati	35. Kansas City, Mo.
6. Baltimore	16. Newark	26. Syracuse	36. Houston
7. Cleveland	17. Minneapolis	27. Birmingham	37. Buffalo
8. St. Louis	18. Boston	28. Rochester	38. Miami
9. Fairbanks	19. Bridgeport	29. El Paso	39. Dallas
10. New York	20. Indianapolis	30. New Haven	40. New Orleans

FIGURE 14–9

The forty most-polluted cities of the United States, according to air-quality tests by the Environmental Protection Agency.

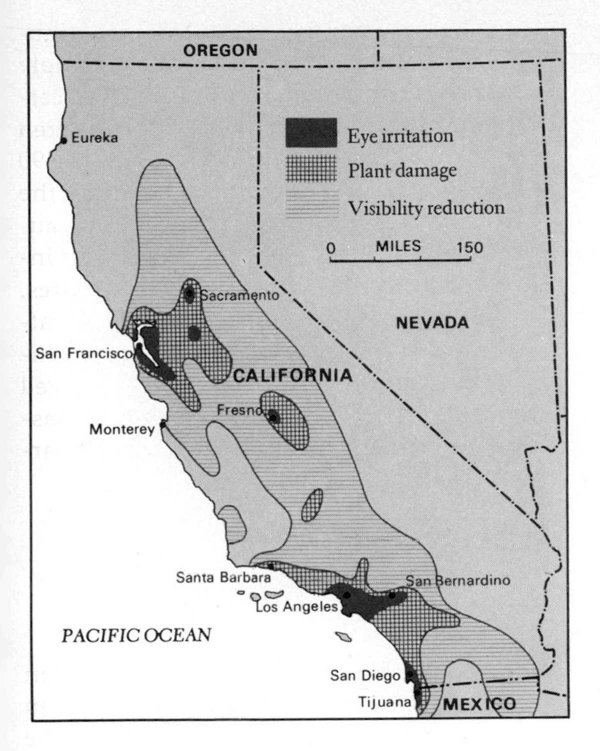

FIGURE 14–10
The extent and some effects of air pollution in California in the early 1960s.

plaguing even remote rural areas where few air pollutants are ever created (Plate IV). Probably most common are the worldwide photochemical air pollutants, or photochemical smogs, created largely by the increasing emissions of nitrogen oxides and hydrocarbons from the engine fuels of transportation systems.

In other instances, and to varying degrees, the rapid increases in urban air pollution in many locations are due to local terrain and weather conditions, which are as yet largely beyond human control (Fig. 14–10 and Plates II, III, and IV). Winter air pollution in the Great Salt Lake basin of Utah, for example, is caused as much by the poor air drainage over the Wasatch mountain range as it is by the high seasonal use of local coals for industrial purposes and home heating. The wind and temperature patterns over the Hudson Valley and neighboring land and water bodies contribute in a large way to the trapping of heavy concentrations of pollutants over New York City for protracted periods. And the cities of Santiago and Lima, on the west coast of South America and backed by the high slopes of the Andes, are also periodically subject to trapped pollution conditions because of the limited ventilation of their situations (Plates II–IV). Tokyo, Hong Kong, and Sydney are each situated amid terrain that resembles, to varying degrees, the mountain-girdled Los Angeles basin, and each is similarly beset by almost constant heavy air pollution.

Water Pollution

The problems created by water pollution are not new. History is replete with accounts of streams and lakes despoiled by culturally induced land erosion and foul-smelling urban sewers, of garbage-laden harbors, diseased water wells, and sludge-choked rivers. The problem has existed for centuries, but it is now progressing at a much faster rate than in the past (Fig. 14–11). The seriousness of its present state is being acknowledged in many parts of the world; there is increasing public interest in additional antipollution laws and in more stringent enforcement of existing water-use regu-

FIGURE 14–11

Pollution within the Great Lakes water system. Left, an aerial view of pollutants flowing into Lake Michigan along the Chicago lakefront. (NASA, Grant Heilman) Right, some effects of water pollution in Pennsylvania, which borders Lake Erie. (Grant Heilman)

lations. Unfortunately the chemistry and mechanics of water pollution are so complicated that there is still wide disagreement among the experts as to its causes and long-range effects.

The "Water Budget"

Water is vital to all the earth's living things; adequate supplies of it have always been deciding factors in the distribution of plant and animal life, including human populations. For most human uses, and some commercial ones as well, the quality of water is as important as is its supply; to be suitable for human consumption, water must be substantially free from brackishness, plant and animal wastes, and bacterial contamination. Such unpolluted, or "fresh," water supplies are known as *potable waters.* Early human migration routes and settlement sites were certainly much influenced by the availability of drinkable water, and today—despite advances in drilling, irrigation, and purification—the location, quality, quantity, ownership, and control of potable waters remain major human concerns.

For all practical considerations, the earth's total estimated water supply, or "water budget," is constant. Moreover, the planet's various water forms—vapor, liquid, and solid—have been maintained in practically constant proportions over thousands of years. Significant variations in their proportions occur only during periods of widespread glaciation or deglaciation, when there is a marked change in the amount of water "locked up" in the continental ice sheets. The constancy of the earth's water forms results from an environmental system called the *hydrologic cycle,* which involves the continuous natural processes of evaporation, movement, condensation, precipitation, runoff, and percolation. But with about 97.2 percent of the world's water composed of oceanic brines, and about 2.2 percent locked up in polar ice caps and glaciers, only a meager 0.6 percent of the water budget consists of continental fresh waters—and most of these are subsurface rather than easily accessible to humans in lakes and rivers.

Major Water Pollutants

"Pure water" is a rare substance. Because of its abilities as a solvent, many dissolved substances are usually present in *natural water.* The most common of these natural pollutants are harmless compounds of such chemicals as sodium, calcium, iron, and magnesium. But humans further pollute the surface and subsurface waters of most of the inhabited world by adding an infinite variety and quantity of alien compounds. Wastes from industrial processes, agricultural processes, and human life processes are all dumped sooner or later onto the land and into the waters, and, either indirectly or directly, find their way into the earth's supply of fresh water.

From Industrial Sources. Industrial chemicals and biological matter often reach streams and lakes and subsurface waters as parts of municipal sewage (Fig. 14–12). The principal contributors of such pollutants include food processing plants, textile mills, pulp and paper plants, iron and steel plants, and synthetic rubber and plastics plants. The effluents of iron and steel plants, an abundant and potent variety of acids, alkalies, and wastes, play a major role in the pollution of streams and lakes in the northeastern United States, northwestern Europe, the Soviet Union, and a score of other large iron and steel centers around the world (Fig. 14–13). Utilizing superheated water, they are also the cause of extensive *thermal pollution* in these waters; today's iron and steel technology uses as much as 12,000 to 14,000 gallons of water per ingot-ton produced, largely for the cooling of high-temperature blast furnaces. As long as the penalties for polluting

FIGURE 14–12
Nondegradable detergent foam covering Conestoga Creek, in Lancaster County, Pennsylvania. (Grant Heilman)

degradable substances are all transformed by natural chemical decomposition, to the limit of the water's regenerating capacity. The chemical nutrients produced in the process are used as food by *algae* and other green water plants—which become the food of microscopic aquatic animals called *zooplankton.* Small fish eat the zooplankton and are eaten in turn by larger fish, and they by still larger ones. The decomposition of dead animals and fish allowed to remain in water begins the *food chain* anew.

But the natural regenerative processes of water can become overburdened. When excessive waste pollutants are dumped into a body of water, maximum quantities of oxygen are used up in their decomposition; the more oxygen thus used, the less the amount available for use by aquatic animals and plants. These animals and plants begin to suffocate and die, adding even more waste pollutants to the water and increasing the depletion of its oxygen content. Not dependent on a supply of oxygen, *anaerobic* bacteria (rather than aerobic) take over the water body and its waste decay and transformation process; they are effective, but emit noxious gases that give off strong odors.

Eutrophication. The addition of nutrients to a water body through cultural activities and the water's natural processes is usually referred to as *eutrophication,* or nutrient enrichment (Fig. 14–15). Heavy eutrophication greatly increases algae growth. Produced beyond the needs of their water's zooplankton, algae will die and become waste themselves. Their increasing decomposition uses up more and more of the water's oxygen. Thus a stream or lake that is artificially stocked with nutrients—as in the runoff of chemical fertilizers or the dumping of industrial detergents—may eventually choke on its own plant growth. It may take thousands of years for a large water body to succumb to this evolutionary process, but a small pond can be transformed into a swamp or marsh within a few decades. Even the downfall of a lake can occur within a human lifetime.

Lake Erie, over 200 miles long, is considered a *dead lake* by some scientists, a century or more of abuse by humans having made it a mammoth cesspool. Human and industrial pollutants have been dumped directly into Lake Erie from all directions: Detroit; Windsor, Ontario; Toledo; Cleveland; Erie; Buffalo; and smaller communities have all contributed their wastes. The Cuyahoga River, which flows into Lake Erie's south side at Cleveland, has at times demonstrated a bacterial count near that of raw sewage; and the Detroit River dumps pollutants into the lake from the whole upper Great Lakes basin. The regeneration of Lake Erie is possible, but it will require the joint efforts of nature and human beings for years or even decades.

FIGURE 14–15

A diagrammatic representation of the eutrophication, or excess fertilization, of large water bodies. The process is caused when heavy pollution promotes a cycle of algae growth and decay.

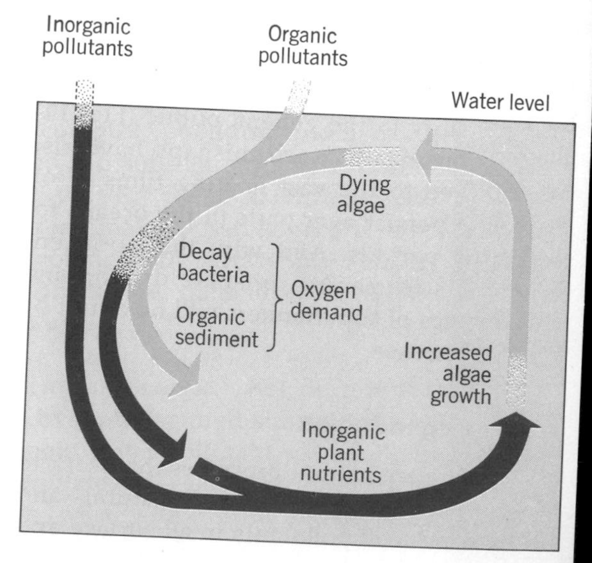

Land Pollution

Humankind's early plant and animal production may have disturbed the upper layers of the cultivated soils and grazing lands, but probably did little to permanently upset their natural processes and formations. The effect of agriculture in recent centuries has been much more harmful; and, in the last three decades, doubling and quadrupling world food demands have placed unparalleled pressures on the world's cultivable soils. Although human understanding of soil exhaustion has progressed markedly in this period, land in many parts of the world is still being overworked and misused.

Throughout most of history, animal manures, night soil, and "green fertilizers" (legumes and grasses plowed under) were the farmer's chief sources of *soil enrichment.* Not until well into the twentieth century did American and European farmers make the first extensive uses of chemical fertilizers and acid neutralizers. As food-production demands have continued to spiral upward, more and better commercial fertilizers have been produced. And along with the wide array of fertilizers, science has provided an equally varied assortment of pesticides and fungicides. Their effects on food production have been phenomenal, but it is becoming clear that they offer unwelcome repercussions as well; some of the new wonder fertilizers and pesticides are now thought to be harmful to human health, both when encountered directly and when ingested through the products and tissue of plants and animals. Excessive use of fertilizers and pesticides is even being discouraged as harmful to the humus layer of soil, for it apparently decreases the layer's natural nutrient production and thus makes the soil more dependent on continuous artificial enrichment.

Agriculture and Mining

Soil erosion has always been a problem in world agriculture and gardening. But the seriousness of the problem has increased greatly since the introduction of deep-plowing a century ago and the marked increase, since the beginning of this century, of clean-row cultivation—the cultivation in even, weeded rows of such crops as cotton, potatoes, and cabbage. Both these techniques generate increases in crop yields, by allowing for better utilization of natural soil moistures, but they also contribute to extensive erosion (Fig. 14–16). All the topsoil of a

FIGURE 14–16

Oddly shaped islands of cultivated land in an extensive area of Kansas once covered by wheat fields. Abetted by noncontour planting and serious droughts, soil erosion has gullied most of the area and made it useless for farming. (Grant Heilman)

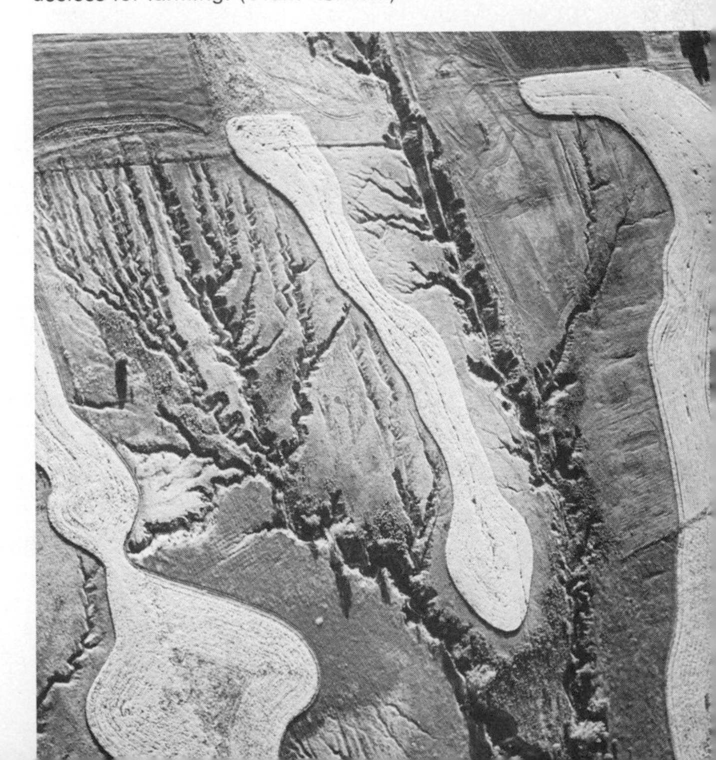

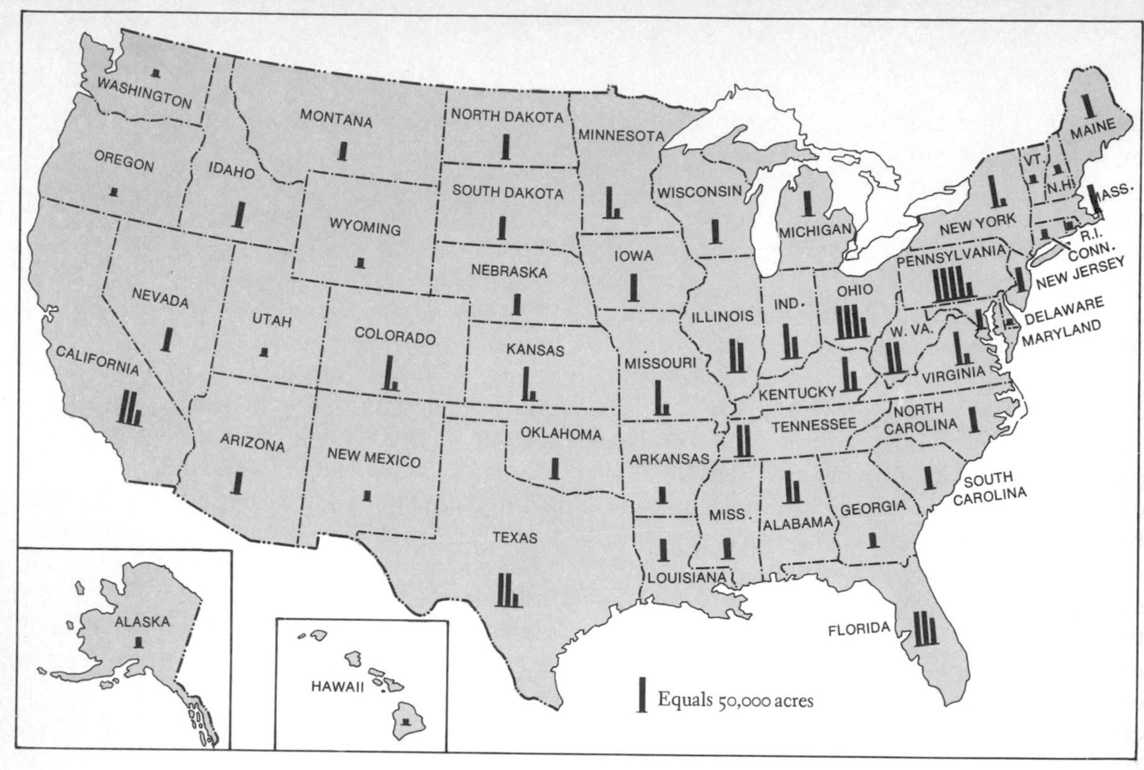

FIGURE 14–17

Above, the extent of strip mining in the United States; by the mid-1970s over 3.5 million acres had been laid waste by surface mining. Below, random absorptions of farmland due to the strip mining of coal, near Appleton City, Missouri. (Grant Heilman)

clean-cultivated field may be washed away by *sheet erosion* within a few years or even a single season. Since the first pioneers made their way across the United States, the fertile topsoils of its agricultural lands have been eroded away in many places by as much as one-third of their original depths—mainly through *eolian erosion,* which carries away soil denuded by excessive cropping and overgrazing (see Chapter 2, section entitled "The Dust Bowl"). Irresponsible farming practices are the basic cause of this and much of the world's soil erosion, but they can now be avoided relatively easily and cheaply.

Ranking second to agriculture as a source of erosion is the extraction of minerals from the earth's surface by means of strip mining, a process that will probably become even more extensive in response to the recent energy crisis. By the beginning of the 1970s, over three and a half million acres of United States land—over 5,000 square miles—had been cut and ridged by strip mining (Fig. 14–17). Strip mining to extract coal accounts for over half of these distressed areas, and the extraction of precious metals, such as gold, and sand and gravel is the cause of most of the rest.

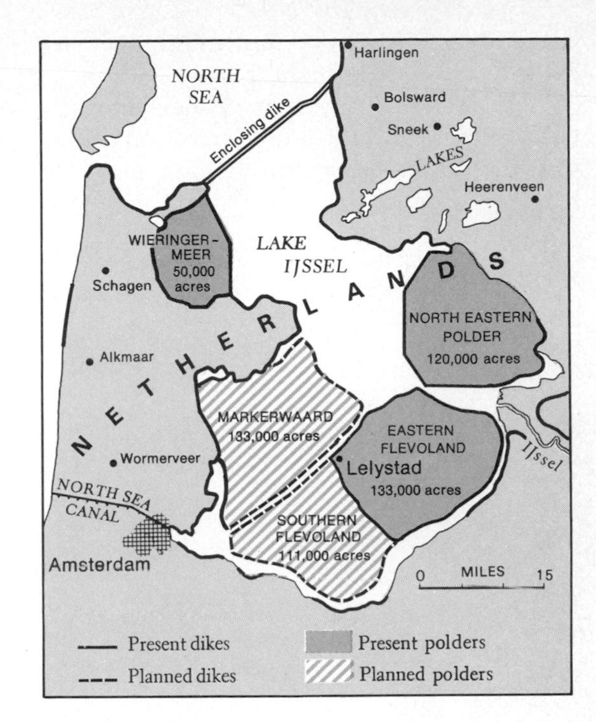

FIGURE 14–18
The diked and drained new lands in the Netherlands.

A Positive Geomorphic Step

As geomorphic agents, with relatively superior skills and highly advanced technologies, people in a developing society can reshape the earth's surface in a variety of ways and with varying degrees of permanence. Some of this geomorphic activity is environmentally constructive, and some is extremely destructive.

One of the most constructive, and extensive, of all artificial landforms is found in the Zuider Zee region of the Netherlands, where the Dutch government has drained the sea water from 950 square miles of land in large sectors called *polders* (Fig. 14–18). The first polder, 50,000 acres facing the North Sea, was drained in 1930. A 20-mile-long dike was then completed across the remaining mouth of the estuary, in 1932, to permanently separate the entire bay area from the sea. More of the Zuider Zee now has been diked and drained, and more is being readied; the deepest part has been converted into Lake IJssel. The second polder, 120,000 acres, was drained in 1942; and the third, 133,000 acres, in 1957. Two remaining areas, one 111,000 acres and the other 133,000 acres, are being readied for draining. After each polder is drained of its sea water, the new lands are worked and flushed extensively for several years under government supervision; when they are suitable for cultivation, they are leased to private farmers.

Comparable landform changes in other parts of the world include the creation of vast agricultural terraces throughout most of south and east Asia—especially in China, Japan, India, Sri Lanka, Nepal, the Philippines, and Indonesia—and the enormous excavations necessary in the building of the Suez and Panama canals (Fig. 14–19).

FIGURE 14–19

One of modern society's major geomorphic feats, the building of a canal across the Isthmus of Panama was both useful to humans and environmentally sound. Above, Panama Canal construction at Culebra Cut in 1904. Below, the Pacific entrance to the canal, through Miraflores Locks, in 1953. (Panama Canal Official Photo)

Other Forms of Pollution

Noise

Another rapidly growing environmental problem in most societies of the world is *noise*. While the dangers of human exposure to intense sound and vibration are usually most severe in highly urbanized, heavily industrialized centers, serious noise problems also exist in suburban, commercial, and entertainment areas. And with the proliferation of television sets, dishwashers, vacuum cleaners, and food blenders, the noise levels in the modern home are beginning to approach those of factories, airports, and discotheques. Prolonged exposure to certain intense noises can cause hearing impairments and, probably, adverse effects on the human nervous system (Fig. 14–20).

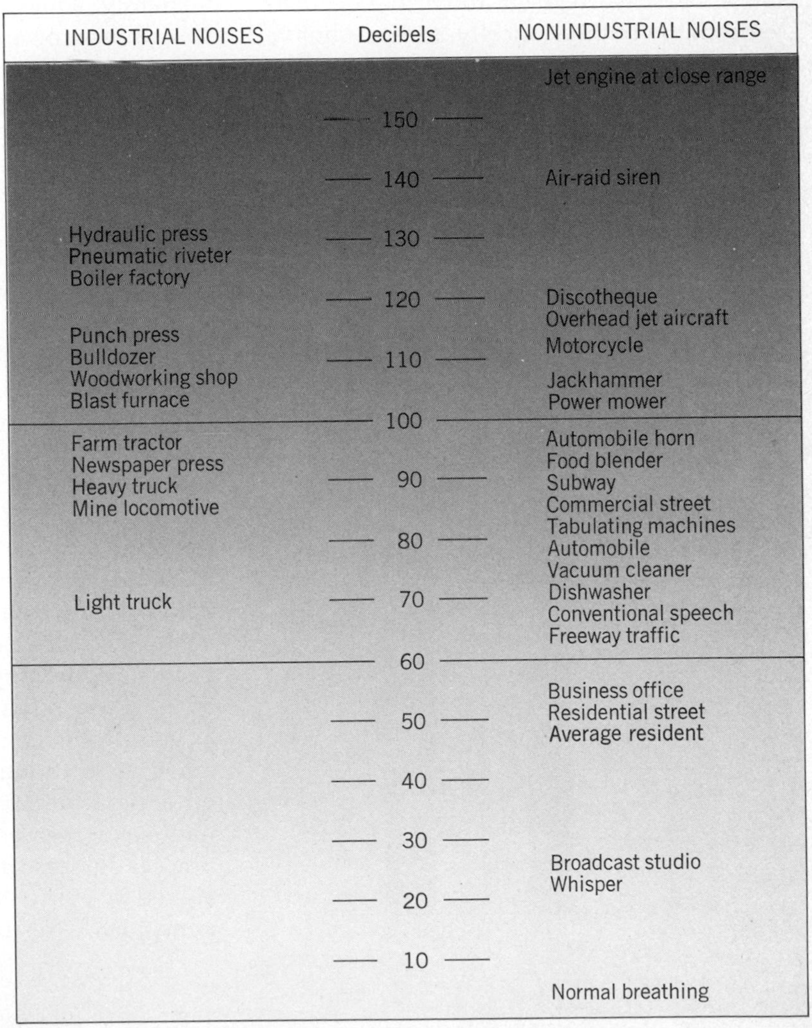

INDUSTRIAL NOISES	Decibels	NONINDUSTRIAL NOISES
		Jet engine at close range
	150	
	140	Air-raid siren
Hydraulic press Pneumatic riveter Boiler factory	130	
	120	Discotheque Overhead jet aircraft
Punch press Bulldozer Woodworking shop Blast furnace	110	Motorcycle Jackhammer Power mower
	100	
Farm tractor Newspaper press Heavy truck Mine locomotive	90	Automobile horn Food blender Subway Commercial street
	80	Tabulating machines Automobile Vacuum cleaner
Light truck	70	Dishwasher Conventional speech Freeway traffic
	60	
	50	Business office Residential street Average resident
	40	
	30	
	20	Broadcast studio Whisper
	10	
		Normal breathing

FIGURE 14–20
A comparison of industrial and nonindustrial noise levels. Such levels are measured in decibels: those below 60 decibels are considered acceptable to human sensibilities; those between 60 and 100 generally annoying; and those above 100 frequently damaging to human health.

Noise levels are expressed in units called *decibels,* which represent the intensity of sounds on a scale from 0 to 160. Least perceptible sounds are low on the scale, and there is a multiplying intensity with each decibel rise. The perceived sound intensity of any noise source at 160 decibels is about four times that of the same noise source at 140 decibels. Most noises that register between 0 and 60 decibels are considered to be of little physical or psychological danger to human beings; those that register between 60 and 100 are generally annoying and at least mildly hazardous if prolonged; and all those noises that register above 100 decibels are probably both physically and psychologically dangerous to most people subjected to them for extended periods.

FIGURE 14–21

Comparative estimates of solid wastes in the United States. (After U.S. Bureau of Solid Waste Management)

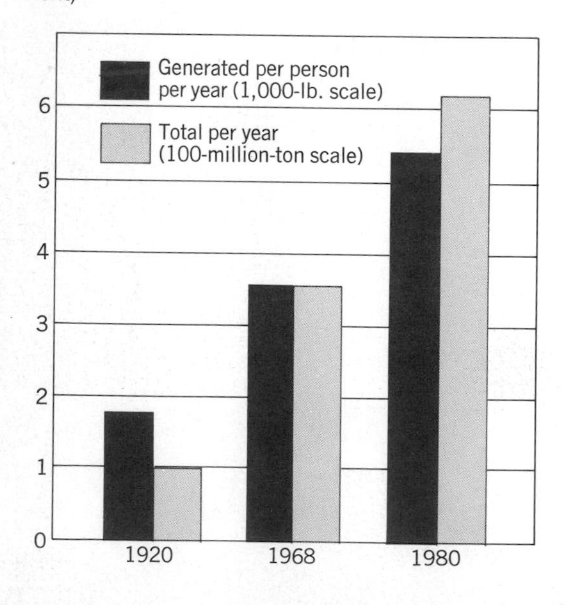

Heavy Metals

Heavy metals, such as mercury and lead, are introduced into the air and water daily, both through the normal processes of nature and the varied activities of human beings. Most of the additions for which humans are responsible are made by means of energy-combustion processes—in factories, power-generation plants, and automobile engines. As solids, liquids, and gases, most heavy metals are highly poisonous to human beings and, consumed in large amounts, can affect the nervous system adversely and permanently. Due in large part to the widespread use of pesticides, these poisonous metals have been diffused throughout large portions of the earth's atmosphere, water, and soil. They are particularly menacing because they collect in the organs and tissue of animals and, unchanged in food preparation, can pass through an entire food chain, with each order in the chain acquiring the accumulations of all preceding orders.

Radiation

Although necessary for all life on earth, *radiation* can also destroy all forms of earthly life. To the radiation that comes directly from the sun, science has added large amounts from radioactive materials—through nuclear fallout, waste from nuclear power plants, and such electronic equipment as X-rays, laser beams, color television, and microwave ovens. Research on the effects of radiation is as yet inconclusive, but exposure to unusual amounts is thought to result in cancer and other harmful changes in animal cells. International agreements banning most testing of nuclear devices in the atmosphere have been signed by many nations of the world, but the atmospheric testing by abstaining nations and all underground testing are without official restrictions. Moreover, even unanimous agreement to the present nuclear treaties would not be a guarantee against large-scale accidental contaminations.

Solid Wastes

Of all the pollutants created in the name of societal progress, *solid wastes* are probably among the most serious environmental threats, and the most difficult to handle. Early societies were not very subtle in disposing of their solid wastes—for the most part bones, bowls, broken tools, and charcoals—but nature was able to cooperate in fairly effective cover-ups, now known as *middenheaps* and considered valuable anthropological records. Modern societies often seem nearly as inept in disposing of their solid wastes, and the composition and volume of these materials sometimes seems to defy nature as well (Fig. 14–21).

Sources and Disposal. Each day the people of the United States throw away great quantities of solid wastes, or trash, their cast-offs reaching a total of over 4 billion tons a year. Fortunately, over 90 percent of this discarded material is associated with agricultural and mining activities and is thus widely dispersed through sparsely populated rural and mountain areas, where, tending to blend with the natural landscape, it offends or threatens relatively few people. The remaining 10 percent of the country's rubbish develops within more heavily populated areas, where it is highly concentrated, highly visible, and a virtual plague on everyone's doorstep—householders, institutions, and businesses alike (Fig. 14–22).

The federal Bureau of Solid Waste Management estimates that household and commercial sources contribute about 6 percent of the country's annual total of solid wastes. Wastepaper represents about one-half of this figure, with bottles, cans, textiles, and lawn rubbish accounting for most of the re-

FIGURE 14–22
While solid wastes are ever becoming a more visible problem of the countryside, the greater portion of every nation's solid wastes, or rubbish, is amassed within heavily populated urban areas. Mounds comparable to the one shown here exist in many urbanized areas of the world, and their varying contents tell much about the interests and socioeconomic levels of different societies. (Bayer from Monkmeyer)

FIGURE 14–23

A shorefront dumping area in New Jersey. Some countries have prohibited such dumps within densely populated areas, but prohibition is futile unless accompanied by effective reclamation and recycling programs. (Bruce Davidson, © 1969 Magnum)

mainder. Industrial solid wastes, though only about 3 percent of the total, represent a much greater threat to the environment than household and commercial wastes.

Until recently, many small commercial and industrial establishments disposed of their solid wastes mainly by burning or merely dumping them in open lots. But federal and state regulations now place heavy restrictions on open burning at dump sites. Sanitary landfill-dumps have been introduced in many progressive communities, but they are only small first steps in a much needed long-range federal solid-waste-management program (Fig. 14–23).

Reclamation and Recycling. There are materials in most solid wastes that can be reclaimed and put to new uses, or recycled. Often among them are such valuable natural resources as iron, copper, and bauxite ores, from slag piles and old tailings; iron and steel, copper, aluminum, and tin, from old machinery, cans, and automobiles; glass, from old bottles and jars; and paper pulp, from old newsprint. An increasing number of less valuable materials is also being recycled in order to avoid the problems of their collection and disposal. With aggravated collection and disposal problems impending, it is likely that even greater importance will have to be given to recycling, especially as it involves the reuse of depletable natural materials, such as lumber or nondegradable substances. Conservation is in style again, and one hopes that societal attitudes toward balanced resource management will continue their present upward trend.

REVIEW AND DISCUSSION

1. Write a short, well-organized essay on the major aspects of environmental damage and conservation attempts during the nineteenth and twentieth centuries.
2. Cite briefly several harmful impacts of air pollution on living beings in either your specific local community or geographical region.
3. Describe *thermal inversions* and indicate where such conditions frequently occur within the United States.
4. Compare and contrast the impacts of air pollutants from the three major sources: transportation devices, power plants, and industry. What are the "atmosphere's self-purification" processes? Discuss one major example.

5. For most human uses, and some commercial ones as well, the quality of water is as important as is its supply. Comment on the relationship of this important fact to water pollution. Discuss the world's "water budget" and its significance in the understanding of water pollution.
6. Discuss the differing distributions and consequences of water pollutants from industrial sources, from agricultural sources, and from human sources, on an international basis.
7. What is meant by *water regeneration?* How is it affected by nondegradable pollutants? What is *eutrophication?* How does this process proceed?
8. Research and write a short essay on one of the world's many polluted water bodies—for example, the Rhine River, Lake Tahoe, the Danube River, the Monongahela River, the Mediterranean Sea, Lake Baikal, the Dead Sea, or the Ganges River. How did this water body become polluted? What is being done to restore it or to keep it from additional harm?
9. What is the status of noise pollution as an ever-growing environmental problem? How widespread is this type of pollution? What serious encounters with it have you had? What were their effects?
10. Of all the pollutants created in the name of societal progress, solid wastes are probably the most aesthetically undesirable. How can solid wastes also do physical damage to the environment?
11. What are the principal sources of solid wastes? What have been the usual means of solid waste disposal in the past, and what may be the best means of disposing of solid wastes in the future? What roles do you think reclamation and recycling can and will play in future environmental protection?
12. Would there be environmental pollution if human beings did not exist? Explain and provide examples to support your argument.

SELECTED REFERENCES

American Chemical Society. *Solid Wastes.* Washington, D.C., 1971.

Andrassy, J. *International Law and the Resources of the Sea.* New York: Columbia University Press, 1970.

Baron, R. A. *The Tyranny of Noise.* New York: St. Martin's, 1971.

Bates, M. "The Human Ecosystem." *Resources and Man.* San Francisco: Freeman, 1969.

Benarde, M. A. *Our Precarious Habitat,* rev. ed. New York: Norton, 1973.

Brady, N. C., ed. *Agriculture and the Quality of Our Environment.* Washington, D.C.: American Association for the Advancement of Science, 1967.

Bryson, R. A., and J. E. Kutzbach. *Air Pollution.* Washington, D.C.: Commission on College Geography, Resource Paper No. 2, Association of American Geographers, 1968.

Bullard, F. M. *Volcanoes in History, in Theory, in Eruption.* Austin: University of Texas Press, 1962.

Carr, D. E. *The Breath of Life.* New York: Norton, 1965.

Clark, J. R. "Thermal Pollution and Aquatic Life." *Scientific American,* Vol. 220 (March 1969), 18–27.

Commoner, B. *The Closing Circle: Nature, Man and Technology.* New York: Knopf, 1971.

Detwyler, T. R., ed. *Man's Impact on Environment: Selected Readings.* New York: McGraw-Hill, 1971.

Doerr, A., and L. Guernsey. "Man as a Geomorphological Agent: The Example of Coal Mining." In F. E. Dohrs and L. M. Sommers, eds., *Physical Geography: Selected Readings.* New York: Crowell, 1967.

Dorst, J. *Before Nature Dies.* Baltimore: Penguin, 1971.

Ehrlich, P., and A. Ehrlich. *Population, Resources, Environment.* San Francisco: Freeman, 1970.

Fenner, D., and J. Klarman. "Power from the Earth." *Environment,* Vol. 13 (December 1971), 19–26, 31–34.

Friedman, W. *The Future of the Oceans.* New York: Braziller, 1971.

"Great Lakes Water Use: Population, Land Use, Power, Waste Disposal, Commodity Flow" (Map). Ottawa: Department of Energy, Mines and Resources, 1971.

Greenwood, N., and J. Edwards. *Human Environments and Natural Systems.* Belmont, Calif.: Wadsworth, 1973.

Hamblim, L. *Pollution—The World Crisis.* New York: Barnes & Noble, 1971.

Hannon, B. M. "Bottles, Cans, Energy." *Environment,* Vol. 14 (March 1972), 11–21.

Hare, F. K. "How Should We Treat the Environment?" *Science,* Vol. 167 (23 January 1970), 352–55.

Hines, L. G. *Environmental Issues: Population, Pollution, and Economics.* New York: Norton, 1973.

Hubschman, J. "Lake Erie: Pollution Abatement, Then What?" *Science,* Vol. 171 (12 February 1971), 536–40.

Kromm, D. E. "Response to Air Pollution in Ljubljana, Yugoslavia." *Annals of the Association of American Geographers,* Vol. 63 (June 1973), 208–17.

Malin, H., and C. Lewicke. "Pollution-Free Power for the Automobile." *Environmental Science and Technology,* Vol. 6 (1972), 512–17.

Marine, G. *America the Raped.* New York: Simon & Schuster, 1969.

Mecklin, J. M. "It's Time to Turn Down All That Noise." *Fortune,* Vol. 80 (October 1969), 130–33, 188–95.

Miller, G. *Replenish the Earth.* Belmont, Calif.: Wadsworth, 1972.

Shepard, P. *Man in the Landscape.* New York: Random House, 1967.

Small, W. E. *Third Pollution: The National Problem of Solid Waste Disposal.* New York: Praeger, 1971.

Steinhart, J., and C. Steinhart. *Blowout: A Case Study of the Santa Barbara Oil Spill.* Belmont, Calif.: Duxbury Press, 1972.

Stern, A., ed. *Air Pollution,* 2nd ed., 3 vols. New York: Academic Press, 1968.

Strobble, M. *Understanding Environmental Pollution.* St. Louis, Mo.: Mosby, 1971.

Turk, A., et al. *Ecology, Pollution and Environment.* Philadelphia: Saunders, 1972.

Warren, C. *Biology and Water Pollution Control.* Philadelphia: Saunders, 1971.

Weinberg, A. "Social Institutions and Nuclear Energy." *Science,* Vol. 177 (March 1972), 27–34.

Wexler, H. "Volcanoes and World Climate." *Scientific American,* Vol. 186 (April 1952), 74–80.

Willrich, T., and N. Hines, eds. *Water Pollution Control and Abatement.* Ames: Iowa State University Press, 1967.

Wilson, C., ed. *Man's Impact on the Global Environment.* Cambridge: Massachusetts Institute of Technology Press, 1970.

Wolozin, H., ed. *The Economics of Air Pollution: A Symposium.* New York: Norton, 1966.

Appendixes

APPENDIX A

Graphic Representations of the Earth

Modern field and laboratory work in geography employs both traditional devices, such as maps and globes, and many recently discovered techniques, such as aerial and color photography and remote sensing—a highly sophisticated system involving the use of high-altitude aircraft or space vehicles. Thus, today's geographer is able to record earth phenomena with much greater facility and speed than his predecessors. The wealth of data gathered by these new means has led to increasingly sophisticated hypotheses concerning the total earth environment.

The Geographic Grid

The *globe* is the closest approximation of the earth yet devised by humankind. A scale model of the entire earth, it is usually small enough to be scanned in a few glances. The ancient Greeks—who, more than 1,600 years before the time of Columbus, postulated that the earth was approximately spherical—were the first to reproduce the earth in the form of a globe. (The oldest existing globe, however, is of German origin and was completed in 1492; it is now in the Germanic Museum at Nurenberg, Germany.)

A glance at any globe reveals a system of intersecting lines. This network consists of the *meridians*—north-south lines connecting the north and south poles—and the *parallels*—east-west lines crossing the meridians at right angles (Fig. A–1).

The meridians are used to measure distances and to designate positions that are east or west of a particular line of reference. Arbitrarily chosen and later confirmed by international agreement, this line of reference is the meridian that passes through Greenwich (London), England, and is designated the *prime meridian*. Measurements that are east or west of the prime meridian are designated by the term *longitude*, and are expressed in degrees and fractions thereof. *East longitude* extends halfway around the earth to the east; *west longitude* extends halfway around the earth to the west. Since all circles are comprised of 360 degrees, the east and west longitudes each encompass 180 degrees (Fig. A–2). For further refinement of measurement and loca-

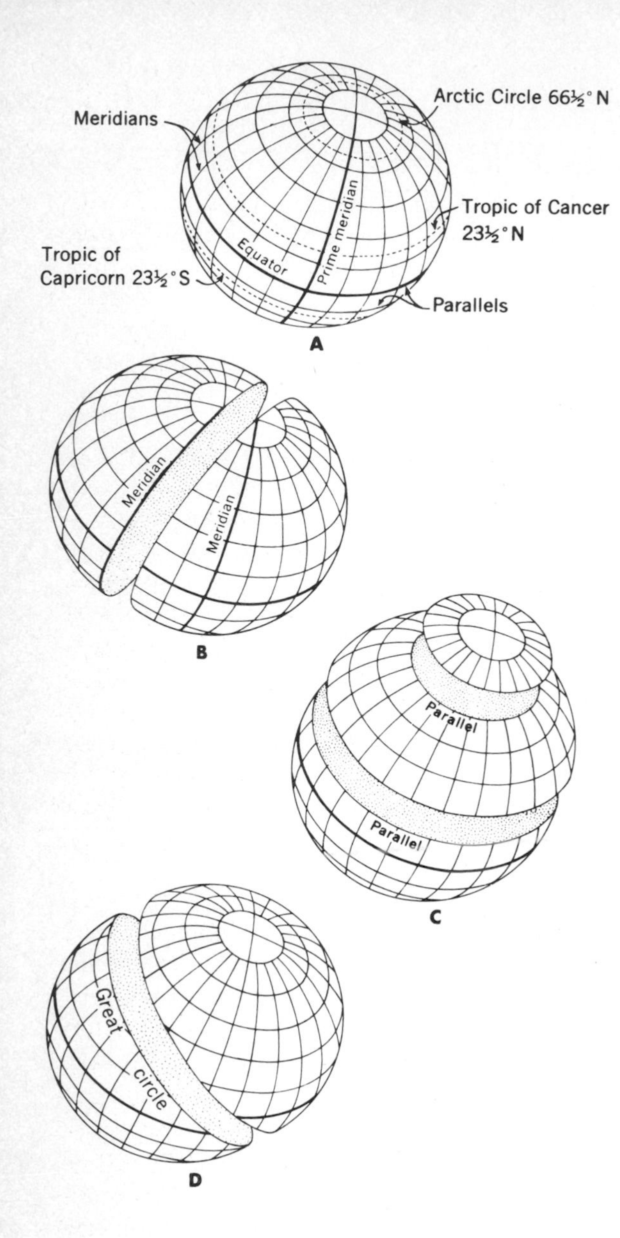

FIGURE A–1

The Globe. A. The most important *lines* on the globe. B. The *meridians.* Any meridian is one half of a great circle. C. The *parallels.* Of these, only the equator is a great circle. D. The *great circles.* Any circle drawn on the earth's surface whose center coincides with the center of the globe is a great circle.

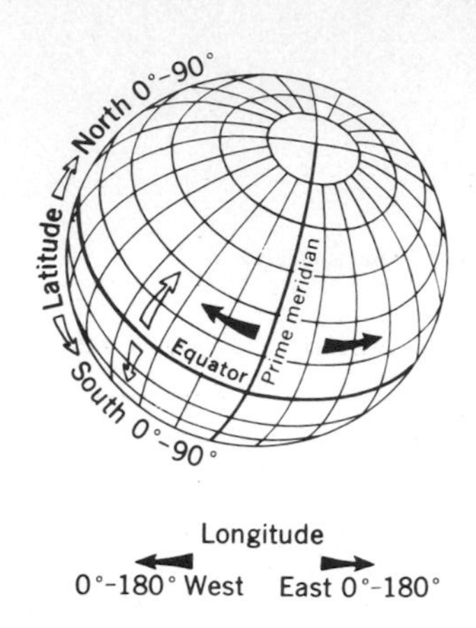

FIGURE A–2

Reference lines. The measurement of *longitude east* and *west* of the prime meridian, and the measurement of *latitude north* and *south* of the equator.

tion, each degree (°) is divided into 60 minutes (′), and each minute may itself be divided into 60 seconds (″). Thus, for example, Colombo, Sri Lanka is said to be located at 79°52′ E and Hilo, the largest island of Hawaii, at 155°01′ W (Plate II).

Parallels are used to measure distances and to designate positions north or south of the equator. These north-south positions, or distances, are called *latitudes,* and, like longitudes, are expressed in degrees, minutes, and seconds. The distance from the equator to either pole is one-fourth of the circumference of a circle, or 90 degrees. Thus, north and south latitudes each encompass 90 degrees (Fig. A–2). In terms of latitude, Colombo, Sri Lanka is located north of the equator at 6°58′ N and Hilo, Hawaii at 19°44′ N.

The full coordinates of Colombo, Sri Lanka are thus 6°58′ N, 79°52′ E; and the full coordinates of Hilo, Hawaii are 19°44′ N, 155°01′ W.

Time Systematization

Standard Time Zones

Time is determined with respect to the average rotation period of the earth on its axis. One rotation period of the earth is designated a day. Since every point on the earth rotates in a 360° circle around the earth's axis once every 24 hours, every point on the earth moves through 15° of the arc every hour of time. This simple space-time relationship provides the working basis for the system of *standard time zones*—a system devised during the last century in response to humanity's urgent need to keep track of time on a global basis.

If the earth rotated in such a fashion as to present the same face—the same half—to the sun at all times (as is almost the case with the planet Mercury), then that half would have constant daylight, and the other half would be in constant darkness. This does not occur, however, because, as noted before, the speed of rotation is such that, for each 24-hour period, the earth presents a steadily changing face to the sun. Thus, there is an infinite succession of mornings, noons, evenings, and midnights, no one of which can possibly occur over the entire earth at the same time. If it is noon in Los Angeles, it is midnight at an *antipodal* (opposite) location somewhere in the Indian Ocean (Plate IX).

Because the earth rotates toward its own east, each morning approaches from the east, as does each noon, sunset, and midnight. Thus, when it is noon in Chicago, it is approximately 1:00 P.M. in New York, but only about 11:00 A.M. in Denver (Fig. A–3). Actually, there are as many noons—that is, noon positions of the sun—as there are longitudinal positions on the earth. Clocks set according to the noon positions of the sun would even show slightly different times for east and west Chicago. Thus, a century ago, with the development of cross-country rail travel in the United States, a more uniform time system became mandatory—time had to be standardized.

In 1883, the railroads of the United States agreed upon a system of standard time zones. In ensuing years, the entire country followed suit. Today, the coterminous United States is divided into four standard time zones established by the Interstate Commerce Commission (Fig. A–3). Each zone is 15° of longitude in width, or, in other words, "one hour of time" in width. All localities within each zone are governed by the time of the zone's *central meridian;* and each zone is one hour earlier than its neighboring zone to its east. The longitudinal boundaries of these zones are irregular lines, for they were drawn so as to avoid splitting major population agglomerations into different zones.

The world standard time zone system is simply an extension of the system developed in the United States. The earth is divided into 24 time zones, each covering 15° of longitude, or one hour of time (Fig. A–4). The "0" time zone is based on the prime meridian, and both to the west and to the east of it are zones consecutively numbered from 1 to 12, indicating the number of hours to be subtracted from (if to the west), or added to (if to the east), *Greenwich Mean Time* in order to get the given zone's *standard time*. The boundaries of these zones also are purposely irregular to avoid, as much as possible, the splitting of major national and economic areas into two or more time zones.

The International Date Line

From the above discussion, it follows that comparisons of the "day" at distant places on the earth can become somewhat complex. One might assume that it is noon on Wednesday in London and that it is midnight on the other side of the earth (at the 180° meridian). But the question is, what midnight?

The answer is relatively simple. By international agreement, each new calendar day begins at midnight at the *International Date Line,* which, in general, follows the course of the 180° meridian: from pole to pole through the central Pacific (Fig. A–4). Every day of the week begins at midnight at the International Date Line, travels westward around the earth, and arrives (and ends) at the date line 24 hours later. Thus, there is a westward procession of the "days," just as there is a westward procession of the "hours" of the day; and, just as it is not the same time all over the earth, it also is not the same calendar day. When crossing the international date

FIGURE A–3

The *standard time zones of the United States.* The meridians shown are the so-called *central,* or *standard,* meridians. Standard time for Alaska was set by Congress as that of the 150°W meridian; but actually, as indicated in Figure A–4, four different times are used.

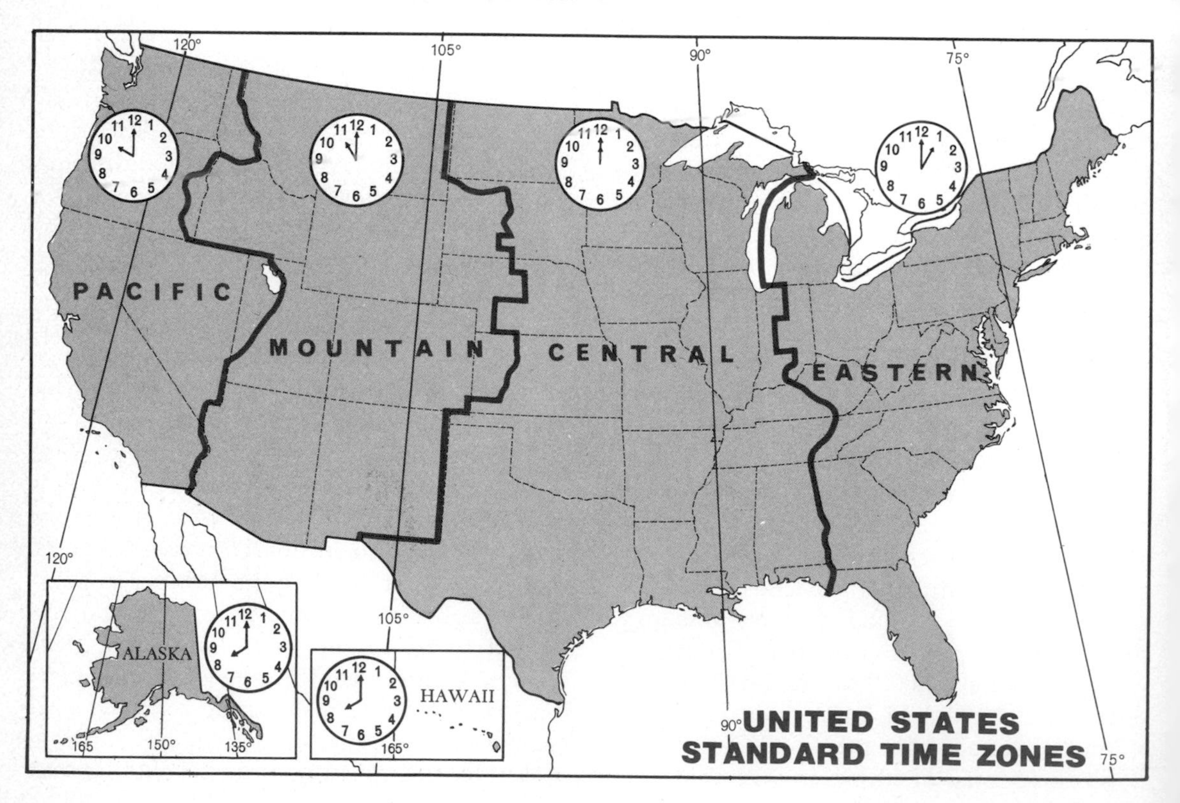

line going westward, a person loses a day—for example, Wednesday morning immediately becomes Thursday, even though only moments have passed. When crossing the date line going eastward, a person repeats the same day—for example, one Wednesday is followed by a second Wednesday (Fig. A–4).

The arbitrary positioning of the International Date Line in the mid-Pacific has proven to be highly convenient, for, in that area, there are few lands and people to be affected by the collision of different days. The slightly irregular trace of the date line is an attempt to place entire island groups to one side or the other of the line for added simplicity and the convenience of the islands' inhabitants.

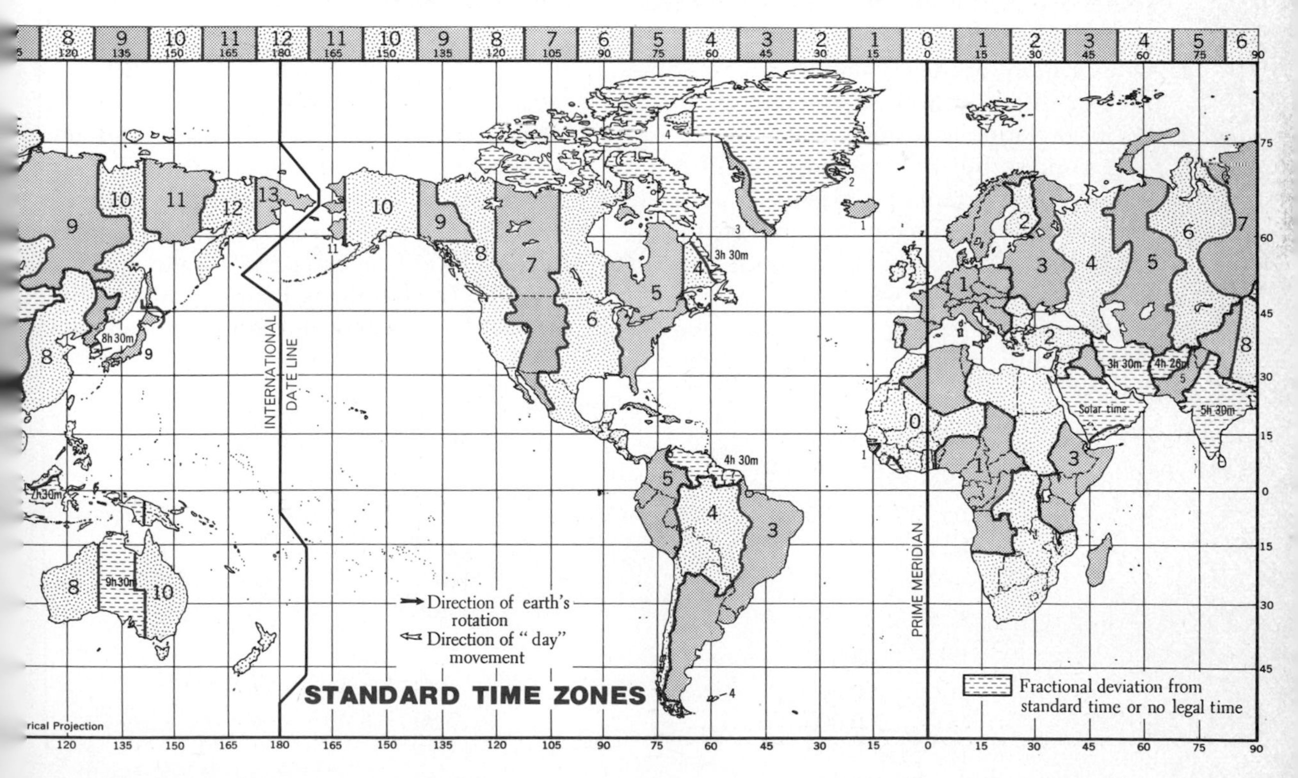

FIGURE A–4

The *standard times zones of the world.* The earth is divided into twenty-four time zones, each covering 15° of longitude. The time of the "0" zone is based on the 0° meridian (located at Greenwich, England), and it is generally adopted as *standard time* for 7°30′ eastward and westward. Each of the zones beyond is *designated* by a number representing the hours to be added to, or subtracted from, that zone's standard time to get *Greenwich Mean Time.* Irregularities in the zonal pattern are due to adjustments made for national and economic convenience.

Maps

Planimetric Maps

Today, maps are virtually infinite in purpose, design, scale, and origin; and they continue to be the principal tools or devices for geographic investigation and exposition. The term *map* derives from the Latin word *mappa,* meaning cover. By today's definition, a map is a scaled representation of all, or some portion, of the earth, constructed on a flat, or plane, surface. The earliest map still in existence is a clay tablet that shows a part of Mesopotamia and that dates from about 2,500 B.C.

For clarity and easy reference, modern maps generally include certain standard components. These are *title, date, legend,* and some identification of the *scale, latitude, longitude, direction,* and *projection* involved. One or more of these components are sometimes omitted, especially when maps appear in close relation to textual materials.

A concise but full *title* that identifies the purpose and extent of a map is an aid in its documentation, cataloguing, and use in subsequent research; this is true both of an intricate map of world vegetation (Plate V) and of a much simpler map of, say, drainage patterns. Many maps also carry a *date* and, sometimes, both the date of the data portrayed and, if different, the date of the map's publication. Very large maps of relatively small areas are frequently self-explanatory in regard to the data they portray, but most maps are of such small size in proportion to their area of coverage that it is necessary to include appropriate symbolization in a *legend* (Plate IV).

Since maps are flattened out globes, or parts of globes, they all have *scale* relationships with the earth, or with whatever segment of it they portray. Several methods, or devices, are commonly used to indicate the scale of a map. One means of scale identification is the *printed word*—as in "one inch equals one mile," or, "one inch equals 315 miles." More frequently, however, the scale of a map is indicated *graphically* by a measured line, graduated to show the relationship between map-distances and corresponding earth-distances (Plate II). This type of graphic, or *bar,* scale is frequently accompanied, or replaced, by a *representative fraction* (RF), or fractional scale—such as 1:24,000. The following listing provides some of the map scales most commonly used on an international basis.

Map scale (RF)	Equivalents on the earth's surface to one inch on the map (in feet and miles)
1: 10,000	833 feet
1: 24,000	2,000
1: 50,000	4,166
1: 62,500	5,208
1: 63,360	1.00 mile
1: 100,000	1.58
1: 500,000	7.89
1: 1,000,000	15.78
1: 50,000,000	789.00

Small- and large-scale maps differ not only in the quantity of detail that can be incorporated but also in the quality of such detail, and in the effectiveness of the generalizations that can or must be made.

A map that indicates latitude and longitude data, either along the margins or by showing parallels and meridians, indicates the precise position of the map area on the face of the earth. This same data may also be a reference to the direction shown on the map, since parallels always extend east-west and meridians north-south. *Direction* is also indicated on some maps by means of a *north arrow,* which may refer to and be marked either *true north* or *magnetic north.* A true north arrow is in alignment with a meridian and thus points to the *geographic* (true) north pole. A magnetic north arrow is in alignment with the flows, or lines, of the magnetic forces that course through the earth. At any point on the earth, the angle between true north and magnetic north is known as the *magnetic declination.* The lines of magnetic force follow wavy paths, not the meridians, and they flow not from geographic pole to geographic pole but between the earth's magnetic poles. They also fluctuate and shift gradually through the years. The *north magnetic pole* is presently at approximately 75° N, 101° W (Fig. A–5); and the *south magnetic pole* is located near the fringe

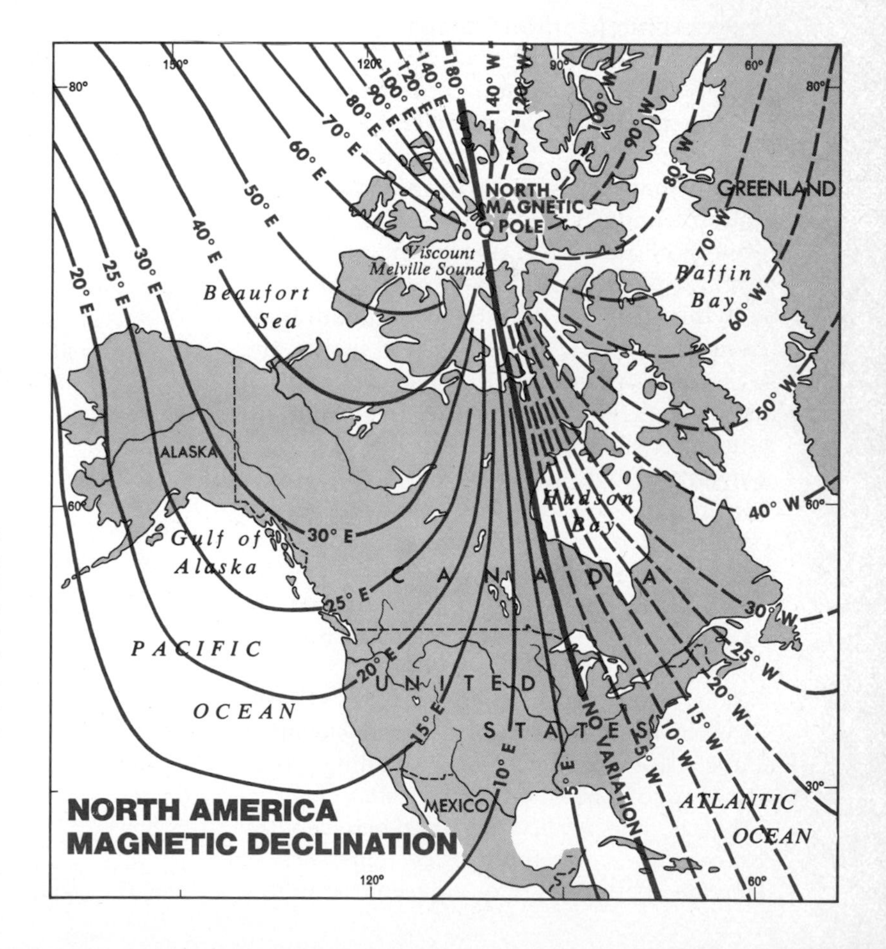

FIGURE A–5

Magnetic declinations (variations of the compass) in part of the Northern Hemisphere. The solid and dashed lines are *isogonic lines* (lines of equal magnetic declination). The heavy line is the *agonic line* (line of no magnetic declination). The dashed lines are in areas where the magnetic compass points west of *true north;* the solid lines are in areas where it points east of true north. The *north magnetic pole* is located approximately at 75° N, 101° W. (Adapted from U.S. Hydrographic Chart No. 1706, for 1965) of Antarctica, south of New Zealand. To allow for corrections of magnetic north to true north, detailed maps showing magnetic declination are prepared by various hydrographic agencies. Each line on such maps is called an *isogonic line;* and all points on a given isogonic line have the same declination. The line along which there is no magnetic declination is called the *agonic line.* In southern California, the magnetic compass points about 15° east of true north, which thus requires a 15° correction to the west. In central Maine, the magnetic compass points about 20° west of true north, which thus requires a 20° correction to the east (Fig. A–5).

Topographic (Contour) Maps

The most useful and most exact of the various relief, or three-dimensional, maps is the *contour map.* In the United States, this type of map is used for the standard topographic maps issued by the United States Geological Survey, Department of the Interior. A portion of such a topographic map, showing part of the Saticoy Quadrangle (contour map 1:24,000) appears inside the back cover of this book. Close examination of this map will show patterns of approximately paralleling fine brown lines bending and curving over its land areas. Each of these fine lines is a *contour line,* and the presence of a sufficient number of them enables one to perceive the third-dimension (depth) of minor landform features. To the trained observer, a contour map takes on the appearance of a relief model.

A contour line connects, or extends through, points that are the same elevation above sea level (Fig. A–6). In reading a contour map, one must know the *contour interval* – that is, the *vertical distance* between the contour lines, or the *vertical difference in elevation* between two adjacent contours.

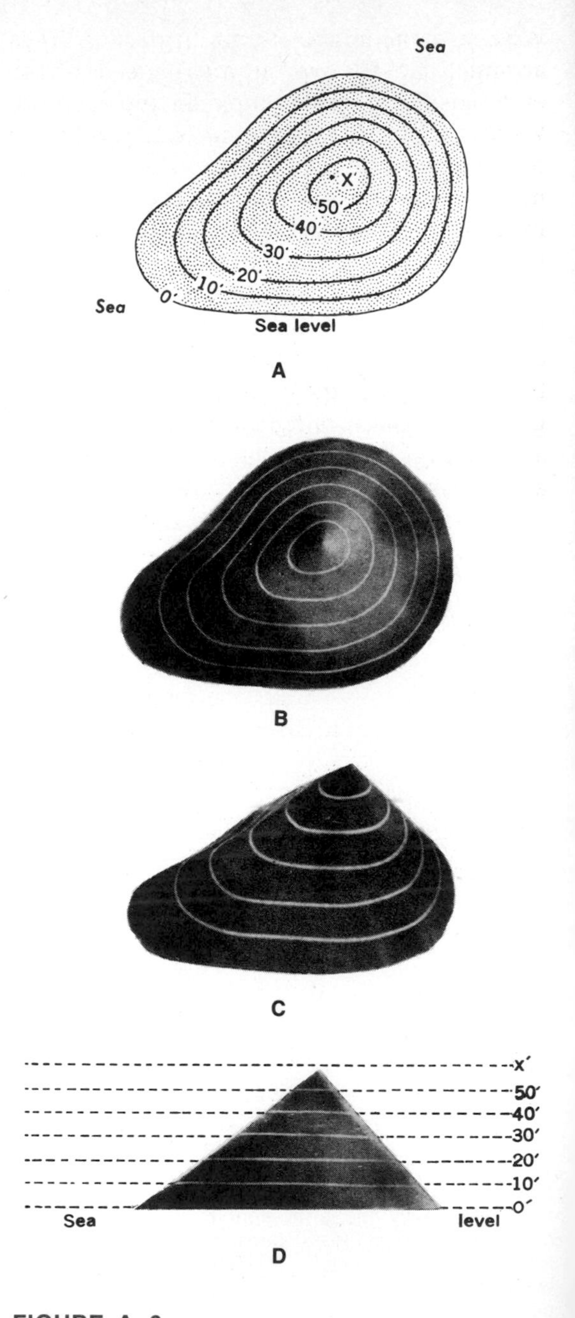

FIGURE A–6

Contour lines as they appear on a *contour map*, and as the same lines appear on views of a model of the same conically-shaped landform. The contour interval used here is 10 feet. A. Contour map. B. Vertical view. C. Oblique view. D. Side view.

With a considerable number of such lines arranged closely over an area, one can generally visualize the varying relief represented. Whatever the nature of the terrain features represented—plains, hills, cliffs, or even mountains—the spacings of their contours indicate their surface *gradients.* (Gradient is the degree of *slope steepness,* as measured in amount of *vertical change* in a given horizontal distance.) The closer the contour lines crowd together, the steeper the gradient; and the wider apart they spread, the flatter the gradient. Contours spread far apart on level areas and crowd together, and even merge, along steep slopes and sheer cliffs.

The stylized contour map in Figure A–7 illustrates several elementary characteristics common to all contour maps, regardless of the scales or terrain types involved. Most contour maps indicate a number of *bench marks,* which are precisely located spots on the earth's surface for which exact elevations have been determined. A permanent bench mark is usually a post that is set into a concrete base and then topped by a small bronze plate. This plate has several important functions. It serves to indicate the area's latitude and longitude, elevation and direction, the plate's date of installation, and the name of the agency that erected it.

FIGURE A–7
A simple and stylized contour map, illustrating the behavior of contours, the types of bench marks, and the determination of elevations and slopes. The contour interval used here is 25 feet. Contour lines wide apart indicate gentle slope. Contour lines close together indicate steep slope.

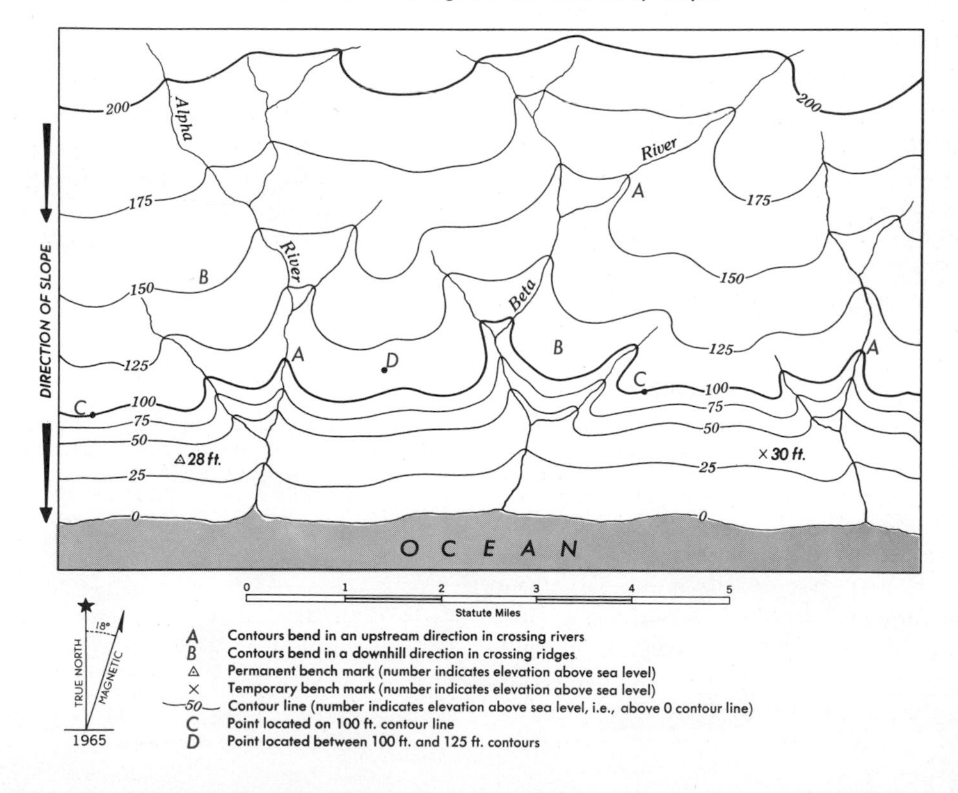

Map Projections

The ideal way to represent the total earth, its components, and its many changing surface aspects is by means of a globe—the only "true representation." But, the globe has several distinct disadvantages, mainly its limited portability, and production and reproduction difficulties. So people depend largely on maps that are flat, and on the projections that make it possible to construct flat maps for extensive areas.

A *map projection* is any orderly realignment of the spherical earth's *grid,* or *graticule,* of meridians and parallels onto a flat or level surface. Every map projection has its own peculiar version of this linear network; some are simple, others more complex and irregular (Fig. A–8). But there is no perfect, or all-purpose, map projection. The construction of every projection involves some distortion of the earth's grid. Each projection thus has its own particular limitations and should be used only for the purposes for which its design is well suited.

The two major properties of map projections (never found together in the same projection) are *equivalence,* or *equal-area,* and *conformality,* or *true shape;* and the two significant secondary properties are *azimuth,* or *true direction,* and *equidistance,* or *true distance.*

Equivalent, or equal-area, map projections portray all of their subject areas in proportion to their true sizes on the earth—but the shapes of these areas are badly distorted. Among such projections are the sinusoidal equal-area projection (Plates II–VIII) and the Mollweide projection (Fig. A–9).

Conformal, or true shape, map projections portray all small features of a small area in their true shapes, but their areas are distorted; one example is the Mercator projection (Fig. A–10). (Unfortunately, no projection can show the true shapes of such large areas as Asia or America—only a globe can do this.) Conformal map projections are designed especially for use in navigation and surveying.

Azimuthal, or true direction, map projections show the correct directions of an area from a stated point on the map. Equidistance, or true distance, map projections show the correct distances of an area from a stated point on the map.

These four groups of map projections include most of the maps in general use today. A few map projections do not fall within any

FIGURE A–8

Map projections. At the left, the relatively simple grid of one type of *cylindrical projection.* In the center, the globe. At the right, the relatively complicated grid of the *sinusoidal projection.*

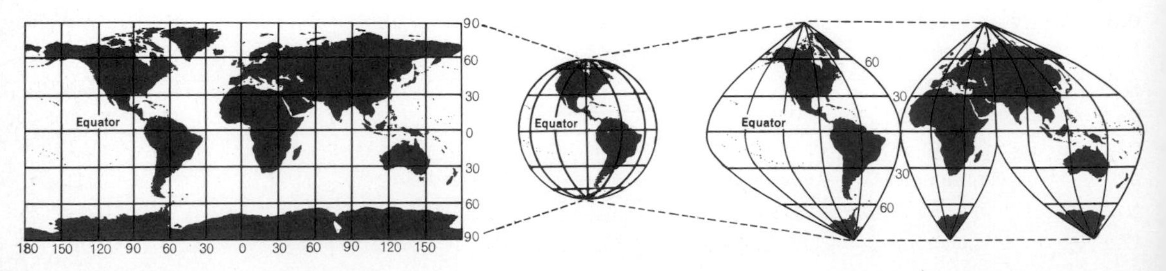

one of these groups; for there are some that distort area, shape, direction, and distance all at the same time—however the degree of distortion may be small in one or all of the properties.

This brief discussion should provide a sufficient basis for an understanding of map projections, why they are necessary, and their major and secondary properties. Further consideration of the above subjects, as well as information on the construction of map projections can be found in most standard textbooks on cartography.

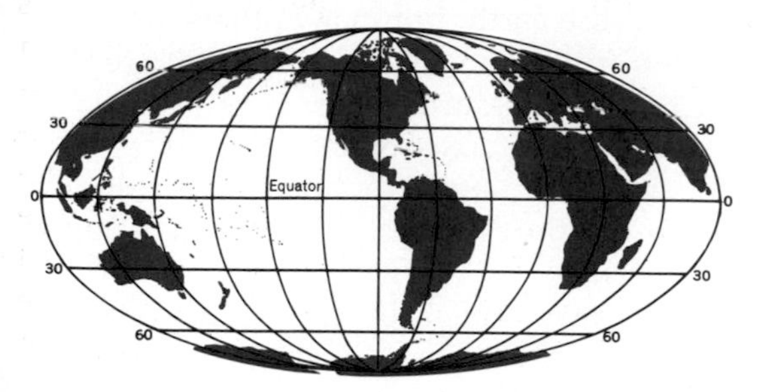

FIGURE A–9
An *equivalent*, or *equal-area, map projection*. The earth as it appears on the Mollweide projection, one of several equal-area projections (compare with Figure A–10).

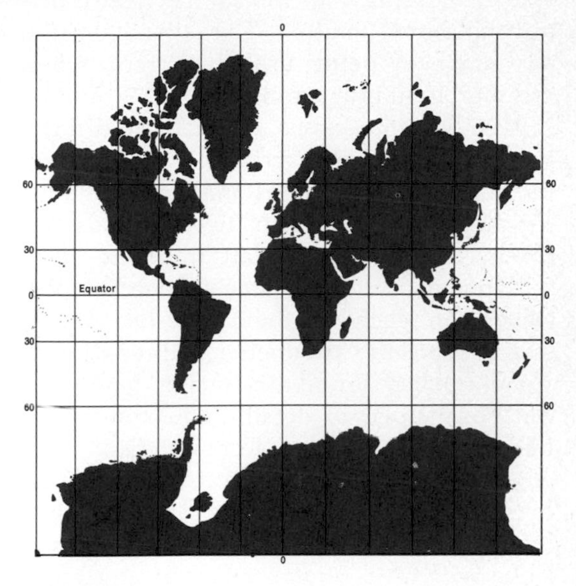

FIGURE A–10
A *conformal projection*. The earth as it appears on the Mercator projection (one of several conformal projections). Note how Greenland and other high-latitude areas assume sizes out of proportion to actual sizes (compare with Figure A–9).

Remote Sensing

Remote sensing is the obtaining and analyzing of an image of a portion of the earth's surface from some position detached from that surface. In a sense, even a photograph taken with a hand-held camera by a person standing on the ground is a form of remote sensing, since the camera is raised above the surface to the height of the photographer's eye. In common usage, however, the term is restricted to images obtained from vehicles detached from the earth and at considerable heights above it.

The simplest and most common form of remote sensing is the *aerial photograph*, or *air photo*, taken by a camera from an airplane. Air photos fall into several classes based upon the kind of film used, the angle photographed, and the altitude of the aircraft.

Black-and-white film is most commonly used, but use is also made of color film (which utilizes the ordinary visible spectrum of light waves), and of both black-and-white and color infra-red film (which utilizes spectra ranging into the shorter heat-transmitting wave lengths). The latter have the advantage of being usable at night, when ordinary light is at a minimum.

Aerial photographs taken from directly overhead are termed *verticals*. They are often used as map substitutes and are invaluable in map-making. When two verticals are taken just moments apart from the same moving airplane, the area of overlap portrayed on each of them can be viewed stereoscopically, producing a three-dimensional image. From this image, landforms can be mapped in detail, and contours can be drawn with great accuracy. Groups of adjacent overlapping verticals can be fitted together to form a *photomosaic*, which, when annotated with place names, becomes a *photo map*. Air photos taken at any angle other than a vertical one are termed *obliques*. Scale varies greatly in an oblique, and only by complex methods of *scale readjustment (resolution)* can such photos be used as a basis for mapping. However, since they have a more realistic appearance to the layman than do verticals, obliques are often preferred for some types of public presentation.

A special type of oblique is produced by *low angle side scanning radar*. Radar beams are emitted from an airplane at right angles to its line of flight and, reflected back in varying strengths from the terrain, are received by a receptor on the same airplane. The receptor records the reflected beams and produces a continuous strip photograph of the landscape to one side of the flight path.

Continuously orbitting satellites and manned space missions are providing us with images of the earth taken from altitudes far greater than those attained by conventional aircraft. Some of these images become available only after the return of the spacecraft to the earth, but many others are transmitted from the vehicle in flight to receptors on the earth's surface. Some of these latter images show areas of cloud cover as they develop from hour to hour, and they are therefore proving quite useful in weather forecasting.

Still very much in its infancy, remote sensing offers great possibilities for the gathering of information without the hardships of field expeditions. Thus, it will be an invaluable aid in the charting of areas that cannot be easily penetrated on foot or safely traversed by low-flying aircraft. Combined with the geographical data obtained on the earth's surface (*ground truths*), the information obtained through remote sensing will make the human capacity for extrapolating environmental data almost endless.

APPENDIX B

One Hundred Contour Maps that Illustrate Specific Landforms and Drainage Conditions

The list of maps given below has been adapted from "A Set of One Hundred Contour Maps That Illustrate Specified Physiographic Features," as issued by the United States Geological Survey, Washington, D.C. The list contains many maps that are already available in college and university libraries or in separate map collections in such institutions. Others may be obtained from the Director of the United States Geological Survey. A few titles are out of print and are not scheduled for reissue.

The following listing is divided into two parts, List A and List B. List A contains contour maps, by name, in alphabetical arrangement. Under the name of each contour map is a list of landforms and drainage features that appear on the map. List B is arranged according to specific landform and drainage features (for example, fiord, hanging valley, terminal moraine, tidal flat, and so on), and the maps on which each feature appears are listed alphabetically by name below the name of the feature.

List A

Antietam, Md.–Va.–W.Va.
- Braided stream
- Dissected peneplain
- Intrenched meander
- Intrenched stream
- Mature topography
- Ridges due to hard rocks
- Valley excavated in soft rocks
- Water gap or narrows
- Wind gap or pass

Aransas Pass, Tex.
- Barrier beach
- Beach, abandoned
- Lagoon or coastal lake
- Offset barrier beach
- Recurved or hooked spit
- Sand dunes
- Tidal marsh
- Truncated headland
- Wave-cut cliff
- Winged headland

Atlanta, Ga.
- Consequent stream (southeast part)
- Dissected peneplain
- Dome of crystalline rock
- Early mature drainage
- Early mature topography
- Intrenched stream (Chattahoochee)
- Migrating divide (between Chattahoochee and Atlantic drainage)
- Monadnock
- Rock sculpture controlled by exfoliation

Avon, Ill.
- Cut-off meander or oxbow lake
- Ground moraine (plain)
- Plain (of ground moraine)
- Stream meander
- Stream prematurely aged by valley filling
- Youthful drainage
- Youthful topography

Ballarat, Calif.–Nev.
- Alluvial slope, possibly including planed rock slope
- Alluvium-filled valley (all valleys)
- Bolson basin (same as above)
- Canyon or gorge
- Mountain not glaciated
- Nonglaciated or V-shaped valley
- Playa
- Rugged or youthful mountain
- Sand dune (Saline and Death valleys)
- Sand wash
- Valley due to dislocation of the rocks

Baton Rouge, La.
- Floodplain
- Lateral drainage diverted by floodplain deposits (Bayou Fountain)
- Natural and artificial levees
- Plain (of erosion)
- Stabilized stream (Mississippi River)
- Stream meander
- Swamp or marsh

Bright Angel, Ariz.
- Canyon or gorge
- Cliff
- Contrast among surfaces formed in different cycles of erosion
- Dissected plateau
- Encroachment of younger upon older drainage
- Hanging valley
- Intrenched stream
- Mesa (Shiva Temple)
- Migrating divide (south of the river)
- Nonglaciated or V-shaped valley
- Old drainage (on plateau)
- Old topography (on plateau)
- Plateau
- Rejuvenated stream
- Rock sculpture controlled by fractures (Bright Angel Creek Valley)
- Rock terrace
- Stream piracy (on plateau)
- Waterfall or rapid
- Youthful drainage
- Youthful topography

Buck Hill, Tex.
- Alluvial slope, possibly including planed rock slope
- Canyon or gorge
- Escarpment
- Mesa
- Plain
- Youthful drainage

Cape Henry, Va.
- Barrier beach
- Drainage blocked by barrier beach

Drowned valley
Estuary (Lynnhaven Bay)
Foreland (Cape Henry)
Sand dune
Swamp or marsh
Wave-formed plain
Youthful drainage
Youthful topography

Capitola, Calif.
Barrier beach
Beach, abandoned (about 200 feet in altitude)
Drainage blocked by barrier beach
Lagoon or coastal lake
Sand dune
Thorofare or tidal runway
Tidal marsh
Wave-cut cliff (at Capitola)
Wave-cut cliff, abandoned (Capitola to Port Watsonville)
Wave-formed terrace (north of Pajaro River)

Castlegate, Utah
Alluvial slope
Canyon or gorge
Dissected alluvial slope
Dissected plateau (Porphyry Bench, etc.)
Early mature topography
Escarpment (Book Cliffs)
Floodplain
Mature topography (northern part)
Nonglaciated or V-shaped valley
Stream meander

Cherry Ridge, Mont.
Cut-off meander or oxbow lake
Esker
Glacial lake and pond
Kame
Kettlehole
Stream meander
Stream prematurely aged by valley filling
Terminal moraine
Underfit stream
Youthful drainage

Chico Landing, Calif.
Alluvial slope (eastern part)
Alluvium-filled valley
Cut-off meander or oxbow lake
Development or shifting of meander
Floodplain
Meander migrating downstream
Natural and artificial levees
Stream meander

Chisos Mountains, Tex.
Alluvial fan or cone (Heath Creek)
Alluvial slope, possibly including planed rock slope
Canyon or gorge
Cuesta
Intrenched meander (Rio Grande)
Intrenched stream (Rio Grande)
Ridges due to hard rocks
Valley excavated in soft rocks

Cincinnati. *See* EAST CINCINNATI

Crater Lake National Park, Oreg.
Caldera (basin of Crater Lake)
Cliff (on rim of caldera)
Contrast among surfaces formed in different cycles of erosion (on rim of caldera)
Crater Lake
Glaciated or trough valley
Mesa
Truncated volcanic cone (ancient Mount Mazama)
Volcanic cone
Youthful drainage

Cross Plains, Wis.
Alluvium-filled valley
Contrast between glaciated and non-glaciated surfaces
Dissected plain
Glacial lake and pond
Ground moraine (on right)
Infantile drainage (on right)
Mature drainage (on left)
Mature topography (on left)
Stream prematurely aged by valley filling
Swamp or marsh
Valley train

Cumberland Island, Ga.
Barrier beach
Beach, abandoned (ridges back of beach)
Offset barrier beach
Recurved or hooked spit
Sand dune
Thorofare or tidal runway
Tidal marsh

Cut Bank, Mont.
Canyon or gorge (Cut Bank Creek)
Dissected alluvial slope (west of Cut Bank Creek)
Escarpment
Glacial lake and pond
Infantile drainage
Kettlehole
Terminal moraine (northeastern part)
Youthful drainage
Youthful topography

Cuyuna, Minn.
Cut-off meander or oxbow lake
Glacial lake and pond
Infantile drainage
Infantile topography
Intrenched stream
Kame
Kettlehole
Lake plain (southeastern part)
Stream meander
Swamp or marsh
Terminal moraine

De Queen, Ark.–Okla.
Consequent stream (southward-flowing streams)
Dissected peneplain (middle part)
Early mature topography (same as above)
Intrenched meander
Intrenched stream
Late mature topography
Mature topography
Ridges due to hard rocks
Subsequent stream (eastward-flowing and westward-flowing streams)
Trellis drainage
Valley excavated in soft rocks
Water gap or narrows

East Cincinnati, Ohio–Ky.
Alluvium-filled valley (Little Miami)
Canyon or gorge
Dissected peneplain
Dissected plateau
Drainage deranged by ice sheet
Early mature drainage
Early mature topography
Floodplain (Little Miami)
Intrenched stream
Meander scar (at Cincinnati and Newport)
Stream terrace (Terrace Park)

Elk Point, S.D.–Neb.–Iowa
Cut-off meander or oxbow lake
Dissected plain
Floodplain
Lateral drainage diverted by floodplain deposits (Big Sioux River)
Mature topography (upland)
Meander, migrating downstream (down to Elk Point)
Plain
Stream meander
Stream prematurely aged by valley filling (Big Sioux River)
Stream terrace
Swamp or marsh
Underfit stream (Big Sioux River)

Erie, Pa.
Beach, abandoned (700- and 740-feet altitude)
Delta
Drainage deranged by ice sheet
Ground moraine (southern part)
Intrenched stream
Lagoon or coastal lake
Recurved or hooked spit
Sand dune
Wave-cut cliff
Wave-formed terrace (about 740-feet altitude)
Youthful drainage

Everett, Pa.
Anticlinal valley (between Evitts-Dunning and Tussey Mountains)
Floodplain
Intrenched meander
Intrenched stream
Ridges due to hard rocks
Ridges due to the erosion of an anticline (Warrior Ridge)
Ridges due to the erosion of a syncline (ridges in Dunning Cove)
Valley excavated in soft rocks
Water gap or narrows (Bedford and Mount Dallas)
Zigzag ridges

Fargo, N.D.–Minn.
Intrenched meander
Intrenched stream
Lake plain (Glacial Lake Agassiz)
Youthful drainage
Youthful topography

Fayetteville, W.Va.
Canyon or gorge
Cliff (top of gorge)
Dissected plateau (northwestern part)
Intrenched meander
Intrenched stream
Mature drainage (northwestern part)
Mature topography (northwestern part)
Plateau
Waterfall or rapid (Kanawha Falls)

Fond du Lac, Wis.
Delta
Drumlin (southeastern part)
Escarpment
Glacial lake and pond
Ground moraine
Infantile drainage
Swamp or marsh
Terminal moraine

Fort Payne, Ala.–Ga.
Anticlinal valley (Big Wills Creek)
Canyon or gorge (Little River Gulf)
Escarpment (east front of Lookout Mountain)
Old drainage (Coosa Valley)
Old topography (Coosa Valley)
Plateau
Ridges due to hard rocks
Stream meander

Genoa, N.Y.
- Delta
- Dissected plateau modified by glaciation
- Ground moraine
- Spit
- Youthful drainage (parallel)
- Youthful topography

Gilbert Peak, Utah–Wyo.
- Comb ridge or cleaver
- Disappearing stream
- Glacial cirque
- Glacial col
- Glacial lake and pond
- Glaciated or trough valley
- Hanging valley
- Kettlehole
- Matterhorn peak
- Mountain modified by alpine glaciation
- Solution basin and sinkhole

Glacier National Park, Mont.
- Alpine or valley glacier
- Canyon or gorge
- Cliff (bordering glacial cirques)
- Comb ridge or cleaver
- Continental or subcontinental divide
- Delta (St. Mary)
- Dissected alluvial slope (Milk River Ridge)
- Fault-line scarp (east front of mountains)
- Glacial cirque
- Glacial lake and pond
- Glaciated or trough valley
- Glacier
- Hanging glacier
- Hanging valley
- Matterhorn peak
- Mountain modified by alpine glaciation
- Mountain range
- Nonglaciated or V-shaped valley (southwestern part)
- Rugged or youthful mountain
- Waterfall or rapid
- Wind gap or pass (Gunsight, Swiftcurrent, etc.)

Grass Creek Basin, Wyo.
- Anticlinal valley (Grass Creek Basin)
- Cuesta
- Dissected alluvial slope
- Mature topography
- Ridges due to hard rocks
- Ridges due to the erosion of an anticline
- Stream terrace (Cottonwood Creek)

Greylock, Mass.–Vt.
- Canyon or gorge
- Escarpment
- Glacial lake and pond
- Ground moraine
- Hanging valley
- Mountain modified by continental glaciation
- Mountain peak, isolated
- Plateau
- Valley excavated in soft rocks

Henderson, Ky.–Ind.
- Alluvium-filled valley
- Development or shifting of meander
- Dissected plain
- Floodplain
- Lateral drainage diverted by floodplain deposits
- Mature topography
- Stream meander
- Underfit stream (Canoe Creek)
- Youthful drainage

Hollow Springs, Tenn.
- Contrast among surfaces formed in different cycles of erosion
- Dip slope (Plateau of the Barrens)
- Dissected plateau
- Early mature drainage
- Early mature topography
- Encroachment of younger upon older drainage
- Migrating divide (western edge of plateau)
- Nonglaciated or V-shaped valley
- Old drainage (on plateau)
- Old topography (on plateau)
- Plateau
- Solution basin and sinkhole

Honomu, Hawaii
- Canyon or gorge
- Consequent stream (all streams)
- Hanging valley
- Rock stack
- Volcanic cone
- Waterfall or rapid
- Wave-cut cliff
- Youthful drainage
- Youthful topography

Hot Springs and vicinity, Ark.
- Dissected peneplain
- Intrenched stream
- Mature topography
- Ridges due to hard rocks
- Ridges due to the erosion of an anticline
- Ridges due to the erosion of a syncline
- Valley excavated in soft rocks
- Water gap or narrows
- Zigzag ridges

Houghton, Mich.
- Delta (Sturgeon and Pilgrim Rivers)
- Drainage deranged by ice sheet
- Glacial lake and pond
- River valley, abandoned
- Sand dune
- Swamp or marsh
- Terminal moraine
- Wave-cut cliff
- Youthful drainage
- Youthful topography

Interlachen, Fla.
- Lake in solution basin
- Lake plain
- Meander scar
- Solution basin and sinkhole
- Stream prematurely aged by valley filling (Oklawaha River)
- Swamp or marsh

Joplin district, Mo.–Kan.
- Alluvial slope
- Dissected plateau
- Early mature topography
- Floodplain
- Intrenched meander
- Intrenched stream
- Waterfall or rapid (Grand Falls)

Kaaterskill, N.Y.
- Beheaded stream (Gooseberry and Schoharie creeks and Beaver Kill)
- Escarpment
- Glacial lake and pond
- Ground moraine
- Late mature topography (lowland)
- Mature topography (upland)
- Migrating divide
- Mountain modified by continental glaciation
- Rock terrace (on escarpment)
- Stream piracy (*see* Beheaded stream)
- Trellis drainage (lowland)
- Wind gap or pass

Kalamazoo, Mich.
- Drainage deranged by ice sheet
- Glacial lake and pond
- Infantile drainage
- Infantile topography
- Kame
- Kettlehole
- Pitted outwash plain
- Stream prematurely aged by valley filling (Kalamazoo River)
- Stream terrace
- Swamp or marsh
- Terminal moraine
- Underfit stream (Kalamazoo River)

Kilmarnock, Va.
- Bay bar
- Beach, abandoned
- Cut-off meander or oxbow lake
- Dissected plain (upland)
- Drowned valley
- Estuary
- Land-tied island or tombolo
- Meander scar (Meachim Creek)
- Recurved or hooked spit
- Tidal marsh
- Wave-formed plain (upland)
- Wave-formed terrace

Kimmswick, Mo.–Ill.
- Contrast among surfaces formed in different cycles of erosion
- Cut-off meander or oxbow lake
- Development or shifting of meander
- Dissected plateau (upland)
- Floodplain
- Lateral drainage diverted by floodplain deposits
- Mature topography (upland)
- Meander scar
- Natural and artificial levees
- Solution basins and sinkhole
- Stream terrace (Meramec River and Sandy Creek)
- Unsymmetrical stream (upland in Missouri)

Knoxville, Iowa
- Cut-off meander or oxbow lake
- Dissected plain
- Early mature drainage
- Early mature topography
- Floodplain
- Stream meander

Lake Placid, N.Y.
- Glacial lake and pond
- Ground moraine (in valley)
- Mountain modified by continental glaciation
- Stream prematurely aged by valley filling (Saranac River)
- Swamp or marsh
- Water gap or narrows (Wilmington Notch)
- Wind gap or pass
- Youthful drainage

Lake Providence, La.
- Cut-off meander or oxbow lake
- Development or shifting of meander
- Floodplain
- Natural and artificial levees
- Swamp or marsh

La Porte, Tex.
- Drowned valley
- Infantile drainage
- Intrenched stream
- Meander scar
- Mud spring (mud volcano), extinct
- Natural and artificial levees

Stabilized stream
Wave-cut cliff
Loveland, Colo.
Artificial lake (for irrigation)
Canyon or gorge
Consequent stream (on plain)
Cut-off meander or oxbow lake
Hogback
Old drainage (on plain)
Old topography (on plain)
Plain
Ridges due to hard rocks
Ridges due to the erosion of an anticline
Ridges due to the erosion of a syncline
Stream meander
Valley excavated in soft rocks
Water gap or narrows
Wind gap or pass
Lower Matanuska Valley, Alaska
Braided stream
Glacial cirque
Glacial lake and pond
Glaciated or trough valley
Glacier
Hanging valley (western part)
Intrenched stream
Kettlehole
Ridges due to the erosion of a syncline
Stream terrace
Terminal moraine
Youthful drainage
Malaga, Wash.
Canyon or gorge
Cliff
Dip slope
Escarpment
Meander scar
Plateau (eastern part)
Rock terrace
Stream terrace (along Columbia River)
Underfit stream (Moses Coulee)
Waterfall or rapid
Manitou, Colo.
Alluvial fan or cone (The Mesa)
Alluvial slope
Artificial lake (municipal water supply)
Canyon or gorge
Contrast among surfaces formed in different cycles of erosion
Glacial cirque
Monadnock (Pikes Peak)
Mountain peak, isolated
Mountain modified by alpine glaciation
Nonglaciated or V-shaped valley
Plain
Swamp or marsh
Terminal moraine
Marysville Buttes and vicinity, Calif.
Alluvial fan or cone (at West Butte and Sutter)
Alluvial slope, possibly including planed rock slope
Alluvium-filled valley
Cut-off meander or oxbow lake
Dissected volcanic cone
Distributary stream
Floodplain
Infantile drainage
Natural and artificial levees
Plain
Stream meander
Swamp or marsh
Wedge segments of volcanic cone, outlined by erosion
Meriden, Conn.
Drainage deranged by ice sheet
Escarpment
Fault-line scarp
Kettlehole
Artificial lake
Rock sculpture controlled by fractures (Hanging Hill, etc.)
Stream terrace
Terminal moraine (Southington)
Valley excavated in soft rocks
Youthful drainage
Mesa Verde National Park, Colo.
Canyon or gorge
Cliff
Consequent stream (southward-flowing)
Contrast among surfaces formed in different cycles of erosion
Dip slope (old plateau surface)
Dissected plateau
Early mature topography
Escarpment (north face of plateau)
Migrating divide
Milton, W.Va.
Dissected peneplain
Dissected plateau
Intrenched meander
Intrenched stream
Mature drainage
Mature topography
Meander migrating downstream (most meanders)
River valley, abandoned (Teays Valley)
Stream piracy (Trace and Sleepy creeks)
Underfit stream (Mud River)
Monadnock, N.H.
Dissected plateau modified by glaciation
Drainage deranged by ice sheet
Glacial lake and pond
Ground moraine
Hanging valley
Monadnock
Mountain peak, isolated
Mountain modified by continental glaciation
Swamp or marsh
Monticello, Ky.
Dissected peneplain
Dissected plateau (southeast corner)
Early mature drainage
Early mature topography
Floodplain
Intrenched meander
Intrenched stream
Mature drainage (southeast corner)
Mature topography (southeast corner)
Meander migrating downstream
Meander scar
Monadnock (Chapel Hill, etc.)
Solution basin and sinkhole
Montpelier, Idaho–Wyo.–Utah
Alluvial slope
Alluvium-filled valley (Thomas Fork Valley)
Bay bar
Consequent stream
Cut-off meander or oxbow lake
Early mature topography
Escarpment
Fault-line scarp
Intrenched meander
Lake plain
Solution basin and sinkhole
Stream meander
Subsequent stream
Superimposed stream
Swamp or marsh
Terrace formed by silted-up lake basin, partly excavated later
Underfit stream (Bear Lake outlet)
Valley due to dislocation of the rocks
Mount Desert, Maine
Fiord (Somes Sound)
Glacial lake and pond
Mountain modified by continental glaciation
Recurved or hooked spit (Meadow Point)
Rock sculpture controlled by fractures
Spit (Hadley Point, etc.)
Swamp or marsh
Mount Hood, Oreg.–Wash.
Alpine or valley glacier
Braided stream
Canyon or gorge
Cliff
Dissected plateau
Glacial cirque
Glacial lake and pond
Glacier
Hanging glacier
Hanging valley
Landslide
Landslide lake
Mature topography
Mountain peak, isolated
Volcanic cone (Mount Hood)
Waterfall or rapid
Mount Mitchell, N.C.
Continental or subcontinental divide (Blue Ridge summit)
Contrast among surfaces formed in different cycles of erosion
Encroachment of younger upon older drainage (Blue Ridge summit)
Intrenched meander (northern part)
Intrenched stream (northern part)
Late mature topography
Migrating divide (Blue Ridge summit)
Mountain not glaciated
Subdued or old mountain
Wind gap or pass (Swannanoa Gap)
Mount Rainier National Park, Wash.
Alpine or valley glacier
Braided stream
Comb ridge or cleaver
Early mature topography
Glacial cirque
Glacial col
Glacial lake and pond
Glaciated or trough valley
Glacier
Hanging glacier
Hanging valley
Matterhorn peak
Medial and lateral moraines
Mesa
Mountain modified by alpine glaciation
Mountain peak, isolated
Nunatak
Rock sculpture controlled by fractures
Volcanic cone
Wedge segments of volcanic cone, outlined by erosion
Wind gap or pass (Chinook Pass)
Mount Riley, N.M.
Alluvial slope, possibly including planed rock slope
Dissected alluvial slope
Mountain peak, isolated

Volcanic cone (western part)
Youthful drainage
Naples, N.Y.
Delta (streams entering Canandaigua Lake)
Dissected plateau, modified by glaciation
Drainage deranged by ice sheet
Glacial lake and pond
Glaciated or trough valley
Ground moraine
Mature topography
River valley, abandoned
Stream prematurely aged by valley filling
Swamp or marsh
Terminal moraine (above Naples)
Underfit streams (Cohocton River and Naples Creek)
Youthful drainage (parallel)
Nashua, Mont.
Alluvium-filled valley
Cut-off meander or oxbow lake
Dissected plain (south of Missouri River)
Floodplain
Glacial lake and pond
Mature topography (south of Missouri River)
Plain
Stream meander
Stream prematurely aged by valley filling (Milk River)
Underfit stream (Milk River)
Youthful drainage (north of Milk River)
Youthful topography (north of Milk River)
Natural Bridge, Va.
Intrenched meander
Intrenched stream
Meander migrating downstream
Mountain not glaciated
Natural bridge
Ridges due to hard rocks
Ridges due to the erosion of a syncline
Solution basin and sinkhole
Synclinal stream
Valley excavated in soft rocks
Niagara River and vicinity, N.Y.
Beach, abandoned (ridge at Dickersonville)
Canyon or gorge
Escarpment (above Lewiston)
Ground moraine
Plain
Waterfall or rapid
Oberlin, Ohio
Beach, abandoned (various "ridges")
Intrenched meander
Intrenched stream
Wave-cut cliff
Youthful topography
Oelwein, Iowa
Contrast between a surface deeply covered with drift and one which has only a slight cover of much older dissected drift
Dissected plain (northeastern part)
Ground moraine (southwestern part)
Mature drainage (northeastern part)
Mature topography (northeastern part)
Plain (southwestern part)
Youthful drainage (southwestern part)
Youthful topography (southwestern part)
Ogallala, Neb.
Braided stream
Lake in dune area
Sand dune
Youthful drainage
Youthful topography
Olancha, Calif.
Alluvial slope (eastern part)
Alluvium-filled valley (eastern part)
Block mountain
Bolson basin (eastern part)
Canyon or gorge
Encroachment of younger upon older drainage
Escarpment (eastern part)
Fault-line scarp (eastern part)
Glacial cirque (northern part)
Glacial lake and pond
Glaciated or trough valley (Kern Canyon)
Hanging valley
Migrating divide (eastern part)
Mountain modified by alpine glaciation (northern part)
Mountain not glaciated (southern part)
Nonglaciated or V-shaped valley
Plateau
Oyster Bay, N.Y.
Bay bar (Bar Beach)
Drowned valley
Land-tied island or tombolo
Recurved or hooked spit (Plum Point)
Spit (Matinecock Point)
Terminal moraine
Wave-cut cliff
Winged headland (Oak Neck)
Palmyra, N.Y.
Beach, abandoned (of Glacial Lake Iroquois)
Drainage deranged by ice sheet
Drumlin
Ground moraine (southern part)
Infantile drainage
Infantile topography
River valley, abandoned (southern part)
Swamp or marsh
Passadumkeag, Maine
Braided stream due to inequalities in its bed
Drainage deranged by ice sheet
Esker ("horsebacks")
Glacial lake and pond
Infantile drainage
Swamp or marsh
Pawpaw, Md.–W.Va.–Pa.
Dissected peneplain (900- to 950-feet altitude)
Intrenched meander
Mature drainage
Mature topography
Meander migrating downstream
Ridges due to hard rocks
Ridges due to the erosion of a syncline (Purslane Mountain and Sideling Hill)
River valley, abandoned (opposite Pawpaw)
Solution basin and sinkhole
Trellis drainage
Valley excavated in soft rocks
Water gap or narrows (at Lineburg)
Peeples, S.C.–Ga.
Cut-off meander or oxbow lake
Floodplain
Plain
Solution basin and sinkhole
Stream meander
Stream prematurely aged by valley filling
Underfit stream
Peever, S.D.–Minn.
Continental or subcontinental divide
Intrenched meander
Glacial lake and pond
Kettlehole
Plain
River valley, abandoned (Glacial Warren River)
Stream meander
Stream terrace
Swamp or marsh
Terminal moraine
Underfit streams (Jim Creek and Minnesota River)
Youthful drainage
Pittsburgh, Pa.
Dissected plateau
Intrenched meander
Intrenched stream
Mature drainage
Mature topography
Migrating divide (main divides all migrating northward)
Nonglaciated or V-shaped valley
River valley, abandoned (in Pittsburgh and back of McKeesport)
Stream meander (Youghiogheny River from Buenavista to Versailles)
Stream terrace
Unsymmetrical stream
Playas, N.M.
Alluvial slope, possibly including planed rock slope
Alluvium-filled valley
Bolson basin (Playas Valley)
Playa (bottom of Playas Lake)
Playa lake
Point Reyes, Calif.
Barrier beach (on Drakes Bay)
Bay bar
Block mountain (Inverness Ridge)
Delta (head of Tomales Bay)
Distributary stream (on delta)
Drainage blocked by barrier beach
Drainage deranged by faulting
Drowned valley (Tomales Bay)
Estuary (Tomales Bay and estuaries off Drakes Bay)
Hanging valley
Intrenched stream (Walker Creek)
Lagoon or coastal lake
Nonglaciated or V-shaped valley
Offset barrier beach
Rock stack
Sand dune
Truncated headland
Valley due to dislocation of the rocks (Tomales Bay and Olema Creek Valley, San Francisco earthquake rift)
Wave-cut cliff
Wave-formed terrace
Port Valdez district, Alaska
Alpine or valley glacier
Bay bar
Braided stream
Delta
Distributary stream
Fiord (Port Valdez)
Glacial outwash plain
Glacier
Hanging glacier
Hanging valley
Ice-dammed glacial lake

Medial and lateral moraines (on glaciers)
Nunatak
Recurved or hooked spit
Tidal flat
Tidewater glacier (Shoup Glacier)
Quincy, Wash.
Cliff
Dip slope
Escarpment
Infantile topography (lake plain)
Lake plain
Stream terrace
Waterfall, abandoned
Ravenswood, W.Va.–Ohio
Dissected peneplain
Floodplain
Intrenched meander
Mature drainage
Mature topography
Meander migrating downstream (Ohio River)
Meander scar (at Letart)
River valley, abandoned (at Racine)
Stream terrace
Redlands, Calif.
Alluvial slope
Alluvium-filled valley
Braided stream (Santa Ana, Wash.)
Canyon or gorge
Contrast among surfaces formed in different cycles of erosion (northern part)
Encroachment of younger upon older drainage (northern part)
Fault-line scarp (front of mountain)
Mountain not glaciated
Nonglaciated or V-shaped valley
Rugged or youthful mountain
Sand wash
Reelfoot Lake, Tenn.–Mo.–Ky.
Delta
Dissected peneplain
Early mature topography
Earthquake lake
Floodplain
Natural and artificial levees
Stream meander
Rehoboth, Del.
Barrier beach
Drowned valley
Estuary
Lagoon or coastal lake
Offset barrier beach
Plain
Sand dune
Solution basin and sinkhole
Thorofare or tidal runway
Tidal marsh
Truncated headland (Rehoboth Beach)
Rochester east, N.Y.
Barrier beach
Bay bar
Beach, abandoned (Ridge Road)
Canyon or gorge
Delta (Irondequoit River)
Drowned valley
Intrenched stream
Kame
Kettlehole
Lagoon or coastal lake
Swamp or marsh
Terminal moraine
Waterfall or rapid (Rochester)
Youthful drainage
Youthful topography
St. Albans, Vt.
Bay bar
Beach, abandoned (east of lake)
Delta (Missisquoi River and Mill Run)
Distributary stream (Missisquoi delta)
Intrenched stream
Natural and artificial levees
Spit (Stephenson Point)
Swamp or marsh
Waterfall or rapid (Highgate Falls)
Wave-formed terrace
St. Croix Dalles, Wis.–Minn.
Canyon or gorge
Glacial lake and pond
Infantile drainage
Infantile topography
Kame
Kettlehole
Stream terrace (at Taylors Falls)
Swamp or marsh
Terminal moraine
Waterfall or rapid (St. Croix Falls)
Seattle, Wash.
Cut-off meander or oxbow lake
Delta
Distributary stream
Floodplain
Glacial lake and pond
Spit
Stream meander
Stream prematurely aged by valley filling
Tidal flat
Wave-cut cliff
Standingstone, Tenn.
Disappearing stream
Dissected peneplain (1,000 and 1,900 feet altitude)
Dissected plateau (edge of plateau)
Intrenched meander
Intrenched stream
Plateau (Cumberland Plateau)
Solution basin and sinkhole
Stockton, Utah
Alluvial fan or cone (of Ophir Creek)
Alluvial fan or cone, abandoned (of Soldier Creek)
Alluvial slope
Alluvium-filled valley
Bay bar, abandoned (north of Stockton)
Beach, abandoned (northeast of Stockton)
Bolson basin (Rush Valley)
Canyon or gorge
Cliff (in Dry Canyon)
Lake plain (all below 5,250 feet)
Mature topography (mountain)
Mountain not glaciated
Nonglaciated or V-shaped valley
Playa (floor of Rush Lake)
Playa lake (Rush Lake)
Recurved or hooked spit, abandoned (Stockton)
Rugged or youthful mountain
Swamp or marsh
Valley due to dislocation of the rocks
Wave-cut cliff, abandoned (northeast of Stockton)
Wave-formed terrace (northeast of Stockton)
Stryker, Mont.
Comb ridge or cleaver
Drumlin
Escarpment
Glacial cirque
Glacial col
Glacial lake and pond
Glaciated or trough valley
Matterhorn peak
Mountain modified by alpine glaciation
Swamp or marsh
Tehama, Calif.
Alluvial fan or cone (Mill and Antelope creeks)
Alluvial slope
Alluvium-filled valley
Canyon or gorge
Contrast among surfaces formed in different cycles of erosion (northeastern part)
Distributary stream (on fans)
Fault-line scarp (base of mountain)
Floodplain
Infantile topography
Nonglaciated or V-shaped valley
Valley due to dislocation of the rocks
Youthful topography
Van Horn, Tex.
Alluvial fan or cone (Victoria Canyon)
Alluvial slope
Alluvium-filled valley
Block mountain (Sierra Diablo)
Bolson basin (Salt Flat)
Canyon or gorge
Cliff (eastern face of Sierra Diablo)
Cuesta (Sierra Diablo)
Escarpment (eastern front of Sierra Diablo)
Fault-line scarp (eastern foot of Sierra Diablo)
Mountain not glaciated
Playa (floor of Salt Lake)
Playa lake (Salt Lake)
Valley due to dislocation of the rocks
Washington and vicinity, D.C.–Md.–Va.
Braided stream
Canyon or gorge
Delta (creeks below Alexandria)
Delta, abandoned
Dissected peneplain (east of Potomac and Anacostia rivers)
Drowned valley
Early mature topography
Estuary (Potomac River)
Fall line (through Washington, from northeast to southwest)
Floodplain
Intrenched meander
Intrenched stream
Meander scar
Rejuvenated stream (Potomac River)
Stream terrace (below Great Falls)
Swamp or marsh
Tidal flat
Waterfall or rapid (Great and Little Falls)
Waukegan, Ill.–Wis.
Extended stream (streams east of Chicago & Northwestern Ry.)
Kame
Kettlehole
Lake plain (higher part of Waukegan and Zion City)
Stream prematurely aged by valley filling
Swamp or marsh
Terminal moraine
Terrace formed by silted-up lake basin, partly excavated later (same as lake plain)
Wave-cut cliff (south of Waukegan)

Wave-cut cliff, abandoned (north of Waukegan)
Wave-formed terrace (lower part of Waukegan and Zion City)
Youthful drainage

White Mountains, Calif.–Nev.
Alluvial fan or cone (Montgomery, Marble, and Milner creeks)
Alluvial slope
Alluvium-filled valley
Bolson basin (Fish Valley)
Contrast among surfaces formed in different cycles of erosion
Distributary stream (on fans)
Early mature topography (mountain)
Fault-line scarp (western foot of White Mountains)
Glacial cirque (summit of White Mountains)
Mountain range
Playa (Fish Valley)
Rugged or youthful mountain
Sand dune
Valley due to dislocation of the rocks (west of White Mountains)

Yosemite Valley, Calif.
Alluvial fan or cone (below Rocky Point)
Alluvial slope
Alluvium-filled valley
Canyon or gorge
Cliff (El Capitan and others)
Contrast among surfaces formed in different cycles of erosion
Distributary stream (on alluvial slope)
Dome of crystalline rock (Half Dome and others)
Glaciated or trough valley
Hanging valley (Bridalveil Valley and others)
Lake plain (bottom of Yosemite Valley)
Roches moutonnées (Liberty Cap and Mount Broderick)
Rock sculpture controlled by fractures
Rock sculpture controlled by exfoliation
Rock slide and talus (foot of cliffs)
Waterfall or rapid

List B

(Asterisks [*] indicate that a particular feature is well shown on the map.)

COASTAL FEATURES

Barrier beach
* Aransas Pass, Tex.
* Cape Henry, Va.
* Capitola, Calif.
* Cumberland Island, Ga.
Oyster Bay, N.Y.
Point Reyes, Calif.
* Rehoboth, Del.
Rochester east, N.Y.

Bay bar
Kilmarnock, Va.
* Montpelier, Idaho–Wyo.–Utah
* Oyster Bay, N.Y.
Point Reyes, Calif.
Port Valdez district, Alaska
* Rochester east, N.Y.
St. Albans, Vt.

Bay bar, abandoned
* Stockton, Utah

Beach, abandoned
* Aransas Pass, Tex.
Capitola, Calif.
Cumberland Island, Ga.
Erie, Pa.
* Kilmarnock, Va.
* Niagara River and vicinity, N.Y.
* Oberlin, Ohio
Palmyra, N.Y.
Rochester east, N.Y.
St. Albans, Vt.
* Stockton, Utah

Delta
See VALLEYS

Drowned valley
* Cape Henry, Va.
* Kilmarnock, Va.
* La Porte, Tex.
Oyster Bay, N.Y.
* Point Reyes, Calif.
* Rehoboth, Del.
Rochester east, N.Y.
* Washington and vicinity, D.C.–Md.–Va.

Estuary
* Cape Henry, Va.
* Kilmarnock, Va.
Point Reyes, Calif.
* Rehoboth, Del.
* Washington and vicinity, D.C.–Md.–Va.

Fiord
* Mount Desert, Maine
* Port Valdez district, Alaska

Foreland
* Cape Henry, Va.

Lagoon or coastal lake
* Aransas Pass, Tex.
Capitola, Calif.
Erie, Pa.
Point Reyes, Calif.
Rehoboth, Del.
Rochester east, N.Y.

Land-tied island or tombolo
Kilmarnock, Va.
Oyster Bay, N.Y.

Offset barrier beach
* Aransas Pass, Tex.
Cumberland Island, Ga.
Point Reyes, Calif.
Rehoboth, Del.

Recurved or hooked spit
Aransas Pass, Tex.
Cumberland Island, Ga.
* Erie, Pa.
Kilmarnock, Va.
Mount Desert, Maine
Oyster Bay, N.Y.
Port Valdez district, Alaska

Recurved or hooked spit, abandoned
* Stockton, Utah

Rock stack
Honomu, Hawaii
* Point Reyes, Calif.

Spit
Genoa, N.Y.
Mount Desert, Maine
* Oyster Bay, N.Y.
St. Albans, Vt.
* Seattle, Wash.

Thorofare or tidal runway
Capitola, Calif.
* Cumberland Island, Ga.
Rehoboth, Del.

Tidal flat
Port Valdez district, Alaska
* Seattle, Wash.
* Washington and vicinity, D.C.–Md.–Va.

Tidal marsh
Aransas Pass, Tex.
Capitola, Calif.
* Cumberland Island, Ga.
Kilmarnock, Va.
Rehoboth, Del.

Truncated headland
Aransas Pass, Tex.
* Point Reyes, Calif.
Rehoboth, Del.

Wave-cut cliff
Aransas Pass, Tex.
Capitola, Calif.
* Erie, Pa.
* Honomu, Hawaii
* Houghton, Mich.
La Porte, Tex.
* Oberlin, Ohio
* Oyster Bay, N.Y.
* Point Reyes, Calif.
Seattle, Wash.
* Waukegan, Ill.–Wis.

Wave-cut cliff, abandoned
* Capitola, Calif.
* Stockton, Utah
* Waukegan, Ill.–Wis.

Wave-formed plain
* Cape Henry, Va.
Kilmarnock, Va.

Wave-formed terrace
Capitola, Calif.
* Erie, Pa.
* Kilmarnock, Va.
Point Reyes, Calif.
* St. Albans, Vt.
* Stockton, Utah
Waukegan, Ill.–Wis.

Winged headland
* Aransas Pass, Tex.
* Oyster Bay, N.Y.

MOUNTAINS

Alluvial fan or cone
See VALLEYS

Alluvial slope
See VALLEYS

Alluvial slope, possibly including planed rock slope
See VALLEYS

Block mountain
Olancha, Calif.
Point Reyes, Calif.
* Van Horn, Tex.

Dissected alluvial slope
See VALLEYS

Dome of crystalline rock
* Atlanta, Ga.
* Yosemite Valley, Calif.

Landslide
See VALLEYS

Matterhorn peak
See FEATURES DUE TO GLACIATION OF MOUNTAINS

Monadnock
See RESULTS OF EROSION IN VARIOUS STAGES OF THE CYCLE

Mountain modified by alpine glaciation
* Gilbert Peak, Utah–Wyo.
* Glacier National Park, Mont.
* Manitou, Colo.

* Mount Rainier National Park, Wash.
Olancha, Calif.
* Stryker, Mont.

Mountain modified by continental glaciation
* Greylock, Mass.–Vt.
* Kaaterskill, N.Y.
* Lake Placid, N.Y.
* Monadnock, N.H.
* Mount Desert, Maine

Mountain not glaciated
* Ballarat, Calif.–Nev.
* Mount Mitchell, N.C.
* Natural Bridge, Va.
* Olancha, Calif.
* Redlands, Calif.
* Stockton, Utah
* Van Horn, Tex.

Mountain peak, isolated
Greylock, Mass.–Vt.
* Manitou, Colo.
* Monadnock, N.H.
* Mount Hood, Oreg.–Wash.
* Mount Rainier National Park, Wash.
* Mount Riley, N.M.

Mountain range
* Glacier National Park, Mont.
* White Mountains, Calif.

Rock slide and talus
See VALLEYS

Rugged or youthful mountain
* Ballarat, Calif.–Nev.
* Glacier National Park, Mont.
* Redlands, Calif.
* Stockton, Utah
White Mountains, Calif.–Nev.

Subdued or old mountains
* Mount Mitchell, N.C.

Volcanic cone
See VOLCANIC FEATURES

Wind gap or pass
Antietam, Md.–Va.–W.Va.
* Glacier National Park, Mont.
* Kaaterskill, N.Y.
Lake Placid, N.Y.
Loveland, Colo.
* Mount Mitchell, N.C.
Mount Rainier National Park, Wash.

RIDGES AND HILLS

Cuesta
* Chisos Mountains, Tex.
* Grass Creek Basin, Wyo.
* Van Horn, Tex.

Hogback
* Loveland, Colo.

Ridges due to hard rocks
* Antietam, Md.–W.Va.–Va.
* Chisos Mountains, Tex.
* De Queen, Ark.–Okla.
* Everett, Pa.
Fort Payne, Ala.–Ga.
Grass Creek Basin, Wyo.
* Hot Springs and vicinity, Ark.
* Loveland, Colo.
Natural Bridge, Va.
Pawpaw, Md.–W.Va.–Pa.

Ridges due to erosion of an anticline
* Everett, Pa.
* Grass Creek Basin, Wyo.
* Hot Springs and vicinity, Ark.
* Loveland, Colo.

Ridges due to erosion of a syncline
* Everett, Pa.
* Hot Springs and vicinity, Ark.
* Loveland, Colo.
* Lower Matanuska Valley, Alaska
* Natural Bridge, Va.
Pawpaw, Md.–W.Va.–Pa.

Sand dune
Aransas Pass, Tex.
Ballarat, Calif.–Nev.
* Cape Henry, Va.
Capitola, Calif.
Cumberland Island, Ga.
Erie, Pa.
Houghton, Mich.
* Ogallala, Neb.
Point Reyes, Calif.
Rehoboth, Del.
White Mountains, Calif.

Zigzag ridges
Everett, Pa.
Hot Springs and vicinity, Ark.

PLATEAUS AND MESAS

Dissected plateau
Bright Angel, Ariz.
* Castlegate, Utah
* East Cincinnati, Ohio–Ky.
* Fayetteville, W.Va.
* Hollow Springs, Tenn.
* Joplin district, Mo.–Kans.
* Kimmswick, Mo.–Ill.
* Mesa Verde National Park, Colo.
* Milton, W.Va.
Monticello, Ky.
Mount Hood, Oreg.–Wash.
* Pittsburgh, Pa.
Standingstone, Tenn.

Dissected plateau modified by glaciation
* Genoa, N.Y.
* Monadnock, N.H.
* Naples, N.Y.

Mesa
Bright Angel, Ariz.
* Buck Hill, Tex.
Crater Lake National Park, Oreg.
Mount Rainier National Park, Wash.

Plateau
Bright Angel, Ariz.
Fayetteville, W.Va.
* Fort Payne, Ala.–Ga.
Greylock, Mass.–Vt.
* Hollow Springs, Tenn.
* Malaga, Wash.
* Olancha, Calif.
* Standingstone, Tenn.

ESCARPMENTS, TERRACES, AND CLIFFS

Cliff
Bright Angel, Ariz.
* Crater Lake, Oreg.
Fayetteville, W.Va.
* Glacier National Park, Mont.
* Malaga, Wash.
* Mesa Verde National Park, Colo.
* Mount Hood, Oreg.–Wash.
* Quincy, Wash.
Stockton, Utah
* Van Horn, Tex.
* Yosemite Valley, Calif.

Escarpment
* Buck Hill, Tex.
* Castlegate, Utah
* Cut Bank, Mont.
* Fond du Lac, Wis.
Fort Payne, Ala.–Ga.
* Greylock, Mass.–Vt.
* Kaaterskill, N.Y.
* Malaga, Wash.
Meriden, Conn.
Mesa Verde National Park, Colo.
Montpelier, Idaho–Wyo.–Utah
* Niagara River and vicinity, N.Y.
* Olancha, Calif.
* Quincy, Wash.
* Stryker, Mont.
* Van Horn, Tex.

Fault-line scarp
* Glacier National Park, Mont.
* Meriden, Conn.
* Montpelier, Idaho–Wyo.–Utah
* Olancha, Calif.
* Redlands, Calif.
* Tehama, Calif.
Van Horn, Tex.
* White Mountains, Calif.–Nev.

Rock terrace
Bright Angel, Ariz.
* Kaaterskill, N.Y.
* Malaga, Wash.

Stream terrace
See VALLEYS

Terrace formed by silted-up lake basin, which later was partly excavated
* Montpelier, Idaho–Wyo.–Utah
* Waukegan, Ill.–Wis.

Wave-cut cliff
See COASTAL FEATURES

Wave-cut cliff, abandoned
See COASTAL FEATURES

Wave-formed terrace
See COASTAL FEATURES

PLAINS

Bolson basin
* Ballarat, Calif.–Nev.
Olancha, Calif.
* Playas, N.M.
Stockton, Utah
Van Horn, Tex.
White Mountains, Calif.–Nev.

Dip slope
* Hollow Springs, Tenn.
Malaga, Wash.
* Mesa Verde National Park, Colo.
Quincy, Wash.

Lake plain
* Cuyuna, Minn.
* Fargo, N.D.–Minn.
* Interlachen, Fla.
* Montpelier, Idaho–Wyo.–Utah
* Quincy, Wash.
Stockton, Utah
* Waukegan, Ill.–Wis.
* Yosemite Valley, Calif.

Plain
* Avon, Ill.
* Baton Rouge, La.
Buck Hill, Tex.
Elk Point, S.D.–Neb.–Iowa
* Loveland, Colo.
* Manitou, Colo.
Marysville Buttes and vicinity, Calif.
Nashua, Mont.
Niagara River and vicinity, N.Y.
Oelwein, Iowa
* Peeples, S.C.–Ga.
* Peever, N.D.–Minn.
* Rehoboth, Del.

Playa
* Ballarat, Calif.–Nev.
* Playas, N.M.
* Stockton, Utah
* Van Horn, Tex.
* White Mountains, Calif.–Nev.

Wave-formed plain
See COASTAL FEATURES

VALLEYS

Alluvial fan or cone
* Chisos Mountains, Tex.
Manitou, Colo.
Marysville Buttes and vicinity, Calif.
* Stockton, Utah
* Tehama, Calif.

* Van Horn, Tex.
* White Mountains, Calif.–Nev.
* Yosemite Valley, Calif.

Alluvial fan or cone, abandoned
* Stockton, Utah

Alluvial slope
Castlegate, Utah
Chico Landing, Calif.
Joplin district, Mo.–Kan.
* Manitou, Colo.
Montpelier, Idaho–Wyo.–Utah
Olancha, Calif.
Redlands, Calif.
* Stockton, Utah
* Tehama, Calif.
* Van Horn, Tex.
* White Mountains, Calif.–Nev.
* Yosemite Valley, Calif.

Alluvial slope, possibly including planed rock slope
* Ballarat, Calif.–Nev.
Buck Hill, Tex.
* Chisos Mountains, Tex.
Marysville Buttes and vicinity, Calif.
* Mount Riley, N.M.
Playas, N.M.

Alluvium-filled valley
* Ballarat, Calif.–Nev.
* Chico Landing, Calif.
Cross Plains, Wis.
East Cincinnati, Ohio–Ky.
Henderson, Ky.–Ind.
Marysville Buttes and vicinity, Calif.
Montpelier, Idaho–Wyo.–Utah
* Nashua, Mont.
Olancha, Calif.
* Playas, N.M.
* Redlands, Calif.
* Stockton, Utah
* Tehama, Calif.
* Van Horn, Tex.
* White Mountains, Calif.–Nev.
* Yosemite Valley, Calif.

Anticlinal valley
Everett, Pa.
* Fort Payne, Ala.–Ga.
* Grass Creek Basin, Wyo.

Canyon or gorge
Ballarat, Calif.–Nev.
Bright Angel, Ariz.
Buck Hill, Tex.
Castlegate, Utah
* Chisos Mountains, Tex.
Cut Bank, Mont.
* East Cincinnati, Ohio–Ky.
* Fayetteville, W.Va.
Fort Payne, Ala.–Ga.
Glacier National Park, Mont.
Greylock, Mass.–Vt.
Honomu, Hawaii
Loveland, Colo.
* Malaga, Wash.
Manitou, Colo.
* Mesa Verde National Park, Colo.
Mount Hood, Oreg.–Wash.
* Niagara River and vicinity, N.Y.
* Olancha, Calif.
Redlands, Calif.
Rochester east, N.Y.
St. Croix Dalles, Wis.–Minn.
Stockton, Utah
Tehama, Calif.
Van Horn, Tex.
* Washington and vicinity, D.C.–Md.–Va.
* Yosemite Valley, Calif.

Delta
Erie, Pa.
Fond du Lac, Wis.
* Genoa, N.Y.
Glacier National Park, Mont.
Houghton, Mich.
* Naples, N.Y.
* Point Reyes, Calif.
* Port Valdez district, Alaska
Reelfoot Lake, Tenn.–Mo.–Ky.
Rochester east, N.Y.
* St. Albans, Vt.
* Seattle, Wash.
* Washington and vicinity, D.C.–Md.–Va.

Delta, abandoned
* Washington and vicinity, D.C.–Md.–Va.

Dissected alluvial slope
* Castlegate, Utah
Cut Bank, Mont.
Glacier National Park, Mont.
* Grass Creek Basin, Wyo.
Mount Riley, N.M.

Drowned valley
See COASTAL FEATURES

Floodplain
Baton Rouge, La.
Castlegate, Utah
* Chico Landing, Calif.
* East Cincinnati, Ohio–Ky.
Elk Point, S.D.–Neb.–Iowa
Everett, Pa.
* Henderson, Ky.–Ind.
Joplin district, Mo.–Kan.
* Kimmswick, Mo.–Ill.
Knoxville, Iowa
* Lake Providence, La.
* Marysville Buttes and vicinity, Calif.
Monticello, Ky.
* Nashua, Mont.
* Peeples, N.C.–Ga.
* Ravenswood, W.Va.–Ohio
* Reelfoot Lake, Tenn.–Mo.–Ky.
Seattle, Wash.
* Tehama, Calif.
* Washington and vicinity, D.C.–Md.–Va.

Glaciated or trough valley (U-shaped)
* Crater Lake National Park, Oreg.
Gilbert Peak, Utah–Wyo.
* Glacier National Park, Mont.
Lower Matanuska Valley, Alaska
* Mount Rainier National Park, Wash.
* Naples, N.Y.
* Olancha, Calif.
* Stryker, Mont.
* Yosemite Valley, Calif.

Hanging valley
Bright Angel, Ariz.
Gilbert Peak, Utah–Wyo.
Glacier National Park, Mont.
Greylock, Mass.–Vt.
* Honomu, Hawaii
Lower Matanuska Valley, Alaska
* Monadnock, N.H.
Mount Hood, Oreg.–Wash.
* Mount Rainier National Park, Wash.
Olancha, Calif.
* Point Reyes, Calif.
Port Valdez district, Alaska
* Yosemite Valley, Calif.

Landslide
* Mount Hood, Oreg.–Wash.

Meander scar
* East Cincinnati, Ohio–Ky.
Interlachen, Fla.
* Kilmarnock, Va.
* Kimmswick, Mo.–Ill.
* La Porte, Tex.
Malaga, Wash.
Monticello, Ky.
* Ravenswood, W.Va.–Ohio
* Washington and vicinity, D.C.–Md.–Va.

Migrating divide
* Atlanta, Ga.
* Grand Canyon National Park (east half), Ariz.
Hollow Springs, Tenn.
* Kaaterskill, N.Y.
* Mesa Verde National Park, Colo.
* Mount Mitchell, N.C.
* Olancha, Calif.
Pittsburgh, Pa.

Natural and artificial levees
* Baton Rouge, La.
Chico Landing, Calif.
Kimmswick, Mo.–Ill.
* Lake Providence, La.
* La Porte, Tex.
Marysville Buttes and vicinity, Calif.
Reelfoot Lake, Tenn.–Mo.–Ky.
* St. Albans, Vt.

Nonglaciated or V-shaped valley
Ballarat, Calif.–Nev.
Bright Angel, Ariz.
Castlegate, Utah
Glacier National Park, Mont.
* Hollow Springs, Tenn.
Manitou, Colo.
Olancha, Calif.
Pittsburgh, Pa.
Point Reyes, Calif.
* Redlands, Calif.
* Stockton, Utah
* Tehama, Calif.

Rock slide and talus
Yosemite Valley, Calif.

River valley, abandoned
* Houghton, Mich.
* Milton, W.Va.
* Naples, N.Y.
* Palmyra, N.Y.
Pawpaw, Md.–W.Va.–Pa.
* Peever, S.D.–Minn.
Pittsburgh, Pa.
Ravenswood, W.Va.–Ohio

Rock terrace
See ESCARPMENTS, TERRACES, AND CLIFFS

Sand wash
Ballarat, Calif.–Nev.
* Redlands, Calif.

Stream terrace
East Cincinnati, Ohio–Ky.
Elk Point, S.D.–Neb.–Iowa
* Grass Creek Basin, Wyo.
Kalamazoo, Mich.
Kimmswick, Mo.–Ill.
Lower Matanuska Valley, Alaska
Malaga, Wash.
Meriden, Conn.
* Peever, S.D.–Minn.
Pittsburgh, Pa.
* Quincy, Wash.
Ravenswood, W.Va.–Ohio
St. Croix Dalles, Wis.–Minn.
* Washington and vicinity, D.C.–Md.–Va.

Valley due to dislocation of the rocks
* Ballarat, Calif.–Nev.
Montpelier, Idaho–Wyo.–Utah
* Point Reyes, Calif.
Stockton, Utah
* Tehama, Calif.
Van Horn, Tex.
White Mountains, Calif.–Nev.

Valley excavated in soft rocks
Antietam, Md.–Va.–W.Va.
Chisos Mountains, Tex.
* De Queen, Ark.–Okla.
* Everett, Pa.
* Greylock, Mass.–Vt.
* Hot Springs and vicinity, Ark.
* Loveland, Colo.
* Meriden, Conn.
Natural Bridge, Va.
* Pawpaw, Md.–W.Va.–Pa.
Water gap or narrows
* Antietam, Md.–Va.–W.Va.
De Queen, Ark.–Okla.
* Everett, Pa.
Hot Springs and vicinity, Ark.
Lake Placid, N.Y.
Loveland, Colo.
Pawpaw, Md.–W.Va.–Pa.

STREAMS

Beheaded stream
* Kaaterskill, N.Y.
Braided stream
Antietam, Md.–Va.–W.Va.
* Lower Matanuska Valley, Alaska
Mount Hood, Oreg.–Wash.
* Mount Rainier National Park, Wash.
* Ogallala, Neb.
* Passadumkeag, Maine
* Port Valdez district, Alaska
* Redlands, Calif.
* Washington and vicinity, D.C.–Md.–Va.
Consequent stream
* Atlanta, Ga.
De Queen, Ark.–Okla.
* Honomu, Hawaii
* Loveland, Colo.
* Mesa Verde National Park, Colo.
Montpelier, Idaho–Wyo.–Utah
Cut-off meander or oxbow lake
Avon, Ill.
Cherry Ridge, Mont.
* Chico Landing, Calif.
Cuyuna, Minn.
Elk Point, S.D.–Neb.–Iowa
Kilmarnock, Va.
Kimmswick, Mo.–Ill.
* Knoxville, Iowa
* Lake Providence, La.
Loveland, Colo.
Marysville Buttes and vicinity, Calif.
Montpelier, Idaho–Wyo.–Utah
* Nashua, Mont.
Peeples, S.C.–Ga.
Seattle, Wash.
Development of shifting meander
* Chico Landing, Calif.
* Henderson, Ky.–Ind.
Kimmswick, Mo.–Ill.
Lake Providence, La.
Disappearing stream
* Gilbert Peak, Utah–Wyo.
* Standingstone, Tenn.
Distributary stream
Marysville Buttes and vicinity, Calif.
* Point Reyes, Calif.
* Port Valdez district, Alaska
* St. Albans, Vt.
* Seattle, Wash.
* Tehama, Calif.
* White Mountains, Calif.–Nev.
* Yosemite Valley, Calif.
Drainage blocked by barrier beach
* Cape Henry, Va.
* Capitola, Calif.
Point Reyes, Calif.
Drainage deranged by faulting
* Point Reyes, Calif.
Drainage deranged by ice sheet
East Cincinnati, Ohio–Ky.
Erie, Pa.
* Houghton, Mich.
* Kalamazoo, Mich.
Meriden, Conn.
Monadnock, N.H.
* Naples, N.Y.
Palmyra, N.Y.
Passadumkeag, Maine
Early mature drainage
Atlanta, Ga.
East Cincinnati, Ohio–Ky.
* Hollow Springs, Tenn.
Knoxville, Iowa
Monticello, Ky.
Encroachment of younger upon older drainage
Bright Angel, Ariz.
* Hollow Springs, Tenn.
Mount Mitchell, N.C.
* Olancha, Calif.
Redlands, Calif.
Extended stream
* Waukegan, Ill.–Wis.
Infantile drainage
Cross Plains, Wis.
* Cut Bank, Mont.
* Cuyuna, Minn.
* Fond du Lac, Wis.
* Kalamazoo, Mich.
La Porte, Tex.
* Marysville Buttes and vicinity, Calif.
* Palmyra, N.Y.
* Passadumkeag, Maine
* St. Croix Dalles, Wis.–Minn.
Intrenched stream
Antietam, Md.–Va.–W.Va.
Atlanta, Ga.
Bright Angel, Ariz.
Chisos Mountains, Tex.
* Cuyuna, Minn.
De Queen, Ark.–Okla.
* East Cincinnati, Ohio–Ky.
Erie, Pa.
* Everett, Pa.
* Fargo, N.D.–Minn.
* Fayetteville, W.Va.
Hot Springs and vicinity, Ark.
Joplin district, Mo.–Kan.
* La Porte, Tex.
Lower Matanuska Valley, Alaska
* Milton, W.Va.
* Monticello, Ky.
Mount Mitchell, N.C.
* Natural Bridge, Va.
* Oberlin, Ohio
Pittsburgh, Pa.
Point Reyes, Calif.
* Rochester east, N.Y.
St. Albans, Vt.
Standingstone, Tenn.
Washington and vicinity, D.C.–Md.–Va.
Lateral drainage diverted by floodplain deposits
Baton Rouge, La.
* Elk Point, S.D.–Neb.–Iowa
Henderson, Ky.–Ind.
Kimmswick, Mo.–Ill.
Mature drainage
Cross Plains, Wis.
* Fayetteville, W.Va.
* Milton, W.Va.
Monticello, Ky.
Oelwein, Iowa
* Pawpaw, Md.–W.Va.–Pa.
* Pittsburgh, Pa.
* Ravenswood, W.Va.–Ohio
Meander migrating downstream
Chico Landing, Calif.
* Elk Point, S.D.–Neb.–Iowa
Milton, W.Va.
* Monticello, Ky.
Natural Bridge, Va.
* Pawpaw, Md.–W.Va.–Pa.
* Ravenswood, W.Va.–Ohio
Old drainage
Bright Angel, Ariz.
* Fort Payne, Ala.–Ga.
* Hollow Springs, Tenn.
* Loveland, Colo.
Rejuvenated stream
Bright Angel, Ariz.
* Washington and vicinity, D.C.–Md.–Va.
Stabilized stream
* Baton Rouge, La.
* La Porte, Tex.
Stream meander
Avon, Ill.
Baton Rouge, La.
Castlegate, Utah
Cherry Ridge, Mont.
* Chico Landing, Calif.
Cuyuna, Minn.
Elk Point, S.D.–Neb.–Iowa
Fort Payne, Ala.–Ga.
* Henderson, Ky.–Ind.
* Knoxville, Iowa
Loveland, Colo.
Marysville Buttes and vicinity, Calif.
Montpelier, Idaho–Wyo.–Utah
* Nashua, Mont.
Peeples, S.C.–Ga.
Peever, S.D.–Minn.
Pittsburgh, Pa.
Reelfoot Lake, Tenn.–Mo.–Ky.
* Seattle, Wash.
Stream piracy
Bright Angel, Ariz.
* Kaaterskill, N.Y.
* Milton, W.Va.
Stream prematurely aged by valley filling
Avon, Ill.
Cherry Ridge, Mont.
Cross Plains, Wis.
Elk Point, S.D.–Neb.–Iowa
Interlachen, Fla.
* Kalamazoo, Mich.
Lake Placid, N.Y.
Naples, N.Y.
* Nashua, Mont.
* Peeples, S.C.–Ga.
* Seattle, Wash.
Waukegan, Ill.–Wis.
Subsequent stream
* De Queen, Ark.–Okla.
Montpelier, Idaho–Wyo.–Utah
Superposed stream
Montpelier, Idaho–Wyo.–Utah
Synclinal stream
* Natural Bridge, Va.
Trellis drainage
* De Queen, Ark.–Okla.
* Kaaterskill, N.Y.
Pawpaw, Md.–W.Va.–Pa.
Underfit stream
Cherry Ridge, Mont.
* Elk Point, S.D.–Neb.–Iowa
Henderson, Ky.–Ind.
* Kalamazoo, Mich.
Malaga, Wash.
Milton, W.Va.

Kaaterskill, N.Y.
Kimmswick, Mo.–Ill.
* Milton, W.Va.
Monticello, Ky.
Mount Hood, Oreg.–Wash.
Naples, N.Y.
Nashua, Mont.
Oelwein, Iowa
* Pawpaw, Md.–W.Va.–Pa.
* Pittsburgh, Pa.
* Ravenswood, W.Va.–Ohio
* Stockton, Utah

Monadnock
* Atlanta, Ga.
Manitou, Colo.
* Monadnock, N.H.
Monticello, Ky.

Old drainage
See STREAMS

Old topography
Bright Angel, Ariz.
* Fort Payne, Ala.–Ga.
* Hollow Springs, Tenn.
* Loveland, Colo.

Prematurely old stream
See STREAMS

Rejuvenated stream
See STREAMS

Rock sculpture controlled by exfoliation
* Atlanta, Ga.
* Yosemite Valley, Calif.

Rock sculpture controlled by fractures
Bright Angel, Ariz.
* Meriden, Conn.
* Mount Desert, Maine
* Mount Rainier National Park, Wash.
* Yosemite Valley, Calif.

Rugged or youthful mountain
See MOUNTAINS

Subdued or old mountain
See MOUNTAINS

Youthful drainage
See STREAMS

Youthful topography
Avon, Ill.
Bright Angel, Ariz.
* Cape Henry, Va.
Cut Bank, Mont.
* Fargo, N.D.–Minn.
* Genoa, N.Y.
* Honomu, Hawaii
* Houghton, Mich.
Nashua, Mont.
* Oberlin, Ohio
Oelwein, Iowa
Ogallala, Neb.
Rochester east, N.Y.
* Tehama, Calif.

GLACIERS

Alpine or valley glacier
Glacier National Park, Mont.
* Mount Hood, Oreg.
* Mount Rainier National Park, Wash.
* Port Valdez district, Alaska

Glacier
* Glacier National Park, Mont.
Lower Matanuska Valley, Alaska
* Mount Hood, Oreg.–Wash.
* Mount Rainier National Park, Wash.
* Port Valdez district, Alaska

Hanging glacier
* Glacier National Park, Mont.
Mount Hood, Oreg.–Wash.
* Mount Rainier National Park, Wash.
* Port Valdez district, Alaska

Tidewater glacier
* Port Valdez district, Alaska

FEATURES DUE TO CONTINENTAL GLACIATION

Contrast between glaciated and nonglaciated surfaces (1), or between a surface deeply covered with drift and one that has only a slight cover of much older dissected drift (2)
* (1) Cross Plains, Wis.
* (2) Oelwein, Iowa

Drainage deranged by ice sheet
See STREAMS

Drumlin
* Fond du Lac, Wis.
* Palmyra, N.Y.
Stryker, Mont.

Esker
Cherry Ridge, Mont.
* Passadumkeag, Maine

Glacial outwash plain
* Port Valdez district, Alaska

Ground moraine
* Avon, Ill.
Cross Plains, Wis.
Erie, Pa.
Fond du Lac, Wis.
Genoa, N.Y.
Greylock, Mass.–Vt.
Kaaterskill, N.Y.
Lake Placid, N.Y.
Monadnock, N.H.
Naples, N.Y.
Niagara Falls and vicinity, N.Y.
Oelwein, Iowa
* Palmyra, N.Y.

Kame
* Cherry Ridge, Mont.
Cuyuna, Minn.
* Kalamazoo east, Mich.
Rochester east, N.Y.
* St. Croix Dalles, Wis.–Minn.
* Waukegan, Ill.–Wis.

Kettlehole
* Cherry Ridge, Mont.
* Cut Bank, Mont.
Cuyuna, Minn.
Gilbert Peak, Utah–Wyo.
* Kalamazoo, Mich.
Lower Matanuska Valley, Alaska
Meriden, Conn.
Peever, S.D.–Minn.
Rochester east, N.Y.
* St. Croix Dalles, Wis.–Minn.
Waukegan, Ill.–Wis.

Mountain modified by continental glaciation
See MOUNTAINS

Pitted outwash plain
* Kalamazoo, Mich.

Terminal moraine
* Cherry Ridge, Mont.
* Cut Bank, Mont.
* Cuyuna, Minn.
Fond du Lac, Wis.
Houghton, Mich.
* Kalamazoo, Mich.
Lower Matanuska Valley, Alaska
Manitou, Colo.
Meriden, Conn.
* Naples, N.Y.
Oyster Bay, N.Y.
Peever, S.D.–Minn.
Rochester east, N.Y.
* St. Croix Dalles, Wis.–Minn.
Waukegan, Ill.–Wis.

Valley train
* Cross Plains, Wis.

FEATURES DUE TO GLACIATION OF MOUNTAINS

Comb ridge or cleaver
* Gilbert Peak, Utah–Wyo.
Glacier National Park, Mont.
* Mount Rainier National Park, Wash.
* Stryker, Mont.

Glacial cirque
* Gilbert Peak, Utah–Wyo.
* Glacier National Park, Mont.
Lower Matanuska Valley, Alaska
* Manitou, Colo.
Mount Hood, Oreg.–Wash.
* Mount Rainier National Park, Wash.
Olancha, Calif.
* Stryker, Mont.
White Mountains, Calif.–Nev.

Glacial col
* Gilbert Peak, Utah–Wyo.
Mount Rainier National Park, Wash.
* Stryker, Mont.

Glaciated or trough valley
See VALLEYS

Matterhorn peak
* Gilbert Peak, Utah–Wyo.
Glacier National Park, Mont.
* Mount Rainier National Park, Wash.
Stryker, Mont.

Medial and lateral moraines
* Mount Rainier National Park, Wash.
* Port Valdez district, Alaska

Mountains modified by alpine glaciation
See MOUNTAINS

Nunatak
* Mount Rainier National Park, Wash.
* Port Valdez district, Alaska

Roches moutonnées
* Yosemite Valley, Calif.

VOLCANIC FEATURES

Caldera
* Crater Lake National Park, Oreg.

Dissected volcanic cone
* Marysville Buttes and vicinity, Calif.

Truncated volcanic cone
* Crater Lake National Park, Oreg.

Volcanic cone
* Crater Lake National Park, Oreg.
* Honomu, Hawaii
* Mount Hood, Oreg.–Wash.
* Mount Rainier National Park, Wash.
* Mount Riley, Ariz.

Wedge segments of volcanic cone outlined by erosion
* Marysville Buttes and vicinity, Calif.
* Mount Rainier National Park, Wash.

SOLUTION FEATURES

Natural bridge
* Natural Bridge, Va.

Solution basin and sinkhole
Gilbert Peak, Utah–Wyo.
Hollow Springs, Tenn.
* Interlachen, Fla.
Kimmswick, Mo.–Ill.
* Monticello, Ky.
Montpelier, Idaho–Wyo.–Utah
Natural Bridge, Va.
Pawpaw, Md.–W.Va.–Pa.
* Peeples, S.C.–Ga.
Rehoboth, Del.
* Standingstone, Tenn.

MISCELLANEOUS

Continental or subcontinental divide
* Glacier National Park, Mont.
* Mount Mitchell, N.C.
Peever, S.D.–Minn.

Fall line
* Washington and vicinity, D.C.–Md.–Va.

Mud spring (mud volcano), extinct
* La Porte, Tex.

APPENDIX C

Climatic Classification and Statistics

The Köppen Classification of Climates

Origin

The most commonly used classification of climates is that devised by a Russian meteorologist, Wladimir Köppen (1846–1940). As early as 1900, Köppen suggested a climatic classification based on plant cover. It was in 1918, however, that he first formulated the actual scheme. In its major aspects, this original plan has not been altered markedly, although there have been many minor modifications. The final form specified by Köppen himself appeared in 1936 as a section of a five-volume work, *Handbook of Climatology*, of which Köppen was one of the authors and an editor.

General Features

The Köppen classification deals with the two climatic elements that are most obviously significant to human beings as they live and make a living on the face of the earth. These two elements are temperature and precipitation. Long-time records of each are available for numerous and widely scattered places on the earth. When these records are analyzed, it is found that certain outstanding combinations regularly occur. For example, some places show a combination of "hot and moist"; others are "hot and dry"; still others are "cold and moist"; and so on, through many temperature and moisture combinations. It is such recognized combinations, or types, that constitute the Köppen system of climatic classification.

This system recognizes five *major* categories. To each of these a letter symbol is given. One of the major categories, for example, is expressed by the letter **A,** which signifies moist climates that are perennially hot. Subdivisions are made within each of the major classes, and each of these in turn is given a letter symbol. One of the subdivisions of the **A** climates is **Af.** The addition of the **f** indicates one special type of **A** climate, one that is not only moist and hot, but moist to the degree that there normally is no season of true drought at any time during the year.

The type of climate, or the *climate-formula,* to use Köppen's designation, is expressed as a combination of two or more letters. The letters are defined both specifically and generally. In the use of the complete system, many detailed differences can be indicated by the addition of as many letter symbols as apply to the measurements at hand. For purposes of a general world survey, the subdivisions of the climatic types usually involve at least a third letter. Thus, a given climate may be symbolized as **Cfa.** The third letter in this instance signifies hot summers, and it, too, is capable of exact, quantitative definition.

Significance of Classification

The Köppen classification of climates is extremely useful, first, because it is a quantitative and wholly objective device for the categorization of observed climatic phenomena. As has been indicated, it makes use of temperature and of precipitation amounts and of temperature and precipitation distribution throughout the year. These are measurable facts that are commonly observed over a large part of the earth's surface. When the Köppen definitions are applied to the observed facts, there is no question as to which category those facts represent. Given a set of climatic statistics, any individual, no matter from what part of the world she or he may come, can arrive at the proper climate-formula for her/his own, or any other, climate. The classification thus provides a sort of international climatic shorthand.

Second, the Köppen classification is an extremely useful geographic tool for world or continental study. The definitions are stated in exact amounts that are capable of accurate representation on maps. It is therefore possible to construct maps on which the areas of different climatic types may be specifically delimited. From these maps, direct comparisons between one part of the world and another, or between one part of a continent and another, may be made. Further, these comparisons are standardized by the quantitative nature of the definitions. On a worldwide or on a continental scale, the patterns disclosed by a map of climatic types are basic to an understanding of the variety that exists. It must here be noted, however, that the classification is of much less value in disclosing significant patterns within small areas.

Third, the Köppen classification is easy to apply. It does not require any special knowledge, either of meteorological details or of mathematical processes. It can, therefore, be used by nonspecialists fully as easily as it can by professional meteorologists.

Finally, the Köppen classification has come into such common use throughout the world that it may now be considered a standard. Nearly all of the many current classifications of climate are variants of the Köppen system. In some variants, names have been substituted for combinations of letter symbols. Thus, it is fairly common to speak of an **Af** climate as a tropical rainforest climate, of a **BW** climate as a desert or saharan climate, and of a **Cs** climate as a Mediterranean climate. In other variants, there have been different definitions proposed for those of Köppen while the main framework remains unaltered. Whether they be descriptive-type variations or definition variations, no one of the alternative classifications has had the worldwide acceptance that has been accorded the Köppen system as a useful geographic tool. This is not to suggest that there are no weaknesses in the Köppen system—for example, the definitions specify types that are far too broad for purposes of climatological research. But, despite difficulties of this sort, the system remains an extremely useful and a widely accepted tool for geographers.

Plan of Classification

Major categories. The five major categories of climate are given the letter symbols **A, B, C, D,** and **E.** Climates that are humid and are always hot, showing no appreciable seasonal changes of temperature, are designated the **A** climates. Dry climates, in which potential evaporation exceeds precipitation, are given the symbol **B.** Humid climates in which there is a definite, though mild, winter are called **C** climates. Humid climates with severe winters use the letter **D.** And climates that have, essentially, no summer season are known as **E** climates. For convenient reference, a tabular summary of general meanings of all symbols is given in Figures C–1 and C–2, in this part of the appendix. In addition, exact definitions for each symbol are given in the second section. The method of determination of the climatic symbol represented by any set of temperature and precipitation statistics is included in the following part of the appendix.

The choice of the particular letters is purely arbitrary. It should be noted that the first letter symbol in the classification is always a capital letter. This is important because, in the definition of subdivisions of the major categories, some of the letters are used again, but they are always written or printed as lower case letters.

The **A, C,** and **D** climates are moist; that is, they are climates in which, regardless of the temperature, there is normally no problem of insufficient precipitation. The **E** climates are so cold that the problem of moisture is not significant. Only in the **B** climates does one find, as a normal condition, the problem of moisture deficiency.

Moisture deficiency, or dryness, must be defined on a sliding scale. Efficiency of precipitation, from the point of view of the cover of vegetation that it produces, is closely related to the rate of evaporation. More water will evaporate at high temperatures than at low ones. If evaporation is high, water available for plants is decreased. Hence, the amount of precipitation that is adequate for plant growth in a cool area is altogether insufficient in an area of constantly high temperatures. Likewise, an area that receives

FIGURE C–1

Twenty-five main climatic types of the Köppen system.

Af	Always hot, always humid climate
Am	Always hot, seasonally excessively humid climate
Aw	Always hot, seasonally droughty climate
BSh	Semiarid, hot climate
BSk	Semiarid, cool or cold climate
BWh	Arid, hot climate
BWk	Arid, cool or cold climate
Cfa	Mild winter, always humid climate with long, hot summers
Cfb	Mild winter, always humid climate with short, warm summers
Cfc	Mild winter, always humid climate with very short, cool summers
Cwa	Mild winter, humid summer climate with long, hot summers
Cwb	Mild winter, humid summer climate with short, warm summers
Csa	Mild winter, humid winter climate with long, hot, droughty summers
Csb	Mild winter, humid winter climate with short, warm, droughty summers
Dfa	Severe winter, always humid climate with long, hot summers
Dwa	Severe winter, humid summer climate with long, hot summers
Dfb	Severe winter, always humid climate with short, warm summers
Dwb	Severe winter, humid summer climate with short, warm summers
Dfc	Severe winter, always humid climate with very short, cool summers
Dwc	Severe winter, humid summer climate with very short, cool summers
Dfd	Severe winter, always humid climate with short summers and excessively cold winters
Dwd	Severe winter, humid summer climate with short summers and excessively cold winters
ET	Polar climate with very short period of plant growth
EF	Polar climate in which plant growth is impossible
H	Undifferentiated mountain climates

most of its precipitation during the warm growing season when the evaporation rate is high must have more water available than an area where most of the precipitation comes during the cooler season when evaporation is low. If evaporation records were kept as frequently and as accurately as are those of temperature and precipitation, it would be possible to determine directly the location of the line along which evaporation equals precipitation. This would permit exact definition of the boundary between climates that are humid and those that are dry. Unfortunately, records of evaporation are comparatively few in number and the points of observation are widely scattered. For these reasons, it has been necessary to introduce into the classification a definition that is not predicated upon a single amount of precipitation, but is in reality a crude estimate of evaporation. This definition makes allowance for temperature, for yearly precipitation, and for seasonal distribution of precipitation.

Subdivisions of the major categories. The **B** climates are subdivided into arid climate, which is represented by the secondary symbol **W,** and semiarid climate, which is represented by the secondary symbol **S.** These are written as capital letters when they are used with **B** and they have no tie with the letters **w** and **s** as those are used with the **A, C,** and **D** types. Beyond the classification as arid or semiarid, a further difference within the **B** climates is made in terms of temperature: those dry climates that are always hot use the symbol **h,** and those that are cold or have a pronounced winter season use the symbol **k.**

In each of the **A, C,** and **D** types the second symbol is one that specifies the nature, the amount, and the distribution of precipitation throughout the year. There are three possibilities so far as the distribution of precipitation during the year is concerned: the precipitation is rather evenly spread throughout the year, is concentrated in the cooler part of the year, or is concentrated in the warmer part of the year. The symbol that indicates an even spread throughout the year is **f;** the one that indicates concentration in the cooler part of the year is **s;** and the one that shows concentration in the warmer part of the year is **w.** Theoretically, any of the three might be used following the first letter of the climate-formula. Actually, **s** occurs so infrequently with either **A** or **D** that it may be omitted so far as those two major categories are concerned. Within the **A** climates, another letter, **m,** is used to indicate an intermediate position between **f** and **w.**

In the **C** and **D** climates a third symbol commonly is used. Since in those two categories the nature of the winter and of the summer have great effect upon the vegetation, the symbols should indicate the nature of each of those seasons. The first letter does this in part; the third letter completes the statement. Three of the letters used in third place characterize the summer and one of them, the winter. Where the summers are long and hot, the symbol **a** is used; for short and warm summers, **b** is used; for very short and cool summers, **c** is the symbol; and where winters are excessively cold and severe, the symbol **d** is used.

Within the climates that, in an ordinary sense, have no summer, **E,** a distinction can be made between areas of practically continuous frost, **F,** and those in which temperatures are high enough to permit the growth of tundra vegetation, **T.**

Summary of Types

Generalized description. In summary, twenty-five climatic types are recognized. These are listed in Figure C–1 in this part

FIGURE C–2
The relationship between the major categories of climate and their subdivisions.

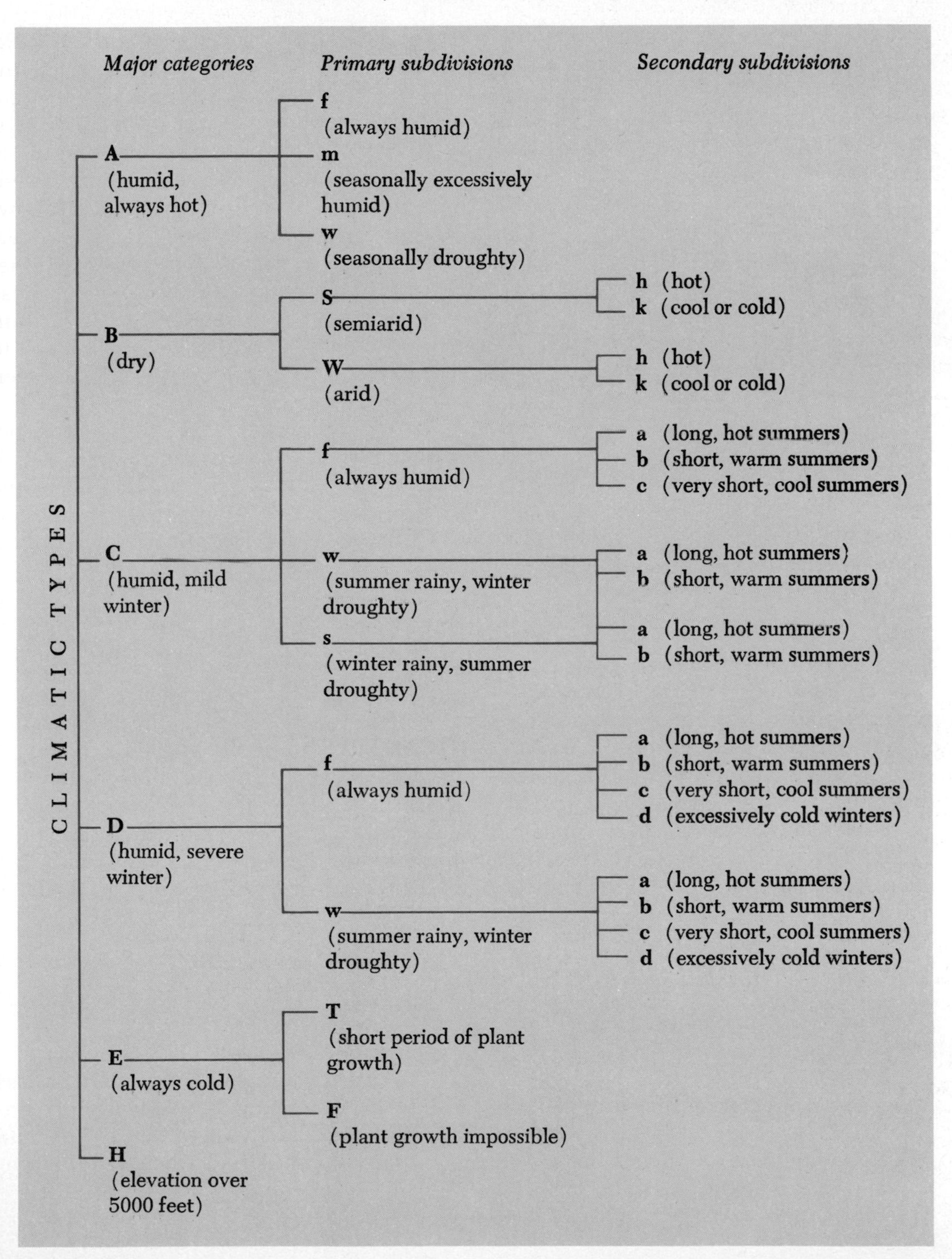

FIGURE C–3

Some climatic type equivalents.

Type name used in this text	*Köppen type*	*Place name example*
Tropical wet	**Af**	Upper Amazon
Tropical monsoon	**Am**	Malabar
Tropical dry-winter	**Aw**	Southern Sudan
Semiarid	**BSh** and **BSk**	Lower Rio Grande and Northern Great Plains
Arid	**BWh** and **BWk**	Sahara and Tarim Basin
Humid subtropical	**Cfa**	Gulf states
Midlatitude marine	**Cfb** and **Cfc**	Northwestern Europe
Subtropical dry-winter and tropical upland	**Cw**	South China and Ethiopia
Mediterranean	**Csa** and **Csb**	Mediterranean countries
Humid continental with hot summers	**Dfa** and **Dwa**	Iowa and North China
Humid continental with warm summers	**Dfb** and **Dwb**	Northern Great Lakes and Northern Manchuria
Subarctic	**Dfc, Dwc, Dfd,** and **Dwd**	Central Canada, Amur Lowland, and Northeastern Siberia
Tundra	**ET**	Northern Alaska
Icecap	**EF**	Antarctica

of the appendix, each with a generalized résumé of its major features. Figure C–2 shows the relationship between the major categories and their subdivisions. It also indicates the relative position of each of the types within the whole scheme.

Equivalent climatic types. It is common practice to refer to climatic types by descriptive names, as has been done in the body of this text, or by names of places in which the climates occur. For convenience, certain of these equivalent names are listed in Figure C–3 of this part of the appendix. It should be pointed out that, even though they are rough equivalents, the descriptive or place-name climatic types lack the uniformity and convenience of detailed, standard definition that the Köppen types possess.

Definitions of Climatic Symbols

Primary Types

A Average temperature of the coolest month 64.4° F. or over.

B Precipitation less in inches than the amount $.44t-8.5$ when it is evenly distributed throughout the year; or less than the amount $.44t-14$ when it is concentrated chiefly in winter; or less than the amount $.44t-3$ when it is concentrated chiefly in summer. In all of these amounts, *t* equals average annual temperature in degrees F. These amounts are worked out in Figure C–4. See note regarding meaning of "concentrated" in Figure C–4.

C Average temperature of the warmest month over 50° F. and of the coldest month between 64.4° F. and 32° F.

D Average temperature of the warmest month over 50° F. and of the coldest month 32° F. and below.

E Average temperature of the warmest month below 50° F.

Secondary Types

As they occur with **A**

f Precipitation of the driest month of the year at least 2.4 inches.

w Precipitation of the driest month of the year less than the amount shown in Figure C–5.

m Precipitation of the driest month of the year less than 2.4 inches, but more than the amount shown in Figure C–5.

As they occur with **B**

W Precipitation for the year less than half the amount shown in the definition for **B.** See Figure C–4.

S Precipitation for the year less than the amount shown in the definition for **B,** but more than half that amount. See Figure C–4.

As they occur with **C** *and* **D**

s Precipitation for the driest month of the summer half of the year less than 1.6 inches and less than 1/3 the amount that falls in the wettest month of the winter half of the year.

w Precipitation of the driest month of the winter half of the year less than 1/10 the amount that falls in the wettest month of the summer half of the year.

f Precipitation not satisfying the definitions for **s** or **w.**

As they occur with **E**

T Average temperature of the warmest month of the year between 50° F. and 32° F.

F Average temperature of the warmest month 32° F. or below.

Tertiary Types

As they occur with **B**

h Average annual temperature 64.4° F. or above.

k Average annual temperature below 64.4° F.

As they occur with **C**

a Average temperature of the warmest month 71.6° F. or above.

b Average temperature of the four warmest months 50° F. or above, and average temperature of the warmest month below 71.6° F.

c Average temperature of from one to three months 50° F. or above, and average temperature of warmest month below 71.6° F.

As they occur with **D**

a Same as with **C.**

b Same as with **C.**

c Same as with **C.**

d Average temperature of the coldest month below −36.4° F. (*Note:* **a, b,** and **c** do not specifically exclude **d,** *but* when **d** can be used, do *not* use **a, b,** or **c.**)

FIGURE C–4

The amount of precipitation (in inches) at the boundary between the **B** climates and the other primary types for each average annual temperature from 32° F. to 90° F., and for the three possible distributions of rain throughout the year.

Average annual temperature	*Precipitation concentrated chiefly in winter* (.44*t* – 14)	*Precipitation evenly distributed throughout the year* (.44*t* – 8.5)	*Precipitation concentrated chiefly in summer* (.44*t* – 3)	*Average annual temperature*	*Precipitation concentrated chiefly in winter* (.44*t* – 14)	*Precipitation evenly distributed throughout the year* (.44*t* – 8.5)	*Precipitation concentrated chiefly in summer* (.44*t* – 3)
32	.08	5.58	11.08	62	13.28	18.78	24.28
33	.52	6.02	11.52	63	13.72	19.22	24.72
34	.96	6.46	11.96	64	14.16	19.66	25.16
35	1.40	6.90	12.40	65	14.60	20.10	25.60
36	1.84	7.34	12.84	66	15.04	20.54	26.04
37	2.28	7.78	13.28	67	15.48	20.98	26.48
38	2.72	8.22	13.72	68	15.92	21.42	26.92
39	3.16	8.66	14.16	69	16.36	21.86	27.36
40	3.60	9.10	14.60	70	16.80	22.30	27.80
41	4.04	9.54	15.04	71	17.24	22.74	28.24
42	4.48	9.98	15.48	72	17.68	23.18	28.68
43	4.92	10.42	15.92	73	18.12	23.62	29.12
44	5.36	10.86	16.36	74	18.56	24.06	29.56
45	5.80	11.30	16.80	75	19.00	24.50	30.00
46	6.24	11.74	17.24	76	19.44	24.94	30.44
47	6.68	12.18	17.68	77	19.88	25.38	30.88
48	7.12	12.62	18.12	78	20.32	25.82	31.32
49	7.56	13.06	18.56	79	20.76	26.26	31.76
50	8.00	13.50	19.00	80	21.20	26.70	32.20
51	8.44	13.94	19.44	81	21.64	27.14	32.64
52	8.88	14.38	19.88	82	22.08	27.58	33.08
53	9.32	14.82	20.32	83	22.52	28.02	33.52
54	9.76	15.26	20.76	84	22.96	28.46	33.96
55	10.20	15.70	21.20	85	23.40	28.90	34.40
56	10.64	16.14	21.64	86	23.84	29.34	34.84
57	11.08	16.58	22.08	87	24.28	29.78	35.28
58	11.52	17.02	22.52	88	24.72	30.22	35.72
59	11.96	17.46	22.96	89	25.16	30.66	36.16
60	12.40	17.90	23.40	90	25.60	31.10	36.60
61	12.84	18.34	23.84				

NOTE: Precipitation is said to be concentrated in summer when 70 percent or more of the average annual amount is received during the warmer six months. It is concentrated in winter when 70 percent or more is received in the cooler six months. If neither season receives 70 percent of the annual total, the precipitation is considered to be evenly distributed. Division of the year into six-month periods is made between March and April on the one hand and between September and October on the other.

FIGURE C–5

The amount of precipitation (in inches) during the driest month of the year along the boundary between **Am** and **Aw** climates for differing amounts of yearly precipitation. If the amount in the driest month is 2.4 inches or more, the symbol **f** is used; if the amount in the driest month is between 2.4 and the amount shown in this table, the symbol **m** is used; and if the amount in the driest month is less than that shown in this table, the symbol **w** is used.

Total yearly precipitation	*Precipitation in driest month*	*Total yearly precipitation*	*Precipitation in driest month*	*Total yearly precipitation*	*Precipitation in driest month*	*Total yearly precipitation*	*Precipitation in driest month*
38.5	2.40	54	1.78	69.5	1.15	85	.54
39	2.38	54.5	1.77	70	1.13	85.5	.51
39.5	2.36	55	1.75	70.5	1.11	86	.50
40	2.34	55.5	1.73	71	1.10	86.5	.48
40.5	2.32	56	1.70	71.5	1.08	87	.46
41	2.30	56.5	1.68	72	1.06	87.5	.44
41.5	2.29	57	1.66	72.5	1.03	88	.42
42	2.26	57.5	1.64	73	1.02	88.5	.40
42.5	2.24	58	1.63	73.5	1.00	89	.37
43	2.22	58.5	1.60	74	.98	89.5	.36
43.5	2.20	59	1.58	74.5	.96	90	.34
44	2.18	59.5	1.56	75	.94	90.5	.32
44.5	2.16	60	1.55	75.5	.92	91	.29
45	2.14	60.5	1.53	76	.90	91.5	.28
45.5	2.12	61	1.51	76.5	.88	92	.26
46	2.10	61.5	1.48	77	.86	92.5	.24
46.5	2.08	62	1.47	77.5	.84	93	.22
47	2.07	62.5	1.45	78	.81	93.5	.20
47.5	2.04	63	1.42	78.5	.80	94	.18
48	2.02	63.5	1.41	79	.78	94.5	.16
48.5	2.00	64	1.38	79.5	.76	95	.14
49	1.98	64.5	1.36	80	.74	95.5	.11
49.5	1.96	65	1.34	80.5	.72	96	.09
50	1.94	65.5	1.33	81	.70	96.5	.07
50.5	1.92	66	1.30	81.5	.68	97	.06
51	1.90	66.5	1.28	82	.66	97.5	.04
51.5	1.88	67	1.26	82.5	.63	98	.02
52	1.86	67.5	1.24	83	.61	98.5	.00
52.5	1.85	68	1.22	83.5	.59		
53	1.82	68.5	1.20	84	.58		
53.5	1.80	69	1.18	84.5	.56		

NOTE: If the total yearly precipitation is 197 inches, there can be two whole months without rain and the symbol m used; if the yearly amount is 295.5 inches, three months may be dry and the symbol m used; etc. The symbol m signifies that the dry period is compensated by excess rain at another time whereas w indicates a lack of compensating amounts. Hence, the dividing amount in the driest month becomes smaller as the total for the year becomes larger.

The Determination of Climatic Types

General Procedure

Primary types. In order to determine the climatic type at a given place, it is necessary to have available average temperatures for each month and for the year, as well as average precipitation for each month and for the year. With these statistics at hand, the actual classification can most easily be accomplished by following the routine procedure stated below. It is essentially a process of elimination.

To determine the first letter of the climate-formula, proceed according to the following questions.

1. Is the average temperature of the warmest month below 50° F.? If it is, the climate is of the **E** type; if it is not, proceed to the second question.
2. Is the amount of precipitation less than the amount specified as the limit of the **B** climates in Figure C–4? If it is less, the climate is of the **B** type; if it is not, proceed to the third question.
3. Is the average temperature of the coolest month 64.4° F. or above? If it is, the climate is of the **A** type; if it is not, proceed to the fourth question.
4. Is the average temperature of the coldest month between 64.4° F. and 32° F.? If it is, the climate is of the **C** type; if it is not, or, in other words, if the average temperature of the coldest month is below 32° F., the climate is of the **D** type.

Thus by following through a sequence of, at most, four questions, it is possible to determine which one of the five primary climatic types is represented by any set of temperature and precipitation statistics.

Secondary types. Once the primary type is determined, similar routines are followed to obtain the second and third letters of the climate-formula. Secondary characteristics of each of the main types are most easily determined by applying the following questions.

1. If the climate is of the **E** type, is the average temperature of the warmest month over 32° F.? If it is, the secondary symbol is **T**, if it is not, the secondary symbol is **F**.
2. If the climate is of the **B** type, is the yearly precipitation over one half the amount specified as the limit of the **B** climates in Figure C–4? If it is, the secondary symbol is **S**; if it is not, the secondary symbol is **W**.
3. If the climate is of the **A** type:
   a. Is the precipitation of the driest month 2.4 inches or more? If it is, the secondary symbol is **f**; if it is not, refer to the next question.
   b. Is the precipitation of the driest month more than the amount specified in Figure C–5? If it is, the secondary symbol is **m**; if it is not, the secondary symbol is **w**.
4. If the climate is of the **C** or **D** type:
   a. Is the driest month of the year in the summer six-month period and is its precipitation less than 1.6 inches? If both conditions are satisfied, proceed to question b; if either is not satisfied, proceed to question c.
   b. Is the precipitation of the driest month of summer less than ⅓ that of the wettest month of winter? If it is, the secondary symbol is **s**; if it is not, the secondary symbol is **f**.

c. Is the precipitation of the driest month of winter less than 1/10 that of the wettest month of summer? If it is, the secondary symbol is **w;** if it is not, the secondary symbol is **f.**

Tertiary types. Tertiary characteristics of each of the **C** and **D** climates are determined most easily as follows. If the climate is of the **C** or **D** type, apply the following questions.

1. Is the average temperature of the coldest month below −36.4° F.? If it is, the symbol is **d;** if it is not, proceed to the next question.
2. Is the average temperature of the warmest month 71.6° F. or above? If it is, the symbol is **a;** if it is not, proceed to the next question.
3. Are the average temperatures of the four warmest months 50° F. or above? If they are, the symbol is **b;** if they are not, the symbol is **c.**

Application of Classification Procedure

First example. The first example is Maracaibo, Venezuela. The statistics are given on the following page.

Beginning with the primary types, we must first ask the question, "Is the average temperature of the warmest month below 50° F.?" The warmest month is August, with an average of 84.4°. This is well above 50° and hence the type cannot be **E.** We have eliminated one of the primary types.

Second, we must ask the question, "Is the amount of precipitation less than the amount specified by definition as the limit of the **B** climates?" It is simplest here to refer to Figure C–4. To use the table, we must know three things about the place in question. These are: (1) the average annual temperature, (2) the average annual precipitation, and (3) the distribution of the precipitation throughout the year—for example, does 70 percent or more of the precipitation come in the winter six months? Does 70 percent or more come in the summer six months? Or does 70 percent come in neither season?

Items 1 and 2 are given in the statistics: Maracaibo has an average annual temperature of 82.3° F. and an average annual precipitation of 18.2 inches. To secure the necessary third item requires the application of simple arithmetic.

The warmer six months at Maracaibo are April through September. During these months, the precipitation in inches is as follows: April, .5; May, 2.4; June, 1.6; July, 1.4; August, 1.3; and September, 3.3. The total for this season is 10.5 inches. Since the total for the year is 18.2 inches and that for the warmer six months is 10.5 inches, the amount that falls in the cooler six months must be 18.2 minus 10.5, or 7.7 inches. The summer, or warmer season, precipitation is 10.5 inches; the winter, or cooler season, precipitation is 7.7 inches. Is either of these amounts equal to or more than 70 percent of the total for the year? Seventy percent of 18.2 is 12.74. Since neither the warmer season precipitation nor that of the cooler season is equal to or greater than 70 percent of the total, the precipitation is considered to be evenly distributed throughout the year.

Now, the three items necessary for the use of Figure C–4 are available: average annual temperature, 82.3° F.; average annual precipitation, 18.2 inches; and evenly distributed precipitation. Consulting Figure C–4 we find in the first column average annual temperatures in whole numbers. The nearest to 82.3 is 82. The third column is headed "Precipitation evenly distributed throughout the year." We follow down this column until

		J	F	M	A	M	J	J	A	S	O	N	D	YEAR
Maracaibo														
	TEMP.	80.7	81.0	81.2	82.5	83.2	83.5	84.1	84.4	83.5	82.0	81.3	81.0	82.3
	PREC.	0.1	0.1	0.3	0.5	2.4	1.6	1.4	1.3	3.3	4.3	2.5	0.4	18.2

we come to the figure on the same line as 82 in the first column. That figure is 27.58. Now we can answer the question, "Is the amount of precipitation less than the amount specified by definition as the limit of the **B** climates?" The precipitation at Maracaibo (18.2 inches) is less than the limit figure (27.58 inches). Hence, Maracaibo has a **B** climate.

We need go no further with the primary types for no set of statistics will fulfill the requirements for more than one. The task is now to determine the secondary symbol.

For the second symbol, only one question is necessary: "Is the yearly precipitation over one half the amount specified in Figure C–4?" The amount specified in Figure C–4 in this instance is 27.58. One half of that is 13.79. Is the annual precipitation at Maracaibo (18.2 inches) over one half the amount (13.79 inches) specified in Figure C–4? It is, so the secondary symbol is **S.** Maracaibo thus has a **BS** climate.

Second example. As a second example of the application of the procedure of classification, the statistics for Iloilo in the Philippine Islands may be examined on the next page.

We follow the same order as in the example of Maracaibo. First, is the average temperature of the warmest month below 50° F.? It is not; therefore the type is not **E.** Is the precipitation for the year less than the amount specified for the limit of the **B** climates? We must know three things to answer this question: (1) the average annual temperature; (2) the average annual precipitation; and (3) the distribution of precipitation throughout the year. The warmer six months at Iloilo is the period of April through September. In that season, 60.8 inches of rain fall. Since 88.7 inches fall during the whole year, the amount in the cooler season is 88.7 minus 60.8, or 17.9 inches. Seventy percent of 88.7 is 62.09. Neither the amount that falls in the warmer season nor that which falls in the cooler season is equal to or more than 62.09. Therefore, the precipitation may be considered to be evenly distributed throughout the year. Referring to Figure C–4, we find 81 in the first column. Opposite it in the third column is 27.14. This represents the amount of precipitation that, under these conditions of temperature, limits the **B** climates. The amount of precipitation at Iloilo is more than 27.14; therefore, the climate at Iloilo is not **B.**

We proceed to the next question. "Is the average temperature of the coolest month 64.4° F. or above?" The coolest month at Iloilo has an average temperature of 78.0° F. Therefore, Iloilo has an **A** climate.

There is a threefold division within the **A** climates. We must now discover into which of these three Iloilo fits. The first question asks "Is the precipitation for the driest month 2.4 inches or more?" The driest month at Iloilo has 1.3 inches of rain. The answer to the question is in the negative and consequently the symbol **f** does not apply. The second question asks "Is the precipitation of the driest month more than the amount specified in Figure C–5?" In order to use that table, we must know the average annual precipitation and the precipitation of the driest month. At Iloilo, the average annual precipitation is 88.7 inches; the precipitation of the driest month is 1.3 inches. On Figure C–5 in the column headed "Total yearly precipitation," the figure 88.5 can be found. This is the figure closest to 88.7. Oppo-

		J	F	M	A	M	J	J	A	S	O	N	D	YEAR
Iloilo														
	TEMP.	78.0	78.5	80.0	80.7	82.3	81.3	80.5	80.7	80.0	80.0	79.7	78.8	81.0
	PREC.	2.4	1.8	1.3	1.6	5.9	10.5	16.3	13.8	12.7	10.3	7.7	4.4	88.7

site it in the column headed "Precipitation of the driest month" is the figure .40. Returning to the question, we can see that the precipitation of the driest month at Iloilo (1.3 inches) is more than the amount specified in Figure C–5 (.40 inch). Therefore, the symbol **m** can be used. The climate at Iloilo is of the **Am** type.

The World Distribution of Climate Types

Since climatic types represent quantitative measurements of temperature and precipitation, it is easy to erect a simplified plan, indicating diagrammatically the place of occurrence of the major types. One such plan is shown in Figure C–6. Each of the corners of the rectangle represents extreme combinations of the two elements, temperature and precipitation. The upper left-hand corner indicates the extreme of cold and dry; the lower left-hand corner indicates the extreme of hot and dry; the upper right-hand corner, cold and wet; and the lower right-hand corner, hot and wet.

The Generalized Continent Diagram

The very simple pattern shown on Figure C–6 takes into account only the factor of quantity for both of the basic climatic elements. To get at the pattern for the world, there must be included some indication of the distribution of land and water, which is so powerful a control of distribution for both temperature and precipitation.

It will be remembered that there are two great "world islands," the Americas on the one hand and the Europe–Asia–Africa land mass on the other. These "world islands" are roughly triangular in broad outline. Their widest extent occurs at about 60° North latitude and they taper into the Southern Hemisphere where both of them pinch out in the middle latitudes. After a short break, land appears again in the circumpolar continent of Antarctica. In the Northern Hemisphere, the continents fray out poleward toward the Arctic Ocean and do not extend to the Pole.

This very generalized distribution of land in each hemisphere is represented diagrammatically in Figure C–7. The outline circle represents the outer edge of a hemisphere. The top-shaped figure upon it represents the outline of land. The continent thus delineated indicates simply land as opposed to water. No attempt is made to show generalized differences of elevation, nor is any attempt made to show detailed position of any actual coastline. The smaller triangular figure is a generalization of the Antarctic continent. The whole figure is called the *generalized continent diagram,* or the *ideal continent.*

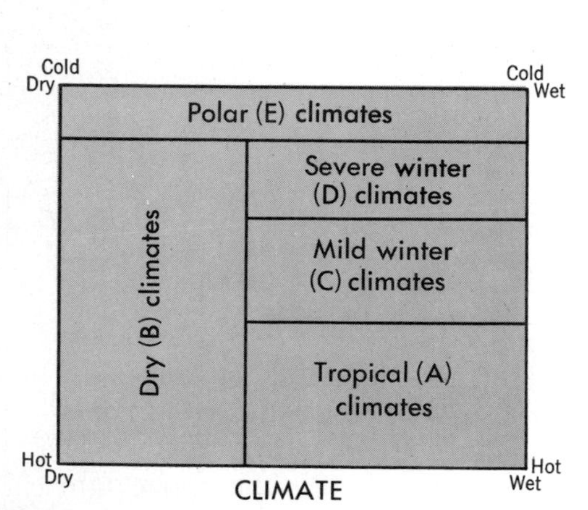

FIGURE C–6

The relationship of temperature and precipitation to major classifications of climate.

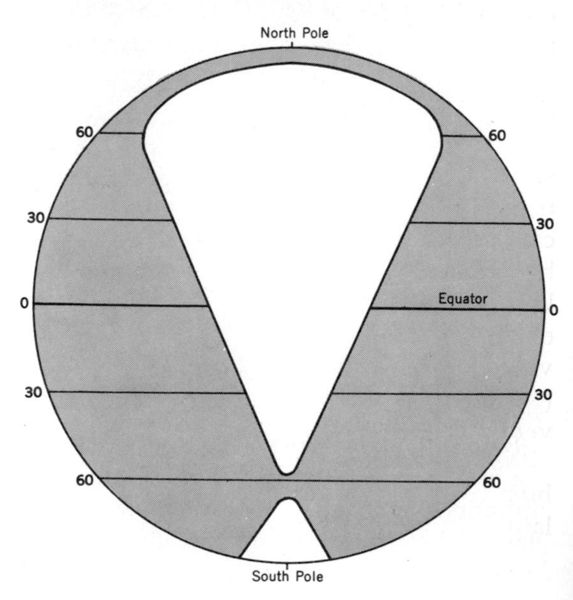

FIGURE C–7

Generalized continent diagram—distribution of land and water.

Distribution of Major Climatic Categories

The zone of greatest heating and of highest precipitation for the world as a whole occurs along the equator, and the greatest cold, whether dry or wet, occurs in the higher latitudes near, but not at, the poles. In the higher latitudes coastal positions are wet and continental positions are comparatively dry. The remaining combination, hot and dry, is found within the tropics extending inland from the continental west coasts. These facts can be placed upon the generalized continent diagram as in Figure C–8.

Now, instead of the pattern that appeared on the rectangular diagram (Fig. C–6), a new pattern can be drawn. A line must appear between the hot-wet and the hot-dry focal points. Lines must be drawn between the hot-dry and the two cold-wet locations. Finally, there must be indicated the limit, in the continental interior, where dryness becomes less significant than coldness. There are thus formed, on the generalized continent diagram, curving loops that extend from the west coasts between about 20° and 30° latitude both North and South into the continental interiors of the higher middle latitudes (Fig. C–9). The areas within the loops are distinguished by their dryness; everything outside, by wetness. The loops, then, outline the areas of dry (**B**) climate. Outside them lie the tropical humid (**A**); the humid mild-winter (**C**); the humid severe-winter (**D**); and the polar (**E**) climates.

Across the polar reaches in both hemispheres perennial cold is so dominant that relative wetness and dryness are not significant. Lines may be drawn (Fig. C–10) in both hemispheres to indicate the equatorward limit of the polar (**E**) climates. The portion of the generalized continent not yet subdivided includes now only tropical humid (**A**); humid mild-winter (**C**); and humid severe-winter (**D**) types.

Temperatures characteristic of the tropical humid (**A**) climates occur only within the low latitudes. On the generalized continent, the

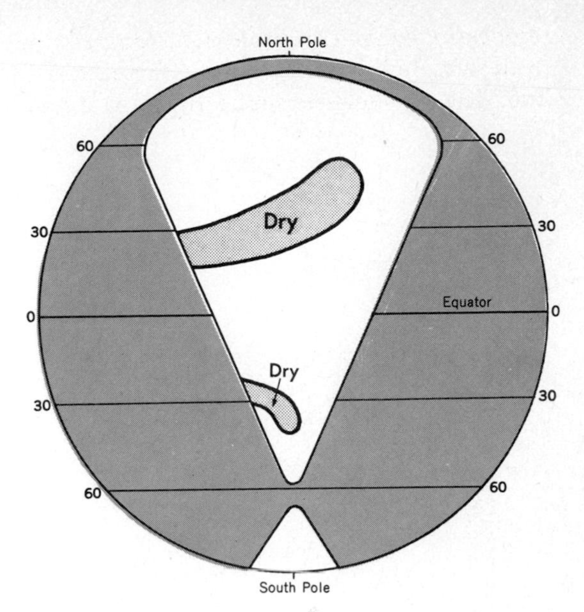

FIGURE C–9
Generalized continent diagram—dry (**B**) climates.

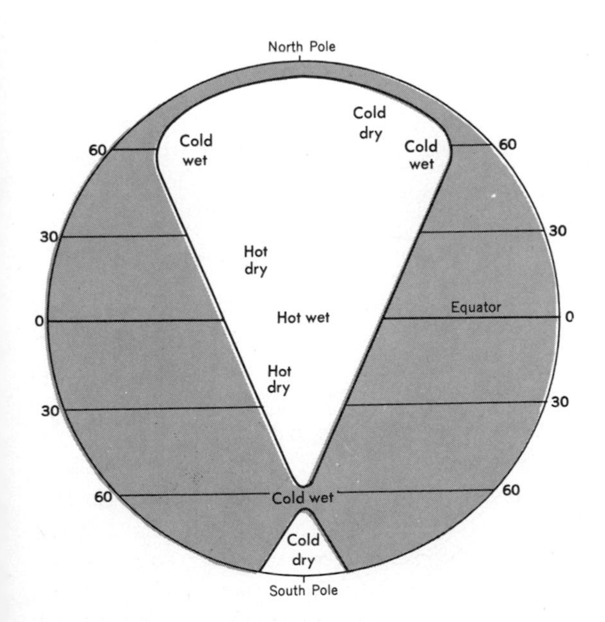

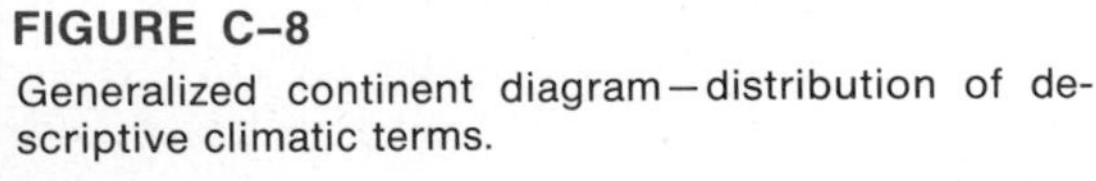
FIGURE C–8
Generalized continent diagram—distribution of descriptive climatic terms.

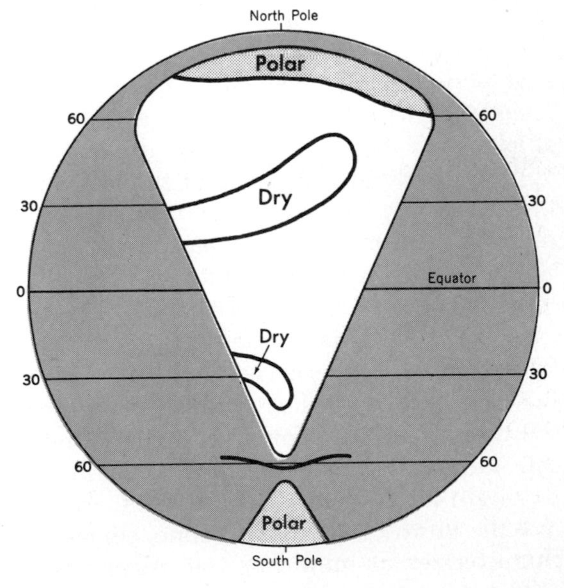

FIGURE C–10
Generalized continent diagram—polar (**E**) and dry (**B**) climates.

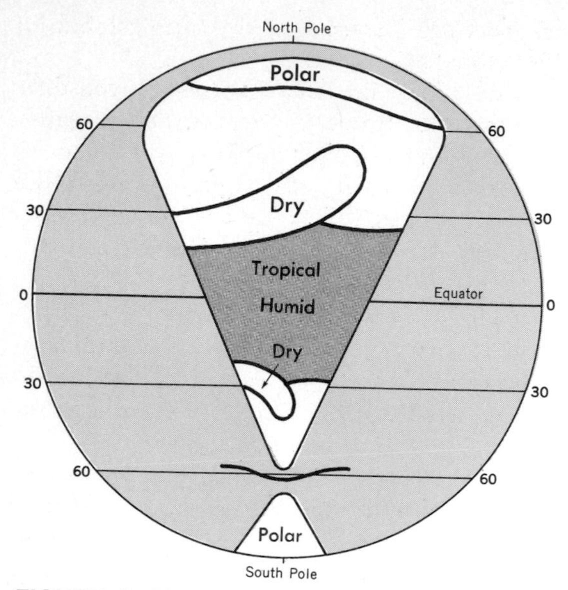

FIGURE C–11

Generalized continent diagram—tropical humid (**A**), polar (**E**), and dry (**B**) climates.

only portion of these latitudes remaining unspecified is that between and east of the two areas of dry (**B**) climate. A line in each hemisphere from the east coast to the dry interior approximately at the edges of the low latitudes provides the remaining boundaries for the tropical humid (**A**) climates (Fig. C–11).

There remain only the humid mild-winter (**C**) and humid severe-winter (**D**) climates to delimit. There is not sufficient continentality in the Southern Hemisphere to allow the development of the humid severe-winter (**D**) type. Hence, it is only necessary to indicate the separation between the two in the Northern Hemisphere. To the east, the west, and the north of the dry (**B**) zone, climate is characterized as humid or wet. As one proceeds away from the equator, coldness increases. Therefore, the boundary between the humid mild-winter (**C**) and the humid severe-winter (**D**) climates must cross the diagram from east to west. Latitude for latitude, west-coast temperatures in the middle latitudes are higher for winter than are those of the east coast. Hence, the limit between the humid mild-winter (**C**) and the humid severe-winter (**D**) climates begins on the west coast at approximately 60° North latitude, curves equatorward across the continent, and leaves the east coast at about 45° North. In the section where it crosses the dry (**B**) climate, its significance is lost, for the climates are dry on either side of it. In that section, the line is omitted from the diagram. The *generalized* distribution of the five major categories of climate as they occur over the land thus appears in Figure C–12.

It must be thoroughly understood that Figure C–12 is not a detailed map of climatic distribution over the world. It is simply a

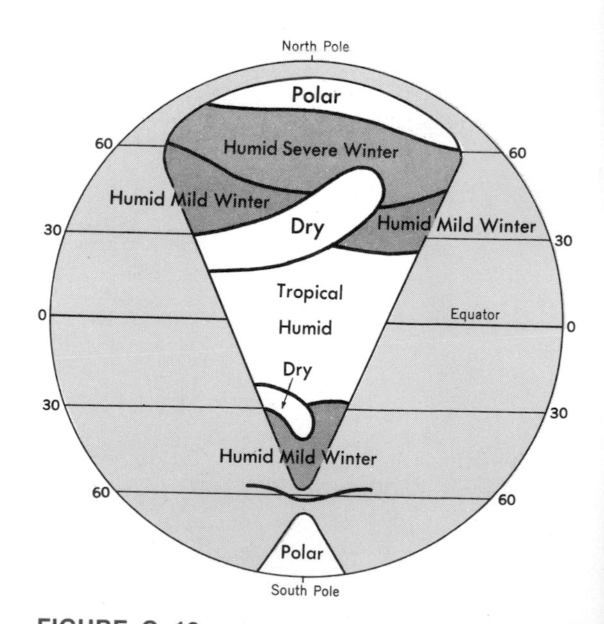

FIGURE C–12

Generalized continent diagram—five major climatic types.

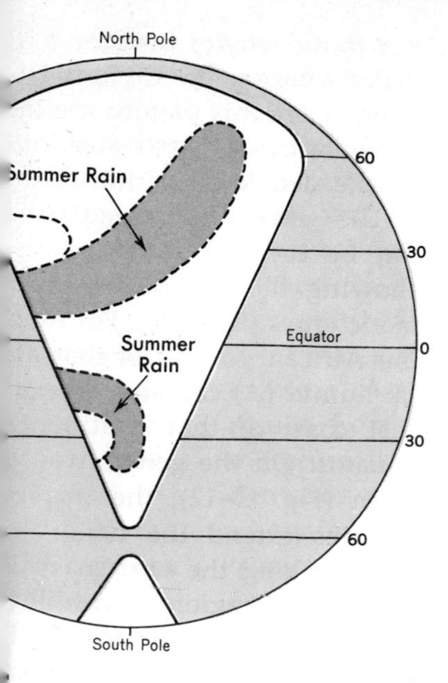

ntinent diagram—areas of summer

terize the tropical humid

irregularities of coastline
d in the Caribbean region
The climatic map (Plate IV)
y (**B**) climate comes to the
e boundary between Mex-
ed States. On the general-
agram (Fig. C–12), the dry
een to be well inland from
he apparent contradiction
fact that the coastline on
ontinent omits the large in-
Gulf of Mexico.
the reason for the lack of
A) climate on the African
of the equator is the actual
astline, this time in relation
s. Prevailing winds are par-
no matter what the season.
tle moisture is carried on-
ndian Ocean. The precipi-
to the tropical humid (**A**)
ore lacking from that source.
upled with the control in-
surface, which was stated
er part of the dryness of the
st is explained and the ab-
humid (**A**) climate is reason-
tood.

es.
have one diagram (Fig. C–12)
generalized pattern of the five
ries of climate: tropical humid
humid mild-winter (**C**), humid
(**D**), and polar (**E**); and one
C–15) showing the general-
f precipitation types. By super-
C–16), we secure the key to
location of the major climatic
generalized continent. As has
reviously, departures from this
primarily from the controls
by irregularities of surface and
coastline position.
bdivision of the humid mild-
d humid severe-winter (**D**) cli-
e on the basis of the character
ers. Without discussing this

ution of Climatic Types

ds inland to form rough half
ure C–13. Within the shaded
tion occurs mainly during
the summer is droughty.
condition, summer rain and
t, is characteristic of truly
tions. As has been explained
re is a tendency toward an

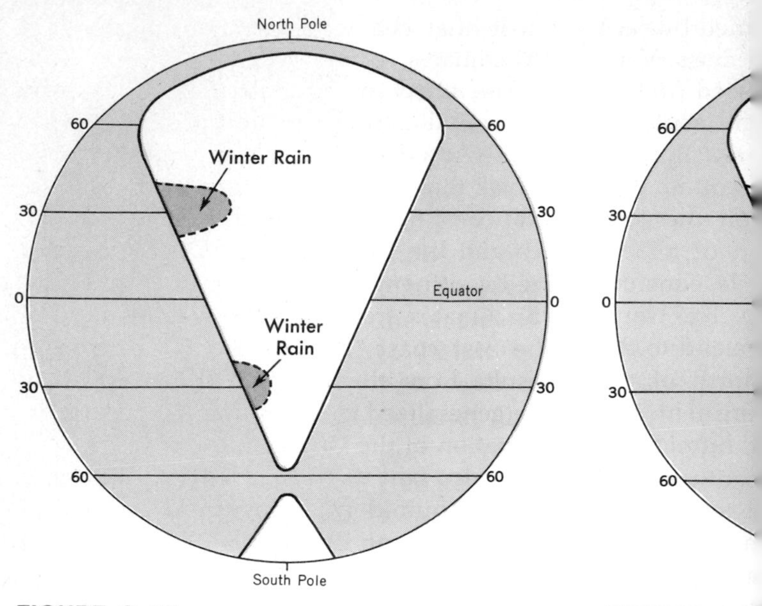

FIGURE C–13
Generalized continent diagram—areas of winter precipitation.

FIGURE C–1
Generalized c
precipitation.

indraft of air into continents in summer and an outdraft in winter. This means that more moisture is available for precipitation in summer than in winter. The indraft is pronounced only when there is a very considerable land mass, but the characteristic tendency toward summer precipitation concentration is everywhere prominent away from coastline locations. Generalization of the regions in which this condition prevails is indicated on Figure C–14.

The only other possibility is that precipitation be relatively evenly spread throughout the year. When the zone of winter concentration (Fig. C–13) and the zone of summer concentration (Fig. C–14) are plotted, that which is left in the area of year-round precipitation. Figure C–15 shows the characteristic location of each of the three possible

rainfall regin
We now
showing the
major catego
(**A**), dry (**B**),
severe-winte
diagram (Fi
ized pattern
position (Fi
the expected
types on the
been stated
design resu
emphasized
by details of
Further su
winter (**C**) a
mates is mac
of the sumr

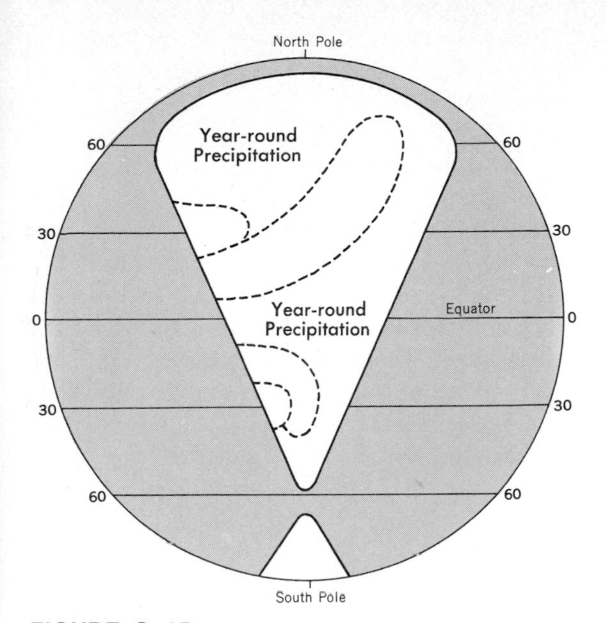

FIGURE C–15
Generalized continent diagram—areas of year-round precipitation.

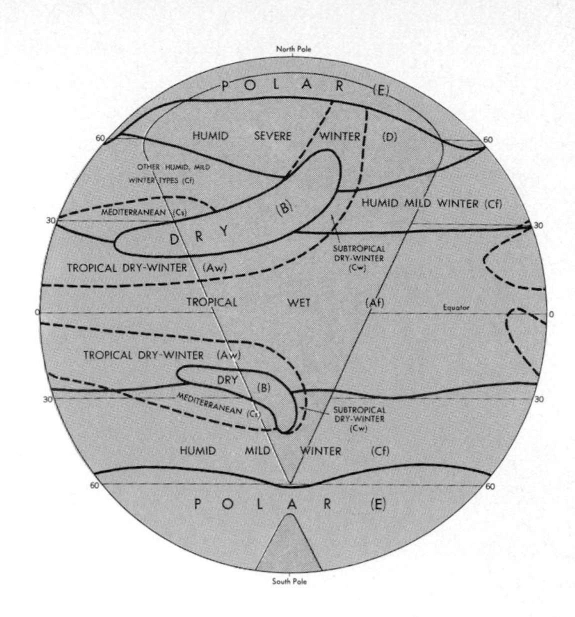

FIGURE C–16
Generalized continent diagram—distribution of climatic types over both land and ocean.

breakdown in detail, the generalized pattern is shown in Figure C–17. In this diagram, part of the Northern Hemisphere portion of the generalized continent has been enlarged so as to show the relationship of the types more clearly.

Climate of the Oceans

Largely because human beings live permanently on the land masses, climate over the oceans is generally neglected. Nevertheless, the same major types are found there as on land. The pattern is shown on Figure C–16.

Two specific features should be noted. First, the area of dry (**B**) climate is greatly reduced over the ocean. This is to be expected, for the existence of dry (**B**) climate is based primarily on two conditions: (1) the absence of large amounts of water vapor in the air and (2) the absence of some means of forcing the condensation of such vapor as is present. Over the oceans both of these deficiencies are generally restricted to sections close to the continental west coasts. Second, the area of humid severe-winter (**D**) climate is likewise very small in proportion to that over the continent. By definition, the humid severe-winter (**D**) climates have very large temperature ranges. In other words, they are continental. As we have seen, the presence of large water bodies tends to reduce temperature range. The result of this tendency is to produce insufficient range over the oceans to satisfy the definition of the humid severe-winter (**D**) type in all except two small areas off the coasts of the higher middle latitudes in the Northern Hemisphere.

Finally, it should be noted that there is a much more regular belted arrangement of climates over the oceans than over the land. This reflects the facts already pointed out in the discussion of temperature and precipitation.

Though climatic maps may show the distribution over the oceans, the boundaries between one type and another are much more broadly generalized than those over the land. Realization of the fact that the climatic types are not confined to the land is essential; so also is the realization of the generalized distribution of those types, for knowledge of their precise distribution over the sea is neither possible nor would it be considered particularly vital.

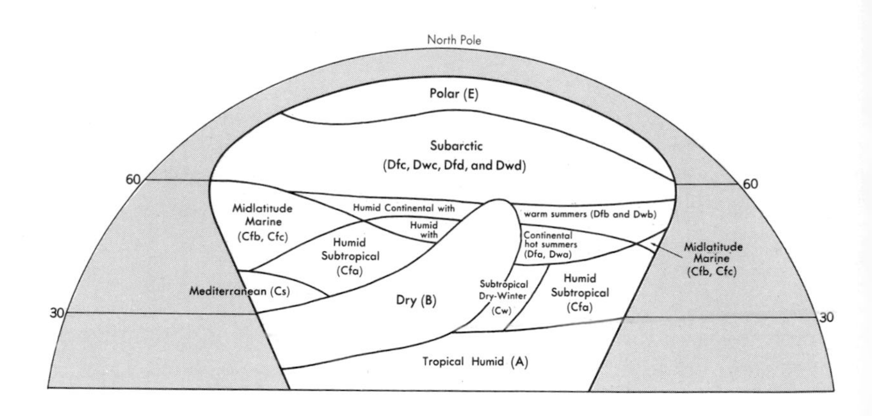

FIGURE C–17

Northern Hemisphere portion of generalized continent diagram, showing in greater detail the distribution pattern of the subdivisions of humid mild-winter (**C**) and humid severe-winter (**D**) climates.

Selected Climatic Statistics

North and Middle America and the Caribbean Area

		J	F	M	A	M	J	J	A	S	O	N	D	YEAR
1. San Juan, Puerto Rico														
	TEMP.	75	75	75	77	79	80	80	80	80	80	78	76	78
	PREC.	4.1	3.0	3.0	4.1	5.2	5.4	5.8	5.9	6.0	5.7	7.0	5.4	60.6
2. Miami, Florida														
	TEMP.	66	68	72	74	79	81	82	82	81	77	73	69	75
	PREC.	3.3	2.5	2.8	3.1	5.9	7.9	7.2	7.3	9.9	9.2	2.4	2.2	63.7
3. Port-au-Prince, Haiti														
	TEMP.	78	78	79	80	81	83	84	84	82	81	80	78	81
	PREC.	1.2	2.7	4.1	6.7	8.0	4.0	2.7	4.8	7.7	7.4	4.2	0.9	54.4
4. Acapulco, Mexico														
	TEMP.	78	78	79	80	83	83	83	81	81	83	80	79	81
	PREC.	0.6	0	0	0	1.7	16.5	6.0	6.3	14.7	5.9	1.9	0.7	54.3
5. Veracruz, Mexico														
	TEMP.	70	71	74	78	81	81	81	81	80	78	75	71	77
	PREC.	1.0	0.6	0.5	0.6	1.7	11.4	13.0	10.7	12.0	5.7	3.1	1.0	61.3
6. Medicine Hat, Canada														
	TEMP.	13	16	28	45	55	63	69	67	56	45	30	19	42
	PREC.	0.6	0.6	0.6	0.7	1.6	2.5	1.8	1.4	1.2	0.6	0.7	0.7	13.0
7. Goodland, Kansas														
	TEMP.	29	33	40	50	59	70	77	74	66	54	41	30	52
	PREC.	0.3	0.6	1.0	1.8	2.5	2.8	2.7	2.5	1.6	1.0	0.6	0.6	18.0
8. San Diego, California														
	TEMP.	54	55	56	58	61	64	67	69	67	63	59	56	61
	PREC.	1.8	2.0	1.5	0.6	0.3	0.1	0.1	0.1	0.1	0.3	0.9	1.8	9.6
9. Mexico City, Mexico														
	TEMP.	54	57	61	63	65	64	62	62	61	59	57	55	60
	PREC.	0.2	0.3	0.5	0.7	1.9	4.1	4.5	4.3	4.1	1.6	0.5	0.3	23.0
10. Phoenix, Arizona														
	TEMP.	50	54	60	67	75	84	90	88	82	70	59	52	69
	PREC.	1.0	0.7	0.6	0.4	0.1	0.1	1.3	1.0	0.7	0.5	0.8	0.7	7.9
11. La Paz, Mexico														
	TEMP.	63	65	68	71	74	78	82	84	82	79	72	66	74
	PREC.	0.2	0.1	0	0	0	0	0.4	1.2	1.4	0.6	0.5	1.1	5.5

		J	F	M	A	M	J	J	A	S	O	N	D	YEAR
12. Cairo, ILLINOIS														
	TEMP.	44	47	57	67	76	85	88	87	81	70	56	46	67
	PREC.	3.9	3.2	4.1	3.9	3.8	3.9	3.0	3.0	2.9	2.8	3.6	3.3	41.4
13. New York, NEW YORK														
	TEMP.	30	31	38	49	59	69	75	73	66	55	44	34	52
	PREC.	3.3	3.3	3.5	3.3	3.5	3.4	4.1	4.4	3.4	3.4	3.6	3.3	42.5
14. Washington, D.C.														
	TEMP.	33	35	42	53	64	73	77	75	68	57	45	36	55
	PREC.	3.1	3.1	3.5	3.3	3.7	3.7	4.3	4.1	3.3	3.1	2.6	3.0	40.8
15. San Antonio, TEXAS														
	TEMP.	53	55	63	69	75	81	83	83	79	70	61	54	69
	PREC.	1.4	1.6	1.8	2.7	3.2	2.7	2.5	2.6	3.5	2.0	2.2	1.8	28.0
16. Montgomery, ALABAMA														
	TEMP.	48	51	58	65	73	80	82	81	76	66	56	49	66
	PREC.	5.1	5.5	6.4	4.3	3.8	4.2	4.7	4.2	2.9	2.4	3.1	4.5	51.1
17. Sitka, ALASKA														
	TEMP.	30	32	35	40	46	51	55	55	51	44	37	33	42
	PREC.	7.1	6.8	5.6	5.4	4.3	3.5	4.0	7.1	9.7	11.7	8.8	7.4	81.4
18. St. Paul, MINNESOTA														
	TEMP.	12	15	28	46	58	67	72	70	61	48	31	19	44
	PREC.	0.9	0.8	1.4	2.4	3.4	4.1	3.5	3.5	3.4	2.0	1.4	1.0	27.8
19. Albany, NEW YORK														
	TEMP.	23	24	33	47	59	68	72	71	63	50	39	28	48
	PREC.	2.6	2.5	2.7	2.7	3.5	4.0	4.1	3.8	3.4	3.4	3.0	2.7	38.4
20. Dubuque, IOWA														
	TEMP.	20	23	35	49	60	70	75	72	64	52	37	25	48
	PREC.	1.5	1.4	2.2	2.6	3.9	4.5	3.7	3.4	4.0	2.5	1.9	1.5	33.1
21. Edmonton, CANADA														
	TEMP.	6	12	22	40	51	57	61	59	50	41	26	14	37
	PREC.	0.8	0.7	0.8	0.9	1.7	3.1	3.2	2.4	1.3	0.7	0.7	0.8	17.1
22. Toronto, CANADA														
	TEMP.	22	23	30	42	53	63	69	67	60	48	37	27	45
	PREC.	2.8	2.7	2.4	2.5	2.8	2.7	2.9	2.8	3.1	2.4	2.8	2.6	32.5
23. Halifax, CANADA														
	TEMP.	23	23	30	39	49	58	65	65	58	49	39	28	44
	PREC.	5.5	4.9	4.9	4.6	4.1	3.7	3.8	4.5	3.6	5.2	5.5	5.3	55.6
24. Duluth, MINNESOTA														
	TEMP.	10	13	24	38	48	58	66	65	57	45	29	17	39
	PREC.	1.0	1.0	1.5	2.1	3.4	4.4	3.9	3.5	3.7	2.7	1.5	1.2	29.9
25. Norman, CANADA														
	TEMP.	−19	−13	−2	19	41	54	59	54	41	24	−1	−15	20
	PREC.	0.4	0.5	0.6	0.5	1.1	1.3	1.8	1.9	1.0	0.8	0.4	0.4	10.7
26. Churchill, CANADA														
	TEMP.	−21	−17	−2	17	30	44	54	52	41	27	6	−11	18
	PREC.	0.6	1.1	1.1	1.0	0.9	2.0	1.8	2.4	2.6	1.3	1.2	0.8	16.8

		J	F	M	A	M	J	J	A	S	O	N	D	YEAR
27. Father Point, CANADA														
	TEMP.	9	10	22	34	44	53	58	56	49	41	30	17	36
	PREC.	2.5	2.4	2.4	1.9	3.0	3.6	2.9	3.2	3.1	3.3	2.9	2.8	34.0
28. Upernivik, WEST GREENLAND														
	TEMP.	−7	−10	−7	6	25	35	41	41	34	25	14	1	17
	PREC.	0.4	0.4	0.6	0.5	0.6	0.6	1.0	1.1	1.0	1.2	1.1	0.5	9.0
29. Angmagssalik, EAST GREENLAND														
	TEMP.	18	16	19	25	34	41	45	43	38	30	23	20	29
	PREC.	3.3	2.0	2.4	2.4	2.4	2.1	1.9	2.4	3.7	5.7	3.3	2.8	34.4

South America

		J	F	M	A	M	J	J	A	S	O	N	D	YEAR
30. Medellín, COLOMBIA														
	TEMP.	71	72	71	71	71	71	71	71	71	69	69	70	71
	PREC.	2.7	3.5	3.3	6.6	7.7	5.5	4.1	4.6	6.2	6.9	5.2	2.5	58.8
31. Santos, BRAZIL														
	TEMP.	76	78	76	73	69	66	66	66	67	69	73	76	71
	PREC.	10.4	9.2	8.1	6.8	6.1	2.4	4.4	4.6	5.6	6.1	7.8	7.0	78.5
32. Manaus, BRAZIL														
	TEMP.	80	80	80	80	80	80	81	82	83	83	82	81	81
	PREC	9.2	9.0	9.6	8.5	7.0	3.6	2.2	1.4	2.0	4.1	5.5	7.7	69.8
33. Rio de Janeiro, BRAZIL														
	TEMP.	79	79	78	75	72	70	69	70	70	72	74	77	74
	PREC.	4.9	4.8	5.2	4.3	3.1	2.3	1.7	1.7	2.6	3.2	4.1	5.4	43.3
34. Coquimbo, CHILE														
	TEMP.	64	64	62	59	57	54	54	54	55	57	59	62	58
	PREC.	0	0	0	0	1.1	1.5	1.1	0.5	0.2	0	0	0	4.4
35. Lima, PERU														
	TEMP.	73	74	74	70	66	63	61	61	61	63	66	70	67
	PREC.	0	0	0	0	0	0.1	0.2	0.4	0.4	0.4	0.2	0.1	1.8
36. Buenos Aires, ARGENTINA														
	TEMP.	74	73	69	61	55	50	49	51	55	60	66	71	61
	PREC.	3.1	2.8	3.9	4.8	2.8	2.0	2.1	2.2	2.9	3.3	4.0	4.0	37.9
37. Puerto Montt, CHILE														
	TEMP.	60	58	56	52	50	46	46	46	47	51	54	57	52
	PREC.	4.6	4.4	5.9	7.4	10.6	10.0	10.8	9.3	6.3	5.5	5.5	5.4	85.7
38. Mar del Plata, ARGENTINA														
	TEMP.	67	66	64	59	53	47	46	47	50	53	59	64	56
	PREC.	2.1	3.1	3.1	3.1	1.9	2.4	2.1	1.8	2.6	2.4	2.6	2.7	29.9
39. Santiago, CHILE														
	TEMP.	69	67	62	57	51	46	46	49	52	57	62	67	57
	PREC.	0	0.1	0.2	0.6	2.4	3.3	2.8	2.1	1.3	0.5	0.2	0.2	13.7
40. Tucumán, ARGENTINA														
	TEMP.	77	75	72	66	60	54	54	57	64	69	73	75	66
	PREC.	6.3	7.5	5.5	3.1	1.2	0.6	0.3	0.5	0.6	2.3	4.2	5.9	38.0

Europe

		J	F	M	A	M	J	J	A	S	O	N	D	YEAR
41. Astrakhan, U.S.S.R.														
	TEMP.	19	21	32	49	64	73	78	75	64	50	38	26	49
	PREC.	0.5	0.3	0.4	0.5	0.6	0.7	0.5	0.5	0.5	0.4	0.4	0.5	5.8

Station		J	F	M	A	M	J	J	A	S	O	N	D	YEAR
42. Turin, ITALY														
	TEMP.	33	37	46	54	61	69	74	73	65	54	43	35	54
	PREC.	2.2	1.6	2.3	4.4	4.8	4.2	2.3	2.6	2.8	3.6	2.6	1.6	35.0
43. Trieste, ITALY														
	TEMP.	39	41	47	54	62	69	74	73	66	58	49	43	56
	PREC.	2.2	2.3	2.8	2.9	3.7	4.1	3.7	3.7	4.0	4.7	3.6	3.4	41.1
44. Paris, FRANCE														
	TEMP.	37	39	43	51	56	62	66	64	59	51	43	37	51
	PREC.	1.4	1.1	1.4	1.5	1.9	2.1	2.0	1.9	1.9	2.1	1.9	1.6	20.8
45. Dublin, IRELAND														
	TEMP.	40	41	42	45	49	55	58	57	54	48	44	41	48
	PREC.	2.2	1.9	1.9	1.9	2.1	2.0	2.6	3.1	2.0	2.6	2.9	2.5	27.7
46. Reykjavik, ICELAND														
	TEMP.	30	30	31	36	43	49	52	51	46	39	34	30	39
	PREC.	3.9	3.3	2.7	2.4	1.9	1.9	1.9	2.0	3.5	3.4	3.7	3.5	34.1
47. Frankfurt am Main, WEST GERMANY														
	TEMP.	32	36	41	49	57	63	66	64	58	49	41	35	49
	PREC.	1.5	1.4	1.7	1.2	1.9	2.2	2.6	2.2	1.9	2.1	1.8	2.0	22.5
48. Edinburgh, SCOTLAND														
	TEMP.	38	39	40	45	49	55	58	57	54	47	42	39	47
	PREC.	2.0	1.8	1.7	1.5	1.9	2.4	2.7	3.0	2.5	2.4	2.3	2.3	26.5
49. Athens, GREECE														
	TEMP.	48	50	53	58	66	74	80	80	73	66	57	52	63
	PREC.	2.1	1.8	1.3	0.9	0.8	0.6	0.3	0.6	0.7	1.4	2.9	2.5	15.9
50. La Coruña, SPAIN														
	TEMP.	49	50	52	55	58	62	64	64	62	57	53	50	57
	PREC.	2.7	3.5	3.2	2.9	2.0	1.4	0.9	1.1	1.7	3.1	3.5	3.5	29.5
51. Granada, SPAIN														
	TEMP.	44	48	52	58	63	70	77	77	70	60	51	44	59
	PREC.	2.0	1.7	2.2	2.1	1.8	0.8	0.1	0.2	1.1	1.8	1.9	1.9	17.6
52. Lisbon, PORTUGAL														
	TEMP.	51	52	55	58	62	67	71	72	69	63	57	51	61
	PREC.	3.4	3.3	3.4	3.1	1.8	0.6	0.2	0.2	1.3	2.4	3.6	4.3	27.6
53. Bucharest, RUMANIA														
	TEMP.	26	31	41	52	62	69	73	72	64	53	40	31	51
	PREC.	1.2	1.1	1.5	1.7	2.3	4.3	2.2	2.2	2.0	1.6	1.6	1.2	22.9
54. Odessa, U.S.S.R.														
	TEMP.	25	28	35	48	59	68	73	71	62	52	41	31	49
	PREC.	0.9	0.7	1.1	1.1	1.3	2.3	2.1	1.2	1.4	1.1	1.6	1.3	16.1
55. Oslo, NORWAY														
	TEMP.	25	26	31	41	51	60	63	60	52	42	33	27	43
	PREC.	1.1	1.1	1.3	1.3	1.7	1.9	2.8	3.5	2.4	2.5	1.9	1.7	23.2
56. Moscow, U.S.S.R.														
	TEMP.	12	15	23	38	53	62	66	63	52	40	28	17	39
	PREC.	1.7	0.9	1.2	1.5	1.9	2.0	2.8	2.9	2.2	1.4	1.6	1.5	21.6

		J	F	M	A	M	J	J	A	S	O	N	D	YEAR
57. Stockholm, Sweden														
	TEMP.	27	26	30	38	48	57	62	59	53	43	35	29	42
	PREC.	1.4	1.3	1.3	1.5	1.5	1.7	2.4	2.9	1.9	1.8	1.9	1.9	21.5
58. Tromsö, Norway														
	TEMP.	26	24	27	32	39	47	52	51	44	36	30	27	36
	PREC.	4.3	4.4	3.1	2.3	1.9	2.2	2.2	2.8	4.8	4.6	4.4	3.8	40.8
59. Archangel, U.S.S.R.														
	TEMP.	7	9	19	30	41	54	60	57	47	35	22	12	33
	PREC.	0.8	0.7	0.8	0.7	1.0	1.5	2.2	2.1	2.0	1.5	1.1	0.8	15.2
60. Vardö, Norway														
	TEMP.	22	21	23	30	35	42	48	48	43	35	28	24	33
	PREC.	2.6	2.8	2.1	1.7	1.4	1.6	1.7	2.0	2.4	2.5	2.5	2.6	25.9

Africa

		J	F	M	A	M	J	J	A	S	O	N	D	YEAR
61. Nouvelle-Anvers, Republic of the Congo														
	TEMP.	79	80	79	78	78	78	77	78	78	77	77	77	78
	PREC.	4.1	3.5	4.1	5.6	6.2	6.1	6.3	6.3	6.3	6.6	2.6	9.3	67.0
62. Freetown, Sierra Leone														
	TEMP.	81	82	82	82	82	80	78	77	79	80	81	81	80
	PREC.	0.6	0.5	1.1	5.4	14.8	21.3	36.8	39.6	32.5	15.2	5.3	1.3	174.4
63. Banana, Republic of the Congo														
	TEMP.	80	81	82	80	79	75	73	73	76	79	80	80	78
	PREC.	2.1	2.3	3.7	6.1	1.9	0	0	0.1	0.1	1.6	5.9	4.7	28.5
64. Mombasa, Kenya														
	TEMP.	80	80	82	81	78	77	75	76	77	78	79	80	79
	PREC.	0.8	0.9	2.3	7.8	13.7	3.6	3.5	2.2	1.9	3.4	5.0	2.2	47.3
65. Kimberley, South Africa														
	TEMP.	76	75	72	64	56	50	51	56	62	68	73	76	65
	PREC.	2.8	3.0	2.8	2.0	0.7	0.5	0.1	0.3	0.2	0.8	2.1	2.9	18.2
66. Gorée, Senegal														
	TEMP.	69	66	68	69	72	78	81	82	82	82	78	72	75
	PREC.	0	0	0	0	0	0.9	3.6	9.9	5.2	0.7	0.1	0	20.4
67. Cairo, Egypt														
	TEMP.	54	57	63	70	77	82	84	83	78	75	66	59	71
	PREC.	0.3	0.2	0.2	0.2	0	0	0	0	0	0.1	0.1	0.2	1.3
68. Swakopmund, Namibia														
	TEMP.	61	63	63	60	61	59	57	55	56	58	59	62	59
	PREC.	0	0.1	0.2	0	0	0	0	0	0	0.1	0	0.2	0.6
69. Cape Town, South Africa														
	TEMP.	69	70	68	63	59	56	55	56	57	61	64	67	62
	PREC.	0.7	0.6	0.9	1.8	3.9	4.4	3.5	3.3	2.2	1.6	1.1	0.8	24.8
70. Algiers, Algeria														
	TEMP.	53	55	58	61	66	71	77	78	75	69	62	56	65
	PREC.	4.2	3.5	3.5	2.3	1.3	0.6	0.1	0.3	1.1	3.1	4.6	5.4	30.0
71. Pretoria, South Africa														
	TEMP.	72	71	68	63	57	53	52	57	63	68	69	71	64
	PREC.	5.5	3.9	3.5	1.1	0.6	0.2	0.1	0.2	1.1	1.8	3.7	4.2	25.9

Asia and Adjacent Islands

	J	F	M	A	M	J	J	A	S	O	N	D	YEAR
72. Legaspi, PHILIPPINES													
TEMP.	78	78	80	82	83	82	81	81	81	81	80	79	81
PREC.	15.2	13.0	7.7	5.4	5.3	8.8	10.1	7.2	10.2	13.0	17.3	18.7	131.9
73. Manokwari, NEW GUINEA													
TEMP.	78	78	79	79	79	79	79	79	79	80	80	79	79
PREC.	10.8	10.7	13.0	11.1	7.8	8.3	6.0	5.3	5.1	4.2	6.6	10.6	99.5
74. Medan, INDONESIA													
TEMP.	75	77	78	79	79	78	78	78	77	77	76	76	77
PREC.	5.4	3.9	4.1	5.2	7.0	5.2	5.2	6.9	8.7	9.8	10.1	8.9	80.4
75. Yap, CAROLINE ISLANDS													
TEMP.	80	80	80	81	81	81	80	80	80	81	81	81	81
PREC.	7.0	6.8	4.9	5.2	9.6	10.8	16.6	16.2	13.4	11.8	10.1	8.9	121.3
76. Djakarta, INDONESIA													
TEMP.	78	78	79	80	80	79	79	79	80	80	79	79	79
PREC.	11.9	13.4	7.9	5.6	4.1	3.7	2.6	1.6	2.8	4.5	5.7	7.6	71.4
77. Moulmein, BURMA													
TEMP.	76	79	84	85	83	80	79	79	80	81	79	76	80
PREC.	0.2	0.2	0.5	2.8	20.0	37.8	45.2	42.8	28.1	8.5	2.2	0.3	188.6
78. Colombo, SRI LANKA													
TEMP.	79	80	81	82	82	80	80	81	80	79	80	79	80
PREC.	3.2	1.9	4.7	11.4	12.1	8.4	4.5	3.8	5.0	14.4	12.5	6.4	88.3
79. Bombay, INDIA													
TEMP.	75	75	78	82	85	82	80	79	79	81	79	76	79
PREC.	0.1	0	0	0.1	0.5	20.6	24.6	14.9	10.9	1.8	0.5	0.1	74.1
80. Tehran, IRAN													
TEMP.	34	42	48	61	71	80	85	83	77	61	51	42	62
PREC.	1.2	0.9	2.4	0.9	0.4	0	0.4	0	0.1	0.1	1.2	1.3	8.9
81. Tashkent, U.S.S.R.													
TEMP.	30	34	47	58	70	77	81	77	67	54	43	36	56
PREC.	1.8	1.4	2.6	2.6	1.1	0.5	0.1	0.1	0.2	1.1	1.4	1.7	14.6
82. New Delhi, INDIA													
TEMP.	58	62	74	86	92	92	86	85	84	79	68	60	77
PREC.	1.0	0.6	0.7	0.3	0.7	3.2	8.4	7.4	4.4	0.4	0.1	0.4	27.6
83. Karachi, PAKISTAN													
TEMP.	65	68	75	81	85	87	84	82	82	80	74	67	78
PREC.	0.6	0.3	0.1	0.1	0	0.4	3.2	1.8	0.7	0	0.2	0.2	7.6
84. Tokyo, JAPAN													
TEMP.	37	38	44	54	62	69	75	78	72	61	50	41	57
PREC.	2.0	2.6	4.3	5.3	5.9	6.3	5.6	4.6	7.5	7.2	4.3	2.3	57.9
85. Shanghai, CHINA													
TEMP.	38	39	46	56	66	73	80	80	73	63	52	42	59
PREC.	2.2	2.3	3.4	3.8	3.7	6.5	5.5	5.9	4.7	3.2	1.7	1.2	44.1
86. Wuhan, CHINA													
TEMP.	39	40	49	61	71	78	84	83	76	65	54	43	62
PREC.	2.1	1.1	2.8	4.8	5.0	7.0	8.6	4.6	2.2	3.9	1.1	0.6	43.8

		J	F	M	A	M	J	J	A	S	O	N	D	YEAR
87. Varanasi, INDIA														
	TEMP.	60	65	77	87	91	89	84	83	83	78	68	60	77
	PREC.	0.7	0.5	0.3	0.1	0.6	5.5	12.5	11.2	6.5	2.2	0.2	0.2	40.5
88. Darjeeling, INDIA														
	TEMP.	40	42	50	56	58	60	62	61	59	55	48	42	53
	PREC.	0.8	1.1	2.0	4.1	7.8	24.2	31.7	26.0	18.3	5.3	0.2	0.2	121.7
89. Peking, CHINA														
	TEMP.	24	29	41	57	68	76	79	77	68	55	39	27	53
	PREC.	0.1	0.2	0.2	0.6	1.4	3.0	9.4	6.3	2.6	0.6	0.3	0.1	24.8
90. Irkutsk, U.S.S.R.														
	TEMP.	−5	1	17	35	48	59	65	60	48	33	13	1	31
	PREC.	0.6	0.5	0.4	0.6	1.2	2.3	2.9	2.4	1.6	0.7	0.6	0.8	14.6

Australia and New Zealand

		J	F	M	A	M	J	J	A	S	O	N	D	YEAR
91. Darwin, AUSTRALIA														
	TEMP.	84	84	84	84	82	79	77	79	83	85	86	85	83
	PREC.	15.2	13.5	9.6	4.1	0.6	0.1	0.1	0.1	0.5	2.0	4.7	9.8	60.3
92. Hall's Creek, AUSTRALIA														
	TEMP.	86	85	83	78	71	66	64	69	76	84	87	87	78
	PREC.	5.7	4.8	3.0	0.8	0.4	0.2	0.2	0.1	0.2	0.6	1.5	3.3	20.8
93. Alice Springs, AUSTRALIA														
	TEMP.	83	82	77	73	60	54	53	58	66	74	79	82	70
	PREC.	1.7	1.7	1.2	0.7	0.6	0.6	0.4	0.4	0.4	0.7	1.0	1.5	10.9
94. Marble Bar, AUSTRALIA														
	TEMP.	93	92	90	84	75	68	67	71	78	84	90	92	82
	PREC.	2.6	3.2	1.9	0.9	0.7	1.1	0.6	0.2	0	0.2	0.4	1.3	13.1
95. Auckland, NEW ZEALAND														
	TEMP.	66	66	64	61	57	54	52	52	55	57	60	63	59
	PREC.	2.6	3.3	3.0	3.5	4.6	5.0	5.0	4.2	3.7	3.6	3.4	2.9	44.8
96. Melbourne, AUSTRALIA														
	TEMP.	68	68	65	59	54	50	49	51	54	58	61	65	59
	PREC.	1.8	1.8	2.1	2.2	2.1	2.0	1.8	1.8	2.4	2.6	2.2	2.2	25.0
97. Hobart, TASMANIA														
	TEMP.	62	62	59	55	51	47	46	48	51	54	57	60	54
	PREC.	1.8	1.6	1.7	1.9	1.8	2.2	2.1	1.8	2.1	2.2	2.4	1.9	23.5
98. Adelaide, AUSTRALIA														
	TEMP.	74	74	70	64	58	54	52	54	57	61	67	71	63
	PREC.	0.7	0.8	1.0	1.7	2.7	3.1	2.6	2.4	2.1	1.7	1.1	0.9	20.8
99. Mackay, AUSTRALIA														
	TEMP.	80	79	77	74	68	64	62	64	68	73	77	79	72
	PREC.	15.0	13.7	15.4	7.3	4.4	2.7	2.3	0.9	1.1	2.4	2.7	7.3	75.2

Indian Ocean

		J	F	M	A	M	J	J	A	S	O	N	D	YEAR
100. KERGUELEN ISLANDS														
	TEMP.	44	45	39	39	35	37	33	34	33	34	39	41	38

Conversion Tables

Conversion of Degrees Fahrenheit (°F) and Degrees Centigrade (°C)

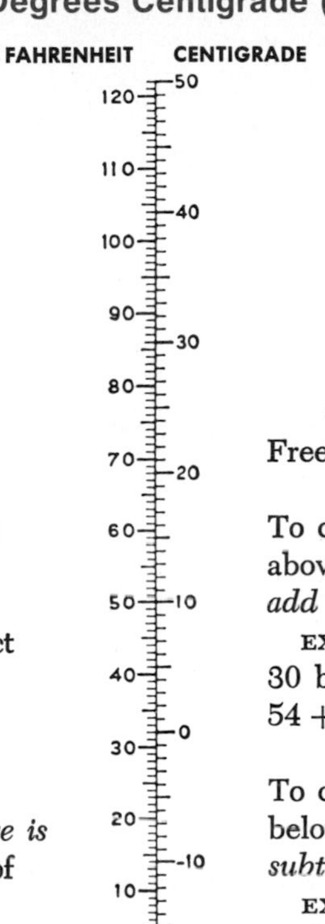

One °F = $\frac{5}{9}$ths °C
Freezing °F = 32° F

To change *°F to °C,* and *if the °F figure is* above *freezing, subtract* 32 and then multiply by $\frac{5}{9}$.

EX. If the °F figure is 50° F, subtract 32 from 50 and then multiply by $\frac{5}{9}$ ($50 - 32 = 18$; $18 \times \frac{5}{9} = \frac{90}{9} = 10$). Thus, 50° F = 10° C.

To change *°F to °C,* and *if the °F figure is* below *freezing,* determine the number of degrees below freezing in °F and then multiply that number by $\frac{5}{9}$.

EX. If the °F figure is −20° F, this means 52 Fahrenheit degrees below freezing. Multiply 52 by $\frac{5}{9}$ ($52 \times \frac{5}{9} = \frac{260}{9} = 28.88$). Thus, −20° F = 28.88° C.

One °C = $\frac{9}{5}$ths °F
Freezing °C = 0° C

To change *°C to °F,* and *if the °C figure is* above *freezing,* multiply by $\frac{9}{5}$ and then *add* 32.

EX. If the °C figure is 30° C, multiply 30 by $\frac{9}{5}$ and add 32 ($30 \times \frac{9}{5} = \frac{270}{5} = 54$; $54 + 32 = 86$). Thus, 30° C = 86° F.

To change *°C to °F,* and *if the °C figure is* below *freezing,* multiply by $\frac{9}{5}$ and then *subtract* 32.

EX. If the °C figure is −20° C, multiply 20 by $\frac{9}{5}$ and subtract 32 ($20 \times \frac{9}{5} = \frac{180}{5} = 36$; $36 - 32 = 4$). Thus, −20° C = −4° F.

COURTESY AMERICAN GEOGRAPHICAL SOCIETY

Conversion of Inches, Millimeters, and Centimeters

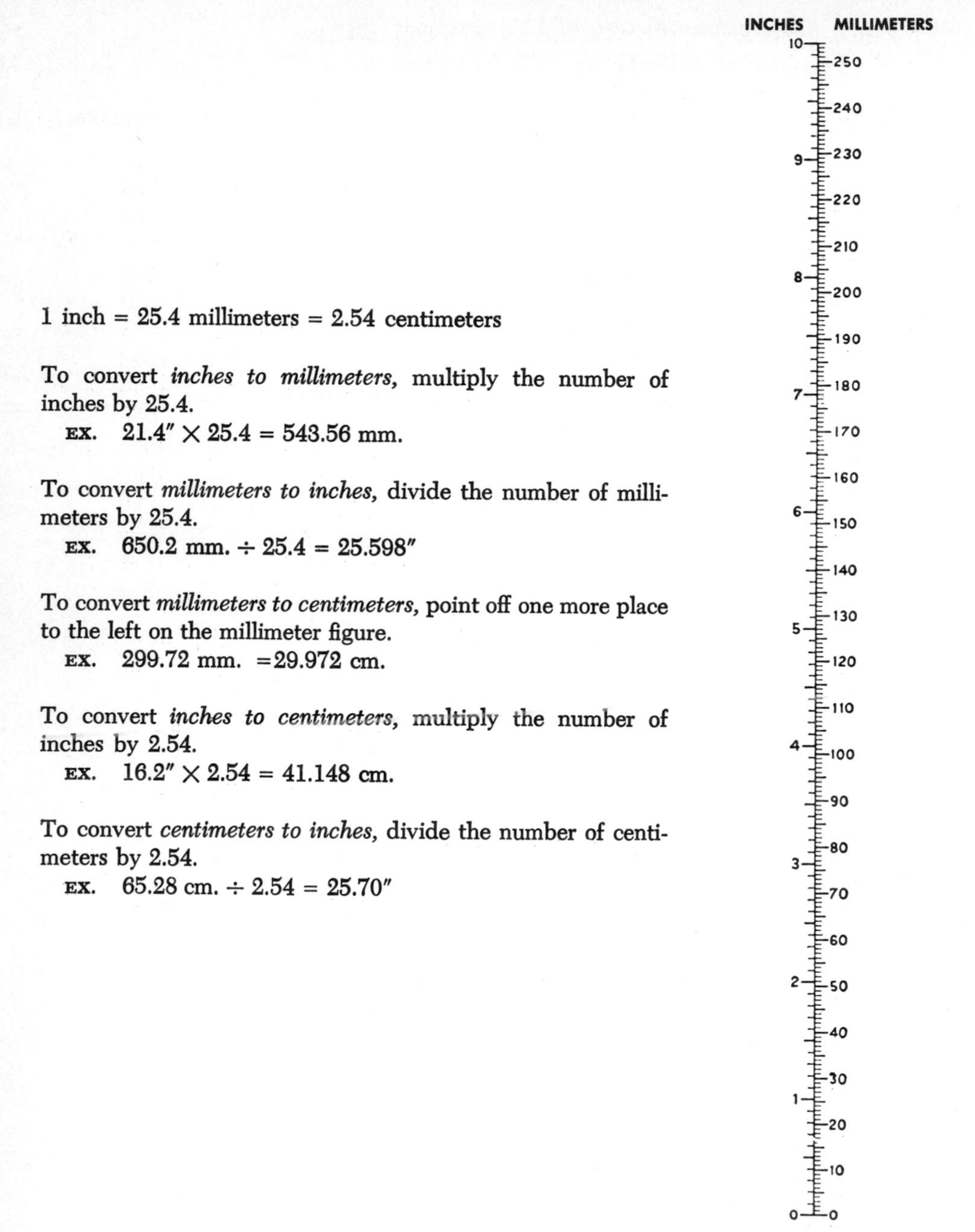

1 inch = 25.4 millimeters = 2.54 centimeters

To convert *inches to millimeters,* multiply the number of inches by 25.4.

EX. 21.4″ × 25.4 = 543.56 mm.

To convert *millimeters to inches,* divide the number of millimeters by 25.4.

EX. 650.2 mm. ÷ 25.4 = 25.598″

To convert *millimeters to centimeters,* point off one more place to the left on the millimeter figure.

EX. 299.72 mm. =29.972 cm.

To convert *inches to centimeters,* multiply the number of inches by 2.54.

EX. 16.2″ × 2.54 = 41.148 cm.

To convert *centimeters to inches,* divide the number of centimeters by 2.54.

EX. 65.28 cm. ÷ 2.54 = 25.70″

COURTESY AMERICAN GEOGRAPHICAL SOCIETY

Comparative Measurement of Atmospheric Pressure

1 *inch of mercury* = 25.4 *millimeters* = 33.86395 *millibars*
1 *millibar* = .02953 *inches* = .75006 *millimeters*

INCHES	MILLIBARS	INCHES	MILLIBARS	INCHES	MILLIBARS
27.0	= 914	28.6	= 968	30.1	= 1019
27.1	= 918	28.7	= 972	30.2	= 1023
27.2	= 921	28.8	= 975	30.3	= 1026
27.3	= 924	28.9	= 979	30.4	= 1030
27.4	= 928	29.0	= 982	30.5	= 1033
27.5	= 931	29.1	= 985	30.6	= 1036
27.6	= 935	29.2	= 989	30.7	= 1040
27.7	= 938	29.3	= 992	30.8	= 1043
27.8	= 941	29.4	= 996	30.9	= 1046
27.9	= 945	29.5	= 999	31.0	= 1050
28.0	= 948	29.6	= 1002	31.1	= 1053
28.1	= 952	29.7	= 1006	31.2	= 1057
28.2	= 955	29.8	= 1009	31.3	= 1060
28.3	= 958	29.9	= 1013	31.4	= 1063
28.4	= 962	30.0	= 1016	31.5	= 1067
28.5	= 965				

Some Linear Measurement Equivalents

1 statute *mile* = 5280 *ft.* = 1.6093 *km.* = 1609.3 *m.*
1 nautical *mile* (international) = 6076.1033 *ft.* = 1.852 *km.* = 1852 *m.*
1 *furlong* = 660 *ft.* = 220 *yds.* = ⅛ statute *mile* = 201.162 *m.*
1 *rod* = 16.5 *ft.* = 5.5 *yds.* = 1/32 statute *mile* = 5.029 *m.*
1 *yard* = 36 *in.* = 3 *ft.* = 1/1760 statute *mile* = .9144 *m.*
1 *foot* = 12 *in.* = 30.48 *cm.* = .3048 *m.*
1 *inch* = 2.54 *cm.* = 25.4 *mm.*

1 *chain* (surveyor's) = 100 *links* = 66 *ft.* = 792 *in.* = 1/80 statute *mile*
1 *fathom* = 8 *spans* = 72 *in.* = 6 *ft.* = 1.8288 *m.*

1 *kilometer* (km.) = 3280.8399 *ft.* = .62137 statute *mile*
1 *meter* (m.) = 39.37 *in.* = 3.28 *ft.*
1 *centimeter* (cm.) = .3937 *in.* = .0328 *ft.*
1 *millimeter* (mm.) = .03937 *in.*

(1 km. = 1000 m.; 1 m. = 100 cm.; 1 cm. = 10 mm.)

Some Area Measurement Equivalents

1 *square mile* (sq. mile) = 640 *acres* = 258.999 *hectares*
1 *acre* = 43,560 *sq. ft.* = .4046 *hectares* = 4046.8564 *sq. m.*
1 *hectare* = 2.471 *acres* = .01 *sq. km.*

1 *square kilometer* (sq. km.) = 247.1054 *acres* = .3861 *sq. miles*
1 *square meter* (sq. m.) = 10.7639 *sq. ft.*

1 *survey township* = 36 *sq. miles* = 36 *survey sections*
1 *survey section* = 1 *sq. mile* = 640 *acres* = 258.999 *hectares*

Some Time Measurement Equivalents

1 *minute* = 60 *seconds*
1 *hour* = 60 *minutes* = 3600 *seconds*
1 *day* = 24 *hours* = 1440 *minutes* = 86,400 *seconds*
1 *year* = 365.2422 *days*

Some Circular Measurement Equivalents

1 *circle* = 360 *degrees* = 4 *quadrants* = 1 *circumference*
1 *quadrant* = 90 *degrees* = ¼ *circle*
1 *degree* = 60 *minutes*
1 *minute* = 60 *seconds*
1 *radian* = 57 *degrees*, 17 *minutes*, 44.8 *seconds* (57°17′44.8″)

Some Longitude—Time Measurement Equivalents

360 *degrees of longitude* = 24 *hours of time* = 1 *day*
15 *degrees of longitude* = 1 *hour*
1 *degree of longitude* = 4 *minutes*
1 *minute of longitude* = 4 *seconds*
1 *second of longitude* = 1⁄15 *second*

Some Weight Measurement Equivalents

1 *short ton* = 2000 *pounds* = 907.18 *kilograms*
1 *long ton* = 2240 *pounds* = 1016.05 *kilograms*
1 *metric ton* = 2204.623 *pounds* = 1000 *kilograms*

1 *pound* = 16 *ounces* = 453.5923 *grams* = .4536 *kilograms*
1 *ounce* = 28.3495 *grams* = 16 *drams*

1 *kilogram* = 35.274 *ounces* = 2.2046 *pounds*
1 *quintal* = 100 *kilograms* = 220.4623 *pounds*
10 *quintals* = 1 *metric ton*

APPENDIX E

Geologic Time Scale

APPENDIX F

The Seventh Approximation (7A): A Classification of Soils

In addition to the great soil groups, the widely accepted classification of soils used in this and most other geography texts, various other classifications have been developed. The basic objectives of all of these classifications are essentially the same, however, despite differences in soil categories and nomenclature. One of the most recent and comprehensive of these classifications is known as the *7th Approximation,* or *7A.* When this new system was presented to the Seventh International Congress of Soil Science in 1960, it had already undergone a series of developmental stages. Since the 1960 presentation represented the seventh stage of its development, the classification has been designated the 7th Approximation.

The 7th Approximation has ten major *orders* of soils, all of which are strictly defined and limited—as are the lower *suborders* and *great groups.* The terminology used is much more rigorous than a simple compounding of geographic names with textural classes. This is because the system attempts to describe the morphology of a soil, as well as to delineate its quantitative dominances of type. The ten orders of soil are: *entisols* (recent); *vertisols* (mixed); *inceptisols* (new surfaces); *aridisols* (desert); *mollisols* (calcified); *spodosols* (podzolized); *alfisols* (pedalfers); *ultisols* (weathered); *oxisols* (laterized); and *histosols* (organic).

There are twenty-nine suborders, within which the oxisols and histosols are as yet undivided. The suborder name is derived by the compounding of one of fifteen prefixes (for example, *aqu* = wet) with the root of one of the order terms; thus, *aquent* stands for a *waterlogged recent soil.* The great groups are designated by adding short suffixes (for example, *hal* = salty) to the suborder names. The great groups are further subdivided into *subgroups, families,* and *series,* according to soil texture, mineral form, and so on.

The 7th Approximation has had some impact on soil science and soil geographers. Reactions to it have been varied, to say the least; but, the genetic directness of 7A may, in time, balance the initial impact caused by its curious nomenclature.

Indexes

Index of Geographical Terms

D

E

F

M

N

O

P

R

General Index

D

E

H

I

N

O

P

Q

R

S

T

U

V

W

X

Y

Z

B 7
C 8
D 9
E 0
F 1
G 2
H 3
I 4
J 5